A NOTE FROM THE AUTHORS

Congratulations on your decision to take the AP Calculus exam! Whether or not you're completing a year-long AP Calculus course, this book can help you prepare for the exam. In it you'll find information about the exam as well as Kaplan's test-taking strategies, a targeted review that highlights important concepts on the exam, and practice tests. Take the diagnostic test to see which subject you should review most, and use the full-length exams to get comfortable with the testing experience. The review chapters include summaries of the most important AP exam topics, so that even if you haven't completed them in class, you won't be surprised on Test Day. Don't miss the strategies for answering the free-response questions: You'll learn how to cover the key points AP graders will want to see.

By studying college-level calculus in high school, you've placed yourself a step ahead of other students. You've developed your critical-thinking and time-management skills, as well as your understanding of sophisticated mathematical proofs and concepts. Now it's time for you to show off what you've learned.

Best of luck,

Tamara Lefcourt Ruby

James Sellers

Lisa Korf

Jeremy Van Horn

Mike Munn

RELATED TITLES

AP Biology

AP Calculus AB & BC

AP Chemistry

AP English Language & Composition

AP English Literature & Composition

AP Environmental Science

AP European History

AP Macroeconomics/Microeconomics

AP Physics B & C

AP Psychology

AP Statistics

AP U.S. Government & Politics

AP U.S. History

AP World History

SAT Premier with CD-Rom

SAT Strategies, Practice, and Review

SAT Subject Test: Biology E/M

SAT Subject Test: Chemistry

SAT Subject Test: Literature

SAT Subject Test: Mathematics Level 1

SAT Subject Test: Mathematics Level 2

SAT Subject Test: Physics

SAT Subject Test: Spanish

SAT Subject Test: U.S. History

SAT Subject Test: World History

AP® CALCULUS AB & BC

2015

AP® CALCULUS AB & BC

2015

Tamara Lefcourt Ruby
James Sellers
Lisa Korf
Jeremy Van Horn
Mike Munn

New York

Published by Kaplan Publishing, a division of Kaplan, Inc.
395 Hudson Street
New York, NY 10014

Printed in the United States of America

10 9 8 7 6 5 4 3 2 1

ISBN-13: 978-1-61865-686-5

TABLE OF CONTENTS

PART ONE: THE BASICS

PART TWO: DIAGNOSTIC TEST

PART THREE: AP CALCULUS REVIEW

**BC content*

**BC content*

**BC content*

**BC content*

PART FOUR: PRACTICE TESTS

**BC content*

ACCESSING YOUR ONLINE COMPANION

Welcome to *AP Calculus AB & BC 2015*!

The online companion provides one-month access to Grockit. Grockit is a comprehensive, personalized platform for AP test-like question practice and review.

Track Your Skill Data: Analyze and track your performance based on content area, difficulty, and time spent on each question.

Target Your Weaknesses: Grockit continually learns as you learn, and targets you with the material you need to work on most.

Study with Peers and Friends: Collaborate and compete with peers or learn from expert instructors.

Register your online companion using these simple steps

1. Go to http://goto.grockit.com/textbook/ap/.

2. Follow the on-screen instructions. Please have a copy of your book available.

Good Luck with your studies!

ABOUT THE AUTHORS

Tamara Lefcourt Ruby holds a PhD in mathematics from the University of Pennsylvania, and is a research associate at Tel Aviv University. She also teaches calculus to students enrolled in The Johns Hopkins University Center for Talented Youth, and taught calculus at Harvard University and the University of Texas for 15 years.

James Sellers is director of undergraduate mathematics at Pennsylvania State University, where he teaches calculus and other math courses.

Lisa Korf is an assistant professor of mathematics at the University of Washington.

Jeremy Van Horn holds a PhD in mathematics from the University of Texas, where he is a math instructor.

Mike Munn earned a BS in math at the University of Notre Dame as well as a master's degree in math from Stony Brook University. After teaching for a brief period, he returned to school and currently is pursuing a PhD in mathematics at the City University of New York.

Jim Gerber is the Head of Mathematics at the American School Foundation of Monterrey in Mexico, where he teaches AP Calculus AB and BC. He is a College Board consultant and has given workshops in Mexico, Guatemala, China, Malaysia, the Philippines and in the US. He has been involved in the AP Calculus reading since 2005.

Kimberly Foltz has taught mathematics at the Indiana Academy for Science, Mathematics, and Humanities in Muncie, Indiana, since 1991. She has served as a College Board consultant, an AP workshop and AP summer institute leader, and a reader and table leader for the AP Calculus exam. In 2009, Kim was selected as the Indiana winner of the Siemens Award for Advanced Placement Teaching.

Bill Scott has been on the math faculty of Phillips Academy in Andover, MA, where he has taught up and down the curriculum from elementary algebra through the honors BC calculus course since 1987. Presently, Bill is the Chair of the Department of Mathematics, Statistics, and Computer Science, and he has served as a workshop leader for the College Board for many years.

Bill Roloff has taught at Lake Park High School in Roselle, IL since 1985. He is National Board Certified and has been an AP Calculus reader and workshop leader since 2001.

Larry Peterson began his AP Calculus career in 1975 and currently teaches at Northridge High School in Layton, Utah. He has been involved with reading the AP Calculus Exam since 1993.

Sergio Stadler has taught AP Calculus at Marist School for 35 years. He has co-authored multiple choice questions for Louis Leithold's Calculus textbook. He has been an AP reader and table leader for the AP Calculus exam, authored TIP's Advanced Placement Calculus AB manual, and been an Atlanta teaching fellow at TIP's Summer Residential Program. He has been a CollegeBoard consultant and an instructor of summer institutes.

Angie Seckar-Martinez has taught math since 1998. She currently teaches Algebra and AP Calculus AB and BC at McCallum High School in Austin, Texas.

KAPLAN PANEL OF AP EXPERTS

Congratulations—you have chosen Kaplan to help you get a top score on your AP exam.

Kaplan understands your goals, and what you're up against—achieving college credit and conquering a tough test—while participating in everything else that high school has to offer.

You expect realistic practice, authoritative advice, and accurate, up-to-the-minute information on the test. And that's exactly what you'll find in this book, as well as every other in the AP series. To help you (and us!) reach these goals, we have sought out leaders in the AP community. Allow us to introduce our expert:

AP Calculus Expert

William Fox has been teaching calculus at the university level for over 20 years. He has been at Francis Marion University in Florence, South Carolina since 1998. He has been a reader and table leader for the AP Calculus AB and BC exams since 1992.

Part One

THE BASICS

CHAPTER 1: INSIDE THE AP CALCULUS EXAM

Advanced Placement exams have been around for half a century. While the format and content have changed over the years, the basic goal of the AP program remains the same: to give high school students a chance to earn college credit or advanced placement. To do this, a student needs to do two things:

- Find a college that accepts AP scores
- Do well enough on the exam

The first part is easy, since the majority of colleges accept AP scores in some form or another. The second part requires a little more effort. If you have worked diligently all year in your coursework, you've laid the groundwork. The next step is familiarizing yourself with the test.

There are actually *two* AP Calculus tests offered to students, a Calculus AB exam and a Calculus BC exam. The BC exam covers some more advanced topics; it's not necessarily a harder test, but its scope is broader. Most students opt to take the AB exam because they find it somewhat easier, but the BC exam is worth more credit at many universities.

Your decision to take the AP Calculus exam involves many factors, but in essence it boils down to a question of choosing between three hours of time spent on the AP Calculus exam and several hours of time spent in a college classroom. Depending on the college, a score of 3, 4, or 5 on the AP Calculus exam can allow you to leap over the introductory courses and jump right into more advanced classes. These advanced classes are usually smaller, more specifically focused, more intellectually stimulating, and simply more interesting than a basic course. If you are concerned solely about fulfilling your mathematics requirement so you can get on with your study of pre-Columbian art, Elizabethan music, or some other area apart from calculus, the AP exam can help you there, too. If you do well on the AP calculus exam, you may never have to take a math class again, depending on the requirements of the college you choose.

INTRODUCTION TO THE AP CALCULUS EXAM

If you're holding this book, chances are you are already gearing up for the AP Calculus exam. Your teacher has spent the year cramming your head full of the math know-how you will need to have at your disposal. But there is more to the AP Calculus exam than math know-how. You have to be able to work around the challenges and pitfalls of the test—and there are many—if you want your score to reflect your abilities. Studying math and preparing for the AP Calculus exam are not the same thing. Rereading your textbook is helpful, but it's not enough.

That's where this book comes in. We'll show you how to marshal your knowledge of calculus and put it to brilliant use on Test Day. We'll explain the ins and outs of the test structure and question format so you won't experience any nasty surprises. We'll even give you test strategies designed specifically for the AP Calculus exam.

Preparing for the AP Calculus exam means doing some extra work. You need to review your text *and* master the material in this book. Is the extra push worth it? If you have any doubts, think of all the interesting things you could be doing in college instead of taking an introductory course that covers only material that you already know.

OVERVIEW OF THE TEST STRUCTURE

The AP Calculus exam consists of two sections. In Section I, you have 105 minutes to answer 45 multiple-choice questions with five answer choices each. This section is worth half of your total score. This is the portion of the test in which the right answer is provided to you as one of five choices. Section II of the exam consists of six "free-response" questions that are worth the other half of your score. The term "free-response" means roughly the same thing as "large, multistep, and involved"; you will use most—probably *all*—of the 90 minutes allotted for Section II answering these six problems. Although these free-response problems are long and made up of multiple parts, they don't usually cover an obscure topic. Instead, they take a fairly basic calculus concept and ask you a *bunch* of questions about it. It's a lot of calculus work, but it's fundamental calculus work.

Both the AB and BC exams have the same format. Everything that is presented on the AB exam is fair game for the BC exam. However, the BC exam also includes a variety of *additional* information that is not covered on the AB exam. This book divides a large body of calculus material into multiple chapters, outlined below. Topics that are unique to the BC exam are marked with an asterisk (*) in this list.

Chapter 3: Calculator Basics

Chapter 4: Graphing with a Calculator

Chapter 5: Equation Solving with a Calculator

Chapter 6: Operations with Functions on a Calculator

NOTE: The AP Calculus exam does not directly test whether you know how to use your calculator properly—it *assumes* that you do. The information in Chapter 3 is included in this book because it is an important part of doing well on the exam.

Chapter 7: Limits

Chapter 8: Asymptotes

Chapter 9: Continuous Functions

Chapter 10: Parametric, Polar, and Vector Functions*

Chapter 11: The Concept of the Derivative

Chapter 12: Computation of Derivatives

Chapter 13: The Derivative at a Point

Chapter 14: The Derivative As a Function

Chapter 15: Second Derivatives

Chapter 16: Applications of Derivatives (including BC content: numerical solution of differential equations using Euler's method and L'Hôpital's Rule)

Chapter 17: Introduction to Integrals

Chapter 18: Applications of Integrals

Chapter 19: Antiderivatives—The Indefinite Integral

Chapter 20: The Fundamental Theorem of Calculus

Chapter 21: Techniques of Antidifferentiation (including BC content: antiderivatives by substitution of variables (including change of limits for definite integrals), parts, and simple partial fractions (nonrepeating linear factors only), and improper integrals (as limits of definite integrals))

Chapter 22: Applications of Antidifferentiation (including BC content: solving logistic differential equations and using them in modeling)

Chapter 23: Numerical Approximations

Chapter 24: The Concept of Series*

Chapter 25: The Properties of Series*

Chapter 26: Taylor Series*

** BC content*

HOW THE EXAM IS SCORED

Scores are based on the number of questions answered correctly. **No points are deducted for wrong answers.** No points are awarded for unanswered questions. Therefore, you should answer every question, even if you have to guess.

When your three-plus hours of testing are up, your exam is sent away for grading. The multiple-choice part is handled by a machine, while qualified readers—current and former calculus teachers and professors—grade your responses to Section II. After an interminable wait, your composite score will arrive. Your results will be placed into one of the following categories, reported on a five-point scale:

5 = Extremely well qualified (to receive college credit or advanced placement)

4 = Well qualified

3 = Qualified

2 = Possibly qualified

1 = No recommendation

Some colleges will give you college credit for a score of 3 or higher, but it's much better to get a 4 or a 5. If you have an idea about which colleges you want to go to, check out their websites or call the admissions office to find their particular rules regarding AP scores.

REGISTRATION AND FEES

You can register for the exam by contacting your guidance counselor or AP coordinator. If your school doesn't administer the exam, contact AP Services for a listing of schools in your area that do. The fee for each AP exam is $89 within the United States, and $119 at schools and testing centers outside of the United States. For students with acute financial need, the College Board offers $26 or $28 credit. In addition, most states offer exam subsidies to cover all or part of the remaining cost for eligible students. To learn about other sources of financial aid, contact your AP coordinator.

ADDITIONAL RESOURCES

For more information on all things AP, contact AP Services:

Phone: (888) 255-5427 or (212) 632-1780
Email: apstudents@info.collegeboard.org
Website: https://apstudent.collegeboard.org/home

CHAPTER 2: STRATEGIES FOR SUCCESS: IT'S NOT ALWAYS HOW MUCH YOU KNOW

There was a time when taking a standardized test was an unusual experience. There was also a time when people used an abacus instead of a calculator. What do these two facts have in common? Both were true a long time ago, but neither one is going to be true again any time soon.

Standardized tests have become common in education today. Following closely on the heels of standardized tests are *strategies* to help you succeed on standardized tests. You are probably familiar with some of the general strategies that help students increase their scores on a standardized exam. Even so, a quick review will help to jog your memory.

GENERAL TEST-TAKING STRATEGIES

1. **Pacing.** Because many tests are timed, proper pacing allows someone to attempt as many questions as possible in the time allotted. Poor pacing causes students to spend too much time on some questions while leaving others untouched because they run out of time before getting a chance to attempt every problem.
2. **Two-Pass System.** Using the two-pass system is one way to help pace yourself better on a test. The key idea is that you don't simply start with the first question and trudge onward from there. Instead, you start at the beginning, but take a first pass through the test answering all the questions that are easy for you. If you encounter a tough problem, you spend only a small amount of time on it and then move on in search of easier questions that might come after that problem. This way, you don't get bogged down on a tough problem when you could be earning points answering later problems that you *do* know. On your second pass, go back through the section and attempt all the tougher problems that you

passed over the first time. You should be able to spend a little more time on them, and this extra time might help you answer the problem. Even if you don't reach an answer, you might be able to employ techniques like the process of elimination to cross out some answer choices and then take a guess.

To make this strategy work for you, you must be organized and careful. You cannot just flip haphazardly through the test, hoping to find problems you can solve. You must go through the test in an orderly way, analyzing problems quickly and deciding how you will handle them. You will also need to mark the questions that you choose to save for later and make sure that you *really do* come back and answer them.

3. **Process of Elimination.** On every multiple-choice test, the correct answer is given to you. The only difficulty lies in spotting the correct answer hidden among incorrect choices. Even so, the multiple-choice format means you don't have to pluck the answer out of the air. Instead, if you can eliminate the answer choices you know are incorrect and only one choice remains, that must be the correct answer.
4. **Patterns and Trends.** The key word here is the *standardized* in "standardized testing." Standardized tests don't change greatly from year to year. Sure, the particular questions won't be the same and different topics will be covered from one administration to the next, but there will also be a lot of overlap from year to year. That's the nature of standardized testing: If the test changed dramatically each time it came out, it would be useless as a tool for comparison. Because of this, certain patterns can be uncovered about any standardized test. Learning about these trends and patterns can help students do better on tests they have not taken previously.
5. **The Right Approach.** Having the right mindset plays a large part in how well people do on a test. Those who are nervous about the exam and hesitant to make guesses often fare much worse than students with an aggressive, confident attitude. Students who start with the first question and struggle methodically forward through each problem don't score as well as students who deal with the easy questions before tackling the harder ones. People who take a test cold have more problems than those who take the time to learn about the test beforehand. In the end, factors like these are what separate people who are good test takers from those who struggle even when they know the material.

These points are valid for every standardized test, but they are quite broad. The rest of this chapter will discuss how these general ideas can be modified to apply specifically to the AP Calculus exam. These test-specific strategies—combined with the factual information covered in your course and in this book's review—are the one-two punch that will help you succeed on this specific test.

HOW THIS BOOK CAN HELP, PLUS SPECIFIC AP CALCULUS EXAM STRATEGIES

Kaplan AP Calculus AB and BC contains precisely the information you will need to ace the test. There's nothing extra in here to waste your time—no pointless review of material that's not on the test, no rah-rah speeches. Just the most potent test-preparation tools available:

1. **Test Strategies Geared Specifically to the AP Calculus Exam**. Many books give the same tired talk about the process of elimination that's been used for every standardized test given in the past 20 years. We're going to talk about the process of elimination as it applies to the AP Calculus exam and *only* to the AP Calculus exam. There are several skills and general strategies that work for this particular test; these will be covered in the next chapter.
2. **A Well-Crafted Review of All the Relevant Subjects**. The best test-taking strategies can't help you get a good score if you don't know the difference between a limit and a derivative. At its core, the AP Calculus exam covers a wide range of calculus topics and it's necessary for you to know that material if you are to do well on the test. However, chances are good that you're already familiar with these subjects, so an exhaustive review is not needed. In fact, it would be a waste of your time. No one wants that, so we've tailored our review section to focus on how the relevant topics typically appear on the exam, and what you need to know to answer the questions correctly. If a topic doesn't come up on the AP Calculus exam, we don't cover it. If it appears on the test, we'll provide you with the facts you need to navigate the problem safely.
3. **Full-Length Practice Tests**. Few things are better than experience when it comes to standardized testing. Taking a practice AP exam gives you an idea of what it's like to answer calculus questions for three hours. Granted, that's not a fun experience, but it is a helpful one. Practice exams give you the opportunity to test and refine your skills, as well as a chance to find out what topics you should spend some additional time studying. And the best part is that it doesn't count! Mistakes you make on our practice exams are mistakes you won't make on the real test.

USING A CALCULATOR ON THE AP CALCULUS EXAM

The AP Calculus exam is split into two sections, and each section is further divided into two parts. This extra division is done because each section has some questions for which you are allowed to use a graphing calculator, as well as some questions for which a graphing calculator is not allowed.

Section I is all multiple-choice questions; you have 105 minutes to complete 45 questions that account for 50 percent of your total score. Section I is divided into two parts, A and B. Part A contains 28 questions that must be completed in 55 minutes; calculators *are not* allowed. Part B contains 17 questions that must be completed in 50 minutes; calculators *are* allowed.

Section II is all free-response questions; you have 90 minutes to complete six questions that account for 50 percent of your total score. Section II is also divided into Part A and Part B. Part A contains two problems that must be completed in 30 minutes; a graphing calculator is required. Part B contains four problems that must be completed in 60 minutes; calculators *are not* allowed.

So graphing calculators are not allowed in Part A of Section I, but you can use them in Part B, while in Section II the situation is reversed (i.e., you can use them in Section II, Part A but not Section II, Part B). It's important for you to know when you *are* permitted to use your graphing calculator and when you are *not.*

When you are working on Section II, you should avoid falling into the graphing-calculator trap of not showing your work or your problem setup. When you solve a problem using a graphing calculator, you *must* show a setup on the answer sheet—show the logic and the method behind the problem, or you won't receive credit! What's more, if a problem's setup involves an integral, say something like $\int_1^3 x^2\, dx$, you need to write this integral, *not* the equivalent calculator-speak, which would be fnInt(x ^ 2, x, 1, 3). Always use standard mathematical notation.

Remember that your calculator is not a cure-all. Because you have a lot of computing power at your disposal, the AP exam you take will demand a lot more, computationally, than AP exams from the pre-graphing-calculator era. In fact, if you don't use your calculator as effectively as possible, this will hurt your score. You are allowed to use the calculator on the exam, but this is not a treat or a prize—some problems on the exam cannot be solved without one.

To succeed on the AP exam you must know calculus *and* your graphing calculator well!

HOW TO APPROACH THE MULTIPLE-CHOICE QUESTIONS

The worst thing that can be said about the multiple-choice questions on the AP Calculus exam is that they count for 50 percent of your total score. Although you might not like multiple-choice questions, there's no denying the fact that it's easier to guess on a multiple-choice question than it is to guess the correct answer to an open-ended question. The answer is always there in front of you for a multiple-choice problem; the trick is to find it in the forest—OK, thicket—of incorrect answers.

Contrast this with Section II of the AP Calculus exam, which accounts for the other 50 percent of your score. In that section, if you don't know how to work a problem, you have to write down what you *do* know and hope that the mythical King of Partial Credit is feeling kindly toward you that day.

Every multiple-choice question on the AP Calculus exam can be described as a "stand-alone" question. A stand-alone question covers a specific topic; the question that follows it hits a different topic. Usually there are between five and 60 words in the question stems and these words provide you with the information you need to answer the problem. Here's a typical question:

> The cost of producing x units of a certain item is $c(x) = 2000 + 8.6x + 0.5x^2$. What is the instantaneous rate of change of c with respect to x when $x = 300$?
>
> (A) 313.6
>
> (B) 308.6
>
> (C) 300.0
>
> (D) 297.2
>
> (E) 200.0

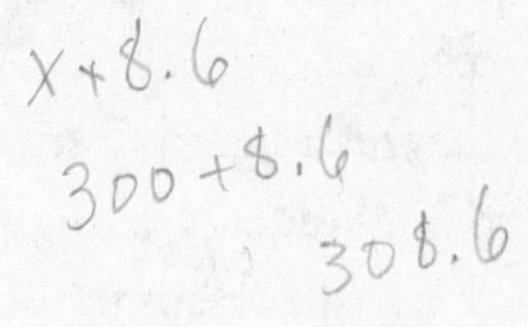

In this problem, you get some information and then you're expected to answer the question. Where this question occurs in the test makes no difference, because there's no patterned order of difficulty in which questions are presented on the AP Calculus exam. Tough questions are scattered between easy and moderately difficult questions.

The stand-alone questions look like a bunch of disconnected calculus questions one after the other and that's just what they are. Because they aren't connected to each other, there's no reason you have to answer these questions in sequential order. A two-pass system, discussed next, should be used here. You can tweak the general idea of the two-pass system and apply it specifically to the AP Calculus exam.

The Two-Pass System on the AP Calculus Exam

If you wanted to, you could take all the AP Calculus questions and arrange them in a spectrum ranging from "fastest to answer" to "slowest to answer." For example, questions involving a graph are often much faster to answer, as long as you can interpret the visual data correctly.

Picking out questions with graphs is not an especially critical way to analyze the exam questions, but some students do no more than that. You should realize that the more advanced your pacing system is, the more time you might have at the end of Section I to answer the questions that you find difficult. To further refine your two-pass abilities, draw up two lists of exam topics. Label one list "Calculus Concepts I Enjoy and Know About" and label the other list "Calculus Concepts That Are Not My Strong Points."

When you get ready to tackle the multiple-choice section, keep these two lists in mind. On your first pass through the section, answer all the questions that deal with concepts you like and know a lot about. If a question covers a subject that's not one of your strong points, skip it and come back on your second pass. The overarching goal is to use the time available to answer the maximum number of questions correctly.

This refinement of the basic two-pass system should give you a clear idea about how to approach the multiple-choice section of the AP Calculus exam. Now that you've got an idea of the correct approach, let's talk a bit about the correct mindset for test-taking.

Comprehensive, Not Sneaky

Some tests are sneakier than others. A sneaky test has questions that are written in convoluted ways; they're designed to trip you up mentally and manipulate your score by using a host of other little tricks. Students taking a sneaky test often have the proper facts, but get many questions wrong because of traps in the questions themselves.

The AP Calculus exam is *not* a sneaky test. It aims to see how much calculus knowledge you have stored in your skull. To do this, it asks a wide range of questions from an even wider range of calculus topics. The exam tries to cover as many different calculus facts as it can, which is why the problems jump from topic to topic. The test works hard to be comprehensive, which means that students who only know one or two calculus topics will soon find themselves struggling.

Understanding these facts about how the test is designed can help you to answer its questions. The AP Calculus exam is comprehensive, not sneaky; it makes questions hard by asking about hard subjects, not by using rhetorical tricks to create hard questions. And you've taken an AP Calculus course, so...

BE INSTINCTUAL!

Trust your instincts when guessing. If you think you know the right answer, chances are that you dimly remember the topic being discussed in your AP course. The test is about knowledge, not traps, so trusting your instincts will help more often than not.

You don't have enough time to think deeply about every tough question, so trusting your instincts can keep you from getting bogged down and wasting precious time on a problem. Some of your educated guesses are likely to be incorrect, but again, the point is not to get a perfect score. A perfect score would certainly be nice, but most people are going to lose at least some points on the AP exam. Your basic goal should be to get as good a score as you can; surviving hard questions by going with your gut feelings can help you to achieve this aim.

On other problems, though, you might have no inkling of what the correct answer should be. In that case, turn to the following key idea.

THINK "GOOD MATH!"

The AP Calculus exam rewards good mathematicians. It covers fundamental topics and you are expected to use logical thinking and good mathematical technique to solve problems. What the test doesn't want is sloppy math or sloppy thinking. It doesn't want answers that are factually incorrect, too extreme to be true, or irrelevant to the topic.

Still, bad math answers invariably appear, because it's a multiple-choice test and you have to have four incorrect answer choices along with the one right answer. So if you don't know how to answer a problem, look at the answer choices and think "Good Math." This may lead you to find some poor answer choices that can be eliminated.

Which of the following gives the derivative of the function $f(x) = x^2$ at the point (2, 4)?

(A) $\lim_{h \to 0} \frac{(x+2)^2 - x^2}{4}$

(B) $\lim_{h \to \infty} \frac{(2+h)^2 - 2^2}{h}$

(C) $\frac{(2+h)^2 - 2^2}{h}$

(D) $\lim_{h \to 0} \frac{(2+h)^2 - 2^2}{h}$

(E) $\lim_{h \to 0} \frac{(4+h)^2 - 4^2}{h}$

Students who recall the definition of a derivative based on limits will have no problem determining that (D) is the correct answer. Even if you don't know how to answer this problem, you can use "Comprehensive, not Sneaky" and "Good Math" to give yourself a chance at guessing the right answer. The easiest answer choice to eliminate is (C), because its solution will contain the letter h, while the problem never mentions this letter as a variable or constant. Choice (A) can also be eliminated, because it defines a function that varies with x as h approaches zero.

Although you may not be able to set up the limit properly, you can attack the problem by actually determining the final answer to the problem and seeing if any of the remaining choices lead to that value. The derivative of the function, $f'(x) = 2x$, and at the point (2, 4), $f'(x) = 4$. Choice (B) factors to $h + 4$, which is infinity as h approaches infinity. Choice (E) factors to $h + 8$, which is 8 when $h = 0$. Finally, choice (D) factors to $h + 4$, which equals 4 when $h = 0$. Because choice (D) is the only choice that provides the correct answer of $f'(x) = 4$, it must be the correct answer.

AP EXPERT TIP

For each multiple-choice question, no points are deducted for wrong answers, but no points are awarded for unanswered questions. Therefore, you should answer every question, even if you have to guess.

HOW TO APPROACH THE FREE-RESPONSE QUESTIONS

The good news is that the topics covered in these questions are usually fairly common calculus topics. You won't see trick questions asking about an obscure subject. However, the fact that a topic is well known doesn't mean the problem will be simple or easy. Most free-response questions are divided into parts—that is, they are stuffed with smaller questions. You won't usually get one broad question like, "Is a Riemann sum ever really happy?" Instead, you'll get an initial setup followed by questions labeled (a), (b), (c), and so on. Expect to write at least one paragraph (or provide multistep equations) for each letter. It's best to clearly label the part you are answering (e.g., [a] or [part a]) and to use each page completely before moving to the second. Putting each part on a single page is not a good idea, because a harried reader might see the end of your part (a) with half a blank page beneath it and think that's the end of your whole answer! You definitely do not want that to happen.

AP EXPERT TIP

Sometimes these parts are related to each other, and sometimes they are independent. Each part will have a labeled area in the answer booklet for you to use.

For the free-response questions, you receive points for responding properly to each sub-question prompt. The more points you score, the better off you are on that question. Going into the details about how points are scored would make your head spin, but in general, the AP Calculus readers have a rubric that works as a blueprint for a good answer. Every subsection of a question has one to five key ideas attached to it. If you write about one of those ideas, you earn yourself a point. Readers always use the same rubric for a question and all questions are evaluated using a rubric. Any reader who reads your exam should, in theory, award you the same number of points on a given question. Readers check and cross-check each other to ensure that each answer is evaluated in the same way. Two readers rarely differ on a score.

There's a limit to how many points you can earn on a single subquestion and there are other complex rules guiding the grading of the AP exams, but it boils down to this: Writing smart things about each question will earn you points on that question.

So don't rush or be unnecessarily terse. You have about 15 minutes for each free-response problem. Use that time to be as precise as you can be for each subquestion. Sometimes doing well on one subquestion earns you enough points to make up for another subquestion that you didn't answer as effectively. When all the points are tallied for that free-response problem, you can still come out strong on total points, even though you didn't ace every single subquestion.

AP EXPERT TIP

At least one question on Part A of Section II will require you to use a graphing calculator.

Finally, because you are allowed to use a calculator on Part A of the Section II, you should *expect* to use it. There's no point in permitting a calculator for part of the exam if they aren't going to make you press some digits in Part A. This isn't

100 percent certain, of course, because nothing on an unknown test ever is. But it is a good guess, like picking the most favored horse to win the race.

If you get a question on a subject you don't know well, things might look grim for that problem. Still, take heart. Quite often, you'll be able to earn at least some points on every question because there will be some subquestion or segment in each one that you know. Remember, the goal is not perfection. If you can ace four of the questions and slug your way to partial credit on the other two, you will put yourself in position to get a good score on the entire test. That's the Big Picture, so don't lose sight of it just because you don't know the answer to one subquestion on Part B.

Be sure to use all the strategies discussed in this chapter when taking the practice exams. Trying out the strategies there will help you become comfortable with them and you should be able to put them to good use on the real exam.

Of course, all the strategies in the world can't save you if you don't know anything about calculus. The next part of the book will help you review the primary concepts and facts that you can expect to encounter on the AP Calculus exam.

STRESS MANAGEMENT

You can beat anxiety the same way you can beat the AP Calculus exam—by knowing what to expect beforehand and developing strategies to deal with it. Following are some time-tested and simple techniques for dealing with stress.

VISUALIZE

Sit in a comfortable chair in a quiet setting. If you wear glasses, take them off. Close your eyes and breathe in a deep, satisfying breath of air. Really fill your lungs until your rib cage is fully expanded and you can't take in any more. Then, exhale the air completely. Imagine you're blowing out a candle with your last little puff of air. Do this two or three more times, filling your lungs to their maximum and emptying them totally. Keep your eyes closed, comfortably but not tightly. Let your body sink deeper into the chair as you become even more comfortable.

Close your eyes and start remembering a real-life situation in which you did well on a test. If you can't come up with one, remember a situation in which you did something that you were really proud of—a genuine accomplishment. Make the memory as detailed as possible. Think about the sights, the sounds, the smells, even the tastes associated with this remembered experience. Remember how confident you felt as you accomplished your goal. Now start thinking about the AP Calculus exam. Keep your thoughts and feelings in line with that prior, successful experience. Don't make comparisons between them. Just imagine taking the upcoming test with the same feelings of confidence and relaxed control.

This exercise is a great way to bring the test down to earth. You should practice this exercise often, especially when you feel burned out on exam preparation. The more you practice it, the more effective the exercise will be for you.

Exercise

Whether it's jogging, walking, biking, mild aerobics, pushups, or a pickup basketball game, physical exercise is a very effective way to stimulate both your mind and body and to improve your ability to think and concentrate. Lots of students get out of the habit of regular exercise when they're prepping for the exam. Also, sedentary people get less oxygen to the blood, and hence to the brain, than active people. You can watch TV fine with a little less oxygen; you just can't think as well.

Any big test is a bit like a race. Finishing the race strong is just as important as being quick early on. If you can't sustain your energy level in the last sections of the exam, you could blow it. Along with a good diet and adequate sleep, exercise is an important part of keeping yourself in fighting shape and thinking clearly for the long haul.

There's another thing that happens when students don't make exercise an integral part of their test preparation. Like any organism in nature, you operate best if all your "energy systems" are in balance. Studying uses a lot of energy, but it's all mental. When you take a study break, do something active. Take a five- to ten-minute exercise break for every 50 or 60 minutes that you study. The physical exertion helps keep your mind and body in sync. This way, when you finish studying for the night and go to bed, you won't lie there unable to sleep because your head is tired while your body wants to run a marathon.

One warning about exercise: It's not a good idea to exercise vigorously right before you go to bed. This could easily cause sleep-onset problems. For the same reason, it's also not a good idea to study right up to bedtime. Make time for a "buffer period" before you go to bed. Take 30 to 60 minutes to take a long hot shower, to meditate, to read a relaxing book, or even watch TV.

Stay Drug Free and Eat Healthy

Using drugs to prepare for or take a big test is not a good idea. Don't take uppers to stay alert. Amphetamines make it hard to retain information. Mild stimulants, such as coffee, cola, or over-the-counter caffeine pills can help you study longer because they keep you awake, but they can also lead to agitation, restlessness, and insomnia. Some people can drink a pot of coffee sludge and sleep like a baby. Others have one cup and start to vibrate. It all depends on your tolerance for caffeine. Remember, a little anxiety is a good thing. The adrenaline that gets pumped into your bloodstream helps you stay alert and think more clearly.

You can also rely on your brain's own endorphins. Endorphins have no side effects and they're free. It just takes some exercise to release them. Running, bicycling, swimming, aerobics, and power walking all cause endorphins to occupy the happy spots in your brain's neural synapses. In addition, exercise develops your mental stamina and increases the oxygen transfer to your brain.

To reduce stress you should eat as healthily as possible. Fruits and vegetables can help reduce stress. Low-fat protein such as fish, skinless poultry, beans, and legumes (like lentils), or whole grains such as brown rice, whole wheat bread and pastas (no bleached flour) are good for your brain-chemistry. Avoid sweet, high-fat snacks. Simple carbohydrates such as sugar can make stress worse, and fatty foods may lower your immunity. Don't eat excessively salty foods either. They can deplete potassium, which you need for nerve functions.

Isometrics

Here's another natural route to relaxation and invigoration. You can do it whenever you get stressed out, including during the test. Close your eyes. Starting with your eyes and—without holding your breath—gradually tighten every muscle in your body (but not to the point of pain) in the following sequence:

- Close your eyes tightly.
- Squeeze your nose and mouth together so that your whole face is scrunched up. (If it makes you self-conscious to do this in the test room, skip the face-scrunching part.)
- Pull your chin into your chest, and pull your shoulders together.
- Tighten your arms to your body, then clench your fists.
- Pull in your stomach.
- Squeeze your thighs and buttocks together, and tighten your calves.
- Stretch your feet, then curl your toes (watch out for cramping in this part).

At this point, every muscle should be tightened. Now, relax your body, one part at a time, in reverse order, starting with your toes. Let the tension drop out of each muscle. The entire process might take five minutes from start to finish (maybe a couple of minutes during the test). This clenching and unclenching exercise will feel silly at first, but if you get good at it, it can help you feel very relaxed.

COUNTDOWN TO THE TEST

It's almost over. Eat a power snack, drink some carrot juice—or whatever makes healthy sense to keep yourself going. Here are Kaplan's strategies for the three days leading up to the test.

Three Days Before the Test

Take a full-length practice test under timed conditions. Use the techniques and strategies you've learned in this book. Approach the test strategically, actively, and confidently.

Two Days Before the Test

Go over the results of your practice test. Don't worry too much about your score, or about whether you got a specific question right or wrong. The practice test doesn't count. But do examine your performance on specific questions with an eye to how you might get through each one faster and better on the test to come.

WARNING: DO NOT take a full practice exam if you have fewer than 48 hours left before the test. Doing so will probably exhaust you and hurt your score on the actual test. Maybe it will help to think of the AP Calculus exam as a marathon. Racers don't run a marathon the day before the real thing—they rest and conserve their energy!

The Night Before the Test

DO NOT STUDY. Get together an "AP Calculus exam kit" containing the following items:

- A calculator with fresh batteries
- A watch
- A few No. 2 pencils
- Erasers
- Photo ID card
- A snack—there are breaks, and you'll probably get hungry

Know exactly where you're going, exactly how you're getting there, and exactly how long it takes to get there. It's probably a good idea to visit your test center sometime before the day of the test, so that you know what to expect—what the rooms are like, how the desks are set up, and so on.

Relax the night before the test. Read a good book, take a long hot shower, watch a movie or something on TV. Get a good night's sleep. Go to bed early and leave yourself extra time in the morning.

The Morning of the Test

- Eat breakfast. Make it something substantial, but not anything too heavy or greasy.
- Don't drink a lot of coffee if you're not used to it. Bathroom breaks cut into your time, and too much caffeine is a bad idea.
- Dress in layers so that you can adjust to the temperature of the test room.
- Read something non-test related. Warm up your brain with a newspaper or a magazine.
- Be sure to get there early. Allow yourself extra time for traffic, mass-transit delays, and/or detours.

During the Test

Don't be shaken. If you find your confidence slipping, remind yourself how well you've prepared. You know the structure of the test; you know the instructions; you've had practice with—and have learned strategies for—every question type.

If something goes really wrong, don't panic. If the test booklet is defective—two pages are stuck together or the ink has run—raise your hand and tell the proctor you need a new book. If you accidentally mis-grid your answer page or put the answers in the wrong section, raise your hand and tell the proctor. He or she might be able to arrange for you to re-grid your test after it's over, when it won't cost you any time.

After the Test

You might walk out of the exam thinking that you blew it. This is a normal reaction. Lots of people—even the highest scorers—feel that way. You tend to remember the questions that stumped you, not the ones that you knew.

We're positive that you will have performed well and scored your best on the exam because you followed the Kaplan strategies outlined in this section. Be confident in your preparation, and celebrate the fact that the AP Calculus exam is soon to be a distant memory.

CHAPTER 3: CALCULATOR BASICS

Calculators are good—that's what professional organizations for math teachers have decided. This is lucky for you, because it means you can use a calculator on the AP exam. In fact, you are *expected* to use a calculator on the AP exam, a *graphing calculator*, to be specific. Graphing calculators have become a standard, essential part of AP Calculus. You should be using yours on a regular basis to become comfortable with it.

There is a huge variety of graphing calculators available on the market and their capabilities have increased as technology has advanced. Some of the newer models are much more powerful than their older, first-generation siblings. It does not really matter which particular brand or type of calculator you have, though you should make certain the calculator meets your needs. What is truly important is that you know how to *use* your calculator.

HOW MUCH COMPUTING POWER DO YOU NEED?

The AP Calculus committee works hard to develop questions that can be solved with any appropriate calculator, but it can't develop tests that are fair to all students if the range of calculator capabilities is too large. To level the playing field, your calculator must meet a minimum performance level to ensure that you can solve all of the problems on the test. Some calculators with more advanced, complicated features are not allowed because the AP Calculus exam committee feels that they would give students who use them an unfair advantage. The trick, then, is to find a calculator that can do everything you need it to do, but not so much that it will be barred from the exam.

The people who write the test assume that your calculator can:

1. Plot a graph in the appropriate viewing window.
2. Find the zeros of a function (a technique often used to solve equations).
3. Calculate the value of the derivative of a function at a point.
4. Calculate the value of a definite integral numerically.

If you can perform these functions with your calculator, then you have enough computational power to solve any problem on the AP exam.

You can't use a non-graphing scientific calculator on the exam—they don't give you enough computing power. You can't use tools that provide too much computing power, either, which means no computers—like laptops—and no mini-computers—including pocket organizers, pen-input devices like PDAs, or anything with a QWERTY keyboard, such as a TI-92 Plus. The AP Course Description has a list of approved calculators; read it and make sure yours is on the list. Then make sure that you *really* know your calculator. Your calculator is your friend, so treat it like one. Use it often. Take it with you. Learn its secrets. Let it help you find answers in class and on tests. (This last one, of course, is something you should only do with your calculator and not a human friend.)

DOES A FANCIER CALCULATOR HELP YOU MORE?

The authors of the AP exam try to write questions that present an equal challenge to students, regardless of what sort of calculators they may have. They try, but they don't always succeed, and it is definitely to your advantage to have the most advanced calculator you can.

For example, suppose you are asked to find the value of c that satisfies the mean value theorem for $f(x) = x \cdot \ln x$ on $[2, 4]$.

Using a less advanced calculator, you need to compute the derivative of f by hand, i.e., you need to use the product rule to compute $f'(x) = \ln x + 1$. Then you need to solve the equation $\frac{4 \cdot \ln 4 - 2 \cdot \ln 2}{4 - 2} = \ln c + 1$. You need a calculator to solve this equation and at this point there are lots of ways to proceed.

More advanced calculators can handle symbolic manipulation, compute the derivative for you, and solve the entire problem in one line of code. For example, to solve this problem using a TI-89, on the home screen type $y = x\ (\ln x)$. Press ENTER, then type F2: *Solve* $(d\ (y\ (x), x, c) = (y\ (4) - y\ (2))/(4 - 2), c$. Press ENTER and your answer appears. With an advanced calculator, you don't need to do any computations yourself.

Distinctions Among Different Types of Calculators

Lower-level calculators cannot perform symbolic manipulations. That is, they cannot multiply or factor polynomials, differentiate or integrate symbolically, or take limits. They evaluate a polynomial at a point; they can evaluate the derivative of a function at the point, but they can't tell you the derivative function. They can compute a numerical approximation of a definite integral, but they can't tell you the exact value. For example, a lower-level calculator will compute $\int_1^3 \frac{1}{\sqrt{x}}\,dx$ as 1.4641016 but it will not give the exact answer $2\sqrt{3} - 2$. It cannot compute antiderivatives such as $\int \frac{1}{\sqrt{x}}\,dx = 2\sqrt{x}$.

The TI-83 Plus is an example of a lower-level graphing calculator. A more powerful graphing calculator can perform symbolic manipulations. It can multiply and factor polynomials, differentiate and integrate symbolically, and take limits of functions. It shows the answer symbolically as an equation, when appropriate. It can show a precise answer with square roots and fractions. The TI-89 is an example of a more powerful model.

There are lots of other calculators that fall into each of these categories. Yours, whatever it is, is just fine as long as you know how to use it. Many calculators use similar key strokes. The important thing is to become familiar with your *own* calculator. If you want to perform an operation and don't know how to do it, refer to your owner's manual. If you've lost yours (have you cleaned your room lately?), most calculator companies have a copy of the instruction manual on their websites.

AP EXPERT TIP

Be sure to use the calculator that you've been using through the school year. The TI-83/84 family have comparable keystrokes, but they are significantly different than the TI-89 calculators. So, it would not be wise to switch to an unfamiliar calculator right before the exam.

USING CALCULATORS WITH THIS BOOK

In subsequent chapters, we'll help you to get familiar with the ways you'll need to use your calculator on the AP exam. We'll give general guidelines for solving problems with any graphing calculator and detailed instructions for solving problems with a TI-83 Plus. The TI-83 is an example of a basic, no-frills, can't-compute-derivatives calculator. Sometimes, when there's a way to solve a problem more elegantly with a more advanced, can-compute-derivatives calculator, we'll also give instructions for the TI-89.

What we won't do is tell you how to get started with your graphing calculator. You should refer to your owner's manual for that basic information. For example, we'll assume that you know how to enter a function into the calculator (e.g., using the [Y=] button on a TI-83) and that you know how to set the viewing window for a graph.

At the beginning of each section, we'll discuss the keys and utilities that are discussed in that section. Then we'll move on to sample problems that illustrate how you would actually use these utilities on the AP exam.

WHEN CAN YOU USE YOUR GRAPHING CALCULATOR?

As mentioned in Chapter 2, the AP exam is divided into two parts: multiple-choice questions and free-response questions. Each chunk is worth 50 percent of your total score. On about one-third of the multiple-choice questions and one-third of the free-response questions, you can use a graphing calculator. You can bring up to two(!) graphing calculators into the exam with you, along with whatever programs they contain. You can use whatever shortcuts you've programmed into your calculator. To maintain the integrity of the test, you can't take any test information out in your calculator.

USING YOUR CALCULATOR

The 45 multiple-choice questions make up Section I of the exam and count for 50 percent of your score; they appear in two separate parts. There are 28 questions in Part A, in which a calculator *is not* allowed, and 17 questions in Part B in which a calculator *is* allowed. You won't necessarily need your graphing calculator for all 17 of the questions in Part B. The six free-response questions make up Section II of the exam and count for the other 50 percent of your score; they also appear in two parts. There are two in Part A, where you are permitted (and will almost definitely need) to use your calculator, and four in Part B, to be completed without a calculator. Let's do the math here. This means that for 32 out of 51 of the questions on the test (more than 60 percent of the exam, or 64 percent of the points) you *cannot* use a calculator. It's not a crutch or a cure-all, even if you have a calculator that takes derivatives, computes integrals, and does your laundry. The bottom line is that you need to know calculus and know it well to succeed on the AP exam.

AP EXPERT TIP

Rounding off too soon can lead to an incorrect answer. Learn how to use the "store" feature of your calculator to store and retrieve intermediate results. Don't lose points by rounding off more than once in the same problem.

ROUNDING ANSWERS ON THE FREE-RESPONSE QUESTIONS

When you're using your calculator (and even when you're not), remember that all numerical answers on the free-response section of the AP exam must either be exact or rounded (or truncated) to three places after the decimal (nearest thousandths) unless otherwise specified. This means that if the correct answer is $\sqrt{2}$, you need to write $\sqrt{2}$ or 1.414. An answer of 1.41 would not receive the final point awarded for the correct answer (although you would probably still get most of the credit for your work). A correct answer that is rounded to the nearest hundredth is counted as a *wrong* answer. If you want to include more decimal places, go right ahead; you won't be penalized. The important thing is not to include *too few*. You should get into the habit *now* of rounding your answers to the nearest thousandth, so that it will be automatic by the time the exam rolls around.

NOT EVERYTHING YOUR CALCULATOR TELLS YOU IS TRUE

A graphing calculator is a tool, and like any tool, it has limitations and quirks. For example, very large numbers will not necessarily be precise. Consider the following example: If we ask a TI-83 Plus to compute $5 \times 10^{12} + 4 - 5 \times 10^{12}$, it will give a correct answer of 4, but if we ask the calculator to compute $5 \times 10^{13} + 4 - 5 \times 10^{13}$—the answer to which is also 4—it will tell us 0.

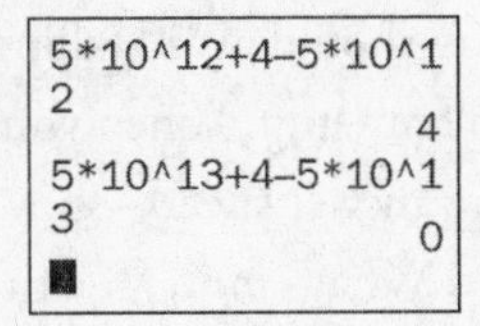

Calculators store a finite number of digits and if we try to add and subtract numbers of very different magnitudes, the small number will slip under the calculator's radar. We must be careful with numbers that are very large or very small, because the calculator will make rounding errors.

Another problem can occur when we try to compute a derivative numerically. Calculators generally use an approximation algorithm that computes a derivative as

$$f'(a) \approx \frac{f(a+h) - f(a-h)}{2h}$$

and samples small values of h. This usually works fine, but in certain situations, it gives an incorrect result. For example, if we ask a TI-83 to compute the derivative of $f(x) = x^{\frac{2}{3}}$ at $x = 0$, it will return the answer 0. But this function is not differentiable at $x = 0$; it has a sharp corner there.

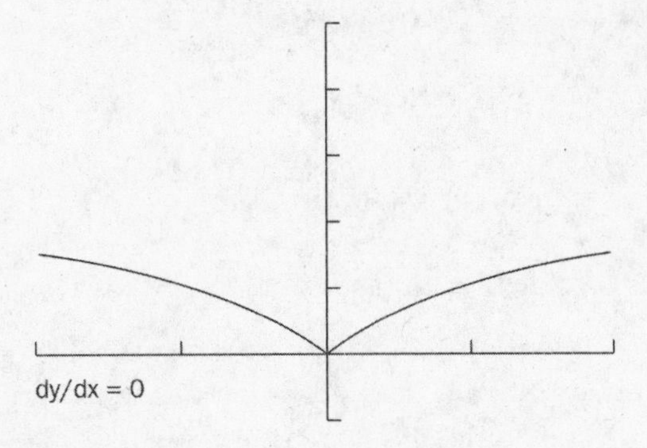

There are other pitfalls to avoid, but some of these will depend on what particular calculator you are using. Knowing your calculator well means knowing its quirks and faults.

CALCULATOR . . . FRIEND OR FOE?

Your graphing calculator has more computing power than ENIAC—one of the first computers, whose vacuum-tube processors filled a huge room. With the rise of more sophisticated graphing calculators, the grading emphasis on the AP exam has shifted from calculations to setup. For example, on a typical three-point problem, two points will be awarded for the setup and only one point for the final answer.

When you are working on Section II you should avoid falling into the graphing-calculator trap of not showing your work or your problem setup. When you solve a problem by using a graphing calculator, you *must* show a setup on the answer sheet—show the logic and the method behind the problem, or you won't receive credit! What's more, if a problem's setup involves an integral, say something like $\int_1^3 x^2\,dx$, you need to write this integral, *not* the equivalent calculator-speak, which would be fnInt(x ^ 2,x,1,3). Always use standard mathematical notation.

Remember that your calculator is not a cure-all. Because you have a lot of computing power at your disposal, the AP exam you take will demand a lot more, computationally, than AP exams from the pre-graphing-calculator era. In fact, if you don't use your calculator as effectively as possible, this will hurt your score. It's not only that you are *allowed* to use the calculator on the exam; some problems *cannot* be solved without one.

To succeed on the AP exam you must know calculus *and* your graphing calculator well!

Part Two

DIAGNOSTIC TEST

AP CALCULUS AB AND BC DIAGNOSTIC TESTS

Take a moment to gauge your readiness for the AP Calculus exam by taking either the AB diagnostic test or the BC diagnostic test, depending on which test you plan to take. The questions in this diagnostic are designed to cover most of the topics you will encounter on the AP Calculus exam. After you take it, you can use the results to give yourself a general idea of the subjects in which you are strong and the topics you need to review more. You can use this information to tailor your approach to the following review chapters. Ideally you'll still have time to read all the chapters, but if you're pressed for time, you can start with the chapters and subjects you really need to work on.

Important note for students planning to take the BC exam: The BC exam consists of both AB and BC material. If you plan to take the BC exam, you should test yourself on **BOTH** diagnostic tests. The BC diagnostic focuses on BC material in an effort to give you the greatest opportunity to test your understanding of BC topics. The real BC exam will not emphasize BC topics as heavily, so you should take the AB diagnostic as well to ensure that you have a good grasp of *both* the AB and the BC material.

Before taking this diagnostic, find a quiet place where you can work uninterrupted for three hours. Time yourself, and take the entire test without interruption—you can always call your friend back *after* you finish. Also, no TV or music. You won't have either luxury while taking the real AP Calculus exam, so you may as well get used to it now. Give yourself 105 minutes to work on the multiple-choice questions. Then, take 10 minutes to read and 90 minutes to answer the free-response questions.

When you've finished, be sure to read the explanations for all questions, even those you answered correctly. Even if you got the problem right, reading the worked solution can give you insights that will prove helpful on the real exam.

Good luck on the diagnostic!

AP Calculus AB Diagnostic Test Answer Grid

To compute your score for this diagnostic test, calculate the number of questions you got right, then divide by 45 to get the percentage of questions that you answered correctly.

The approximate score range is as follows:

5 = 80–100% (extremely well qualified)

4 = 60–79% (well qualified)

3 = 50–59% (qualified)

2 = 40–49% (possibly qualified)

1 = 0–39% (no recommendation)

A score of 49% is a 2, so you can definitely do better. If your score is low, keep on studying to improve your chances of getting credit for the AP Calculus AB exam.

1. Ⓐ Ⓑ Ⓒ Ⓓ Ⓔ	16. Ⓐ Ⓑ Ⓒ Ⓓ Ⓔ	31. Ⓐ Ⓑ Ⓒ Ⓓ Ⓔ
2. Ⓐ Ⓑ Ⓒ Ⓓ Ⓔ	17. Ⓐ Ⓑ Ⓒ Ⓓ Ⓔ	32. Ⓐ Ⓑ Ⓒ Ⓓ Ⓔ
3. Ⓐ Ⓑ Ⓒ Ⓓ Ⓔ	18. Ⓐ Ⓑ Ⓒ Ⓓ Ⓔ	33. Ⓐ Ⓑ Ⓒ Ⓓ Ⓔ
4. Ⓐ Ⓑ Ⓒ Ⓓ Ⓔ	19. Ⓐ Ⓑ Ⓒ Ⓓ Ⓔ	34. Ⓐ Ⓑ Ⓒ Ⓓ Ⓔ
5. Ⓐ Ⓑ Ⓒ Ⓓ Ⓔ	20. Ⓐ Ⓑ Ⓒ Ⓓ Ⓔ	35. Ⓐ Ⓑ Ⓒ Ⓓ Ⓔ
6. Ⓐ Ⓑ Ⓒ Ⓓ Ⓔ	21. Ⓐ Ⓑ Ⓒ Ⓓ Ⓔ	36. Ⓐ Ⓑ Ⓒ Ⓓ Ⓔ
7. Ⓐ Ⓑ Ⓒ Ⓓ Ⓔ	22. Ⓐ Ⓑ Ⓒ Ⓓ Ⓔ	37. Ⓐ Ⓑ Ⓒ Ⓓ Ⓔ
8. Ⓐ Ⓑ Ⓒ Ⓓ Ⓔ	23. Ⓐ Ⓑ Ⓒ Ⓓ Ⓔ	38. Ⓐ Ⓑ Ⓒ Ⓓ Ⓔ
9. Ⓐ Ⓑ Ⓒ Ⓓ Ⓔ	24. Ⓐ Ⓑ Ⓒ Ⓓ Ⓔ	39. Ⓐ Ⓑ Ⓒ Ⓓ Ⓔ
10. Ⓐ Ⓑ Ⓒ Ⓓ Ⓔ	25. Ⓐ Ⓑ Ⓒ Ⓓ Ⓔ	40. Ⓐ Ⓑ Ⓒ Ⓓ Ⓔ
11. Ⓐ Ⓑ Ⓒ Ⓓ Ⓔ	26. Ⓐ Ⓑ Ⓒ Ⓓ Ⓔ	41. Ⓐ Ⓑ Ⓒ Ⓓ Ⓔ
12. Ⓐ Ⓑ Ⓒ Ⓓ Ⓔ	27. Ⓐ Ⓑ Ⓒ Ⓓ Ⓔ	42. Ⓐ Ⓑ Ⓒ Ⓓ Ⓔ
13. Ⓐ Ⓑ Ⓒ Ⓓ Ⓔ	28. Ⓐ Ⓑ Ⓒ Ⓓ Ⓔ	43. Ⓐ Ⓑ Ⓒ Ⓓ Ⓔ
14. Ⓐ Ⓑ Ⓒ Ⓓ Ⓔ	29. Ⓐ Ⓑ Ⓒ Ⓓ Ⓔ	44. Ⓐ Ⓑ Ⓒ Ⓓ Ⓔ
15. Ⓐ Ⓑ Ⓒ Ⓓ Ⓔ	30. Ⓐ Ⓑ Ⓒ Ⓓ Ⓔ	45. Ⓐ Ⓑ Ⓒ Ⓓ Ⓔ

AP CALCULUS AB DIAGNOSTIC TEST

Directions: Solve the following problems, using available space for scratchwork. Do not use a calculator unless a question instructs you to do so. After examining the form of the choices, decide which one is the best of the choices given and fill in the corresponding oval on the answer sheet. No credit will be given for anything written in the test book. Do not spend too much time on any one problem.

Note: For the actual test, no scrap paper is provided.

In this test:

(1) The domain of a function f is the set of all real numbers x for which $f(x)$ is a real number, unless otherwise specified.

(2) The inverse of a trigonometric function f may be indicated using the inverse function notation f^{-1} or with the prefix "arc" (e.g., $\sin^{-1}x = \arcsin x$).

1. Evaluate $\lim\limits_{h \to \frac{1}{2}} \dfrac{e^h - \sqrt{e}}{h - \frac{1}{2}}$.

(A) e^2

(B) $e^2 - 1$

(C) $\sqrt{e}$

(D) e

(E) The limit does not exist.

2. $\int_0^{\frac{\pi}{2}} \sqrt{\sin x}\cos x \; dx =$

(A) 0

(B) 1

(C) $\sqrt{\pi}$

(D) $\dfrac{2}{3}$

(E) $\dfrac{2\pi}{3}$

3. The graph of $y = x^3 + 21x^2 - x + 1$ is concave down for

(A) $x < -7$

(B) $-7 < x < 7$

(C) all x

(D) $x > 7$

(E) $x < -7$ and $x > 7$

4. If $F(x) = \dfrac{x^3 + x^5}{\ln(x^2)}$, then $F'(\sqrt{e}) =$

(A) $e^2(3e + 1)$

(B) $e(3e + 1)$

(C) $2e$

(D) $\sqrt{e}(e+1)$

(E) $3e^2 - 1$

GO ON TO THE NEXT PAGE

5. What are the values for which the function $f(x) = \frac{2}{3}x^3 - x^2 - 4x + 3$ is increasing?
   (A) $-1 < x < 2$
   (B) $x < -1$
   (C) $x < -1$ and $x > 2$
   (D) $0 < x < 2$
   (E) $x > -1$

6. If $f(x) = \cos^2(\sin(2x))$, then $f'\left(\frac{\pi}{8}\right) =$
   (A) $1+\sqrt{2}\cos\left(\frac{\sqrt{2}}{2}\right)$
   (B) $\sqrt{2}\sin\left(\frac{\sqrt{2}}{2}\right)$
   (C) $\cos\left(\sin\left(\frac{\sqrt{2}}{2}\right)\right)$
   (D) $-2\sqrt{2}\cos\left(\frac{\sqrt{2}}{2}\right)\sin\left(\frac{\sqrt{2}}{2}\right)$
   (E) $-\sqrt{2}\cos\left(\frac{\sqrt{2}}{2}\right)$

7. The graph of f is given below. Which of the following statements is true about f?

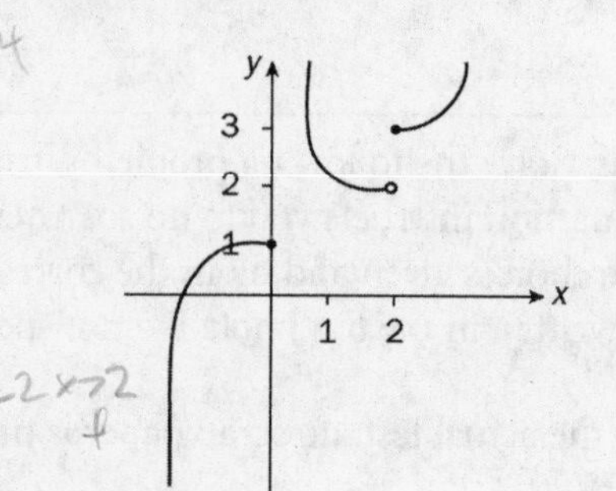

   (A) $\lim_{x\to 0} f(x) = 1$
   (B) $\lim_{x\to 0^-} f(x) = 2$
   (C) $\lim_{x\to 2^+} f(x) = 3$
   (D) $\lim_{x\to 2} f(x) = 2$
   (E) $\lim_{x\to 2^-} f(x) = \text{undefined}$

8. Evaluate $\lim_{x\to 0} \frac{\sin^2(4x)}{x^2}$.
   (A) $\frac{1}{4}$
   (B) 0
   (C) 16
   (D) 4
   (E) The limit does not exist.

GO ON TO THE NEXT PAGE

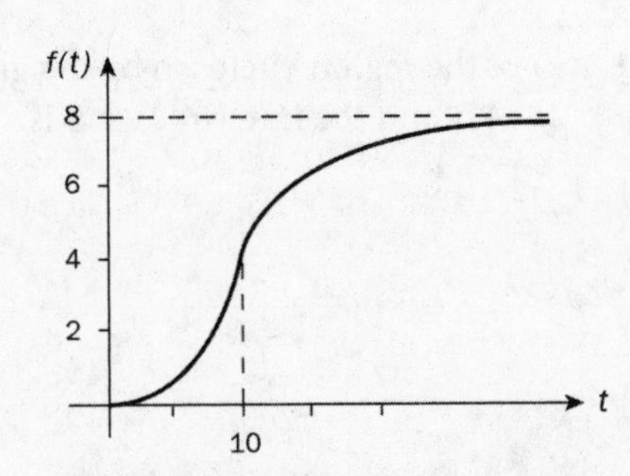

9. The graph of f is given above. Which graph below could represent the graph of f'?

(A)
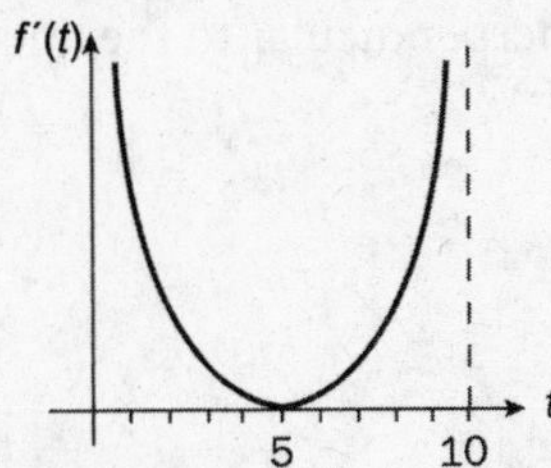

(B)
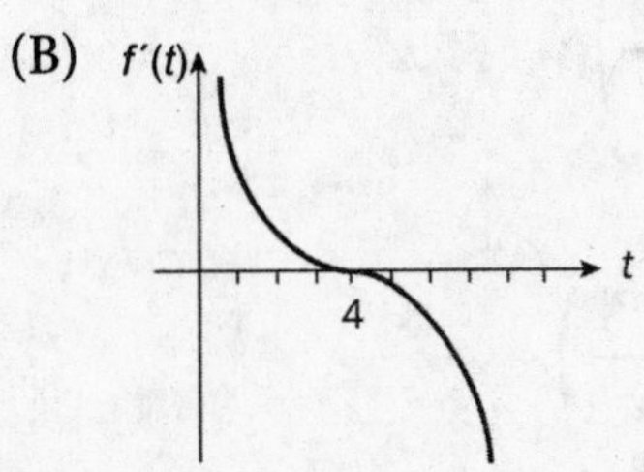

(C)
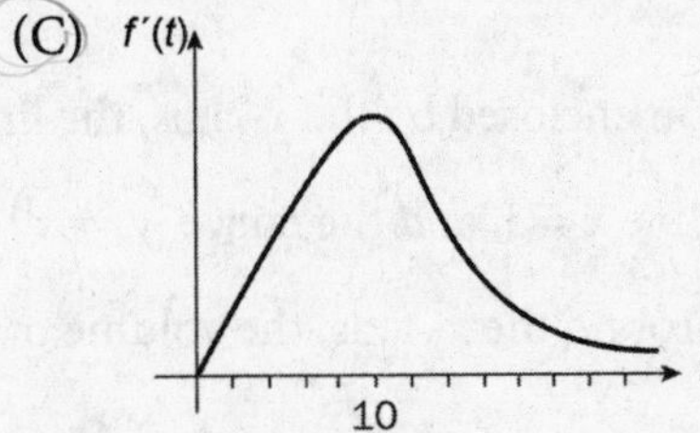

(D)
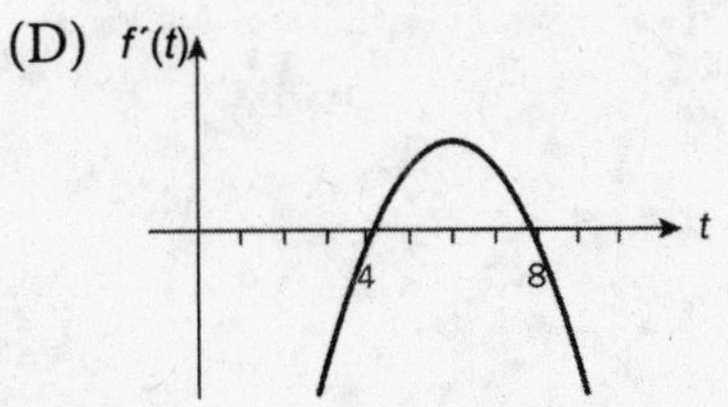

(E)
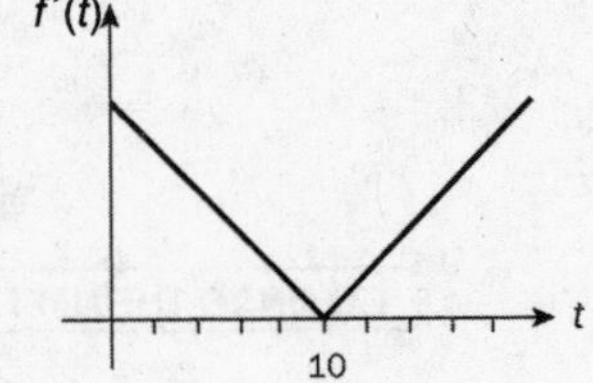

10. Evaluate $\lim\limits_{x\to\infty} \dfrac{x^3+3x^5+x}{2x^5+3x^2+5}$.

(A) 0

(B) $-\dfrac{4}{3}$

(C) $\dfrac{2}{3}$

(D) $\dfrac{3}{2}$

(E) The limit does not exist.

11. Evaluate $\lim\limits_{x\to 1} \dfrac{2x^2+x-3}{1-x^2}$.

(A) $-\dfrac{5}{2}$

(B) 0

(C) $\dfrac{4}{3}$

(D) $-\dfrac{3}{2}$

(E) The limit does not exist.

12. The solution to the differential equation $\dfrac{dy}{dx} = \dfrac{x^2}{y^3}$, where $y(3) = 3$ is

(A) $y = \sqrt[4]{\frac{3}{4}x^3 - 45}$

(B) $y = \sqrt[4]{\frac{4}{3}x^3 + 45}$

(C) $y = \sqrt[4]{\frac{4}{3}x^4 + 5}$

(D) $y = \sqrt[4]{\frac{4}{3}x^4} + \sqrt[4]{45}$

(E) $y = \sqrt[4]{\frac{4}{3}x^4 + 45}$

13. The slope of the tangent to the curve $2x^3y^2 - 5x^2y = 18$ at the point (1,2) is

(A) $-\dfrac{4}{5}$

(B) $-\dfrac{3}{2}$

(C) 0

(D) $-\dfrac{4}{3}$

(E) -1

GO ON TO THE NEXT PAGE

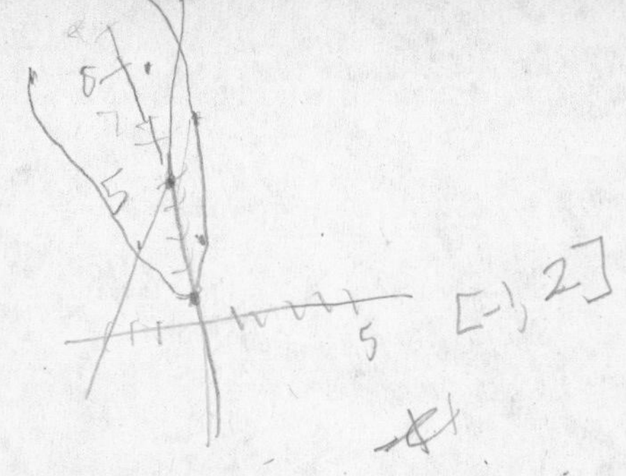

14. Which of the following is a slope field for the differential equation $\frac{dy}{dx} = \frac{2x^3}{y}$?

(A)

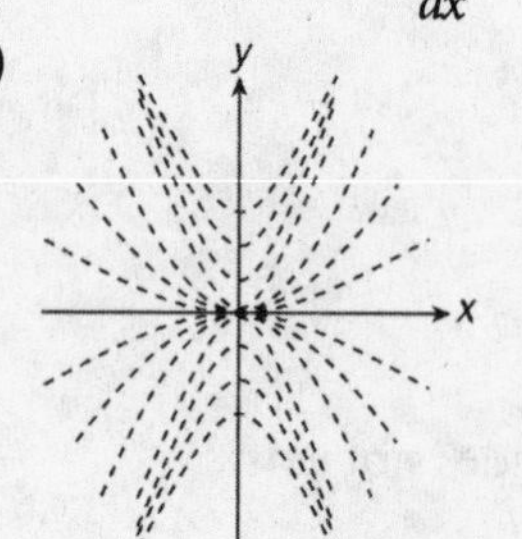

(B)

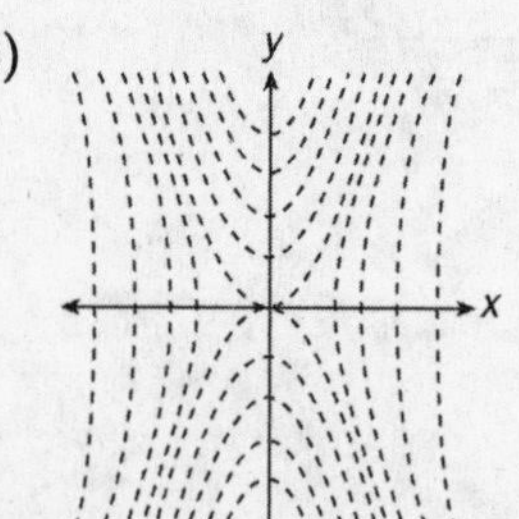

(C)

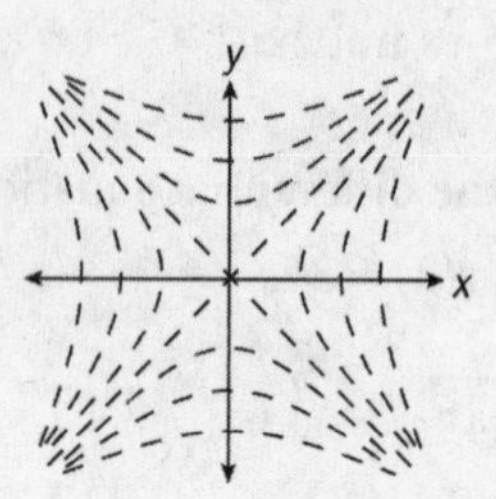

(D)

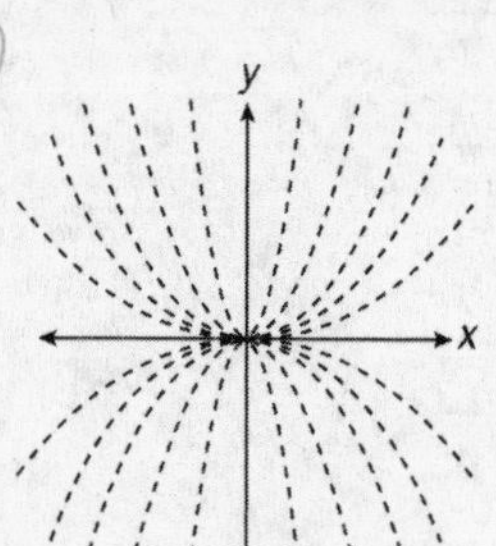

(E)

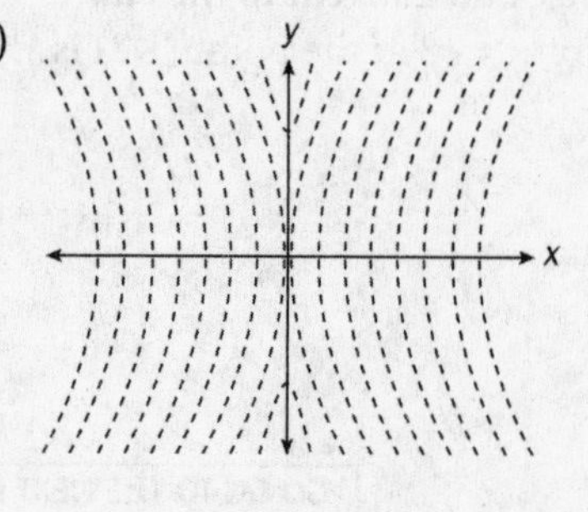

15. The area of the region enclosed by the graph of $y = 2x^2 + 1$ and the line $y = 2x + 5$ is

(A) 5
(B) 3
(C) $\frac{7}{2}$
(D) 9
(E) 1

16. At what point on the graph of $y = 2x^{\frac{3}{2}}$ is the tangent line perpendicular to the line $2x + 3y = 6$?

(A) (0, 0)
(B) $\left(\frac{1}{16}, \frac{1}{32}\right)$
(C) $\left(\frac{3}{2}, \frac{3\sqrt{6}}{2}\right)$
(D) $\left(\frac{1}{4}, \frac{1}{4}\right)$
(E) $\left(\frac{3}{16}, \frac{3\sqrt{3}}{32}\right)$

17. If the region enclosed by the x–axis, the line $x = 1$, the line $x = 3$, and the curve $y = x^{\frac{3}{2}}$ is revolved around the x–axis, the volume of the solid is

(A) 20π
(B) $\frac{25}{2}\pi$
(C) 21π
(D) 4π
(E) $\frac{100}{3}\pi$

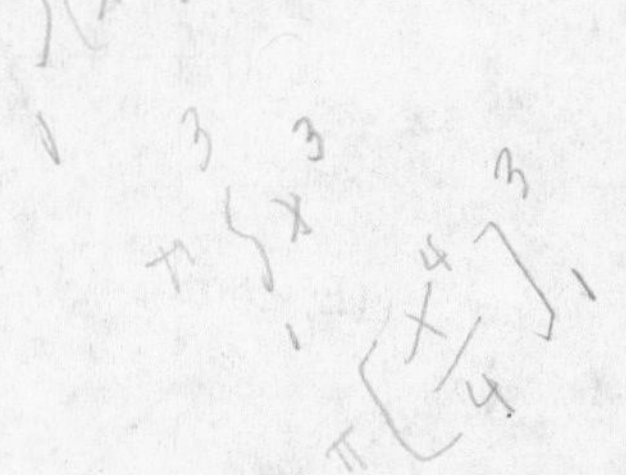

GO ON TO THE NEXT PAGE

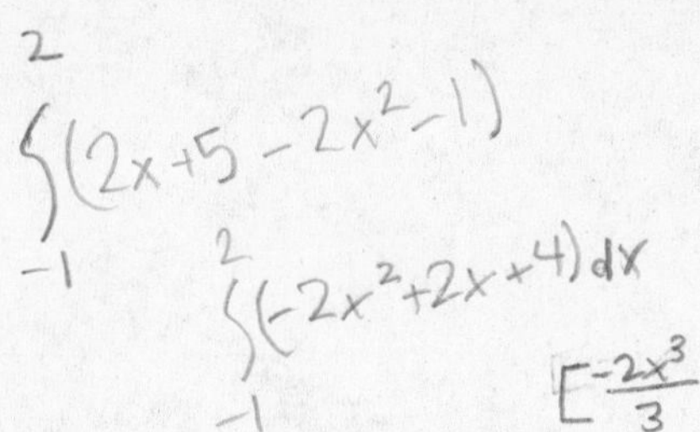

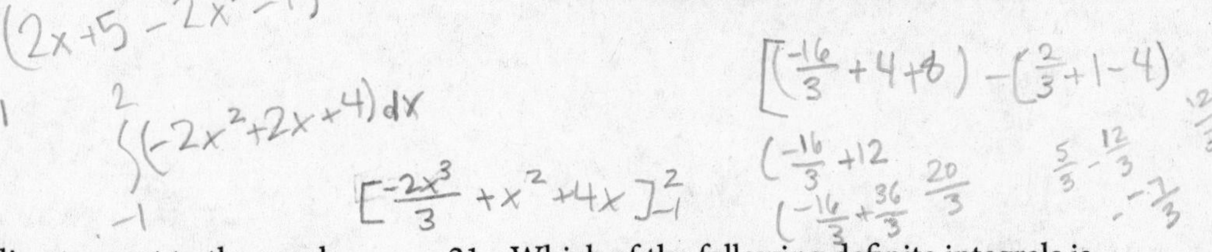

18. An equation of the line tangent to the graph of $y=2\sin\left(2x+\frac{3\pi}{4}\right)$ at $x=\frac{\pi}{8}$ is

(A) $y-\frac{\pi}{2}=-4x$

(B) $y+\frac{\pi}{4}=\frac{3}{4}x$

(C) $y=4x+\frac{\pi}{2}$

(D) $y+\frac{\pi}{2}=4x$

(E) $2y=-4x+\pi$

19. Find the derivative of $f(x)=\int_0^{x^2}\cos(t^2)\,dt$.

(A) $\cos(x^4)$

(B) $\cos(x^2)$

(C) $2x\cos(x^2)$

(D) $x\cos(x^4)$

(E) $2x\cos(x^4)$

20. Suppose that f is a continuous function and differentiable everywhere. Suppose also that $f(0) = 1, f(5) = -4, f(-5) = -3$. Which of the following statements must be true about f?

I. f has exactly two zeros.
II. f has at least two zeros.
III. f must have a zero between 0 and −5.
IV. There is not enough information to determine anything about the zeros of f.

(A) I only

(B) II only

(C) I and III only

(D) II and III only

(E) IV

21. Which of the following definite integrals is equivalent to $\int_0^2 f(x)dx$?

I. $\int_2^4 f(x-2)dx$

II. $\int_{-2}^0 f(x-2)dx$

III. $\int_{-3}^{-1} f(x+3)dx$

(A) I only

(B) II only

(C) III only

(D) I and II only

(E) I and III only

22. Let $f(x) = x^2 - 4x + e^{2x-1}$. Find the equation of the tangent line to f at the point where $x=\frac{1}{2}$.

(A) $y=-2x+\frac{1}{4}$

(B) $y=-x-\frac{1}{4}$

(C) $y=-2x-\frac{3}{4}$

(D) $y=-x+\frac{5}{4}$

(E) $y=x+\frac{5}{4}$

23. Let $f(x)=\int_{-2}^{x}(3t^2-5t+1)dt$. Then $f'(2)$ =

(A) −2

(B) 0

(C) 3

(D) 4

(E) 6

GO ON TO THE NEXT PAGE

24. Let $f(t) = -t^3 + 6t^2 - 8t - 3$. Then the instantaneous rate of change of f at $t = -1$ is

(A) −27
(B) −26
(C) −23
(D) 6
(E) 12

Suppose f is a continuous, differentiable function. Selected values of f are shown in the table below. Use the table to answer questions 25–27.

x	−4	−1	0	3	7
$f(x)$	6	−5	−3	6	3

25. What is the average rate of change of f on the interval $[-1,7]$?

(A) 1
(B) 3
(C) $\frac{1}{3}$
(D) 0
(E) 8

26. In which interval does the Mean Value Theorem guarantee at least one value c, such that $f'(c) = 0$?

(A) $-1 < x < 7$
(B) $-4 < x < 7$
(C) $-1 < x < 3$
(D) $0 < x < 7$
(E) $-4 < x < 3$

27. $\int_{-4}^{7} f(x)dx$ is approximated by a left Riemann sum using four subintervals. Which of the following is the value of the left Riemann sum?

(A) 28
(B) 31
(C) 44
(D) 56
(E) 77

Question 28 requires a calculator.

28. The position of a particle is given by $s(t) = \frac{1}{3}t^3 - 4t^2 + 12t + 1$ for $0 \le t \le 10$. On what interval(s) is the particle slowing down?

(A) $0 \le t \le 2$
(B) $2 \le t \le 6$
(C) $0 \le t \le 4$
(D) $0 \le t \le 2 \cup 4 \le t \le 6$
(E) $2 \le t \le 6 \cup 6 \le t \le 10$

29. Find the slope of the tangent line to $3x^2y - 7xy^3 + 3 = y$ at $(2,1)$.

(A) $-\frac{5}{31}$
(B) $\frac{5}{31}$
(C) $\frac{8}{31}$
(D) 0
(E) $-\frac{7}{31}$

Question 30 requires a calculator.

30. How many points of inflection does the graph of $f(x) = 2^{\sin x}$ have on the interval $0 \le x \le 10$?

(A) 1
(B) 2
(C) 3
(D) 4
(E) 5

GO ON TO THE NEXT PAGE

31. The graph of f is shown below. $\int_0^8 f(x)dx =$

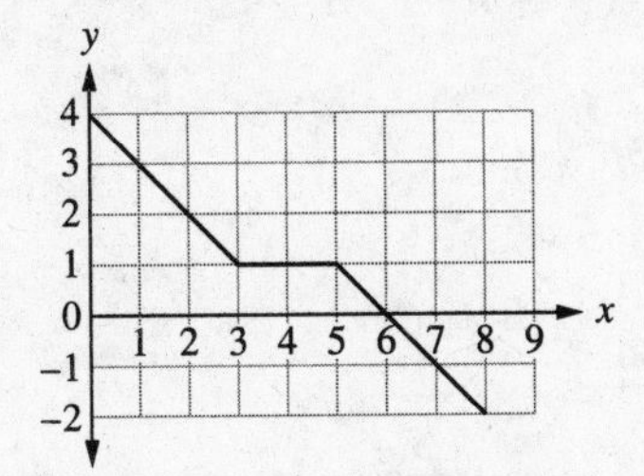

(A) 11

(B) 10

(C) 9.5

(D) 8

(E) 7.5

Question 32 requires a calculator.

32. For $t \geq 0$, the height of a particle is given by $h(t) = 3\sin(4t) - 2t + 1$. What is the average height of the particle on the interval $1 \leq t \leq 4$?

(A) −0.792

(B) −1.531

(C) −3.924

(D) −5.632

(E) −11.772

33. Where on the interval $0 \leq x \leq 2\pi$ does the graph of $f(x) = \cos x - \frac{1}{4}x^2 + 3x$ change from concave up to concave down?

(A) $x = \frac{1}{3}\pi$ only

(B) $x = \frac{4}{3}\pi$ only

(C) $x = \pi$ only

(D) $x = \frac{2}{3}\pi$ and $\frac{4}{3}\pi$

(E) $x = \frac{2}{3}\pi$ only

34. Let $f'(x) = x + 3 + 2\sin x$. How many points of inflection does the graph of f have on the domain $[-3,6]$?

(A) 0

(B) 1

(C) 2

(D) 3

(E) 4

35. The volume of a cube is decreasing at the rate of 750 cm^3/min. When the length of one edge of the cube is 5 cm, how fast is the area of one face of the cube changing?

(A) −130 cm^2/min

(B) −100 cm^2/min

(C) −30 cm^2/min

(D) 40 cm^2/min

(E) 50 cm^2/min

Question 36 requires a calculator.

36. The velocity of a particle is given by $v(t) = 2t - 3\cos(t^2) - 3$, for $t \geq 0$. If the position of the particle at time $t = 2$ is −4, then what is the position of the particle at time $t = 3$?

(A) −7.082

(B) −3.057

(C) −2.724

(D) 1.276

(E) 9.418

GO ON TO THE NEXT PAGE

37. $\lim_{t \to 2} \frac{\ln t - \ln 2}{t-2} =$

(A) 0
(B) $\frac{1}{2}$
(C) ln 2
(D) 2
(E) The limit does not exist.

Question 38 requires a calculator.

38. The derivative of a function *f* is given by $f'(x) = 2^{-x} - 3\cos(x^2)$. On the interval $-3 < x < 3$, at which values of *x* does *f* have a local maximum?

(A) $x \approx -0.937$ and $x \approx 1.194$
(B) $x \approx -1.835$ and $x \approx 1.767$
(C) $x \approx -2.454$ and $x \approx 2.508$
(D) $x \approx 0.441$ and $x \approx 2.508$
(E) $x \approx -0.937$ and $x \approx 2.188$

Question 39 requires a calculator.

39. The base of a solid region is bordered in the first quadrant by the *x*-axis, the *y*-axis, and the line $y = 3 - \frac{1}{2}x$. Cross-sections perpendicular to the *x*-axis are squares. Find the volume of the solid.

(A) 3
(B) 6
(C) 9
(D) 15
(E) 18

40. $\lim_{n \to \infty} \sum_{k=1}^{n} \left(-1 + \frac{4}{n}k\right)^2 \cdot \frac{4}{n} =$

(A) $\frac{26}{3}$
(B) 3
(C) 0
(D) $\frac{28}{3}$
(E) The limit does not exist.

Question 41 requires a calculator.

41. Let $F(x) = \int_{-1}^{x^2} \left(\ln(t+1) - t^2 + \sin t\right) dt$. $F''(1) \approx$

(A) −1.919
(B) −2.098
(C) −2.770
(D) −2.991
(E) −3.256

Question 42 requires a calculator.

42. The initial population of a colony of flies is 12. Ten days later, the population is 60. The population of flies grows at a rate $\frac{dy}{dt} = ky$. Find the value of *k*.

(A) $k \approx 1.609$
(B) $k \approx 1.208$
(C) $k \approx 0.843$
(D) $k \approx 0.179$
(E) $k \approx 0.161$

GO ON TO THE NEXT PAGE

43. $\int_{-1}^{4} f(3x-1)dx$ is equivalent to:

(A) $\frac{1}{3}\int_{-4}^{11} f(u)du$

(B) $3\int_{-4}^{11} f(u)du$

(C) $\int_{-4}^{11} f(u)du$

(D) $\frac{1}{3}\int_{-1}^{4} f(u)du$

(E) $\frac{1}{3}\int_{-1}^{4} f(u-1)du$

44. If f is continuous on the interval $[-2,6]$, then which of the following must be true?

I. $f'(c) = 0$ for some c in $(-2,6)$

II. f has a minimum somewhere in $[-2,6]$

III. $f(c) = 0$ for some c in $(-2,6)$

(A) I only
(B) II only
(C) III only
(D) I and II only
(E) II and III only

45. Let $a < b < c < d$ and

$\int_{a}^{b} f(x)dx = 12, \int_{a}^{c} f(x)dx = -5$ and

$\int_{b}^{d} f(x)dx = -9$. Then $\int_{c}^{d} f(x)dx =$

(A) −2
(B) 2
(C) 7
(D) 8
(E) 26

GO ON TO THE NEXT PAGE

FREE-RESPONSE QUESTIONS

Directions: Solve the following problems, using available space for scratchwork. Show how you arrived at your answer.

You may wish to look over the problems before starting to work on them; on the actual test, it is not expected that everyone will be able to complete all parts of all problems. All problems are given equal weight, but the individual parts of a particular problem are not necessarily given equal weight. You should not spend too much time on any one problem.

- You should write out all your work for each part. On the actual test, you will do this in the space provided in the test booklet. Be sure to write clearly and legibly. If you make a mistake, you can save time by crossing it out rather than trying to erase it. Erased or crossed-out work will not be graded.
- Show all your work. Clearly label any functions, graphs, tables, or other objects that you use. On the actual exam, you will be graded on the correctness and completeness of your methods as well as your answers. Answers without any supporting work may not receive credit.
- Justifications (i.e., the request that you "justify your answer") require that you give mathematical (non-calculator) reasons.
- Work must be expressed in standard mathematical notation, not calculator syntax.
- Unless otherwise specified, answers (numeric or algebraic) need not be simplified.
- If you use decimal approximations in calculations, the readers of the actual exam will grade you on accuracy. Unless otherwise specified, your final answers should be accurate to three places after the decimal point.
- Unless otherwise specified, the domain of function f is the set of all real numbers x for which $f(x)$ is a real number.

GO ON TO THE NEXT PAGE

PART A

Time: 30 Minutes
3 Problems

A GRAPHING CALCULATOR IS ALLOWED ON THIS PORTION OF THE EXAM

1. A particle moves along the x–axis so that its velocity at any time $t > 0$ is given by $v(t) = 5t^2 - 4t + 7$. The position of the particle, $x(t)$, is 8 for $t = 3$.

   (a) Write a polynomial for the position of the particle at any time $t \geq 0$.

   (b) Find the total distance traveled by the particle from time $t = 0$ until time $t = 2$.

   (c) Does the particle achieve a minimum velocity? And if so what is the position of the particle at this time?

2. Let R be the region in the first quadrant bordered by the y-axis and the graphs of $f(x) = 3\ln(x + 1)$ and $g(x) = \dfrac{1}{x+1} + 3$.

   (a) Find the area of R.

   (b) Find the volume if R is rotated around the x-axis.

   (c) The vertical line $x = k$ divides R into two regions of equal area. Set up, but do not solve, an integral equation that finds the value of k.

3. An online retailer has a warehouse that receives packages that are later shipped out to customers. The warehouse is open 18 hours per day. On one particular day, packages are delivered to the warehouse at a rate of $D(t) = 300\sqrt{t}$ packages per hour. Packages are shipped out at a rate of

   $S(t) = 60t + 300\sin\left(\dfrac{\pi}{6}t\right) + 300$ packages per

   hour. For both functions, $0 \leq t \leq 18$, where t is measured in hours. At the beginning of the workday, the warehouse already has 4000 packages waiting to be shipped out.

   (a) What is the rate of change of the number of packages in the warehouse at time $t = 4$?

   (b) To the nearest whole number, evaluate $\int_3^{12} S(t)\,dt$. What is the meaning of $\int_3^{12} S(t)\,dt$?

   (c) To the nearest whole number, how many packages are in the warehouse at the end of the 18-hour day?

   (d) Over the 18-hour day, to the nearest whole number, what is the maximum number of packages in the warehouse and at what time did that occur?

GO ON TO THE NEXT PAGE

PART B

Time: 60 Minutes
3 Problems

NO CALCULATOR IS ALLOWED ON THIS PORTION OF THE EXAM

Note: If you have extra time, you can go back and work on Part A of Section II, but you cannot use a calculator to complete your work at this time.

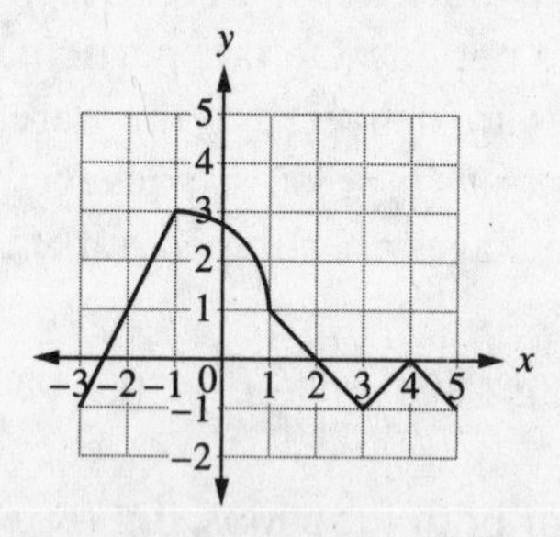

4. The graph above is the graph of $y = f(x)$, which consists of line segments and a quarter-circle centered at $(-1,1)$. Let $g(x) = \int_1^x f(t)dt$.
   (a) Find $g(-1)$ and $g(5)$.
   (b) Find the equation of the tangent line to g at the point where $x = -1$.
   (c) Find all values in $-3 < x < 5$ where g has a horizontal tangent line. Determine if each value is a local maximum, local minimum, or neither. Justify your answer.
   (d) d. Find the x-coordinate of all points of inflection of g. Justify your answer.

5. Let $\frac{dy}{dt} = t(3 - y)$. One solution to this differential equation, $y = f(t)$, passes through the point (4, 2).
   (a) Find the equation of the tangent line to $y = f(t)$ at (4,2) and use it to approximate the value of $f(5)$.
   (b) Find an expression for $y = f(t)$ by solving the differential equation $\frac{dy}{dt} = t(3 - y)$ with the initial condition $f(4) = 2$.
   (c) The claim is made that the graph below is a different solution curve, $y = g(t)$, to $\frac{dy}{dt} = t(3 - y)$. Is this possible? Explain why or why not.

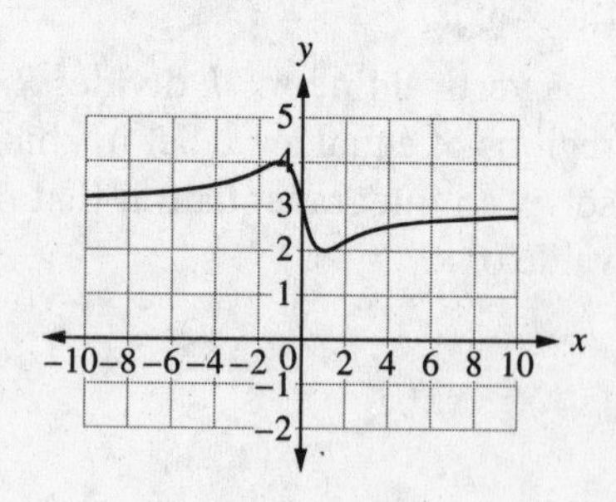

GO ON TO THE NEXT PAGE

6. Suppose f is continuous, decreasing, and concave down and has values shown in the table below.

x	0	3	6	9	12
$f(x)$	17	16	14	11	6

(a) Use the data to estimate $f'(9)$. Show the computations that lead to your answer.

(b) Use the data to evaluate $\int_0^{12} f'(x)dx$. Show the computations that lead to your answer.

(c) Use a right Riemann sum with four subintervals to find the average value of f over the interval $0 \le x \le 12$. Is your estimate an overestimate or underestimate? Explain.

(d) The tangent line to f at $x = 6$ is used to approximate $f(7)$. The secant line from $x = 6$ to $x = 9$ is also used to approximate $f(7)$. Which calculation is an under approximation of $f(7)$ and which is an over approximation of $f(7)$? Justify your answer.

GO ON TO THE NEXT PAGE

GO ON TO THE NEXT PAGE

IF YOU FINISH BEFORE TIME IS CALLED, YOU MAY CHECK YOUR WORK ON THIS SECTION ONLY. DO NOT TURN TO ANY OTHER SECTION IN THE TEST.

AP CALCULUS AB DIAGNOSTIC TEST ANSWER KEY

1. C
2. D
3. A
4. B
5. C
6. D
7. C
8. C
9. C
10. D
11. A
12. B
13. D
14. B
15. D
16. D
17. A
18. A
19. E
20. D
21. E
22. B
23. C
24. C
25. A
26. E
27. A
28. D
29. B
30. D
31. D
32. C
33. B
34. D
35. B
36. C
37. B
38. E
39. E
40. D
41. C
42. E
43. A
44. B
45. D

ANSWERS AND EXPLANATIONS

1. C

This problem is a direct application of the definition of the derivative. Remember: $f'(a) = \lim_{h \to a} \frac{f(h) - f(a)}{h - a}$.

Here $f(x) = e^x$ and $a = \frac{1}{2}$. We know that $f'\left(\frac{1}{2}\right) = \frac{d}{dx}e^x\Big|_{x=\frac{1}{2}} = e^{\frac{1}{2}} = \sqrt{e}$ and $f'\left(\frac{1}{2}\right) = \lim_{h \to 0} \frac{e^h - \sqrt{e}}{h - \frac{1}{2}}$. So the result follows.

2. D

Let $u = \sin x$; $du = \cos x \, dx$. We can rewrite the integral as:

$$\int_0^{\frac{\pi}{2}} \sqrt{\sin x} \cos x \, dx = \int u^{\frac{1}{2}} du = \frac{2}{3}u^{\frac{3}{2}} = \frac{2}{3}(\sin x)^{\frac{3}{2}}\Big|_0^{\frac{\pi}{2}} = \frac{2}{3}(1)^{\frac{3}{2}} - \frac{2}{3}(0)^{\frac{3}{2}} = \frac{2}{3}.$$

3. A

The function f is concave down for all x such that $f''(x) < 0$. By direct calculation, $y' = 3x^2 + 42x - 1$ and $y'' = 6x + 42$. So f is concave down when $6x + 42 < 0$, that is, when $x < -7$.

4. B

To calculate the derivative of a quotient we must use the quotient rule:

$\left[\frac{f}{g}\right]' = \frac{gf' - fg'}{g^2}$. With $f(x) = x^3 + x^5$ and $g(x) = \ln(x^2)$,

we get $F'(x) = \frac{[\ln(x^2)][3x^2 + 5x^4] - [x^3 + x^5][\frac{1}{x^2} \cdot 2x]}{[\ln(x^2)]^2}$.

And since $\ln(e) = 1$, we get, after much reducing, $F'(\sqrt{e}) = e(3e + 1)$.

5. C

The function f is increasing for all x such that $f'(x) > 0$. By direct calculation, $f'(x) = 2x^2 - 2x - 4 = 2(x^2 - x - 2) = 2(x - 2)(x + 1)$. Therefore, $f'(x) > 0$ when $(x - 2)$ and $(x + 1)$ are either both positive or both negative. This occurs when $x < -1$ and $x > 2$.

6. D

In this problem we have to use the chain rule three times. Perhaps it helps to rewrite f as $f(x) = [\cos(\sin(2x))]^2$. Then by applying the chain rule (three times) we get $f'(x) = 2[\cos(\sin(2x))] \times [-\sin(\sin(2x))] \times \cos(2x) \times 2 = -4\cos(\sin(2x))\sin(\sin(2x))\cos(2x)$. Now by plugging in the value $x = \frac{\pi}{8}$ and using the fact that $\sin\frac{\pi}{4} = \cos\frac{\pi}{4} = \frac{\sqrt{2}}{2}$, we get $f'\left(\frac{\pi}{8}\right) = -2\sqrt{2}\cos\left(\frac{\sqrt{2}}{2}\right)\sin\left(\frac{\sqrt{2}}{2}\right)$.

7. C

We can easily calculate each limit and check which graphs are plausible. Choice (C) is correct because as x approaches 2 from the right, $f(x)$ approaches 3. The limit in (A) is undefined because the left and right side limits are not equal. For (B), by examining the graph, the limit of f as x approaches 0 from the left is 1, not 2. For (D), the left hand and right hand limits are not equal so the limit does not exist. For (E), the left hand limit of the function is defined. In fact, we get $\lim_{x \to 2^-} f(x) = 2$.

8. C

Students are expected to know that $\lim_{x \to 0} \frac{\sin x}{x} = 1$. For this problem, the limit can be rewritten:

$\lim_{x \to 0} \frac{\sin^2(4x)}{x^2} = \lim_{x \to 0}\left(\frac{\sin 4x}{x}\right)^2 = \lim_{x \to 0}\left(4 \times \frac{\sin 4x}{4x}\right)^2 = 16 \times \lim_{x \to 0}\left(\frac{\sin 4x}{4x}\right)^2 = 16$. Note that as $x \to 0$, $4x \to 0$, so the limit is still 1.

9. C

Notice that the graph of f has a point of inflection at $x = 10$; i.e., it changes concavity at $x = 10$. Points of inflection can only occur when $f'' = 0$ or is undefined. Since f'' is the rate of change of f', the slope of the tangent line to the graph of f' must be 0 at the point of inflection. We can therefore eliminate all but (C).

10. D

Because the degree of the numerator polynomial is equal to the degree of the denominator polynomial, the limit as x approaches infinity is simply the ratio of the coefficients of the highest-order terms. Note that

$$\lim_{x \to \infty} \frac{x^3 + 3x^5 + x}{2x^5 + 3x^2 + 5} = \lim_{x \to \infty} \frac{\frac{x^3}{x^5} + \frac{3x^5}{x^5} + \frac{x}{x^5}}{\frac{2x^5}{x^5} + \frac{3x^2}{x^5} + \frac{5}{x^5}} = \frac{0 + 3 + 0}{2 + 0 + 0} = \frac{3}{2}.$$

11. A

The limit is indeterminate because if we plug in the value $x = 1$ we get $\frac{0}{0}$. However, by factoring the numerator and the denominator we can reduce the argument and easily plug in the value $x = 1$ to evaluate the limit. See that $\lim_{x \to 1} \frac{2x^2 + x - 3}{1 - x^2} = \lim_{x \to 1} \frac{(2x + 3)(x - 1)}{(1 - x)(1 + x)} = \lim_{x \to 1} \frac{-(2x + 3)}{1 + x} = \frac{-5}{2}$.

12. B

First, cross-multiply and integrate both sides to get $\int y^3 dy = \int x^2 dx$. Thus $\frac{y^4}{4} = \frac{x^3}{3} + C$ and so, solving for y, we get $y = \sqrt[4]{\frac{4}{3}x^3 + C}$. Now using the initial condition we can figure out what C should be. We know that $3 = \sqrt[4]{\frac{4}{3}3^3 + C}$; solving for C we get $C = 45$. Therefore, we can conclude that $y = \sqrt[4]{\frac{4}{3}x^3 + 45}$.

13. D

Differentiate the expression term by term, keeping in mind the product rule and attaching a term dx or dy each time we differentiate x or y, respectively. Then solve for $\frac{dy}{dx}$. We get

$$(2x^3y^2) - (5x^2y) = 18$$

$$(4x^3ydy + 6x^2y^2dx) - (10xydx + 5x^2dy) = 0$$

$$4x^3ydy - 5x^2dy = 10xydx - 6x^2y^2dx$$

$$(4x^3y - 5x^2)dy = (10xy - 6x^2y^2)dx$$

$$\frac{dy}{dx} = \frac{(10xy - 6x^2y^2)}{(4x^3y - 5x^2)}.$$

Now evaluate $\frac{dy}{dx}$ at the point (1,2) to get the slope of the tangent line to the curve. We get

$$\left.\frac{dy}{dx}\right|_{(1,2)} = \frac{[10(1)(2) - 6(2)^2(1)^2]}{[4(2)(1)^3 - 5(1)^2]} = -\frac{4}{3}.$$

14. B

Cross-multiply and integrate to get $\int y\, dy = \int 2x^3 dx$. Thus, $\frac{y^2}{2} = \frac{x^4}{2} + C$. Solving for y, we see that $y = \pm\sqrt{x^4 + C}$. For example, when $C = 0$, we have $y = \pm x^2$. These integral curves are given in (B).

15. D

To calculate the area between the curves $y = 2x^2 + 1$ and $y = 2x + 5$, we must evaluate the integral $\int_a^b (2x + 5) - (2x^2 + 1)dx$. To determine which values to use for a and b as the limits of the integral, we calculate the x values where the two curves intersect. Solve $2x^2 + 1 = 2x + 5$ by factoring to get $x = 2$ and $x = -1$. Set $a = -1$, $b = 2$. The enclosed area, A, is therefore given by the equation

$$A = \int_{-1}^{2} (2x + 5) - (2x^2 + 1)\, dx = \int_{-1}^{2} -2x^2 + 2x + 4\, dx = \left[\frac{-2x^3}{3} + x^2 + 4x\right]_{-1}^{2} = -6 + 15 = 9.$$

16. D

Because $y = 2x^{\frac{3}{2}}$, we get $y' = 3\sqrt{x}$.

So the slope of the tangent at x is $3\sqrt{x}$. $2x + 3y = 6$ implies $y = \frac{-2}{3}x + 2$, so the slope of the line given is $\frac{-2}{3}$. Recall that perpendicular lines have negative reciprocal slopes. Thus we must solve the equation $3\sqrt{x} = \frac{3}{2}$ for x. We get $x = \frac{1}{4}$. Therefore $\left(\frac{1}{4}, \frac{1}{4}\right)$ is the point where the slope of the tangent is perpendicular to the line $2x + 3y = 6$.

17. A

The equation to use is $V = \int_a^b \pi[f(x)]^2\,dx$, where $f(x)$ = radius of a cross section given by the function and a and b are the left and right bounds of the surface of revolution. The $\pi[f(x)]^2$ term measures the area of a cross-sectional disk and the $\int_a^b dx$ term adds them up to give the entire volume of the solid. So,

$$V = \int_1^3 \pi[x^{\frac{3}{2}}]^2\,dx = \int_1^3 \pi x^3\,dx = \left.\frac{\pi x^4}{4}\right|_1^3 = \pi\left[\frac{3^4}{4} - \frac{1}{4}\right] = 20\pi$$

18. A

To get the equation of the tangent line, we need to know the slope of the tangent line and a point on the tangent line. To calculate the slope of the tangent at $x = \frac{\pi}{8}$, find $y'\left(\frac{\pi}{8}\right)$.

Using the chain rule, $y'\left(\frac{\pi}{8}\right) = 2\cos\left(2x + \frac{3\pi}{4}\right) \cdot 2 = 4\cos\left(2x + \frac{3\pi}{4}\right) = 4\cos\left(\frac{\pi}{4} + \frac{3\pi}{4}\right) = -4$. Because the line we want is tangent to the curve $y = 2\sin\left(2x + \frac{3\pi}{4}\right)$, $\left(\frac{\pi}{8}, y\left(\frac{\pi}{8}\right)\right) = \left(\frac{\pi}{8}, 0\right)$ is a point on the tangent line. Using the point–slope form we conclude that the equation of the tangent line is $y - \frac{\pi}{2} = -4x$.

19. E

Set $f(x) = \int_0^x \cos(t^2)dt$. Therefore, $f(x^2) = \int_0^{x^2} \cos(t^2)\,dt$. From the Fundamental Theorem of Calculus we know that $f'(x) = \cos(x^2)$. Now applying the chain rule we see that $\frac{d}{dx}\left(\int_0^{x^2} \cos(t^2)\,dt\right) = \frac{d}{dx} f(x^2) =$

$f'(x^2) \cdot 2x = \cos((x^2)^2) \cdot 2x = 2x\cos(x^4)$.

20. D

This question is a direct application of the intermediate value theorem, which states that if f is a continuous function on the closed interval $[a,b]$ and d is a real number between $f(a)$ and $f(b)$, then there exists a c in $[a,b]$ such that $f(c) = d$. Because $f(0) = 1$, $f(5) = -4$, by the Intermediate Value Theorem there clearly exists a c_1 between 0 and 5 such that $f(c_1) = 0$. Similarly, because $f(0) = 1$, $f(-5) = -3$, there exists a c_2 between 0 and -5 such that $f(c_2) = 0$. Therefore, c_1 and c_2 are at least two zeros of f and by their location, (D) must be the correct answer.

21. E

In I, the function is shifted right 2 units, so the limits of integration are shifted right two units as well. In II, the function is shifted right two units, but the limits of integration are shifted two units left. In III, the function is shifted left three units and the limits of integration are also shifted left three units.

22. B

In order to find the equation of a line, we need the slope and a point the line passes through. $f\left(\frac{1}{2}\right)=-\frac{3}{4}$, so the point is $\left(\frac{1}{2},-\frac{3}{4}\right)$. $f'(x)=2x-4+2e^{2x-1}$, so $m=f''\left(\frac{1}{2}\right)=-1$. $y+\frac{3}{4}=-1\left(x-\frac{1}{2}\right)\Rightarrow y=-x-\frac{1}{4}$.

23. C

Using the Fundamental Theorem of Calculus, $f'(x)=3x^2-5x+1\Rightarrow f'(3)$.

24. C

Instantaneous rate of change of a function at a point is the slope of the tangent line at that point. Therefore, $f'(t)=-3t^2+12t+8\Rightarrow f'(-1)=-23$.

25. A

Average rate of change is simply the slope between two points: $\frac{f(7)-f(-1)}{7-(-1)}=\frac{3-(-5)}{8}=\frac{8}{8}=1$

26. E

The only interval where the average rate of change equals 0 is on [–4,3].

27. A

$\int_{-4}^{7} f(x)\,dx \approx 6\cdot 3+-5\cdot 1+-3\cdot 3+6\cdot 4=28.$

28. D

Sketching the graphs of $v(t)=s'(t)=t^2-8t+12$ and $a(t)=v'(t)=s''(t)=2t-8$ shows that the velocity and acceleration have opposite signs on the intervals $0\le t\le 2\cup 4\le t\le 6$.

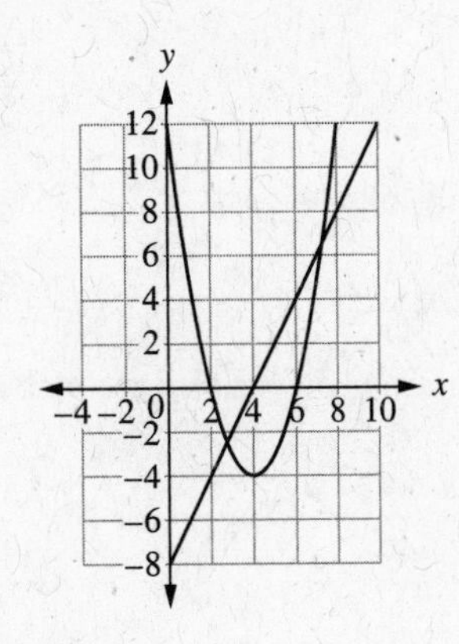

29. B

Using implicit differentiation, $3x^2 \cdot y' + 6xy - 21xy^2 y' - 7y^3 = y' \Rightarrow y' = \dfrac{6xy - 7y^3}{-3x^2 + 21xy^2 + 1}$. so

$$m = \left.\frac{6xy - 7y^3}{-3x^2 + 21xy^2 + 1}\right|_{(x,y)=(2,1)} = \frac{5}{31}.$$

30. D

Graphing $f'(x) = 2^{\sin x} \cdot \ln 2 \cdot \cos x$ on $0 \le x \le 10$ shows 4 direction changes (relative extrema). Therefore f has 4 points of inflection.

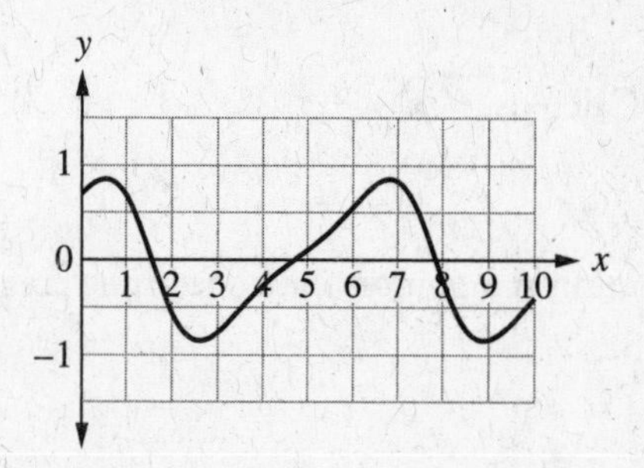

31. D

Adding the signed areas of the various regions defined by the function and the x-axis,
$\int_0^8 f(x)\,dx = \frac{1}{2}(4+1)\cdot 3 + 2 + \frac{1}{2} - 2 = 8.$

32. C

The average value of a function f on an interval $[a,b]$, is defined by: $\frac{1}{b-a}\int_a^b f(x)\,dx$.

Therefore, $\frac{1}{3}\int_1^4 h(t)\,dt \approx -3.924$.

33. B

$f(x) = \cos x - \frac{1}{4}x^2 + 3x \Rightarrow f'(x) = -\sin x - \frac{1}{2}x + 3 \Rightarrow f''(x) = -\cos x - \frac{1}{2}$. $f''(x) = 0$ at $x = \frac{2}{3}\pi$ and $x = \frac{4}{3}\pi$,

but f'' changes from positive to negative only at $x = \frac{4}{3}\pi$.

34. D

Looking at the graph of $y = f'(x)$ on the domain $[-3,6]$, there are 3 changes in direction. Therefore the graph of f has 3 points of inflection on $[-3,6]$.

35. B

Let $V = s^3$. Then $\dfrac{dV}{dt} = 3s^2\dfrac{ds}{dt}$. When $s = 5$, then $-750 = 3\cdot 5^2\cdot\dfrac{ds}{dt}$, so $\dfrac{ds}{dt} = -10$ cm/min.

The area of one face of the cube is $A = s^2$. So $\dfrac{dA}{dt} = 2s\dfrac{ds}{dt} = 2\cdot 5\cdot -10 = -100\ \text{cm}^2/\text{min}$.

36. C

We calculate the position to be the initial position (–4) plus the change from the initial position. Therefore, $s(3) = -4 + \int_2^3 v(t)\,dt \approx -2.724$.

37. B

This problem is the definition of the derivative of $f(t) = \ln t$ at $t = 2$.

Therefore, $\lim\limits_{t\to 2}\dfrac{\ln t - \ln 2}{t-2} = \dfrac{d}{dt}(\ln t)\Big|_{t=2} = \dfrac{1}{t}\Big|_{t=2} = \dfrac{1}{2}$.

38. E

Looking at the graph of f', the only places where f' crosses the x-axis and changes from positive to negative are at $x \approx -0.937$ and $x \approx 2.188$.

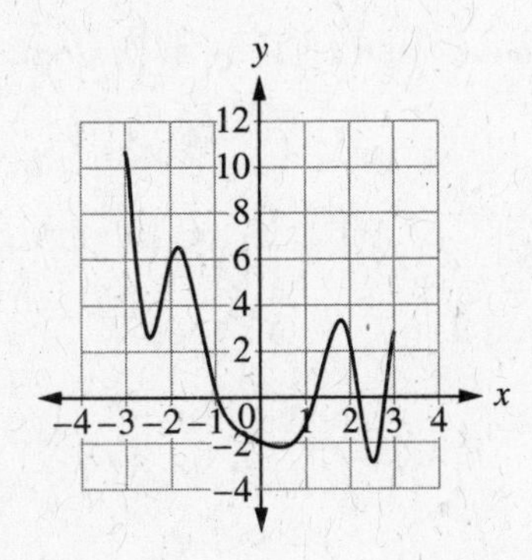

39. E

$\int_0^6 \left(3-\tfrac{1}{2}x\right)^2 dx = 18$.

40. D

$-1+\dfrac{4}{n}k$ represents the values of x. $\dfrac{4}{n}$ represents Δx, the width of each rectangle. The limits of integration begin at -1 and once k reaches the value of n, $\dfrac{4}{n}k = \dfrac{4}{n}n = 4$. So the upper limit of integration is 4 units to the right of -1.

Therefore, $\lim\limits_{n\to\infty}\sum\limits_{k=1}^{n}\left(-1+\dfrac{4}{n}k\right)^2\cdot\dfrac{4}{n} = \int_{-1}^{3} x^2\,dx = \dfrac{28}{3}$.

41. C

Using the Fundamental Theorem of Calculus along with the Chain Rule,

$$F(x)=\int_{-1}^{x^2}\left(\ln(t+1)-t^2+\sin t\right)dt \Rightarrow F'(x)=\left(\ln\left(x^2+1\right)-x^4+\sin\left(x^2\right)\right)\cdot 2x.$$

$$\frac{d}{dx}\left(F'(x)\right)\bigg|_{x=1} \approx -2.770.$$

42. E

$$\frac{dy}{dt}=ky \Rightarrow y=Ae^{kt}.$$

$$A=12, t=10, y=60 \Rightarrow 60=12e^{k\cdot 10} \Rightarrow 5=e^{10k} \Rightarrow k=\frac{\ln 5}{10}\approx 0.161.$$

43. A

Let $u = 3x - 1 \Rightarrow du = 3dx \Rightarrow dx = \frac{1}{3}du$. Limits of integration become $u = 3(-1) - 1 = -4$ and $u = 3(4) - 1 = 11$.
So $\int_{-1}^{4} f(3x-1)\,dx = \frac{1}{3}\int_{-4}^{11} f(u)\,du.$

44. B

Of the three choices, the only one known for certain is II by the Extreme Value Theorem.

45. D

$$\int_c^d f(x)\,dx=\int_c^a f(x)\,dx+\int_a^b f(x)\,dx+\int_b^d f(x)\,dx=-(-5)+12+(-9)=8.$$

FREE–RESPONSE

1. (a) The position of the particle is given by

$x(t) = \int v(t)dt = \int 5t^2 - 4t + 7dt = \frac{5}{3}t^3 - 2t^2 + 7t + C$. Solve for C by substituting $x(3) = 8$: $8 = \frac{5}{3}(27) - 2(9) + 21 + C = 45 - 18 + 21 + C = 48 + C$. Therefore, $C = -40$ and the position of the particle can be written

$$x(t) = \frac{5}{3}t^3 - 2t^2 + 7t - 40.$$

(b) Total distance traveled is given by $\int_0^2 |v(t)|\,dt$.

$v(t)$ is positive everywhere, so this is just

$$x(2) - x(0) = \left(\frac{5}{3}2^3 - 2\cdot 2^2 + 7\cdot 2 - 40\right) - \left(\frac{5}{3}0^3 - 2\cdot 0^2 + 7\cdot 0 - 40\right) = \frac{40}{3} - 8 + 14 = \frac{58}{3}.$$

(c) We find the extrema by setting $v'(t) = 0$. $v'(t) = 10t - 4 = 0$ when $t = \frac{2}{5}$. Note that $v''(t) = 10 > 0$, so this value is a minimum.

The position is, then, $x\left(\frac{2}{5}\right) = \frac{5}{3}\left(\frac{8}{125}\right) - 2\left(\frac{4}{25}\right) + \frac{14}{5} - 40 = -37\frac{31}{75}$.

Note: Any question that asks you to compute total distance is asking for $\int |v(s)|\,ds$, where v is the velocity. Our problem is easier, since $|v(s)| = v(s)$ for our function. For part (c), it isn't enough to just compute the zero of v'. You also need to explain, however briefly, why v actually attains a minimum (and doesn't just continue to decrease to $-\infty$). In the solution, this is because we stated that v is positive everywhere, and so can't go to $-\infty$.

2. (a) Solving $f(x) = g(x)$, we get $x \approx 2.034$.

$$A = \int_0^{2.034} (g(x) - f(x))\,dx \approx 3.212.$$

(b) Volume of a representative slice $= \pi\left((g(x))^2 - (f(x))^2\right)\cdot \Delta x$.

$$V = \int_0^{2.034} \pi\left((g(x))^2 - (f(x))^2\right)dx \approx 50.265.$$

(c) $\int_0^k (g(x) - f(x))\,dx = \frac{1}{2}\cdot 3.212$ OR $\int_0^k (g(x) - f(x))\,dx = \int_k^{2.034} (g(x) - f(x))\,dx$.

3. (a) The units should be "packages per hour." Both $S(t)$ and $D(t)$ are measured in packages per hour. Therefore, $D(4) - S(4) \approx -199.808$ packages per hour.

(b) $\int_3^{12} S(t)\,dt \approx 6177.042.$

Between 3 and 12 hours after the warehouse opened, approximately 6177 packages were shipped out.

(c) The number of packages in the warehouse at the end of the day is the initial amount (4000) plus the total change in the number of packages: $4000 + \int_0^{18} (D(t) - S(t))\,dt \approx 3008$

(d) Let $p(x) = 4000 + \int_0^x (D(t) - S(t))\,dt.$

$p'(x) = D(x) - S(x) = 0 \Rightarrow x \approx 5.526$ or $x \approx 12.111.$

x	$p(x)$
0	4000
5.526	2895.783
12.111	4394.908
18	3007.591

The maximum number of packages is approximately 4395 at time $t \approx 12.111$ hours.

4. (a) $g(-1)$ represents the area between the function and the x-axis between $x = 1$ and $x = -1$.

Therefore, $g(-1) = -\left(2 + \frac{1}{4}\pi \cdot 2^2\right) = -2 - \pi.$

Similarly, $g(5) = \frac{1}{2} - 1 - \frac{1}{2} = -1.$

(b) $g'(x) = f'(x) \Rightarrow m = g'(-1) = f(-1) = 3.$
$y - (-2 - \pi) = 3(x + 1) \Rightarrow y + 2 + \pi = 3x + 3 \Rightarrow y = 3x + 1 - \pi.$

(c) $g'(x) = f(x) = 0 \Rightarrow x = -2.5,\ x = 2,\ x = 4.$

A local minimum occurs when $x = -2.5$ because g' changes from negative to positive.

A local maximum occurs when $x = 2$ because g' changes from positive to negative.

Neither a local maximum nor a local minimum occurs when $x = 4$ because g' does not change sign.

(d) Inflection points occur when $x = -1$ and $x = 4$ because g' changes from increasing to decreasing (relative maxima). An inflection point also occurs when $x = 3$ because g' changes from decreasing to increasing (relative minima).

5. (a) $m = 4(3 - 2) = 4 \Rightarrow y - 2 = 4(t - 4) \Rightarrow y = 4t - 14$
$f(5) \approx 4.5 - 14 = 6$

(b) Solve by separating the variables. Be careful with the constant C.

$$\frac{dy}{dt} = t(3-y) \Rightarrow \frac{dy}{3-y} = t \cdot dt$$

$$\int \frac{1}{3-y}\,dy = \int t\,dt$$

$$-\ln|3-y| = \tfrac{1}{2}t^2 + C$$

$$-\ln 1 = \tfrac{1}{2}\cdot 4^2 + C$$

$$-8 = C$$

$$-\ln|3-y| = \tfrac{1}{2}t^2 - 8$$

$$\ln|3-y| = -\tfrac{1}{2}t^2 + 8$$

$$3 - y = e^{-\frac{1}{2}t^2+8}$$

$$y = 3 - e^{-\frac{1}{2}t^2+8}$$

(c) This graph cannot be the solution curve $y = g(t)$ because $\frac{dy}{dt} = 0$ when $t = 0$. This graph clearly does not have zero slope when $t = 0$.

6. (a) $f'(9) \approx \frac{6-14}{12-6} = -\frac{8}{6} = -\frac{4}{3}$ (Be sure to show the difference quotient.)

OR $f'(9) \approx \frac{6-11}{12-9} = -\frac{5}{3}$.

OR $f'(9) \approx \frac{11-14}{9-6} = -\frac{3}{3} = -1$.

(b) Using the evaluation part of the Fundamental Theorem of Calculus,

$$\int_0^{12} f'(x)\,dx = f(x)\Big|_0^{12} = f(12) - f(0) = 6 - 17 = -11.$$

(c) $\frac{1}{12-0}\int_0^{12} f(x)\,dx \approx \frac{1}{12}[16\cdot 3 + 14\cdot 3 + 11\cdot 3 + 6\cdot 3] = \frac{141}{12} = \frac{47}{4}$.

For a decreasing function, a right Riemann sum is an underestimate of the integral. Therefore, $\frac{141}{12}$ underestimates the average value of f.

(d) Because f is concave down, the tangent line approximation of $f(7)$ will be an over approximation, and the secant line approximation of $f(7)$ will be an under approximation.

AP CALCULUS AB DIAGNOSTIC TEST CORRELATION CHART

Use the results of your test to determine which topics you should spend the most time reviewing.

Multiple Choice Question	Review Chapter: Subsection of Chapter
1	Chapter 12: Exponential and Logarithmic Functions
2	Chapter 21: When F is Complicated: The Substitution Game
3	Chapter 15: Concavity and the Sign of f''
4	Chapter 12: Rules for Computing Derivatives
5	Chapter 14: Increasing and Decreasing Behavior of f and the Sign of f'
6	Chapter 12: Rules for Computing Derivatives
7	Chapter 7: Basic Definition and Understanding Limits Graphically
8	Chapter 7: Some Important Limits
9	Chapter 16: Graphing Using Derivatives
10	Chapter 7: Evaluating Limits Algebraically
11	Chapter 7: Evaluating Limits Algebraically
12	Chapter 16: Differential Equations
13	Chapter 12: Implicit Differentiation
14	Chapter 16: Differential Equations
15	Chapter 18: Area Between Two Curves
16	Chapter 13: The Slope of a Curve at a Point
17	Chapter 18: Volumes of Solids
18	Chapter 13: The Slope of a Curve at a Point
19	Chapter 20: What's So Fundamental? The Connection Between Integrals and Derivatives
20	Chapter 9: Properties of Continuous Functions
21	Chapter 17: Properties of the Definite Integral
22	Chapter 13: The Slope of a Curve at a Point
23	Chapter 20: What's So Fundamental? The Connection Between Integrals and Derivatives
24	Chapter 13: The Slope of a Curve at a Point
25	Chapter 11: A Geometric Interpretation of the Derivative: Slope
26	Chapter 14: The Mean Value Theorem
27	Chapter 23: Approximating the Area Under the Curve Using Riemann Sums
28	Chapter 16: Interpreting the Derivative As a Rate of Change
29	Chapter 12: Implicit Differentiation
30	Chapter 15: Inflection Points
31	Chapter 17: Finding the Area Under a Curve
32	Chapter 18: Average Value of a Function
33	Chapter 15: Concavity and the Sign of f'' Chapter 15: Inflection Points
34	Chapter 14: Corresponding Characteristics of the Graphs of f and f' Chapter 15: Inflection Points
35	Chapter 16: Related Rates

Multiple Choice Question	Review Chapter: Subsection of Chapter
36	Chapter 18: Distance Traveled by a Particle Along a Line
37	Chapter 11: The Concept of the Derivative
38	Chapter 14: Critical Points and Local Extrema
39	Chapter 18: Volumes of Solids
40	Chapter 17: Introducing the Definite Integral
41	Chapter 20: What's So Fundamental? The Connection Between Integrals and Derivatives
42	Chapter 22: Separable Differential Equations
43	Chapter 21: When *F* Is Complicated: The Substitution Game
44	Chapter 9: Properties of Continuous Functions
45	Chapter 17: Properties of the Definite Integral

Free Response Question	Review Chapter: Subsection of Chapter
1(a)	Chapter 22: Accumulation Functions and Antiderivatives
1(b)	Chapter 18: Distance Traveled by a Particle
1(c)	Chapter 16: Optimization
2(a)	Chapter 18: Area Bounded by Curves
2(b)	Chapter 18: Solids of Revolution
2(c)	Chapter 17: Properties of the Definite Integral
3(a)	Chapter 18: Accumulated Change Chapter 22: Accumulation Functions and Antiderivatives
3(b)	Chapter 22: Accumulation Functions and Antiderivatives
3(c)	Chapter 22: Accumulation Functions and Antiderivatives
3(d)	Chapter 14: Critical Points and Local Extrema
4(a)	Chapter 17: Introducing the Definite Integral
4(b)	Chapter 23: Approximating the Area Under Functions Given Graphically or Numerically
4(c)	Chapter 14: Critical Points and Local Extrema
4(d)	Chapter 15: Inflection Points
5(a)	Chapter 13: The Slope of a Curve at a Point Chapter 16: Differential Equations
5(b)	Chapter 22: Separable Differential Equations
5(c)	Chapter 16: Differential Equations Chapter 22: Differential Equations
6(a)	Chapter 13: Approximating Derivatives Using Tables and Graphs
6(b)	Chapter 20: What's So Fundamental? The Connection Between Integrals and Derivatives
6(c)	Chapter 23: Approximating the Area Under the Curve Using Riemann Sums
6(d)	Chapter 11: A Geometric Interpretation: The Derivative As a Slope Chapter 13: Approximating Derivatives Using Tables and Graphs

AP Calculus BC Diagnostic Test Answer Grid

To compute your score for this diagnostic test, calculate the number of questions you got right, then divide by 45 to get the percentage of questions that you answered correctly.

The approximate score range is as follows:

5 = 80–100% (extremely well qualified)

4 = 60–79% (well qualified)

3 = 50–59% (qualified)

2 = 40–49% (possibly qualified)

1 = 0–39% (no recommendation)

A score of 49% is a 2, so you can definitely do better. If your score is low, keep on studying to improve your chances of getting credit for the AP Calculus AB exam.

1. Ⓐ Ⓑ Ⓒ Ⓓ Ⓔ
2. Ⓐ Ⓑ Ⓒ Ⓓ Ⓔ
3. Ⓐ Ⓑ Ⓒ Ⓓ Ⓔ
4. Ⓐ Ⓑ Ⓒ Ⓓ Ⓔ
5. Ⓐ Ⓑ Ⓒ Ⓓ Ⓔ
6. Ⓐ Ⓑ Ⓒ Ⓓ Ⓔ
7. Ⓐ Ⓑ Ⓒ Ⓓ Ⓔ
8. Ⓐ Ⓑ Ⓒ Ⓓ Ⓔ
9. Ⓐ Ⓑ Ⓒ Ⓓ Ⓔ
10. Ⓐ Ⓑ Ⓒ Ⓓ Ⓔ
11. Ⓐ Ⓑ Ⓒ Ⓓ Ⓔ
12. Ⓐ Ⓑ Ⓒ Ⓓ Ⓔ
13. Ⓐ Ⓑ Ⓒ Ⓓ Ⓔ
14. Ⓐ Ⓑ Ⓒ Ⓓ Ⓔ
15. Ⓐ Ⓑ Ⓒ Ⓓ Ⓔ
16. Ⓐ Ⓑ Ⓒ Ⓓ Ⓔ
17. Ⓐ Ⓑ Ⓒ Ⓓ Ⓔ
18. Ⓐ Ⓑ Ⓒ Ⓓ Ⓔ
19. Ⓐ Ⓑ Ⓒ Ⓓ Ⓔ
20. Ⓐ Ⓑ Ⓒ Ⓓ Ⓔ
21. Ⓐ Ⓑ Ⓒ Ⓓ Ⓔ
22. Ⓐ Ⓑ Ⓒ Ⓓ Ⓔ
23. Ⓐ Ⓑ Ⓒ Ⓓ Ⓔ
24. Ⓐ Ⓑ Ⓒ Ⓓ Ⓔ
25. Ⓐ Ⓑ Ⓒ Ⓓ Ⓔ
26. Ⓐ Ⓑ Ⓒ Ⓓ Ⓔ
27. Ⓐ Ⓑ Ⓒ Ⓓ Ⓔ
28. Ⓐ Ⓑ Ⓒ Ⓓ Ⓔ
29. Ⓐ Ⓑ Ⓒ Ⓓ Ⓔ
30. Ⓐ Ⓑ Ⓒ Ⓓ Ⓔ
31. Ⓐ Ⓑ Ⓒ Ⓓ Ⓔ
32. Ⓐ Ⓑ Ⓒ Ⓓ Ⓔ
33. Ⓐ Ⓑ Ⓒ Ⓓ Ⓔ
34. Ⓐ Ⓑ Ⓒ Ⓓ Ⓔ
35. Ⓐ Ⓑ Ⓒ Ⓓ Ⓔ
36. Ⓐ Ⓑ Ⓒ Ⓓ Ⓔ
37. Ⓐ Ⓑ Ⓒ Ⓓ Ⓔ
38. Ⓐ Ⓑ Ⓒ Ⓓ Ⓔ
39. Ⓐ Ⓑ Ⓒ Ⓓ Ⓔ
40. Ⓐ Ⓑ Ⓒ Ⓓ Ⓔ
41. Ⓐ Ⓑ Ⓒ Ⓓ Ⓔ
42. Ⓐ Ⓑ Ⓒ Ⓓ Ⓔ
43. Ⓐ Ⓑ Ⓒ Ⓓ Ⓔ
44. Ⓐ Ⓑ Ⓒ Ⓓ Ⓔ
45. Ⓐ Ⓑ Ⓒ Ⓓ Ⓔ

AP CALCULUS BC DIAGNOSTIC TEST

Directions: Solve the following problems, using available space for scratchwork. Do not use a calculator unless a question instructs you to do so. After examining the form of the choices, decide which one is the best of the choices given and fill in the corresponding oval on the answer sheet. No credit will be given for anything written in the test book. Do not spend too much time on any one problem.

Note: For the actual test, no scrap paper is provided.

In this test:

(1) The domain of a function f is the set of all real numbers x for which $f(x)$ is a real number, unless otherwise specified.

(2) The inverse of a trigonometric function f may be indicated using the inverse function notation f^{-1} or with the prefix "arc" (e.g., $\sin^{-1}x = \arcsin x$).

1. Evaluate $\lim_{x\to 0} \dfrac{\tan^{-1} 2x}{x}$

(A) 0
(B) 1
(C) 2
(D) 3
(E) The limit does not exist.

2. Let $f(x) = 3x^2 + 1$ and $g(x) = x^3 - 9x$. For how many values of a is the line tangent to f at $x = a$ parallel to the line tangent to g at $x = a$?

(A) 0
(B) 1
(C) 2
(D) 3
(E) Not enough information to determine

3. Which of the following statements is true about the series below?

$$\sum_{n=1}^{\infty} \frac{(\sin n)(n+1)}{(2n+3)^4}$$

(A) The series converges by the alternating series test.
(B) The series converges absolutely because the series of absolute values is a p–series with $p = 4$.
(C) We can conclude by the comparison test—comparing the series of absolute values with the series $\sum_{n=1}^{\infty} \frac{1}{n^3}$—that the given series converges absolutely.
(D) We can conclude by the limit comparison test—comparing the series of absolute values with the series $\sum_{n=1}^{\infty} \frac{1}{n^3}$—that the series converges absolutely.
(E) We can conclude by the limit comparison test—comparing the series of absolute values with the series $\sum_{n=1}^{\infty} \frac{1}{n^4}$—that the given series does *not* converge absolutely.

GO ON TO THE NEXT PAGE

4. At time t the position of a particle on the xy plane is given by $x(t) = e^t$, $y(t) = e^t \sin t$. Which of the following integrals gives the distance traveled by the particle from time $t = 0$ to time $t = 1$?

(A) $\int_0^1 \sqrt{e^t(1+\sin t+\cos t)}\, dt$

(B) $\sqrt{2}\int_0^1 e^t \sqrt{1+\sin t\cos t}\, dt$

(C) $\int_0^1 e^t \sqrt{(1+\sin t+\cos t)}\, dt$

(D) $\int_0^1 e^{\frac{t}{2}} \sqrt{(1+\sin t+\cos t)}\, dt$

(E) $\sqrt{2}\int_0^1 e^t\, dt$

5. The figure below shows the portions of the graphs of the ray $\theta = \frac{\pi}{4}$ and the curve $r = 2\cos\theta + \sin\theta$ that lie in the first quadrant. Which of the following integrals expresses the area of the shaded region R?

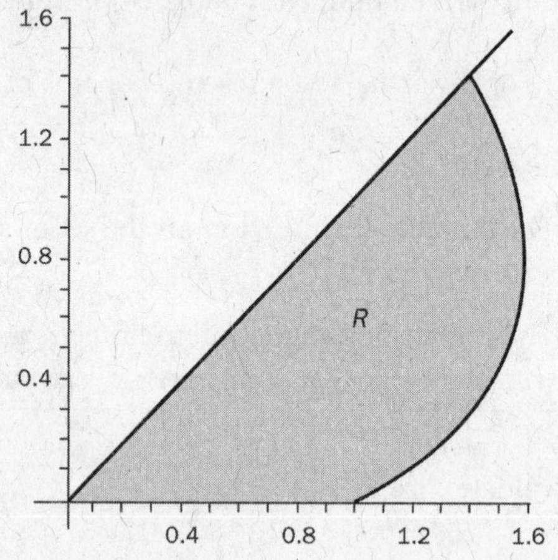

(A) $\int_0^{1.6} \left(\frac{\pi}{4} - r\right) d\theta$

(B) $\int_0^{\frac{\pi}{4}} r\, d\theta$

(C) $\int_0^{\frac{\pi}{4}} r^2\, d\theta$

(D) $\int_0^{\frac{\pi}{4}} \frac{r^2}{2}\, d\theta$

(E) $\int_0^{1.6} \left(\frac{\pi}{4} - \frac{r^2}{2}\right) d\theta$

GO ON TO THE NEXT PAGE

6. Evaluate $\int_1^e (\ln x)^2\,dx$ (hint: use integration by parts twice).

   (A) e
   (B) $e-2$
   (C) $\frac{1}{3}(e^3-1)$
   (D) 1
   (E) The integral diverges.

Question 7 requires a calculator.

7. Let $f(x) = x^3 - 3x^2 + 3x + 7$. Which of the following statements is true?

   (A) f has a relative extremum at $x = 1$ and no inflection points.
   (B) f is everywhere increasing and does not change concavity.
   (C) f has no relative extrema but has an inflection point at $x = 1$.
   (D) f has a relative maximum and an inflection point at $x = 1$.
   (E) f has a relative minimum and an inflection point at $x = 1$.

8. If $f(x) = e^{\sin(3x)}$ then $f'(0) =$

   (A) 1
   (B) -1
   (C) 3
   (D) e
   (E) e^3

9. A population is modeled by the logistic differential equation $\frac{dP}{dt} = \frac{P}{3}\left(1-\frac{P}{8}\right)$. If $P(0) = 2$, for what value of P is the population growing the fastest?

   (A) 8
   (B) 3
   (C) 4
   (D) 0
   (E) 24

10. Evaluate the improper integral $\int_1^{+\infty} e^{-3x}\,dx$.

   (A) $\frac{1}{e^3}$
   (B) The integral diverges to ∞.
   (C) $\frac{1}{3e^3}$
   (D) The integral diverges to $-\infty$.
   (E) $-\frac{1}{3}\left(1-\frac{1}{e^3}\right)$

11. Consider the curve given parametrically by $x = \sin\left(\frac{\pi}{2}t\right)$, $y = 2t^3$. The equation of the line tangent to the curve at the point where $y = \frac{1}{4}$ is

   (A) $y-\frac{1}{4}=\frac{3}{2}x$
   (B) $y-\frac{1}{4}=\frac{3}{2}\left(x-\frac{\sqrt{2}}{2}\right)$
   (C) $y-\frac{1}{4}=\frac{3\sqrt{2}}{\pi}x$
   (D) $y-\frac{1}{4}=\frac{3\sqrt{2}}{\pi}\left(x-\frac{\sqrt{2}}{2}\right)$
   (E) Not enough information to determine

GO ON TO THE NEXT PAGE

Question 12 requires a calculator.

12. The region S in the first quadrant is bounded by the curves $y = 1$, $y = \sin x$, and the y–axis. If S is revolved around the x–axis, the volume of the resulting solid of revolution is

(A) $\frac{\pi^2}{2} - \pi \approx 1.793$

(B) $\frac{\pi}{4} \approx 0.785$

(C) $\frac{\pi^2}{4} \approx 2.467$

(D) $\frac{3\pi^2}{4} - 2\pi \approx 1.119$

(E) $\frac{3\pi}{4} - 2 \approx 0.356$

13. Evaluate $\int \frac{7}{(x-3)(x+4)}\,dx$.

(A) $\ln\left|\frac{x-3}{x+4}\right| + C$

(B) $\ln|(x-3)(x+4)| + C$

(C) $\ln\left|\frac{x+4}{x-3}\right| + C$

(D) $7\ln\left|\frac{x-3}{x+4}\right| + C$

(E) $7\ln\left|\frac{x+4}{x-3}\right| + C$

14. The graph of f'' is shown below. Which of the following could be the graph of f?

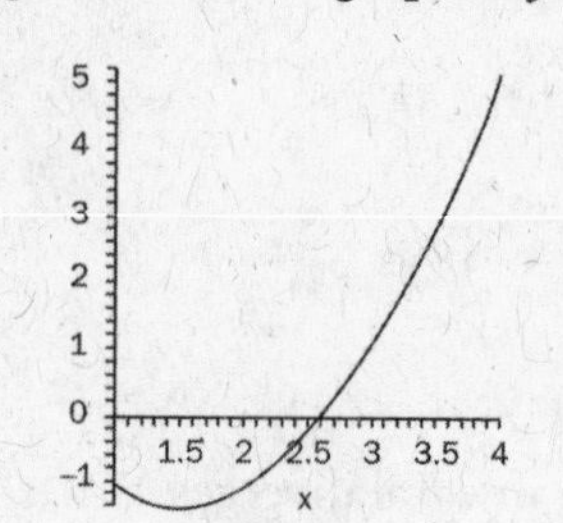

(A)

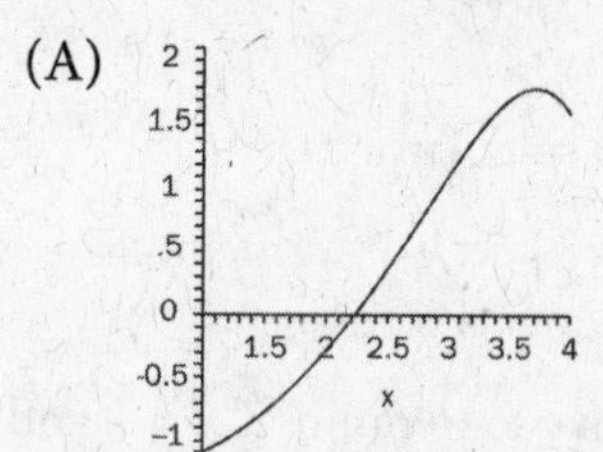

(B)

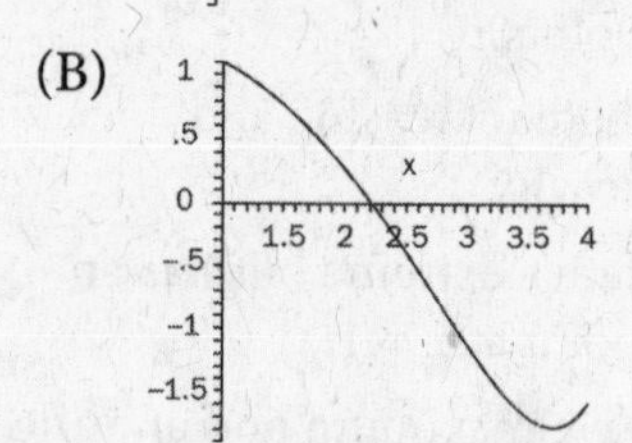

(C)

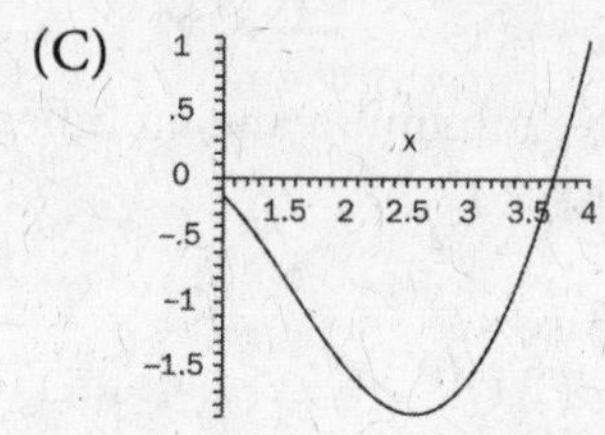

(D)

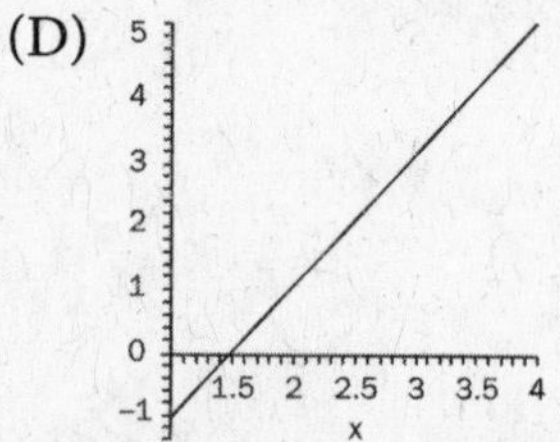

(E)

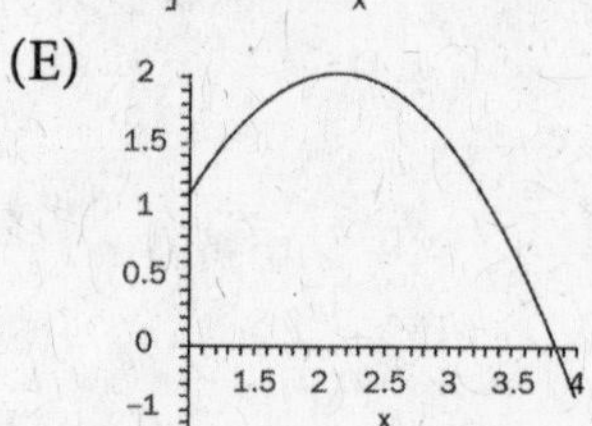

GO ON TO THE NEXT PAGE

15. Suppose $f(x)$ is continuous, but not differentiable at the point $x = 2$ and $f(2) = 3$. Which of the following statements must be true about f?

I. $\lim_{x\to 2^+} f(x) = \lim_{x\to 2^-} f(x)$

II. $\lim_{x\to 2^+} f(x) = 3$

III. $\lim_{x\to 2} \frac{f(x)-3}{x-2}$ does not exist.

(A) I only
(B) I and II only
(C) III only
(D) II and III only
(E) I, II, and III

16. Consider the curve defined by the equation $2x - \cos y = y$. What is $\frac{d^2y}{dx^2}$ at the point $\left(\frac{1}{2}, 0\right)$?

(A) 0
(B) 1
(C) 2
(D) 3
(E) 4

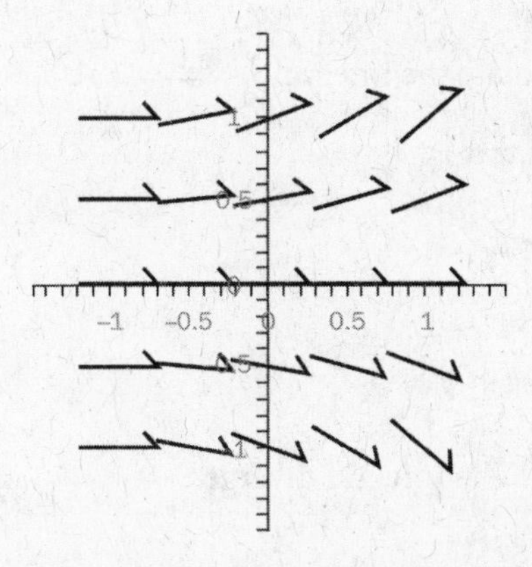

17. The slope field for the differential equation $\frac{dy}{dx} = \frac{y}{f(x)}$ is shown above. Which of the following could be $f(x)$?

(A) $x - 2$
(B) $(x - 2)^2$
(C) x
(D) $-x^2$
(E) $\sqrt{x}$

18. The function $g(x) = \int_1^{x^2} e^t \tan^{-1} t\, dt$. What is $\frac{d}{dx}(g(x))$ evaluated at $x = 1$?

(A) 0
(B) 1
(C) $-\frac{\pi}{4e}$
(D) $\frac{\pi \cdot e}{4}$
(E) $\frac{\pi \cdot e}{2}$

GO ON TO THE NEXT PAGE

19. Evaluate the integral $\int_1^e \frac{(\ln x^2)^2}{3x}\,dx$.

(A) $\frac{8}{3}$

(B) $\frac{4}{9}$

(C) 4

(D) $\frac{4}{3e}$

(E) The integral diverges.

20. Suppose $f(x)$ and $g(x)$ are continuous functions, and for some c in the interval (0, 1), $f(c) = g(c)$, but $f'(c) \neq g'(c)$. Let h be the function defined by

$$h(x)=\begin{cases} f(x) & \text{for } 0\le x\le c \\ g(x) & \text{for } c< x\le 1 \end{cases}$$

Then the average value of h on the closed interval [0, 1] is

(A) $\int_0^1 \frac{f(x)+g(x)}{2}$

(B) $\int_0^c f(x)\,dx + \int_c^1 g(x)\,dx$

(C) $\frac{1}{c}\int_0^c f(x)\,dx + \frac{1}{1-c}\int_c^1 g(x)\,dx$

(D) $\frac{1}{2}\left(\frac{1}{c}\int_0^c f(x)\,dx + \frac{1}{1-c}\int_c^1 g(x)\,dx\right)$

(E) The average value cannot be computed because h is not differentiable at c.

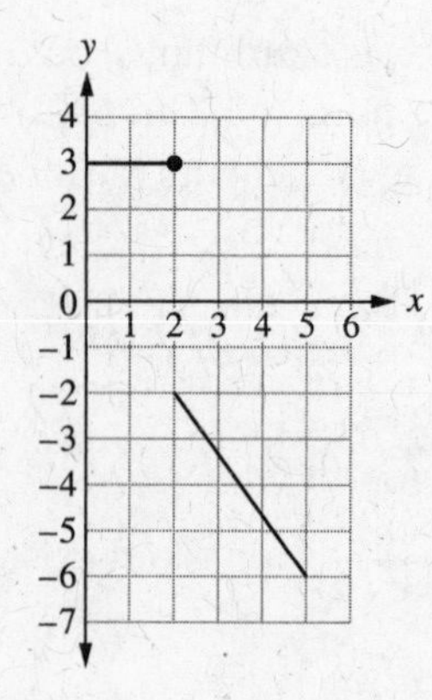

21. The graph of f is shown above for $0 \le x \le 5$. What is the value of $\int_0^5 f(x)\,dx$?

(A) −6

(B) 0

(C) 6

(D) 12

(E) 18

22. The Maclaurin series for the function f is given by $f(x)=\sum_0^\infty \left(\frac{x}{3}\right)^n$. What is the value of $f(-2)$?

(A) −3

(B) $-\frac{3}{5}$

(C) $-\frac{2}{3}$

(D) $\frac{3}{5}$

(E) $\frac{5}{3}$

GO ON TO THE NEXT PAGE

23. $\int x\cos(2x)\,dx =$

(A) $\frac{1}{2}x\sin(2x)-\frac{1}{4}\cos(2x)+C$

(B) $\frac{1}{2}x\sin(2x)+\frac{1}{4}\cos(2x)+C$

(C) $\frac{1}{2}x\sin(2x)-\frac{1}{2}\cos(2x)+C$

(D) $\frac{1}{2}x\sin(2x)+\frac{1}{2}\cos(2x)+C$

(E) $2x\sin(2x)-4\cos(2x)+C$

Question 24 requires a calculator.

24. Water is being chlorinated at a rate $w(t)$ where $w(t)$ is measured in gallons per minute and t is measured in minutes. At time $t = 7$ minutes, 80 gallons of water had been treated. Selected values for $w(t)$ are shown in the table below. Using a left Riemann sum with the three intervals indicated by the table, approximate the number gallons that have been treated by time $t = 19$ minutes.

t(minutes)	7	10	15	19
$W(t)$ (gallons per minute)	14.2	16.3	20.4	19.4

(A) 197.1

(B) 205.7

(C) 228.5

(D) 285.7

(E) 308.5

25. $\int \frac{2dx}{x^2-3x-4} =$

(A) $\frac{2}{5}\ln\left|\frac{x+1}{x-4}\right|+C$

(B) $\frac{2}{5}\ln|(x-4)(x+1)|+C$

(C) $\frac{2}{5}\ln\left|\frac{x-4}{x+1}\right|+C$

(D) $\frac{4}{5}\ln|x-4|-\frac{2}{5}\ln|x+1|+C$

(E) $2\ln\left|\frac{x-4}{x+1}\right|+C$

26. The points (−4, −2) and (1, 6) are on the graph of $y = f(x)$ that satisfies the differential equation $\frac{dy}{dx}=x+y^2$. Which of the following must be true?

(A) (1, 6) is a local maximum of f.

(B) (1, 6) is a point of inflection of the graph of f.

(C) (−4, −2) is a local minimum of f.

(D) (−4, −2) is a local maximum of f.

(E) (−4, −2) is a point of inflection of the graph of f.

27. What is the radius of convergence of the series $\sum_{n=0}^{\infty}\frac{(x-3)^{2n}}{5^n}$?

(A) 5

(B) $2\sqrt{5}$

(C) $\sqrt{5}$

(D) $\frac{\sqrt{5}}{2}$

(E) 0

GO ON TO THE NEXT PAGE

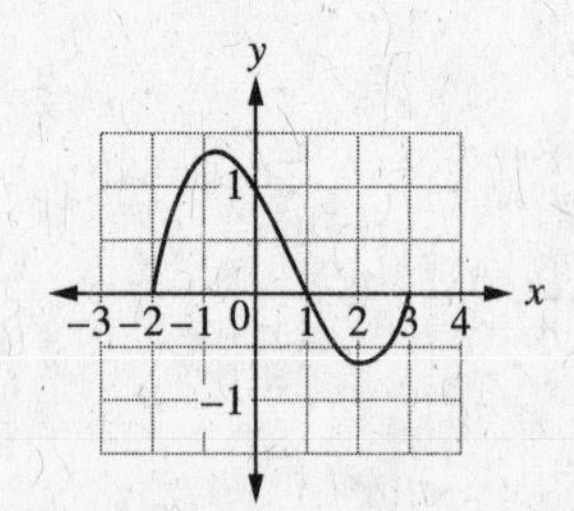

28. The graph of $y = f(x)$ is shown above. If $g(x) = \int_{-1}^{x} f(x)\,dx$, which of the following is true?

(A) $g'(1) < g(1) < g(0)$
(B) $g(0) < g(1) < g'(1)$
(C) $g(1) < g(0) < g'(1)$
(D) $g'(1) < g(0) < g(1)$
(E) $g(1) < g'(1) < g(0)$

Question 29 requires a calculator.

29. Let $y = f(x)$ be the solution to the differential equation $\frac{dy}{dx} = x + y$ with the initial condition $f(2) = 6$. What is the approximation for $f(3)$ using Euler's method with two steps of equal length beginning at $x = 2$?

(A) $\left(3, \frac{13}{4}\right)$
(B) $\left(3, \frac{65}{4}\right)$
(C) $(3,17)$
(D) $\left(3, \frac{61}{2}\right)$
(E) $(3,74)$

30. Which of the following functions can be represented by the power series $1 - \frac{x^2}{3} + \frac{x^4}{5} - \frac{x^6}{7} + \ldots + (-1)^n \frac{x^{2n}}{2n+1} + \ldots$ for $|x| \le 1$?

(A) $\text{Arc}\tan x$
(B) $\frac{\text{Arc}\tan x}{x}$
(C) $\frac{\sin x}{x}$
(D) $\sin x$
(E) $\frac{e^x - 1}{x}$

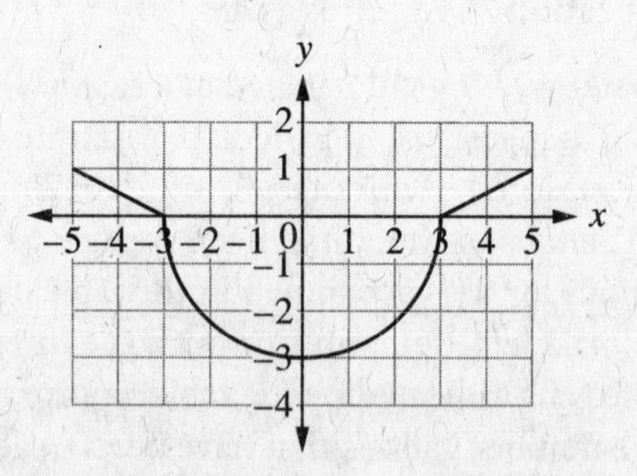

31. The graph of $f'(x)$, the derivative of f, consists of two line segments and a semicircle, as shown in the figure above. If $f(3) = 2$, then $f(-5) =$

(A) $1 - 9\pi$
(B) $1 - \frac{9\pi}{2}$
(C) $\frac{9\pi}{2} - 1$
(D) $\frac{9\pi}{2} + 1$
(E) $9\pi - 1$

GO ON TO THE NEXT PAGE

32. What is the $\lim_{x \to 3} \frac{\int_3^x \cos t\, dt}{x^2 - 9}$?

(A) $\frac{\sin 3}{9}$

(B) $-\frac{\sin 3}{6}$

(C) $\frac{\cos 3}{6}$

(D) $\frac{\sin 3}{6}$

(E) $\frac{\cos 3}{2}$

33. A function f has derivatives of all orders for all real numbers. It is known that $f(2) = 4, f'(2) = -3, f''(2) = 3$, and $f'''(2) = -1$. Which of the following statements are necessarily true?

I. $f(1)$ can be determined exactly from the given information.

II. The Taylor series for f at $x = 2$ converges to $f(x)$ for all values of x.

III. The third-degree Taylor polynomial for f at $x = 2$ can be determined from the given information.

(A) I only
(B) II only
(C) III only
(D) I and II only
(E) II and III only

34. If $P(t)$ is the size of a population at time t, which of the following differential equations describes quadratic growth in the size of the population?

(A) $\frac{dP}{dt} = 50$

(B) $\frac{dP}{dt} = 50t$

(C) $\frac{dP}{dt} = 25t^2$

(D) $\frac{dP}{dt} = 50P$

(E) $\frac{dP}{dt} = 25P^2$

35. Let r be a polar curve described by $r = 1 + 3\cos\theta$. What is the slope of the line tangent to the graph of r at $\theta = \frac{\pi}{2}$?

(A) -3

(B) $-\frac{1}{3}$

(C) 0

(D) $\frac{1}{3}$

(E) 3

36. For what values of p will both series $\sum_{n=1}^{\infty} \frac{1}{n^{p+3}}$ and $\sum_{n=1}^{\infty} \left(\frac{p}{3}\right)^n$ converge?

(A) $-2 < p < 3$ only.
(B) $-2 < p < 1$ only.
(C) $2 < p < 3$ only.
(D) $p < -2$ or $p > 3$ only.
(E) There are no values of p for which both series converge.

GO ON TO THE NEXT PAGE

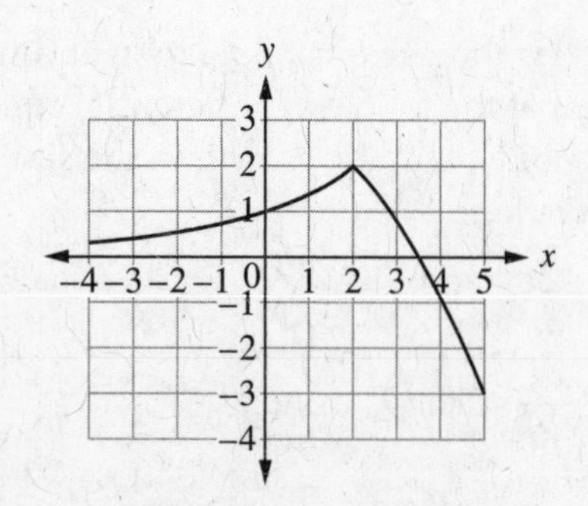

37. The graph of $g(x)$ is shown above on the interval $-4 \le x \le 4$. Which of the following statements is false?

(A) g is continuous at $x = 2$.

(B) g has a critical point at $x = 2$.

(C) g has an absolute maximum at $x = 2$.

(D) g changes concavity at $x = 2$.

(E) g is differentiable at $x = 2$.

Question 38 requires a calculator.

38. Let $f(x) = e^{-x}$ and $g(x) = x^3$. For what values of x is the rate of change of $f(x)$ less than the rate of change of $g(x)$?

(A) All real values of x.

(B) (0.632, 2.484)

(C) $(-\infty, 0.632)$ and $(2.484, \infty)$

(D) (−0.910, 0.459)

(E) (−0.910, 2.484) and (0.459, 0.632)

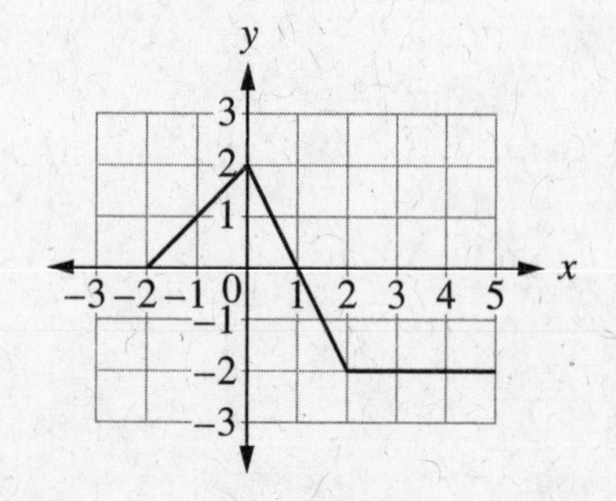

39. The graph of a piecewise linear function $f(x)$ is shown above. What is the value of $\int_{-2}^{5}(2f(x)+4)\,dx$?

(A) −4

(B) 0

(C) 10

(D) 20

(E) 24

40. Let f be a function with derivatives of all orders for $x > 0$ such that $f(4) = 8$, $f'(4) = -2$, $f''(4) = -6$, and $f'''(4) = 24$. Which of the following is the third-degree Taylor polynomial for f about $x = 4$?

(A) $8-2x-6x^2+24x^3$

(B) $8-2x-3x^2+4x^3$

(C) $8-2(x-4)-3(x-4)^2+4(x-4)^3$

(D) $8-2(x-4)-3(x-4)^2+8(x-4)^3$

(E) $8-2(x-4)-6(x-4)^2+24(x-4)^3$

GO ON TO THE NEXT PAGE

41. Suppose that $f'(x) < 0$ for all real numbers x and $\int_0^4 f(t)\,dt = 0$. Which of the following could be a table of values for the function f?

(A)

x	y
0	5
3	4
4	1

(B)

x	y
0	3
3	5
4	−4

(C)

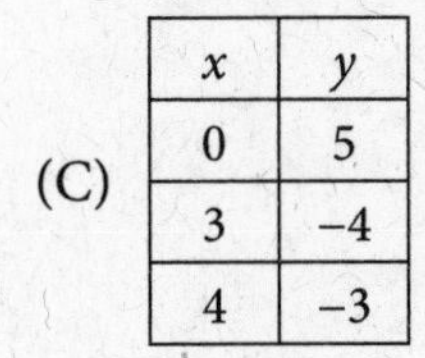

x	y
0	5
3	−4
4	−3

(D)

x	y
0	1
3	1
4	−1

(E)

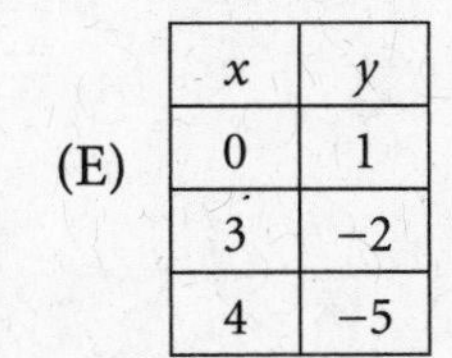

x	y
0	1
3	−2
4	−5

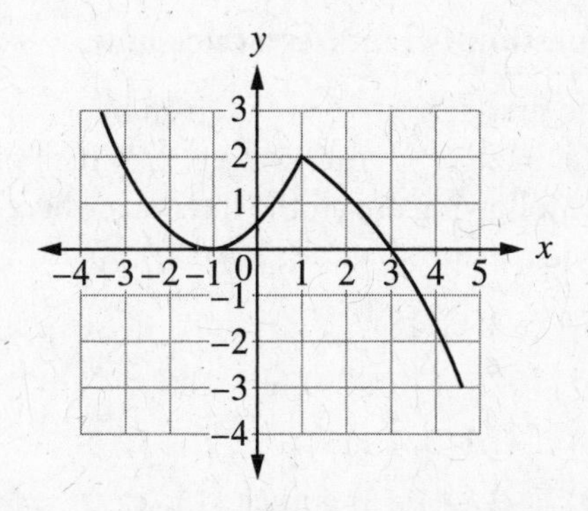

42. The graph of $f'(x)$, the derivative of f, is shown above. Which of the following statements must be true?

I. f has a relative minimum at $x = -1$.

II. f has a relative maximum at $x = 1$.

III. f is concave upward for $-1 < x < 1$.

(A) I only
(B) II only
(C) III only
(D) I and II only
(E) II and III only

43. Let f be a function that is continuous at $x = -2$. Which of the following must be true?

(A) $f(-2) < \lim_{x \to -2} f(x)$
(B) $f'(-2)$ exists
(C) $f'(x)$ is negative for all $x < -2$ and positive for all $x > -2$
(D) $\lim_{x \to -2^-} f(x) \neq \lim_{x \to -2^+} f(x)$
(E) $\lim_{x \to -2} f(x) = f(-2)$

GO ON TO THE NEXT PAGE

Question 44 requires a calculator.

44. The first derivative of f is given by $f'(x) = 2^{(x^3 - 4x)}$ for $-2 \le x \le 2$. Which of the following are all the intervals where f is concave up?

(A) $(-2, 0.118)$
(B) $(-2, -1)$ and $(0.118, 2)$
(C) $(-1.115, 1.115)$
(D) $(-2, -1.115)$ and $(1.115, 2)$
(E) $(-2, 0.451)$

45. Let f and g be differentiable functions for all x. Selected values for $f(x)$, $g(x)$, $f'(x)$, and $g'(x)$ are given in the table below. If $h(x) = f(g(x))$, what is the value of $h'(2)$?

x	$f(x)$	$g(x)$	$f'(x)$	$g'(x)$
1	0	2	4	6
2	−1	3	0	−5
3	6	−2	4	−1
4	−2	−1	2	2

(A) −20
(B) −4
(C) 0
(D) 4
(E) 5

GO ON TO THE NEXT PAGE

Free-Response Questions

Directions: Solve the following problems, using available space for scratchwork. Show how you arrived at your answer.

You may wish to look over the problems before starting to work on them; on the actual test, it is not expected that everyone will be able to complete all parts of all problems. All problems are given equal weight, but the individual parts of a particular problem are not necessarily given equal weight. You should not spend too much time on any one problem.

- You should write out all your work for each part. On the actual test, you will do this in the space provided in the test booklet. Be sure to write clearly and legibly. If you make a mistake, you can save time by crossing it out rather than trying to erase it. Erased or crossed-out work will not be graded.
- Show all your work. Clearly label any functions, graphs, tables, or other objects that you use. On the actual exam, you will be graded on the correctness and completeness of your methods as well as your answers. Answers without any supporting work may not receive credit.
- Justifications (i.e., the request that you "justify your answer") require that you give mathematical (non-calculator) reasons.
- Work must be expressed in standard mathematical notation, not calculator syntax.
- Unless otherwise specified, answers (numeric or algebraic) need not be simplified.
- If you use decimal approximations in calculations, the readers of the actual exam will grade you on accuracy. Unless otherwise specified, your final answers should be accurate to three places after the decimal point.
- Unless otherwise specified, the domain of function f is the set of all real numbers x for which $f(x)$ is a real number.

GO ON TO THE NEXT PAGE

PART A

Time: 30 Minutes
2 Problems

A GRAPHING CALCULATOR IS ALLOWED ON THIS PORTION OF THE EXAM

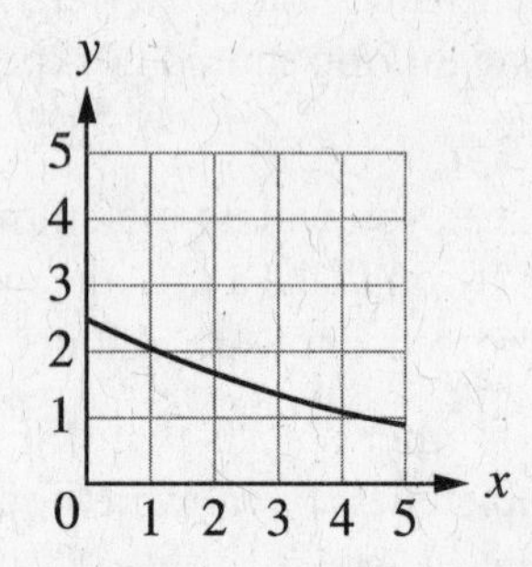

1. Let R be the region bounded by the graphs of $y = e^{-\frac{x}{2}}$, $y = 5$, $x = \ln 2$, and the y-axis as shown in the figure above.

   (a) Find the area of R.

   (b) The region R is revolved about the line $y = 6$ to form a solid. Find the volume of the generated solid.

   (c) The region R is the base of a solid. For this solid, each cross-section perpendicular to the x-axis is a quarter circle. Find the volume of this solid.

t(hours)	0	2	4	5	7
$W(t)$ (gallons)	6	10	8.5	6	2.2

2. The amount of water in a small storage tank is given by a differentiable function $W(t)$ for $0 \le t \le 7$, where t is measured in hours and $W(t)$ is measured in gallons. Selected values for $W(t)$ are shown in the table above.

   (a) Use the values in the table to estimate $W'(3)$ Use this estimate to explain the meaning of $W'(3)$ in the context of this problem.

   (b) Using trapezoidal sums as indicated by the data in the table, estimate the average number of gallons of water in the tank during the 7 hours.

   (c) Is there a time t, $0 \le t \le 7$, when $W'(t) = 0$? If so, explain why. If not, explain why not.

   (d) $W(t)$ can be modeled by the function $G(t) = 6 + 4\sin\left(\frac{\pi t}{5}\right)$. On average, how many gallons of water are in the tank from $0 \le t \le 7$?

GO ON TO THE NEXT PAGE

PART B

Time: 60 Minutes
4 Problems

NO CALCULATOR IS ALLOWED ON THIS PORTION OF THE EXAM

Note: If you have extra time, you can go back and work on Part A of Section II, but you cannot use a calculator to complete your work at this time.

3. Consider the differential equation $\frac{dy}{dx} = -\frac{x}{2y}$.
   (a) On the axes provided sketch the slope field for the given differential equation at the 12 points indicated.

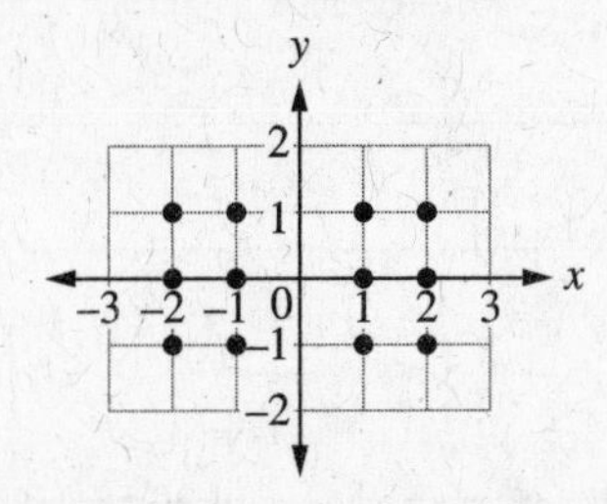

   (b) Let $y = f(x)$ be the particular solution to the given differential equation with the initial condition $f(-1) = 1$. Is f concave upward when $x = -1$? Explain.
   (c) Find the particular solution to $y = f(x)$ for the given differential equation with the initial condition $f(-1) = 1$

4. A vector describing the position of a particle moving in the xy-plane has position $(x(t),y(t))$ at any time t. The position of the particle at time $t = 1$ is $(8,3)$, and the velocity vector for time $t > 0$ is given by $\left\langle 5+\frac{3}{t^2}, 4-\frac{1}{t^2}\right\rangle$.
   (a) Find the acceleration vector at time $t = 3$.
   (b) Find the position vector at time $t = 3$.
   (c) For what time t, if any, for $t > 0$, does the line tangent to the path of the particle have a slope of -6?
   (d) The particle approaches a line as $t \to \infty$. Find the slope of this line. Show the work that leads to your conclusion.

5. Let $f(x) = xe^{-x}$ and $G(x) = \int_1^x f(t)\,dt$.
   (a) Find $\int_1^\infty f(x)\,dx$.
   (b) Write an equation for the line tangent to the graph of $G(x)$ at $x = 2$.
   (c) Evaluate $G'(x^2)$ at $x = 2$.

6. The function f has a Maclaurin series that converges to $f(x)$ for all x in the interval of convergence. The coefficient of the x^n term in the series is given by $\frac{(-2)^n}{(n+1)}$ for $n > 1$.
   (a) Write a formula for the nth derivative of f at $x = 0$, $f^{(n)}(0)$.
   (b) Find the values of x for which the Maclaurin series for f converges. Indicate whether the convergence is absolute or conditional. Show the work that leads to your answer.
   (c) Use the fifth-degree Maclaurin polynomial for f to approximate $f(0.09)$.
   (d) Show that the error in using the fifth-degree Maclaurin polynomial to approximate $f(0.09)$ is less than 10^{-5}.

GO ON TO THE NEXT PAGE

GO ON TO THE NEXT PAGE

IF YOU FINISH BEFORE TIME IS CALLED, YOU MAY CHECK YOUR WORK ON THIS SECTION ONLY. DO NOT TURN TO ANY OTHER SECTION IN THE TEST.

AP CALCULUS BC DIAGNOSTIC TEST ANSWER KEY

1. C
2. C
3. C
4. B
5. D
6. B
7. C
8. C
9. C
10. C
11. D
12. C
13. A
14. B
15. E
16. E
17. B
18. E
19. B
20. B
21. A
22. D
23. B
24. D
25. A
26. C
27. C
28. D
29. B
30. B
31. D
32. C
33. C
34. B
35. E
36. A
37. E
38. A
39. D
40. C
41. E
42. C
43. E
44. D
45. A

ANSWERS AND EXPLANATIONS

1. C

We first try to evaluate this limit by substituting the value $x = 0$ into the expression. We get the indeterminate form $\frac{\text{"0"}}{0}$, telling us we need to "try something else." The limit satisfies the hypotheses of L'Hôpital's Rule, so we instead compute the equivalent limit $\lim_{x \to 0} \frac{(\tan^{-1} 2x)'}{x'} = \lim_{x \to 0} \frac{2}{1 + 4x^2}$, which we evaluate by substituting the value $x = 0$ into the expression $\frac{2}{1 + 4x^2}$, giving us 2.

2. C

The lines tangent to f and g at $x = a$ are parallel if their slopes are equal. The slope of the tangent line at a point is the derivative of the function evaluated at that point. Combining these ideas tells us that the solution to this question is the number of solutions to the equation $f'(a) = g'(a)$. We compute these derivatives: $f'(x) = 6x$; $g'(x) = 3x^2 - 9$ and find the solutions to the equation $6a = 3a^2 - 9 \rightarrow a^2 - 2a - 3 = 0 \rightarrow a = 3, -1$. We see that the equation has two solutions.

Note: Because we're not asked to find the values for a, we don't actually need to solve the equation. We could get away with computing the discriminant of the equation to determine the number of solutions. However, this equation is pretty straightforward and solving it is less complicated than computing the discriminant.

3. C

This is a sneaky series to work with because of the sin n in the numerator. The series does take on both positive and negative values, but it does not alternate in a regular "+ – + – + – . . . " pattern. It is thus not an alternating series and we can eliminate answer (A).

The series cannot be written in the form $\sum_{n=1}^{\infty} \frac{1}{n^p}$, so it is not a p-series. Therefore we can eliminate answer (B). We can, however, compare the series of absolute values $\sum_{n=1}^{\infty} \left| \frac{(\sin n)(n+1)}{(2n+3)^4} \right|$ to the p-series $\sum_{n=1}^{\infty} \frac{1}{n^3}$, since

$$\left| \frac{(\sin n)(n+1)}{(2n+3)^4} \right| \leq \left| \frac{1 \cdot (n+1)}{(2n+3)^4} \right| \leq \frac{1}{n^3}.$$

Because the general term of the series of absolute values is smaller than the general term $\frac{1}{n^3}$ (the series is smaller than a series we know converges), we can conclude that the series of absolute values $\sum_{n=1}^{\infty} \left| \frac{(\sin n)(n+1)}{(2n+3)^4} \right|$ converges.

This means the original series, $\sum_{n=1}^{\infty} \frac{(\sin n)(n+1)}{(2n+3)^4}$, converges absolutely.

Note: The limit comparison test, suggested in answer (D), will not work because the limit of the ratios of the general terms of the series does not exist (sin n in the numerator). The limit comparison test would work if we compared the series of absolute values with convergent p-series $\sum_{n=1}^{\infty} \frac{1}{n^2}$.

4. B

The distance traveled by the particle from time $t = 0$
to $t = 1$ is the length of the curve traced by the particle over this interval, i.e., the arc length. Because

the particle's position is given parametrically, we use the parametric form of the equation for arc length:

$\int_{t=a}^{t=b} \sqrt{\left(\frac{dx}{dt}\right)^2 + \left(\frac{dy}{dt}\right)^2}\, dt$. Compute the derivatives $\frac{dx}{dt} = e^t$; $\frac{dy}{dt} = e^t \sin t + e^t \cos t$, then substitute the derivatives and starting and ending times into the formula for arc length: $\int_{t=0}^{t=1} \sqrt{(e^t)^2 + (e^t \sin t + e^t \cos t)^2}\, dt$. We simplify the integrand to find the answer $\sqrt{2}\int_0^1 e^t \sqrt{1 + \sin t \cos t}\, dt$.

5. D

The area of a simple region enclosed by a polar curve r between two rays $\theta = \alpha$ and $\theta = \beta$ is $\int_{\alpha}^{\beta} \frac{r^2}{2}\, d\theta$.

In our case, the simple polar region R is enclosed by r and the rays $\theta = 0$ and $\theta = \frac{\pi}{4}$; therefore, the integral in answer (D) expresses the area of R.

6. B

First we set up $u = (\ln x)^2$ and $dv = dx$. Using integration by parts, the integral becomes $[x(\ln x)^2]_1^e - 2\int_1^e \ln x\, dx$
We again use inte-gration by parts to compute the integral in the resulting expression, this time

setting $u = \ln x$ and $dv = dx$, to get $\int_1^e \ln x\, dx = [x \ln x - x]_1^e$. We combine the information to find $\int_1^e (\ln x)^2\, dx = [x(\ln x)^2 - 2(x \ln x - x)]_1^e$. We substitute the limits of integration into the expression and simplify to get the answer $e - 2$.

7. C

The easiest way to answer this question is to graph the function and use the graph to determine which statement is correct. On a graph, we see that the graph has no relative extrema, but does have an inflection point. We can use the trace function to determine that the inflection point occurs at $x = 1$, so answer (C) is correct.

We can also answer this question without a graphing calculator. We compute the first derivative $f'(x) = 3x^2 - 6x + 3$. We solve the equation $0 = 3x^2 - 6x + 3 = 3(x - 1)^2 \rightarrow x = 1$. This means the point $x = 1$ is a critical point. We can use the First or Second Derivative Test to determine whether the function has a relative extremum at this critical point, or if we look carefully at the derivative, we can rewrite it as $f'(x) = 3(x - 1)^2$, so $f'(x) \geq 0$ everywhere. Therefore the function is always increasing, so it has no relative extrema and we can eliminate answers (A), (D), and (E). Next, we determine where the function changes concavity. We compute $f''(x) = 6x - 6$ and solve the equation $0 = 6x - 6 \rightarrow x = 1$ to see that this point is a candidate for an inflection point, i.e., the function might change concavity at $x = 1$. A sign analysis determines that $f''(x) > 0$, for $x > 1$, so the function is concave up on this interval; $f''(x) < 0$, $x < 1$, so the function is concave down on this interval. We conclude that $x = 1$ is an inflection point; i.e., the function changes concavity at this point, so the correct answer is (C).

8. C

This problem is an application of the double–zoom–whammy chain rule, which means we have to apply the chain rule twice to compute the derivative. We do this and find $f'(x) = 3\cos 3x \cdot e^{\sin(3x)}$. We evaluate the derivative at $x = 0$ and simplify to find $f'(0) = 3 \cdot \cos(0) \cdot e^{\sin(0)} = 3$.

9. C

The quickest way to solve this problem is to recognize that the population growth is modeled by a logistic differential equation, $\frac{dy}{dt} = rP\left(1 - \frac{P}{K}\right)$, and that the carrying capacity (maximum population) is K and the rate of growth is greatest when $P = \frac{K}{2}$. In this case $K = 8$, so the population is growing the fastest when $P = 4$.

Note: A logistic differential equation can take many forms. If the form that shows up on the exam is not the same as the form you learned in class, you may not be able to apply quick rules to answer questions. All of the forms are separable differential equations, so you can always tackle questions about logistic growth by solving the differential equation and proceeding from there. It may not be elegant, but it will work.

10. C

To compute this improper integral, rewrite the integral as the limit of a definite integral and compute the resulting definite integral and its limit to find $\int_1^{+\infty} e^{-3x}\,dx = \lim_{c\to\infty} \int_1^c e^{-3x}\,dx = \lim_{c\to\infty} \left[-\frac{1}{3}e^{-3x}\right]_1^c = \frac{1}{3e^3}$.

11. D

To determine the equation of the tangent line, we need to find the x and y coordinates of a point on the line and the slope, which is the derivative $\frac{dy}{dx}$ of the function at the point where $y = \frac{1}{4}$. We'll use this point, $y = \frac{1}{4}$, as our point on the line. Next, we need to find its x coordinate. Substitute into the definition $y = 2t^3$ and solve for t to find $t = \frac{1}{2}$. Substitute this value for t into the definition $x = \sin\left(\frac{\pi}{2}t\right)$ and simplify to find $x = \frac{\sqrt{2}}{2}$.

Now we've found the points we need: $\left(\frac{\sqrt{2}}{2}, \frac{1}{4}\right)$. Next, find the slope. Because the curve is defined parametrically, use the formula $\frac{dy}{dx} = \frac{dy/dt}{dx/dt}$ and evaluate this derivative at the value of t corresponding to $x = \frac{\sqrt{2}}{2}$, $y = \frac{1}{4}$ which was computed above as $t = \frac{1}{2}$. Therefore, we need to find $\left.\frac{dy}{dx}\right|_{\frac{1}{4},\frac{\sqrt{2}}{2}} = \left.\frac{dy/dt}{dx/dt}\right|_{t=\frac{1}{2}}$. Compute $\frac{dy}{dt} = 6t^2$ and $\frac{dx}{dt} = \frac{\pi}{2}\cos\frac{\pi}{2}t$, now evaluate these derivatives at $t = \frac{1}{2}$:

$$\left.\frac{dy}{dx}\right|_{\frac{1}{4},\frac{\sqrt{2}}{2}} = \left.\frac{dy/dt}{dx/dt}\right|_{t=\frac{1}{2}} = \left.\frac{6t^2}{\frac{\pi}{2}\cos\frac{\pi}{2}t}\right|_{t=\frac{1}{2}} = \frac{3\sqrt{2}}{\pi}.$$

Now we have point $(x_0, y_0) = \left(\frac{1}{4}, \frac{\sqrt{2}}{2}\right)$ and slope $m = \frac{3\sqrt{2}}{\pi}$ of the line and we can use the point-slope form of the line equation, $y - y_0 = m(x - x_0)$ to find the equation of the line tangent to the curve:

$$y - \frac{1}{4} = \frac{3\sqrt{2}}{\pi}\left(x - \frac{\sqrt{2}}{2}\right).$$

12. C

If we draw a quick sketch of the region, we see that the typical cross section is a washer, whose outer radius is $R = 1$ and whose inner radius is $\sin x$. The boundaries of the region are $x = 0$ and $x = \frac{\pi}{2}$. Set up the integral that expresses the volume of the solid, $\int_0^{\frac{\pi}{2}} (\pi \cdot 1^2 - \pi\sin^2 x)\,dx$, simplify and use a trig identity to reduce the problem to an integral

$$\int_0^{\frac{\pi}{2}} (\pi \cdot 1^2 - \pi\sin^2 x)\,dx = \pi\int_0^{\frac{\pi}{2}} (1 - \sin^2 x)\,dx$$

$$\underset{\text{trig identity}}{=} \pi\int_0^{\frac{\pi}{2}} \left(1 - \frac{1}{2}(1 - \cos 2x)\right)dx = \pi\int_0^{\frac{\pi}{2}} \left(\frac{1}{2} + \frac{1}{2}\cos 2x\right)dx = \frac{\pi^2}{4} \approx 2.467$$

13. A

To compute this integral, we use the technique of partial fractions. Rewrite the integrand as $\frac{7}{(x-3)(x+4)} = \frac{A}{(x-3)} + \frac{B}{(x+4)}$, simplify, and substitute values for x into the simplified expression to find $A = 1$ and $B = -1$. Rewrite the integrand as $\int \frac{7}{(x-3)(x+4)}\,dx = \int \frac{1}{(x-3)} + \frac{-1}{(x+4)}\,dx$ and compute these basic antiderivatives to find

$$\int \frac{1}{(x-3)} + \frac{-1}{(x+4)}\,dx = \ln|x-3| - \ln|x+4| + C.$$

Apply properties of the logarithm to rewrite the answer as $\ln\left|\frac{x-3}{x+4}\right| + C$.

14. B

Using the relationship between the second derivative and concavity, we can read the concavity of the function f from the given graph of f''; f is concave up on the intervals where f'' is positive, and f is concave down on the intervals where f'' is negative. The function f changes concavity at the point where the graph of f'' crosses the x–axis. On the graph of f'', we see that $f'' < 0$ for x less than approximately 2.6 and $f'' > 0$ for values of x greater than approximately 2.6. Therefore, the correct answer will show a graph that is concave down on $x < 2.6$ and concave up on $x > 2.6$, with a change in concavity only at the point $x \approx 2.6$. Looking at the graphs in the answer choices, we see that only the graph in answer (B) satisfies these conditions.

15. E

Because $f(x)$ is continuous at $x = 2$ and $f(2) = 3$, it follows from the definition of continuity that $\lim_{x \to 2} f(x) = 3$. It follows from the definition of the limit that $\lim_{x \to 2^+} f(x) = \lim_{x \to 2^-} f(x) = \lim_{x \to 2} f(x)$; therefore statements I and II are correct and we can eliminate answer choices (A), (C), and (D). Because f is not differentiable at $x = 2$, it follows from the definition of the derivative at a point that the limit in III does not exist. Therefore, statement III is also correct. All three choices are correct, so the answer is (E).

16. E

We use implicit differentiation to compute this second derivative. We first differentiate implicitly to find the equation that defines the first derivative $\frac{dy}{dx}$; $2 + \frac{dy}{dx}\sin y = \frac{dy}{dx}$. Now differentiate this equation implicitly to find the equation that defines the second derivative

$\frac{d^2y}{dx^2}$; $\frac{d^2y}{dx^2}\sin y + \left(\frac{dy}{dx}\right)^2 \cos y = \frac{d^2y}{dx^2}$. Evaluate the second derivative at the point $\left(\frac{1}{2}, 0\right)$. To do this, first compute $\frac{dy}{dx}$ by evaluating each term in the equation defining $\frac{dy}{dx}$ at $\left(\frac{1}{2}, 0\right)$ and solving the resulting expression for $\frac{dy}{dx}$. $2 + \frac{dy}{dx}\sin(0) = \frac{dy}{dx} \Rightarrow \frac{dy}{dx} = 2$. We use this information to solve for $\frac{d^2y}{dx^2}$.

$$\frac{d^2y}{dx^2}\sin(0) + (2)^2\cos(0) = \frac{d^2y}{dx^2} \Rightarrow \frac{d^2y}{dx^2} = 4.$$

17. B

Answer this question by looking at the arrows defining the slope field and use process of elimination. Each arrow indicates the slope at a point. In the second quadrant, the arrows point upward, indicating that the slopes at the points in the portion of the second quadrant shown in the graph, given by $\frac{y}{f(x)}$, are positive. Because y is positive in the second quadrant, $f(x)$ must be a function that is positive in the portion of the second quadrant shown. We can thus eliminate answers (A) and (C), because the functions for these answers are negative for $x < 0$ and we can eliminate (D) because the function $-x^2$ is negative everywhere. We can also eliminate answer (E) because the slope field shown in the graph is defined for $x < 0$; $\sqrt{x}$ is not defined for negative values of x

so the slope field $\frac{dy}{dx} = \frac{y}{\sqrt{x}}$ is not defined for $x < 0$ and so cannot be the slope field shown in the graph. In the fourth quadrant, the arrows point downward, indicating that the slope at the points shown in the fourth quadrant is negative. Because y is negative in the fourth quadrant, the function $f(x)$ must be positive at the points shown on the graph. The function $(x-2)^2$ is always positive, so answer (B) is correct.

18. E

To compute this derivative we use the second Fundamental Theorem of Calculus and the chain rule to compute $\frac{d}{dx}(g(x)) = 2x \cdot e^{x^2} \tan^{-1} x^2$. Evaluate this derivative at $x = 1$; $2 \cdot 1 \cdot e^{1^2} \tan^{-1} 1^2 = \frac{\pi \cdot e}{2}$.

19. B

To compute this definite integral, make the u–substitution $u = \ln x^2$. Then $du = \frac{2}{x}$; now rewrite the limits of integration as u–limits: $x = 1 \rightarrow u = 0$;

$x = e \rightarrow u = 2$. The integral becomes $\frac{1}{6}\int_0^2 u^2 du$. Now compute: $\frac{1}{6}\int_0^2 u^2 du = \frac{1}{6} \cdot \frac{u^3}{3}\Big|_0^2 = \frac{4}{9}$.

20. B

The average value of a continuous function h on a closed interval $[a, b]$ is $\frac{1}{b-a}\int_a^b h(x)\,dx$.

Because f and g are continuous and $f(c) = g(c)$, our function h is continuous on the interval $[0, 1]$. The average value is given by $\frac{1}{1-0}\int_0^1 h(x)\,dx$. Expressing the definite integral in terms of f and g we get

$$\frac{1}{1-0}\left(\int_0^c f(x)\,dx + \int_c^1 g(x)\,dx\right) = \int_0^c f(x)\,dx + \int_c^1 g(x)\,dx.$$

21. A

When dealing with discontinuous functions, it is best to separate the problems into separate sections and find the area of each region. Finally, add the answers together to get the final total.

Area of region of (0, 2) is 6

Area of region (2, 5) is –12

Sum is 6 + –12 = –6

22. D

When dealing with series, try to determine the type of series shown. Then you can apply the tools associated with that series type.

This is a geometric series with first term of 1 and a common ratio of $\frac{-2}{3}$.

$$f(2) = \sum_{n=0}^{\infty}\left(\frac{-2}{3}\right)^n$$
$$S = \frac{a_0}{1-r} = \frac{1}{1-\frac{-2}{3}} = \frac{3}{5}$$

23. B

Use integration by parts to solve this problems.

Let

$$u = x \Rightarrow du = dx$$

$$dv = \cos(2x)dx \quad v = \frac{1}{2}\sin(2x)$$

$$\int x\cos(2x)\,dx = \frac{1}{2}x\sin(2x) - \int\frac{1}{2}\sin(2x)$$

$$\Rightarrow \frac{1}{2}x\sin(2x) + \frac{1}{4}\cos(2x) + C$$

24. D

A common error on accumulation problems is to forget to add the initial condition to the sum determined by the definite integral.

$$G(19) = 80 + \int_7^{19} w(t)\ dt$$
$$G(19) \doteq 80 + 3(14.2) + 5(16.3) + 4(20.4) = 285.7$$

25. A

When u-substitution doesn't work and there is a quadratic expression in the denominator, try separation by partial fractions.

$$\frac{2}{x^2 - 3x - 4} = \frac{2}{(x+1)(x-4)}$$

Multiply both sides by the common denominator.

$2 = A(x+1) + B(x-4)$

When $x = -1$,

$$2 = A(0) + B(5) \Rightarrow B = -\frac{2}{5}.$$

When $x = 4$,

$$2 = 5A + B(0) \Rightarrow A = \frac{2}{5}.$$

So,

$$\int \frac{2}{(x+1(x-4)} dx = \int \left(\frac{\frac{2}{5}}{x+1} - \frac{\frac{2}{5}}{x-4} \right) dx$$

$$\Rightarrow \frac{2}{5} \int \frac{1}{x+1} - \frac{1}{x-4} dx$$

$$= \frac{2}{5} \left(\ln|x+1| - \ln|x-4| \right) + C = \frac{2}{5} \ln\left| \frac{x+1}{x-4} \right| + C$$

26. C

Relative extrema can be justified using the second derivative test. You must first verify that the first derivative has a critical point. Then, use the second derivative test.

$\frac{dy}{dx} = 0$ at (−4, −2) which implies a critical point.

$$\frac{d^2y}{dx^2} = 1 + 2y \left. \frac{dy}{dx} \right|_{(-4,-2)} = 1 \Rightarrow$$

Since the second derivative is positive, we know our critical point is a relative minimum.

27. C

Using the ratio test

$$\lim_{n \to \infty} \left| \frac{(x-3)^{2(n+1)}}{5^{n+1}} \cdot \frac{5^n}{(x-3)^{2n}} \right| = \lim_{n \to \infty} \left| \frac{(x-3)^2}{5} \right| < 1$$

$$\Rightarrow -1 < \frac{(x-3)^2}{5} < 1 \quad -5 < (x-3)^2 < 5$$

$$\Rightarrow -\sqrt{5} < x-3 < \sqrt{5}$$

The diameter of convergence is $2\sqrt{5}$, so the radius of convergence is $\sqrt{5}$.

Remember that finding the radius of convergence does not require a test on the endpoints of the interval.

28. D

$g'(1) = f(1) = 0$

$g(0)$ is the area under the curve from $x = -1$ to $x = 0$ which is strictly a positive value.

$g(1)$ is the area under the curve from $x = -1$ to $x = 1$ which is a bigger value than $g(0)$.

29. B

Original x	Original y	$dy = (x + y)dx$	Next x	Next y
2	6	$(2+6)\frac{1}{2}=4$	2.5	10
2.5	10	$\left(\frac{5}{2}+10\right)\frac{1}{2}=\frac{25}{4}$	3	$\frac{65}{4}$

30. B

You are expected to know the Taylor series expansion formulas for seven basic functions: sine, cosine, $e^x, \frac{1}{1-x}, \frac{1}{1+x}$, ln x, and arctan x.

This is the power series expansion for $\frac{Arc\tan x}{x}$.

31. D

Don't forget to add the initial condition.

$$f(-5) = 2 + \int_3^{-5} f'(x)\,dx = 2 + \frac{9\pi}{2} - 1 = \frac{9\pi}{2} + 1$$

32. C

$$\lim_{x\to 3} \frac{\int_3^x \cos t\,dt}{x^2 - 9} = \frac{0}{0}.$$

Use L'Hôpital's rule.

$$\lim_{x\to 3} \frac{\frac{d}{dx}\int_3^x \int \cos t\,dt}{\frac{d}{dx}\left(x^2 - 9\right)} = \lim_{x\to 3} \frac{\cos x}{2x} = \frac{\cos 3}{6}.$$

33. C

I is false because the information given is only concerning the values of the function and its derivatives at $x = 2$. II is not necessarily true because no information about the remainder is given. III is true because we have all the information needed to write the third-degree Taylor polynomial for f about $x = 2$.

34. B

If a function grows quadratically, its derivative must be linear.

35. E

Tangent slopes in polar format are done parametrically.

$$x = r\cos\theta \Rightarrow x = (1+3\cos\theta)\cos\theta$$

$$\frac{dx}{d\theta} = (1+3\cos\theta)(-\sin\theta)+\cos\theta(-3\sin\theta)$$

$$y = r\sin\theta \Rightarrow y = (1+3\cos\theta)\sin\theta$$

$$\frac{dy}{d\theta} = (1+3\cos\theta)(\cos\theta)+\sin\theta(-3\sin\theta)$$

$$\frac{dy}{dx} = \frac{\frac{dy}{d\theta}}{\frac{dx}{d\theta}} = \left.\frac{(1+3\cos\theta)(\cos\theta)+\sin\theta(-3\sin\theta)}{(1+3\cos\theta)(-\sin\theta)+\cos\theta(-3\sin\theta)}\right|_{\theta=\frac{\pi}{2}} = 3$$

36. A

Solve the first series using the p-series rule. The second series is a geometric series.

$$\sum_{n=1}^{\infty}\frac{1}{n^{p+3}} \Rightarrow p+3>1 \rightarrow p>-2 \text{ and}$$

$$\sum_{n=1}^{\infty}\left(\frac{p}{3}\right)^n \Rightarrow -1<\frac{p}{3}<1 \rightarrow -2<p<3$$

37. E

A *continuous* function is not differentiable at a corner or a cusp—any "sharp point."

38. A

Rate of change involves the first derivative of the function.

$f'(x) < g'(x)$ for all values of x.

39. D

When integrating from left to right, regions above the x-axis have positively signed area. Regions below the x-axis have negatively signed area.

$$\int_{-2}^{5}(2f(x)+4)dx = 2\int_{-2}^{5}f(x)\,dx + \int_{-2}^{5}4\,dx$$
$$= 2(-4)+28 = 20$$

40. C

This deals with the basic formula for a Taylor series centered at $x = a$.

$$f(x) = \frac{f(a)}{0!}(x-a)^0 + \frac{f'(a)}{1!}(x-a)^1 + \frac{f''(a)}{2!}(x-a)^2 + \frac{f'''(a)}{3!}(x-a)^3 + \ldots + \frac{f^{(n)}}{n!}(x-a)^n + \ldots$$

A Taylor polynomial has a finite number of terms. The series formula is truncated at that term.

$$T_3(4) = f(4) + \frac{f'(4)}{1!}(x-4)^1 + \frac{f''(4)}{2!}(x-4)^2 + \frac{f'''(4)}{3!}(x-4)^3$$

41. E

If $f'(x) < 0$, then f is always decreasing. So the y-values in the table should be decreasing. Also, if the value of the definite integral is zero, then there must be some positive values and then negative values for y. Only e meets both conditions.

42. C

When a function is concave upward, $f''(x) > 0$. This means that the graph of the first derivative is increasing. The graph of $f'(x)$ as shown increases only on the interval $-1 < x < 1$.

43. E

If a function is continuous, then $\lim_{x \to a} f(x) = f(a)$. Only option (E) matches this definition.

44. D

A function is concave up when $f''(x) > 0$. If $f'(x) = 2^{x^3-4x}$, then $f''(x) = 2^{x^3-4x}(\ln 2)(3x^2 - 4)$, which is positive on the intervals $(-2, -1.115)$ and $(1.115, 2)$.

45. A

This problem involves the application of the chain rule.

$$h(x) = f(g(x))$$
$$\Rightarrow h'(x) = f'(g(x))\,g'(x)$$
$$\Rightarrow h'(2) = f'(g(2))\,g'(2) = f'(3)g'(2) = (4)(-5) = -20$$

FREE-RESPONSE ANSWERS

1. (a) $A = \int_0^{\ln 2} \left(5 - e^{-\frac{x}{2}}\right) dx = 2.8799$

(b) $V = \pi \left(\int_0^{\ln 2} \left(6 - e^{-\frac{x}{2}}\right)^2 - (6-5)^2 \right) dx = 55.702 \text{ or } 55.703$

(c) $V = \frac{\pi}{4} \int_0^{\ln 2} \left(5 - e^{-\frac{x}{2}}\right)^2 dx = 9.401 \text{ or } 9.402$

2. (a) $W'(3) \doteq \frac{W(4) - W(2)}{4-2} = \frac{8.5-10}{2} = -0.75$

At time $t = 3$ the amount of water in the tank is decreasing at a rate of -0.75 gallons per hour.

(b) $\textit{Average} = \frac{1}{7} \int_0^7 W(t)\, dt$

$\int_0^7 W(t)\, dt$

$\doteq 2\left(\frac{1}{2}\right)(6+10) + 2\left(\frac{1}{2}\right)(10+8.5) + 1\left(\frac{1}{2}\right)(8.5+6) + 2\left(\frac{1}{2}\right)(6+2.2)$

$= 49.95$

$\textit{Average} = \frac{49.95}{7} = 7.135 \text{ or } 7.136$

(c) Yes, since $W(5) = W(0) = 6$ and W is differentiable, Rolle's Theorem guarantees there will be at least one point where $W'(t) = 0$.

(d) $\textit{Average} = \frac{1}{7} \int_0^7 W(t)\, dt$

$= \frac{1}{7} \int_0^7 \left(6 + 4\sin\left(\frac{\pi t}{5}\right)\right) dt = 7.190$

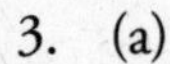

3. (a)

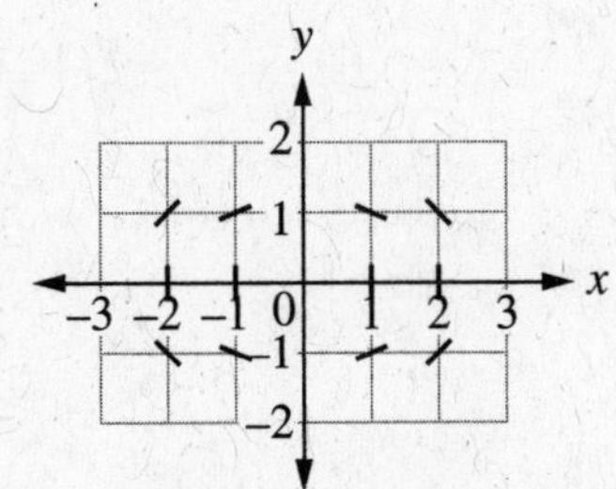

(b)

$$y'' = \frac{2y(-1) - x \bullet 2\frac{dy}{dx}}{(2y)^2}$$

$$y'|_{(-1,1)} = -\frac{-1}{2(1)} = \frac{1}{2}$$

$$y''|_{(-1,1)} = \frac{2(1)(-1) - (-1)(2)\left(\frac{1}{2}\right)}{(2(1))^2} < 0$$

No, since $y'' < 0$ on the interval containing $(-1,1)$, $f(x)$ must be concave down.

(c)

$$\frac{dy}{dx} = \frac{-x}{2y}$$

$$2y\,dy = -x\,dx$$

$$\int 2y\,dy = \int -x\,dx$$

$$y^2 = -\frac{x^2}{2} + C \quad at\ (-1,1)$$

$$1 = -\frac{1}{2} + C \Rightarrow C = \frac{3}{2}$$

$$y^2 = -\frac{x^2}{2} + \frac{3}{2}$$

$$y = \pm\sqrt{-\frac{x^2}{2} + \frac{3}{2}}$$

choose positive branch

$$f(x) = \sqrt{-\frac{x^2}{2} + \frac{3}{2}}$$

4. (a) $\vec{a}(3) = \vec{v}'(3) = \left\langle \frac{-6}{t^3}, \frac{2}{t^3} \right\rangle\Big|_{t=3} = \left\langle \frac{-2}{9}, \frac{2}{27} \right\rangle$

(b)

$$x = 8 + \int_1^3 \left(5 + \frac{3}{t^2}\right) dt = 8 + \left(5t - \frac{3}{t}\right)\Big|_1^3 = 20$$

$$y = -3 + \int_1^3 \left(4 - \frac{1}{t^2}\right) dt = 3 + \left(4t + \frac{1}{t}\right)\Big|_1^3 = \frac{31}{3}$$

(c)

$$\frac{dy}{dx} = \frac{\frac{dy}{dt}}{\frac{dx}{dt}} = \frac{4 - \frac{1}{t^2}}{5 + \frac{3}{t^2}} = \frac{4t^2 - 1}{5t^2 + 3}$$

$$\frac{4t^2 - 1}{5t^2 + 3} = -6 \Rightarrow 4t^2 - 1 = -30t^2 - 18$$

$$\Rightarrow 34t^2 = -17 \Rightarrow t^2 = -\frac{17}{34}$$

no solution

(d)

$$\lim_{t \to \infty} \frac{4 - \frac{1}{t^2}}{5 + \frac{3}{t^2}} = \frac{4}{5}$$

5. (a) $\int_1^{\infty} xe^{-x}\ dx = \lim_{b \to \infty} \int_1^b xe^{-x}\ dx$

Integration by parts

$$\int_1^b xe^{-x}\ dx = -xe^{-x} - e^{-x}\Big|_1^b$$

$$\Rightarrow -e^{-x}(x+1)\Big|_1^b$$

$$= -e^{-b}(b+1) - (-e^{-1}(1+1)) = -e^{-b}(b+1) + 2e^{-1}$$

$$\lim_{b \to \infty} -e^{-b}(b+1) + 2e^{-1} = 2e^{-1}$$

(b) $G(2) = \int_1^2 xe^{-x}\ dx = 2e^{-1} - 3e^{-2}$

$$\left(2, 2e^{-1} - 3e^{-2}\right) = (x, y)$$

$$m = G'(2) = f(2) = 2e^{-2}$$

$$T: y - \left(2e^{-1} - 3e^{-2}\right) = 2e^{-2}(x - 2)$$

(c) $G'(x^2) = f(x^2)(2x)$

$= x^2 e^{-x^2} \cdot 2x\Big|_{x=2} = 16e^{-4}$

6. (a) The coefficient on the x^n term in a Maclaurin series is given by $\frac{f^{(n)}(0)}{n!}$. Set this expression equal to $\frac{(-2)^n}{(n+1)}$ and solve for $f^{(n)}(0)$: $f^{(n)}(0) = \frac{(-2)^n \cdot n!}{(n+1)}$.

(b) Find the interval of convergence for a power series $\sum_{n=0}^{\infty} a_n x^n$ by first applying the ratio test to the series and solving the resulting inequality. Therefore, solve $\lim_{n\to\infty}\left|\frac{a_{n+1}}{a_n}x\right| < 1$. Now solve the inequality

$$\lim_{n\to\infty}\left|\frac{\frac{(-2)^{n+1}}{(n+1+1)}}{\frac{(-2)^n}{(n+1)}}x\right| = \lim_{n\to\infty}\left|\frac{2^{n+1}}{(n+2)} \cdot \frac{(n+1)}{2^n}x\right| = |2x| < 1$$

to find that the series converges on $-\frac{1}{2} < x < \frac{1}{2}$. The ratio test does not give a conclusive answer about the convergence at the endpoints of the interval, so analyze these points separately.

For $x = \frac{1}{2}$, the series is $\sum_{n=0}^{\infty}\frac{(-2)^n}{(n+1)}\cdot\left(\frac{1}{2}\right)^n = \sum_{n=0}^{\infty}\frac{(-1)^n}{(n+1)}$. This converges conditionally by the alternating series test, but not absolutely, since the series of absolute values $\sum_{n=0}^{\infty}\frac{1}{(n+1)}$ diverges by the integral test.

For $x = -\frac{1}{2}$, the series is $\sum_{n=0}^{\infty}\frac{(-2)^n}{(n+1)}\cdot\left(-\frac{1}{2}\right)^n = \sum_{n=0}^{\infty}\frac{1}{(n+1)}$, which, as above, diverges by the integral test.

The interval of convergence is therefore $-\frac{1}{2} < x \leq \frac{1}{2}$.

↑ converges conditionally

By the ratio test the convergence is absolute on the interior of the interval; the only point where the series converges conditionally is the right endpoint $x = \frac{1}{2}$.

(c) Using the formula given for the coefficient on the degree n term, we compute the fifth-degree Maclaurin polynomial for f: $1 - x + \frac{4}{3}x^2 - 2x^3 + \frac{16}{5}x^4 - \frac{16}{3}x^5$, so

$f(0.09) \approx 1 - (0.09) + \frac{4}{3}(0.09)^2 - 2(0.09)^3 + \frac{16}{5}(0.09)^4 - \frac{16}{3}(0.09)^5 \approx 0.9195.$

(d) We do not need to use the Lagrange Error Bound to calculate a bound on the error because the terms of the series alternate uniformly "+-+-+-"... In an alternating series, the error is less than the absolute value of the next term, i.e., error $\leq \left|\frac{(-2)^6}{6+1}(0.09)^6\right| < 0.000005 < 10^{-5}$.

AP CALCULUS BC DIAGNOSTIC TEST CORRELATION CHART

Use the results of your test to determine which sections you should spend the most time reviewing.

Multiple Choice Question	Review Chapter: Subsection of Chapter
1	Chapter 16: L'Hôpital's Rule
2	Chapter 13: The Slope of a Curve at a Point
3	Chapter 25: Comparing Series to Test for Convergence or Divergence
4	Chapter 10: Parametric Functions Chapter 16: Interpreting the Derivative As a Rate of Change
5	Chapter 10: Polar Functions Chapter 19: Antiderivatives—The Indefinite Integral
6	Chapter 21: Integration by Parts
7	Chapter 14: Critical Points and Local Extrema Chapter 15: Inflection Points
8	Chapter 12: Derivatives of Trigonometric Functions Chapter 12: Chain Rule Chapter 12: Exponential and Logarithmic Functions
9	Chapter 22: Solving Logistic Differential Equations and Using Them in Modeling
10	Chapter 21: Antiderivatives by Improper Integrals
11	Chapter 10: Parametric Functions Chapter 13: The Slope of a Curve at a Point
12	Chapter 18: Solids of Revolution
13	Chapter 21: Antiderivatives by Simple Partial Fractions
14	Chapter 14: Corresponding Characteristics of the Graphs of f and f' Chapter 15: Inflection Points
15	Chapter 9: Continuity and Limits Chapter 11: The Relationship Between Differentiability and Continuity
16	Chapter 12: Implicit Differentiation
17	Chapter 16: Differential Equations
18	Chapter 20: What's So Fundamental? The Connection Between Integrals and Derivatives
19	Chapter 21: When F Is Complicated: The Substitution Game
20	Chapter 17: Properties of the Definite Integral Chapter 18: Average Value of a Function
21	Chapter 21: Antiderivatives by Improper Integrals
22	Chapter 25: The Geometric Series with Applications
23	Chapter 21: Integration by Parts
24	Chapter 23: Approximating the Area Under the Curve Using Riemann Sums
25	Chapter 21: Antiderivatives by Simple Partial Fractions
26	Chapter 14: Corresponding Characteristics of the Graphs of f and f'
27	Chapter 26: From Series to Power Series
28	Chapter 20: Interpreting the Fundamental Theorems Graphically

Multiple Choice Question	Review Chapter: Subsection of Chapter
29	Chapter 16: Differential Equations Via Euler's Method
30	Chapter 26: From Series to Power Series
31	Chapter 14: Corresponding Characteristics of the Graphs of f and f'
32	Chapter 16: L'Hôpital's Rule
33	Chapter 26: Taylor Polynomial Approximation
34	Chapter 22: Separable Differential Equations
35	Chapter 12: Derivatives of Parametric, Polar and Vector Functions
36	Chapter 25: The Integral Test and the Convergence of the p-Series
37	Chapter 11: The Relationship Between Differentiability and Continuity
38	Chapter 14: Equations Involving Derivatives: Translating Between Verbal Descriptions and Equations Involving Derivatives
39	Chapter 20: Interpreting the Fundamental Theorems Graphically
40	Chapter 26: Taylor Polynomial Approximation
41	Chapter 23: Approximating the Area Under Functions Given Graphically or Numerically
42	Chapter 14: Equations Involving Derivatives: Translating Between Verbal Descriptions and Equations Involving Derivatives Chapter 15: Inflection Points
43	Chapter 11: The Relationship Between Differentiability and Continuity
44	Chapter 15: Concavity and the Sign of f'
45	Chapter 12: Chain Rule

AP CALCULUS BC DIAGNOSTIC TEST CORRELATION CHART

Free Response Question	Review Chapter: Subsection of Chapter
1(a)	Chapter 18: Area Between Two Curves
1(b)	Chapter 18: Solids of Revolution
1(c)	Chapter 18: Volumes of Solids
2(a)	Chapter 13: The Slope of a Curve at a Point
2(b)	Chapter 18: Average Value of a Function Chapter 23: Trapezoidal Approximation
2(c)	Chapter 14: Mean Value Theorem
2(d)	Chapter 18: Average Value of a Function
3(a)	Chapter 16: Differential Equations
3(b)	Chapter 12: Implicit Differentiation Chapter 15: Concavity and the Sign of f'
3(c)	Chapter 22: Separable Differential Equations
4(a)	Chapter 16: Interpreting the Derivative as a Rate of Change
4(b)	Chapter 16: Interpreting the Derivative as a Rate of Change
4(c)	Chapter 12: Derivatives of Parametric, Polar, and Vector Functions
4(d)	Chapter 8: Limits Approaching Infinity
5(a)	Chapter 21: Integration by Parts Chapter 21: Antiderivatives by Improper Integrals
5(b)	Chapter 20: What's So Fundamental? The Connection Between Integrals and Derivatives
5(c)	Chapter 12: Chain Rule Chapter 20: What's So Fundamental? The Connection Between Integrals and Derivatives
6(a)	Chapter 26: Taylor Polynomial Approximation
6(b)	Chapter 26: From Series to Power Series
6(c)	Chapter 26: Taylor Polynomial Approximation
6(d)	Chapter 26: How Close Is Close Enough? The Lagrange Error Bound

Part Three

AP CALCULUS REVIEW

CHAPTER 4: GRAPHING WITH A CALCULATOR

IF YOU LEARN ONLY TWO THINGS IN THIS CHAPTER . . .

1. To graph a function, your calculator needs to know the formula for the function and the viewing window, or part of the graph you want to see.
2. Entering more than one function allows you to see more than one graph on one set of axes, and is important for determining where curves intersect.

USING YOUR CALCULATOR TO GRAPH A FUNCTION

There are many situations on the AP exam in which it is helpful to see a graph of a function. A graph can help us visualize the function or figure out what equation or equations we need to solve. Usually graphing a function (or functions) is a precursor to doing something else with the function(s)—like solving an equation involving the function(s), finding the intersection point(s) of two or more curves, finding the area between two curves, or finding the volume of a solid of revolution.

To graph a function, we need to tell the calculator two things: the formula for the function and the viewing window, or the part of the graph we want to see. Your calculator should have a screen for entering functions and a screen for setting the graph window. There is also a screen to help you zoom in or out on specific parts of the graph. Your calculator probably also has a default viewing window; the graph will show in this window if you do not set a specific viewing window.

TI-83: We enter functions on the Y= screen. We can set the viewing window using the WINDOW screen, where we enter the minimum and maximum *x*- and *y*-values. For instance, if we set:

```
WINDOW
Xmin=-2
Xmax=4
Xscl=1
Ymin=-5
Ymax=10
Yscl=1
Xres=1
```

then when we press GRAPH, the x-axis will be drawn with values from –2 to 4 and the y-axis will be drawn with values from –5 to 10.

We can also set the viewing window using the ZOOM key. There are different options for how to zoom in on the graph.

Example:

Graph $y = x^2 + 20$

We start by entering this function and viewing it in the standard window.

TI-83:

Press	To
Y= X, T, θ, n x^2 + 20 ZOOM 6	Enter $y_1 = x^2 + 20$ on the Y= and view the graph

What do you see? On the TI-83 you don't see anything. The standard viewing window for the TI-83 is [–10, 10] × [–10, 10] and this graph doesn't live in this window. (Why? Did you say that this is a parabola that opens up and its vertex is at (0, 20)? Good for you.)

If we know what window we want to view, it makes sense to set the window manually. For this function, the window [–10, 10] × [19, 35] gives a nice picture of the overall graph.

TI-83:

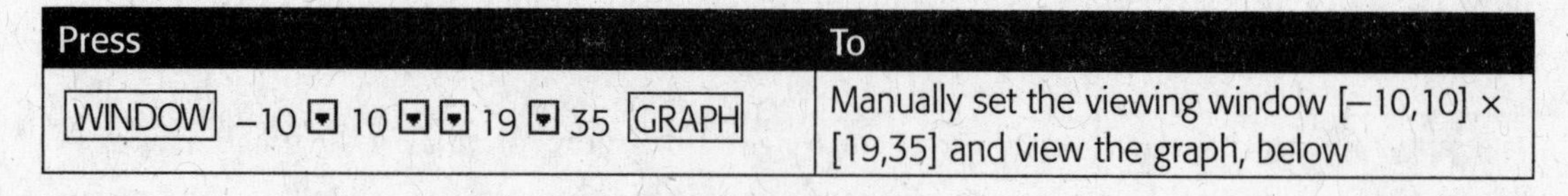

Press	To
WINDOW –10 ▼ 10 ▼ ▼ 19 ▼ 35 GRAPH	Manually set the viewing window [–10,10] × [19,35] and view the graph, below

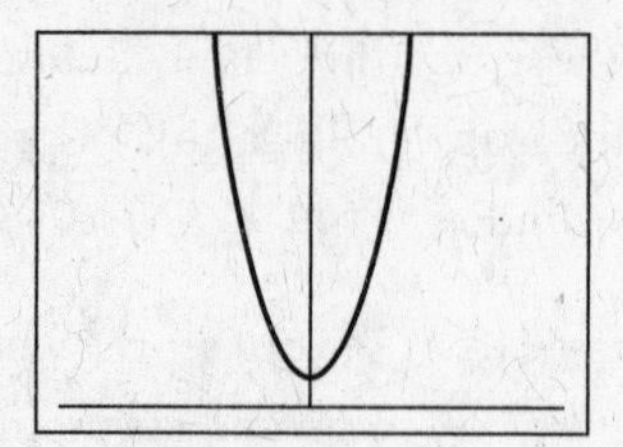

Which Window Is Best?

The window of the graph is not an exact science. It must be large enough to cover the important parts of the graph and small enough so that the image of these parts is as large as possible. The window used in each example or question in the text is only one possibility. Other windows will often work just fine as well.

Zoom

If we don't know what window to view, the zoom menu is a good place to start. There should be a "best fit" option on the zoom menu that will give you a snapshot of the graph that includes an appropriate viewing window.

TI-83:

Press	To
ZOOM 0	Select option **0:ZoomFit** from the ZOOM menu which will automatically draw the graph in an appropriate window

We can also use the calculator's zoom menu to enlarge a specific piece of a graph. If we want to zoom in on the vertex of the parabola in the example above, we can move the cursor to the vertex, using the arrow keys and then use the "zoom in" option on the zoom menu to enlarge the graph near the vertex.

TI-83:

Press	To
ZOOM 2 Arrow buttons	Select option **2:ZoomIn** from the ZOOM. The graph will then be redrawn and we can position the cursor
Arrow buttons ▲ ▼ ▶ ◀	Position the cursor over the point to blow up
ENTER	Zoom in

and

Press	To
ENTER	Zoom in some more

and

Press	To
ZOOM 0	Return to best-fit option **0:ZoomFit**

If we zoom in repeatedly on the vertex, the function starts to look like a straight line. Why does this happen? Did you say the smooth functions look like straight lines locally?

It's a good idea to experiment with the features on your zoom menu. If you aren't a zoom expert, take some time now to learn about this aspect of your calculator.

GRAPHING MORE THAN ONE FUNCTION

On your calculator's screen for entering functions, you should be able to enter more than one function—i.e., you should be able to see the graphs of multiple functions on the same set of axes. We do this, for instance, when we want to see where two curves intersect. There are many reasons we could want this information. For example, we might want to visualize the solution to a system of two equations in two unknowns (the solutions to this system will be the points where the graphs of the two functions intersect). We may be trying to set up an integral to find the area between two curves.

GRAPHING IN ACTION ON THE AP EXAM

Point of Intersection of Two Curves

Questions on the AP exam that require graphing for more than just discovery purposes (i.e., getting a general idea of what to do) usually involve finding the intersection of two curves. On the exam, we'd then use this information to do something like compute the area between two curves or find the volume of a solid of revolution.

Example:

We are given the functions $y_1 = \cos^{-1} x$ and $y_2 = 2x^2$

Region C is bounded by the y-axis and the graphs of y_1 and y_2.
Region D is bounded by the x-axis and the graphs of y_1 and y_2.

(a) Find the x coordinate of the point of intersection of y_1 and y_2.

(b) Find the area of D.

Solution:

We'll solve just part (a) here, to illustrate how a graphing utility can be used to find the intersection of two curves. We can solve part (b) with a calculator, using commands that we discuss in Chapter 6. This problem appears again in the questions at the end of Chapter 6.

To graph trig functions, we need to have our calculator in radian mode. It's also helpful to have the calculator set for floating decimal points. This means that a decimal will display the way we usually see it. For example, the number 5.67 will not appear with extra decimal places (e.g., 5.67000000), the way it would if we fixed the number of decimal points.

TI-83/TI-89:

Press	To
MODE ▼	Open the MODE menu and go to options for decimal appearance
Arrow buttons ► ◄ ENTER	Highlight **FLOAT** and select this option
▼	Go to radian/degree option
Arrow buttons ► ◄ ENTER	Highlight **RADIAN** and select this option
2nd MODE	To QUIT the MODE menu

If we don't have any idea about where these two curves live on the axes and where they intersect, the standard viewing window is a good place to start. We can then position the cursor over the point where the two curves intersect and zoom in. Once we see where the two curves intersect, we can refine the viewing window using the window menu.

TI-83:

Press	To
Y =	Access the Y = screen to input functions
$\cos^{-1}$ X,T,θ,n	Enter the function $Y_1 = \cos^{-1} x$
▼	Scroll down to Y_2
2 X,T,θ,n x^2	Enter the function $Y_2 = 2x^2$
ZOOM 6	To select the **6:Z Standard** from the ZOOM menu and view the graph in the standard [−10,10] × [−10,10] window
Arrow buttons ▲ ▼ ► ◄	Position the cursor near the intersection point
ZOOM 2 ENTER	Zoom in near the intersection point

When the graph is showing, we can zoom in again and/or use the arrow keys to get an idea of the parameters of the graph. When we move the cursor with the arrow keys, the *x*- and *y*-coordinates of the cursor show on the screen. Once we explore the graph, we decide on an appropriate viewing window. The window [−.2,1.5] × [−.2,2] makes sense.

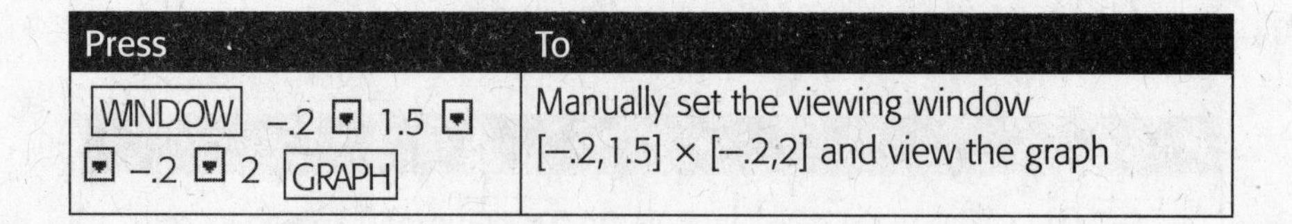

Press	To
WINDOW −.2 ▾ 1.5 ▾ ▾ −.2 ▾ 2 GRAPH	Manually set the viewing window [−.2,1.5] × [−.2,2] and view the graph

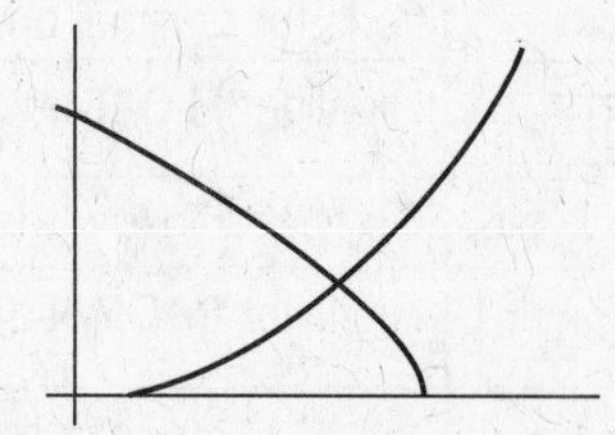

Once the graph is showing,

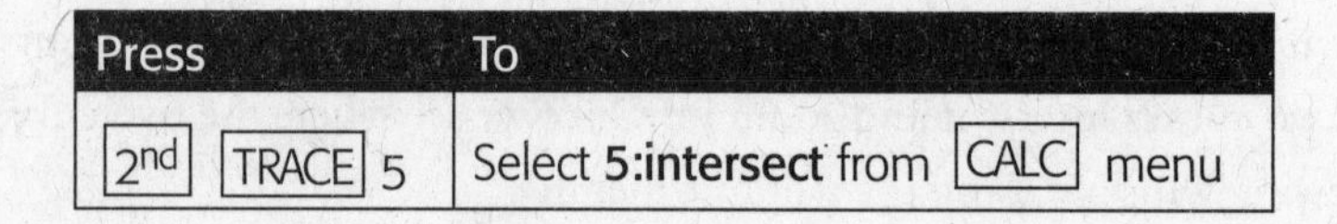

Press	To
2nd TRACE 5	Select **5:intersect** from CALC menu

We are now prompted to choose points on each of the curves and a starting guess. Since the cursor is close to the point of the intersection, we can skip these steps.

Press	To
ENTER ENTER ENTER	Skip prompts for first curve, second curve, and guess

The cursor always starts in the middle of the screen.

The coordinates of the intersection point appear. The point is 0.65463568.

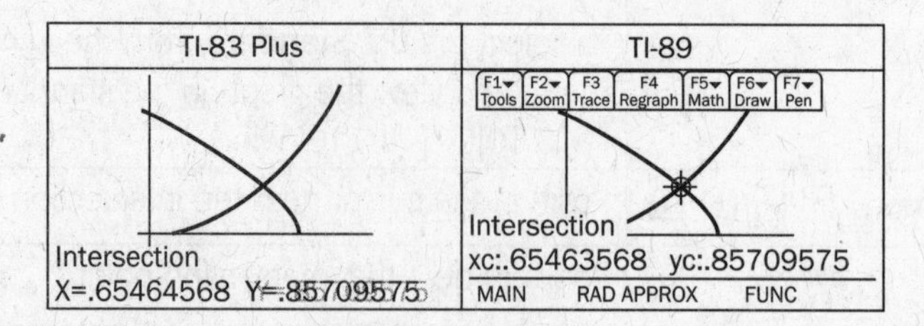

Note that this value is now stored in variable x. It can be saved by returning immediately to the main menu and storing it in memory A.

Press	To
2nd MODE	QUIT the graph and return to main screen
X,T,θ,n STO ALPHA MATH ENTER	Store the contents of the *x*-coordinate of the intersection point in memory location A

We can recall this value later, if we want to use it.

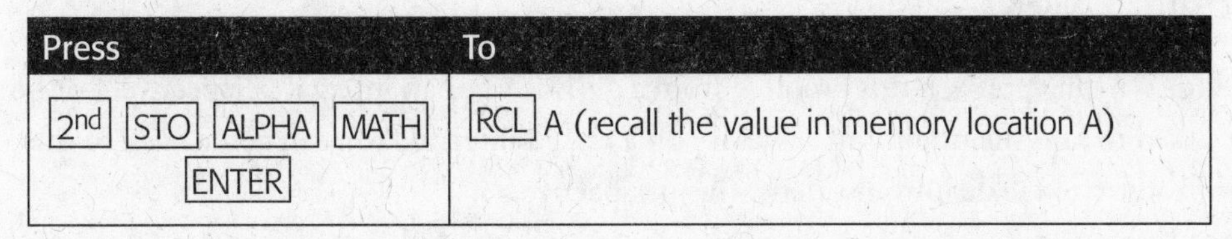

Press	To
2nd STO ALPHA MATH ENTER	RCL A (recall the value in memory location A)

TI-89: Most of the commands for this problem on the TI-89 are the same as the TI-83 with a few exceptions.

1. The CALC command on the TI-89 refers to calculus commands, not graphing commands. To find the intersection of two curves on the TI-89, you need to have the graph screen open. Then press F5 Math and the Intersect command appears.
2. When executing the Intersect command, move the cursor close to the point of intersection. When they ask for first curve, press ENTER and second curve, press ENTER. Then you are asked for a lower bound. Toggle to the left of the point of intersection, press ENTER.

For the upper bound, toggle to the right, press ENTER. The value of the point of intersection, 0.65463568 is automatically stored in *xc* and can be retrieved in the main screen.

Note: On the TI-89 all parentheses must be closed!

Finding the Area Under a Curve—AKA Definite Integral

A calculator is an excellent tool for understanding problems such as "Given a function that describes the flow of people into a museum, how many people have entered the park by a certain time?" or "Given the rate at which a tank is filling, find the volume of liquid in the tank by a certain time." Problems like these are called accumulation problems because some quantity is accumulating.

Example:

At a fuel depot, the rate of change of the gas in a tank from time $t = 0$ to $t = 5$ hours is given by the function $y = t(t-2)(t-3)(t-5)^2$. Let $f(t)$ represent the amount of gas in the tank at time t. At $t = 0$ there are 100 gallons of gas in the tank.

(a) What is the relationship between y and $f(t)$?
(b) For what value of t is the number of gallons of gas in the tank a minimum? Justify your answer.
(c) For what value of t is the number of gallons of gas in the tank a maximum? Justify your answer.

Solution:

(a) To even get started with this problem you need to answer and understand this part. If y is the rate of change in the amount of gas in the tank and $f(t)$ is the amount of gas in the tank at time t, then y is the rate of change of f—otherwise known as the derivative of f. That is, $y = f'(t)$.

Now that we know this we can use our calculator and the Fundamental Theorem of Calculus to answer parts (b) and (c).

(b) and (c) We can approach this problem from a derivative or an integral perspective—or both. We are asked to find the minimum value of *f* on a closed interval. You can come back to this problem and try to tackle it from a derivative perspective.

Using the Fundamental Theorem of Calculus, we know that the amount of fuel in the tank at time *t* is given by the area under the curve from zero to *t*. We can graph the function to get a handle on the area.

We graph the function $y = x(x - 2)(x - 3)(x - 5)^2$ on a reasonable window, and try to qualitatively judge where the area is a minimum and where it is a maximum. The window [0,5] × [−10,50] works well.

Press	To
[Y =]	Access the [Y =] screen to input functions
[CLEAR]	Clear the function in Y_1, if necessary
[▼] [CLEAR]	Clear the function in Y_2, if necessary Repeat this step to clear any remaining functions
[▲]	To move cursor back to Y_1. Repeat, if necessary
[X,T,θ,n] ([X,T,θ,n] −2) ([X,T,θ,n] −3) ([X,T,θ,n] −5) [x^2]	Enter the function $Y_1 = x(x - 2)(x - 3)(x - 5)^2$
[WINDOW] 0 [▼] 5 [▼] [▼] −10 [▼] 50 [GRAPH]	Manually set the viewing window [0,5] – [−10,50] and view the graph

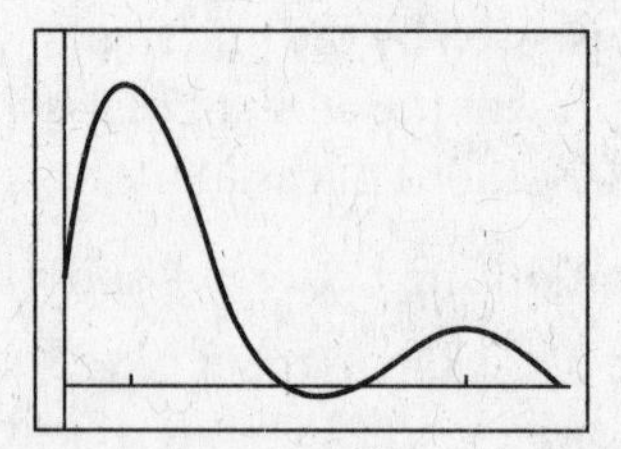

The net accumulation of gas in the tank from zero to *t* is given by the area under the curve from 0 to *t* (where the area under the *x*-axis is considered negative). Since most of the area lies above the *x*-axis, for any time *t*, the net accumulation of gas in the tank will be positive—that is, there is always *more* gas in the tank than at the starting time. This shows the number of gallons of gas in the tank is at a minimum at $t = 0$ and is at a maximum at $t = 5$, without doing any calculus!

REVIEW QUESTIONS

1. If the graph of the derivative of y is described by $\frac{dy}{dx} = (x-1)(x-2)(x-3)^2 - .22$, how many critical points does the graph of y have on [0, 4]?

   (A) 1
   (B) 2
   (C) 3
   (D) 4
   (E) 5

2. If $g(x) = x^{\frac{2}{3}}$, then which of the following must be true?

   I. The function $g(x)$ is defined for all x.
   II. The function $g(x)$ is differentiable for all x.
   III. The function $g(x)$ has a minimum at $x = 0$.

   (A) II only
   (B) III only
   (C) I and II only
   (D) I and III only
   (E) I, II, and III

3. The derivative of f is given by $f'(x) = (x-2)(x-4)(x-6)(x-8)$. At what values of x will f have a relative maximum?

   (A) 5
   (B) 4 and 6
   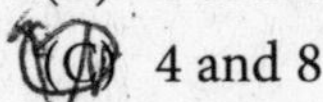
   (C) 4 and 8
   (D) 2 and 4
   (E) 2 and 6

4. Suppose $f(x)$ is a differentiable function for all x and $f(x) < 0$ for all x. If $g'(x) = f(x)(x+2)^2(x-1)^3(x-3)$, which statement is correct?

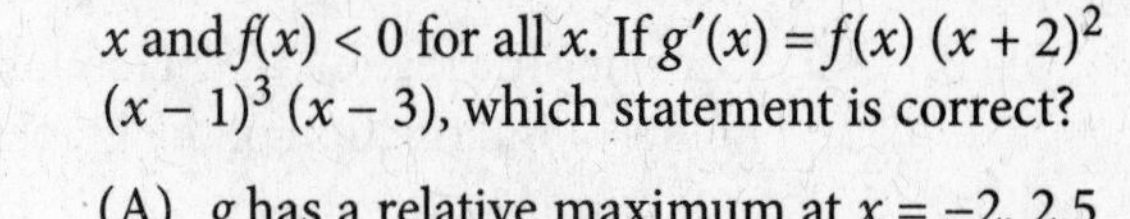

   (A) g has a relative maximum at $x = -2, 2.5$ and a relative minimum at $x = -1$.
   (B) g has a relative maximum at $x = -2, 1, 2.5$ and a relative minimum at $x = -1$.
   (C) g has a relative maximum at $x = 1$ and a relative minimum at $x = 3$.
   (D) g has a relative maximum at $x = 3$ and a relative minimum at $x = 1$.
   (E) g has a relative maximum at $x = 4$ and a relative minimum at $x = 2$.

5. Let g be the function the derivative of which is given in problem 4. How many inflection points does g have?

   (A) 0
   (B) 1
   (C) 2
   (D) 3
   (E) 4

6. Which of the following statements about the function $y = x^3(3-x)$ is true?

   I. The function has an absolute maximum.
   II. The function has an absolute minimum.
   III. The function has a relative minimum.

   (A) I only
   (B) II only
   (C) III only
   (D) I and II
   (E) I and III

7. Find the value of $x > 0$ for which the graphs of $f(x) = x \ln x$ and $g(x) = x^3(3-x)$ have parallel tangent lines.

   (A) 1.768
   (B) 1.983
   (C) 2.067
   (D) 2.155
   (E) 2.398

Free-Response Question

8. Consider the region, R, enclosed by the graphs of $f(x) = \cos(x - 0.4)$ and $g(x) = x^2$. Find the points of intersection of the graphs of f and g.

ANSWERS AND EXPLANATIONS

1. B

We graph the derivative function in an appropriate window, say, $[0,4] \times [-1,1]$ and count how many times the curve crosses the *x*-axis. See first two frames below. At first, it looks like the graph crosses the *x*-axis three times or maybe four. What is happening between 2 and 3? If we zoom in twice on that point and expand the image, it is clear that the curve does not cross the *x*-axis at that point. It only crosses the *x*-axis twice, yielding two critical points.

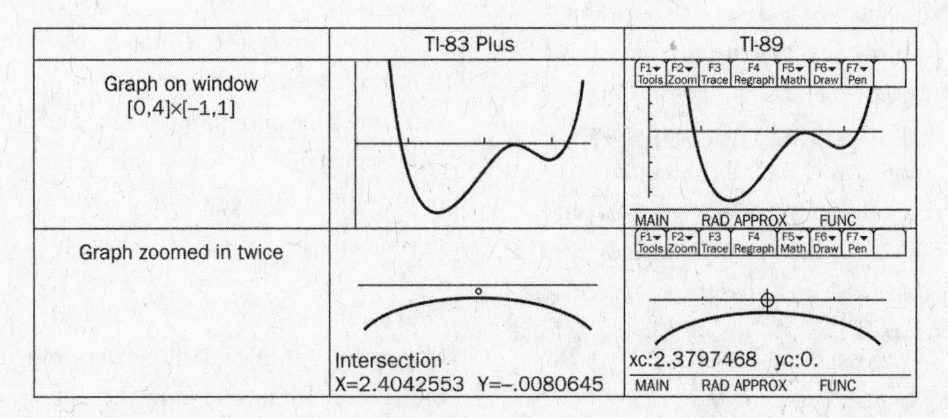

2. D

Solving without a calculator: The function g is defined for all values of x, so statement I is true.

We next compute the derivative:

$$g'(x) = \frac{2}{3}x^{-\frac{1}{3}} = \frac{2}{3\sqrt[3]{x}}.$$

The derivative is not defined at $x = 0$. Therefore, g is not differentiable at $x = 0$, so II is not true.

This is a good place to throw in some test strategy. After verifying that statement I is true and statement II is not true, we can eliminate answer choices (A), (B), (C), and (E). We're done without even checking whether statement III is true!

To show that g has a minimum at $x = 0$ compute $g'(-1) = -1$ and $g'(1) = 1$. The slope of g, $g'(x)$ changes from negative to positive as the x values increase across 0. Therefore, at $x = 0$ the graph of g has a minimum and it happens to be an absolute minimum.

Solving with a calculator: We graph g on an appropriate window, say $[-4,4] \times [-.5,2.5]$, and examine both the function and the graph. The function is defined for all values of x as shown above (statement II is true).

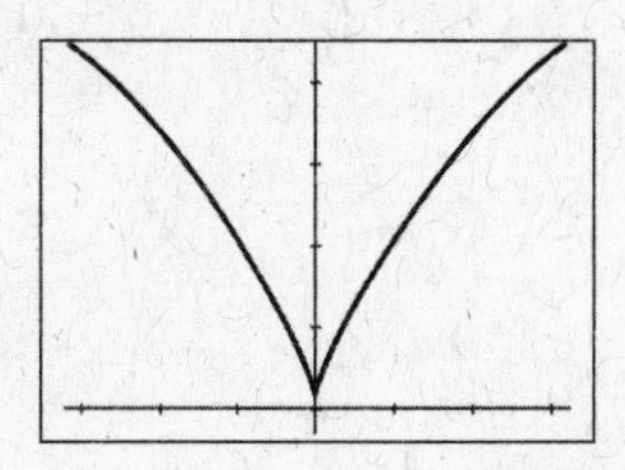

By inspection the graph has a minimum at $x = 0$, (III is true) and a sharp point at $x = 0$ is a point where the derivative does not exist (statement II is not true). Choice (D) is correct.

3. E

We can answer this question with the use of a calculator. A relative maximum occurs when the graph of the derivative changes from positive to negative. When the graph of the derivative is positive, the slope of the function is positive and the function is going up. When the graph of the derivative is negative, the slope of the function is negative and the function is going down. At the point where the slope first goes up and then down, the graph has a relative maximum point. The graph of the derivative goes from plus to minus at only $x = 2$ and $x = 6$. We graph this function on an appropriate window, say $[0,10] \times [-20,20]$ and observe that the sign of the function changes from positive to negative at $x = 2$ and $x = 6$. The correct answer is therefore (E).

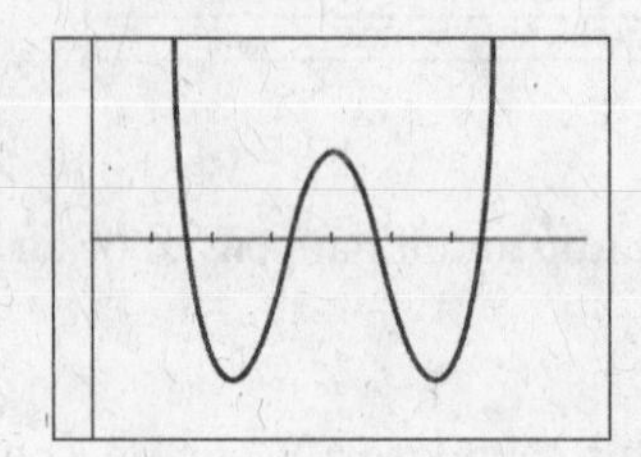

4. D

Calculator Solution: Think of a simple function that is negative for all x, for example: $f(x) = -1$. Now we can think of $g'(x)$ as $(-1)\,(x+2)^2\,(x-1)^3\,(x-3)$ in a window such as $[-4, 4] \times [-40, 44]$.

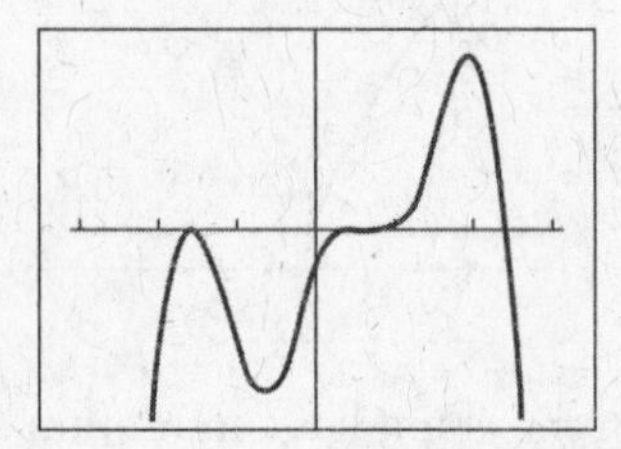

From the graph, we see that there is a relative maximum at $x = 3$, because the graph of the derivative changes from positive to negative at that point. There is a relative minimum at $x = 1$, because the graph of the derivative changes from negative to positive at that point. At $x = -2$ and 2.5, there may be a maximum or minimum of $g'(x)$ but not of $g(x)$.

5. D

A point of inflection occurs when a function changes concavity, that is, when the second derivative changes sign. The second derivative is the derivative of the derivative, so when the graph of the derivative changes from increasing to decreasing or vice versa, an inflection point occurs. The inflection points correspond to the relative maximums and minimums on the graph of g' that we created for problem 4.

6. A

We can answer this question by viewing the graph of the function in a window that is large enough to show all of the turns of the function, say $[-3,5] \times [-10,10]$.

TI-83: In the Y= screen, enter $Y_1 = x^3(3 - x)$. Set the window $[-3,5] \times [-10,10]$ and press GRAPH.

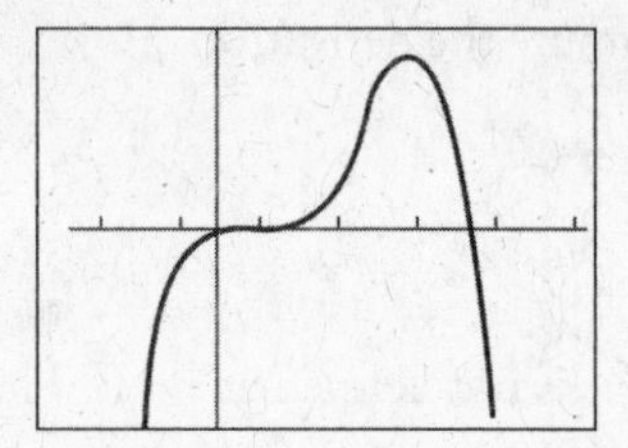

From the graph, we see that the function has an absolute maximum and does not have an absolute or a relative minimum. The correct answer is (A).

7. D

We will be solving this problem using a graphing utility. We'll see a similar problem in the next chapter that we will solve with the solver utility.

Two lines are parallel when they have equal slopes. The slope of the tangent line is the derivative of the function. Therefore, to solve this problem, we need to solve the equation $f'(x) = g'(x)$.

We first compute the derivatives:

$$f(x) = x\ln x \quad \Rightarrow \quad f'(x) = x \cdot \frac{1}{x} + 1 \cdot \ln x$$
$$= 1 + \ln x = 1 + \ln x$$
$$g(x) = x^3(3 - x) = 3x^3 - x^4 \quad \Rightarrow \quad g'(x)$$
$$= 9x^2 - 4x^3$$

We now graph these two derivative functions on the same set of axes and view them in an appropriate window, say [–0.1,–3] × [–8,8]. We can then use the calculator's utility for finding the intersection of two curves.

TI-83 or TI-89: Use the CALC (TI-83) or F5 (TI-89) function and find the intersection of the curves.

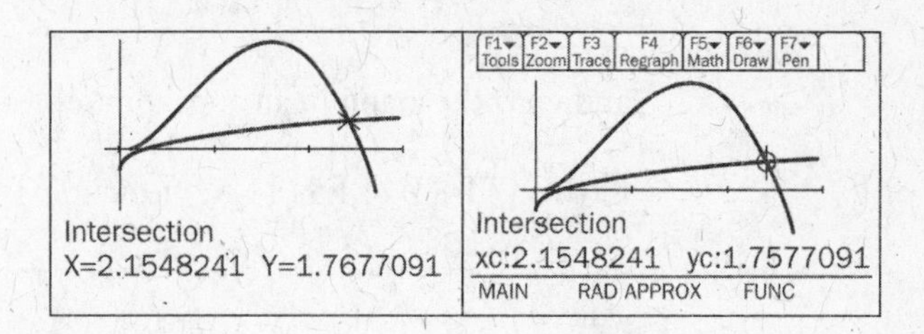

f'g+fg'
(1)(lnx)+(x)(1/x)
lnx+1

x^3(3-x)
3x^3-x^4
9x^2-4x^3

FREE-RESPONSE ANSWER

8. Before graphing trig functions, we must make sure that our calculator is in radian mode. We then draw these two curves on the same set of axes and use the calculator's utility to find the points where the two curves intersect.

TI-83 and TI-89: On the Y= screen enter $Y_1 = \cos(x - 0.4)$ and in Y_2 enter $Y_2 = x^2$ and view the graph in an appropriate window like $\left[-\frac{\pi}{2}, \frac{\pi}{2}\right] \times [-0.1, 1.2]$. Use the CALC (TI-83) or F5 (TI-89) function and find the intersection of the two curves.

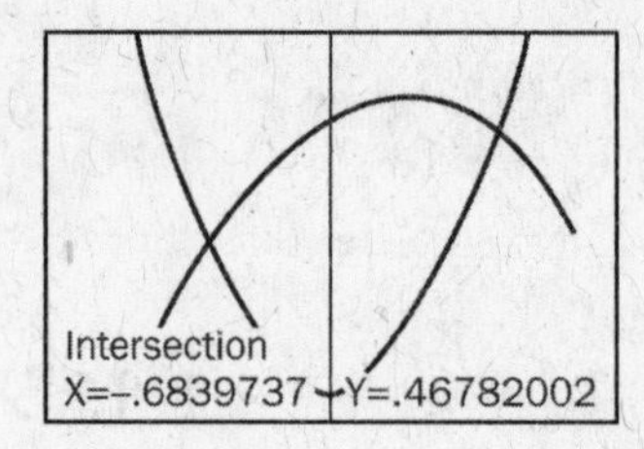

The two points of intersection are as follows: $x = -.683, f(x) = g(x) = .467$ and $x = .929, f(x) = g(x) = .863$ (alternatively, the first solution can be $x = -.684, f(x) = g(x) = .468$).

If we were going on to answer additional questions about these curves, we could store the coordinates of the intersection points in the calculator's memory.

On an actual AP exam, this problem would have several more parts, asking for things like a volume of a solid of revolution or an area between two curves. We'll see this problem again later in this book, where we'll address its additional parts.

CHAPTER 5: EQUATION SOLVING WITH A CALCULATOR

IF YOU LEARN ONLY THREE THINGS IN THIS CHAPTER . . .

1. Once an equation has been set equal to zero, solutions of equations are the points where the graph crosses the *x*-axis.
2. A more sophisticated calculator may save you time on the exam if it is able to take over at an earlier point in the problem.
3. Graphing calculators make solving volumes of solids of revolution and slice problems much easier, although on the AP exam, the problem setup is often more important than the correct solution.

A graphing calculator is an excellent tool for solving equations. Two utilities are available on most graphing calculators that can be used to find numerical solutions—finding the zero of a graph and the solver utility. More sophisticated calculators, like the TI-89, can even solve an equation for a letter, a variable or a constant, but that isn't necessary on the AP exam, so we won't discuss it here.

One way to solve an algebraic equation is to set it equal to zero; this is the plan of attack we use when solving most equations by hand—quadratic equations, trig equations, etc. By setting equations to zero, we can also use a graphing calculator to find the solution. Any graphing calculator can find a numerical solution to any equation that is set to zero. More sophisticated models don't need to have the equation set to zero to solve it.

SOLVING EQUATIONS BY FINDING THE ZERO OF A GRAPH

Once we have an algebraic equation set to zero, we can enter the corresponding function into the calculator and graph it. For example, if we want to solve the equation $x^3 - 9x + 3 = 4x^2$, we first set it equal to zero: $x^3 - 4x^2 - 9x + 3 = 0$, then graph the function $y = x^3 - 4x^2 - 9x + 3$ in an appropriate window.

The solutions of the equation are the points where the graph crosses the x-axis. Graphing calculators have a utility that allows us to find these points, by focusing in around each point where the graph crosses the axis. We have to find the coordinate of each point of intersection separately.

TI-83: Enter the function $y = x^3 - 4x^2 - 9x + 3$ into the [Y=] screen and set an appropriate window. The window [–5, 7] × [–50, 50] gives a nice overall picture of the graph. On the [CALC] ([2nd] [TRACE]) menu choose the "zero" utility by using the arrow keys or pressing [2]. We are prompted for a left bound. Move the cursor to the left of the left-most zero. Press [ENTER]. We are prompted for a right bound. Move the cursor just to the right of this zero. Be careful not the move the cursor too far to the right—you want to trap just one zero between the left bound and the right bound. After the cursor is positioned to the right of the zero, press [ENTER]. We are prompted for a guess. We can skip this step by pressing [ENTER] again. The cursor appears over the zero and the coordinate $x = -1.826545$ appears. On the exam, we round the answer to $x = -1.827$.

We repeat this procedure for the other two zeros by setting left- and right-bounds on either side of each of these zeros. We find that the other zeros occur at $x = .2970$ and $x = 5.530$.

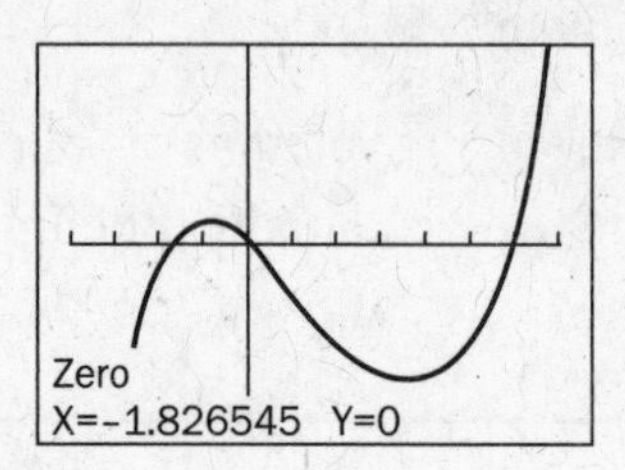

Finding the zero of the graph is the best way to solve equations—it has the fewest quirks and what you see is what you get. It's important, though, to make sure that the function is being viewed in an appropriate window.

By graphing an equation, we see what a function looks like and we get visual data about the number and locations of the solutions. Maybe the curve crosses the x-axis several times, indicating several solutions, but maybe it doesn't cross the x-axis at all.

GRAPHING CALCULATORS DON'T LIKE GRAPHS THAT DON'T CROSS THE AXIS

Because of the algorithm that some calculators use to find the zeros of a graph, the calculator may have trouble finding the coordinates of a zero that just touches the graph and doesn't cross it.

For example, on the TI-83, if we try to solve $x^2 = 0$ by graphing $y = x^2$ and finding the zero of the graph, it will return an error message: "ERR: NO SIGN CHNG." Ummm…what? Even someone who slept through most of algebra knows that 0 is a solution to this equation. But the calculator won't even take it on. The bottom line is, there is no replacement for knowing calculus and you need to know your calculator well enough to know its quirks.

Become skillful with the graphing method before using the solver method that comes next. Any equation that can be solved with the solver utility can also be solved using the zero utility.

The example cited above is an example of a polynomial function. Polynomial functions are often used in calculus problems, so it is helpful to know two of their important properties when graphing them on the calculator:

The degree of a polynomial indicates the maximum number of real roots. For example, if you want to graph $y = x^4 - 16x^3 - 7x^2 + 490x$, a fourth-degree polynomial, you should be sure that your window is expanded far enough to accommodate the possibility that four roots will appear. In the first graph below, created on a TI-84 calculator, a ZOOM Standard window [–10,10] by [10,10] was used. This graph is not very useful because only two roots are showing and it is not possible to tell where the function increases and decreases. In the second graph, ZOOMFit was used. Here, only three of the four roots are showing. From this graph it is clear that Xmax must be increased to show all four roots. The window of the third graph, showing all four roots, is [–10,20] by [–2200,1500].

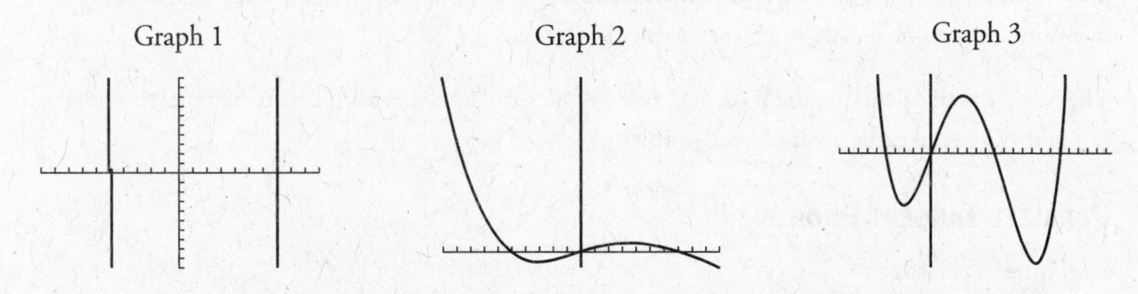

The "end behavior" of a polynomial function is also determined by its degree. In our example above, because the degree 4 is an even number, both ends of the graph should either point up or point down (think of the parabolas $y = x^2$ and $y = -x^2$). If the degree of a polynomial is odd, then one end will point up and one end will point down (think of the basic cubic functions $y = x^3$ and $y = -x^3$). So, by knowing about end behavior, we see that Graph 2 above must be incomplete.

SOLVER UTILITY

Graphing calculators have a specific utility, often called a solver or an equation solver, to solve equations. On older models, this solver may only be able to solve equations that are set to zero. The TI-83 is an example of a calculator like this. Newer, fancier models, such as the TI-89, can handle more complicated equations. A plain-vanilla model will give you a numeric approximation for an answer after you supply a starting guess; a newer model can give you a specific answer.

TI-83: The solver is found in the MATH menu and requires that equations be set equal to zero.

Be careful! The solver may seem like the most straightforward way to solve equations, but, on basic calculator models (like the TI-83/84), it is very quirky. Consider solving equations graphically, instead!

TI-89: The solver function is found in F2 and is quite flexible and fast. The equation need not be set equal to zero.

Regardless of your calculator of choice, make sure you graph any equation in order to put the output from a solver program in context. The AP testmakers will be ready with trap answer choices for students who forget to do so!

EQUATION SOLVING IN ACTION ON THE AP EXAM

Many problems on the AP exam require you to understand a problem conceptually, set it up, and then solve it. Thanks to graphing calculators, we are sometimes spared the need to slog through the computations once we've set the problem up. But keep in mind that the people grading the exam know that it's easier to solve complex calculus problems these days; as a result, they weight the problem setup more heavily when they grade the exam.

The fancier your calculator, the more toil you will be spared on the exam because a fancier model can take over at an earlier point in the problem.

PARALLEL TANGENT PROBLEM

Example:

Find the value of $x > 0$ for which the graphs of $f(x) = e^{3x}$ and $g(x) = \ln\sqrt{x}$ have parallel tangent lines.

Solution:

We start by rephrasing this question in terms of something we can compute—derivatives. Two lines are parallel if their slopes are equal. The slope of the tangent line is the derivative. In the language of calculus, we are being asked to find the points where the derivatives of the two functions are equal. That is, we need to solve the equation $f'(x) = g'(x)$.

More sophisticated calculators, such as the TI-89, can solve equations involving derivatives. In the solver ([F2]) we enter:

TI-89:

```
F1 Tools | F2 Algebra | F3 Calc | F4 Other | F5 PrgmIO | F6 Clean up

■solve(d/dx(e^(3·x)) = d/dx(1n(√x ▸
                         x=.1172445704
solve(d(e^(3x),x)=d(1n(√(...
MAIN     RAD APPROX     FUNC     1/30
```

On a basic model like the TI-83 or TI-84, we need to compute the derivatives by hand. On a TI-89, the calculator can compute the derivatives for us:

$$f(x) = e^{3x} \quad \Rightarrow \quad f'(x) = 3e^{3x}$$

$$g(x) = \ln\sqrt{x} = \frac{1}{2}\ln x \quad \Rightarrow \quad g'(x) = \frac{1}{2x}$$

We can now have the equation we need to solve:

$$3e^{3x} = \frac{1}{2x}$$

This problem cannot be solved using paper and pencil. However, it is not difficult with a graphing calculator. One way to proceed is to set the equation equal to zero and use the solver utility. Of course, it's never a bad idea to graph the function first, in this case $y = 3e^{3x} - \frac{1}{2x}$, to get an idea of what we're looking at. After we graph the function, we have an idea of where the solution will be.

TI-83: Enter the equation $0 = 3e^{3x} - \frac{1}{2x}$ in the equation solver. From our graph of the function, it appears that the solution is somewhere around 0.1, so we can use that as our starting guess. We then press [SOLVE] ([ALPHA] [ENTER]).

```
EQUATION SOLVER          | 3e^(3X)-1/(2X)=0
eqn:0=3e^(3X)-1/         | X=.1172445704■...
(2X)■                    | bound={-1E99,1...
```

The solution is $x \approx 0.117$.

VOLUME OF A SOLID OF REVOLUTION

Volumes of solids of revolution and slice problems become much easier on a graphing calculator. When the calculator performs all of the algebraic and numeric operations, our only tough job is to set them up. But the folks who write the AP exam know this, and in recent years much more credit has been given for properly setting up these problems than for finding their solutions.

Example:

Consider the region, R, bounded by the graphs of the functions $f(x) = 2\sin(\pi x)$ and $g(x) = x(x - 1)$.

(a) Find the area of the region R.

(b) For what value of k would the area over [0, k] be $\frac{1}{3}$ of the area of region R?

(c) Suppose the base of a solid is the region R. If cross sections of the solid perpendicular to the x-axis are semi-circles, find the volume of the solid.

(d) Find the volume of the solid generated when the region R is revolved about the horizontal line $y = 4$.

Solution:

At this point in this example, we'll solve only part (a) to see how to use the solver to find the points where the two curves intersect.

(a) To get started, we need to determine where the two curves intersect. That is, we need to solve $2\sin(\pi x) = x(x - 1)$. We graph the functions $f(x) = 2\sin(\pi x)$and $g(x) = x(x - 1)$ on the same set of axes. We can then zoom in on the appropriate region of the graph. To find the points where the curves intersect we can use the solver utility, the zero utility, or we can go back to a tool from Chapter 4 and use the intersect utility to find where the curves intersect.

TI-83: On the [Y=] screen enter the functions $y_1 = 2\sin(\pi x)$ and $y_2 = x(x - 1)$. Set an appropriate viewing window; the window [-1,3] × [3,3] works nicely. Now we'll use the intersect utility on the [CALC] ([2nd] [TRACE]) menu. Select intersect from the menu. Use the arrow to move the cursor to a point near the first intersection point on the first curve. Press [ENTER]. Use the arrows to move the cursor to a point on the second curve near this point. Press [ENTER]. We are prompted for a guess. Enter a guess or skip this step and just press [ENTER] again. The coordinates of the intersection point, (0, 0) appear. Repeat this procedure for the other intersection point to find the coordinates (1, 0).

REVIEW QUESTIONS

1. Find the value of x for which the functions $f(x) = ex^2$ and $g(x) = 2 - (x + 4)^2$ have parallel tangent lines.

   (A) −2.586
   (B) −1.228
   (C) −1.029
   (D) −0.259
   (E) −0.541

2. If $f(x) = \frac{e^x}{x}$ what is the value of c guaranteed by applying the Mean Value Theorem to f on the interval [1, 5]?

   (A) 3.389
   (B) 3.497
   (C) 3.482
   (D) 3.504
   (E) 3.612

3. For which value of x on the interval $\left[\frac{\pi}{4}, \frac{\pi}{2}\right]$ is the slope of the line tangent to the curve $f(x) = \tan(e^x)$ equal to 7?

   (A) 1.203
   (B) 1.297
   (C) 1.301
   (D) 1.306
   (E) 1.354

4. If the derivative of f is given by $f'(x) = 8x^3 - e^x$, then the graph of f has a minimum point at $x =$

   (A) 0.5894
   (B) 0.5967
   (C) 0.6001
   (D) 0.6134
   (E) 0.6295

5. Which of the following is the equation of the line tangent to the graph of $f(x) = e^{3x}$ at the point when the slope of this tangent line is 8?

   (A) $y = \frac{8}{3}x - 0.026$
   (B) $y = 8x + 0.051$
   (C) $y = 8x - 0.051$
   (D) $y = 8x - 0.026$
   (E) $y = 8x + 0.026$

Free-Response Question

6. Given the function $f(x) = \sin(e^x)$

(a) Find the smallest positive value of x such that $f(x) = 0$.

(b) Find the slope of the line tangent to the curve at the point from part (a).

(c) Find the equation of the line tangent to the curve at the point from part (a).

ANSWERS AND EXPLANATIONS

1. C

Two lines are parallel when they have equal slopes. The slope of the tangent line is the derivative. Therefore, to solve this problem we need to solve the equation $f'(x) = g'(x)$.

On a basic graphing calculator, we first need to compute the derivatives manually:

$$f(x) = e^{x^2} \quad \Rightarrow \quad f'(x) = 2x \cdot e^{x^2}$$

$$g(x) = 2 - (x+4)^2 \quad \Rightarrow \quad g'(x) = -2(x+4)$$

Using the derivatives above, we then need to solve the equation

$$2x \cdot e^{x2} = -2(x+4)$$

On a basic calculator, we need to set this equation equal to zero and then solve it using the calculator solver utility:

$$0 = 2x \cdot e^{x2} + 2(x+4)$$

As usual, it's never a bad idea to graph the related function $y = 2x \cdot e^{x2} + 2(x+4)$ to get an idea of what we're looking for.

TI-83: Go to the solver on the MATH menu. Once in the solver, press the up arrow to get to the equation. Enter the equation. Press ENTER. Enter a guess if you have one, or press SOLVE (ALPHA ENTER). The solution is shown in the right screen, $x = -1.029$.

EQUATION SOLVER eqn:0=2Xe^(X²)+2 (X+4)▮	2Xe^(X²)+2(X+...=0 ▪X=-1.029436676... bound={-1E99,1... ▪left-rt=0

A more sophisticated calculator will compute the derivative for us. We can ask it to solve the equation and specify the range in which we want it to search for the solution.

$$\frac{d}{dx}\left(e^{x^2}\right) = \frac{d}{dx}\left(2 - (x+4)^2\right)$$

TI-89: The solver function is found in F2. Because the answer choices for x lie between −3 and 0, it is wise to limit the range of x. At the end of the equation, type the vertical line "when" or "such that" (found under the equal sign) and $x > -3$ and $x < 0$. The word "and" is found in CATALOG.

```
F1 Tools | F2 Algebra | F3 Calc | F4 Other | F5 PrgmIO | F6 Clean up

▪solve(d/dx(e^x²) = d/dx(2-(x▸
                     x=-1.02943667606
...+4)^2,x),x)|x>-3 and x<0
MAIN      RAD APPROX     FUNC     1/30
```

We thus find the solution $x = -1.029$.

2. B

You should know the Mean Value Theorem by name—it is referred to on the exam.

The Mean Value Theorem states that if f is continuous on [a, b] and differentiable on the open interval (a, b), then there exists a c on (a, b), such that

$$f'(c) = \frac{f(b) - f(a)}{b - a}$$

We are asked to find this c. Even with a basic graphing calculator, there is a more sophisticated way to approach this problem, which we'll see in the next chapter. For now, we'll use just a basic method for solving equations. We'll revisit this question in the next chapter.

Before we begin, we should graph the function to see whether it looks continuous and differentiable on the interval in which we are interested, say on the window [0,6] × [2,15].

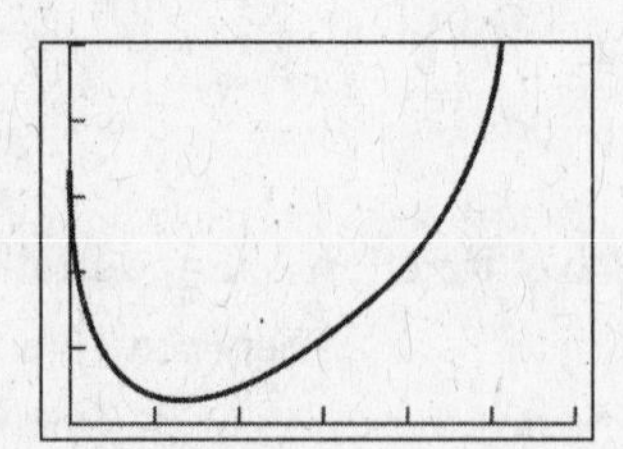

To solve this problem on a basic graphing calculator, we compute:

$$f'(x) = \frac{x \cdot e^x - e^x \cdot 1}{x^2} = e^x \cdot \frac{x-1}{x^2}$$

We evaluate $f(5)$ and $f(1)$.

$$f(5) = \frac{e^5}{5} \approx 29.6826$$

$$f(1) = \frac{e}{1} \approx 2.7183$$

Now we solve the equation:

$$e^c \cdot \frac{c-1}{c^2} = \frac{f(5) - f(1)}{5-1} = 6.741$$

On the calculator, we need to solve the equation

$$0 = e^x \cdot \frac{x-1}{x^2} - 6.741$$

We can use the solver or see where the graph of the related function

$y = e^x \cdot \dfrac{x-1}{x^2} - 6.741$ crosses the x-axis.

TI-83: On the Y= screen enter the function $y = e^x \cdot \dfrac{x-1}{x^2} - 6.741$. The zoom standard window (ZOOM Z Standard) gives a good picture of the graph. On the CALC (2nd TRACE) menu, choose the zero utility. Move the cursor to the left of the point where the curve crosses the axis. Press ENTER. Now move the cursor to the right of where the graph crosses the axis. Press ENTER. We are prompted for a guess. To skip this step, press ENTER again.

The solution $x = 3.4969...$ appears, so the correct answer choice is (B).

3. E

The solution to this problem is the solution to the equation $f'(x) = 7$.

On a basic graphing calculator we need to compute the derivative of $f(x) = \tan(e^x)$ manually. We find

$$f'(x) = e^x \cdot \sec^2(e^x)$$

Next we use the calculator to solve the equation

$$e^x \cdot \sec^2(e^x) = 7 \quad \underset{\text{set the equation equal to zero}}{\Rightarrow} \quad e^x \cdot \sec^2(e^x) - 7 = 0$$

by using the solver utility or another method discussed in this chapter to find the solution to this equation on the interval $\left[\frac{\pi}{4}, \frac{\pi}{2}\right]$.

TI-83: Use the solver utility on the MATH menu. On the solver screen, use the arrow to move up to the eqn:= line. Because $\sec x = \dfrac{1}{\cos x}$, we write $(\cos x)^{-2}$ instead of $\sec^2 x$. Enter the expression $e^x \cdot (\cos(e^x))^{-2} - 7$. Press ENTER. Because we want to find the solution on the interval $\left[\frac{\pi}{4}, \frac{\pi}{2}\right]$, enter an initial value between $\frac{\pi}{4}$ and $\frac{\pi}{2}$, say 1.5.

Press SOLVE (ALPHA ENTER).

EQUATION SOLVER eqn:0=e^(X)(cos(e^(X)))^-2-7	e^(X)(cos(e^(...=0 X=1.3541656874... bound={-1E99,1...

TI-89: In the main screen type: *Solve*(*d*(*tan*(*e*^(*x*)), *x*)=7, *x*) | $x > \pi/4$ and $x < \pi/2$.

The "and" is found in CATALOG or it can be typed using the alpha-lock and the space symbol (above the minus sign).

```
F1 Tools | F2 Algebra | F3 Calc | F4 Other | F5 PrgmIO | F6 Clean up
■solve(d/dx(tan(e^x))=7,x)|▸
                    x=1.354165687
...tan(e^(x)),x)=7,x)|x>π/4...
MAIN    RAD APPROX    FUNC    1/30
```

Using either calculator, the solution $x = 1.354\ldots$ appears, so the correct answer is (E).

4. D

A function has a minimum where the sign of the derivative changes from negative to positive. To solve this problem, we graph the derivative $f'(x) = 8x^3 - e^x$ on an appropriate window—a window that includes all of the answer choices makes sense. We then use the zero utility to find the point where the graph crosses the axis changing from negative to positive as *x*.

TI-83: On the Y= screen enter the function $Y_1 = 8x^3 - e^x$. View the graph in the window $[0,1] \times [-5,5]$. Use the zero utility on the CALC (2nd TRACE) menu. Use the arrow keys to move the cursor to the left of the zero. Press ENTER. Use the arrow keys to move the cursor to the right of the zero. Press ENTER. We are prompted for a guess. To skip this step press ENTER again. The solution $x = .6134$ appears, so the correct answer is (D).

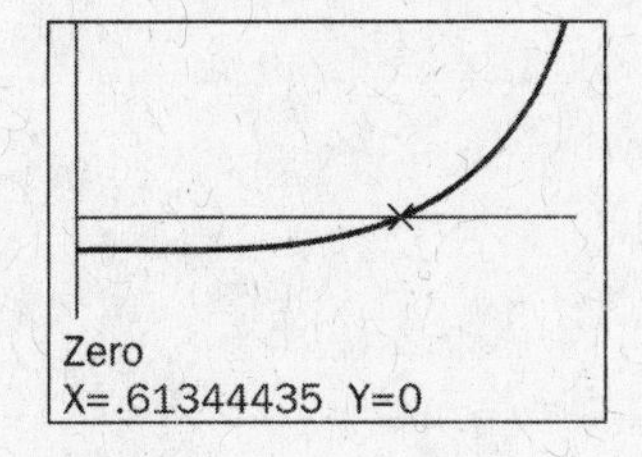

5. B

The slope of the tangent line is the derivative, so we need to find the value of *x* for which $f'(x) = 8$. We then find the *y*-value corresponding to this *x*-value and use this information to find the equation of the tangent line.

On a basic calculator, we need to compute the derivative manually:

$$f'(x) = 3e^3x$$

We then need to solve the equation

$3e^{3x} = 8 \Rightarrow 3e^{3x} - 8 = 0$, using one of the methods discussed in this section.

TI-83: We can use the solver utility. Select the solver utility from the MATH menu.

Use the arrow to move to the **eqn:0**= line. Enter the expression $3e^{3x} - 8$. Press ENTER SOLVE (ALPHA ENTER). The value $x = .3269\ldots$ appears.

EQUATION SOLVER eqn:0=e^(3X)–8	3e^(3X)–8=0 X=.32694308433... bound={0,10}

On a more advanced calculator, we do not need to compute the derivative manually; we can have the calculator compute it as part of the equation we are solving. We now need to find the *y*-value that corresponds to $x = .3269\ldots$ by evaluating the original function at this point. We find $y = 2.667$. We can use the point-slope form of the equation of a line to find:

$$y - 2.667 = 8(x - .3269) \Rightarrow y = 8x + 0.051$$

The correct answer is (B).

FREE-RESPONSE ANSWER

6. (a) To find the smallest positive value of x for which $f(x) = 0$, we need to solve the equation $0 = \sin(e^x)$. Let's first graph the function, making sure our calculator is in radian mode (remember!), to get a sense of what the function $f(x) = \sin(e^x)$ looks like. Because we're looking for a small positive value of x and because the value of this function lies between -1 and 1, the window $[0,3] \times [-1,1]$ makes sense.

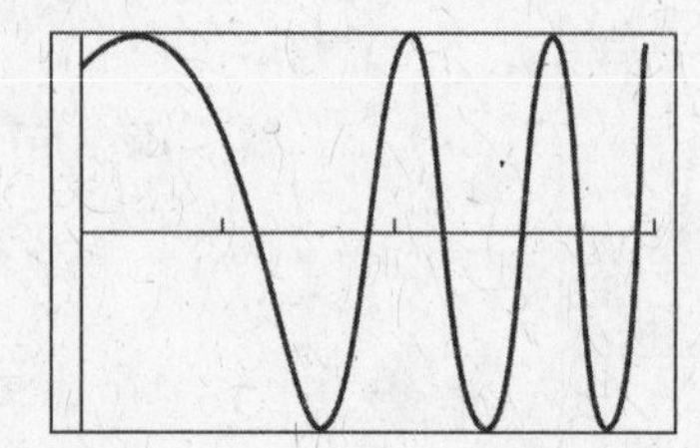

Now we use one of the methods discussed in this chapter to find the coordinate of the point furthest to the left on the graph where the curve crosses the axis. Since we already have the graph displayed, it makes sense to use the zero utility to find the point.

TI-83: Press CALC) (2nd TRACE) by pressing 2. Press ENTER. The graph appears. Use the arrows to move the cursor to the left of the left-most zero. Press ENTER. Move the cursor to the right of this zero. Press ENTER. We are prompted for a guess. To skip this step press ENTER. The coordinate of the zero, $x = 1.1447\ldots$ appears. The (rounded) solution to part a is $x = 1.145$.

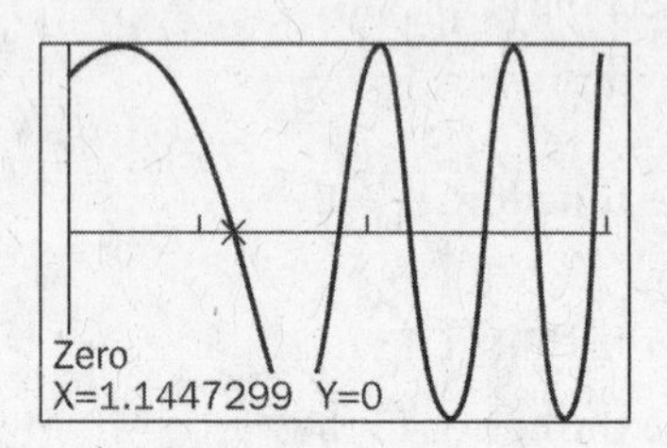

(b) To find the slope of the tangent line at $x = 1.145$, we evaluate the derivative of f at this point. We compute:

$$f(x) = \sin(e^x)$$
$$f'(x) = e^x \cos(e^x)$$
$$f'(1.145) = e^{1.145} \cos(e^{1.145}) = -3.142$$

(c) We'll find the equation of the tangent line using point-slope form. We need the slope of the line, which we have from part (b), and a point on the line, which we have from part (a) (because this is the point where the curve crosses the axis, the y-coordinate of this point is zero).

$y - 0 = -3.142(x - 1.145)$

We don't need to simplify algebra, so it's fine to leave the equation in this form.

To verify that our equation is correct, we can graph the function and the tangent line on the same set of axes, adjusting the window (if necessary) to get a good picture. The window [0,3] × [−2,2] works well.

TI-83: Go to the Y= screen and add the equation $Y_2 = -3.142(x - 1.145)$.

Press GRAPH.

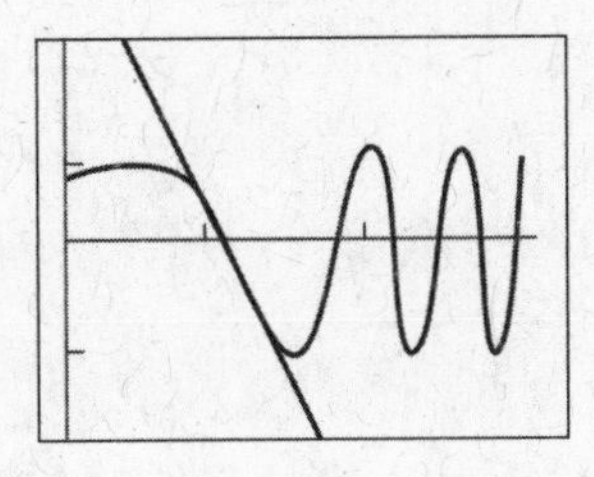

CHAPTER 6: OPERATIONS WITH FUNCTIONS ON A CALCULATOR

IF YOU LEARN ONLY TWO THINGS IN THIS CHAPTER . . .

1. All graphing calculators approved for use on the AP exam are able to evaluate a derivative at a point and compute a definite integral; make sure you are familiar with these utilities on your calculator.
2. Questions on speed and total distance are much easier with a graphing calculator, and often appear on the AP exam.

Graphing calculators can do a lot of nifty things. Even a basic model can do things like compute a definite integral, evaluate a derivative at a point, and evaluate a function at a point.

A basic model can tell you that if $f(x) = x^2$, then $f'(2) = 6$, but it can't tell you that $f'(x) = 2x$.

It can tell you that $\int_{1}^{\frac{\pi}{2}} \cos x \, dx = 1$, but it can't tell you that $\int \cos x \, dx = \sin x + c$. More sophisticated models like the TI-89 *can* carry out abstract, symbolic computations. An advanced graphing calculator will compute $(x^2)' = 2x$ and $\int \cos x \, dx = \sin x + c$.

In this chapter, we'll highlight some of these features and talk about how they might come into play on the exam. Sometimes there is more than one way to accomplish the same task; we'll point that out, too. While we're describing these features, we'll work with basic functions—things we could easily evaluate in our heads. That way, we can focus on the technique and not the math.

EVALUATING A FUNCTION

We saw in the previous chapters that we often need to determine the value of a function at a specific point while working a problem. For example, suppose we want to evaluate the function $f(x) = x^2 + 1$ at $x = 3$. One way to approach this task is to ignore the powerful capabilities of the calculator and use just the regular, non-algebraic computing power of your calculator (or, for that matter, your brain). Type $3^2 + 1$ ENTER on the main screen. The answer 10 appears.

Fancier techniques involve entering the function into the Y= screen.

TI-83: On the Y= screen enter the function $Y_1 = x^2 + 1$.

USING A GRAPH TO EVALUATE A FUNCTION

Another way to evaluate a function at a point uses a graphing utility. In Chapter 4 we saw that a graphing calculator can find the coordinates of a zero of the graph. The calculator also has a utility to find the *y*-value of a given *x*-coordinate. To do this we graph the function and view it in a window that includes the point we are interested in. We then use a calculator utility (usually called something like "value") to find the information we want.

TI-83: View the graph of the function in a reasonable window, say [0,4] × [0,20]. On the CALC (2nd TRACE) menu choose value by pressing 1 or ENTER. We are prompted for an *x*-value. Type (3) ENTER. The answer $y = 10$ appears at the bottom of the screen.

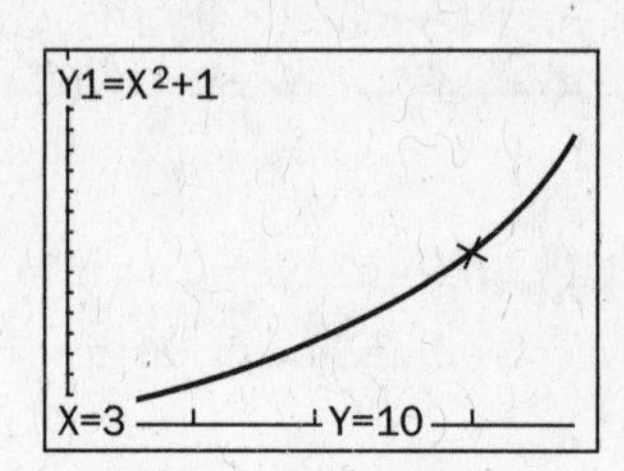

Y-VARS TECHNIQUE

Most graphing calculators have a utility that allows you to evaluate a function that is entered in the Y= screen at a point.

TI-83: Return to the Main Menu by pressing QUIT if needed. Press VARS. Use the right arrow to move the cursor over Y-VARS. Press ENTER and when Y_1 is highlighted, press ENTER again. Now Y_1 will appear on the main menu. Type (3) ENTER and 10 will appear.

Caution: If you try to type Y1 on the key pad, you will not get the function evaluated at the point. You would get the contents of Y multiplied by 1. To evaluate this expression using the Y-VARS technique, you must use the VARS menu!

TABLE TECHNIQUE

Graphing calculators have a utility that allows us to create a table of x- and y-values. Generally, we have the option of entering a starting x-value and taking the remaining values in regular increments, or entering the individual x-values for which we want to know the corresponding y-values. This second option is especially useful when we are trying to evaluate a limit computationally; we can have the function evaluated at, say, $x = 0.1, 0.01$, and 0.001.

TI-83: Press [TBLSET] ([2nd] [WINDOW]). We are prompted for a starting value. In TblStart = enter 3. In ΔTbl = , enter 1. **Indpnt: Auto**, and **Depend: Auto**. Now press [TABLE] ([2nd] [GRAPH]). We see the answer for $x = 3$, and also for $x = 4, 5, 6, \ldots$

TABLE SETUP
TblStart=3
ΔTbl=1
Indpnt: Auto Ask
Depend: Auto Ask

X	Y1	
3	10	
4	17	
5	26	
6	37	
7	50	
8	65	
9	82	
X=3		

If we had another function in Y_2, we would see that function evaluated at 3, 4, 5, 6,…and so forth.

Example:

$$\text{Evaluate } \lim_{x \to 0^+} \frac{\ln x}{\log_{10} x} \text{ computationally.}$$

Solution:

To evaluate this limit we will use the table utility to create a table of values that are approaching zero, entering the x-values manually this time.

TI-83: On the [Y=] screen enter $y_1 = \frac{\ln x}{\log x}$. Press [TBLSET] ([2nd] [WINDOW]). We are prompted for a starting value. In **TblStart** = enter 3. In **ΔTbl** = , enter 1. Use the right arrow key to move the **Indpnt: to Ask**. Keep the setting **Depend: Auto**. Now press [TABLE] ([2nd] [GRAPH]). Press 0.1 [ENTER]. Press 0.01 [ENTER]. Press 0.001 [ENTER].

X	Y1	
.1	2.3026	
.01	2.3026	
X=		

The y-values appear for all the x-values…all of the y-values are the same! Is this a mistake?

Take a minute to set the standard viewing window ([ZOOM] Z Standard) and view the graph of the function by pressing [GRAPH]. Hey, the graph is a horizontal line! What's going on?

"Aha," says the excellent calculus student (that's you, right?).

$$\frac{\ln x}{\log x} = \frac{\ln x}{\frac{\ln x}{\ln 10}} = \ln 10 \approx 2.3026$$

This is a funky way of writing the constant function $y = \ln 10$.

More Sophisticated Calculators

More sophisticated calculators are a bit like computers, so we just enter a simple line of code to tell the calculator to evaluate the function after we enter the function.

TI-89: Enter $y = x^2 + 1$ in y_1. Return to main menu. Type $y_1(3)$ and ENTER. Or type $y_1(x) \mid x = 3$ and ENTER and 10 will appear. This technique is so simple that it is the main technique used.

EVALUATING A DERIVATIVE AT A POINT

All of the graphing calculators allowed on the exam will evaluate a derivative at a point. Check your owner's manual to see exactly where to find this utility on your calculator. To use this utility, we generally have to tell the calculator the function, the variable, and the point at which we want the derivative evaluated.

TI-83: The utility to evaluate a derivative at a point is called **nDeriv** and it is found on the MATH menu. Press 8 or scroll with the arrow keys to find it. The syntax for the nDeriv command is nDeriv(*function, variable, value*).

For example, to evaluate the derivative of $f(x) = x^2 + 1$ at $x = 3$ we go to the nDeriv utility by pressing MATH 8. The prompt "nDeriv" appears.

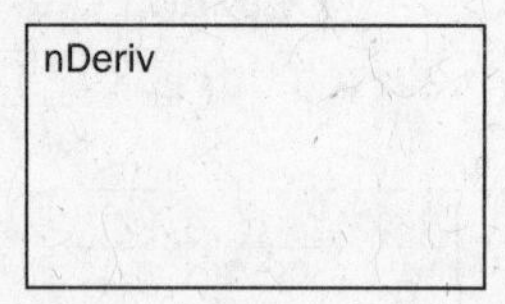

Enter $(x^2 + 1, x, 3)$ ENTER. The answer 6 appears.

nDeriv(▮

EVALUATING A DERIVATIVE USING Y-VARS

We can also recall a function stored on the [Y=] menu and evaluate its derivative at a point using the Y-VARS technique discussed previously. This is useful when we are doing numerous things with a single function, such as graphing it, finding its zeros, etc.

TI-83: On the [Y=] screen enter the function $y_1 = x^2 + 1$. Go to the nDeriv utility by pressing [MATH] 8. The nDeriv prompt screen appears. To recall the stored function, press [VARS]. Press the right arrow key to highlight Y-VARS. Press 1 or [ENTER] to select the function option. Press 1 or [ENTER] to select the Y_1 option. The nDeriv screen appears with Y_1 filled in: nDeriv (Y_1.

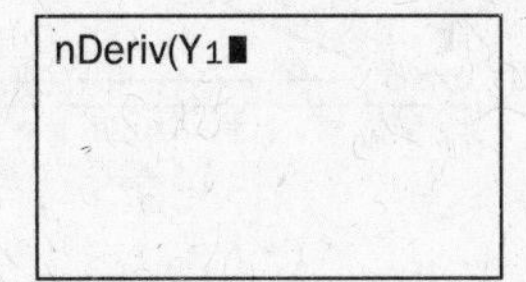

Careful now. The syntax of the nDeriv command is nDeriv (*function,variable,value*). We need to enter a function in the first slot, so we continue by filling in nDeriv ($Y_1(X), X, 3$).

The answer 6 appears.

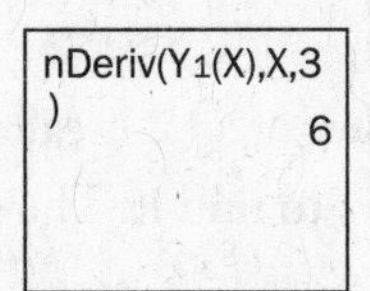

If we try to fill in nDeriv ($Y_1(3)$), an error message will appear.

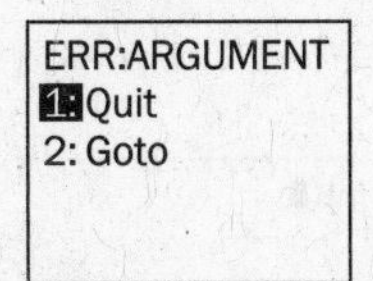

USING A GRAPHING UTILITY TO EVALUATE A DERIVATIVE AT A POINT

Some calculators will allow you to compute the derivative at a point using a graphing utility.

For example, to evaluate the derivative of $y = x^2 + 1$ at $x = 3$ this way, we first graph the function $y = x^2 + 1$ on a window that includes $x = 3$. We then use the appropriate graphing utility (check your instruction manual to find the utility for your calculator) to evaluate the derivative at $x = 3$. We are prompted for the point.

TI-83: On the Y= screen enter the function $y_1 = x^2 + 1$. Graph the function on an appropriate window, say $[-1, 5] \times [0, 30]$. Press CALC (2nd TRACE) 6 (for the *dy/dx* option). We can use the arrow keys to move along the curve; the *x*- and *y*-values appear at the bottom of the screen. Alternatively, we can enter a value for *x*. Press 3 ENTER. The value of the derivative appears: $dy/dx = 6$.

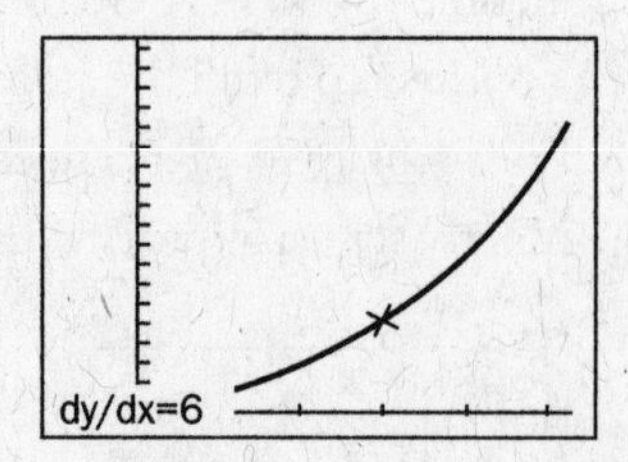

COMPUTING A DEFINITE INTEGRAL

All of the calculators allowed on the exam will compute a definite integral as well. Once again, check your owner's manual to see exactly where to find this utility on your calculator. To compute a definite integral, we generally have to tell the calculator the function, the variable, and the lower and upper limits of integration.

TI-83: The utility for computing a definite integral is called **fnInt** and it is found on the MATH menu. Press 9 or scroll with the arrow keys to find it. The syntax for the fnInt command is fnint(*function,variable,lower,upper*).

For example, to compute the definite integral of $\int_0^4 x^2 + 1\, dx$, we go to the fnInt utility by pressing MATH 9. The prompt "fnInt" appears.

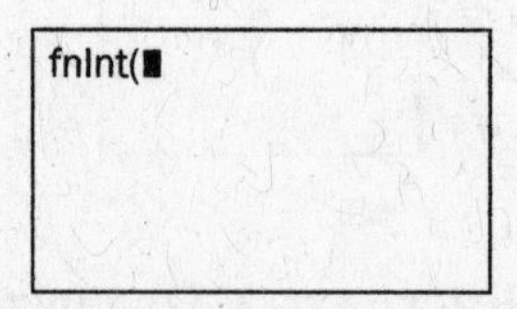

Enter ($x^2 + 1$, *x*,0,4) ENTER. The answer 25.33333 appears.

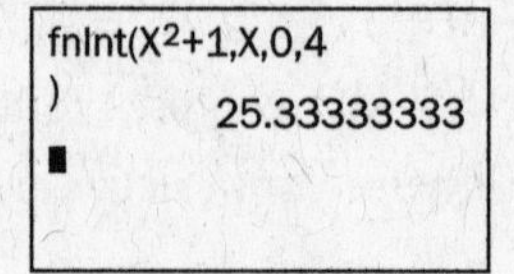

EVALUATING A DEFINITE INTEGRAL USING Y-VARS

As with nDeriv, we can recall a function stored on the [Y=] menu and evaluate a definite integral using the Y-VARS technique to recall the function.

TI-83: On the [Y=] screen, enter the function $y_1 = x^2 + 1$. Go to the fnInt utility by pressing [MATH] 9. The fnInt prompt screen appears. To recall the stored function, press [VARS]. Press the right arrow key to highlight Y-VARS. Press 1 or [ENTER] to select the function option. Press 1 or [ENTER] to select the Y_1 option. The fnInt screen appears with Y_1 filled in: fnInt(Y_1.

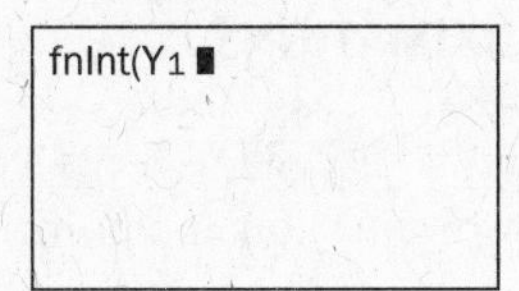

We continue by filling in: fnInt($Y_1(X)$, X, 0, 4).

The answer 25.33333 appears.

USING A GRAPHING UTILITY TO EVALUATE A DEFINITE INTEGRAL

Some calculators will allow you to view the definite integral as the area under the curve and compute the definite integral using a graphing utility.

For example, to evaluate $\int_0^4 x^2 + 1\,dx$ this way, we first graph the function $y = x^2 + 1$ on a window that includes the limits of integration. We then use the appropriate graphing utility (check your owner's manual) to compute the definite integral. We are prompted for upper and lower limits of integration.

TI-83: On the [Y=] screen enter the function $y_1 = x^2 + 1$. Graph the function on an appropriate window, say $[-1, 5] \times [0, 30]$. Press [CALC] ([2nd] [TRACE]) 7 (for the $\int f(x)\,dx$ option). We are prompted for a lower limit. Press 0 [ENTER]. We are prompted for an upper limit. Press 4 [ENTER]. The area is swept out on the graph and the answer $\int f(x)\,dx = 25.33333....$ appears at the bottom of the screen.

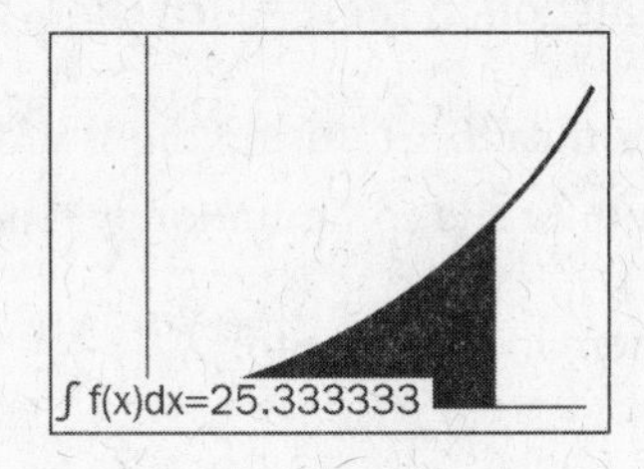

AP EXPERT TIP

Caution: This last method is reliable for many of the functions you will encounter, but the algorithm your calculator uses for the fnInt command is more accurate.

OTHER UTILITIES

Graphing calculators can also compute other interesting information about functions. For instance, most calculators can find the maximum and minimum values of a function on an interval (where appropriate). Experiment with your calculator and check out the owner's manual to see what other helpful utilities your calculator possesses.

FUNCTION OPERATIONS IN ACTION ON THE AP EXAM

RIEMANN SUM OR TRAPEZOIDAL APPROXIMATIONS

Often questions about approximating the area under a curve using Riemann sums pop up on the AP exam. The Y-VARS technique of evaluating a function at a point is particularly helpful in these often messy calculations.

Example:

Consider the function $y = \dfrac{x^2}{\sqrt{x+1}}$.
Find the approximation of the area under the curve on [4, 12] dividing the interval into four partitions and using:

(a) Riemann Sum Approximation (using right endpoints as the sampling value in each interval $x = 6, 8, 10, 12$)
(b) Trapezoidal Approximation

Solution:

(a) and (b): First we recall that the formula for a Riemann Sum Approximation with four partitions is $R_4 = \dfrac{b-a}{4}\left[f(x_1) + f(x_2) + f(x_3) + f(x_4)\right]$, where x_1, x_2, x_3, x_4, are, in this case, the right endpoints—6, 8, 10, 12. We recall that the formula for Trapezoidal Approximation with four partitions is: $T_4 = \dfrac{b-a}{2 \cdot 4}\left[f(x_0) + 2f(x_1) + 2f(x_2) + 2f(x_3) + f(x_4)\right]$ where, in our case, x_0, x_1, x_2, x_3, x_4 are 4, 6, 8, 10, 12.

We can use the Y-VARS technique to avoid a lot of the computation.

On all calculators, we start by typing the function in the [Y=] screen. Type $y_1 = \dfrac{x^2}{\sqrt{x+1}}$.

TI-83: On the main screen we enter the formula for R_4, using the Y-VARS technique. Remember, we must enter $Y_1(k)$ from the Y-VARS menu, or the computation will not work!

The calculator screen appears below. We then do the same for T_4.

TI-89: We type it as we would write it:

Riemann Sum Approximation: $R_4 = (12 - 4) \div 4 \times (Y_1(6) + Y_1(8) + Y_1(10) + Y_1(12))$

Trapezoidal Approximation:
$T_4 = (12 - 4) \div (2 \cdot 4) \times (Y_1(4) + 2Y_1(6) + 2Y_1(8) + 2Y_1(10) + Y_1(12))$

TI-83 Riemann Sum	TI-83 Trapezoidal Sums	TI-89 Both Sums
(12-4)/4*(Y1(6)+ Y1(8)+Y1(10)+Y1 (12)) 210.0592059 ∎	(12-4)/(2*4)*(Y1 (4)+2Y1(6)+2Y1(8)+2Y1(10)+Y1(12)) 177.2762093 ∎	F1 Tools F2 Algebra F3 Calc F4 Other F5 PrgmIO F6 Clean up ∎ $\frac{12-4}{4}$·(y1(6) + y1(8) + y1(1▸ [answer] 210.05921 ∎ $\frac{12-4}{2\cdot4}$·(y1(4) + 2·y1(6) + 2▸ 177.27621 (12-4)/(2*4)*(y1(4)+2*y1(... MAIN RAD APPROX FUNC 2/30

"Particle Along a Line" Problems: Speed and Total Distance

Questions on speed and total distance now appear regularly on the calculator portion of the AP Calculus exams. Both topics are much easier with a graphing calculator.

Speed is the magnitude of velocity, $\|v\ (t)\|$.

Example:

A particle moves along the y-axis with a velocity at time $t \geq 0$ given by $v(t) = \sqrt{t+2} \cdot \cos\sqrt{t}$. At time $t = 0$, the particle is at $y = 3$.

(a) Find the acceleration of the particle at $t = 4$. Is the speed increasing or decreasing? Justify your answer.

(b) Find all times on the interval (0,5) when the particle changes direction.

(c) Find the total distance traveled by the particle over the interval $0 \leq t \leq 5$.

(d) What is the position of the particle at $t = 5$?

Solution:

(a) The acceleration of the particle is the derivative of velocity—a messy job to do by hand for this function—but we can have the calculator compute $a(4) = v'(4)$.

TI-83 and TI-89 Screens: The first calculation is the velocity at $t = 4$, $v(4)$ or $(Y_1(4))$ and the second is its derivative. Notice that with a calculator, the variable x is typed for t.

```
Plot1 Plot2 Plot3
\Y1=√(X+2 )*cos*(√(X))
\Y2=
\Y3=
\Y4=
\Y5=
\Y6=
```

```
Y1(4)
        -1.019347408
nDeriv(Y1(X),X,4)
        -.6417742928
```

```
F1 Tools F2 Zoom F3 Edit F4 ✓ F5 All F6 Style
PLOTS
✓y1=√(x+2)·cos(√x)
y2=
y3=
y4=
y5=
y6=
y7=
y7(x)=
MAIN   RAD APPROX   FUNC
```

```
F1 Tools F2 Algebra F3 Calc F4 Other F5 PrgmIO F6 Clean up
■ y1(4)          -1.01934740761
■ d/dx (y1(x)) |x=4
                 -.641774297338
d(y1(x),x) |x=4
MAIN   RAD APPROX   FUNC   2/30
```

The key punches that are used on the calculator *should not* appear in your exam answers. You are likely to lose credit if something like nDeriv($Y_1(X),X,4$) shows up on your answer sheets. The answer on the exam should be written as:

$v(4) = -1.019 < 0$

$a(4) = v'(4) = -0.642 < 0$

Because both the acceleration and the velocity are negative, the speed is increasing.

(b) The particle changes direction when $v(t)$ changes sign. We can see when this happens by graphing $v(t)$.

TI-83: The function is already entered in the Y= screen. Now view the graph on the appropriate window and use the zero utility to find the coordinate of the point where the velocity changes sign.

The answer on the exam should appear as below with an explanation! The person grading the exam doesn't see your calculator screen. They must see some justification for your claim that v changes sign, for instance by computing $v(2)$ and $v(3)$ to show that the sign of v changes.

The particle changes direction when velocity changes from positive to negative at $t \approx 2.467$. We can confirm this computationally.

For $t < 2.467$, try $t = 2$, $v(2) \approx 0.312 > 0$

For $t > 2.467$, try $t = 3$, $v(3) \approx -0.359 < 0$

(c) Position is the antiderivative of velocity. Total distance is the sum of the distance going up plus the distance going down. Both are taken as positive quantities and do not cancel each other out. Before graphing calculators existed, solving a problem like this involved a lot of tough algebra and often a lot of computational mistakes as well. With the calculator, it is now much easier. We only need to integrate the absolute value of the velocity function over the interval.

The absolute value function (abs) is found in the catalog.

TI-83/TI-89:

On the AP exam the answer needs to show the setup:

$$\int_0^5 |v(t)|\, dt = 4.016 \text{ or } \int_0^5 \left|\sqrt{t+2}\cdot\cos\sqrt{t}\right| dt = 4.016$$

Either of the setups shown here is fine. However, there should be no mention of Y_1 or fnInt.

(d) The position of the particle at $t = 5$ is the initial position ($y(0) = 3$) plus the change in position from time $t = 0$ to $t = 5$.

$$y(5) = y(0) + \int_0^5 v(t)\, dt = 3 + \int_0^5 v(t)\, dt \approx 2.779$$

TI-83/TI-89:

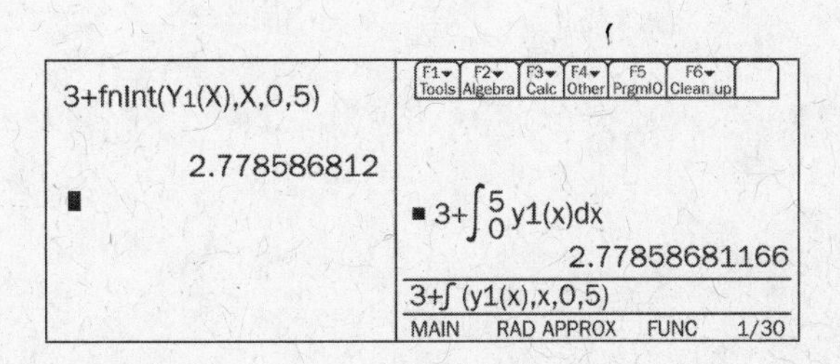

Another way to interpret this question is to think of it as an application of the (drumroll please…) **Fundamental Theorem of Calculus**. In general terms, this theorem can be written as $\int_a^b f'(x)\, dx = f(b) - f(a)$. An **FTC Equation** of this type is often used multiple times in each AP exam, so it is very important to recognize when to use it.

In this problem, $v(t)$, the velocity function, represents $f'(x)$, which can be thought of as a generic rate-of-change function. The position function, $y(t)$, is $f(x)$, an antiderivative of the velocity (rate-of-change) function.

We are looking for $y(5)$, the position of the particle at $t = 5$, so we can set up an FTC Equation:

$$\int_0^5 v(t)\,dt = y(5) - y(0).$$

Of the three quantities represented in this equation, the first can be found with your calculator (using fnInt), the second is the unknown quantity, and the third, $y(0)$, is given. So, by substitution, $-0.2214131883 = y(5) - 3$, and this leads to $y(5) = 2.779$.

REVIEW QUESTIONS

1. If $f(x) = \sqrt{e^x + \tan x}$, then $f'(1) =$

   (A) 1.386
   (B) 1.486
   (C) 1.498
   (D) 2.068
   (E) 2.168

2. A particle is moving along the x-axis with a velocity of $v(t) = 2\sin t + \cos t - 2$ for $t \geq 0$ in minutes. What is the total distance traveled after five minutes?

   (A) −9.526
   (B) 8.024
   (C) 8.415
   (D) 9.817
   (E) 9.526

3. A particle is moving along the y-axis, according to the equation $y(t) = \ln t - 2\sin t$ for $0 < t \leq \frac{\pi}{2}$. Find the position of the particle at the instant that the acceleration is zero.

   (A) −1.661
   (B) 1.277
   (C) 8.415
   (D) 9.817
   (E) 9.526

4. Suppose $G(x)$ is an antiderivative of $\frac{\cos x}{e^x}$ and $G(-2) = -1$. Then, $G(5)=$

   (A) −3.485
   (B) −1.002
   (C) 0.371
   (D) 1.847
   (E) 0.818

5. The Trapezoidal Approximation of the definite integral, $\int_0^9 \ln\left(1 + \sqrt{x}\right)dx$, with $n = 3$ is

   (A) 8.809
   (B) 8.837
   (C) 8.940
   (D) 8.953
   (E) 8.979

6. Find the minimum value of the function $f(x) = \ln(x) + \sin(x)$ on the interval $\left[\frac{\pi}{4}, \pi\right]$.

   (A) 0.4655
   (B) 1.145
   (C) 1.6055
   (D) 0
   (E) 0.6817

FREE-RESPONSE QUESTIONS

7. We are given the functions $y_1 = \cos^{-1}x$ and $y_2 = 2x^2$. Region C is bounded by the y-axis and the graphs of y_1 and y_2. Region D is bounded by the x-axis and the graphs of y_1 and y_2.

   (a) Find the x-coordinate of the point of intersection of y_1 and y_2.

   (b) Find the area of region D.

8. Consider the area enclosed by the graphs of $f(x) = \sin x$ and $g(x) = \left(\frac{1}{2}x - \frac{1}{2}\right)^2$. Find the value of K so that the vertical line $x = K$ divides this region into two regions of equal area.

ANSWERS AND EXPLANATIONS

1. B

We use the calculator's utility for evaluating a derivative at a point. Before calculating the function, make sure your calculator is in radian mode.

TI-83/TI-89:

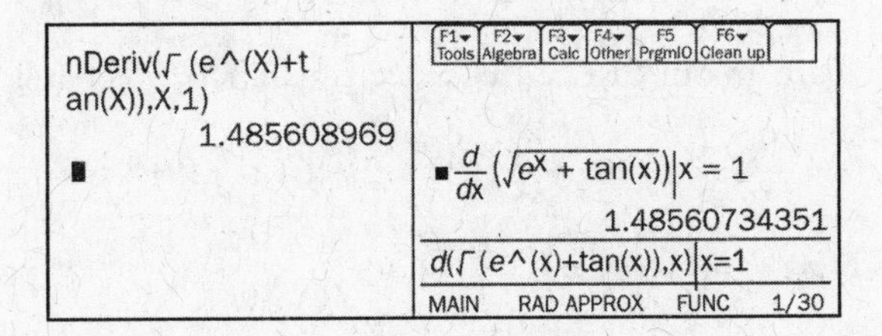

Notice that the syntax is different for the TI-83 and the TI-89. If you forget to include the vertical "when" bar, the TI-89 will return an expression rather than a number.

$$f'(1) \approx 1.486$$

2. D

The total distance traveled is given by the definite integral of the speed function (the absolute value of the velocity function.) That is, $\int_0^5 |v(t)|\,dt$. We use our calculator's utility for computing a definite integral. Your calculator will take the absolute value of a function; check your owner's manual about how to do this on your calculator.

TI-83: The absolute value function is found on the MATH menu or in the catalog.

```
fnInt(abs(2sin(X)+
cos(X)-2),X,0,5)
                9.817072058
```

On the exam, we show the setup:

$$\int_0^5 |v(t)|\,dt = 9.817$$

Remember—the setup should not make any mention of fnInt!

3. A

Acceleration is the second derivative of position, so we need to solve the equation $a(t) = y''(t) = 0$. We then need to evaluate y at the solution of this equation.

Working with a basic model, we need to compute $y'(t) = v(t)$ and $y''(t) = a(t)$ manually:

$$y'(t) = v(t) = \frac{1}{t} - 2\cos t$$

$$y''(t) = a(t) = -\frac{1}{t^2} + 2\sin t$$

We then use one of the methods described in Chapter 5 to solve $a(t) = y''(t) = 0$.

TI-83: We'll use the graphing utility to find the zero of the graph. Enter the function $Y_1 = -\frac{1}{t^2} + 2\sin t$ on the Y= screen. View the graph in an appropriate window. Use the zero utility on the CALC (2nd TRACE) menu to find the coordinates of the zero.

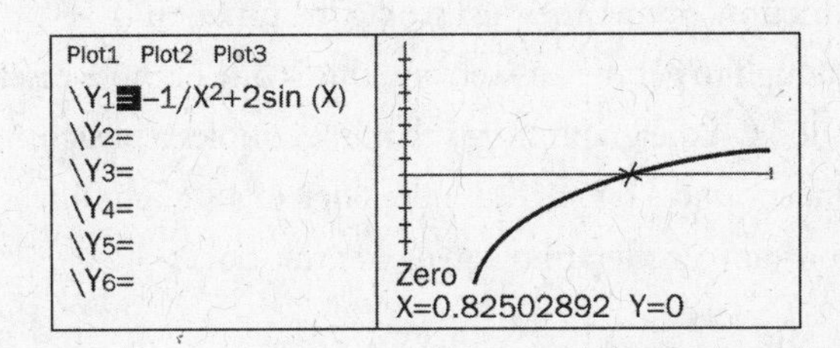

Now immediately press QUIT (2nd MODE) to return to the main screen. Enter $\ln(x) - 2\sin x$ ENTER. The calculator evaluates this function at the x-value of the zero we just found.

```
1n(X)-2sin(X)
        -1.661471651
```

TI-89: Enter: solve
$(d(\ln(t) - 2 * \sin(t), 2) = 0, t) \,|t>0$ and $t<1$.

Press ENTER.

Then simplify the entry line to read: $\ln(t) - 2 * \sin(t)|$ and "grab" the t value from the previous answer by highlighting it and pressing ENTER. Press ENTER again to perform the operation.

$$y(t) = \ln t - 2\sin t \quad \Rightarrow \quad y'(t) = \frac{1}{t} - 2\cos t$$
$$\Rightarrow \quad y''(t) = -\frac{1}{t^2} + 2\sin t$$

```
F1 Tools | F2 Algebra | F3 Calc | F4 Other | F5 PrgmIO | F6 Clean up
■ solve( d²/dt² (ln(t)-2·sin(t▸
                    t = .825028924015
■ ln(t)-2·sin(t)|t = .82502▸
                    -1.66147165091
...sin(t)|t=.82502892401501
MAIN    RAD APPROX    FUNC    2/30
```

We find $y(t) = -1.661$. This result means that the particle lies below the x-axis at the instant that the acceleration is zero.

Alternate method: Start by graphing the original function. Sometimes just seeing the function is enough to get the answer, and this is one of those cases. A look at the graph shows that it resembles a straight line around –1.7. Because the slope of a straight line is constant, acceleration equals 0 at this point.

4. E

The antiderivative of $\frac{\cos x}{e^x}$ is $\int \frac{\cos x}{e^x} dx$. By the Fundamental Theorem of Calculus, if $G'(x) = g(x)$, then $\int_a^b g(x)dx = G(b) - G(a)$, so $\int_{-2}^{5} \frac{\cos x}{e^x} dx = G(5) - G(-2)$.

We can use our graphing calculator to compute:

$$\int_{-2}^{5} \frac{\cos x}{e^x} dx \approx 1.818.$$

We are given that $G(-2) = -1$, therefore we can solve

$$1.818 \approx G(5) + 1 \quad \Rightarrow \quad G(5) \approx 0.818$$

TI-83: We use fnInt$(\cos x/e^x, x, -2, 5)$ to find

$$\int_{-2}^{5} \frac{\cos x}{e^x} dx \approx 1.818.$$

5. A

The formula for the Trapezoidal Approximation with three partitions is

$$T = \frac{b-a}{2 \cdot 3}\left[f(x_0) + 2f(x_1) + 2f(x_2) + f(x_3)\right],$$

where, in our case, x_0, x_1, x_2, x_3 are 0, 3, 6, 9. That is,

$$T = \frac{9-0}{2 \cdot 3}\left[f(0) + 2f(3) + 2f(6) + f(9)\right],$$ where $f(x) = \ln\left(1 + \sqrt{x}\right)$.

We can enter $\ln\left(1 + \sqrt{x}\right)$ into the Y= screen and use the Y-VARS utility on our graphing calculator to perform this computation.

TI-83/TI-89:

```
(9/6)(Y1(0)+2Y1(3)
+2Y1(6)+Y1(9))
        8.809278116
```

```
F1 Tools | F2 Algebra | F3 Calc | F4 Other | F5 PrgmIO | F6 Clean up
■ 9/6·(y1(0)+2·y1(3)+2·y1▸
                  8.80927811629
(9/6)*(y1(0)+2*y1(3)+2*y1...
MAIN    RAD APPROX    FUNC    1/30
```

We find $T = 8.809$.

6. A

Let's look at a solution to this problem that does not involve taking any derivatives. We know from the Intermediate Value Theorem that a continuous function attains a maximum and a minimum value on a closed interval. The only question is...where?

We can graph this function on the window $\left[\frac{\pi}{4}, \pi\right] \times [0,2]$. We can tell from looking at the graph that the minimum value of this function occurs at the left endpoint. We can use the value utility to find the y-value of the left endpoint.

TI-83: On the Y= screen enter $Y_1 = \ln(x) + \sin(x)$ view the graph in the window $\left[\frac{\pi}{4}, \pi\right] \times [0, 2]$. On the CALC (2nd TRACE) menu, scroll down to **1:value** and press ENTER. We are prompted for an x-value. Enter $\frac{\pi}{4}$ or .7854 $\left(\frac{\pi}{4} \approx .7854\right)$ Press ENTER. The y-value .4655. . . appears at the bottom of the screen, corresponding to answer choice (A).

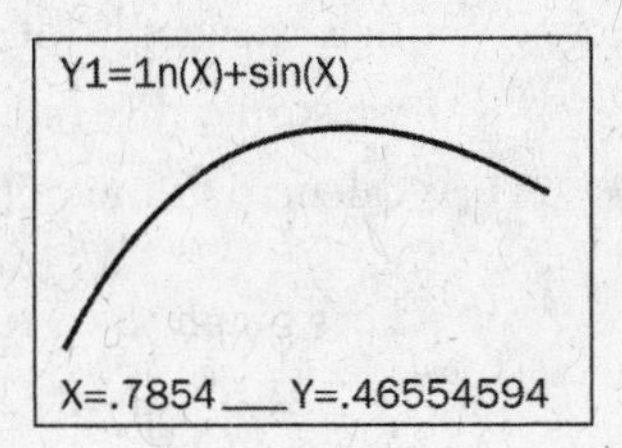

FREE-RESPONSE ANSWERS

7. (a) We saw this question earlier in Chapter 4, where we used the intersection utility to find that the graphs of $y_1 = \cos^{-1} x$ and $y_2 = 2x^2$ intersect at the point $x = 0.65463568$, and we stored this value in the calculator's memory. If the value is no longer stored in the calculator's memory, you can store it now, if you want to save yourself some typing.

   Let's graph these functions once more on the same set of axes, using the window $[-.2, 1.5] \times [-.2, 2]$:

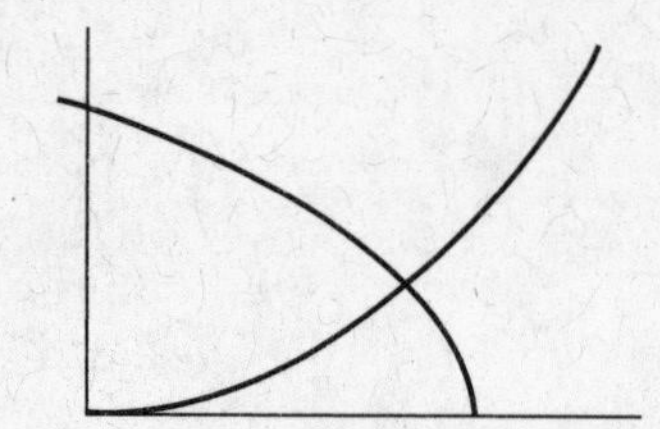

   (b) To find the area of region D, we first need to determine which curve lies on top. We then need to determine the point at which each of the curves intersects the x-axis. We can use the calculator's utility to trace the graph to determine which function is on top.

TI-83: After graphing the functions, press TRACE. A cursor appears on one of the curves, and the equation for the curve on which the cursor is sitting is displayed at the top of the screen. If the cursor is positioned near the intersection of the two curves, use the arrow keys to move the cursor so it is clearly positioned on only one of the curves.

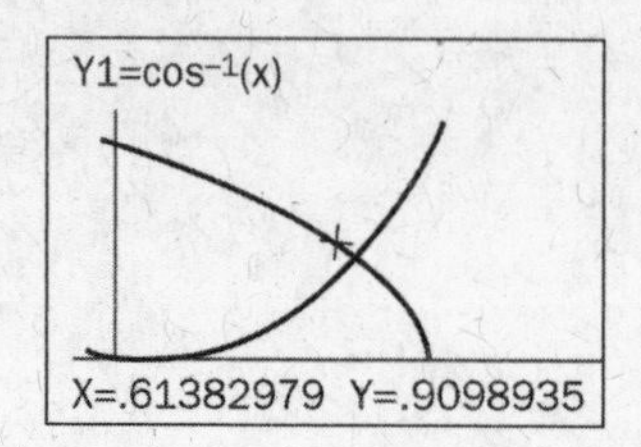

We see that the curve $y_1 = \cos^{-1} x$ lies on top to the left of the point of intersection; $y_2 = 2x^2$ lies on top to the right of the point of intersection. Therefore the left curve bounding region D is $y_2 = 2x^2$ and the right curve bounding region D is $y_1 = \cos^{-1} x$.

We know that the curve $y_2 = 2x^2$ intersects the x-axis at $x = 0$ and the curve $y_1 = \cos^{-1} x$ intersects the x-axis at $x = \cos 0 = 1$. Using this information, we determine that the area of region D is the sum of two integrals; one from 0 to the point where the curves intersect and one from the point where the curves intersect to 1. Let's call this point A. The integral should be written on the exam as:

$$Area_D = \int_0^A 2x^2 dx + \int_A^1 \cos^{-1} x\, dx,$$

where $A = 0.65463568$, the intersection point we found earlier.

```
X→A
            .65463568
fnInt (2X²,X,0, .6
5463568)+fnInt(c
os⁻¹(X), X, .654635
68,1)
           .3818880375
```

```
F1 Tools | F2 Algebra | F3 Calc | F4 Other | F5 PrgmIO | F6 Clean up
■ ∫_0^xc (2 · x2)dx + ∫_xc^1 cos-1(x)c▸
■ ∫_0^xc y2(x)dx + ∫_xc^1 y1(x)dx   ▸
                 .381887554576
...,x,0,xc+∫(y1(x),x,xc,1)
MAIN   RAD APPROX   FUNC   2/30
```

TI-83: Let's look at another way to find the fnInt command—using the catalog. Press CATALOG 2nd 0). Scroll down to fnInt and press ENTER. A blank fnInt screen appears. Enter the command: fnInt($2X^2$, X, 0, 0.65463568) + fnInt($\cos^{-1}(X), X$,0.65463568,1). The answer .38188...appears.

If you've saved the intersection point $x = 0.65463568$ in memory, you can save yourself some typing *and* eliminate the chance of mis-entering the coordinate by using RCL (2nd STO>) A (ALPHA MATH) ENTER.

TI-89: In the main menu, the fnInt command appears in the CALC menu and is printed as an integral sign. The rest of the commands are the same as the TI-83. At the end, type a lowercase *xc* instead of A. After pressing ENTER, what appears on the screen is an equation that looks just like it would if it were written on your paper. This is called "Pretty Print."

Remember that on the exam, you *must* show the setup for the problem and the setup has to look like people-calculus, not calculator-calculus. For example, you are likely to receive credit for this setup:

$$\int_0^A (2x^2)dx$$

However, no one should expect credit for this setup:

fnInt($2x^2$, x, 0, *RCL* A)

8. To solve this problem, we begin by finding where these two curves intersect. We graph the curves on the same set of axes and use the intersect utility to find the coordinates of the points where they intersect. We can then save the coordinates of the intersection points in the calculator memory.

TI-83: On the Y= screen enter $Y_1 = \sin x$ and in Y2 enter $Y_2 = .5(x - .5)^2$. View the graph on a reasonable window, say $[-1, 5] \times [-1, 3]$.

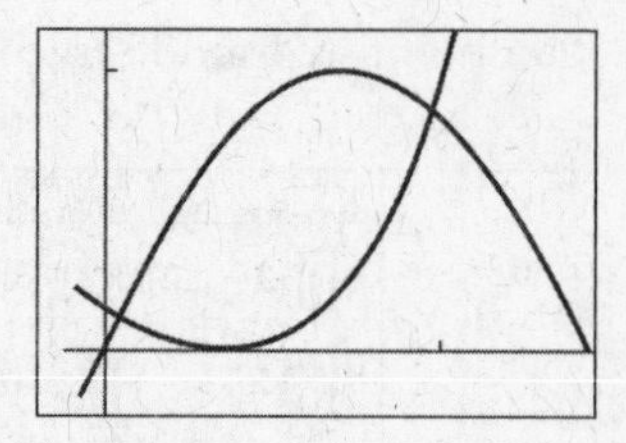

Use the zero utility on the CALC (2nd TRACE) menu to find their points of intersection. Store the *x*-coordinate of the left-hand intersection point (0.0858610092) in A and the *x*-coordinate of the right-hand point (1.880223325) in B.

We now need to compute half the area under the curve. To do this we compute the area between the curves using the calculator's utility to compute a definite integral and then take half of this area. We can then store this data to use again.

TI-83: In the main screen, compute the value of the integral and store it in memory C:

0.5 * fnInt (*Y*1(*X*) – *Y*2(*X*), *X*, *RCL A* ENTER, *RCL B* ENTER) ENTER STO C ENTER

We find that $C = 0.425638\ldots$—half the area between the curves. Next, we need to find the number K for which $\int_A^K \sin(x) - 0.5(x - 0.5)^2 \; dx = C$. This is a great opportunity for the solver—we can use it to solve for the limit of integration in the definite integral.

The solver utility can be used to solve for a limit of integration in a definite integral.

TI-83: We use the data we've stored in memory and we use the solver to solve for the limit of integration *K*. In the MATH menu open Solver. In **eqn:0=**, enter fnInt(*Y*1(*X*) – *Y*2(*X*),*X*,*RCL A* ENTER,*K*) –*RCL C* ENTER.

EQUATION SOLVER eqn:0=fnInt(Y1(X)–Y2 (X),X,.0858610092,K) –.4253830907	fnInt(Y1(X)–Y...=0 X=0 ▪K=1.0028481338... bound={0,2} ▪left–rt=0

TI-89: In the main screen enter SOLVE

$\left(\int (y1(x) - y2(x), x, a, k) = .5 * \int (y1(x) - y2(x), x, a, b)\right).$

Add bounds (0, 2), according to the answer choices.

```
F1 Tools | F2 Algebra | F3 Calc | F4 Other | F5 PrgmIO | F6 Clean up
■ xc → a        .085861009162
■ xc → b     1.880022332455
■ solve(∫ₐᵏ(y1(x)–y2(x))dx=▸
              k = 1.00284813381
...x),x,k,b),k)|k>0 and k<2
MAIN    RAD APPROX    FUNC    3/30
```

The solution is 1.003.

CHAPTER 7: LIMITS

IF YOU LEARN ONLY THREE THINGS IN THIS CHAPTER . . .

1. Limits describe what happens as x approaches a certain number, but do not tell what happens when x equals that number.
2. There are two main simplification strategies for evaluating limits: factoring and conjugation.
3. The following limit is useful in working with limits involving trigonometric functions: $\lim_{x \to 0} \frac{\sin x}{x} = 1$.

Limits pop up all the time in day-to-day life. When you reach the limit on your credit card you can't use it again until you pay off part of the balance. Athletes push themselves to the limit. If you push your parents to the limit of their patience, well...just don't go there.

Limits are an important concept in mathematics, too. They are a way of describing the way a function behaves near a point (but not *at* the point).

In this chapter, we'll work with finite limits, which are limits that approach a finite point. We'll also restrict our attention to limits that, if they exist, equal a finite number. We can also talk about limits approaching infinity and limits that equal infinity; we'll see those in Chapter 8.

BASIC DEFINITIONS AND UNDERSTANDING LIMITS GRAPHICALLY

INFORMAL DEFINITION OF A LIMIT

Suppose $f(x)$ is a function and c is any real number. If the values of $f(x)$ get close to a number L as x gets close to c, then we say that the limit of $f(x)$ exists at $x = c$ and is equal to L. We can also say that the limit of $f(x)$ as x approaches c is L. We write $\lim_{x \to c} f(x) = L$.

To be mathematically precise, we need to clarify what we mean when we say $f(x)$ gets close to L as x gets close to c. The formal definition of the limit, found in most standard calculus texts, makes these ideas precise, but, for our computational purposes, the informal definition is fine.

Limits talk about what happens as *x approaches* a number. The limit doesn't talk about, and doesn't care about, what happens when *x* actually *equals* that number.

EVALUATING LIMITS GRAPHICALLY

Consider the graphs of the three functions $f(x)$, $g(x)$, and $h(x)$ shown below.

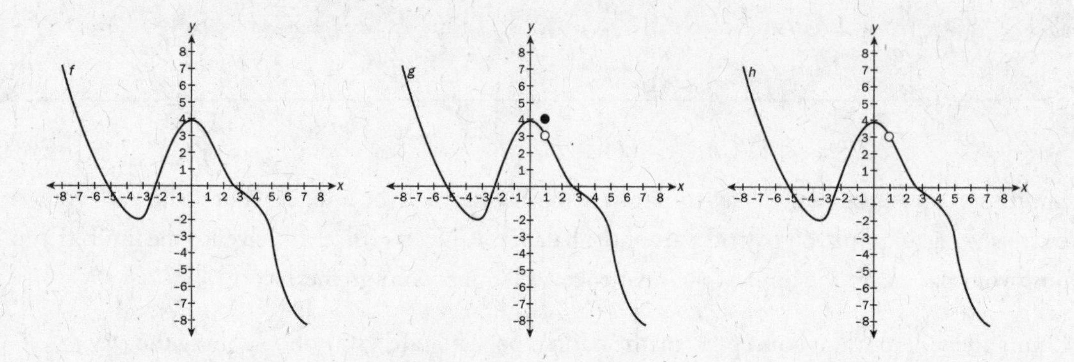

The behavior of each function *at* the point $x = 1$ is different: $f(1) = 3$, $g(1) = 4$, and h is not defined at the point $x = 1$. But on all three graphs, as x approaches 1 the value of the function is getting close to 3, so the limit of all three functions as $x \to 1$ is 3.

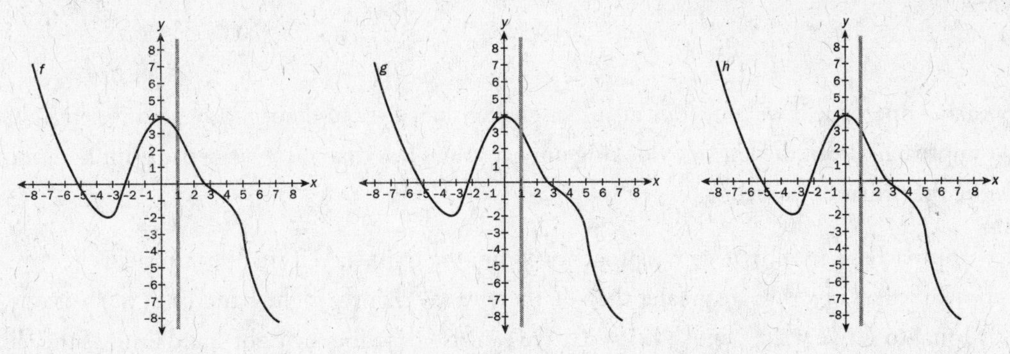

A good way to think about this is to block out the part of each graph where $x = 1$. Without the point $x = 1$, the three graphs appear identical and it looks for all the world like each function is heading for 3 as x approaches 1. This is exactly the information that the limit provides; it tells us to ignore the behavior at the point and only consider behavior near the point. Using limit notation we write:

$$\lim_{x \to 1} f(x) = 3$$

$$\lim_{x \to 1} g(x) = 3$$

$$\lim_{x \to 1} h(x) = 3$$

The graph of a new function $f(x)$ is shown below. From the graph we see that as x approaches 1, the values of the function do not approach a single value. When we consider only values of x that are greater than 1, the values of f approach 2. When we consider only values of x that are less than 1, the values of f approach −2. In this case we say that the limit does not exist.

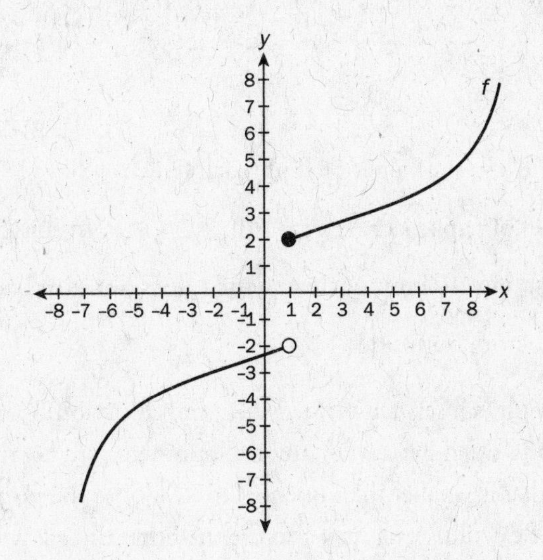

One-Sided Limits

The example above illustrates an important point. When we examine the behavior of a function near a point c, sometimes the function approaches one value from the right side of c—that is, it approaches one value if we only look at values greater than c—and it approaches a different value as we approach c from the left side (looking only at values less than c). We can use limit notation to describe this situation as well.

If $f(x)$ approaches a number L as x approaches c from the right (i.e., if we only consider the function at values of x that are greater than c), then we say that the right-hand limit of f exists at c and is equal to L. We write $\lim_{x \to c^+} f(x) = L$. We can also describe the right-hand limit as the "limit from above."

Similarly, if $f(x)$ approaches a number L as x approaches c from the left (i.e., if we only consider the function at values of x that are less than c), then we say that the left-hand limit of f exists at c and is equal to L. We write $\lim_{x \to c^-} f(x) = L$. We can also describe the left-hand limit as the "limit from below."

Evaluating One-Sided Limits Graphically

Consider the graph of the function $f(x)$ shown below.

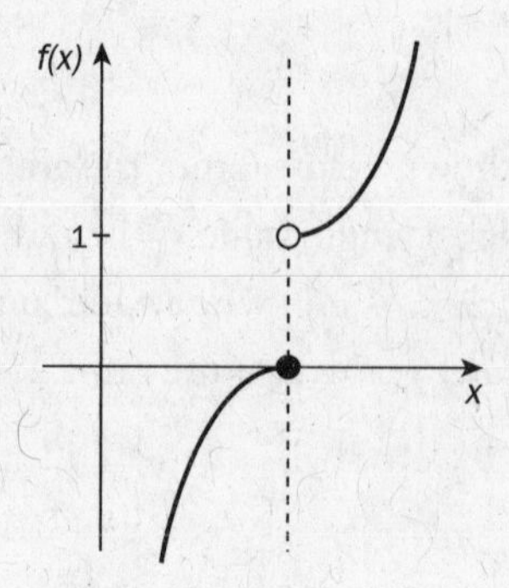

As x approaches 1 from the left, the values of f approach 0, i.e., $\lim_{x \to 1^-} f(x) = 0$. As x approaches 1 from the right, the values of f approach 1, i.e., $\lim_{x \to 1^+} f(x) = 1$. In this case both the left-hand and right-hand limits exist, but *the* limit of the function does not exist because the function is not approaching a single value from both sides.

The greatest integer function (sometimes termed the 'floor' function) is an excellent example of a situation that involves one-sided limits. For any real number x, we say that the greatest integer in x is the largest integer that is smaller than or equal to x. We use the notation $[x]$ to denote the greatest integer in x. Some calculus texts may also use the notation $\lfloor x \rfloor$, which is just as valid. However, note that [x] is the notation used by the test-designers, and hence is the version we will use throughout this section. If x *is* an integer, then the greatest integer in x is x itself. If we place x on a number line, then $[x]$ is the first integer we hit when we move left along the number line. For example, $[4.3] = 4$, $[5] = 5$, $[-4.5] = -5$. (Did you get that? When we move *left* from -4.5, the first integer we hit is -5, *not* -4.)

If we draw a graph of the greatest integer function, we see that it looks like steps.

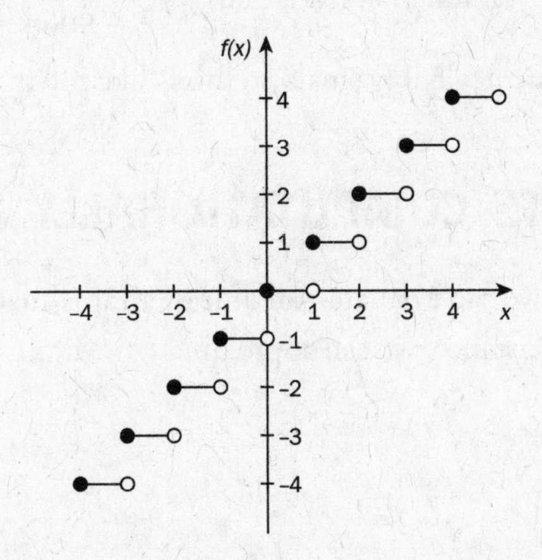

Notice that the ends of the steps are integer values along the x-axis. The left end of each step is included in the step, indicated by a closed circle on the graph. The right end of the step is an open circle, indicating that the right end of the step is *not* included in that step. For instance, there is a step from $x = 2$ to $x = 3$ shown on the graph. For all values $2 \le x < 3$, $[x] = 2$, but at the right endpoint $x = 3$, the value of the function is 3. The graph jumps up at that point. We use the graph to evaluate the left- and right-hand limits.

$$\lim_{x \to 3+} [x] = 3$$
$$\lim_{x \to 3-} [x] = 2$$

The limit of the function does not exist at $x = 3$, because the function is not approaching a single value. Notice that the actual value of [3] does not come into play when we are determining the limit.

We concluded that the limit of the greatest integer function did not exist at $x = 3$ by realizing that as we approached 3 from different directions, the function approached two different values. This is a common reason that a limit can fail to exist.

Theorem

The limit of a function exists at a point if and only if the left- and right-hand limits exist and are equal. The limit is the value of the (equal) one-sided limits. We write:

$$\lim_{x \to c} f(x) = L \text{ iff } \lim_{x \to c+} f(x) = L = \lim_{x \to c-} f(x).$$

We have looked at limits as some numerical value. What about when $x \to \infty$? Obviously, functions like $x^n (n > 0)$, e^x, and $\ln x$ approach infinity. Functions like $\frac{1}{x^n}, e^{-x}$, and $\frac{1}{\ln x}$ approach $\frac{1}{\infty}$ which is zero. What about trigonometric functions like $\lim_{x \to \infty} \sin x$? This limit does not exist, as the function will vary between −1 and 1. What is $\lim_{x \to \infty} \frac{x + 9000}{x^2 - 9000}$? We should recognize that, as $x \to \infty$ the 9000 values become negligibly small, so this is like solving $\lim_{x \to \infty} \frac{x}{x^2} = \lim_{x \to \infty} \frac{1}{x} = 0$.

EVALUATING LIMITS COMPUTATIONALLY

When we don't have a graph, we can evaluate—or at least approximate—a limit using computational methods. For instance, we can approximate the limit

$$\lim_{x \to -2} \frac{x^2 - 4}{x^2 + 5x + 6}$$

by evaluating the expression at values of x that are near −2. The limit doesn't care about what happens *at* $x = -2$. That's a good thing, because if we try to evaluate the expression at $x = -2$, we get $\frac{\text{“0”}}{0}$ which does not provide any information. Instead, we use information about the value of the expression at values of x *near* −2:

x	−1.9	−1.99	−1.999	−2.001	−2.01	−2.1
$\frac{x^2 - 4}{x^2 + 5x + 6}$	−3.545	−3.950	−3.995	−4.005	−4.050	−4.556

Analyzing the data, we can say fairly confidently that as x approaches −2, the value of the expression is approaching −4, that is

$$\lim_{x \to -2} \frac{x^2 - 4}{x^2 + 5x + 6} = -4$$

EVALUATING LIMITS ALGEBRAICALLY

When we are asked to find the limit of a function that is given algebraically (i.e., with a formula), we can also use algebraic methods to find the limit.

Let's begin with some basic facts about limits. You can verify these results by making a table of values.

- $\lim_{x \to c} a = a$
- $\lim_{x \to c} x = c$

We build up from these basic facts by using rules that allow us to algebraically manipulate limits.

THEOREM: OPERATIONS WITH LIMITS

Suppose the limit as $x \to c$ of $f(x)$ and $g(x)$ exist, say

$$\lim_{x \to c} f(x) = L$$
$$\lim_{x \to c} g(x) = M$$

then:

	Rule (formula)	Rule in words
1)	$\lim_{x \to c}[f(x) \pm g(x)] = \lim_{x \to c} f(x) \pm \lim_{x \to c} g(x) = L \pm M$	The limit of the sum (or difference) is the sum (or difference) of the limits
2)	$\lim_{x \to c}[f(x) \cdot g(x)] = \left(\lim_{x \to c} f(x)\right)\left(\lim_{x \to c} g(x)\right) = L \cdot M$	The limit of the product is the product of the limits
3)	$\lim_{x \to c} \frac{f(x)}{g(x)} = \frac{\lim_{x \to c} f(x)}{\lim_{x \to c} g(x)} = \frac{L}{M}$, for $M \neq 0$	The limit of the quotients is the quotient of the limits
4)	$\lim_{x \to c} \sqrt[n]{f(x)} = \sqrt[n]{\lim_{x \to c} f(x)} = \sqrt[n]{L}$, for $L > 0$, if n is even)	The limit of the roots is the root of the limit

These rules are also true for one-sided limits, $x \to c^+$ and $x \to c^-$.

Using our basic limit facts and rules, we can now use algebraic methods to solve many types of limit problems.

When we are asked to compute $\lim_{x \to c} f(x)$, the first thing we do is try to plug the value c into the function. If we get an answer that's a real number, we are done. This basically follows from the rules we just reviewed.

For example, $\lim_{x \to 3} x^2 = 9$, because if we evaluate the function x^2 at the point $x = 3$ we get 9. But if we try to compute $\lim_{x \to 1} \frac{x^2 - 1}{x - 1}$ by plugging $x = 1$ into the function, we obtain $\frac{0}{0}$. We call this an "indeterminate form," because we don't get any information from this expression. To evaluate the limit, we need to simplify the problem using algebra. We can simplify $\frac{x^2 - 1}{x - 1}$ by factoring:

$$\frac{x^2 - 1}{x - 1} = \frac{(x + 1)\cancel{(x - 1)}}{\cancel{(x - 1)}}$$

As long as we stay away from the actual value $x = 1$, the functions $\frac{x^2 - 1}{x - 1}$ and $x + 1$ are identical. But the limit doesn't care about the behavior of the function at $x = 1$; therefore

$$\lim_{x \to 1} \frac{x^2 - 1}{x - 1} = \lim_{x \to 1} (x + 1).$$

We can evaluate $\lim_{x \to 1} (x + 1)$ by plugging in the value $x = 1$.

We find that

$$\lim_{x \to 1} \frac{x^2 - 1}{x - 1} = \lim_{x \to 1} \frac{(x + 1)\cancel{(x - 1)}}{\cancel{(x - 1)}} = \lim_{x \to 1} (x + 1) = 2$$

To gain some insight as to why some answers to limit problems are obtained by simply "plugging in the number," and why some require rewriting the given function, we can look at the graphs of these functions to validate these strategies.

For $\lim_{x \to 3} x^2 = 9$, we know that $y = x^2$ is a parabola which is a special type of polynomial function. An important property of all polynomials is that they are **continuous** for all values of x. Informally, this means that we can draw the graph of any polynomial function without lifting our pencil from the paper. Mathematically, this translates to a limit statement: A function f is continuous at $x = c$ if and only if $\lim_{x \to c} f(x) = f(c)$. So, for our example, when $f(x) = x^2$ and $c = 3$, then $\lim_{x \to 3} x^2 = f(3) = 3^2 = 9$.

Thus, for functions continuous at a given point $x = c$, we can find the limit as x approaches c by finding the function value, $f(c)$.

For $\lim_{x \to 1} \frac{x^2 - 1}{x - 1}$, the function $y = \frac{x^2 - 1}{x - 1}$ is not continuous at $x = 1$, so $f(1)$ will not give us the limit. As stated above, $f(1)$ gives $\frac{0}{0}$, an indeterminate expression. Graphically, this means that the function has a "hole" at $x = 1$. In other words, we have to lift our pencil to draw the graph. By reducing the function, we see that it represents the linear function $y = x + 1$ with a "hole" at the point (1, 2). So graphically, we can see that as x approaches 1, $f(x)$ approaches 2.

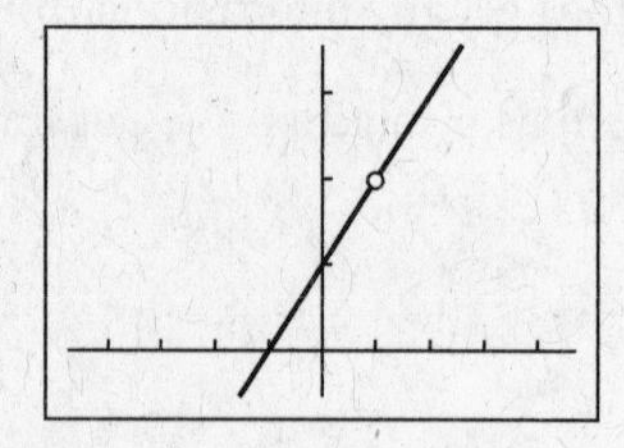

If we need to simplify a limit to evaluate it, there are generally two main simplification strategies that will get us through most problems: factoring and conjugating.

Factoring worked in the example above. Limits that have sums or differences involving square roots are good candidates for conjugation. The conjugate of an expression of the form $a + b$ is the expression $a - b$, and, in certain situations, multiplying by the conjugate simplifies things. For example, to evaluate the limit $\lim_{x\to 4}\frac{4(x-4)}{\sqrt{x}-2}$, we first try to plug in $x = 4$, and we get the indeterminate form $\frac{0}{0}$. We multiply both the numerator and the denominator by the conjugate of the denominator; this doesn't change the limit. After conjugating and simplifying, we are left with a limit we can evaluate by plugging in

$$\lim_{x\to 4}\frac{4(x-4)}{\sqrt{x}-2} = \lim_{x\to 4}\frac{4(x-4)}{\sqrt{x}-2}\cdot\frac{\sqrt{x}+2}{\sqrt{x}+2} = \lim_{x\to 4}\frac{4\cancel{(x-4)}\cdot(\sqrt{x}+2)}{\cancel{x-4}} \underset{\text{evaluate by plugging in}}{=} 16$$

It is worth noting that this function can also be reduced by factoring. In the numerator, the quantity $(x - 4)$ can be thought of as the difference of two perfect squares. As such, it may be factored as the product $(\sqrt{x}-2)(\sqrt{x}+2)$. So now we have an alternate solution to the problem above:

$$\lim_{x\to 4}\frac{4(x-4)}{\sqrt{x}-2} = \lim_{x\to 4}\frac{4(\sqrt{x}-2)(\sqrt{x}+2)}{\sqrt{x}-2} = \lim_{x\to 4}4(\sqrt{x}+2) = 16.$$

SOME IMPORTANT LIMITS

There are two special limits that pop up frequently in calculus and in life, so they're worth mentioning here. We won't show the proofs.

LIMITS INVOLVING TRIGONOMETRIC FUNCTIONS

When we need to work with limits involving trigonometric functions, we usually rely on the following limit.

Theorem:

$$\lim_{x\to 0}\frac{\sin x}{x} = 1.$$

This is a useful piece of information and it is only true when we are expressing x in radians, *not* in degrees. If we tried to evaluate this limit by plugging in we'd get the indeterminate form $\frac{0}{0}$—the algebraic manipulations of factoring and conjugating won't help us here. We can verify that this limit is correct by plugging in smaller and smaller values for x and evaluating $\frac{\sin x}{x}$ at these values of x. If we do this—and make sure our calculators are in radian mode—we'll see that the answers are all close to 1.

Factoring and conjugating generally won't be helpful in computing limits involving trig functions. This sine limit is usually what we use when we need to compute limits involving trig; we try to make bundles of

$$\frac{\text{sine of something approaching zero}}{\text{that } exact \text{ same something}}$$

For instance, we can use our rule for the limit of $\frac{\sin x}{x}$ to evaluate $\lim_{x\to 0}\frac{\sin\frac{x}{2}}{\frac{x}{2}}$. Because the limit is in the form "sine of something approaching zero over that exact same something," the limit is 1. But, to evaluate $\lim_{x\to 0}\frac{\sin\frac{x}{2}}{x}$, we need to do a little work first. To apply the sine limit to this example, the denominator must be $\frac{x}{2}$. We accomplish this by multiplying both the numerator and denominator by $\frac{1}{2}$:

$$\lim_{x\to 0}\frac{\sin\frac{x}{2}}{x} = \lim_{x\to 0}\frac{\frac{1}{2}}{\frac{1}{2}}\cdot\frac{\sin\frac{x}{2}}{x} = \lim_{x\to 0}\frac{\frac{1}{2}\sin\frac{x}{2}}{\frac{x}{2}} = \frac{1}{2}\overbrace{\lim_{x\to 0}\frac{\sin\frac{x}{2}}{\frac{x}{2}}}^{1} = \frac{1}{2}$$

As with all limits, the first thing to try with trig limits is plugging in. If that doesn't work, we try to make bundles of $\frac{\sin \text{“blah”}}{\text{“blah”}}$.

For example, to evaluate $\lim_{x\to 0}\frac{\tan 2x}{x}$, we first try to plug in, getting the indeterminate form $\frac{0}{0}$. Next, we use algebra and trig manipulations to set ourselves up to use the sine limit:

$$\lim_{x\to 0}\frac{\tan 2x}{x} = \lim_{x\to 0}\frac{\frac{\sin 2x}{\cos 2x}}{x}$$

$$= \lim_{x\to 0}\frac{\sin 2x}{x\cos 2x} = \lim_{x\to 0}\frac{2}{2}\cdot\frac{\sin 2x}{x\cos 2x} = \lim_{x\to 0}\overbrace{\frac{\sin 2x}{2x}}^{1}\cdot\frac{2}{\cos 2x} = 2$$

multiply numerator and denominator by 2 to get a $2x$ in the denominator

rearrange

evaluate the limit by plugging in

The limit $\lim_{x \to 0} \frac{\cos x - 1}{x}$ can be evaluated by cleverly using the sine limit:

$$\lim_{x \to 0} \frac{\cos x - 1}{x} = \lim_{x \to 0} \frac{\cos x - 1}{x} \cdot \frac{\cos x + 1}{\cos x + 1} = \lim_{x \to 0} \frac{\cos^2 x - 1}{x(\cos x + 1)} \underset{\text{trig identity}}{=} \lim_{x \to 0} \frac{-\sin^2 x}{x(\cos x + 1)} =$$

$$\underset{\text{rearrange}}{=} \lim_{x \to 0} \overbrace{\frac{\sin x}{x}}^{1} \cdot \frac{-\sin x}{\cos x + 1} \underset{\text{evaluate by plugging in}}{=} \frac{0}{1+1} = 0$$

This limit is used to compute the derivatives of the sine and cosine functions.

The sine and cosine limits calculated above are among the most important limits that you need to remember. Therefore, we repeat our conclusions below:

$$\lim_{x \to 0} \frac{\sin x}{x} = 1$$

$$\lim_{x \to 0} \frac{\cos x - 1}{x} = 0$$

With regard to the special trig limit, $\lim_{x \to 0} \frac{\sin x}{x} = 1$, some graphical insight can provide an important level of understanding of this idea. Let's graph the functions $y = \sin x$ and $y = x$ on the same screen in the window [–4.7, 4.7] by [–3.1, 3.1].

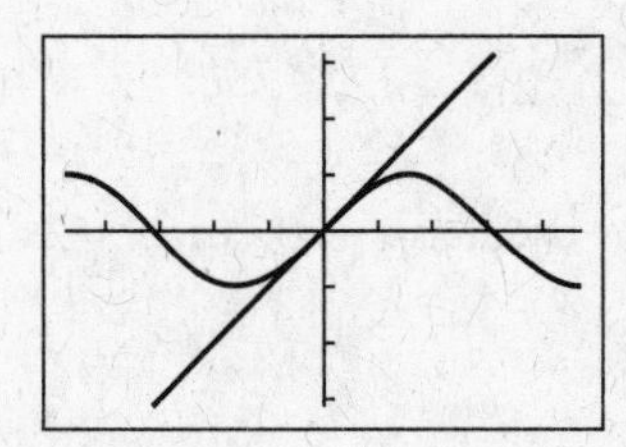

From the graph, we now see that as x gets close to 0 from both sides, the values of $\sin x$ and the values of x are virtually identical because the graphs appear to coincide in the vicinity of $x = 0$ (the graphs actually coincide only at the origin). This observation leads us to assert the following: As $x \to 0$, $\sin x \approx x$. Therefore, it makes sense that if two quantities approach each other in value, then the ratio of these two quantities must approach 1.

This idea can be extended to expressions involving multiples of x. If we graph both $y = \sin 2x$ and $y = 2x$ on the same screen, we will again see the two functions virtually coinciding in the vicinity of $x = 0$.

This leads us to a generalization of a statement made above:

As $x \to 0$, $\sin kx \approx kx$.

Now, what about tangent and cosine? By looking at the pairs of graphs $y = \tan x$ and $y = x$, (Graph 1) and at $y = \cos x$ and $y = x$ (Graph 2), we see that $\tan x$ is well approximated by $y = x$ near $x = 0$, but $\cos x$ is not.

Graph 1

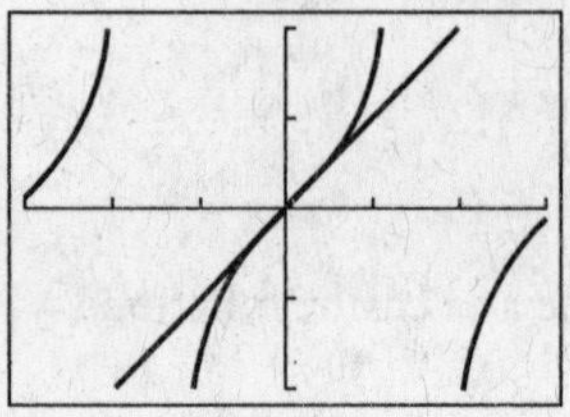

Graph 2

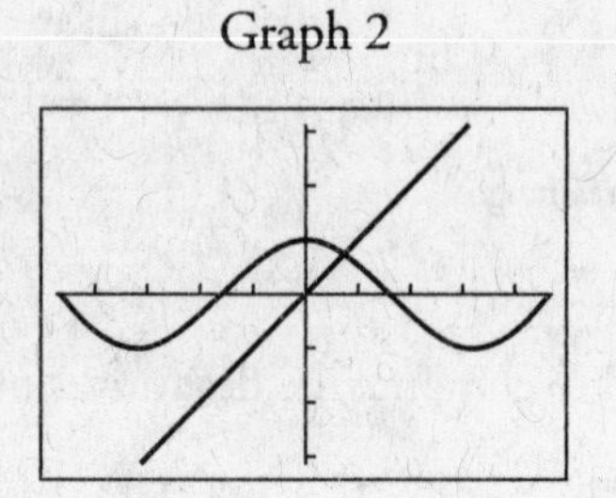

Thus, to summarize,

As $x \to 0$, $\sin kx \approx kx$.

As $x \to 0$, $\tan kx \approx kx$, where k is a constant.

Using these statements we can now revisit some limit problems demonstrated earlier, avoiding the algebraic manipulations and more difficult substitutions:

$\lim_{x\to 0} \frac{\sin\frac{x}{2}}{x} = \lim_{x\to 0} \frac{\frac{x}{2}}{x} = \frac{1}{2}$. Note: We were able to go from step 1 to step 2 because as $x \to 0$, $\sin\frac{x}{2} \approx \frac{x}{2}$.

$\lim_{x\to 0} \frac{\tan 2x}{x} = \lim_{x\to 0} \frac{2x}{x} = 2$. Here, we used, as $x \to 0$, $\tan 2x \approx 2x$.

Introducing . . . *e*

The number e is a constant that pops up when we talk about growth, such as the way bacteria colonies and investments grow. There are many ways to arrive at a definition of this constant, but one way involves limits.

We define the constant e to be:

$$\lim_{h\to 0} (1 + h)^{\frac{1}{h}} = e$$

If we try to evaluate this limit by plugging in we get the indeterminate form "1^∞." One person could look at this expression and say, "well, it's something slightly bigger than 1 to the infinity-th power, so the limit must be infinity." Another person could say "it's 1 to a power, and 1 to any power is 1, so the limit must be 1." Who is correct? Well, whenever this kind of tug-of-war between 1 and infinity happens, the answer usually involves *e*. But what is *e*?

We can get an estimate of *e* by using a calculator to evaluate the expression $(1 + h)^{\frac{1}{h}}$ for smaller and smaller values of h. As we plug in progressively smaller values of h into the expression, we find that $e \approx 2.71828$. Using computation to approximate limits is a perfectly acceptable procedure, so we'll leave it at that. The constant *e* is a number around 2.718.

Let's rephrase the definition of *e* in a more informal way, to expand the possibilities of working with this limit. We can think of *e* as

$$\left(1 + \text{something approaching zero}\right)^{\text{precise reciprocal power}}$$

Thinking of *e* this way lets us look for *e* in other limits. For example, $e = \lim_{h \to 0} (1 + 2h)^{\frac{1}{2h}}$, $e = \lim_{h \to 0} \left(1 - \frac{h}{2}\right)^{-\frac{2}{h}}$, etc.

We can use this information and our rules for manipulating limits to compute limits that involve *e*.

For example:

$$\lim_{h \to 0} (1 + 2h)^{\frac{1}{h}} = \lim_{h \to 0} \left[(1 + 2h)^{\frac{1}{h}}\right]^{\frac{2}{2}} = \lim_{h \to 0} \left[\cancelto{e}{(1 + 2h)^{\frac{1}{2h}}}\right]^{\frac{2}{1}} = e^2$$

REVIEW QUESTIONS

1. $\lim_{x\to -1} \dfrac{2x^2 - x - 3}{x^2 - 2x - 3} =$

(A) -2

(B) -1.25

(C) 0

(D) 1.25

(E) The limit does not exist.

2. $\lim_{x\to 0} \dfrac{\tan 4x}{6x} =$

(A) $\dfrac{1}{3}$

(B) $\dfrac{2}{3}$

(C) 0

(D) $-\dfrac{2}{3}$

(E) The limit does not exist.

3. $\lim_{x\to -1} \cos(\pi x) =$

(A) π

(B) 1

(C) 0

(D) -1

(E) The limit does not exist.

4. $\lim_{x\to 0} \dfrac{\frac{1}{x+4} - \frac{1}{4}}{x} =3$

(A) $-\dfrac{1}{16}$

(B) $-\dfrac{1}{4}$

(C) 0

(D) $\dfrac{1}{4}$

(E) The limit does not exist.

5. $\lim_{x\to 0^+} \dfrac{\cos x}{x} =$

(A) $-\infty$

(B) -1

(C) 0

(D) 1

(E) ∞

6. $\lim_{x\to a} \dfrac{x^2 - 2ax + a^2}{x - a} =$

(A) $-\infty$

(B) a

(C) 0

(D) ∞

(E) The limit does not exist.

7. $\lim_{x\to \infty} \dfrac{x^3 - 3x^2 + 2x - 1}{2^x - 1} =$

(A) -1

(B) 0

(C) 1

(D) ∞

(E) The limit does not exist.

8. $\lim_{x\to -\infty} \dfrac{e^x + 2x^3}{6 - 3x^3} =$

(A) $-\infty$

(B) $-\dfrac{2}{3}$

(C) 0

(D) $\dfrac{2}{3}$

(E) ∞

9. Let $f(x)=\begin{cases}\dfrac{\sin x}{x}; x\geq 0\\ \cos x; x<0\end{cases}$ Which of the following statements about $f(x)$ is true?

I. $\lim_{x\to 0^+} f(x)=1$

II. $\lim_{x\to 0^-} f(x)=1$

III. $\lim_{x\to 0} f(x)=1$

(A) None of these statements are true
(B) I only
(C) II only
(D) I and II only
(E) I, II and III

10. Consider the graph of the function shown below. Which of the following statements is true?

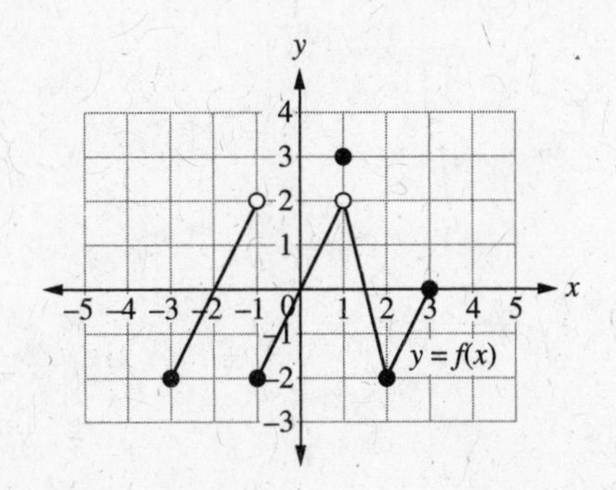

(A) $\lim_{x\to -3^+} f(x)=0$

(B) $\lim_{x\to -1^-} f(x)=2$

(C) $\lim_{x\to -1^+} f(x)$ does not exist.

(D) $\lim_{x\to 1} f(x)$ does not exist.

(E) $\lim_{x\to 1} f(x)=3$

11. $\lim_{x\to\infty}\dfrac{\cos x}{x}=$

(A) -1
(B) 0
(C) 1
(D) ∞
(E) The limit does not exist.

12. $\lim_{x\to\infty}\dfrac{x}{\cos x}=$

(A) -1
(B) 0
(C) 1
(D) ∞
(E) The limit does not exist.

13. $\lim_{x\to -\infty}\dfrac{\ln|x|+\pi}{x}=$

(A) $-\infty$
(B) 0
(C) e
(D) ∞
(E) The limit does not exist.

14. $\lim\limits_{x\to 1}\dfrac{\sqrt{x}-1}{x-1}=$

(A) 0

(B) $\dfrac{1}{2}$

(C) 1

(D) ∞

(E) The limit does not exist.

15. $\lim\limits_{x\to 3}\dfrac{\dfrac{1}{x-1}-\dfrac{1}{2}}{x-3}=$

(A) $-\dfrac{1}{4}$

(B) $-\dfrac{1}{2}$

(C) 0

(D) $\dfrac{1}{4}$

(E) $\dfrac{1}{2}$

FREE-RESPONSE QUESTIONS

16. Refer to the graph below to determine the following. If any limit does not exist, explain why.

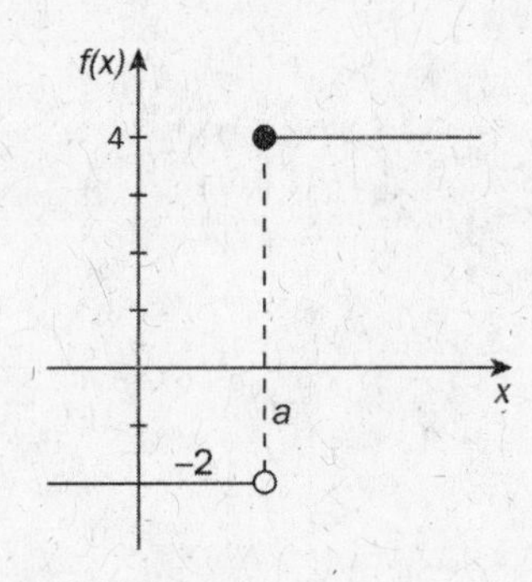

(a) $\lim_{x \to a^-} f(x) =$

(b) $\lim_{x \to a^+} f(x) =$

(c) $\lim_{x \to a} f(x) =$

(d) $f(a) =$

(e) Does $f'(a)$ exist?

17. Refer to the graph below to determine the following. If any limit does not exist, explain why.

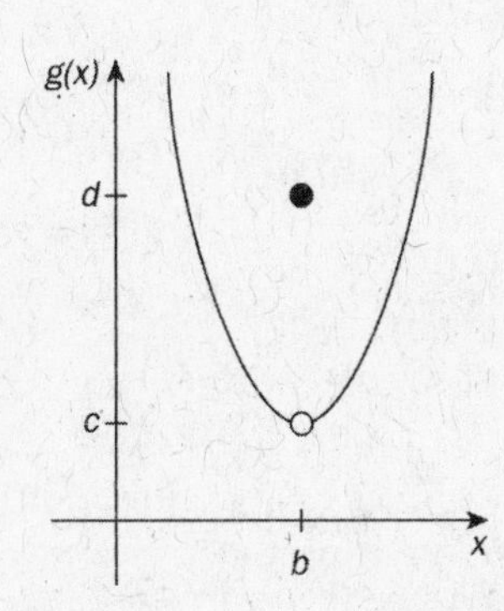

(a) $\lim_{x \to b^-} g(x) =$

(b) $\lim_{x \to b^+} g(x) =$

(c) $\lim_{x \to b} g(x) =$

(d) $g(b) =$

(e) Does $g'(b)$ exist?

18. Find the following limits. If the limit does not exist, explain why.

(a) $\lim_{x\to 5} \frac{|x|}{x}$

(b) $\lim_{x\to 5} \frac{|x-5|}{x-5}$

19. For $y = f(x)$, where $f(x) = \begin{cases} -x, & x < -1 \\ x+1, & x \geq -1 \end{cases}$, find

(a) $\lim_{x\to -1^-} f(x)$

(b) $\lim_{x\to -1^+} f(x)$

(c) $\lim_{x\to -1} f(x)$

20. For $y = f(x)$, where $f(x) = \begin{cases} 1-|x|, & x \leq 1 \\ \cos\left(\frac{\pi}{2}x\right), & x > 1 \end{cases}$, find

(a) $\lim_{x\to 1^-} f(x)$

(b) $\lim_{x\to 1^+} f(x)$

(c) $\lim_{x\to 1} f(x)$

ANSWERS AND EXPLANATIONS

1. D

If we try to plug into this limit, we get the indeterminate form $\frac{0}{0}$. We use the rules for manipulating limits to obtain:

$$\lim_{x\to-1}\frac{2x^2-x-3}{x^2-2x-3} = \lim_{x\to-1}\frac{(2x-3)\cancel{(x+1)}}{(x-3)\cancel{(x+1)}}$$
$$= \lim_{x\to-1}\frac{2x-3}{x-3}$$
$$= \lim_{x\to-1}\frac{2(-1)-3}{-1-3}$$
$$= \frac{-5}{-4} = \frac{5}{4} = 1.25$$

2. B

We first try to plug into this function; we obtain the indeterminate form $\frac{0}{0}$. Now we use our rules for manipulating limits and the limit $\lim\limits_{x\to0}\frac{\sin x}{x} = 1$ to obtain:

$$\lim_{x\to0}\frac{\tan 4x}{6x} \underset{\text{definition of } \tan x}{=} \lim_{x\to0}\frac{\sin 4x}{6x\cos 4x}$$
$$= \frac{1}{6}\lim_{x\to0}\frac{1}{x}\cdot\frac{\sin 4x}{\cos 4x}$$
$$= \frac{1}{6}\lim_{x\to0}\frac{1}{\cos 4x}\cdot\frac{\sin 4x}{x}$$
$$= \frac{1}{6}\lim_{x\to0}\frac{1}{\cos 4x}\cdot\frac{\sin 4x}{x}\cdot\frac{4}{4}$$
$$= \frac{1}{6}\lim_{x\to0}\frac{4}{\cos 4x}\cdot\frac{\sin 4x}{4x}$$
$$= \frac{1}{6}\lim_{x\to0}\frac{4}{\cos 4x}\cdot\cancelto{1}{\lim_{x\to0}\frac{\sin 4x}{4x}}$$
$$= \frac{1}{6}\cdot4\cdot1 = \frac{2}{3}$$

3. D

When we try to evaluate this limit by plugging in, we get an answer that makes sense.

$$\lim_{x\to-1}\cos(\pi x) = \lim_{x\to-1}\cos(\pi(-1))$$
$$= \lim_{x\to-1}\cos(-\pi)$$
$$= -1$$

This problem is a good example of why you should always begin trying to evaluate a limit by plugging in. You can save yourself time and work by cutting out the algebraic manipulation.

4. A

We first try to evaluate this limit by plugging in; we obtain the indeterminate form $\frac{0}{0}$. We use the rules for manipulating limits to obtain:

$$\lim_{x\to0}\frac{\frac{1}{x+4}-\frac{1}{4}}{x} = \lim_{x\to0}\frac{\frac{4-(x+4)}{4(x+4)}}{\frac{x}{1}}$$
$$= \lim_{x\to0}\frac{4-(x+4)}{4x(x+4)}$$
$$= \lim_{x\to0}\frac{\cancel{4}-x-\cancel{4}}{4x(x+4)}$$
$$= \lim_{x\to0}\frac{-\cancel{x}}{4\cancel{x}(x+4)}$$
$$= \lim_{x\to0}\frac{-1}{4(x+4)} = -\frac{1}{16}$$

5. E

As x approaches 0 from the right, the numerator approaches 1 and the denominator approaches zero but through positive values. Thus, the fraction goes to 1 over an increasingly small but positive number, which is an increasingly large positive number. Thus, the limit grows to infinity.

6. C

$$\lim_{x\to a}\frac{x^2-2ax+a^2}{x-a}=\lim_{x\to a}\frac{(x-a)(x-a)}{x-a}=\lim_{x\to a}(x-a)=0$$

7. B

As the limit is approaching infinity, we consider the dominant terms in the numerator and denominator. x^3 dominates in the polynomial in the numerator, and 2^x dominates in the denominator. Since exponential functions grow faster than power functions, the denominator grows to infinity faster than the numerator, and the fraction approaches zero.

8. B

As x is approaching negative infinity, e^x approaches zero, so $2x^3$ dominates in the numerator while $-3x^3$ dominates in the denominator. Thus, since the dominant terms in both numerator and denominator are the same power, limit is the ratio of the coefficients.

9. E

$\lim_{x\to 0^+} f(x)=\lim_{x\to 0^+}\frac{\sin x}{x}=1$, thus I is true.

$\lim_{x\to 0^-} f(x)=\lim_{x\to 0^-}\cos x=1$, thus II is true.

Since I and II are true, III is true by definition.

10. B

$\lim_{x\to -3^+} f(x)=-2$ ∴ (A) is false.

$\lim_{x\to -1^-} f(x)=2$ ∴ (B) is true.

$\lim_{x\to -1^+} f(x)=-2$ ∴ (C) is false.

$\lim_{x\to 1} f(x)=2$ ∴ (D) and (E) are false.

11. B

Since the numerator is confined to values between ±1, and the denominator grows to infinity, the fraction gets closer to 0 as x goes towards infinity.

12. E

The numerator grows towards infinity, and is divided by numbers which are constrained between ±1. This will not approach a finite value. The graph of the curve is shown below and confirms that the limit does not exist.

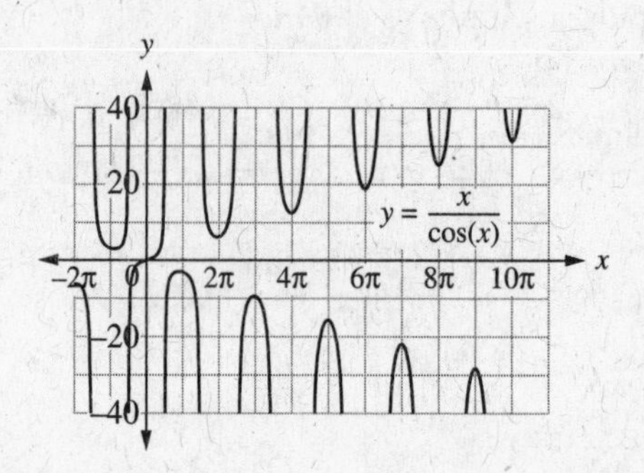

13. B

In the numerator, the dominant term is $\ln|x|$. The x in the denominator grows faster than this, and therefore the denominator will dominate the numerator and the function will approach zero as x grows to negative infinity.

14. B

$$\lim_{x\to 1}\frac{\sqrt{x}-1}{x-1}=\lim_{x\to 1}\frac{\sqrt{x}-1}{x-1}\cdot\frac{\sqrt{x}+1}{\sqrt{x}+1}$$

$$=\lim_{x\to 1}\frac{x-1}{(x-1)(\sqrt{x}+1)}=\lim_{x\to 1}\frac{1}{\sqrt{x}+1}=\frac{1}{2}$$

15. A

$$\lim_{x\to 3}\frac{\frac{1}{x-1}-\frac{1}{2}}{x-3}=\lim_{x\to 3}\frac{\frac{2}{2(x-1)}-\frac{1(x-1)}{2(x-1)}}{x-3}$$

$$=\lim_{x\to 3}\frac{\frac{2-x+1}{2(x-1)}}{\frac{(x-3)}{1}}=\lim_{x\to 3}\frac{3-x}{2(x-1)}\cdot\frac{1}{x-3}=\lim_{x\to 3}\frac{-1}{2(x-1)}=-\frac{1}{4}$$

FREE-RESPONSE ANSWERS

16. (a) $\lim_{x\to a^-} f(x) = -2$

(b) $\lim_{x\to a^+} f(x) = 4$

(c) Because the limit from the left (–2) is not equal to the limit from the right (4), the limit does not exist, i.e., $\lim_{x\to a} f(x)$ does not exist.

(d) $f(a) = 4$

(e) $f'(a)$ does not exist. The function must be continuous at a point to be differentiable there.

17. (a) $\lim_{x\to b^-} g(x) = c$

(b) $\lim_{x\to b^+} g(x) = c$

(c) Because the limit from the left (c) equals, the limit from the right (c), the limit exists and equals c.

$\lim_{x\to b} g(x) = c$

(d) $g(b) = d$

(e) $g'(b)$ does not exist. See answer in 5(e).

18. (a) At $x = 5$, the absolute value of x is equal to x itself. Therefore $\lim_{x\to 5} \frac{|x|}{x} = \frac{x}{x} = 1.$

(b) To evaluate this limit we must check the left- and right-hand limits separately, because the absolute value function $|x - 5|$ changes at $x = 5$. That is

$$|x-5| = \begin{cases} x-5, & x > 5 \\ -(x-5), & x \le 5 \end{cases}$$

$$\lim_{x\to 5^-} \frac{|x-5|}{x-5} = \lim_{x\to 5^-} \frac{-(x-5)}{x-5} = -1 \quad \text{and} \quad \lim_{x\to 5^+} \frac{|x-5|}{x-5} = \lim_{x\to 5^+} \frac{x-5}{x-5} = 1$$

Because the limit from the left is not equal to the limit from the right, the limit does not exist.

This problem is really the same as $\lim_{x\to 0} \frac{|x|}{x}$, which also does not exist.

19. (a) To the left of 1, the function is defined by $-x$, therefore

$$\lim_{x\to -1^-} f(x) = \lim_{x\to -1^-} (-x) = -(-1) = 1$$

(b) To the right of 1, the function is defined by $x + 1$, therefore

$$\lim_{x \to -1^+} f(x) = \lim_{x \to -1^+} (x+1) = 0$$

(c) Because the limit from the left (+1) is not equal to the limit from the right (0), the limit $\lim_{x \to -1} f(x)$ does not exist.

20. (a) To the left of 0 this function is given by the formula $1 - |x|$, so the limit is

$$\lim_{x \to 1^-} f(x) = \lim_{x \to 1^-} 1 - |x| = \lim_{x \to 1^-} 1 - |1| = 0$$

(b) To the right of 0 the function is given by the formula $\cos\left(\frac{\pi}{2}x\right)$, so the limit is

$$\lim_{x \to 1^+} f(x) = \lim_{x \to 1^+} \cos\left(\frac{\pi}{2}x\right) = \lim_{x \to 1^+} \cos\left(\frac{\pi}{2} \cdot 1\right) = \cos\frac{\pi}{2} = 0$$

(c) Even though this function is defined by different formulas to the left and right of $x = 1$, the limit from the left (0) *does* equal the limit from the right (0). Therefore the limit does exist and equals 0, i.e., $\lim_{x \to 1} f(x) = 0$.

CHAPTER 8: ASYMPTOTES

IF YOU LEARN ONLY FOUR THINGS IN THIS CHAPTER . . .

1. The limit of $f(x)$ as x approaches c is infinite if:

   $f(x)$ increases without bound as x approaches c (as x approaches c from the left or right, respectively). We write $\lim_{x\to c} f(x) = \infty$, ($\lim_{x\to c^+} f(x) = \infty$, or $\lim_{x\to c^-} = \infty$, respectively).

   OR

   $f(x)$ decreases without bound as x approaches c (x approaches c from the left or right, respectively). We write $\lim_{x\to c} f(x) = -\infty$, ($\lim_{x\to c^+} f(x) = -\infty$, or $\lim_{x\to c^-} = -\infty$, respectively).
2. A function can have zero, one, or many vertical asymptotes.
3. Don't assume that a function has a vertical asymptote at every point where the denominator is zero.
4. Functions that become infinite as x approaches infinity can grow at different rates; generally, polynomials of higher degree grow faster than polynomials of lower degree.

In this chapter we'll take a look at two different kinds of limits that concern infinity. One kind involves computing a limit as x approaches a finite number and the answer has to do with infinity. We call these infinite limits. The other kind involves taking a limit as x approaches infinity. We'll begin with infinite limits.

INFINITE LIMITS

Let's start by examining two limits: $\lim_{x \to 0} \frac{1}{x^2}$ and $\lim_{x \to 0} \frac{1}{x}$. In both cases, if we try to plug in we get $\frac{1}{0}$. What does this mean? Well, we're not sure. Let's take a look at these limits computationally by making tables of values.

$\lim_{x \to 0} \frac{1}{x^2}$:

First we approach zero from the right:

x	0.1	0.01	0.001	0.0001
$\frac{1}{x^2}$	100	10,000	1,000,000	100,000,000

Now, we approach zero from the left:

x	−0.1	−0.01	−0.001	−0.0001
$\frac{1}{x^2}$	100	10,000	1,000,000	100,000,000

Does this function look like it is approaching a value as x approaches zero? Well, not exactly. As x approaches zero, the value of the function gets larger and larger. It is blowing up toward infinity. We say this limit is infinite, or write $\lim_{x \to 0} \frac{1}{x^2} = \infty$.

Now, we'll try the second limit.

$\lim_{x \to 0} \frac{1}{x}$:

We approach zero from the right:

x	0.1	0.01	0.001	0.0001
$\frac{1}{x}$	10	100	1,000	10,000

Now, we approach zero from the left:

x	−0.1	−0.01	−0.001	−0.0001
$\frac{1}{x}$	−10	−100	−1,000	−10,000

This is not the same situation as the previous limit. In the first limit, from both the left and right sides, the function $\frac{1}{x^2}$ was approaching infinity as $x \to 0$. In the second case, the function $\frac{1}{x}$ approached infinity from the right side of zero and approached *negative* infinity from the left side of zero. This limit does not exist and we write $\lim_{x \to 0} \frac{1}{x}$ DNE.

Now we'll make the idea of infinite limits a bit more precise. We say that the limit of $f(x)$ as x approaches c (a one-sided limit as x approaches c from the left or right) is infinite if either one of the following conditions is satisfied:

- $f(x)$ increases without bound as x approaches c (as x approaches c from the left or right, respectively). We write $\lim_{x \to c} f(x) = \infty$, ($\lim_{x \to c^+} f(x) = \infty$, or $\lim_{x \to c^-} f(x) = \infty$, respectively).

or

- $f(x)$ decreases without bound as x approaches c (x approaches c from the left or right, respectively). We write $\lim_{x \to c} f(x) = -\infty$, ($\lim_{x \to c^+} f(x) = -\infty$, or $\lim_{x \to c^-} f(x) = -\infty$, respectively).

When we try to plug into a limit and get "$\frac{\text{something}}{0}$" as long as "something" $\neq 0$, it means that infinity is coming into play. The limit will either be infinity, negative infinity, or will not exist (because the one-sided limits are positive infinity and negative infinity).

How do we know whether the limit is positive infinity, is negative infinity, or does not exist? Evaluating the limit numerically, as we just did in the previous examples, is one option. Another option is to analyze the signs of the numerator and the denominator. Because the denominator is approaching zero and the numerator is not, we know that the function is blowing up—toward positive infinity or negative infinity—the only question is the sign. Let's take another look at our two examples $\lim_{x \to 0} \frac{1}{x^2}$ and $\lim_{x \to 0} \frac{1}{x}$.

We'll again start with $\lim_{x \to 0} \frac{1}{x^2}$ and examine the left-hand and right-hand limits separately.

As $x \to 0^+$, the numerator is fixed, positive, and the denominator is a square so it is also positive. The denominator is approaching zero, so the limit is positive infinity. As $x \to 0^-$, the numerator is fixed, positive, and the denominator is also positive. The denominator is approaching zero, so the limit is again positive infinity. Because the left-hand and right-hand limits are both positive infinity, we conclude that $\lim_{x \to 0} \frac{1}{x^2} = +\infty$.

AP EXPERT TIP

When we say that the limit of a function is positive or negative infinity, this does not mean that the limit exists. We use the notation $\lim_{x \to c} f(x) = \infty$ to describe a particular way that the limit does not exist. The limit does not exist *because* the function's values are approaching positive or negative infinity. In this case, you should write $\lim = \infty$ or $\lim = -\infty$, not lim DNE.

Now let's look at $\lim_{x \to 0} \frac{1}{x}$ and again examine the left-hand and right-hand limits separately.

As $x \to 0^+$, the numerator is fixed, positive, and the denominator is also positive, because $x > 0$. The denominator is approaching 0, so the limit is positive infinity. As $x \to 0^-$ the numerator is a fixed, positive, but the denominator is negative, because $x < 0$. The denominator is approaching zero, but this time the limit is negative infinity. Because the left-hand and right-hand limits are not the same, we conclude $\lim_{x \to 0} \frac{1}{x}$ does not exist and write DNE.

VERTICAL ASYMPTOTES—UNDERSTANDING INFINITE LIMITS GRAPHICALLY

What does the graph of a function look like near a point where the limit is infinite? We saw above that infinite limits occur when the denominator is approaching zero and the numerator is not approaching zero. This can only occur at a point where the function is undefined.

Looking again at the function $f(x) = \frac{1}{x^2}$, we observe that it is undefined at $x = 0$. We can use the information provided by the limit $\lim_{x \to 0} \frac{1}{x^2} = \infty$, computed above, to draw a graph of the function near the point $x = 0$:

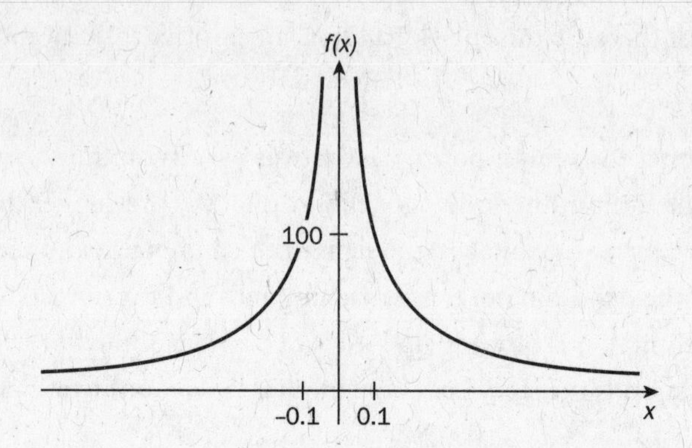

When the limit is infinite (or at least one of the two one-sided limits is infinite) at a point $x = c$, we say that the function has a vertical asymptote at $x = c$. When we graph a function, we indicate a vertical asymptote by drawing a dotted line at the location of the vertical asymptote, the line $x = c$. A function can behave four ways near a vertical asymptote:

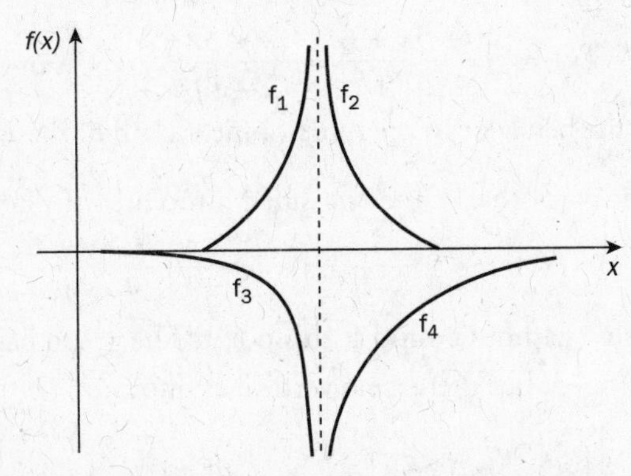

A function can have zero, one, or many vertical asymptotes. Because vertical asymptotes occur where the function increases without bound, a good place to check for vertical asymptotes are points where the function is undefined.

For example, the function $f(x) = \dfrac{x-1}{x^2-2} = \dfrac{x-1}{(x+\sqrt{2})(x-\sqrt{2})}$ has a vertical asymptote at $x = +\sqrt{2}$ and $x = -\sqrt{2}$, because as x approaches either of these values, the function looks like $\dfrac{\text{a number not close to zero}}{\text{a number close to zero}}$. More precisely, by computing a table of values or examining the signs of the numerator and denominator, we find:

$$\lim_{x\to\sqrt{2}^+} \frac{x-1}{x^2-2} = \infty \quad \text{and} \quad \lim_{x\to\sqrt{2}^-} \frac{x-1}{x^2-2} = -\infty$$

Because the one-sided limits are different, we say that $\lim_{x\to\sqrt{2}} \dfrac{x-1}{x^2-2}$ DNE, but because the one-sided limits are infinite, the function has a vertical asymptote at $x = +\sqrt{2}$.

Similarly, because

$$\lim_{x\to-\sqrt{2}^+} \frac{x-1}{x^2-2} = \infty \quad \text{and} \quad \lim_{x\to-\sqrt{2}^-} \frac{x-1}{x^2-2} = -\infty$$

the function has a vertical asymptote at $x = -\sqrt{2}$.

Be Careful Looking for Vertical Asymptotes

Many people mistakenly assume that a function has a vertical asymptote at every point where the denominator is zero. This is *not* true. Just because a function with a denominator is undefined at a point where the denominator equals zero, it does not necessarily have a vertical asymptote there; it may only have a hole. The most comprehensive way to determine whether a function has a vertical asymptote at a point is to evaluate the one-sided limits at that point.

Don't assume that $f(x)$ has a vertical asymptote at $x = c$ because f has a denominator that is equal to zero at c. Make sure that at least one of the one-sided limits $\lim_{x\to c^+} f(x)$, $\lim_{x\to c^-} f(x)$ is infinite!

For example, the function $f(x) = \dfrac{x^2 + 5x + 6}{x^2 - 9} = \dfrac{(x + 3)(x + 2)}{(x + 3)(x - 3)}$ is undefined at both $x = 3$ and $x = -3$. If we examine the behavior of the function near both of these points, we see that the function is blowing up near $x = 3$, that is, the one-sided limits are infinite, so the function has a vertical asymptote at this point. On the other hand, $\lim\limits_{x \to -3} \dfrac{x^2 + 5x + 6}{x^2 - 9} = \dfrac{1}{6}$, so the function is not increasing or decreasing without bound at this point. The graph has a hole, because the function is not defined at $x = -3$, but there is no vertical asymptote:

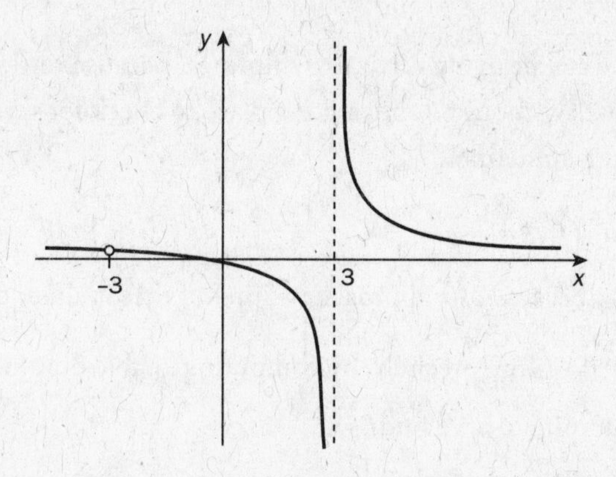

When finding limits, it is important to make a distinction between a function that is UNDEFINED, and one that is INDETERMINATE. If f is undefined at a point $x = c$, then $f(c)$ is $\dfrac{\text{nonzero}}{0}$, and f has a vertical asymptote at $x = c$. Thus, $\lim\limits_{x \to c} f(x)$ is likely to be $+\infty$, $-\infty$, or DNE. However, if $g(x)$ is indeterminate at $x = c$, then $g(c)$ is $\dfrac{0}{0}$, and g has a hole at $x = c$. Thus, $\lim\limits_{x \to c} g(x)$ is likely to exist as a finite number after you rewrite the function g in reduced form.

In the next chapter, you'll learn about types of discontinuities. You can use that knowledge to help you determine whether or not a function with a zero in the denominator has a vertical asymptote. Simply, if the zero in the denominator can be eliminated through factoring and canceling out with the numerator, you're dealing with a point discontinuity and NOT a vertical asymptote. If the zero in the denominator can't be simplified away, you're likely dealing with a vertical asymptote. This simple 'cancellation test' can be much quicker than fully evaluating a limit at the point in question!

Vertical Asymptotes Where There Is No Denominator

A function without a denominator can also have vertical asymptotes. The function $f(x) = \tan x$ is one example—although not the best example, since we can rewrite $f(x) = \tan x$ as a function *with* a denominator, $f(x) = \dfrac{\sin x}{\cos x}$. This function blows up wherever $\cos x = 0$, because at these points

$\sin x = \pm 1$; that is the one-sided limits are infinite. Therefore, $f(x) = \tan x$ has vertical asymptotes at $x = \frac{\pi}{2} + n\pi$, n is an integer.

The function $f(x) = e^{\frac{1}{x^2}}$ is another example of a function without a denominator that has a vertical asymptote; $\lim_{x \to 0} e^{\frac{1}{x^2}} = +\infty$, so this function has a vertical asymptote at $x = 0$.

LIMITS APPROACHING INFINITY

We also gain insight into a function when we see what happens to its value when x gets very, very large (i.e., when $x \to \infty$) and when x gets very, very small (i.e., when $x \to -\infty$). We use the notation $\lim_{x \to \infty} f(x)$ and $\lim_{x \to -\infty} f(x)$ to ask these questions.

As we take arbitrarily large values of x (or, respectively, arbitrarily large and negative values of x), the values of $f(x)$ sometimes approach a finite number L. We say that as x approaches infinity (or negative infinity), $f(x)$ approaches L, and write $\lim_{x \to \infty} f(x) = L$ (or $\lim_{x \to -\infty} f(x) = L$).

We can find these limits computationally: for example, if we can compute $\lim_{x \to \infty} \frac{x+1}{x-3}$ by creating a table of values.

x	10	100	1,000	1,000,000
$\frac{x+1}{x-3}$	$\frac{11}{7} = 1.571$	$\frac{101}{97} = 1.041$	$\frac{1001}{997} = 1.004$	$\frac{1,000,001}{999,997} = 1.000004$

From the table, it appears that $\lim_{x \to \infty} \frac{x+1}{x-3} = 1$.

We can also compute this limit algebraically, using our rules for manipulating limits:

$$\lim_{x \to \infty} \frac{x+1}{x-3} = \lim_{x \to \infty} \frac{\cancel{x}\left(1 + \cancel{\frac{1}{x}}^{\,0}\right)}{\cancel{x}\left(1 - \cancel{\frac{3}{x}}_{\,0}\right)} = 1$$

To compute this limit algebraically, we used the distributive law to pull out the highest power of x that appeared in each of the numerator and the denominator. We then used the rules for manipulating limits to compute.

Let's try this with some other functions:

$$\lim_{x\to-\infty} \frac{x^3+2x-9}{x^2+4x+2} = \lim_{x\to-\infty} \frac{\left(x^{3}\left(1+\frac{2}{x^2}-\frac{9}{x^3}\right)\right)}{\left(x^{2}\left(1+\frac{4}{x}+\frac{2}{x^2}\right)\right)} = \lim_{x\to-\infty} x = -\infty$$

We can also think about this limit even more informally. For the function $\frac{x^3+2x-9}{x^2+4x+2}$, as x gets very, very large and negative, the x^3 term is controlling the behavior of the function. The $2x$ and the 9 are insignificant. For example, even for $x = -1{,}000$, $x^3 = -1{,}000{,}000{,}000$, while $2x = -2{,}000$. This means that $x^3 + 2x - 9 = -1{,}000{,}002{,}009$.

In terms of order of magnitude, this number is not significantly different from $x^3 = -1{,}000{,}000{,}000$. The term of highest degree, the x^3 term, is controlling the behavior of the numerator. For very large, negative values of x, the numerator of the function "looks like" x^3. Similarly, for very large and negative values of x the denominator "looks like" x^2, because x^2 is the term of highest degree in the denominator.

As $x \to -\infty$, the function $\frac{x^3+2x-9}{x^2+4x+2}$ behaves like the function $\frac{x^3}{x^2} = x$. Therefore

$\lim_{x\to-\infty} \frac{x^3+2x-9}{x^2+4x+2} = \lim_{x\to-\infty} \frac{x^3}{x^2} = \lim_{x\to-\infty} x = -\infty$, as we computed previously.

We can perform the same steps on an expression with a radical. Consider the following function:

$$f(x) = \frac{2x+5}{\sqrt{7x^2-6x-1}}$$

Now suppose we wanted to evaluate $\lim_{x\to\infty} f(x)$. It's easy to be put off by such a complicated looking function! We can make our lives easier by putting everything under the radical. Remember that $\sqrt{x^2} = x$. Let's rewrite our function as follows:

$$f(x) = \frac{\sqrt{(2x+5)^2}}{\sqrt{7x^2-6x-1}} = \sqrt{\frac{(2x+5)^2}{7x^2-6x-1}}$$

Aha! With everything under the radical, we can evaluate the limit just like we did previously.

$$\lim_{x\to\infty} f(x) = \sqrt{\frac{(4x^2+20x+25)\frac{1}{x^2}}{(7x^2-6x-1)\frac{1}{x^2}}}$$

As we distribute the $\frac{1}{x^2}$ through, we see that our function quickly becomes much simpler.

$$\lim_{x\to\infty} f(x) = \sqrt{\frac{4+\frac{20}{x}+\frac{25}{x^2}}{7-\frac{6}{x}-\frac{1}{x^2}}}$$

As we take that limit, we see that our result simplifies to:

$$\lim_{x\to\infty} f(x) = \sqrt{\frac{4+0+0}{7+0+0}} = \sqrt{\frac{4}{7}}$$

While it may be necessary to demonstrate all of the steps in showing the limit of a polynomial fraction as $x \to \infty$ on a free-response question, keep in mind you can use some shortcuts in the multiple choice section. In the case of a function that consists of the quotient of two polynomials, remember the following rules of thumb:

As $x \to \infty$:

1. If the degree of the numerator is larger than the degree of the denominator, the limit increases without bound ($=\infty$)
2. If the degree of the numerator is smaller than the degree of the denominator, the limit is zero
3. If the degrees are equal, the limit is equal to the ratio of the leading coefficient of the numerator to the leading coefficient of the denominator.

The same rules generally apply when $x \to -\infty$. However, you will also need to consider the sign of the function as well when applying rules one and three. If the degree of either the numerator or the denominator is odd, your limit will be negative. Remember if both are negative, the negative signs will cancel out and your overall limit will be positive.

Not All Functions Have Finite Limits Approaching Infinity

Some functions do not approach a finite value when $x \to \infty$ or $x \to -\infty$. Polynomials become infinite as x gets very large and positive or very large and negative. For example, $\lim_{x\to\infty} x^3 + 5x - 9 = \infty$.

Periodic functions, such as $\sin x$, do not have limits approaching infinity because they oscillate. That is, $\lim_{x\to\infty} \sin x$ DNE.

And, some functions, particularly exponential functions, can approach a finite value as $x \to -\infty$, and grow infinitely large as $x \to +\infty$. For example:

$$\lim_{x\to-\infty} e^x = 0 \text{ and } \lim_{x\to+\infty} e^x = +\infty$$

HORIZONTAL ASYMPTOTES

The behavior of a function as $x \to \infty$ and $x \to -\infty$ shows up on the graph of the function as well. If $\lim_{x\to\infty} f(x) = L$ or $\lim_{x\to-\infty} f(x) = L$, we say that the line $y = L$ is a horizontal asymptote for the graph. If $\lim_{x\to\infty} f(x) = L$, as $x \to \infty$, the graph will follow the line $y = L$. We indicate a horizontal asymptote on a graph by drawing a dotted line. To determine whether the graph lies above or below the horizontal asymptote, we can substitute large values of x into the function and see if $f(x)$ is slightly smaller or slightly larger than L, and then draw the graph accordingly.

A function can, in principle, have different horizontal asymptotes as *x* approaches infinity and as *x* approaches negative infinity. However, in reality such functions are complicated constructions and rarely appear on the AP exam.

USING ASYMPTOTES TO SKETCH A GRAPH

Let's go back to our earlier example, $f(x) = \frac{1}{x^2}$, and use the vertical and horizontal asymptotes to get a rough sketch.

We saw earlier that this function has a vertical asymptote $x = 0$ and that the function is blowing up toward positive infinity from both directions as $x \to 0$.

We can compute that $\lim_{x\to\pm\infty} \frac{1}{x^2} = 0$, so this function has a horizontal asymptote at $y = 0$. If we plug very large positive values or very large negative values for x into f, we see that the function is slightly bigger than zero as x approaches positive and negative infinity. This tells us that the graph lies above the horizontal asymptote. We can add this information to our earlier graph.

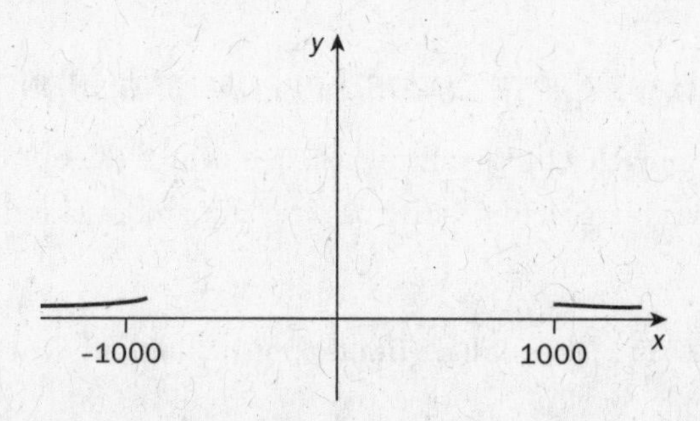

Now we connect the pieces of the graph, and, voila, we have a rough sketch of the function $f(x) = \frac{1}{x^2}$.

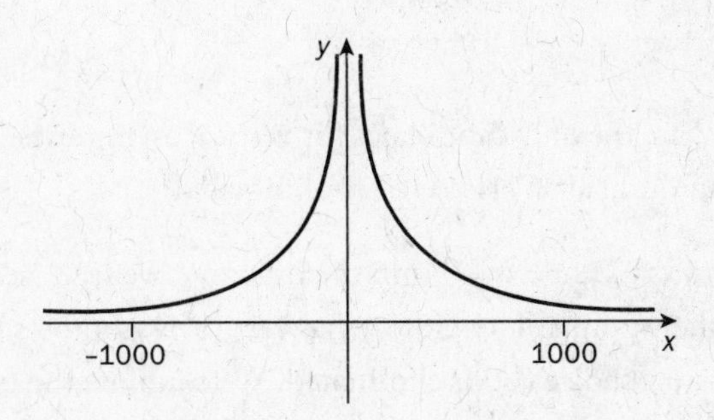

Example:

Which of the following could be the graph of the function
$f(x) = \dfrac{x^2 + 2}{(x + 3)(x - 2)}$?

(A)

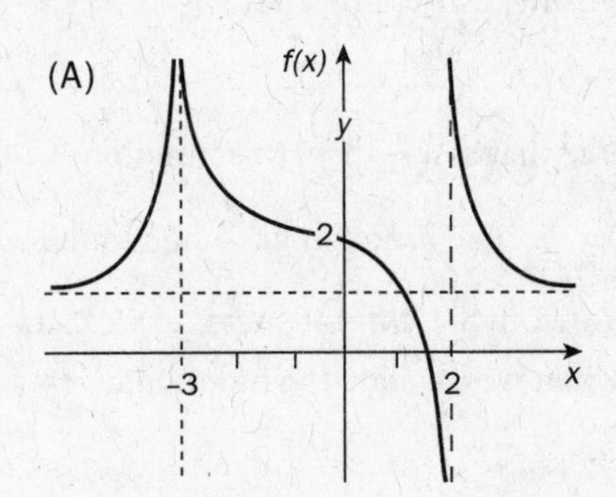

(B)

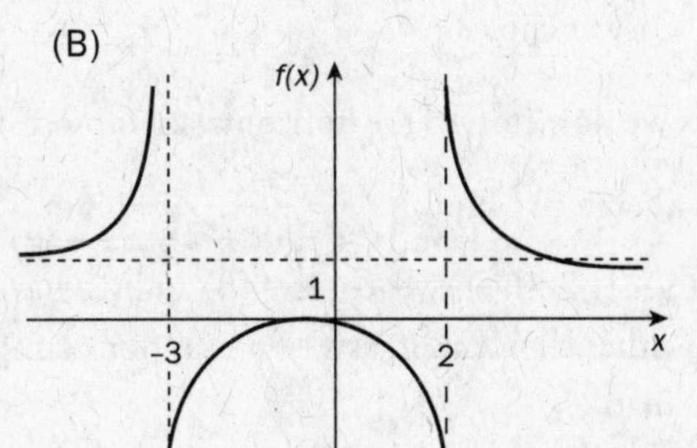

(C)

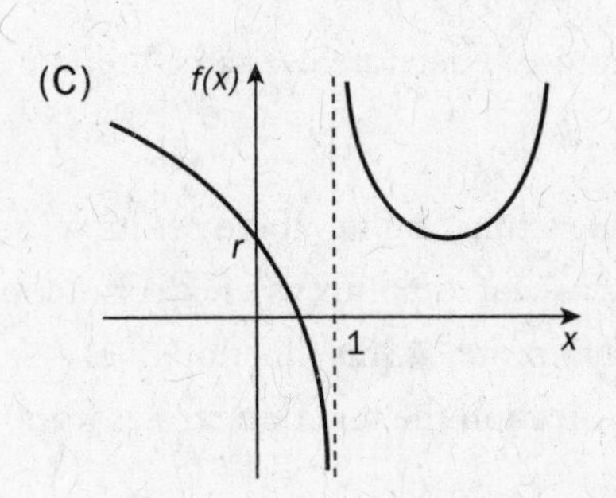

(D)

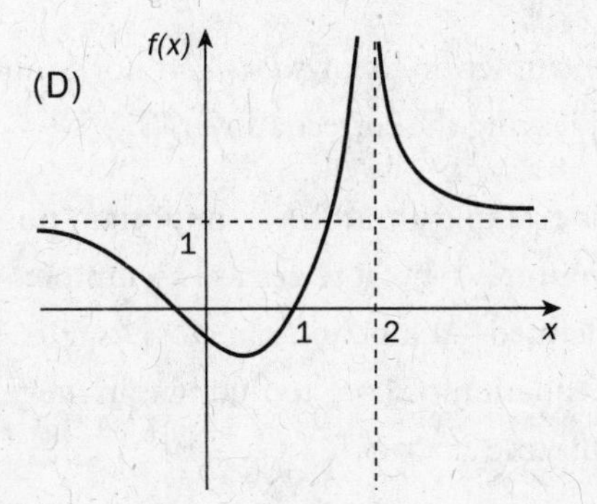

(E)

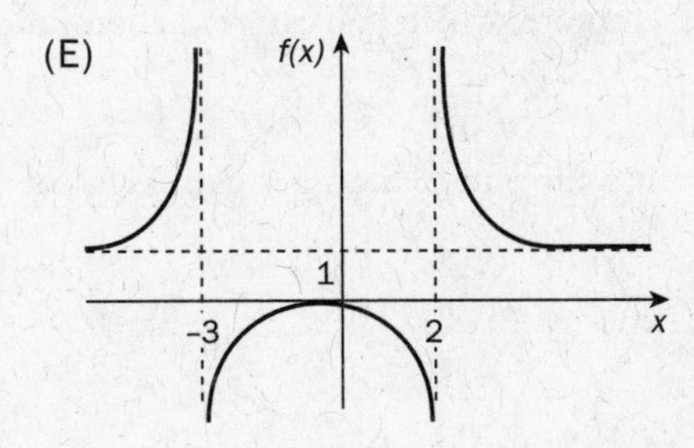

Solution: B

The lines $x = -3$ and $x = 2$ are the only candidates for vertical asymptotes. Answer choice (C), which has a vertical asymptote at $x = 1$, is therefore eliminated.

By plugging values close to $x = 2$ and $x = -3$ into the function, we see that the function is blowing up at both of these points. Therefore, f has vertical asymptotes both at $x = 2$ and $x = -3$, so answer choice (D) is eliminated. We compute the limits $\lim\limits_{x \to 2^+} \frac{x^2 + 2}{(x + 3)(x - 2)} = \infty$, $\lim\limits_{x \to 2^-} \frac{x^2 + 2}{(x + 3)(x - 2)} = -\infty$ to determine the behavior of the function near 2; we compute the limits $\lim\limits_{x \to -3^+} \frac{x^2 + 2}{(x + 3)(x - 2)} = -\infty$, $\lim\limits_{x \to -3^-} \frac{x^2 + 2}{(x + 3)(x - 2)} = +\infty$ to determine the behavior near 3. On the graph in answer choice (A), both the limit from the left and the limit from the right approach *positive* infinity as x approaches -3, so we can eliminate this answer choice.

We now turn to the horizontal asymptote. We see that as $x \to \pm\infty$, this function "looks like" $\frac{x^2}{x \cdot x}$, therefore $\lim\limits_{x \to \pm\infty} \frac{x^2 + 2}{(x + 3)(x - 2)} = \lim\limits_{x \to \pm\infty} \frac{x^2}{x^2} = 1$. The function has a horizontal asymptote at $y = 1$. Both answer choices (B) and (E) have a horizontal asymptote here, so we cannot eliminate either choice. But, as $x \to \infty$, say, for example when we evaluate the function at $x = 1{,}000$, we see that:

$$f(1{,}000) = \frac{1{,}000^2 + 2}{(1{,}000 + 3)(1{,}000 - 2)} = \frac{1{,}000{,}002}{1{,}000{,}994} < 1$$

so the function follows *below* its horizontal asymptote. Therefore, we can eliminate answer choice (E), leaving the correct answer, (B).

This example illustrates an important point. While a function never crosses its vertical asymptotes—because vertical asymptotes can only occur at points where the function is undefined—a function can cross its horizontal asymptote, as this function does. We can solve the equation $f(x) = 1$ to find that in the positive direction the function crosses its horizontal asymptote at $x = 8$.

A function can also have a tilted or curved asymptote, which you can read about in most standard calculus texts.

Vertical and horizontal asymptotes are tools that allow us to draw a rough sketch of certain functions.

RATES OF GROWTH

Earlier, we evaluated limits of a function approaching positive or negative infinity by seeing what the function "looked like" for very large and very small values of x (i.e., very large, negative values of x). Though we didn't mention it then, we were touching on an important point regarding the rates at which functions grow.

Functions that become infinite as x approaches infinity can grow at different rates. We can compare the functions $f(x) = x$ and $g(x) = x^2$ and $h(x) = x^4$, either by comparing tables of values for f, g, and h for large values of x, or by comparing the graphs of these functions as x gets large. If we make this comparison, we see that $g(x) = x^2$ is growing much faster than $f(x) = x$, and that $h(x) = x^4$ is growing the fastest.

In general, polynomials of higher degree grow faster than polynomials of lower degree; in fact, for $q > p > 0$, x^q grows faster than x^p, for any real numbers p, q.

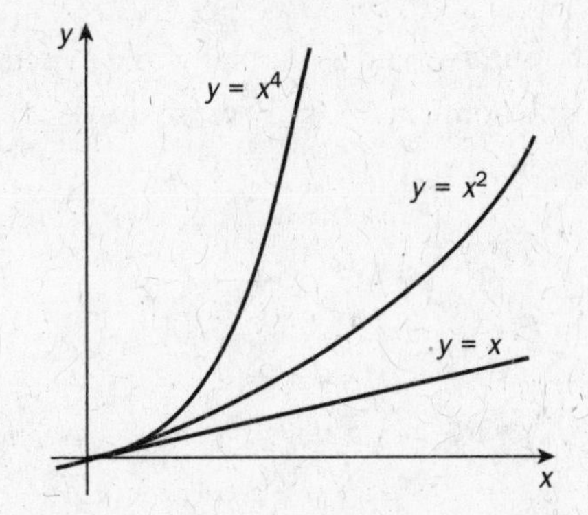

Exponential functions grow faster than polynomials; we can see this on the graphs of x^4 and e^x drawn on the same set of axes below.

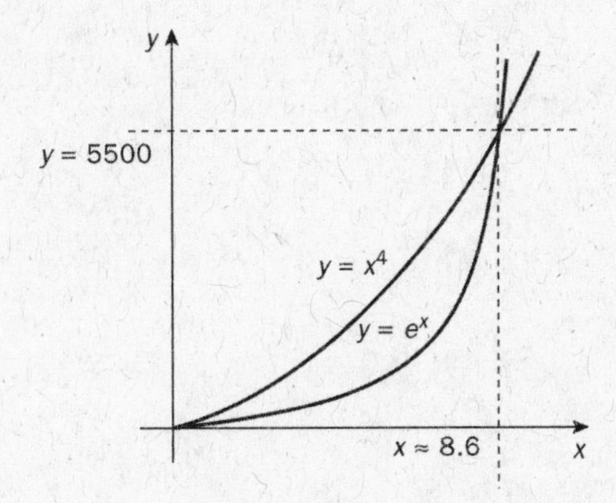

There are general rules about rates of growth. This list of function types is arranged from slowest to fastest rate of growth:

- Logarithmic functions
- x^r, in order of increasing r, $r > 0$
- Exponential functions

If a function is a sum of different types of functions—logarithmic, exponential, polynomial—its growth rate is determined by its fastest growing component. For example, the function $f(x) = 2^x + x^3$ grows at the rate of 2^x. We say that this function grows exponentially, or that it has an exponential rate of growth. Similarly, a function can have a logarithmic or polynomial rate of growth.

We can use information about rates of growth to compute limits approaching infinity. For example, consider $\lim_{x \to \infty} \frac{\sin x + \log x}{x^2 + \log x} = 0$. The numerator is growing logarithmically—notice that because $\sin x$ does not approach infinity, it does not play a role in determining the growth rate; the denominator is growing at the polynomial rate x^2. Because the denominator is growing faster than the numerator, this limit equals zero.

REVIEW QUESTIONS

1. Which of the following functions does not have a horizontal asymptote?

   (A) $f(x) = \dfrac{\sqrt{x^2+3}}{2x-1}$

   (B) $f(x) = \sin \dfrac{1}{x}$

   (C) $f(x) = \dfrac{x}{x+1} \tan \dfrac{1}{x}$

   (D) $f(x) = x \cos \dfrac{1}{x}$

   (E) $f(x) = \dfrac{\sqrt{2x^2+1}}{x-3} \sin \dfrac{1}{x}$

2. Suppose that $g(x) = \sin^2 x \cdot \sqrt{(x^6+4)}$, and $\lim\limits_{x\to\infty} \dfrac{g(x)}{f(x)} = 0$. Which of the following functions could be f?

   (A) x

   (B) x^2

   (C) x^3

   (D) x^4

   (E) $\ln x$

3. $\lim\limits_{x\to\infty} \dfrac{\sin \frac{1}{x}}{\frac{1}{x}} =$

   (A) 0

   (B) 1

   (C) −1

   (D) ∞

   (E) The limit does not exist.

4. $\lim\limits_{x\to\infty} \dfrac{\left(1+\frac{1}{x}\right)^x \cdot x}{\sqrt{x^2+1}} =$

   (A) e

   (B) $2e$

   (C) 0

   (D) ∞

   (E) The limit does not exist.

5. Which of the following functions has a vertical asymptote at $x = 4$?

   (A) $\dfrac{x+5}{x^2-4}$

   (B) $\dfrac{x^2-16}{x-4}$

   (C) $\dfrac{4x}{x+1}$

   (D) $\dfrac{x+6}{x^2-7x+12}$

   (E) None of the above

6. The graph of the function $f(x)$ is shown below:

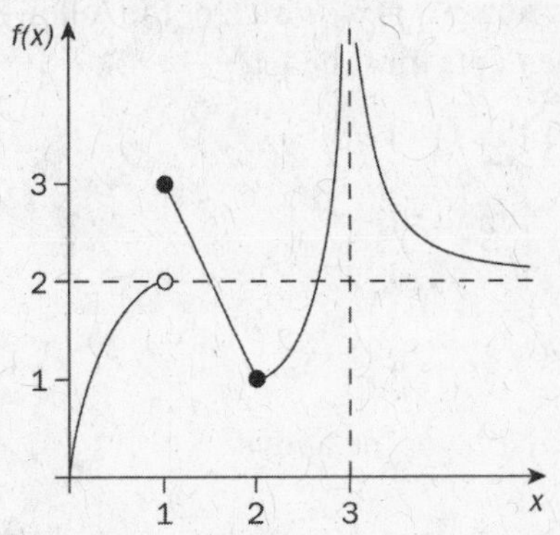

   Which of the following statements is true about f?

   I. f is undefined at $x = 1$.

   II. f is defined but not continuous at $x = 2$.

   III. f is defined and continuous at $x = 3$.

   (A) Only I

   (B) Only II

   (C) I and II

   (D) I and III

   (E) None of the statements are true.

7. Which of the following statements are true for the function $f(x)=\dfrac{2x^3+3x+1}{2^x}$

I. $f(x)$ has a horizontal asymptote of $y = 1$
II. $f(x)$ has a horizontal asymptote of $y = 0$
III. $f(x)$ has a vertical asymptote of $x = 0$

(A) I only
(B) II only
(C) III only
(D) I and III only
(E) II and III only

8. If $y = a$ is a horizontal asymptote of the function $y = f(x)$, which of the following statements must be true?

(A) $\lim\limits_{x\to\infty} f(x)=a$
(B) $\lim\limits_{x\to a} f(x)=\infty$
(C) $f(a) = 0$
(D) $f(x)\neq a \forall x \in \mathbb{R}$
(E) None of the above.

9. Consider the function: $f(x)=\dfrac{x^2-5x+6}{x^2-4}$.
Which of the following statements is true?

I. $f(x)$ has a vertical asymptote of $x = 2$
II. $f(x)$ has a vertical asymptote of $x = -2$
III. $f(x)$ has a horizontal asymptote of $y = 1$

(A) I only
(B) II only
(C) I and III only
(D) II and III only
(E) I, II and III

10. Which of the following functions has both a vertical and horizontal asymptote?

(A) $f(x)=\dfrac{1}{1+e^{-x}}$
(B) $f(x) = \tan x$
(C) $f(x)=\dfrac{x}{x^2+2}$
(D) $f(x)=\dfrac{x}{x^2-2}$
(E) $f(x)=\dfrac{x^2+2}{x}$

11. The function $f(x)=\begin{cases}\dfrac{x^2+2x+3}{x^2-1}; x\geq 0\\ \dfrac{x}{e^x}; x<0\end{cases}$

has which of the following asymptotes?

(A) $y = 0$ only.
(B) $y = 1$ only.
(C) $y = 1, x = 1$ only.
(D) $y = 1, x = \pm 1$ only.
(E) $y = 0, y = 1, x = \pm 1$.

12. Which of the following equations has $y = -2$ as an asymptote?

(A) $f(x)=\dfrac{-2x^3}{x^2+3x-1}$
(B) $f(x)=\dfrac{-2x^3}{x^4+3x-1}$
(C) $f(x) = -2ex$
(D) $f(x)=\dfrac{e^x-2x^2+3}{x^2+3x-1}$
(E) $f(x) = \ln x - 2$

13. If the function $f(x)=\dfrac{-ax^3+bx^2+cx+d}{e^x-wx^3+w}$ has a horizontal asymptote of $y=2$ and a vertical asymptote of $x=0$, then $w-a=$

(A) -1

(B) 0

(C) 1

(D) ∞

(E) The limit does not exist.

14. The function $f(x)=\dfrac{2x^3+3x^2+4x-5}{x^2+1}$ has which of the following asymptotes?

(A) $y=2$ only.

(B) $x=-1$ only.

(C) $y=3$ only.

(D) $y=2$, $x=-1$ only.

(E) None of the above

15. For a certain function $f(x)$, the following information is known to be true:

- $\lim_{x\to\infty} f(x)=-3$
- $\lim_{x\to-\infty} f(x)=-\infty$
- $\lim_{x\to-3^-} f(x)=-\infty$
- $\lim_{x\to-3^+} f(x)=\infty$
- $f(-3)$ does not exist.

Which of the following statements must be true?

I. $f(x)$ has a horizontal asymptote of $y=-3$.

II. $f(x)$ has a vertical asymptote of $x=-3$.

III. $f(x)$ *could be* a continuous function.

(A) I only

(B) II only

(C) I and II only

(D) I and III only

(E) I, II and III

Free-Response Questions

16. Consider the following criteria:

   I. $f(1) = 3$

   II. $\lim\limits_{x \to 1^+} = 2$

   III. $\lim\limits_{x \to 1^-} = 3$

   IV. f has a vertical asymptote at $x = 1$.

   V. f has a jump discontinuity at $x = 1$. (Hint: see Chapter 9 for a review of discontinuities)

   VI. f is continuous at $x = 1$.

   (a) What is the maximum number of the criteria listed above that can simultaneously be satisfied by a function f?

   (b) On the axes provided, draw a graph of a function that satisfies the maximum number of criteria simultaneously.

   (c) What is the minimum number of functions needed to satisfy all of the criteria?

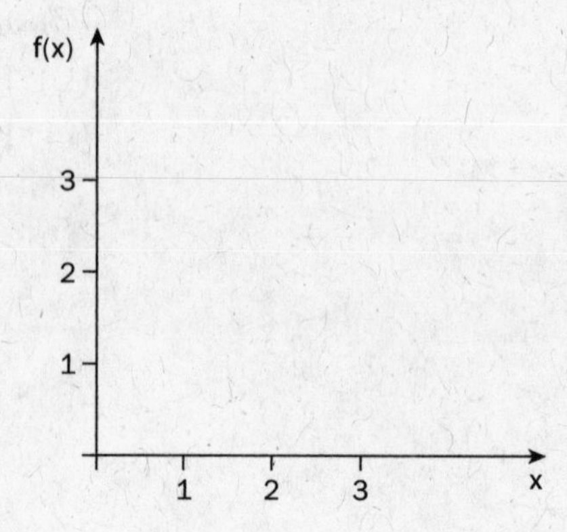

17. Does the function $f(x) = \left(1 + \dfrac{2}{3x}\right)^x$ have a horizontal asymptote? If so, find its equation. If not, explain why.

ANSWERS AND EXPLANATIONS

1. D

To determine whether a function has a horizontal asymptote, we compute the limit of the function as x approaches infinity and negative infinity. If one of the limits exists and is finite, then the function has a horizontal asymptote.

(A) $$\lim_{x\to\pm\infty}\frac{\sqrt{x^2+3}}{2x-1} = \lim_{x\to\pm\infty}\frac{x\sqrt{1+\frac{3}{x^2}}}{x\left(2-\frac{1}{x}\right)} = \frac{1}{2};$$

therefore this function has horizontal asymptote at $y=\frac{1}{2}$

(B) $\lim_{x\to\pm\infty}\sin\frac{1}{x}=0$ this function has a horizontal asymptote at $y = 0$.

(C) $\lim_{x\to\pm\infty}\frac{x}{x+1}\tan\frac{1}{x} = \lim_{x\to\pm\infty}\frac{x}{x+1}\cdot\lim_{x\to\pm\infty}\tan\frac{1}{x}$ $= 0$, the function has a horizontal asymptote at $y = 0$.

(D) $\lim_{x\to+\infty} x\cos\frac{1}{x}=\infty$; $\lim_{x\to-\infty} x\cos\frac{1}{x}=-\infty$.

Because these limits are not finite, the function does not have a horizontal asymptote.

(E) We use rules for manipulating limits to rewrite:

$$\lim_{x\to\pm\infty}\frac{\sqrt{2x^2+1}}{x-3}\sin\frac{1}{x} = \lim_{x\to\pm\infty}\frac{\sqrt{2x^2+1}}{x-3}\cdot\lim_{x\to\pm\infty}\sin\frac{1}{x}$$

We compute each limit separately:

$$\lim_{x\to\pm\infty}\frac{\sqrt{2x^2+1}}{x-3} = \lim_{x\to\pm\infty}\frac{x\sqrt{2+\frac{1}{x^2}}}{x\left(1-\frac{3}{x}\right)} = \sqrt{2}$$

$$\lim_{x\to\pm\infty}\sin\frac{1}{x}=0$$

We conclude that

$$\lim_{x\to\pm\infty}\frac{\sqrt{2x^2+1}}{x-3}\sin\frac{1}{x}=\sqrt{2}\cdot 0 = 0.$$

Therefore, this function has a horizontal asymptote at $y = 0$.

2. D

This question is about the relative rates of growth for g and f. Because the limit is zero, f must be growing at a faster rate than g. Because the sine function oscillates, it does not affect the rate of growth of g relative to f. The growth of g is determined by the $\sqrt{x^6+4}$ term, which, in turn, is determined by the growth of $\sqrt{x^6}=x^3$. Therefore, g is growing at the rate of a third-degree polynomial. Of all the answer choices, only x^4—answer choice (D)—grows faster than x^3, therefore it is the only possibility.

3. B

This limit is a twist on the limit $\lim_{x\to 0}\frac{\sin x}{x}=1$ from Chapter 7. Notice that if $x\to\infty$, then $\frac{1}{x}\to 0$; therefore this limit is just a fancier way of writing the limit from the previous chapter.

4. A

We first rewrite this limit as:

$$\lim_{x\to\infty}\frac{1+\frac{1}{x}^{x}\cdot x}{\sqrt{x^2+1}} = \lim_{x\to\infty}1+\frac{1}{x}^{x}\cdot\lim_{x\to\infty}\frac{x}{\sqrt{x^2+1}}$$

The first limit is a twist on the limit from Chapter 7: $\lim_{h\to 0}(1+h)^{\frac{1}{h}}=e$; as $x\to\infty$, $\frac{1}{x}\to 0$, so the first limit $\lim_{x\to\infty}1+\frac{1}{x}^{x}$ is equal to e.

To compute the second limit we use the rules for manipulating limits:

$$\lim_{x\to\infty}\frac{x}{\sqrt{x^2+1}} = \lim_{x\to\infty}\frac{\not{x}}{\not{x}\sqrt{1+\frac{1}{x^2}}} = 1$$

Combining the two limits, we find

$$\lim_{x\to\infty}\frac{\left(1+\frac{1}{x}\right)^x\cdot x}{\sqrt{x^2+1}} = \lim_{x\to\infty}\left(1+\frac{1}{x}\right)^x\cdot\lim_{x\to\infty}\frac{x}{\sqrt{x^2+1}}$$
$$= e\cdot 1 = e$$

5. D

We can eliminate answer choice (A) immediately; the only candidates for the vertical asymptotes of this function are the points where the denominator equals zero—$x = \pm 2$.

We can also eliminate answer choice (C), because the only candidate for a vertical asymptote of this function is $x = -1$.

In answer choice (B), $x = 4$ *is* a candidate for a vertical asymptote because the function is not defined at this point. To determine whether this point is in fact a vertical asymptote, we compute the limit:

$$\lim_{x\to 4}\frac{x^2-16}{x-4} = \lim_{x\to 4}\frac{(x+4)\cancel{(x-4)}}{\cancel{x-4}} = 8$$

Because the limit is finite, the function *does not* have a vertical asymptote at $x = 4$.

In answer choice (D), we rewrite the function: $\frac{x+6}{x^2-7x+12} = \frac{x+6}{(x-3)(x-4)}$ and see that $x =$ 4 *is* a candidate for a vertical asymptote because the function is not defined at this point. We compute the limit

$$\lim_{x\to 4^+}\frac{x+6}{(x-3)(x-4)} = \frac{10}{1\cdot\text{something small and positive}}$$
$$= +\infty$$

$$\lim_{x\to 4^-}\frac{x+6}{(x-3)(x-4)} = \frac{10}{1\cdot\text{something small and negative}}$$
$$= -\infty$$

Because the one-sided limits are infinite, this function has a vertical asymptote at $x = 4$ and the correct answer is (D).

6. E

Looking on the graph, we see that f is defined but not continuous at $x = 1$; f is defined and continuous at $x = 2$, and f is neither defined nor continuous at $x = 3$. Therefore, statements I, II, and III are all false and the answer is (E).

7. B

In questions like these, treat options I, II and III as true false statements.

I – False. The dominant term in the numerator is $2x^3$, while in the denominator it is 2^x. This makes the end behavior model $y = \frac{2x^3}{2^x}$. Since the denominator dominates the numerator for large values of x, $\lim_{x\to\infty} f(x) = 0$ and therefore $f(x)$ has a horizontal asymptote of $y = 0$. Meanwhile, $\lim_{x\to-\infty} f(x) = -\infty$ since 2^x approaches 0 as x approaches negative infinity. Thus, $f(x)$ has only one asymptote, and it is $y = 0$.

II – True. As explained above.

III – False. For a function to have a vertical asymptote, the denominator must go to zero. In this case, when x is zero, the denominator is 1, and therefore there is no vertical asymptote at $x = 0$. Thus, only statement II is true.

8. E

If $y = a$ is a horizontal asymptote of the function $y = f(x)$, either $\lim_{x\to\infty} f(x) = a$ or $\lim_{x\to-\infty} f(x) = a$. Since either of these is sufficient for a horizontal asymptote at $y = a$, choice a might be true, but does not have to be true. Choice b is one of the conditions for a vertical asymptote, but has no relevance for horizontal asymptotes. Option c is discussing the behavior of the function at x = a, and since horizontal asymptotes are concerned with the function's behavior as it approaches either positive or negative infinity, we have no information about what is happening at x = a, so again, this might be true but does not have to be. Option d implies that the function never crosses the horizontal line $y = a$. It is a common misconception that a function cannot cross it's horizontal asymptotes, but this is not true. Consider, for example, the function shown below:

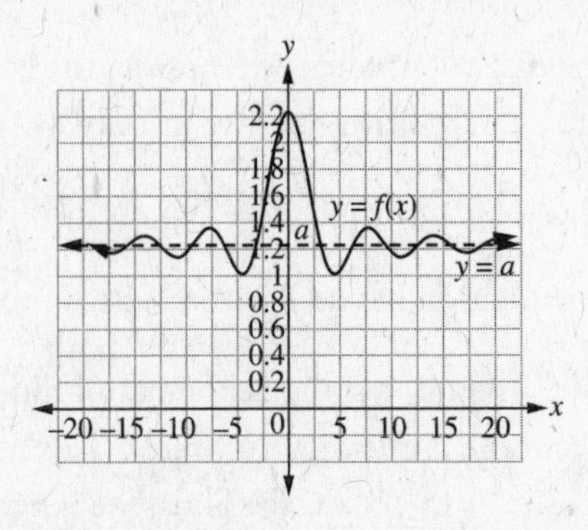

This is an example of a function that has an asymptote at $y = a$ but crosses this line infinitely many times. Thus, all of the first 4 statements *could* be true, but none of them must be.

9. D

In questions like these, treat options I, II and III as true false statements.

I – False. If we substitute x = 2 into the function, $f(x) \to \frac{0}{0}$ which indicates a probable hole at x = 2, and not an asymptote. Taking the limit we see:

$$\lim_{x\to 2}\frac{x^2-5x+6}{x^2-4} = \lim_{x\to 2}\frac{(x-3)(x-2)}{(x+2)(x-2)} = \lim_{x\to 2}\frac{(x-3)}{(x+2)} = -\frac{1}{4}$$

This indicates the presence of a hole and not a vertical asymptote at = 2.

II – True. If we substitute x = –2 into the function, $f(x) \to \frac{20}{0}$ which indicates a vertical asymptote at x = –2.

III – True. To check for horizontal asymptotes, we consider the limits as x goes to infinity or negative infinity. In this case: In this case: $\lim_{x\to\infty} f(x) = 1$ which indicates that $y = 1$ is a horizontal asymptote. Therefore, II and III are both true, and the correct solution is d.

10. D

(A) – the denominator never equals zero since $e^{-x} > 0 \forall x \in \mathbb{R}$. Therefore the function has no vertical asymptotes and so a is wrong.

(B) – tan x has vertical asymptotes, but no horizontal asymptotes, so b is wrong.

(C) – the denominator never equals zero since $x^2 > 0 \forall x \in \mathbb{R}$. Therefore the function has no vertical asymptotes and so c is wrong.

(E) – $\lim_{x\to\infty} f(x) = \infty$, $\lim_{x\to-\infty} f(x) = -\infty$. Therefore, $f(x)$ has no horizontal asymptotes and e is wrong.

(D) – is the correct answer. $\lim_{x\to\infty} f(x) = 0 \quad y = 0$ is a horizontal asymptote.

If we substitute $x = \pm\sqrt{2}$ into the function, $f(x) \to \frac{4}{0}$, indicating vertical asymptotes at $x = \pm\sqrt{2}$.

11. C

For horizontal asymptotes, we consider the behavior as x approaches positive/negative infinity. Since this is a piecewise function, we must take care to be certain that we are using the correct piece of the function in each case.

$$\lim_{x\to\infty} f(x) = \lim_{x\to\infty} \frac{x^2+2x+3}{x^2-1} = 1 \qquad y=1$$

is a horizontal asymptote.

$$\lim_{x\to-\infty} f(x) = \lim_{x\to-\infty} \frac{x}{e^x} = -\infty \therefore$$

there is no horizontal asymptote going to the left.

For vertical asymptotes, we consider cases where the denominator of a rational function is zero. Since e^x never equals zero, we consider only the top piece of the piecewise function.

$x^2-1=0$ when $x=\pm1$. Since this function only exists when $x \geq 0$, the only possibility is $x=1$. If we substitute $x=1$ into the function, $f(x)\to\frac{6}{0}$ which indicates a vertical asymptote at $x=1$. Thus, the only asymptotes are $y=1$ and $x=1$.

12. D

For horizontal asymptotes, we consider the behavior as x approaches positive/negative infinity. In choice d, we observe that an exponential function is involved. This is important because exponential functions exhibit very different behavior at each end. To the right, e^x grows to infinity, but to the left it goes to zero and is therefore dominated by the $-2x^2$. Thus:

$$\lim_{x\to-\infty} f(x) = \lim_{x\to-\infty} \frac{e^x-2x^2+3}{x^2+3x-1} = \lim_{x\to-\infty} \frac{-2x^2}{x^2} = -2 \qquad y=-2$$

is a horizontal asymptote.

13. C

For horizontal asymptotes, we consider the behavior as x approaches positive/negative infinity.

$\lim_{x\to\infty} f(x) = \lim_{x\to\infty} \frac{-ax^3}{e^x} = 0 \quad y=0$ is a horizontal asymptote. Therefore, we must look to the left to find the desired asymptote of $y=2$.

$\lim_{x\to-\infty} f(x) = \lim_{x\to-\infty} \frac{-ax^3}{-wx^3} = \frac{a}{w}$. For an asymptote of $y=2$, $\frac{a}{w}=2$.

If f has a vertical asymptote at $x=0$, then the denominator must be zero when $x=0$.

$$e^0 - w(0)^3 + w = 0$$

Thus: $1+w=0$

$$w=-1$$

but:

$$\frac{a}{w}=2 \to \frac{a}{-1}=2 \to a=-2$$

So: $w-a=-1-(-2)=1$

14. E

For horizontal asymptotes, we consider the behavior as x approaches positive/negative infinity.

$$\lim_{x\to\infty} f(x)=\infty,\ \lim_{x\to-\infty} f(x)=-\infty$$

Thus, the function has no horizontal asymptotes.

For vertical asymptotes, the denominator must be zero (and the numerator non-zero). This never happens as $x^2+1>0\,\forall x\in\mathbb{R}$. Thus, the function has no vertical asymptotes.

15. C

$\lim_{x\to\infty} f(x)=-3$ is sufficient to guarantee a horizontal asymptote of $y=-3$, thus statement I is true.

$\lim_{x\to-3^-} f(x)=-\infty$ and $\lim_{x\to-3^+} f(x)=\infty$ both guarantee a vertical asymptote of $x=-3$, thus statement II is true.

Since $f(-3)$ does not exist, $f(x)$ is not continuous, thus statement III is false.

FREE-RESPONSE ANSWERS

16. (a) Answer choices IV, V, and VI are mutually exclusive. A function cannot be continuous at a point, have a jump discontinuity at the same point, and have an asymptote at that point. It is possible for a function with a jump discontinuity, however, to satisfy I, II, and III. Therefore, criteria I, II, III, and V can be satisfied simultaneously and, the maximum number of criteria that can be satisfied simultaneously is four.

(b) (Answers will vary)

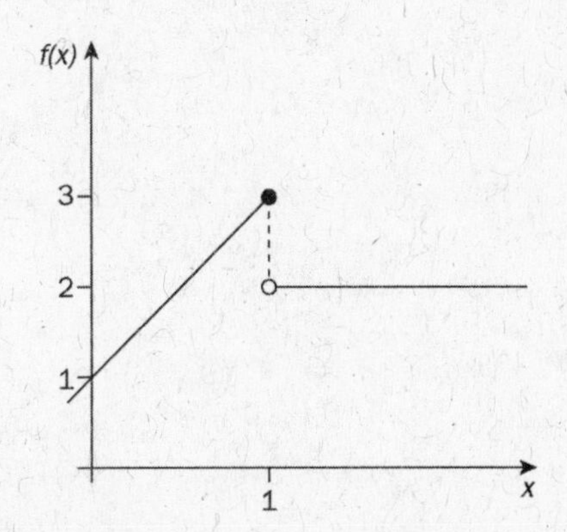

(c) We saw in part (a) that criteria I, II, III, and V can be simultaneously satisfied by a single function. Criteria IV, V, and VI are mutually exclusive. Therefore, two additional functions are needed—one to satisfy criteria IV and one to satisfy criteria VI. The minimum number of functions needed to satisfy all six criteria is, therefore, three.

17. To determine whether the function has a horizontal asymptote, we compute the limits as $x \to \infty$ and $x \to -\infty$. The limit $\left(\lim_{x\to\infty} 1 + \frac{2}{3x}\right)^x$ is a twist on the limit that defined e in Chapter 7; as $x \to \infty$, $\frac{1}{x}$ approaches zero. Therefore, we can manipulate this limit to be of the form $(1 + \text{something approaching zero})^{\text{precise reciprocal power}}$. We write:

$$\lim_{x\to\pm\infty}\left(1+\frac{2}{3x}\right)^x = \lim_{x\to\infty}\left[\left(1+\frac{2}{3x}\right)^x\right]^{\frac{3}{2}\cdot\frac{2}{3}} = \lim_{x\to\infty}\left[\left(1+\frac{2}{3x}\right)^{\frac{3}{2}x}\right]^{\frac{2}{3}} = e^{\frac{2}{3}}$$

The limit $x \to \pm\infty$ is finite, equal to $e^{\frac{2}{3}}$. Therefore, this function does have a horizontal asymptote, and its equation is $y = e^{\frac{2}{3}}$.

CHAPTER 9: CONTINUOUS FUNCTIONS

IF YOU LEARN ONLY FOUR THINGS IN THIS CHAPTER . . .

1. A function is continuous at $x = c$ if:

   $f(x)$ is defined at $x = c$

   $\lim_{x \to c} f(x)$ exists

   $\lim_{x \to c} f(x) = f(c)$

2. There are three types of discontinuities: point discontinuities, jump discontinuities, and vertical asymptotes.
3. **The Intermediate Value Theorem:** If f is continuous on the closed interval $[a, b]$, then for any number k between $f(a)$ and $f(b)$, there is some $c \in [a, b]$ with $f(c) = k$. That is, f takes on every value between $f(a)$ and $f(b)$ on the interval $[a, b]$.
4. **The Extreme Value Theorem:** If f is continuous on the closed interval $[a, b]$, then f takes on a maximum and a minimum value on $[a, b]$.

CONTINUITY AND LIMITS

Continuous means uninterrupted, either in extent, time, sequence, or substance. In mathematics, we say that a function is continuous if it has no holes or breaks.

One way to think about continuity is that it means the function doesn't give us any surprises. If the function looks like it is heading for a certain value, then it is. If the values of x are close together, then the values of $f(x)$ will be close together, too.

We can also think about continuity graphically: A function is continuous if we can draw it without lifting up our pencil. What does this mean?

Well, the situations in which we *do* need to lift our pencil up to draw a function are the following:

- The function is not defined at the point (because the function has a hole or a vertical asymptote).
- The function is defined at the point, but the limit of the function does not exist.
- The function is defined and the limit exists, but the values of the function and limit are different.

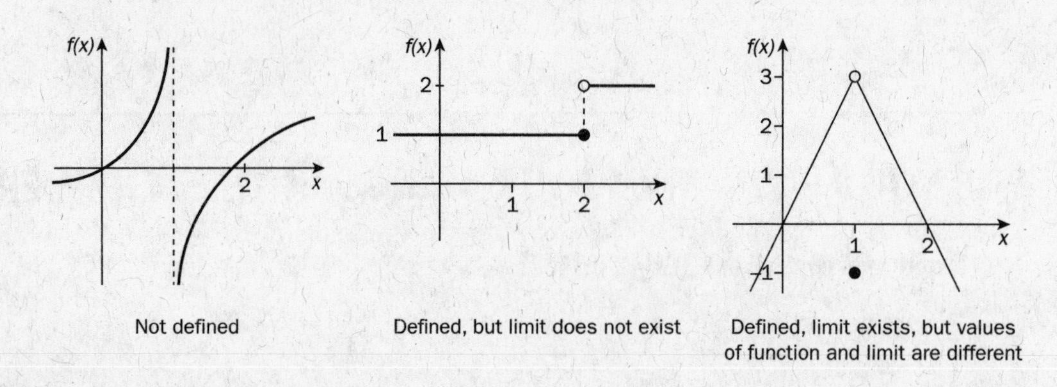

Not defined   Defined, but limit does not exist   Defined, limit exists, but values of function and limit are different

We can use this list and the graphs to make a more precise definition of continuity: A function $f(x)$ is continuous at the point $x = c$ if:

(a) $f(x)$ is defined at $x = c$

(b) $\lim_{x \to c} f(x)$ exists

(c) $\lim_{x \to c} f(x) = f(c)$

If a function is not continuous at $x = c$, we say it is discontinuous there, or has a discontinuity at $x = c$.

If a function is continuous at all x in the interval $[a, b]$, we say that f is continuous on the interval $[a, b]$. The same definition holds for open or half-open intervals.

Going back to our graphs above, we can see that the first graph is not continuous at $x = 1$ because it is not defined there. The second function is not continuous at $x = 2$ because $\lim_{x \to 2^+} f(x) = 2$, while $\lim_{x \to 2^-} f(x) = 1$. The third function is not continuous at $x = 1$ because, although $\lim_{x \to 1} f(x) = 3$, this is not the value of the function at $x = 1$; $f(1) = -1$.

Example:

Is the function defined by $f(x)=\begin{cases}\sqrt{x-3}, & 3\le x<7\\ x-5, & x\ge 7\end{cases}$ continuous at

(a) $x = 4$?

(b) $x = 7$?

Justify your answers.

Solution:

(a) At $x = 4$, the function is defined on both sides by the same formula, $\sqrt{x-3}$. The graph of $f(x) = \sqrt{x-3}$ does not have any holes (on its domain), so the function is continuous at $x = 4$. This graphical analysis confirms the analytical computation $\lim_{x\to 4}\sqrt{x-3} = 1 = f(4)$. Since the function is defined at this point and the value of the function is equal to the value of the limit, the function is continuous at $x = 4$.

(b) The point $x = 7$ is a point to which we must pay special attention; the definition of the function changes at this point, so it is possible that the left- and right-hand limits will not be equal. But, we compute:

$$\lim_{x\to 7^-}\sqrt{x-3} = 2$$

$$\lim_{x\to 7^+} x - 5 = 2$$

Because the left- and right-hand limits are equal, we determine that $\lim_{x\to 7} f(x) = 2$. Using the definition of the function, we see that, at $x = 7$, the function is defined by the formula $x - 5$. Therefore $f(7) = 7 - 5 = 2$. Because $\lim_{x\to 7} f(x) = 2 = f(7)$, we conclude that the function is continuous at $x = 7$. We could also determine this by looking at the graph of the function:

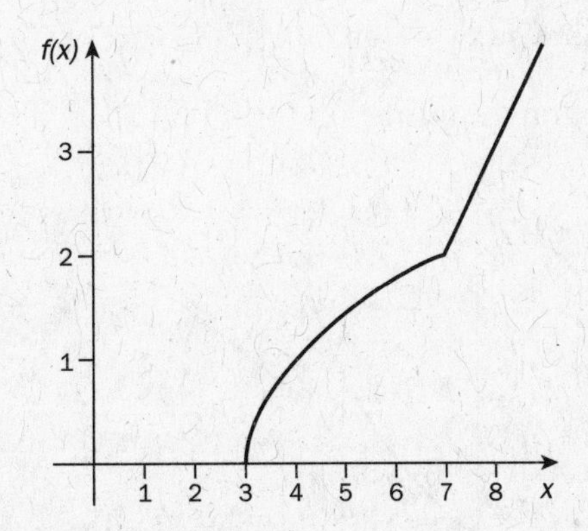

The graph of the function has no holes or breaks at $x = 7$, so the function is continuous there.

Types of Discontinuities

There are three types of discontinuities: point discontinuities, jump discontinuities, and vertical asymptotes. In some calculus texts, point discontinuities are often termed removable discontinuities, while jump discontinuities and vertical asymptotes are lumped together as nonremovable discontinuities.

We say that a function has a point discontinuity at $x = c$ if $\lim_{x\to c} f(x)$ exists, but $\lim_{x\to c} f(x) \neq f(c)$. The function $f(x) = \frac{x^2 - 1}{x - 1}$ has a point discontinuity at $x = 1$. The function is not defined at $x = 1$, because the denominator is equal to zero at that point, but $\lim_{x\to 1} \frac{x^2 - 1}{x - 1} = \lim_{x\to 1} x + 1 = 2$. The graph of the function has a hole at the point $x = 1$. Notice that whenever the zero in the denominator can be eliminated by factoring and canceling out terms, the discontinuity is removable (as it is "removed" from the function).

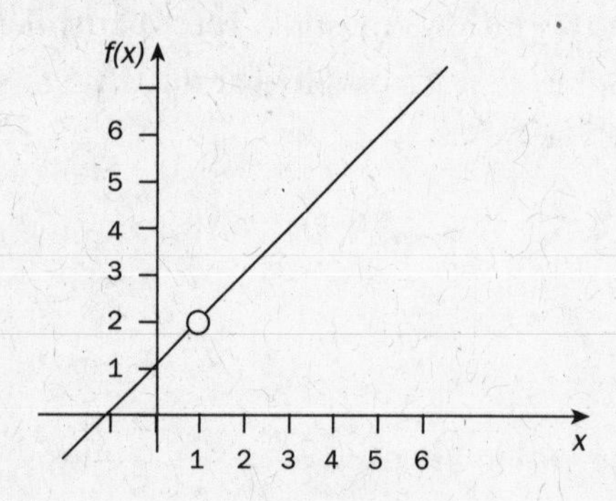

We say a function has a jump discontinuity at $x = c$ if the left- and right-hand limits are finite, but $\lim_{x\to c^+} f(x) \neq \lim_{x\to c^-} f(x)$. The piecewise defined function $f(x) = \begin{matrix} x, & x > 1 \\ 2x^2, & x \leq 1 \end{matrix}$ has a jump discontinuity at $x = 1$, since $\lim_{x\to 1^+} f(x) = \lim_{x\to 1^+} x = 1$ and $\lim_{x\to 1^-} f(x) = \lim_{x\to 1^-} 2x^2 = 2$. We can also see the jump discontinuity on the graph.

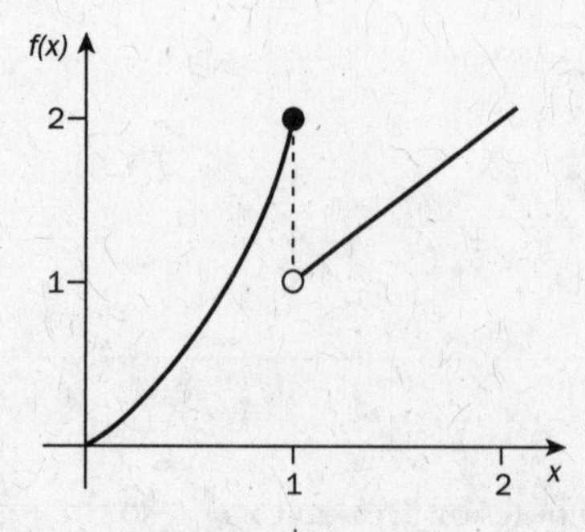

The third type of discontinuity is a vertical asymptote. As we discussed in Chapter 8, a function has an asymptote at $x = c$, when at least one of the right- and left-hand limits of the function are infinite there.

BUILDING CONTINUOUS FUNCTIONS

Let's start by observing that some well-known functions are continuous: $f(x) = c$, $f(x) = x$. Starting with these two basic functions, we can use the definition of continuity and the properties of limits to build a library of continuous functions.

THEOREM

If the functions f and g are continuous at $x = c$, then:

- $f \pm g$ is continuous at $x = c$
- $f \cdot g$ is continuous at $x = c$
- $\frac{f}{g}$ is continuous at $x = c$ if $g(c) \neq 0$; it is discontinuous if $g(c) = 0$

THEOREM

If g is continuous at $x = c$ and f is continuous at $g(c)$, then the composition $f \circ g$ is continuous at $x = c$.

It follows from these theorems that polynomials are continuous everywhere and rational functions are continuous everywhere that their denominators do not equal zero.

PROPERTIES OF CONTINUOUS FUNCTIONS

Continuous functions have some nice properties. The fact that we can draw continuous functions without lifting up our pencil leads us to several conclusions about the behavior of continuous functions.

THE INTERMEDIATE VALUE THEOREM

If f is continuous on the closed interval $[a, b]$, then for any number k between $f(a)$ and $f(b)$, there is some $c \in [a, b]$ with $f(c) = k$. That is, f takes on every value between $f(a)$ and $f(b)$ on the interval $[a, b]$.

We can illustrate this theorem with a picture. For example, suppose we are given that f is continuous on the interval $[1, 5]$, and we are told that $f(1) = 6$ and $f(5) = -3$. We can put this information on a graph. Because the function is continuous on the interval $[1, 5]$, we can draw the graph of the function between 1 and 5 without lifting up our pencil. That means the function must cross any horizontal line between $y = 6$ and $y = -3$, say $y = 3$, at least once. That is, there is at least one value of x between 1 and 5 where the value of the function is 3. We don't know what this value of x is, or how many times on the interval the function takes on the value, but we know that for at least one point $1 \le c \le 5$, we have $f(c) = 3$.

Several possibilities are shown below.

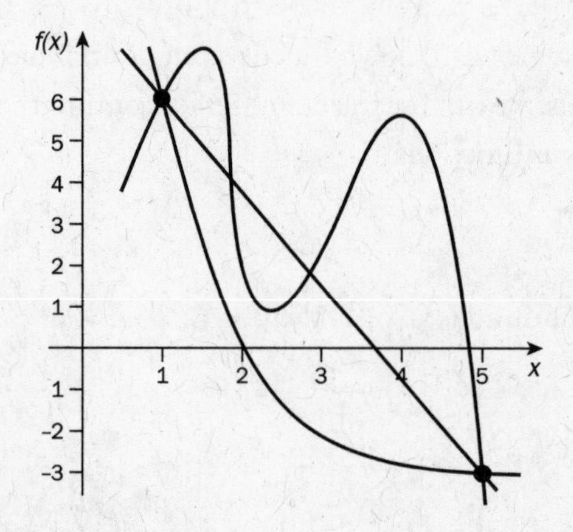

Notice that the assumption that f is continuous on $[a, b]$ is necessary. For example, the function shown in the graph below is not continuous. It has a jump discontinuity at $x = 1$, and we can see from the graph that the function approaches 3 from the right, but does not take on the value of 3.

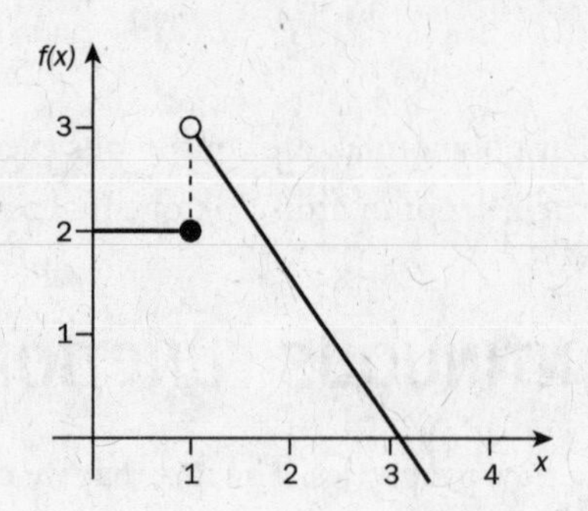

The intermediate value theorem can be used to approximate zeros of a function—points at which the graph of a function crosses the *x*-axis.

THEOREM

If f is continuous on the closed interval $[a, b]$, and $f(a)$ and $f(b)$ have opposite signs, then f has a zero between a and b, that is, for some point $a \leq c \leq b, f(c) = 0$.

For example, the polynomial $f(x) = -x^3 - 2x + 9$ has at least one real root, since $f(0) = 9$ and $f(10) = -1{,}011$. A calculator can be used to find a decimal approximation of this root.

A common question type on the free-response portion of the AP exam is to give students a function in the form of a table of values that describes some "real world" situation. If it is given (as it usually is) that the function is continuous on the given interval, then a question about the Intermediate Value Theorem is likely to appear. To illustrate, using a generic example, let f be a Continuous function on the interval $0 \leq x \leq 50$ with values shown in the following table. (We should take a minute to note here that if the function was not continuous, i.e., if it consisted of only six discrete points, then the Intermediate Value Theorem and most of the other theorems about functions on a closed interval that you learn in this course don't apply.)

x	0	5	20	30	45	50
$f(x)$	−10	−8	0	12	15	18

A typical question might be: On the interval $0 \le x \le 50$, must there be a value of x for which $f(x) = 1$?

Answer: Yes. Because f is continuous on $0 \le x \le 50$, $f(20) = 0$, and $f(30) = 12$, the Intermediate Value Theorem guarantees that there exists at least one x on the interval $20 < x < 30$ for which $f(x) = 1$.

The Extreme Value Theorem

The Extreme Value Theorem describes another property of continuous functions on closed intervals.

If f is continuous on the closed interval $[a, b]$, then f takes on a maximum and a minimum value on $[a, b]$.

We can use this theorem to know there is a most cost-effective pigpen, or maximum profits and minimum costs.

The assumption that f is continuous on the *closed* interval $[a,b]$ is necessary. For example, the function $f(x) = x$ does not attain a maximum or a minimum value on the open interval (1,2). The maximum and minimum *want* to occur at the endpoints of this interval—but they are not included in the domain we are examining.

The function $f(x) = \frac{1}{x}$ has a vertical asymptote at $x = 0$, so it does not obtain a maximum or minimum value on any interval that includes zero.

The Extreme Value Theorem tells us when we are *guaranteed* to find maximum and minimum values for a function. But even a function that does not meet the criteria of the theorem *could* attain a maximum and/or a minimum value on an interval. For example, the function $f(x) = x^2$ obtains a minimum value on the open interval (−2, 1) and, in fact, on any open interval that includes zero, because the vertex of this upward-opening parabola is its low point.

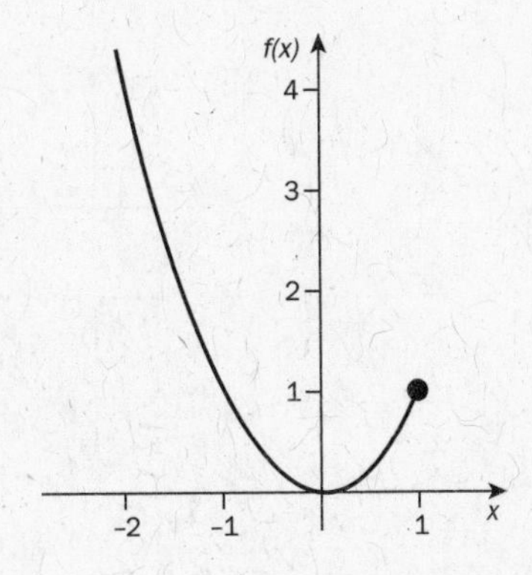

REVIEW QUESTIONS

1. For which value of k is the following function continuous at $x = 4$?

$$f(x) = \begin{cases} \sin\frac{\pi}{x}, & x \le 4 \\ k\sqrt{\frac{x}{2}}, & x > 4 \end{cases}$$

(A) $k = 2$
(B) $k = 1$
(C) $k = -1$
(D) $k = \frac{1}{2}$
(E) $k = -\frac{1}{2}$

2. Suppose $f(a) = 2$ and $f(b) = -3$. Which of the following statements is always true?

(A) If f is continuous on $[a, b]$ then the maximum value of f on $[a, b]$ is less than 2.
(B) f takes on a maximum value on $[a, b]$.
(C) If f is continuous, then f has a zero on $[a, b]$.
(D) If f is continuous on $[a, b]$, then f may take on a minimum value on $[a, b]$.
(E) If f is not continuous on $[a, b]$, then f does not take on a minimum value on $[a, b]$.

3. Suppose that f is not continuous at $x = 1$, f is defined at $x = 1$, and $\lim_{x\to 1} f(x) = L$, where L is finite. Which of the following could be the graph of f?

(A)
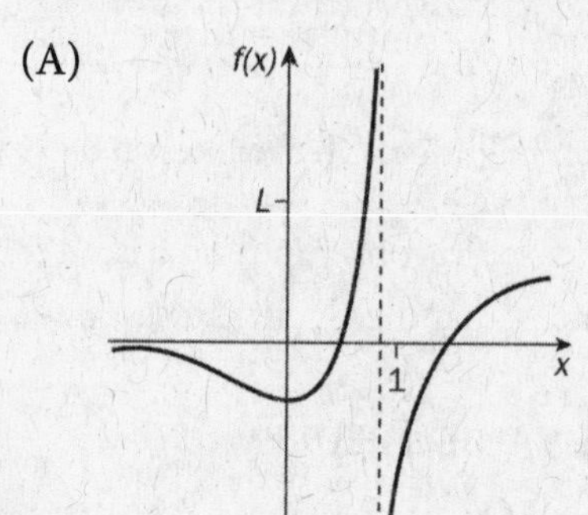

(B)
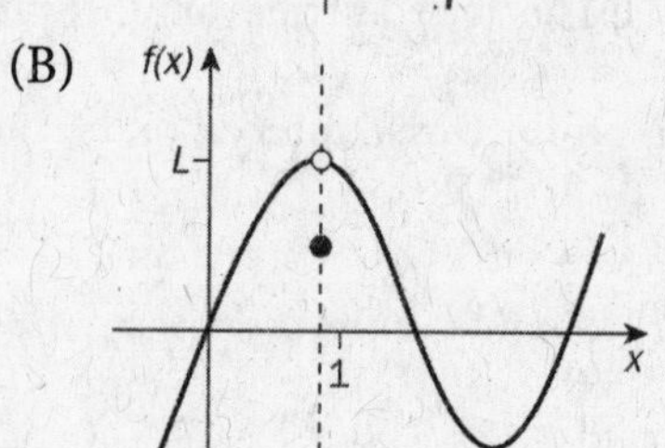

(C)
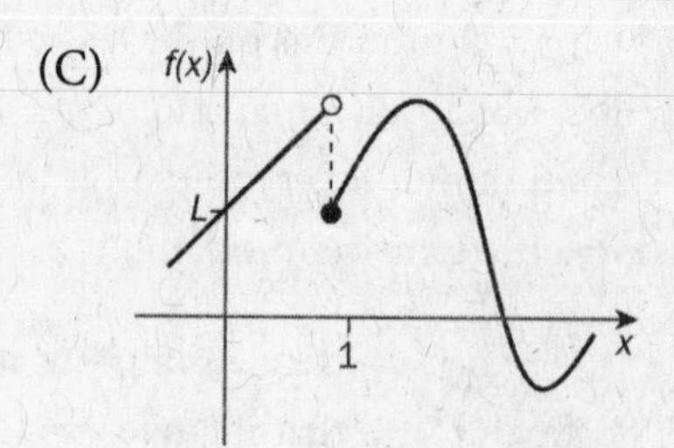

(D)
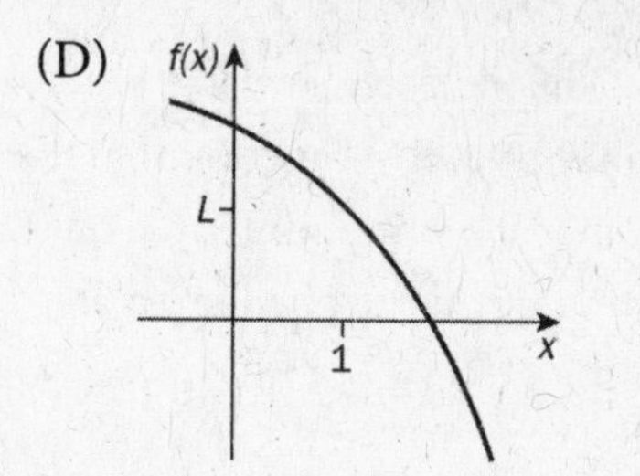

(E)
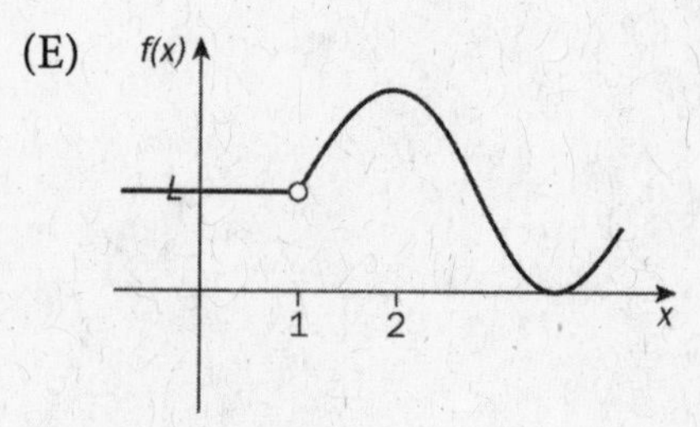

4. For what value of k will the function $f(x) = \dfrac{x^2 - (k+2)x + 6}{x - k}$ have a point discontinuity at $x = k$?

(A) $k = -1$
(B) $k = 0$
(C) $k = 1$
(D) $k = 2$
(E) $k = 3$

5. Suppose f is continuous at $x = c$, $\lim\limits_{x\to c^+} f(x) = s + t + 1$, $\lim\limits_{x\to c^-} f(x) = s - t + 2$, and $f(c) = 2t$. Which of the following statements is true?

(A) $s = -\frac{1}{2}$, $t = \frac{1}{2}$
(B) $s = 1$, $t = \frac{1}{2}$
(C) $s = -\frac{2}{3}$, $t = \frac{1}{3}$
(D) There is not enough information to determine the values of s and t.
(E) None of the statements above is true.

6. Suppose f is continuous on the closed interval $[0, 4]$ and suppose $f(0) = 1, f(1) = 2, f(2) = 0, f(3) = -3, f(4) = 3$. Which of the following statements about the zeros of f on $[0, 4]$ is always true?

(A) f has exactly one zero on $[0, 4]$.
(B) f has more than one zero on $[0, 4]$.
(C) f has more than two zeros on $[0, 4]$.
(D) f has exactly two zeros $[0, 4]$.
(E) None of the statements above is true.

7. Let $f(x)$ be a continuous function for all x. If $f(x) = \dfrac{2x^2 - 2}{x^2 - 3x + 2}$ when $x^2 - 3x + 2 \neq 0$ then what is $f(1)$?

(A) -4
(B) -2
(C) 0
(D) 2
(E) 4

8. Consider the function $y = f(x)$ shown below. Which of the following statements is true?

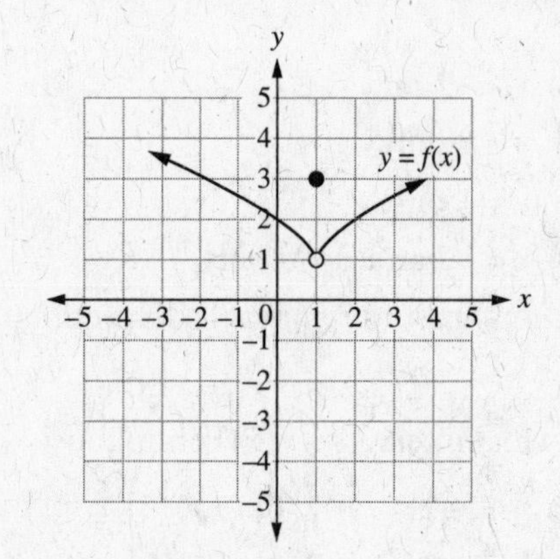

(A) $\lim\limits_{x\to 1} f(x) = 3$
(B) $f(1) = 1$
(C) $f(x)$ is continuous for all x.
(D) $\lim\limits_{x\to 1} f(x) = f(1)$
(E) None of the above

Questions 9 through 12 are based on the function $f(x)$ shown in the graph below:

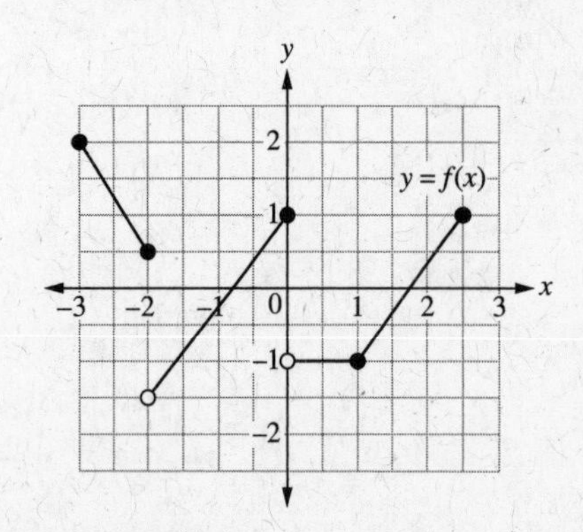

9. The function $f(x)$ has a removable discontinuity at:
   (A) $x = -2$ only
   (B) $x = 0$ only
   (C) $x = 1$ only
   (D) $x = -2$ and $x = 0$ only
   (E) $f(x)$ has no removable discontinuities.

10. On what intervals is $f(x)$ continuous?
    (A) $[-3, -2] \cup [-2, 0] \cup [0, 3]$
    (B) $[-3, -2] \cup [-2, 0] \cup [0, 3]$
    (C) $[-3, -2] \cup [-2, 0] \cup [0, 3]$
    (D) $[-3, -2] \cup [-2, 0] \cup [0, 3]$
    (E) $[-3, -2] \cup [-2, 0] \cup [0, 1] \cup [1, 3]$

11. Where is the function continuous but not differentiable?
    (A) $x = -2$ only
    (B) $x = 0$ only
    (C) $x = 1$ only
    (D) $x = -2$ and $x = 0$ only
    (E) $x = -2$ and $x = 0$ and $x = 1$ only

12. The function has a jump discontinuity at:
    (A) $x = -2$ only
    (B) $x = 0$ only
    (C) $x = 1$ only
    (D) $x = -2$ and $x = 0$ only
    (E) $f(x)$ has no jump discontinuities.

13. Let

$$f(x)=\begin{cases} \dfrac{e^x-1}{x}; x \neq 0 \\ k \quad ; x=0 \end{cases}.$$

$f(x)$ is continuous at $x = 0$ if $k =$

(A) 0
(B) 1
(C) e-1
(D) e
(E) None of the above

14. Consider the function

$$f(x)=\begin{cases} \dfrac{-2x(x+1)(x-2)}{x^2-x-2}; x\neq -1, x\neq 2 \\ 6\ ;x=-1 \\ -4\ ;x=2 \end{cases}.$$

$f(x)$ is continuous everywhere except at:

(A) x = −1 only
(B) x = 2 only
(C) x = −1 and x = 2
(D) x = 0, x = −1 and x = 2.
(E) $f(x)$ is continuous for all real values of x.

15. Consider the function

$$f(x)=\begin{cases} \dfrac{\sin x}{x}; x\neq 0 \\ 1\quad ;x=0 \end{cases}.$$

Which of the following is/are true?

I. $\lim_{x\to 0} f(x)$ exists
II. $f(0)$ exists
III. $f(0)$ is continuous for all real values of x.

(A) I only
(B) II only
(C) III only
(D) I and II only
(E) I, II and III

FREE-RESPONSE QUESTIONS

16. The graph of a function f is shown below and describes the position of a particle as it moves along the y-axis.

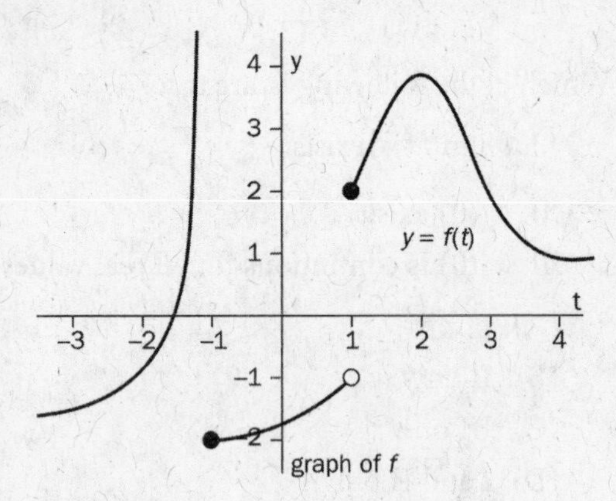

(a) Describe the movement of the particle on the interval $[-1, 2]$.

(b) Assume $f(t)>1$ for $t > 4$. Describe the movement of the particle as t approaches infinity.

(c) Can we use the Intermediate Value Theorem on the interval $[-3, 0]$ to show that f has a zero in that interval? On the interval $[0, 3]$? Explain your reasoning.

17. Let $f(x) = \dfrac{\tan x}{\tan 2x}$. Find the points of discontinuity of f on $[0, 2\pi]$ and determine whether each discontinuity is a point discontinuity, a jump discontinuity, or a vertical asymptote (hint: use the trigonometric identities $\sin 2x = 2\sin x\cos x$ and $\cos 2x = \cos^2 x - \sin^2 x$).

ANSWERS AND EXPLANATIONS

1. D

The function $f(x)$ will be continuous at $x = 4$ if the limit exists at $x = 4$ and is equal to the value of the function at that point. At $x = 4$, the function is defined by the formula $\sin\frac{\pi}{x}$, therefore $f(4) = \sin\frac{\pi}{4} = \frac{\sqrt{2}}{2}$. In addition, $\lim_{x\to 4^-} f(x) = \lim_{x\to 4^-} \sin\frac{\pi}{4} = \frac{\sqrt{2}}{2}$. Thus f will be continuous at $x = 4$ if $\lim_{x\to 4^+} f(x) = \lim_{x\to 4^+} k\sqrt{\frac{x}{2}} = \frac{\sqrt{2}}{2}$.

Using the rules for manipulating limits, we compute that $\lim_{x\to 4^+} k\sqrt{\frac{x}{2}} = k\sqrt{2}$. Therefore, to find k, we solve:

$$k\sqrt{2} = \frac{\sqrt{2}}{2} \Rightarrow k = \frac{1}{2}$$

2. C

Choice (C) is true. By the Intermediate Value Theorem, if a function is continuous on the closed interval $[a, b]$, and $f(a)$ and $f(b)$ have opposite signs, then the function must cross the x-axis on the interval $[a, b]$, that is, f has a zero on $[a, b]$. Choice (A) is not true; the Extreme Value Theorem guarantees that a function that is continuous on a closed interval *does* achieve a maximum value on that interval, but it makes no conclusion about the value of the maximum. Choice (B) is false; the Extreme Value Theorem guarantees that if a function is *continuous* on a closed interval, then it takes on a maximum value. Without continuity, we are not guaranteed that the function will attain a maximum value. Choice (D) is false; the Extreme Value Theorem *guarantees* that the function will take on a minimum value in this case, since f is continuous on a closed interval. Choice (E) is false; the Extreme Value Theorem tells the conditions under which f is *guaranteed* to take on a maximum and minimum value, but if the hypotheses of the theorem are not met, we do not know one way or the other.

3. B

Begin by eliminating the graphs for which f is continuous at $x = 1$. This will eliminate choice (D). Next, eliminate graphs for which f is not defined at $x = 1$. This eliminates choices (A) and (E). Of the two remaining graphs, choose the one for which $\lim_{x\to 1^+} f = \lim_{x\to 1^-} f$

4. E

The function will have a point discontinuity at $x = k$ if the limit exists and is finite at $x = k$, but the value of the limit is not the value of the function at $x = k$. The function is not defined at $x = k$ because the denominator is equal to zero at this point. For the function to have a jump discontinuity here, the limit $\lim_{x\to k} \frac{x^2 - (k+2)x + 5}{x - k}$ must exist and be finite. The easiest way to proceed at this point is to reverse-engineer the answer. We can try all the answer choices, and see which one produces a finite value for the limit.

When $k = -1$, the limit becomes $\lim_{x\to -1} \frac{x^2 - (-1+2)x + 6}{x - (-1)} = \lim_{x\to -1} \frac{x^2 - x + 6}{x + 1}$. This limit is infinite.

When $k = 0$, the limit becomes $\lim_{x\to 0} \frac{x^2 - (0+2)x + 6}{x - (0)} = \lim_{x\to 0} \frac{x^2 - 2x + 6}{x}$. This limit is infinite.

When $k = 1$, the limit becomes $\lim_{x\to 1} \frac{x^2 - (1+2)x + 6}{x - (1)} = \lim_{x\to 1} \frac{x^2 - 3x + 6}{x - 1}$. This limit is infinite.

When $k = 2$, the limit becomes $\lim_{x\to 2} \frac{x^2 - (2+2)x + 6}{x - (2)} = \lim_{x\to 2} \frac{x^2 - 4x + 6}{x - 2}$. This limit is infinite.

When $k = 3$, the limit becomes $\lim_{x\to 3} \frac{x^2 - (3+2)x + 6}{x - (3)} = \lim_{x\to 3} \frac{x^2 - 5x + 6}{x - 3} = \lim_{x\to 3} \frac{\cancel{(x-3)}(x-2)}{\cancel{x-3}} = 1$.

This limit is finite, so answer choice (E) is correct.

5. A

Because f is continuous at $x = c$, the three values $\lim_{x\to c^+} f(x)$, $\lim_{x\to c^-} f(x)$, and $f(c)$ must be equal. That is, $s + t + 1 = s - t + 2 = 2t$. We use this information to set up a system of two equations in two unknowns which we then solve for s and t:

$$\left.\begin{aligned} s + t + 1 &= 2t \\ s - t + 2 &= 2t \end{aligned}\right\} \Rightarrow \left.\begin{aligned} s &= t - 1 \\ s &= 3t - 2 \end{aligned}\right\} \Rightarrow t - 1 = 3t - 2 \Rightarrow 2t = 1 \Rightarrow t = \frac{1}{2},\ s = -\frac{1}{2}$$

6. B

Because f is continuous on the closed interval [0, 4], it follows from the Intermediate Value Theorem and the data given that f has a zero between $x = 1$ and $x = 3$. We are given that $f(2) = 0$, so we are not guaranteed that there will be another zero between $x = 1$ and $x = 3$; this may be the only one. From the given data, we also know that f has a zero between $x = 3$ and $x = 4$. Therefore, the Intermediate Value Theorem guarantees that there are at least two zeros of f on the interval [0, 4], that is, f has more than one zero on [0, 4]—Answer (B).

7. A

In order for $f(x)$ to be continuous at x = 1, $\lim_{x\to 1} f(x) = f(1)$.

$$\lim_{x\to 1} f(x) = \lim_{x\to 1} \frac{2x^2 - 2}{x^2 - 3x + 2} = \lim_{x\to 1} \frac{2(x+1)(x-1)}{(x-2)(x-1)} = \lim_{x\to 1} \frac{2(x+1)}{(x-2)} = \frac{4}{-1} = -4$$

8. E

$\lim_{x\to 1} f(x) = 1$, so choice (A) is incorrect.

$f(1) = 3$, so choice (B) is incorrect.

The previous two statements prove that choice (D) is incorrect.

We see that the function has a displaced point at x = 1, so choice (C) is incorrect.

Thus, choice (E) is the only correct option.

9. E

The only discontinuities that are removable are holes and displaced points — this function has neither. $f(x)$ has 2 jump discontinuities, at $x = -2$ and $x = 0$ but neither of these are removable.

10. D

The function is continuous everywhere except at x = −2 and x = 0, so those are the endpoints we need to consider on the intervals of continuity. At x = −2, the function is continuous from the left, but has a hole when we approach from the right. Therefore we include x = 2 in the interval from the left [−3, −2] but not in the interval from the right [−2, 0].

Likewise, at x = 0, the function is continuous from the left, but has a hole when we approach from the right. Therefore we include x = 0 in the interval from the left [−2, 0] but not in the interval from the right [0, 3]. Since there is no discontinuity at x = 1, we do not concern ourselves with that point when considering intervals of continuity.

11. C

The function is continuous at x = 1 because $\lim_{x\to 1^-} f(x) = \lim_{x\to 1^+} f(x) = f(1) = -1$. The function is not differentiable at x = 1 because it has a corner there (ie: the slopes on either side are finite, but the left hand slope does not equal the right hand slope).

12. D

A jump discontinuity occurs at a value x if the left and right-hand limits are finite but not equal. In this case, that occurs at $x = -2$ and $x = 0$ only as:

$$\lim_{x\to -2^-} f(x) = \frac{1}{2} \text{ while } \lim_{x\to -2^+} f(x) = -\frac{3}{2} \text{ and } \lim_{x\to -0^-} f(x) = 1 \text{ while } \lim_{x\to -0^+} f(x) = -1$$

13. B

In order for $f(x)$ to be continuous at x = 0, $\lim_{x\to 0} f(x) = f(0)$. Substituting x = 0 into f(x) yields 0/0. We can therefore use L'Hopital's rule to compute the limit:

$$\lim_{x\to 0} f(x) = \lim_{x\to 0} \frac{e^x - 1}{x} = \lim_{x\to 0} e^x = e^0 = 1.$$

Then, for continuity, $f(0) = k$ must also equal 1.

14. E

$$f(x)=\begin{cases} \dfrac{-2x(x+1)(x-2)}{(x+1)(x-2)}; x\neq -1, x\neq 2 \\ 2\ ; x=-1 \\ -4 ; x=2 \end{cases}$$

In order for $f(x)$ to be continuous at x = –1:

$$\lim_{x\to -1} f(x)=f(-1)$$

$$\lim_{x\to -1} f(x)=\lim_{x\to -1}\frac{-2x(x+1)(x-2)}{(x+1)(x-2)}=\lim_{x\to -1}(-2x)=-2(-1)=2=f(-1)$$

$\therefore f(x)$ is continuous at x = –1

In order for $f(x)$ to be continuous at x = 2:

$$\lim_{x\to 2} f(x)=f(2)$$

$$\lim_{x\to 2} f(x)=\lim_{x\to 2}\frac{-2x(x+1)(x-2)}{(x+1)(x-2)}=\lim_{x\to 2}(-2x)=-2(2)=-4=f(2)$$

$\therefore f(x)$ is continuous at x = 2

$\therefore f(x)$ is continuous for all real values of x.

15. E

In order for $f(x)$ to be continuous at x = 0, $\lim_{x\to 0} f(x)=f(0)$.

In this case: substituting x = 0 into f(x) yields 0/0. We can therefore use L'Hopital's rule to compute the limit:

$$\lim_{x\to 0} f(x)=\lim_{x\to 0}\frac{\sin x}{x}=\lim_{x\to 0}\frac{\cos x}{1}=\cos(0)=1.$$

Thus, statement 1 is true. Since we are told that $f(0)$ = 1, we know that $f(0)$ exists and so statement II is true. And since $\lim_{x\to 0} f(x)=1$ and $f(0)$ = 1, we know that $\lim_{x\to 0} f(x)=f(0)$ and thus $f(x)$ is continuous at x = 0 which makes statement III true.

Free-Response Answers

16. (a) At $t = -1$, the particle is at $y = -2$. As t approaches 1, the particle moves in a positive direction along the y-axis. At $t = 1$, the particle disappears and reappears at $y = 2$.

(b) As t approaches infinity, the particle moves in a negative direction along the y-axis, ever more slowly toward, but never reaching, $y = 1$.

(c) The Intermediate Value Theorem does not apply to the interval [–3, 0], nor to the interval [0, 3], because the function f is not continuous on either region. The function f has discontinuities at $t = -1$ and $t = 1$.

Strive to use complete sentences when answering a free-response question. For parts (a) and (b), do your best to explain the behavior of a discontinuity and an asymptote in your own words. For part (c), remember that the fundamental statement about the IVT is that the function must be continuous. It doesn't matter that the function actually has a zero in the interval [–3, 0]. Because the function is not continuous on that interval, the IVT does not apply.

17. We use the definition of the tangent function to rewrite f as:

$$f(x) = \frac{\tan x}{\tan 2x} = \frac{\frac{\sin x}{\cos x}}{\frac{\sin 2x}{\cos 2x}} = \frac{\sin x \cos 2x}{\cos x \sin 2x}$$

$$\underset{\text{trig identity}}{=} \frac{\sin x \cdot (\cos^2 x - \sin^2 x)}{\cos x \cdot 2\sin x \cdot \cos x} = \frac{\sin x(\cos x + \sin x)(\cos x - \sin x)}{2\sin x \cdot \cos^2 x}$$

This function has discontinuities at all points where the denominator is equal to zero. On $[0, 2\pi]$, that is, where

$$\cos x = 0 \Rightarrow x = \frac{\pi}{2}, \frac{3\pi}{2}$$
$$\sin x = 0 \Rightarrow x = 0, \pi, 2\pi$$

Combining the two sets of zeros, we see that f is discontinuous at $x = 0, \frac{\pi}{2}, \pi, \frac{3\pi}{2}, 2\pi$.

To determine what type of discontinuity occurs at each point, we need to evaluate the limit as x approaches each of these points.

$$\lim_{x\to\frac{\pi}{2}} f(x) = \lim_{x\to\frac{\pi}{2}} \frac{\cancel{\sin x}(\cos x + \sin x)(\cos x - \sin x)}{2\cancel{\sin x} \cdot \cos^2 x} = -\infty$$

$$\lim_{x\to\frac{3\pi}{2}} f(x) = \lim_{x\to\frac{3\pi}{2}} \frac{\cancel{\sin x}(\cos x + \sin x)(\cos x - \sin x)}{2\cancel{\sin x} \cdot \cos^2 x} = -\infty$$

$$\lim_{x\to 0} f(x) = \lim_{x\to 0} \frac{\cancel{\sin x}(\cos x + \sin x)(\cos x - \sin x)}{2\cancel{\sin x} \cdot \cos^2 x} = \frac{1}{2}$$

$$\lim_{x\to \pi} f(x) = \lim_{x\to \pi} \frac{\cancel{\sin x}(\cos x + \sin x)(\cos x - \sin x)}{2\cancel{\sin x} \cdot \cos^2 x} = \frac{1}{2}$$

$$\lim_{x\to 2\pi} f(x) = \lim_{x\to 2\pi} \frac{\cancel{\sin x}(\cos x + \sin x)(\cos x - \sin x)}{2\cancel{\sin x} \cdot \cos^2 x} = \frac{1}{2}$$

The function has vertical asymptotes at each discontinuity where the limit is infinite, i.e., where $x = \frac{\pi}{2}, \frac{3\pi}{2}$. The function has a point discontinuity at each discontinuity where the limit exists and is finite, i.e., where $x = 0, \pi, 2\pi$.

CHAPTER 10: PARAMETRIC, POLAR, AND VECTOR FUNCTIONS*

IF YOU LEARN ONLY THREE THINGS IN THIS CHAPTER . . .

1. Parametric equations take one real number (t) and return two real numbers (x and y).
2. By using the variables r and θ, polar equations can elegantly describe complicated curves in the x-y coordinate plane.
3. Vector functions are often used in real-world scenarios and often take one real input (time) and return three outputs (x, y, and z coordinates or directions).

There are many different ways to represent a function. So far we have considered a function as a rule in which a dependent variable is controlled by an independent variable such as $f(x)$ or $h(t)$. Of course, these can't be the only types of functions running around a math course; when describing planar curves, other types of functions are better suited for representing these curves. In this chapter we'll look at parametric and polar functions, which are often used to simplify the expression of planar curves. We will also define vector functions, that is, a function that takes a real number and gives back a vector instead of another real number.

BC students, understanding this chapter is critical! Many of the questions on the BC exam take the same calculus concepts that are covered on the AB exam, but apply them to parametric, polar, or vector functions. If at all possible, learn to *think* in these new coordinate systems by Test Day. Don't miss a question that tests a simple concept, just because it uses one of these strange-looking functions.

**BC content*

PARAMETRIC FUNCTIONS

Consider the planar curve below.

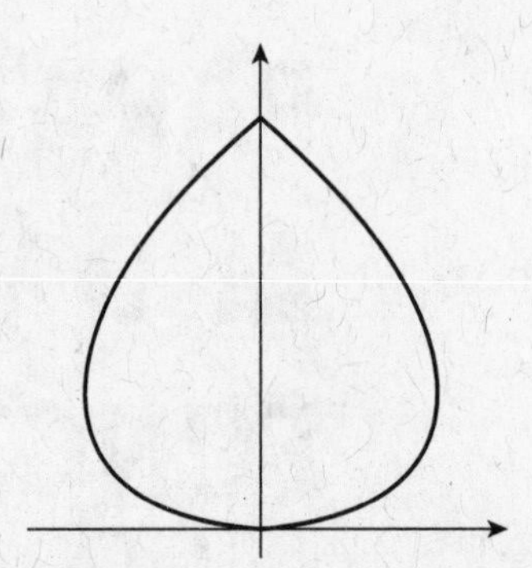

Though the picture on the previous page looks simple enough, we notice some immediate problems. This can't be described by a function we are used to working with because the graph clearly fails the vertical line test. To write an equation to represent this planar curve, we introduce a new variable t and write the x and y coordinates of the curve as functions of the new variable t. That is, we write $x = f(t)$ and $y = g(t)$.

Each value t determines a point $(x, y) = (f(t), g(t))$ on the curve and as t changes, so do $f(t)$ and $g(t)$. For example, the planar curve is defined by the function $f(t) = 5 \sin t$ and $g(t) = t^2$ where $-\pi \le t \le \pi$. To see why this is true, let's revert back to the first days of pre-calculus and vary t to create points to plot on the x–y axis.

t	$x = f(t) = 5 \sin t$	$y = g(t) - t^2$
$-\pi$	0	π^2
$\frac{-\pi}{2}$	–5	$\frac{\pi^2}{4}$
0	0	0
$\frac{\pi}{2}$	5	$\frac{\pi^2}{4}$
π	0	π^2

Plotting these points, we get:

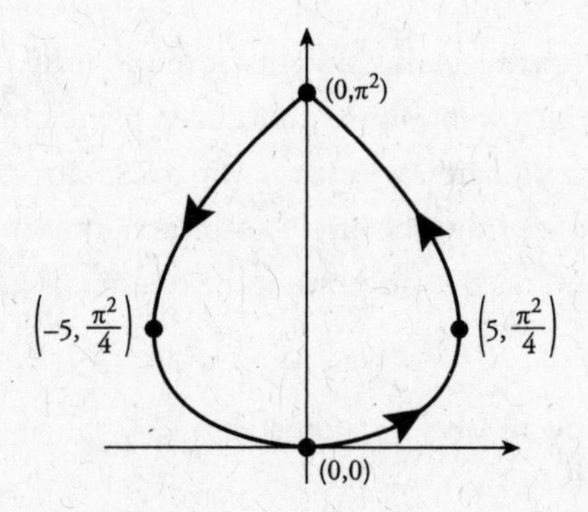

Furthermore, the arrow on the curve indicates the direction of the change of t. Because we have $-\pi \leq t \leq \pi$, we let t move from $-\pi$ to π and indicate this direction of movement with an arrow.

Using the same technique, let's plot the graph of another parametric function. Suppose we are given $x = f(t) = t + 4$ and $y = g(t) = t^2$.

To plot this graph, we make a chart as before.

t	$x = f(t) = t + 4$	$y = g(t) = t^2$
−2	2	4
−1	3	1
0	4	0
1	5	1
2	6	4

We then graph the corresponding ordered pairs (x, y) to get:

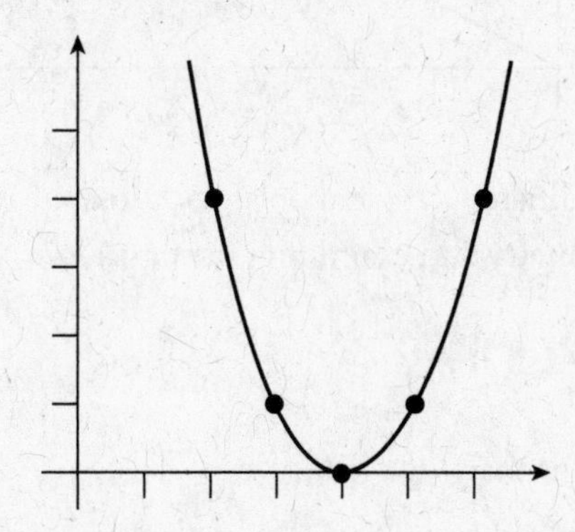

Now let's look at another type of planar curves.

POLAR FUNCTIONS

Polar coordinates provide a different labeling system for the coordinate plane that is more useful in many situations. As you know, any point on the coordinate plane can be identified by an ordered pair (x, y). When using polar coordinates, we use the variables r and θ to label points in the plane. The value of r denotes the distance from the origin and the value of θ denotes the angle of rotation from the positive x-axis to the point. By convention, the origin is denoted by the point (0, 0) in polar coordinates.

The picture below illustrates how to label the point $A = (2\frac{1}{2}, \frac{\pi}{4})$ in polar coordinates.

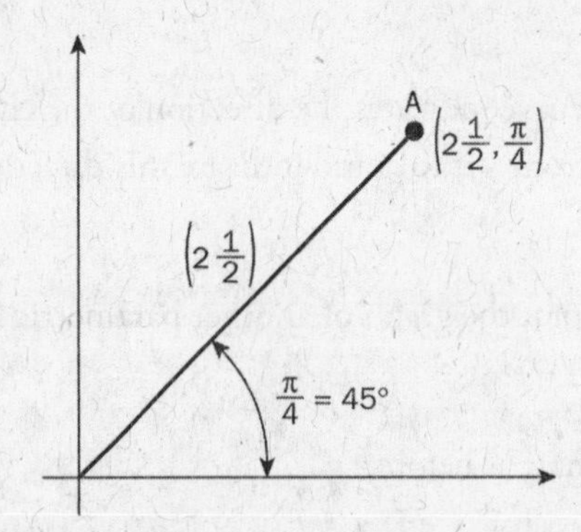

As one would expect, polar and Cartesian coordinates are very closely related. Any point in the plane can be written in either polar or Cartesian coordinates. Consider the picture below.

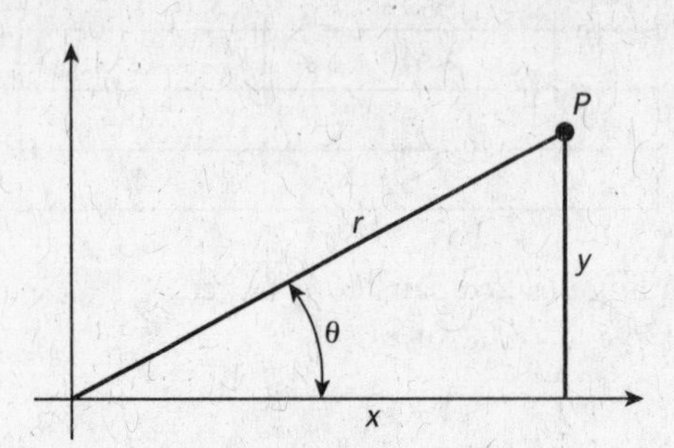

The point P has Cartesian coordinates (x, y) and polar coordinates (r, θ). Recalling the definitions of sine and cosine from trigonometry and examining the figure, we can easily see that

$$\cos\theta = \frac{x}{r} \quad \text{and} \quad \sin\theta = \frac{y}{r},$$

which (solving for x and y) tells us that

$$x = r\cos\theta \text{ and } y = r\sin\theta$$

Thus we can write the Cartesian coordinates as a function of two variables, r and θ.

Much like parametric functions, polar functions provide a convenient means of representing planar curves. We can write $f(r, \theta) = (x(r, \theta), y(r, \theta))$.

This may seem like a lot of extra work for nothing, but it's definitely worth it in certain cases. The use of polar coordinates greatly simplifies the expression of some planar curves. Consider the equation for the unit circle in rectangular coordinates. We have $x^2 + y^2 = 1$, which we then solve for y to get $y = +\sqrt{1 - x^2}$ (the equation for the upper half) and $y = -\sqrt{1 - x^2}$ (the equation for the bottom half). Graphing these two functions simultaneously gives us a planar curve that is the unit circle.

Now that seems like a little too much work just to get an expression for the simplest planar curve ever. What's more, the expression doesn't look all that appealing, either. This is a perfect opportunity to let polar coordinates simplify the situation. To express the unit circle in polar coordinates, we write

$$f(r, \theta) = (\cos\theta, \sin\theta) \text{ for } -\pi \leq \theta \leq \pi, \text{ or } r = 1.$$

As we let θ vary, our curve sweeps out the unit circle, similar to the behavior of parametric curves. In general, the equation for a circle centered at the origin of radius r is given by $f(r, \theta) = (r\cos\theta, r\sin\theta)$ for $-\pi \leq \theta \leq \pi$.

In general, polar functions can be written with r as a function of θ, that is $r = f(\theta)$.

For example, let's sketch the curve given by $r = 1 + \sin\theta$. First, we let θ vary from $0 \leq \theta \leq 2\pi$ and calculate the corresponding values of r.

θ	$r = 1 + \sin\theta$	(θ, r)
0	1	(0,1)
$\frac{\pi}{4}$	$1 + \frac{\sqrt{2}}{2}$	$\left(\frac{\pi}{4}, 1 + \frac{\sqrt{2}}{2}\right)$
$\frac{\pi}{2}$	2	$\left(\frac{\pi}{2}, 2\right)$
$\frac{3\pi}{4}$	$1 + \frac{\sqrt{2}}{2}$	$\left(\frac{3\pi}{4}, 1 + \frac{\sqrt{2}}{2}\right)$
π	1	$(\pi, 1)$
$\frac{3\pi}{2}$	0	$\left(\frac{3\pi}{2}, 0\right)$

We then plot these ordered pairs to graph the function. Graphing the ordered pairs from the right column we can sketch the graph.

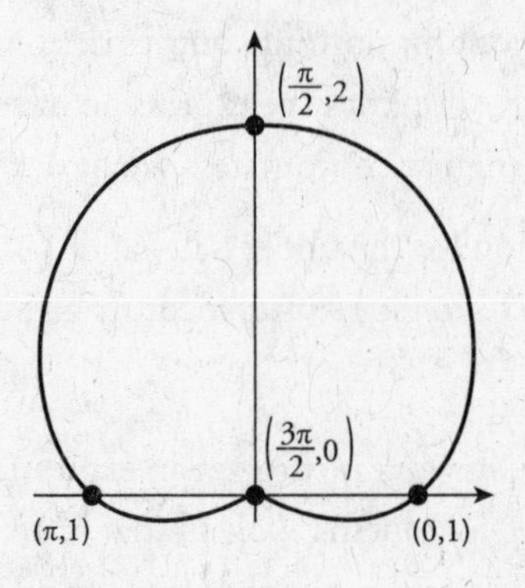

This function is called a cardioid because the graph resembles a heart.

When graphing polar curves in polar mode on the calculator, be sure that the window settings for θ_{min} and θ_{max} will accommodate your curve's period. Either 0 to π or 0 to 2π will suffice for most curves (circles, cardioids, limaçons, and roses). Also, to minimize the distortion of polar graphs on the TI-83/84 calculator, choose your settings so that the ratio $[X_{min}, X_{max}] : [Y_{min}, Y_{max}] = 3 : 2$. See the graphs below to illustrate:

Nice curve: [−3, 3] by [−2, 2]

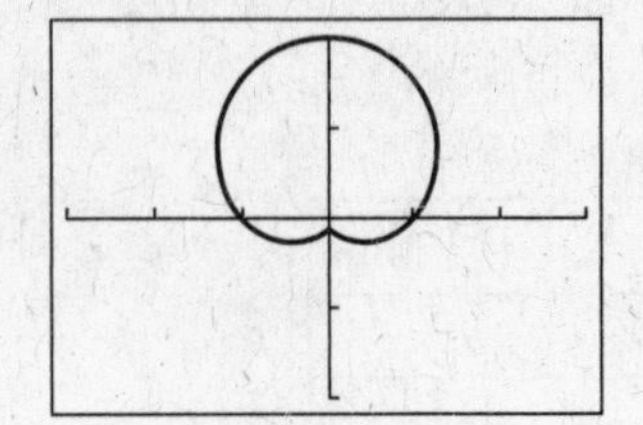

Distorted curve: [−2, 2] by [−2, 2]

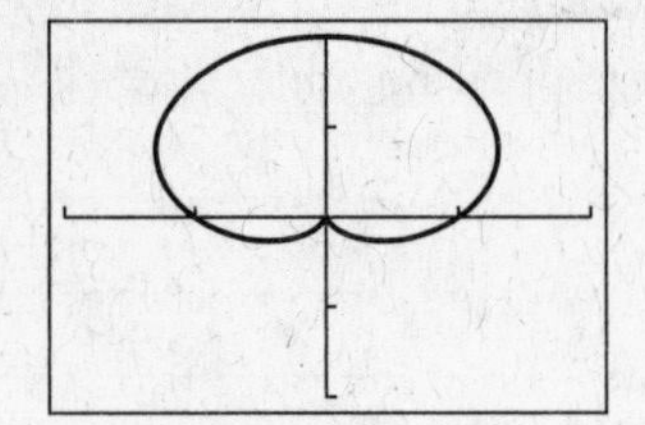

Additionally, in polar mode, the default θ_{step} = .1308996...represents the number $\frac{\pi}{24}$. This number works well in the TRACE mode because it allows you to trace to points on your polar curve that are multiples of $\frac{\pi}{24}$. These include those important and popular trig inputs such as $\frac{\pi}{6}$, $\frac{\pi}{4}$, $\frac{\pi}{3}$, and $\frac{\pi}{2}$.

VECTOR FUNCTIONS

In previous chapters, we dealt primarily with functions that took one real number and returned one real number. Parametric equations take one real number and return two real numbers, while polar equations take two real numbers and return two real numbers. Now we will look at vector functions.

Vector functions take one real number and return three real numbers. These functions are usually written in the form $v(t) = (v_1(t), v_2(t), v_3(t))$. For example, the function $v(t) = \left(3t^2 + 5t, e^t + 1, \frac{\ln t}{2}\right)$ is a vector function. It's like three equations for the price of one!

Suppose we are given the vector-valued function $v(t) = (t^3 - 3, 2t - 1, t - t^2)$. Then when $t = 2$, we have the corresponding vector $v(2) = (5, 3, -2)$. But what does this mean?

Think of a vector as a direction in 3-space. Each component tells you where the vector is pointing. The first component tells you how far to move in the x-direction, the second component tells you how far to move in the y-direction, and the last component tells you how far to move in the z-direction. For the vector in question, start at the origin and move 5 units on the x-axis, 3 units on the y-axis, and −2 units on the z-axis. We get something like

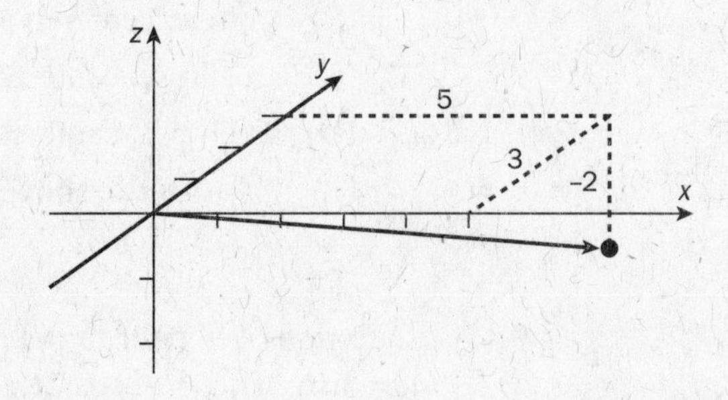

Whereas parametric and polar functions could be used to give equations of curves lying on the plane, vector functions can be used to give an equation of a curve moving through 3-space. However, vector functions are also important for a variety of other reasons. Many problems in physics and engineering are greatly simplified with the use of vector functions. We'll do more work with vector functions in upcoming chapters.

REVIEW QUESTIONS

1. Which curve below is represented by the parametric equations below?

   $x = 3\sin t$ and $y = t^2$ where $-\pi \leq t \leq \pi$

   (A)

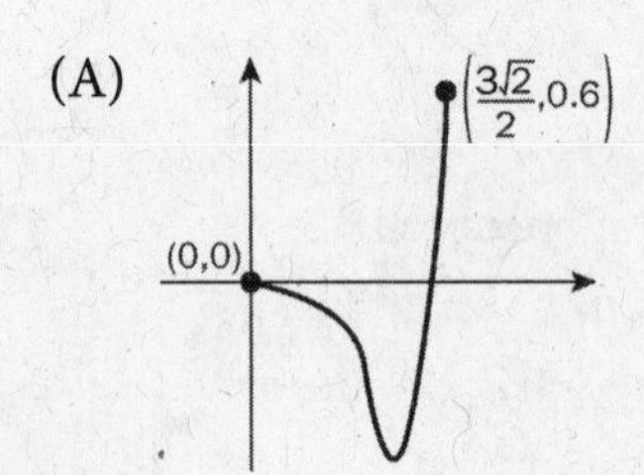

   (B)

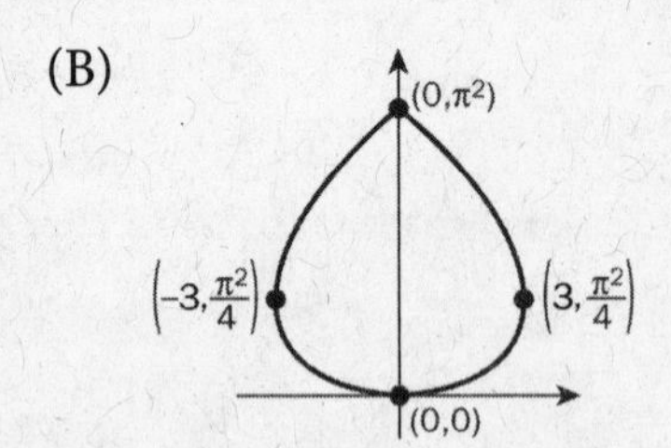

   (C)

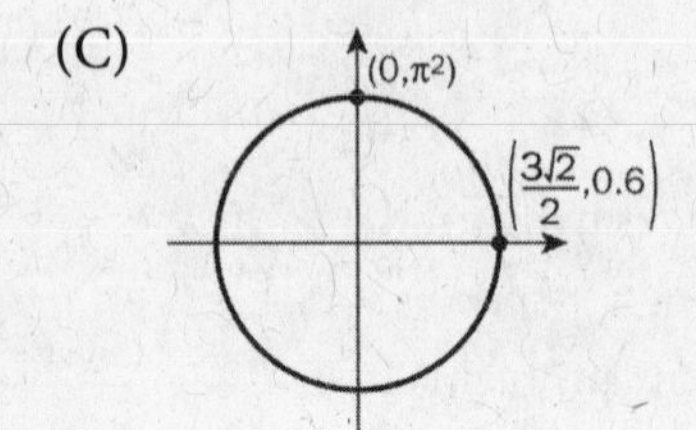

   (D)

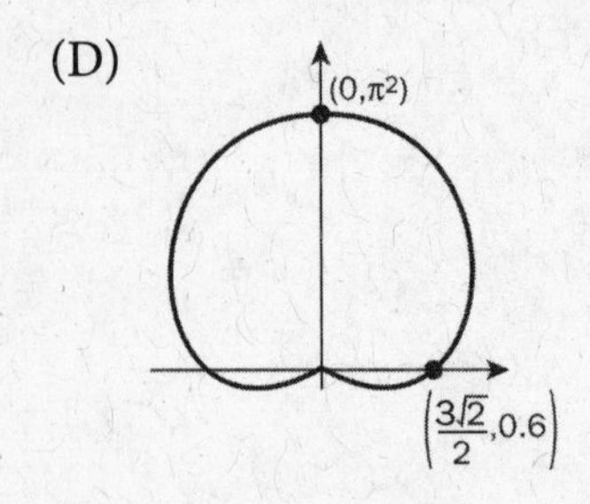

   (E)

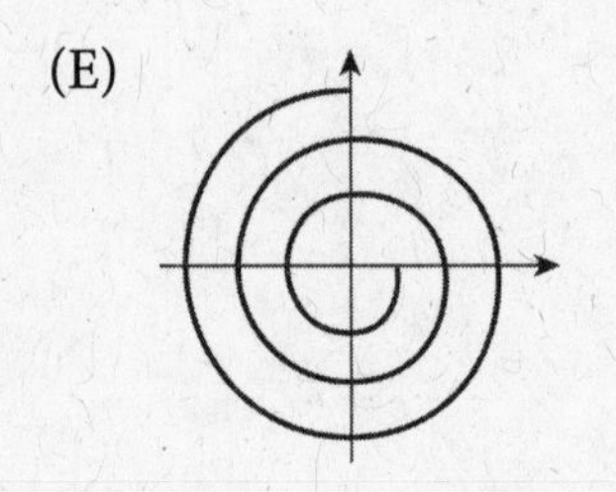

2. Eliminating the parameter in the equations $x = t^2 - t$, $y = t + 1$ yields which of the following equations?

   (A) $x = y^2 - y$
   (B) $x = y^2 - 3y$
   (C) $x = y^2 - 3y - 2$
   (D) $x = y^2 - 3y + 2$
   (E) None of the above

3. The graph of the parametric equations $x = 2\sin t$, $y = 3\cos t$ is:

   (A) a circle, centered at the origin
   (B) a circle, centered at (2,3)
   (C) an ellipse, centered at the origin
   (D) an ellipse, centered at (2,3)
   (E) None of the above

Questions 4–6 refer to the parametric equations: $x = 3\sin(2t)$, $y = -2\cos(2t)$:

4. The graph of the parametric equations is:

   (A) a circle, centered at the origin
   (B) a circle, centered at (3,2)
   (C) an ellipse centered at the origin
   (D) an ellipse centered at (3,2)
   (E) None of the above

5. What range for the parameter will result in the curve being traced exactly once?

   (A) $0 \leq t \leq \frac{\pi}{2}$
   (B) $0 \leq t \leq \pi$
   (C) $0 \leq t \leq 2\pi$
   (D) $0 \leq t \leq 3\pi$
   (E) $0 \leq t \leq 4\pi$

6. Assuming that t increases from 0, which of the following statements is true?

   (A) The curve begins on the positive y-axis and travels in a clockwise direction.

   (B) The curve begins on the positive y-axis and travels in an anti-clockwise direction.

   (C) The curve begins on the negative y-axis and travels in a clockwise direction.

   (D) The curve begins on the negative y-axis and travels in an anti-clockwise direction.

   (E) None of the above

7. Which of the following Cartesian coordinates corresponds to the polar coordinate $\left(-2, \frac{4\pi}{3}\right)$?

   (A) $\left(1, \sqrt{3}\right)$

   (B) $\left(-1, -\sqrt{3}\right)$

   (C) $\left(-2, 240\right)$

   (D) $\left(1, -\sqrt{3}\right)$

   (E) $\left(-1, -\sqrt{3}\right)$

8. Which of the following polar coordinates corresponds to the Cartesian coordinate $(-1,1)$?

   (A) $\left(\sqrt{2}, -\frac{\pi}{4}\right)$

   (B) $\left(1, \frac{3\pi}{4}\right)$

   (C) $\left(-1, \frac{3\pi}{4}\right)$

   (D) $\left(-\sqrt{2}, \frac{3\pi}{4}\right)$

   (E) $\left(\sqrt{2}, \frac{3\pi}{4}\right)$

9. Which of the following is a Cartesian equivalent to the polar equation: $2r = 1 + 3r\cos\theta$?

   (A) $2y^2 - 7x^2 - 6x - 1 = 0$

   (B) $2y^2 - 7x^2 - 1 = 0$

   (C) $4y^2 - 5x^2 - 1 = 0$

   (D) $4y^2 + 5x^2 - 6x - 1 = 0$

   (E) $4y^2 - 5x^2 - 6x - 1 = 0$

10. Which of the following is a polar equation for the circle centered at (4,0) with radius 4?

    (A) $r = 8\cos\theta$

    (B) $r = 4\cos\theta$

    (C) $r = 8\sin\theta$

    (D) $r = 4\sin\theta$

    (E) $r = -4\sin\theta$

11. Which of the following is the graph of $r = 2 - 2\sin\theta$?

(A)

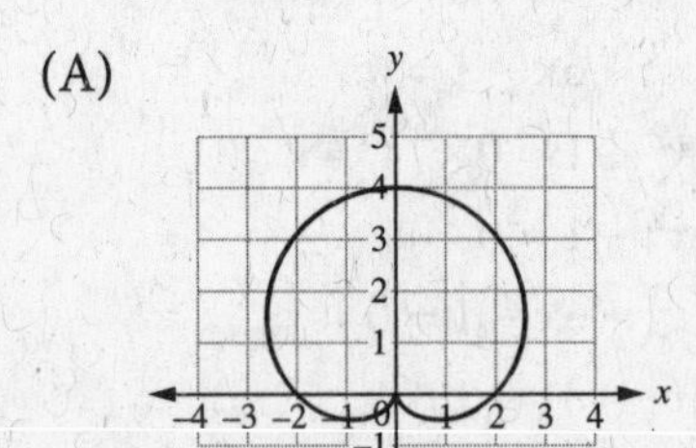

(B)

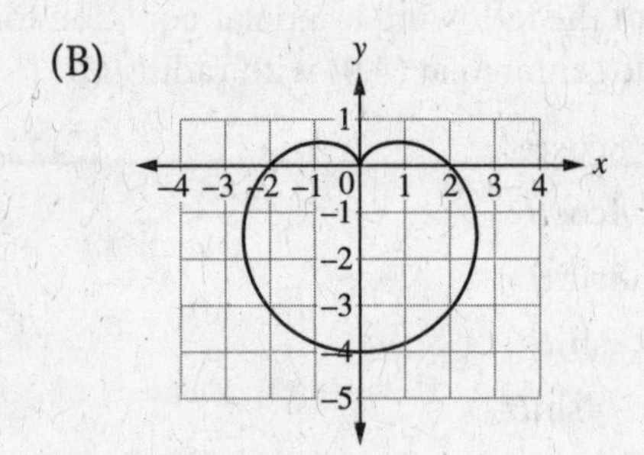

(C)

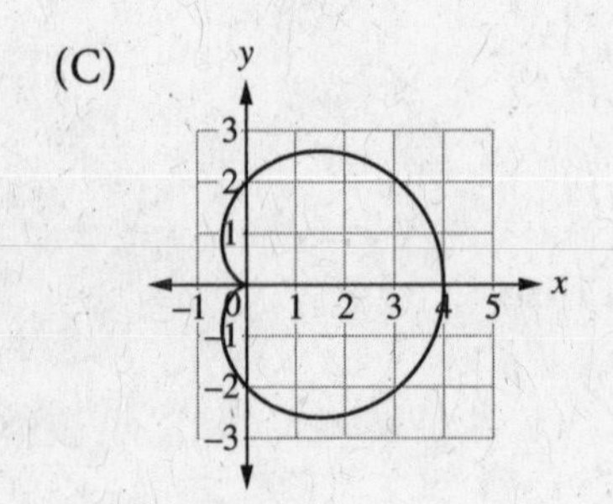

(D)

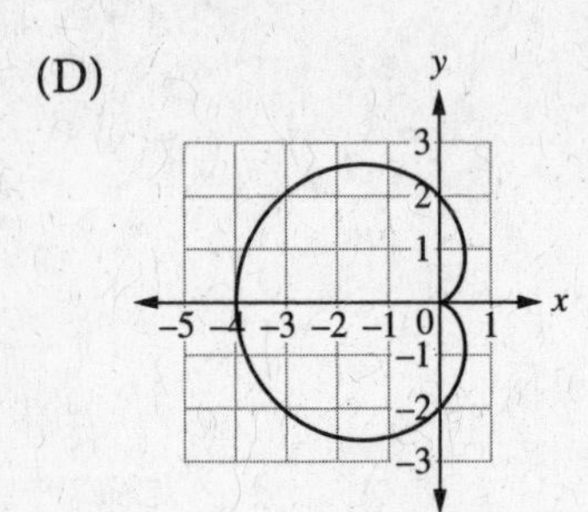

(E)

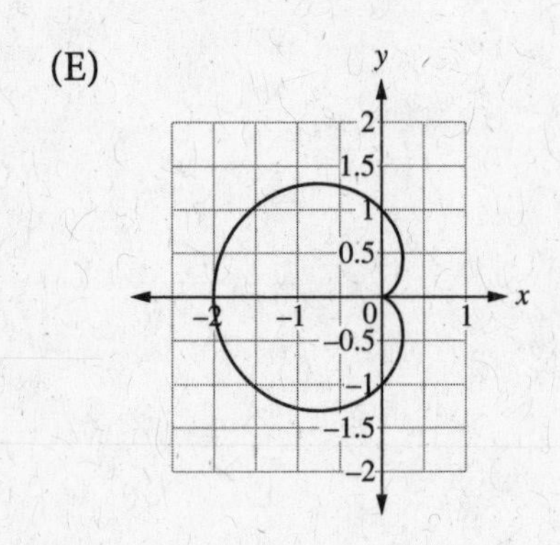

12. State the component form and length of the vector v with initial point $A(2,-1)$ and terminal point $B(-1,3)$.

(A) $v <3,-4>, |v| = 5$

(B) $v <-3,4>, |v| = 5$

(C) $v <3,-4>, |v| \sqrt{7}$

(D) $v <-3,4>, |v| \sqrt{7}$

(E) None of the above

13. Consider the vectors $a = \langle -2,3 \rangle$ and $b = \langle 1,-4 \rangle$. Find $|-3a - 2b.|$

(A) $\sqrt{17}$

(B) 5

(C) $\sqrt{65}$

(D) $\sqrt{305}$

(E) $\sqrt{353}$

14. Consider the vectors $a = \langle -2,3 \rangle$ and $b = \langle 1,-4 \rangle$. Find $a \cdot b$.

(A) −14

(B) −10

(C) 10

(D) 14

(E) $\sqrt{221}$

15. Find the angle between the vectors $a = \left\langle -\sqrt{3}, 1 \right\rangle$ and $b = \left\langle -3, -\sqrt{3} \right\rangle$.

(A) $\dfrac{\pi}{6}$

(B) $\dfrac{\pi}{4}$

(C) $\dfrac{\pi}{3}$

(D) $\dfrac{\pi}{2}$

(E) $\dfrac{2\pi}{3}$

FREE-RESPONSE QUESTIONS

16. Consider the parametrically defined curve given by:

$$\left.\begin{aligned} x(t) &= 2\sin^2 \frac{t}{3} \\ y(t) &= \cos \frac{t}{3} \end{aligned}\right\} 0 \le t \le 2\pi$$

   (a) Sketch the graph of the curve, labeling the direction of motion.

   (b) Find a Cartesian equation for a curve that contains the parametric equation.

   (c) Determine the range of the parameter that will trace the curve exactly once.

17. Consider the graph of the limaçon given by: $r = 1 - 2\sin\theta$.

   (a) Sketch the graph of this curve.

   (b) What is the shortest interval for θ that will produce the graph?

   (c) What range of θ; $0 \le \theta \le 2\pi$ represents the inner loop of the limaçon?

   (d) Find a Cartesian equivalent for this polar curve.

ANSWERS AND EXPLANATIONS

1. B

To graph this parametric equation, we need to set up a table of values and plot some test points. We have

t	$x = 3\sin t$	$y = t^2$
$-\pi$	0	$\pi^2 \approx 9.8$
$\frac{-\pi}{2}$	-3	$\frac{\pi^2}{4} \approx 2.4$
$\frac{-\pi}{4}$	$\frac{-3\sqrt{2}}{2}$	$\frac{\pi^2}{16} \approx 0.6$
0	0	0
$\frac{\pi}{4}$	$\frac{3\sqrt{2}}{2}$	$\frac{\pi^2}{16} \approx 0.6$
$\frac{\pi}{2}$	3	$\frac{\pi^2}{4} \approx 2.4$
π	0	$\pi^2 \approx 9.8$

When we plot these points on the *x–y* axis we get the curve.

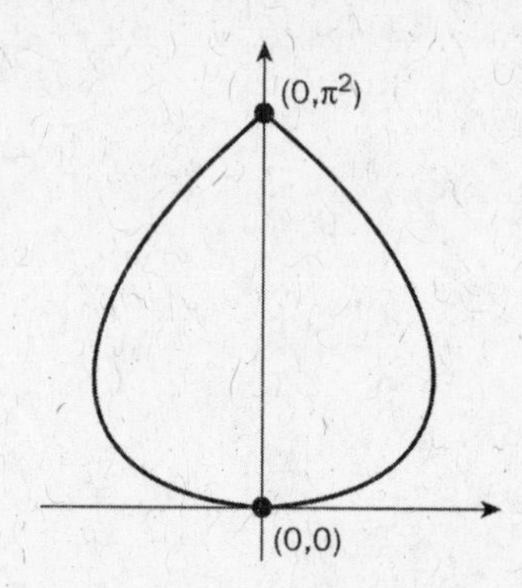

Clearly the answer is (B).

2. D

Eliminating the parameter yields:

$$=t+1 \Rightarrow t = y-1$$
$$=t^2 - t$$
$$=(y-1)^2 - (y-1)$$
$$= y^2 - 2y + 1 - y + 1$$
$$= y^2 - 3y + 2$$

3. C

$$x = 2\sin t \Rightarrow \frac{x}{2} = \sin t$$

$$y = 3\cos t \Rightarrow \frac{y}{3} = \cos t$$

$$\therefore \left(\frac{x}{2}\right)^2 + \left(\frac{y}{3}\right)^2 = \sin^2 t + \cos^2 t = 1$$

$$\frac{x^2}{4} + \frac{y^2}{9} = 1 \rightarrow \textit{ellipse, center } (0,0)$$

4. C

$$x = 3\sin 2t \Rightarrow \frac{x}{3} = \sin 2t$$

$$y = -2\cos 2t \Rightarrow -\frac{y}{2} = \cos 2t$$

$$\therefore \left(\frac{x}{3}\right)^2 + \left(-\frac{y}{2}\right)^2 = \sin^2 2t + \cos^2 2t = 1$$

$$\frac{x^2}{9} + \frac{y^2}{4} = 1 \rightarrow \textit{ellipse, center } (0,0)$$

5. B

Since the period of both $x = 3\sin(2t)$ and $y = -2\cos(2t)$ is $\frac{2\pi}{2} = \pi$, we know that both the x and y coordinates will repeat every π radians. Thus, the desired range for t is: $0 \le t \le \pi$.

We can confirm this by making a table of values:

t	$x = 3\sin(2t)$	$y = -2\cos(2t)$
0	0	−2
$\frac{\pi}{4}$	3	0
$\frac{2\pi}{4} = \frac{\pi}{2}$	0	2
$\frac{3\pi}{4}$	−3	0
$\frac{4\pi}{4} = \pi$	0	−2

6. D

To see the starting point and direction of motion, we set up a table of values and plot the points on a Cartesian plane.

t	$x = 3\sin(2t)$	$y = -2\cos(2t)$
0	0	−2
$\frac{\pi}{4}$	3	0
$\frac{2\pi}{4} = \frac{\pi}{2}$	0	2
$\frac{3\pi}{4}$	−3	0
$\frac{4\pi}{4} = \pi$	0	−2

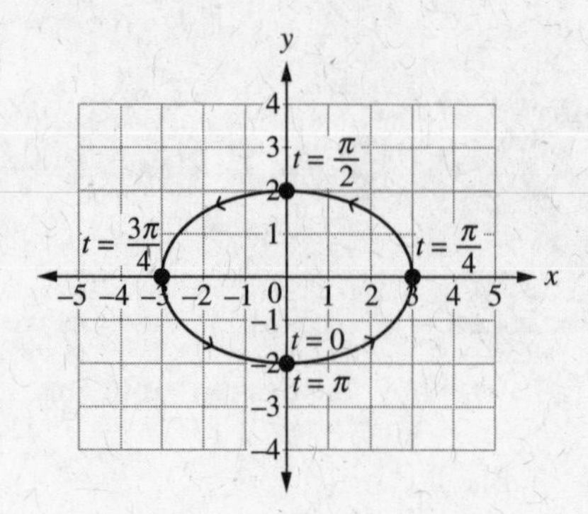

7. A

We are given that $(r, \theta) = \left(-2, \frac{4\pi}{3}\right)$. Thus: r = −2 and $\theta = \frac{4\pi}{3}$. To convert from polar coordinates into Cartesian coordinates we use the conversion equations: $x = r\cos\theta$ and $y = r\sin\theta$.

$$x = r\cos\theta = -2\cos\frac{4\pi}{3} = -2\left(-\frac{1}{2}\right) = 1$$

So:

$$y = r\sin\theta = -2\sin\frac{4\pi}{3} = -2\left(-\frac{\sqrt{3}}{2}\right) = \sqrt{3}$$

and

$$(r, \theta) = \left(-2, \frac{4\pi}{3}\right) \rightarrow (x, y) = \left(1, \sqrt{3}\right)$$

8. E

We are given that $(x,y) = (-1,1)$. Thus: $x = -1$ and $y = 1$. To convert from Cartesian coordinates into polar coordinates we use the conversion equations: $r = \sqrt{x^2 + y^2}$ and $\theta = \tan^{-1}\left(\frac{y}{x}\right)$.

So:

$$r = \sqrt{x^2 + y^2} = \sqrt{(-1)^2 + 1^2} = \sqrt{2}$$

and

$$\theta = \tan^{-1} \frac{-1}{1} = -\frac{\pi}{4} + n\pi; n \in \mathbb{Z}.$$

Since our Cartesian coordinate, $(-1,1)$ is in the second quadrant, $\theta = -\frac{\pi}{4} + \pi = \frac{3\pi}{4}$. Thus, an equivalent polar coordinate is given by $\left(\sqrt{2}, \frac{3\pi}{4}\right)$.

9. E

We begin by making use of the two known facts: $x = r\cos\theta$ and $r = \sqrt{x^2 + y^2}$. Substituting these into the given equation yields:

$$2r = 1 + 3r\cos\theta$$

$$2\sqrt{x^2 + y^2} = 1 + 3x$$

Squaring both sides of this equation gives:

$$\left(2\sqrt{x^2 + y^2}\right)^2 = (1 + 3x)^2$$

$$4\left(x^2 + y^2\right) = 1 + 6x + 9x^2$$

$$4x^2 + 4y^2 = 1 + 6x + 9x^2$$

$$\therefore 4y^2 - 5x^2 - 6x - 1 = 0$$

10. A

We begin by recalling that the standard form of a circle, centered at (h,k) with radius r is:
$(x - h)^2 + (y - k)^2 = r^2$

Thus, the Cartesian equation for the given circle is: $(x - 4)^2 + y^2 = 16$. Expanding and simplifying yields:

$$x^2 - 8x + 16 + y^2 = 16$$

$$x^2 + y^2 = 8x$$

$$\text{but: } x^2 + y^2 = r^2 \text{ and } x = r\cos\theta$$

$$\therefore r^2 = 8r\cos\theta$$

$$r = 8\cos\theta$$

11. B

We recognize that all curves of the form $r = a \pm a\sin\theta$ are vertically oriented cardioids, which eliminate choices (C), (D), and (E) (all of which are horizontally oriented). Plugging in a couple of values for θ allows us to choose between (A) and (B). For example, when $\theta = \frac{\pi}{6}, r = 2 - 2\sin\left(\frac{\pi}{6}\right) = 2 - 2\left(\frac{1}{2}\right) = 1$, which rules out choice (A) and leaves (B) as the only remaining option.

Alternatively, we could plot a table of values for the curve and use this to identify the correct graph.

θ	$r = 2 - 2\sin\theta$	(θ,r)
0	2	(0,2)
$\frac{\pi}{6}$	1	$\left(\frac{\pi}{6},1\right)$
$\frac{2\pi}{4} = \frac{\pi}{2}$	0	$\left(\frac{\pi}{2},0\right)$
$\frac{5\pi}{6}$	1	$\left(\frac{5\pi}{6},1\right)$
$\frac{4\pi}{4} = \pi$	2	$(\pi,2)$
$\frac{3\pi}{2}$	4	$\left(\frac{3\pi}{2},4\right)$

12. B

Consider the graphical representation of the vector, shown below:

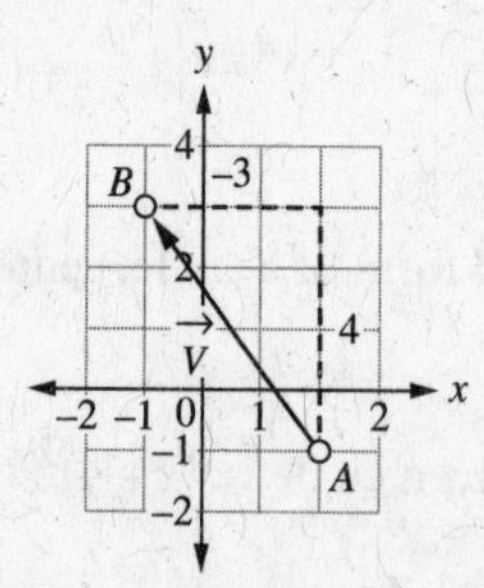

We can see that in going from A to B, we go −3 in the x (horizontal) direction and +4 in the y (vertical) direction. Thus, the vector representation of v is $\langle -3,4 \rangle$. To find the length of the vector, we use Pythagorus:

$$|v| = \sqrt{(-3)^2 + 4^2} = \sqrt{25} = 5.$$

So option B is correct.

13. A

We begin by finding $-3a - 2b$:

$$\begin{aligned}-3a-2b &= -3\langle -2,3\rangle - 2\langle 1,-4\rangle \\ &= \langle 6,-9\rangle - \langle 2,-8\rangle \\ &= \langle 4,-1\rangle\end{aligned}$$

To find the magnitude of this vector, we use the Pythagorean theorem:

$$\left|\langle 4,-1\rangle\right| = \sqrt{(4)^2+(-1)^2} = \sqrt{17}$$

14. A

$a \cdot b = \langle -2,3\rangle \cdot \langle 1,-4\rangle = (-2)(1) + (3)(-4) = -2 - 12 = -14$

15. C

The angle θ between the vectors a and b is given by:

$$\theta = \cos^{-1}\frac{a\cdot b}{|a||b|}$$

$$a\cdot b = \left\langle -\sqrt{3},1\right\rangle\cdot\left\langle -3,-\sqrt{3}\right\rangle = \left(-\sqrt{3}\right)(-3)+(1)\left(-\sqrt{3}\right) = 3\sqrt{3}-\sqrt{3} = 2\sqrt{3}$$

$$|a| = \sqrt{\left(-\sqrt{3}\right)^2+1^2} = \sqrt{4} = 2$$

$$|b| = \sqrt{(-3)^2+\left(-\sqrt{3}\right)^2} = \sqrt{12} = 2\sqrt{3}$$

$$\therefore \theta = \cos^{-1}\frac{a\cdot b}{|a||b|} = \cos^{-1}\left(\frac{2\sqrt{3}}{2\left(2\sqrt{3}\right)}\right) = \cos^{-1}\frac{1}{2} = \frac{\pi}{3}$$

Free-Response Answers

16. (a) To assist us in sketching the graph, we can generate a table of values for these parametric equations:

t	$x(t)=2\sin^2\left(\frac{t}{3}\right)$	$y(t)=\cos\left(\frac{t}{3}\right)$
0	0	1
$\frac{\pi}{2}$	$\frac{1}{2}$	$\frac{\sqrt{3}}{2}$
π	$\frac{3}{2}$	$\frac{1}{2}$
$\frac{3\pi}{2}$	2	0
2π	$\frac{3}{2}$	$-\frac{1}{2}$

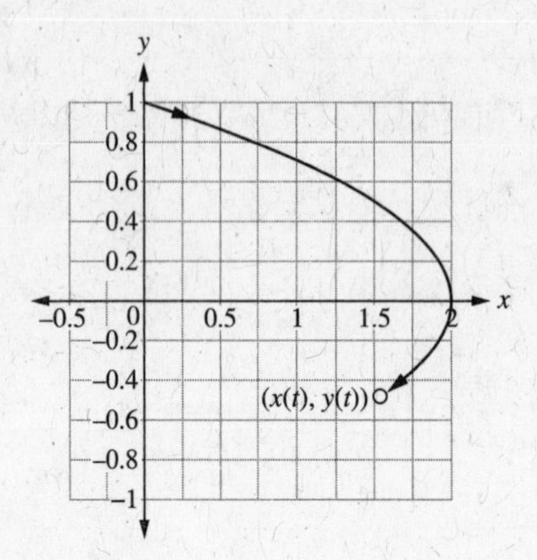

(b) $x(t)=2\sin^2\left(\frac{t}{3}\right)$ $\quad y(t)=\cos\left(\frac{t}{3}\right)$

$\therefore \frac{x}{2}=\sin^2\left(\frac{t}{3}\right)$ $\quad \therefore y^2=\cos^2\left(\frac{t}{3}\right)$

$So: \frac{x}{2}+y^2=\sin^2\left(\frac{t}{3}\right)+\cos^2\left(\frac{t}{3}\right)=1$

$\frac{x}{2}+y^2=1$

$\frac{x}{2}=-y^2+1$

$x=-2y^2+2 \rightarrow x-2=-2y^2$

This is a horizontally oriented parabola with vertex (2,0), which agrees with the table of values we generated previously.

(c) We begin by considering $x(t)=2\sin^2\left(\frac{t}{3}\right)\geq 0\,\forall t\in\mathbb{R}$. Since x is never less than 0, we need to determine at what value of t, x(t) = 0.

$$2\sin^2\left(\frac{t}{3}\right)=0$$

$$\sin^2\left(\frac{t}{3}\right)=0$$

$$\sin\left(\frac{t}{3}\right)=0$$

$$\frac{t}{3}=0,\pi,2\pi\ldots\Rightarrow t=0,3\pi,6\pi\ldots$$

So, we can see that x *may* repeat after t = 3π. Of course, it is not enough that x repeats, we must also confirm that the y-values are repeating. We can confirm that the parabola will repeat after t = 3π by extending our table from part a:

t	$x(t)=2\sin^2\left(\frac{t}{3}\right)$	$y(t)=\cos\left(\frac{t}{3}\right)$
0	0	1
$\frac{\pi}{2}$	$\frac{1}{2}$	$\frac{\sqrt{3}}{2}$
π	$\frac{3}{2}$	$\frac{1}{2}$
$\frac{3\pi}{2}$	2	0
2π	$\frac{1}{2}$	$-\frac{1}{2}$
$\frac{5\pi}{2}$	$\frac{1}{2}$	$-\frac{\sqrt{3}}{2}$
3π	0	-1
$\frac{7\pi}{2}$	$\frac{1}{2}$	$-\frac{\sqrt{3}}{2}$
4π	$\frac{3}{2}$	$-\frac{1}{2}$

Thus, the curve is traced out completely if $0\leq t\leq 3\pi$

17. (a) We can generate a table of values for this polar equation to assist us in sketching the curve:

θ	$r = 1 - 2\sin\theta$	(r, θ)
0	1	(1, 0)
$\frac{\pi}{2}$	−1	$\left(-1, \frac{\pi}{2}\right)$
π	1	$(1, \pi)$
$\frac{3\pi}{2}$	3	$\left(3, \frac{3\pi}{2}\right)$
2π	1	$(1, 2\pi)$

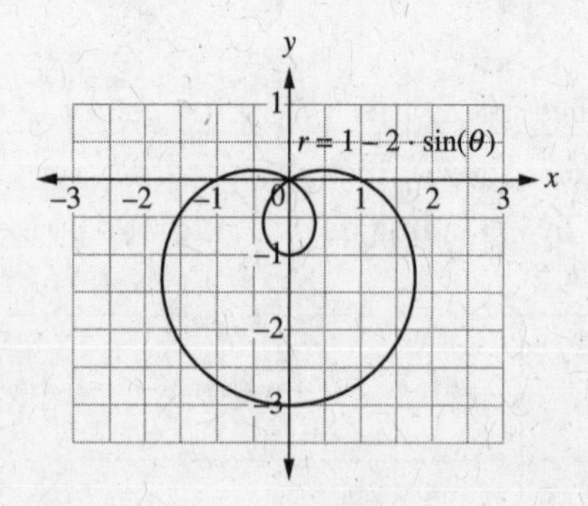

(b) From the table above, we can see that the graph will repeat every 2π radians. Thus, any interval of length 2π will be sufficient to produce one full cycle of the curve.

(c) To find the range of values for which the curve passes through the inner loop, we first find where the graph is at the pole: i.e., where does $r = 1 - 2\sin\theta = 0$.

$$1 - 2\sin\theta = 0$$
$$2\sin\theta = 1$$
$$\sin\theta = \frac{1}{2}$$

This happens an infinite number of times, but only twice on the specified interval $0 \le \theta \le 2\pi$

$$\therefore \theta = \frac{\pi}{6}, \frac{5\pi}{6}$$

We can see from the chart above that at $\frac{\pi}{2}$ the curve is on the inner loop. Since $\frac{\pi}{6} < \frac{\pi}{2} < \frac{5\pi}{6}$, we can deduce that the inner loop is created when $\frac{\pi}{6} \le \theta \le \frac{5\pi}{6}$.

(d) $r = 1 - 2\sin\theta$.

To convert this into Cartesian coordinates, we begin by recalling that
$y = r\sin\theta \Rightarrow \sin\theta = \frac{y}{r}$

Making this substitution, our equation becomes: $r = 1 - 2\left(\frac{y}{r}\right)$

Multiplying both sides of the equation by r yields: $r^2 = r - 2y$

Now, we make use of the fact that $x^2 + y^2 = r^2 \rightarrow r = \sqrt{x^2 + y^2}$ which allows us to transforms our previous equation completely into x and y:

$$r^2 = r - 2y$$
$$x^2 + y^2 = \sqrt{x^2 + y^2} - 2y$$
$$x^2 + y^2 + 2y = \sqrt{x^2 + y^2}$$
$$\left(x^2 + y^2 + 2y\right)^2 = x^2 + y^2$$

CHAPTER 11: THE CONCEPT OF THE DERIVATIVE

IF YOU LEARN ONLY THREE THINGS IN THIS CHAPTER . . .

1. Derivatives can be understood physically as a rate of change and graphically as the slope of a curve.
2. The derivative of f at x_0 is the instantaneous rate of change of f at x_0, or the limit of the difference quotient.
3. If the limit in the definition of the derivative does not exist, then f is not differentiable at x_0 and f does not have a tangent line at this point.

Derivatives are one of the two main concepts in calculus (the other one is integrals, which we'll discuss in subsequent chapters). Derivatives can be understood physically as a rate of change and graphically as the slope of a curve. These ideas are connected via the relationship between slopes and rates of changes.

A PHYSICAL INTERPRETATION OF THE DERIVATIVE: SPEED

To understand derivatives, let's start with a physical concept we understand intuitively—speed. If a car travels 100 miles in two hours, its average speed over that time period is:

$$\frac{100 \text{ miles}}{2 \text{ hours}} = 50 \text{ mph}$$

In general, the average speed of an object traveling during some time period is the change in the object's position during that time divided by the change in the time, or the rate at which the position is changing. We use Δ (i.e., delta) to denote "change in" and write:

$$\text{average speed} = \frac{\text{change in position}}{\text{change in time}} = \frac{\text{position}}{\text{time}}$$

We use the function $p(t)$ to denote the position of the object at time t. If we're measuring the average speed of the object from time t_0 to time t_1, we can write:

$$\underset{(\text{from time } t_0 \text{ to time } t_1)}{\text{average speed}} = \frac{p(t_1) - p(t_0)}{t_1 - t_0}$$

or

$$\underset{(\text{from time } t_0 \text{ to time } t_1)}{\text{average speed}} = \frac{\Delta p}{\Delta t},$$

where we understand from the context that Δt refers to the change over the time period t_0 to t_1.

As the time interval gets smaller and smaller, t_0 and t_1 get closer and closer together. When this happens, we're finding the average speed between time t_0 and a time that is just a little tiny bit after t_0. We write $t_1 = t_0 + \Delta t$, where Δt is a number close to zero. We can find the average speed over a time interval Δt of one minute, one second, 0.1 seconds, etc. It is not necessary for t_1 to be greater than t_0, i.e., for the second time to be after the first. If t_1 is an earlier time, Δt is negative. This is fine. When we replace t_1 with $t_0 + \Delta t$, the formula for average speed becomes:

$$\text{average velocity} = \frac{p(t_0 + \Delta t) - p(t_0)}{(t_0 + \Delta t) - t_0} = \frac{p(t_0 + \Delta t) - p(t_0)}{\Delta t}$$

The change in time Δt can be negative, so Δt can approach zero from both directions. As the time interval gets close to zero, the average speed over the interval becomes the velocity *at* the point t_0, called the "instantaneous speed." We must say "Δt approaches 0" instead of just setting $\Delta t = 0$ in the formula for average speed. If we try to set $\Delta t = 0$ in this formula, we get an undefined expression:

$$\frac{p(t_0 + 0) - p(t_0)}{0} = \frac{p(t_0) - p(t_0)}{0} = \frac{0}{0}$$

When we talk about Δt approaching zero, we've jumped into the realm of limits and we write the definition of the instantaneous speed at time t_0 as:

$$v = \lim_{\Delta t \to 0} \frac{p(t_0 + \Delta t) - p(t_0)}{\Delta t}$$

Speed is the rate of change of position with respect to time. We can talk about the rate of change of other functions as well. For example, in economics, the cost of producing a product is a function of the number of units produced. The rate of change of this function is called the "marginal cost of production." In medicine, the amount of medication in the bloodstream is a function of the time that has elapsed since the medication was administered. The rate of change of this function is called the "absorption rate."

We can extend our computations of velocity as a rate of change by considering the rate of change of a function f at the point x_0 with respect to x_0. The average rate of change of $f(x)$ over the interval x_0 to $x_0 + \Delta x$ is:

$$\frac{\Delta f}{\Delta x} = \frac{f(x_0 + \Delta x) - f(x_0)}{(x_0 + \Delta x) - x_0} = \frac{f(x_0 + \Delta x) - f(x_0)}{\Delta x}$$

This ratio is called the difference quotient. Do not confuse average rate of change with an average value. Sometimes in "the difference quotient" the expression Δx is replaced by h, where h, like Δx, is a number close to zero. In this case, the difference quotient becomes $\frac{f(x_0 + h) - f(x_0)}{h}$. To find the instantaneous rate of change of f at time x_0, we take the limit of the difference quotient

$$\lim_{\Delta x \to 0} \frac{\Delta f}{\Delta x} = \lim_{\Delta x \to 0} \frac{f(x_0 + \Delta x) - f(x_0)}{\Delta x}$$

or

$$\lim_{h \to 0} \frac{f(x_0 + h) - f(x_0)}{h}$$

"The instantaneous rate of change of f" is quite a mouthful, so we replace this with the word "derivative." The derivative of f at x_0 is the instantaneous rate of change of f at x_0, or the limit of the difference quotient.

If the function is linear, that is, $f(x) = mx + b$, then the difference quotient is the familiar slope of the line. For instance, if $f(x) = 4x - 5$, then the difference quotient between the point x_0 and $x_1 = x_0 + h$ is:

$$\frac{[4(x_0 + h) - 5] - [4x_0 - 5]}{h} = \frac{\cancel{4x_0} + 4h - \cancel{5} - \cancel{4x_0} + \cancel{5}}{h} = \frac{4\cancel{h}}{\cancel{h}} = 4,$$ which is the slope of the line.

A line is a function where the difference quotient between any two points is constant—equal to the slope m. The rate of change of a line is constant, and the derivative of a linear function at any point is equal to its slope.

A GEOMETRIC INTERPRETATION: THE DERIVATIVE AS A SLOPE

We can interpret the definition of the derivative as an instantaneous rate of change geometrically, on a graph. The rate of change of y between two points x_0 and x_1 is given by the difference quotient $\frac{f(x_1) - f(x_0)}{x_1 - x_0}$. This difference quotient is the slope of the line between the points $(x_0, f(x_0))$ and $(x_1, f(x_1))$. The line between two points on a graph is called the "secant line" between the points. If we draw the graph of $y = f(x)$ and the secant line between the points, we can visualize this slope.

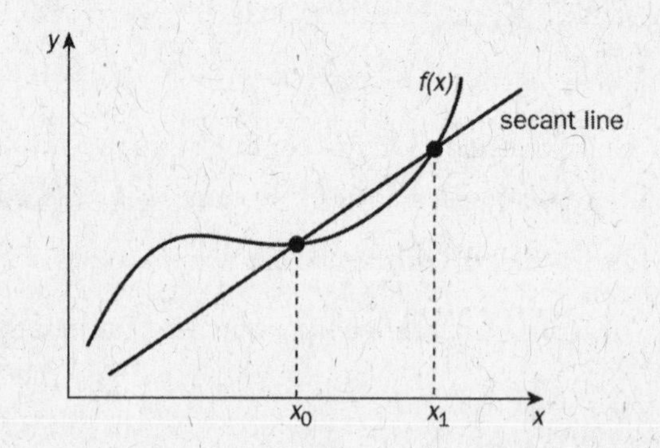

As the points x_0 and x_1 get closer and closer together, it is hard to distinguish between the two points x_0 and x_1, or, using our earlier notation, x_0 and $x_0 + \Delta x$. As $\Delta x \to 0$, the points run together. If the function f is smooth at x_0, that is, if it doesn't have a sharp turn at this point, then, as $\Delta x \to 0$, the secant line between the points becomes tangent to the curve, i.e., it is the line that best approximates the direction of the curve. We call this the tangent line to the graph at the point x_0. The slope of the tangent line to the graph at x_0 is the limit of the slopes of the secant lines between x_0 and points close to x_0—i.e., $x_0 + \Delta x$, for Δx close to zero. We denote the slope of this line by m_{tan}, where it is understood that we are referring to the slope at the point x_0. This slope is given by the limit of the difference quotient.

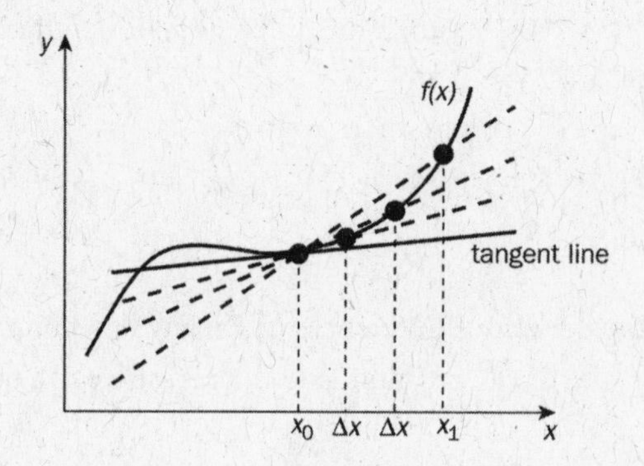

$$m_{\tan} = \lim_{\Delta x \to 0} \frac{\Delta f}{\Delta x} = \lim_{\Delta x \to 0} \frac{f(x_0 + \Delta x) - f(x_0)}{\Delta x}$$

or

$$m_{\tan} = \lim_{h \to 0} \frac{f(x_0 + h) - f(x_0)}{h}$$

This last limit statement is just one of several ways of representing the slope of a tangent line. Suppose, in the first diagram on the previous page (182), we rename x_0 and x_1 with the letters a and b respectively. This means that $b - a$ represents h and that $b \to a$ means $h \to 0$. So, by substitution, we have an equivalent representation of the derivative of the function f: $m_{\tan} = \lim_{b \to a} \frac{f(b) - f(a)}{b - a}$. But, this time, the given point at which the tangent line is drawn is $x = a$, instead of $x = x_0$. You may be thinking, "Isn't one limit statement for this idea enough to remember?" Because this concept is one of the BIG IDEAS of calculus, recognizing a derivative when you see one (even if it is well disguised), will turn some intrinsically difficult multiple-choice questions into easy ones.

As we saw previously, if the function is a line, then the slope between the points is constant. Because the slope of the secant line between any two points on a linear function $y = mx + b$ is equal to m, the limit of the slopes is also m. The tangent line to a linear function at any point is just the line itself.

THE RELATIONSHIP BETWEEN DIFFERENTIABILITY AND CONTINUITY

If the limit in the definition of the derivative exists, we say that f is differentiable at x_0.

When Is a Function Differentiable at x_0?

If the limit in the definition of the derivative does not exist, then f is not differentiable at x_0, and f does not have a tangent line at this point. This can happen for several reasons. Consider the definition of the derivative:

$$\lim_{\Delta x \to 0} \frac{f(x_0 + \Delta x) - f(x_0)}{\Delta x}$$

(1) If f is not defined at x_0, $f(x_0)$ does not exist, so f is not differentiable at x_0.

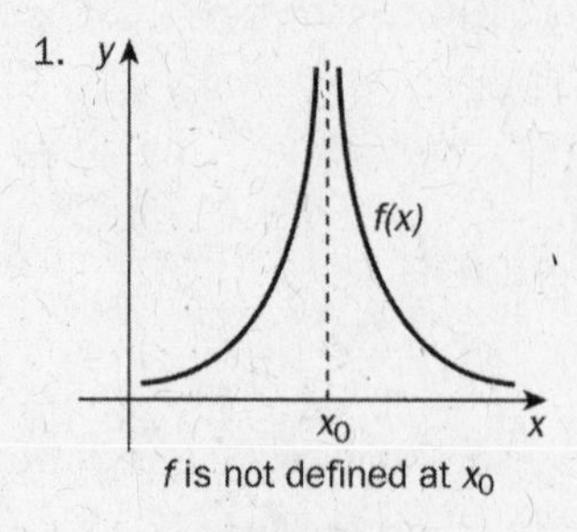

f is not defined at x_0

(2) If f is defined but not continuous at x_0, then $\lim_{\Delta x \to 0} f(x_0 + \Delta x) \neq f(x_0)$ and the above limit is not defined.

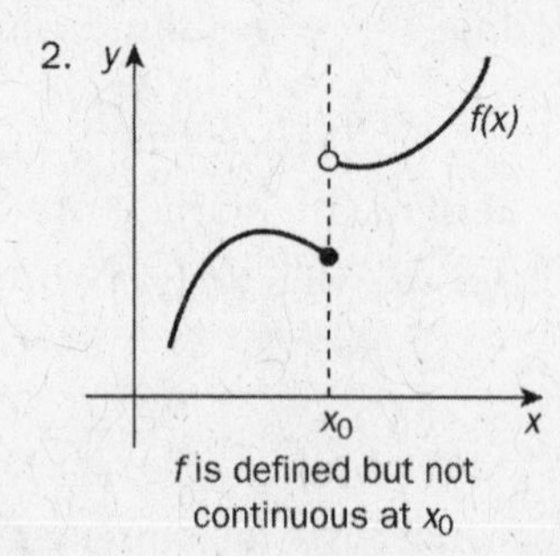

f is defined but not continuous at x_0

(3) Even if f is defined and continuous at x_0, the limit can still fail to exist. This happens when f has a sharp turn at x_0. For example, the absolute value function $f(x) = |x|$ is not differentiable at $x = 0$, even though it is defined and continuous there.

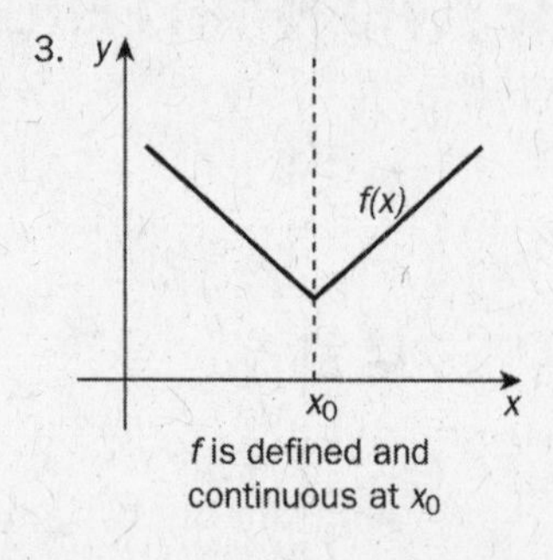

f is defined and continuous at x_0

How Are Differentiability and Continuity Related?

We saw previously that a continuous function may not be differentiable. For example, the function $y = |x|$ is defined and continuous for all real numbers, but it is not differentiable at $x = 0$.

Continuity does not imply differentiability. The reverse, however, is true. If a function is differentiable at a point x_0, it is continuous at this point, i.e., *differentiability implies continuity.* This follows from the definition of differentiability, but we won't go into the details here.

A visual memory aid for this idea is something we can call a "CD dartboard." If you throw a dart and it lands in **D**, then it also lands in **C**.

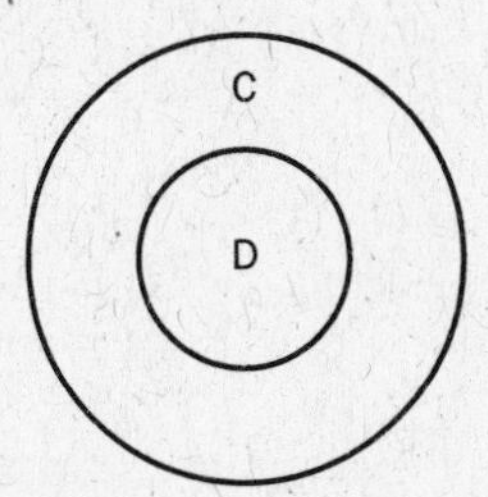

However, if a dart lands in **C**, it doesn't necessarily land in **D**.

Mathematically, if a function is **D**ifferentiable at $x = a$, then the function is **C**ontinuous at $x = a$. Using the language of formal logic that you may have studied in previous courses, we can further conclude: If D $\Rightarrow$ C is true, then its contrapositive ~C $\Rightarrow$ ~D is true, but the converse, C $\rightarrow$ D is not necessarily true.

REVIEW QUESTIONS

1. Below is the graph of the function $y = f(x)$, line l passing through the points $(a, f(a))$ and $(b, f(b))$ on the graph of f, and line m which is tangent to the graph of f at $(a, f(a))$.

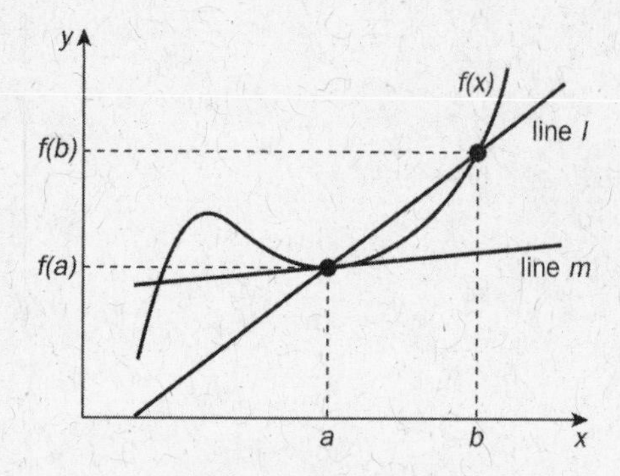

Which of the following statements is true?

I. Line m is the derivative of f at $x = a$.

II. The slope of line l is the average rate of change of f from a to b.

III. The slope of line m is given by $\lim_{\Delta x \to 0} \frac{f(a + \Delta x) - f(a)}{\Delta x}$.

(A) I only
(B) III only
(C) I and III
(D) II and III
(E) All of the above

2. Suppose $f(x)$ is not differentiable at $x = a$. Which of the following must be true?

(A) $f(x)$ is not defined at $x = a$ but could be continuous at $x = a$.
(B) $f(x)$ is not continuous at $x = a$ but could be defined at $x = a$.
(C) $f(x)$ is defined at $x = a$ but is not continuous at $x = a$.
(D) $f(x)$ is continuous and defined at $x = a$.
(E) $f(x)$ could be defined and continuous at $x = a$.

3. Suppose the derivative of the function $f(x)$ exists at $x = a$. Which of the following expressions is NOT equal to the derivative of f at a?

(A) $\lim_{h \to 0} \frac{f(a + h) - f(a)}{h}$

(B) $\lim_{h \to 0} \frac{f(a) - f(a - h)}{h}$

(C) $\lim_{h \to 0} \frac{f(a - h) - f(a)}{h}$

(D) $\lim_{h \to 0} \frac{f(a) - f(a + h)}{-h}$

(E) $\lim_{b \to a} \frac{f(a) - f(b)}{a - b}$

4. If $f'(6) = -2$, which of the following must be true?

I. f is decreasing at $x = 6$
II. f is continuous at $x = 6$
III. f is concave up at $x = 6$

(A) I and II only
(B) I and III only
(C) II and III only
(D) I, II and III
(E) None of the above

5. Which of the following statements is true for the function $y = f(x)$ whose graph is shown below?

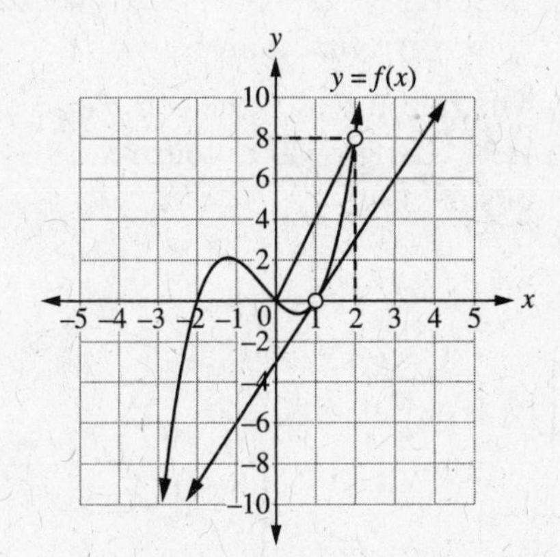

(A) The instantaneous rate of change of f at x = 1 is greater than the average rate of change of f on the interval [0,2].

(B) $f\,(1) < \dfrac{f(2) - f(0)}{2}$

(C) $f'(1) = 0$

(D) The average rate of change of f on [0,2] is 8.

(E) None of the above

6. Which of the following statements must be true for the differentiable function $f(x)$?

I. The average rate of change of $f(x)$ is given by $f'(x)$
II. $\lim\limits_{x \to a} f(x) = f(a) \forall a \in \mathbb{R}$
III. $\lim\limits_{x \to a} \dfrac{f(x) - f(a)}{x - a}$ exists for all real values of a.

(A) I and II only
(B) I and III only
(C) II and III only
(D) I, II and III
(E) None of the above

7. Suppose that the position of a particle is given by the differentiable function $x(t)$. Which of the following statements must be true?

I. $\lim\limits_{h \to 0} \dfrac{x(t+h) - x(t)}{h}$ gives the velocity of the particle at any time t.
II. $\lim\limits_{t \to 3} \dfrac{x(t) - x(3)}{t - 3}$ gives the velocity of the particle at time t = 3.
III. $\dfrac{x(4) + x(1)}{2}$ gives the average velocity on the interval [1,4]

(A) I and II only
(B) I and III only
(C) II and III only
(D) I, II and III
(E) None of the above

8. Which of the following statements about the derivative of f at x_0 is false?

   (A) The derivative of f at x_0 is the instantaneous rate of change of f at x_0.
   (B) The derivative of f at x_0 is the limit of a difference quotient.
   (C) The derivative of f at x_0 is the slope of the tangent to f at x_0.
   (D) The derivative of f at x_0 is the limit of the average rate of change of f on an interval as the length of the interval approaches zero.
   (E) All of the previous statements are true.

9. Which of the following expressions would NOT give the slope of the differentiable function f at a?

   (A) $\lim\limits_{h\to 0}\frac{f(a+h)-f(a)}{h}$
   (B) $\lim\limits_{x\to a}\frac{f(x)-f(a)}{x-a}$
   (C) $\lim\limits_{x\to a}\frac{f(x)-f(a)}{h}$
   (D) $\lim\limits_{h\to 0}\frac{f(a+h)-f(a-h)}{2h}$
   (E) $\lim\limits_{x\to a}\frac{f(a)-f(x)}{a-x}$

10. Consider the function $c(t)$ representing the costs incurred by a painting company since the company's first year of business, where c is measured in $\frac{\$}{m^2}$ and t is measured in years since the company was founded. $c'(4)$ represents:

   (A) The cost, in $\frac{\$}{m^2}$ that the company had four years after it was founded.
   (B) The total cost, in \$, that the company incurred during it's fourth year of business.
   (C) The total cost, in \$, that the company incurred in it's first four years of business.
   (D) The average rate of change in costs in the company's first four years.
   (E) The rate at which the company's costs were changing after four years of business.

11. Consider the function $r(t)$ representing the flow of water past a bridge, where r is measured in $\frac{m^3}{\text{min}}$ and t is measured in hours since midnight. If $r'(5) = 3$ then:

(A) 3 m^3 of water must have flowed past the bridge by 5:00 a.m.

(B) 3 m^3 of water must have flowed past the bridge between 5:00 and 6:00 a.m.

(C) water is flowing past the bridge at a rate of $3\frac{m^3}{\text{min}}$ at 5:00 a.m.

(D) the volume of water flowing past the bridge is increasing by $3\frac{\frac{m^3}{\text{min}}}{hr}$ at 5:00 a.m.

(E) the volume of water flowing past the bridge at 6:00 a.m. will be greater than that which is flowing past the bridge at 5:00 a.m.

12. Consider the differentiable function $f(x)$. Some known values of f are given in the table below:

x	2.0	2.2	2.4	2.6
$f(x)$	3	4	6	9

Use the given values to estimate $f'(2.1)$.

(A) 0.5

(B) 3

(C) 3.5

(D) 4

(E) 5

13. Consider the differentiable function $y = h(t)$, which represents the height of coffee in a coffee pot, t minutes after the coffee begins to brew. Assume that it takes 5 minutes to brew a pot of coffee, that the pot is initially empty, and that no coffee is taken out of the pot while the coffee is brewing. Which of the following statements must true?

I. $h'(t) \geq 0 \forall t \in (0,5)$

II. $h'(2)$ represents the slope of the secant line of $y = h(t)$ on $0 \leq t \leq 2$

III. $h'(2)$ represents the rate at which the height of coffee in the pot is rising at $t = 2$ minutes.

(A) I only

(B) I and II only

(C) I and III only

(D) II and III only

(E) I, II and III

14. Consider the function $Q(t)$, representing the temperature, measured in °C, for a city where t is measured in hours since midnight. If the average rate of change in Q on the interval [5,10] is $2\frac{^\circ C}{hr}$ then

(A) the temperature at 10:00 a.m. was 20°C.

(B) the temperature increased by 10°C between 5:00 a.m. and 10:00 a.m.

(C) the temperature was increasing at 5:00 a.m.

(D) $Q'(5) = 2$

(E) the coldest temperature between midnight and 10:00 a.m. occurred before 5:00 a.m.

15. Which of the following statements are true about $\lim_{x \to 2} \frac{\sqrt{x+2}-2}{x-2}$?

I. $\lim_{x \to 2} \frac{\sqrt{x+2}-2}{x-2}$ represents the slope of the tangent to $f(x) = \sqrt{x+2}$ at any given value of x.

II. $\lim_{x \to 2} \frac{\sqrt{x+2}-2}{x-2} = \frac{1}{4}$

III. $\lim_{x \to 2} \frac{\sqrt{x+2}-2}{x-2}$ represents $f'(2)$ if $f(x) = \sqrt{x+2}$

(A) I only

(B) I and II only

(C) I and III only

(D) II and III only

(E) I, II and III

Free-Response Questions

16. A ball is thrown up in the air from the ground. The height of the ball above the ground at time t is given by the function $h(t)$.

    (a) What is the average velocity of the ball over its entire trajectory—i.e., from the moment it is thrown until it hits the ground?

    (b) Use the characterization of the derivative as the slope of the tangent line to explain why the instantaneous velocity of the ball must be zero at some point before it hits the ground.

17. For each part of the question, sketch a function that satisfies the given criteria or explain why no function exists that satisfies the given conditions.

    (a) f is continuous and differentiable everywhere.

    (b) f is differentiable everywhere but discontinuous at $x = 1$.

    (c) f is defined everywhere but not differentiable at $x = 1$.

18. Suppose that the derivative of the function $f(x)$ is 1 at the point $x = a$. Use the characterization of the derivative as the slope of the tangent line to explain why $f(b) > f(a)$ for $b > a$ and b sufficiently close to a.

19. The height of a particle above the ground at time t is given by the graph below.

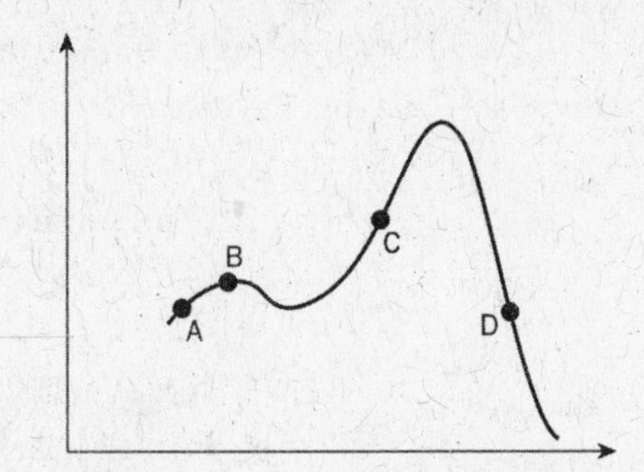

    (a) Between which two labeled points on the graph is the particle's average velocity the least?

    (b) At which of the labeled points on the graph is the particle's instantaneous velocity the greatest?

    (c) At which of the labeled points on the graph is the particle changing directions?

    (d) At which of the labeled points on the graph is the particle's instantaneous velocity zero?

ANSWERS AND EXPLANATIONS

1. D

Statement II is true, because the average rate of change between two points is given by the slope of the secant line between them. Statement I is false because the derivative is the *slope* of the tangent line, not the line itself. Statement III is true because the slope of the tangent line is the limit of the difference quotient giving the slope of the secant line between point a and a point very close to a, i.e., $a + \Delta x$.

Let's pause here to talk about test strategy. Say that the first thing you do is determine that statement I is incorrect, because you've drilled into your brain that the derivative is the slope of the tangent line. The very next thing you should do is cross off answer choices (A), (C), and (E). Even if you have no idea whether statements II and III are correct, you can guess between the two remaining answer choices, (B) and (D), and you have a 50 percent chance of guessing correctly. Those are pretty good odds for guessing on a test.

2. E

Choice (E) must be true. There are functions that fail to be differentiable at a point, but are nonetheless defined and continuous there. The function $y = |x|$ at $x = 0$ is such an example. Choice (A) is always false. From the definition of continuity, a function must be defined at a point to be continuous there. Choice (B) could be false. A function can be continuous and defined at a point where it is not differentiable. For example, the function $y = |x|$ is not differentiable at $x = 0$, but is continuous and defined there. Choice (C) could be false. A function can be defined and continuous at a point where it is not differentiable. Once again, the function $y = |x|$ is not differentiable at $x = 0$, but is continuous and defined there. Choice (D) could be false. A function is not differentiable at a point where it is not continuous or not defined.

For example, the function $y = \frac{1}{x}$ at $x = 0$ is not differentiable at $x = 0$ because it is not defined at that point.

3. C

This question shows that there are many different ways of expressing the limit of the difference quotient to get equivalent expressions for the derivative. Choice (A) is our familiar definition of the derivative; it is the limit of the difference quotient of the function between a and $a + h$. Choice (B) represents the limit of the difference quotient between a and $a - h$. If we let $g = -h$ in the expression and rearrange terms, we get

$$\lim_{h\to 0}\frac{f(a) - f(a-h)}{h} = \lim_{-g\to 0}\frac{f(a) - f(a+g)}{-g}$$

$$= \lim_{-g\to 0}\frac{f(a+g) - f(a)}{g}.$$

We can replace $-g \to 0$ by $g \to 0$ because, as g approaches zero from both directions, $-g$ is also approaching zero from both directions. When we do this, we get our familiar definition $\lim_{g\to 0}\frac{f(a+g) - f(a)}{g}$.

Choice (C) is the additive inverse of the expression in choice (B). Because the expression in choice (B) is the derivative, the expression in choice (C) is the additive inverse, or negative of the derivative. To show this more explicitly, we make the same substitution that we made previously, and let $g = -h$. Then, after substituting, rewriting, and using arithmetic properties of limits, the expression in choice (C) becomes

$$\lim_{h\to 0}\frac{f(a-h) - f(a)}{h} = \lim_{-g\to 0}\frac{f(a+g) - f(a)}{-g}$$

$$= \lim_{g\to 0}\frac{f(a+g) - f(a)}{-g}$$

$$= -\lim_{g\to 0}\frac{f(a+g) - f(a)}{g}.$$

This is the additive inverse, or negative of our familiar definition of the derivative.

Using arithmetical properties of the limit, we can rewrite the expression in choice (D) as:

$$\lim_{h\to 0}\frac{f(a)-f(a+h)}{-h} = \lim_{h\to 0}\frac{-f(a+h)+f(a)}{-h}$$
$$= \lim_{h\to 0}\frac{-[f(a+h)-f(a)]}{-h}$$
$$= \lim_{h\to 0}\frac{f(a+h)-f(a)}{h}$$

which is, once again, our familiar definition of the derivative.

If we write b as $a+h$, then h approaches 0 as $b \to a$. We use arithmetic properties of the limit to rewrite the expression in choice (E) as:

$$\lim_{b\to a}\frac{f(a)-f(b)}{a-b} = \lim_{h\to 0}\frac{f(a)-f(a+h)}{a-(a+h)}$$
$$= \lim_{h\to 0}\frac{f(a)-f(a+h)}{-h}$$
$$= \lim_{h\to 0}\frac{f(a+h)-f(a)}{h}$$

once again getting our familiar definition of the derivative.

4. A

$f'(6) = -2$ means that f has a slope of -2 at $x = 6$. A negative slope means that the function is decreasing, so option I is true.

Also, we know that f is differentiable at $x = 6$ since we are told that the derivative at $x = 6$ is -2. Since differentiability implies continuity, option II must also be true.

Concavity is determined by the second derivative of f. Since we do not know anything about f'', we do not know if $f''(6) > 0$. Thus, option III *might* be true, but does not have to be.

Thus, choice (A) is correct.

5. B

The slope of the tangent line to f at $x = 1$ is the instantaneous rate of change of f at $x = 1$, which by inspection from the graph can be seen to be 3. Thus, $f'(1) = 3$.

The average rate of change of f on [0,2] is the slope of the secant line on the interval, given by: $\frac{f(2)-f(0)}{2} = \frac{8-0}{2} = 4$ which is greater than $f'(1)$ – therefore choice (A) is incorrect, and since $f'(1) = 3 < \frac{f(2)-f(0)}{2} = \frac{8-0}{2} = 4$ choice (B) is correct.

Since $f'(1) = 3$ we can see that choice (C) is incorrect.

As proven earlier, the average rate of change of f on [0,2] is 4, so choice (D) is incorrect.

6. C

We consider each of the three statements individually:

I. $f'(x)$ is the instantaneous rate of change of $f(x)$, which may equal the average rate of change but does not have to. Statement I is false.

II. $\lim_{x\to a} f(x) = f(a)$ implies that f is continuous at $x = a$. Since f is a differentiable function, it must be continuous for all values of x, and therefore statement II is true.

III. $\lim_{x\to a}\frac{f(x)-f(a)}{x-a} = f'(a)$, and since f is a differentiable function, $f'(a)$ must exist for all values of x. Thus, statement III is true.

So choice (C) is correct.

7. A

We consider each of the three statements individually:

I. $\lim_{h \to 0} \frac{x(t+h)-x(t)}{h} = x'(t) = \frac{dx}{dt} = v(t)$ – so statement I is true.

II. $\lim_{t \to 3} \frac{x(t)-x(3)}{t-3} = x'(3) = v(3)$ – so statement II is true.

III. The average velocity is given by $\frac{x(4)-x(1)}{4-1} = \frac{x(4)-x(1)}{3}$, so statement III is false.

8. E

Each of these statements is a different representation of a derivative.

9. C

We begin by recognizing that the slope of the differentiable function *f* at *a* is the same as saying the derivative of *f* at a. There are many ways to represent a derivative. Choices (A) and (B) are the two familiar forms that come in most textbooks, and so both of these are correct.

We know that a derivative is obtained at *a* by obtaining a secant slope and allowing the run to collapse to zero at the point *a*. In choice (D), the secant line is obtained by choosing a point a distance h on either side of a, which makes the slope $\frac{f(a+h)-f(a-h)}{2h}$ so choice (D) also works. Choice (E) is correct because $\frac{f(x)-f(a)}{x-a} = \frac{f(a)-f(x)}{a-x}$. Choice (C) is the only one that does NOT work, because the numerator is the change in y values but, based on those y-values there is indication that the change in x-values would equal h.

10. E

One way to look at this kind of question is to begin by considering the units of measurement of the answer. Since the derivative is equivalent to the slope of the tangent line, we consider the slope of a line (ie: rise/run) where the vertical axis is measured in $\frac{\$}{m^2}$ and the horizontal axis is measured in years. Such a slope would have units of measurement $\frac{\frac{\$}{m^2}}{year}$, which rules out choices (A), (B) and (C).

The average rate of change of cost would given by the slope of a secant line for the function *c*(*t*) on the closed interval [0,4], but we are given $c'(4)$, which is the slope of the tangent line at t = 4; therefore choice (D) is incorrect. We know that $c'(4)$ is the slope of the tangent line at t = 4, which in turn is the rate of change of *c*(*t*) at t=4. Thus, choice (E) is the only correct option.

11. D

We begin by considering the units of measurement of $r'(5)$. Since $r'(5)$ is the slope of the tangent to $r(t)$ at t = 5 hours, we consider the units of slope (ie: rise/run), which in this case gives $\frac{\frac{m^3}{\min}}{hr}$. That rules out choices (A), (B) and (C).

Since $r'(5)$ represents the instantaneous rate of change in r at exactly t = 5, choice (D) is correct.

Choice (E) is incorrect as it is talking about a change over an interval. We know that the flow is increasing at 5:00 a.m. but we do not know that this will continue to be the case for the entire hour. It is possible that sometime after 5:00 the rate at which the water is flowing begins to diminish, and the volume of water flowing past the bridge at t = 6 in this case could be less than that at t = 5. Thus, choice (D) is the only one that is correct.

12. E

Since we don't have the actual function, it is not possible to calculate an exact value for the derivative at 2.1. We know that $f'(2.1)$ is the slope of the tangent to f at $x = 2.1$. Since we cannot get this with the given information, we estimate the slope of the tangent with the slope of the closest secant line: in this case, the secant on the closed interval [2, 2.2]. The slope of this secant line is:

$$\frac{f(2.2)-f(2)}{2.2-2}=\frac{4-3}{0.2}=\frac{1}{0.2}=5$$

So the correct answer is (E).

13. C

We consider each of the three statements individually:

I. $h'(t)$ represents the rate of change in the height of coffee at any time t. Since it takes 5 minutes for the coffee to brew, and no coffee is removed from the pot during this time, we know that the height $h(t)$ cannot be decreasing at any time in the 5-minute interval. Thus, $h'(t) \geq 0$ on that time interval and statement I is true.

II. $h'(2)$ represents the slope of the tangent line to $y = h(t)$ at $t = 2$. While the slope of tangent line at $t = 2$ might be the same as the slope of the secant line of $y = h(t)$ on $0 \leq t \leq 2$, this is not necessarily true, and thus statement II does not have to be true, therefore we treat the statement as false.

III. $h'(t)$ represents the rate of change in the height of coffee at any time t. Thus, $h'(2)$ represents the rate at which the height of coffee in the pot is rising after 2 minutes and statement III is true. Therefore, the answer is (C).

14. B

Average rate of change is the slope of the secant line on a given interval. In this case, the average rate of change is given by: $\frac{Q(10)-Q(5)}{10-5}$

Since we are given that the average rate of change of Q is 2, we can equate these:

$$\frac{Q(10)-Q(5)}{10-5}=2$$

$$\therefore Q(10)-Q(5)=10$$

Since Q(t) represents the temperature at time t, we can see that the temperature must have increased by 10 degrees between t = 5 and t = 10, so choice (B) is correct.

Looking at the other choices:

(A) we do not know what the temperature was at 5:00 a.m., and so it is impossible to know what the temperature was at 10:00, only that it had increased by 10°*C*.

(C) We only know that the temperature increased by a total of 10°*C* between 5:00 a.m. and 10:00 a.m. but this does not necessarily imply that the temperature was increasing at all times on that interval. For example, the temperature may have dropped from 5:00 to 6:00 and then increased from 6:00 to 10:00 – as long as the temperature at 10:00 was 10°*C* higher than the temperature at 5:00.

(D) $Q'(5)$ represents the instantaneous rate of change of temperature at 5:00 a.m., and while this might be the same as the average rate of change from 5:00 to 10:00 (i.e.: the slope of the tangent might be the same as the slope of the secant) they do not have to be.

(E) Again, all we know is that the temperature increased by a total of $10°C$ between 5:00 a.m. and 10:00 – we cannot infer any information about what happened before 5:00 a.m.

15. D

We need to recognize $\lim\limits_{x \to 2} \frac{\sqrt{x+2}-2}{x-2}$ as a form of one of the expressions for the definition of a derivative at a point:

$$f'(a) = \lim_{x \to a} \frac{f(x) - f(a)}{x - a}.$$

These expressions are equivalent if we substitute $f(x) = \sqrt{x+2}$ and $a = 2$. Therefore:

$\lim\limits_{x \to 2} \frac{\sqrt{x+2}-2}{x-2} = f. (2)$ if $f(x) = \sqrt{x+2}$, making statement III true.

This represents the slope of the tangent to $f(x) = \sqrt{x+2}$ at the specific location where x=2. It does not give the slope for any other location, so statement I is false.

To check statement II, we simply evaluate the given limit:

$$\lim_{x \to 2} \frac{\sqrt{x+2}-2}{x-2} \quad \text{(multiply by the conjugate)}$$

$$= \lim_{x \to 2} \frac{\sqrt{x+2}-2}{x-2} \cdot \frac{\sqrt{x+2}+2}{\sqrt{x+2}+2}$$

$$= \lim_{x \to 2} \frac{x+2-4}{(x-2)\left(\sqrt{x+2}+2\right)}$$

$$= \lim_{x \to 2} \frac{\cancel{(x-2)}}{\cancel{(x-2)}\left(\sqrt{x+2}+2\right)}$$

$$= \lim_{x \to 2} \frac{1}{\left(\sqrt{x+2}+2\right)}$$

$$= \frac{1}{\sqrt{4}+2} = \frac{1}{4}$$

Thus, statement II is true and we choose (D).

FREE-RESPONSE ANSWERS

16. (a) The average velocity of the ball over its entire trajectory is given by the change in its height divided by the total time the ball is in the air. Because the ball starts and ends its journey at the same height (i.e., ground level) the change in height over the entire trajectory is zero. Therefore, no matter how long the ball is in the air, the average velocity of the ball is zero:

$$\text{average velocity} = \frac{\Delta\,\text{height}}{\Delta\,\text{time}} = \frac{0}{\Delta\,\text{time}} = 0$$

(b) We draw a sketch of the ball's trajectory. From experience, we know that the ball follows a smooth trajectory; it does not suddenly change direction. The ball reaches its maximum height at some time a. We label this point on the graph. Because the ball's trajectory is smooth, we can draw the line tangent to the graph at the point $t = a$. This tangent line is horizontal. Therefore its slope is zero and the instantaneous velocity of the ball at time $t = a$ is zero.

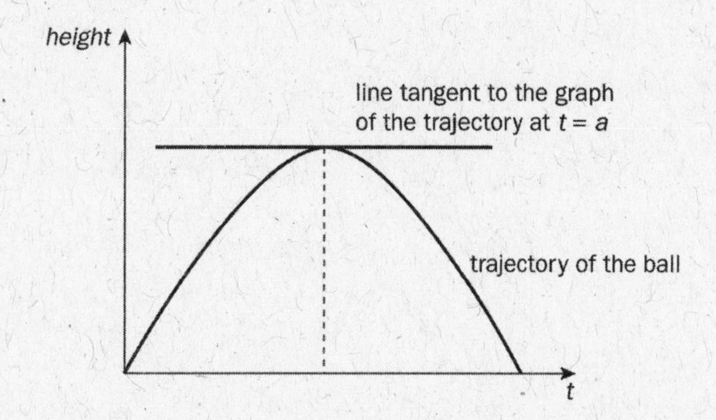

17 (a) Many functions are continuous and differentiable everywhere. For example, any line is continuous and differentiable everywhere.

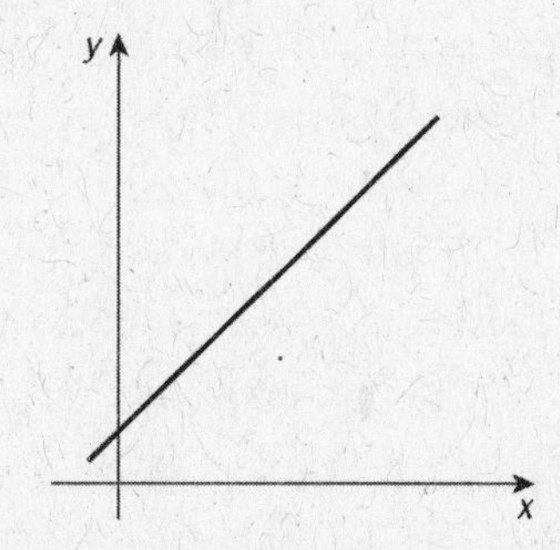

(b) There is no function satisfying these criteria. Because differentiability implies continuity, a function that is not continuous at $x = 1$ cannot be differentiable there.

(c) If a function has a sharp corner at $x = 1$, it is not differentiable at this point. The function $y = |x = 1|$ is such a function.

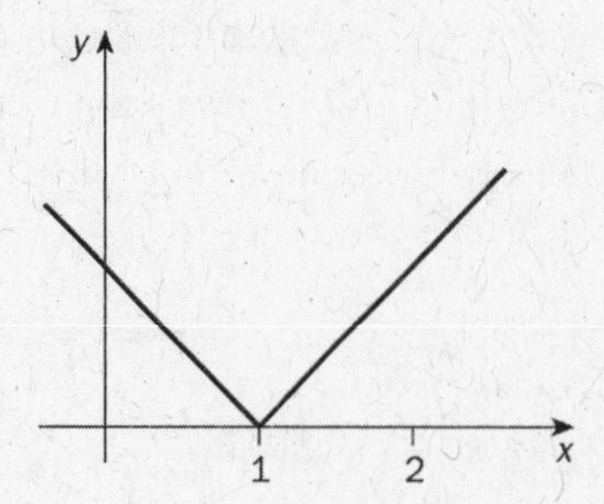

18. Because the derivative of $f(x)$ at $x = a$ is 1, the slope of the line tangent to the curve $f(x)$ at $x = a$ is 1. The slope of the tangent line is the limit of the slope of the secant lines. Therefore, for any number b that is close enough to a, the slope between $(a, f(a))$ and $(b, f(b))$ is close to 1. That is $\frac{f(b) - f(a)}{b - a} \approx 1$. Because $b > a$, the denominator is positive. Therefore the numerator must also be positive, so $f(b) > f(a)$.

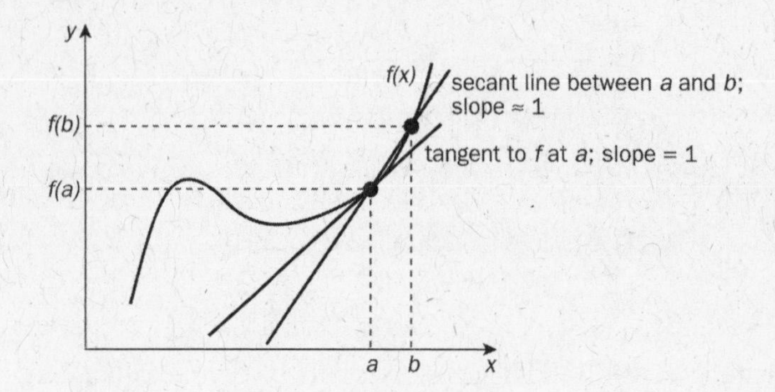

19 (a) The particle's average velocity between two points is given by the slope of the secant line between the two points. Because the lines connecting some pairs of points have negative slope, the smallest average velocity occurs between one of these pairs. Therefore we only need to consider pairs of points where the slope of the line connecting them is negative. We draw lines between points B and D and points C and D. By inspection, the slope of the line between C and D is smaller—i.e., more negative—than the slope of the line between points B and D. Therefore the average velocity of the particle between points C and D is smallest.

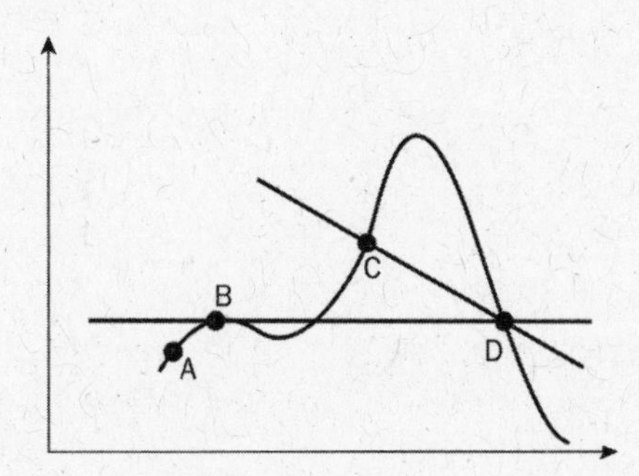

(b) The particle's instantaneous velocity is greatest at the point where the slope of the tangent line is greatest. We draw the line tangent to the curve at each of the labeled points. By inspection, the slope of the line tangent to the curve at point C is the greatest. Therefore the particle's instantaneous velocity is greatest at point C. Remember that veolicty has both magnitude and direction, so a negative velocity is always less than a positive velocity.

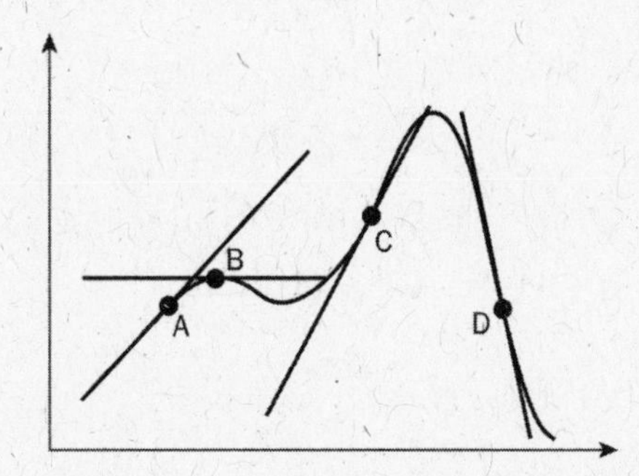

(c) The particle changes direction at point B. Before this point the height of the particle is increasing, so the particle is moving away from the ground. After this point, the height of the particle is decreasing, so the particle is moving toward the ground. Therefore the particle changes direction at point B.

(d) The instantaneous velocity of the particle is given by the slope of the tangent line. The slope of the line tangent to the curve at point B is zero, therefore the instantaneous velocity of the particle at this point is zero. (See the graph in part (b).)

CHAPTER 12: COMPUTATION OF DERIVATIVES

IF YOU LEARN ONLY FIVE THINGS IN THIS CHAPTER . . .

1. **The Power Rule:** If n is an integer, then $\frac{d}{dx}(x^n) = nx^{n-1}$.
2. **The Constant Multiple Rule:** $\frac{d}{dx}[cf(x)] = c \cdot \frac{d}{dx}(f(x))$.
3. **The Sum and Difference Rule:** If f and g are differentiable functions with derivatives f' and g' respectively, then $[f(x) \pm g(x)]' = f'(x) \pm g'(x)$.
4. **The Product Rule:** If f and g are differentiable at a point x, then their product is differentiable at x and the derivative is $[f(x)g(x)]' = f'(x)g(x) + f(x)g'(x)$.
5. **The Quotient Rule:** If f and g are differentiable at a point x and $g(x) \neq 0$, then their quotient is differentiable at x and the derivative is $\left[\frac{f(x)}{g(x)}\right]' = \frac{f'(x)g(x) - f(x)g'(x)}{[g(x)]^2}$.

Later in this chapter, we'll see that it is possible to use the definition of the derivative as the limit of the difference quotient not only to compute the derivative of a function at a point, but also to find a general formula for the derivative function $f'(x)$. We'll compute the derivative of linear functions $f(x) = mx + b$ and find that the derivative of a linear function is always its slope m. We'll also compute the derivative of $f(x) = x^2$ and find that $(x^2)' = \lim_{h \to 0} \frac{(x+h)^2 - x^2}{h} = 2x$. Using this formula, we can, for instance, find the derivative of x^2 at $x = 7$ by evaluating $f'(7) = 2 \cdot 7 = 14$.

In this chapter we will find formulas for the derivatives of different types of functions and find rules for computing the derivatives of sums, products, quotients, and composites of functions. Let's start by finding the derivative of a constant function $f(x) = c$. First we apply the definition of the derivative, $\frac{d}{dx}(f(x)) = \lim_{h \to 0} \frac{f(x+h) - f(x)}{h}$.

DERIVATIVE OF A CONSTANT FUNCTION

Because f is a constant function, both $f(x+h)$ and $f(x)$ equal c. That is,

$\frac{d}{dx}(f(x)) = \lim_{h \to 0} \frac{c-c}{h} = \lim_{h \to 0} 0 = 0$. We see that the derivative equals zero.

Theorem: The Derivative of a Constant Function

For any constant function $f(x) = c$, $\frac{d}{dx}(f(x)) = \frac{d}{dx}(c) = 0$.

DERIVATIVE OF x^n

First let's look at the linear function $f(x) = mx + b$. Any linear function can be written in this form, i.e., the slope-y-intercept form, where m is the slope of the line and b is its y-intercept.

In the following chapter, we'll see that the derivative of a line is its slope, thus $(mx + b)' = m$. We'll also compute the derivative of $f(x) = x^2$ and find that $(x^2)' = 2x$. Right now, we'll use the definition of the derivative to find a formula for the derivative of $f(x) = x^3$:

$$\begin{aligned}(x^3)' &= \lim_{h \to 0} \frac{(x+h)^3 - x^3}{h} = \lim_{h \to 0} \frac{x^3 + 3x^2h + 3xh^2 + h^3 - x^3}{h} \\ &= \lim_{h \to 0} \frac{3x^2h + 3xh^2 + h^3}{h} = \lim_{h \to 0} \frac{(3x^2 + 3xh + h^2)\cancel{h}}{\cancel{h}} \\ &= 3x^2 + 3x \cdot 0 + 0^2 = 3x^2\end{aligned}$$

We find that $(x^3)' = 3x^2$.

Let's keep going, and find a formula for the derivative of $f(x) = x^4$:

$$\begin{aligned}(x^4)' &= \lim_{h \to 0} \frac{(x+h)^4 - x^4}{h} = \lim_{h \to 0} \frac{x^4 + 4x^3h + 6x^2h^2 + 4xh^3 + h^4 - x^4}{h} \\ &= \lim_{h \to 0} \frac{(4x^3 + 6x^2h + 4xh^2 + h^3)\cancel{h}}{\cancel{h}} \\ &= 4x^3 + 6x^2 \cdot 0 + 4x \cdot 0^2 + 0^3 = 4x^3\end{aligned}$$

as well as a formula for the derivative of $f(x) = x^5$:

$$\begin{aligned}(x^5)' &= \lim_{h \to 0} \frac{(x+h)^5 - x^5}{h} = \lim_{h \to 0} \frac{x^5 + 5x^4h + 10x^3h^2 + 10x^2h^3 + 5xh^4 + h^5 - x^4}{h} \\ &= \lim_{h \to 0} \frac{(5x^4 + 10x^3h + 10x^2h^2 + 5xh^3 + h^4)\cancel{h}}{\cancel{h}} \\ &= 5x^4\end{aligned}$$

Let's stop for moment and see what we have so far.

$f(x)$	$f'(x)$
$1 = x^0$	0
$x = x^1$	1
x^2	$2x$
x^3	$3x^2$
x^4	$4x^3$
x^5	$5x^4$

A pattern is developing, one which suggests that, for $n \geq 0$, $(x^n)' = nx^{n-1}$. This is, in fact, true, and you can see a proof using the binomial theorem in any standard calculus book. This formula is true for $n < 0$ as well. What does this mean? Well, if $n < 0$, then we set $-m = n$. Therefore, $m > 0$. By the definition of negative exponents, $x^{-m} = \dfrac{1}{x^m}$.

Let's compute the derivative of $f(x) = x^{-2} = \dfrac{1}{x^2}$ using the definition of the derivative and make sure that the formula we get follows the rule above.

$$(x^{-2})' = \left(\frac{1}{x^2}\right)' = \lim_{h \to 0} \frac{\dfrac{1}{(x+h)^2} - \dfrac{1}{x^2}}{h} = \lim_{h \to 0} \frac{\dfrac{x^2 - (x+h)^2}{(x+h)^2 x^2}}{h}$$

$$= \lim_{h \to 0} \frac{x^2 - (x^2 + 2xh + h^2)}{(x+h)^2 x^2 h} = \lim_{h \to 0} \frac{(-2x - h)\cancel{h}}{(x+h)^2 x^2 \cancel{h}}$$

$$= \frac{-2x}{x^4} = -2x^{-3}$$

The pattern we observed when the exponent n is a non-negative integer appears to hold for negative integers as well. This is correct; the proof can be found in any standard calculus text. We thus have the following rule.

Theorem: The Power Rule

If n is an integer, then $\dfrac{d}{dx}(x^n) = nx^{n-1}$.

Or, in words, to take the derivative of a power function, subtract one from the exponent, and multiply the entire function by the old exponent.

Remember, that the power rule doesn't work when we have an exponent of 0. In that case, we have a constant, and as we've previously learned, the derivative of a constant is just 0.

DERIVATIVE OF $\sqrt{x}$

$$\frac{d}{dx}\left(\sqrt{x}\right) = \lim_{h\to 0}\frac{\sqrt{x+h}-\sqrt{x}}{h} = \lim_{h\to 0}\frac{\sqrt{x+h}-\sqrt{x}}{h}\cdot\frac{\sqrt{x+h}+\sqrt{x}}{\sqrt{x+h}+\sqrt{x}}$$

$$= \lim_{h\to 0}\frac{(x+h)-x}{h(\sqrt{x+h}+\sqrt{x})} = \lim_{h\to 0}\frac{\cancel{h}}{\cancel{h}(\sqrt{x+h}+\sqrt{x})}$$

$$= \frac{1}{2\sqrt{x}}$$

Theorem: The Derivative of $\sqrt{x}$

$$\frac{d}{dx}\left(\sqrt{x}\right) = \frac{1}{2\sqrt{x}}$$

As the above theorem indicates, the derivative of $\sqrt{x}$ is simply a special case of the power rule. Remember $\sqrt{x} = x^{\frac{1}{2}}$. We can apply the power rule — subtract one from the exponent and multiply by the old exponent — to get the above result. Remember that $x^{-\frac{1}{2}} = \frac{1}{x^{\frac{1}{2}}}$.

RULES FOR COMPUTING DERIVATIVES

Now that we have a formula for computing the derivative of x^n for any integer, we'd like to extend this rule to find derivatives of any monomial cx^n as well as any polynomial $a_nx^n + a_{n-1}x^{n-1}\ldots a_1x + a_0$.

Constant Multiple Rule

Suppose that $f(x)$ is a differentiable function with derivative $f'(x)$ and suppose c is a constant. According to the definition of the derivative, $[(cf)(x)]' = \lim_{h\to 0}\frac{(cf)(x+h)-(cf)(x)}{h}$.

Using the definition of a constant multiple times a function, we can rewrite that limit as $[(cf)(x)]' = \lim_{h\to 0}\frac{c\cdot f(x+h)-c\cdot f(x)}{h}$.

Now using the rules from previous chapters on manipulating limits, we can manipulate the limit above to compute the derivative:

$$[(cf)(x)]' = \lim_{h\to 0}\frac{c\cdot[f(x+h)-f(x)]}{h} = c\cdot\lim_{h\to 0}\frac{[f(x+h)-f(x)]}{h} = c\cdot f'(x).$$

Theorem: The Constant Multiple Rule

$$\frac{d}{dx}[cf(x)] = c\cdot\frac{d}{dx}(f(x)).$$

Now we can compute the derivative of any monomial. For instance:

$(4x^7)' \underset{\text{constant multiple rule}}{=} 4 \cdot (x^7)' \underset{\text{power rule}}{=} 4 \cdot 7x^6 = 28x^6$, or, in an example using Leibniz notation,

$$\frac{d}{dx}\left(\frac{5}{x^3}\right) \underset{\text{constant multiple rule}}{=} 5 \cdot \frac{d}{dx}\left(\frac{1}{x^3}\right) = 5 \cdot \frac{d}{dx}(x^{-3}) \underset{\text{power rule}}{=} 5(-3x^{-4}) = -15x^{-4}$$

$$= -\frac{15}{x^4}.$$

Now we'll develop more rules for computing derivatives. After the next rule, we will be able to quickly compute the derivative of a general polynomial.

Sum and Difference Rule

Suppose that $f(x)$ and $g(x)$ are differentiable functions with derivatives $f'(x)$ and $g'(x)$. According to the definition of the derivative, $[(f+g)(x)]' = \lim_{h \to 0} \frac{(f+g)(x+h) - (f+g)(x)}{h}$. Using the definition of a sum of functions we can rewrite that limit as:

$$\begin{aligned} [(f+g)(x)]' &= \lim_{h \to 0} \frac{[f(x+h) + g(x+h)] - [f(x) - g(x)]}{h} \\ &= \lim_{h \to 0} \left[\frac{f(x+h) - f(x)}{h} + \frac{g(x+h) - g(x)}{h} \right] \end{aligned}$$

By using the rules you already learned on manipulating limits, we can rewrite the limit above as:

$$\begin{aligned} [(f+g)(x)]' &= \lim_{h \to 0} \frac{f(x+h) - f(x)}{h} + \lim_{h \to 0} \frac{g(x+h) - g(x)}{h} \\ &= f(x) + g(x) \end{aligned}$$

A parallel argument works to find a formula for the derivative of $[(f-g)(x)]'$. We have the following theorem.

Theorem: The Sum and Difference Rule

If f and g are differentiable functions with derivatives f' and g' respectively, then $[f(x) \pm g(x)]' = f'(x) \pm g'(x)$.

Derivatives of Polynomials

Using the constant multiple rule and the sum and difference rule, we can now compute the derivative of any polynomial. For example,

AP EXPERT TIP

Translated into plain English, here's the symbolic rule: A constant factor can be moved through a derivative sign.

AP EXPERT TIP

As we move forward through the derivative rules, you will see that the same notion does not apply to products and quotients. In other words, it is NOT true that the derivative of a product/quotient is the product/quotient of the derivatives.

$$\frac{d}{dx}(4x^3 - 5x + 4) \underset{\text{sum and difference rule}}{=} \frac{d}{dx}(4x^3) - \frac{d}{dx}(5x) + \frac{d}{dx}(4)$$

$$\underset{\text{constant multiple rule}}{=} 4 \cdot \frac{d}{dx}(x^3) - 5 \cdot \frac{d}{dx}(x) + \frac{d}{dx}(4)$$

$$\underset{\substack{\text{power rule} \\ \text{derivative of a constant function}}}{=} 4 \cdot 3x^2 - 5 \cdot 1 + 0 = 12x^2 - 5$$

We can use the following procedure to compute the derivative of any polynomial. Notice in the example that it is acceptable for the coefficients to be fractions or decimals.

Step	Procedure	Example
1	We are given a polynomial to differentiate.	$\frac{d}{dx}\left(8x^4 - 1.2x^2 + \frac{2}{3}x + \pi\right)$
2	Use the sum and difference rule to break the derivative into a sum and difference of derivatives.	$\frac{d}{dx}(8x^4) - \frac{d}{dx}(1.2x^2) + \frac{d}{dx}\left(\frac{2}{3}x\right) + \frac{d}{dx}(\pi)$
3	Use the constant multiple rule to pull the constant out of each derivative.	$8\frac{d}{dx}(x^4) - 1.2\frac{d}{dx}(x^2) + \frac{2}{3}\frac{d}{dx}(x) + \frac{d}{dx}(\pi)$
4	Use the power rule and constant rule to differentiate each term.	$8 \cdot 4x^3 - 1.2 \cdot 2x + \frac{2}{3} + 0$
5	Simplify.	$32x^3 - 2.4x + \frac{2}{3}$

Note: Remember that π is a constant, so its derivative is zero.

AP EXPERT TIP

As the rules get more complicated, rewriting them in verbal form will help you to remember them. We can rewrite the product rule as follows: *The derivative of the product of two functions is the first times the prime of the second, plus the second times the prime of the first.*

Product Rule and Quotient Rule

Now we'll move on to rules for computing the derivatives of more complicated functions. We won't write the proofs here—you can find them in any standard calculus text.

To find the derivative of a product of two functions, we use the following rule.

Theorem: The Product Rule

If f and g are differentiable at a point x, then their product is differentiable at x and the derivative is $[f(x)g(x)]' = f(x)g'(x) + g(x)f'(x)$.

For example, we can find the derivative of the function $f(x) = \left(x^3 + 7\right)\left(5x^2 - \frac{3}{2}\right)$ in two ways. First, we'll compute the derivative using the product rule:

$$\begin{aligned}\left[(x^3 + 7)\left(5x^2 - \frac{3}{2}\right)\right]' &= (x^3 + 7)'\left(5x^2 - \frac{3}{2}\right) + (x^3 + 7)\left(5x^2 - \frac{3}{2}\right)' \\ &= 3x^2\left(5x^2 - \frac{3}{2}\right) + (x^3 + 7)(10x) \\ &= 15x^4 - \frac{9}{2}x^2 + 10x^4 + 70x \\ &= 25x^4 - \frac{9}{2}x^2 + 70x\end{aligned}$$

We can also compute the derivative of f by multiplying and then computing the derivative of the resulting polynomial.

$$\begin{aligned}\left[(x^3 + 7)\left(5x^2 - \frac{3}{2}\right)\right]' &= \left(5x^5 - \frac{3}{2}x^3 + 35x^2 - \frac{21}{2}\right)' \\ &= 25x^4 - \frac{9}{2}x^2 + 70x\end{aligned}$$

Both methods give the same answer, which is what we expect. In general, if you can multiply something out or do some other algebraic manipulation to make the problem simpler *before* computing the derivative, it's a good idea to go ahead and do it.

Sometimes, though, you can't avoid using the product rule. We'll see some examples a bit later.

To compute the derivative of a quotient of functions, we use another theorem.

Theorem: The Quotient Rule

If f and g are differentiable at a point x and $g(x) \neq 0$, then their quotient is differentiable at x and the derivative is $\left[\frac{f(x)}{g(x)}\right]' = \frac{f'(x)g(x) - f(x)g'(x)}{[g(x)]^2}$. The quotient rule can easily be derived given the product rule and power rule. First, note that $\left[\frac{f(x)}{g(x)}\right]'$ can be rewritten as:

AP EXPERT TIP

The more complicated quotient rule can be most easily memorized with a little poetry: *Low-de-high minus high-de-low, draw the line and square below.*

The denominator (*low*) multiplies the derivative (*de*) of the numerator (*high*). Then we subtract the numerator (*high*) times the derivative (*de*) of the denominator (*low*). Finally, the difference of these two products is divided by (*draw the line*) the square of the denominator (*square below*).

$$
\begin{aligned}
[f(x)[g(x)^{-1}]]' &= f(x)[g(x)^{-1}]' + [g(x)^{-1}]f'(x) \\
&= f(x)[-1 \cdot [g(x)]^{-2} \cdot g'(x)] + [g(x)^{-1}]f'(x) \\
&= \frac{-f(x)g'(x)}{g^2(x)} + \frac{f'(x)}{g(x)} = \frac{g(x)f'(x) - f(x)g'(x)}{g^2(x)}
\end{aligned}
$$

In particular, we can find the derivative of the reciprocal of a function:

$$
\begin{aligned}
\left[\frac{1}{f(x)}\right]' &= \frac{(1)' \cdot f(x) - 1 \cdot f'(x)}{[f(x)]^2} = \frac{0 \cdot f(x) - f'(x)}{[f(x)]^2} \\
&= \frac{-f'(x)}{[f(x)]^2}
\end{aligned}
$$

DERIVATIVES OF TRIGONOMETRIC FUNCTIONS

We want to build up a library of functions and their derivatives. Let's continue by finding the derivatives of the six basic trigonometric functions.

$$
\begin{aligned}
\frac{d}{dx}(\sin x) &= \lim_{h\to 0} \frac{\sin(x+h) - \sin x}{h} \\
&= \lim_{h\to 0} \frac{\sin x \cos h + \cos x \sin h - \sin x}{h} \\
&= \lim_{h\to 0} \frac{\sin x(\cos h - 1) + \cos x \sin h}{h} \\
&= \sin x \cdot \lim_{h\to 0} \frac{(\cos h - 1)}{h} + \cos x \cdot \lim_{h\to 0} \frac{\sin h}{h} \\
&= \sin x \cdot 0 + \cos x \cdot 1
\end{aligned}
$$

(Remember that $\lim_{h\to 0} \frac{(\cos h - 1)}{h} = 0$ and $\lim_{h\to 0} \frac{\sin h}{h} = 1$.)

We find that $\frac{d}{dx}(\sin x) = \cos x$. We can also use the definition of the derivative to find that $\frac{d}{dx}(\cos x) = -\sin x$.

We can use the quotient rule to find the derivative of $\tan x$ by rewriting $\tan x$ as $\frac{\sin x}{\cos x}$:

$$
\begin{aligned}
(\tan x)' &= \left(\frac{\sin x}{\cos x}\right)' = \frac{(\sin x)' \cos x - \sin x(\cos x)'}{\cos^2 x} \\
&= \frac{\cos x(\cos x) - \sin x(-\sin x)}{\cos^2 x} = \frac{\cos^2 x + \sin^2 x}{\cos^2 x} \\
&= \frac{1}{\cos^2 x} = \sec^2 x
\end{aligned}
$$

Because $\cot x = \frac{\cos x}{\sin x}$, $\sec x = \frac{1}{\cos x}$, and $\csc x = \frac{1}{\sin x}$, we can use the quotient rule, and the special case of the quotient rule for finding the derivative of the reciprocal to find the derivatives of these trigonometric functions. It is useful to organize the derivatives of the six basic trigonometric functions in a table.

$f(x)$	$f'(x)$
$\sin x$	$\cos x$
$\cos x$	$-\sin x$
$\tan x$	$\sec^2 x$
$\cot x$	$-\csc^2 x$
$\sec x$	$\sec x \tan x$
$\csc x$	$-\csc x \cot x$

CHAIN RULE

Although the derivative of any function can theoretically be computed using the definition of the derivative as the limit of the difference quotient, in reality these limits can be very difficult to compute. The more complicated a function is, the harder it is to work with. The chain rule is one way of simplifying the process of taking derivatives of complicated functions.

Let's start by motivating this idea with some examples. Suppose we want to find the derivative of $f(x) = (x^3 + 4)^2$. We can expand this binomial to rewrite as $f(x) = x^6 + 8x^3 + 16$. The resulting expression is a polynomial written in a manageable form and we know how to use the sum and difference rules, the constant multiple rule, and the rule for finding the derivative of a constant to take its derivative. We find that $f'(x) = 6x^5 + 24x^2$.

Similarly, we can find the derivative of the function $f(x) = \cos\left(x + \frac{\pi}{4}\right)$ by using trig identities and the product rule:

$$f(x) = \cos\left(x + \frac{\pi}{4}\right) = \cos x \cos\frac{\pi}{4} - \sin x \sin\frac{\pi}{4}$$

$$\begin{aligned} f\ (x) &= \cos\frac{\pi}{4}(\cos x)' - \sin\frac{\pi}{4}(\sin x)' \\ &= \cos\frac{\pi}{4}(-\sin x) - \sin\frac{\pi}{4}(\cos x) \\ &= -\left(\sin x \cos\frac{\pi}{4} + \cos x \sin\frac{\pi}{4}\right) \\ &= -\sin\left(x + \frac{\pi}{4}\right) \end{aligned}$$

But what if the function we're starting with is $f(x) = (x^2 + 4x + 7)^{14}$, or something equally unpleasant to expand? It would be time consuming and annoying to expand f to compute its derivative.

The chain rule, as its name implies, is a chain of steps for computing the derivative of complicated functions. We use "complicated" to mean "a composite of functions." Using this rule, we can break down $f(x) = (x^2 + 4x + 7)^{14}$ into $g(x) = x^2 + 4x + 7$ and $h(x) = x^{14}$, then $f(x) = h(g(x))$. The chain rule is a rule for finding the derivatives of composite functions.

Theorem: The Chain Rule

If g is differentiable at x and h is differentiable at $g(x)$, then $h(g(x))$ is differentiable at x and its derivative is $[h(g(x))]' = \underbrace{h'(g(x))}_{\text{the derivative of the outside function evaluated at the inside function}} \cdot \underbrace{g'(x)}_{\text{the derivative of the inside function}}$.

As usual, this proof can be found in any standard calculus text.

In our example, $g(x) = x^2 + 4x + 7$ and $h(x) = x^{14}$. We compute the derivatives:

$$h'(x) = 14x^{13}$$
$$h'(g(x)) = 14(x^2 + 4x + 7)^{13}$$
$$g'(x) = 2x + 4$$

and plug this information into the chain rule to find that:

$$f'(x) = [(x^2 + 4x + 7)^{14}]' = \underbrace{14(x^2 + 4x + 7)^{13}}_{\text{the derivative of the outside function evaluated at the inside function}} \cdot \underbrace{(2x + 4)}_{\text{the derivative of the inside function}}$$

What about the function $f(x) = \sin(x^2)$? Without the chain rule, we would have to return to the definition of the derivative to compute the derivative of this function. But we can use the chain rule to find the derivative of $f(x) = \sin x^2$. We write $g(x) = x^2$, $h(x) = \sin x$, $f(x) = h(g(x))$.

$$(\sin x^2)' = \underbrace{\cos x^2}_{\text{the derivative of the outside function evaluated at the inside function}} \cdot \underbrace{2x}_{\text{the derivative of the inside function}}$$
$$= 2x \cos x^2$$

Going back to our original example, by expanding the function $f(x) = (x^3 + 4)^2$, we found the derivative $f'(x) = 6x^5 + 24x^2$. Now let's compute the derivative using the chain rule. In this case, we'll take $g(x) = x^3 + 4$ and $h(x) = x^2$. Using the chain rule, we find:

$$f'(x) = [(x^3 + 4)^2]' = \underbrace{2(x^3 + 4)^1}_{\substack{\text{the derivative of the}\\ \text{outside function evaluated}\\ \text{at the inside function}}} \cdot \underbrace{3x^2}_{\substack{\text{the derivative}\\ \text{of the inside}\\ \text{function}}}$$

$$= 6x^5 + 24x^2$$

We see that our answers agree, which is what we expect.

Suppose $f(x) = t(s(r(x)))$. We apply the chain rule once with $h(x) = t(x)$ and $g(x) = s(r(x))$. We find that $f'(x) = \underbrace{t'(s(r(x)))}_{\substack{\text{the derivative of the}\\ \text{outside function evaluated}\\ \text{at the inside function}}} \cdot \underbrace{s(r(x))'}_{\substack{\text{the derivative}\\ \text{of the inside}\\ \text{function}}}$

Now we apply the chain rule again; this time with $h(x) = s(x)$ and $g(x) = r(x)$:

$$s(r(x))' = \underbrace{s'(r(x))}_{\substack{\text{the derivative of the}\\ \text{outside function evaluated}\\ \text{at the inside function}}} \cdot \underbrace{r'(x)}_{\substack{\text{the derivative}\\ \text{of the inside}\\ \text{function}}}$$

Combining the two pieces, we find that

$$f'(x) = \underbrace{t'(s(r(x)))}_{\substack{\text{the derivative of the}\\ \text{outside function evaluated}\\ \text{at the middle function}}} \cdot \underbrace{s'(r(x))}_{\substack{\text{the derivative of the}\\ \text{middle function evaluated}\\ \text{at the inside function}}} \cdot \underbrace{r'(x)}_{\substack{\text{the derivative}\\ \text{of the inside}\\ \text{function}}}$$

For example, we can compute the derivative of $f(x) = \sqrt{\sin x^2}$ by iterating applications of the chain rule. We take as our three functions $\sqrt{x}$, $\sin x$, and x^2. We compute the derivative

$$(\sqrt{\sin x^2})' = \underbrace{\frac{1}{2\sqrt{\sin x^2}}}_{\substack{\text{the derivative of the}\\ \text{square root function evaluated}\\ \text{at } \sin x^2}} \cdot \underbrace{\cos x^2}_{\substack{\text{the derivative of}\\ \sin x \text{ evaluated}\\ \text{at } x^2}} \cdot \underbrace{2x}_{\substack{\text{the derivative}\\ \text{of } x^2}}$$

$$= \frac{2x\cos x^2}{2\sqrt{\sin x^2}} = \frac{x\cos x^2}{\sqrt{\sin x^2}}$$

Remember that especially with trig functions, your answer might look very different from all of the answer choices for a multiple-choice question, yet still be correct. For instance, we can rewrite our answer here as:

$$\frac{x\cos x^2}{\sqrt{\sin x^2}} = x\sqrt{\cot x^2 \cos x^2}$$

The chain rule can be iterated any number of times. For example, to find the derivative of $f(x) = \sin^2(\sqrt{x^2+3})$, we use the four functions x^2, $\sin x$, $\sqrt{x}$, and $x^2 + 3$. We iterate the chain rule three times.

$$\left[\sin^2\left(\sqrt{x^2+3}\right)\right]' = 2\sin\left(\sqrt{x^2+3}\right) \cdot \cos\left(\sqrt{x^2+3}\right) \cdot \frac{1}{2\sqrt{x^2+3}} \cdot 2x$$

$2\sin\left(\sqrt{x^2+3}\right)$: the derivative of x^2 evaluated at $\sin\left(\sqrt{x^2+3}\right)$

$\cos\left(\sqrt{x^2+3}\right)$: the derivative of $\sin x$ evaluated at $\left(\sqrt{x^2+3}\right)$

$\frac{1}{2\sqrt{x^2+3}}$: the derivative of $\sqrt{x}$ evaluated at x^2+3

$2x$: the derivative of x^2+3

$$= \frac{2x\sin\left(\sqrt{x^2+3}\right)\cos\left(\sqrt{x^2+3}\right)}{\sqrt{x^2+3}}$$

$$= \frac{x\sin 2\sqrt{x^2+3}}{\sqrt{x^2+3}}$$

A common source of confusion when applying the chain rule is understanding the difference between $f(x) = \sin^2 x$ and $g(x) = \sin x^2$. For function f, the squaring operation applies to $\sin x$. That is, function f can be rewritten as $f(x) = (\sin x)^2$ (this is how you would enter it in your calculator).

To find $f'(x)$, the outer function is x^2 and the inner function is x.

Thus, $f'(x) = 2\sin x \cdot \cos x$

$2\sin x$: the derivative of x^2 evaluated at $\sin x$

$\cos x$: the derivative of $\sin x$

On the other hand, for function g, the squaring operation applies to the argument x (on the AP exam, this function would most likely appear as $\sin(x^2)$). So, to find $g'(x)$, the outer function is $\sin x$ and the inner function is x^2.

Thus, $g'(x) = \cos(x^2) \cdot 2x$

$\cos(x^2)$: the derivative of $\sin x$ evaluated at x^2

$2x$: the derivative of x^2

EXPONENTIAL AND LOGARITHMIC FUNCTIONS

Before working through this section, you might want to do a quick review of inverse functions, exponential functions, and logarithmic functions in a pre-calculus text.

EXPONENTIAL FUNCTIONS

There are many ways to build up the theory of exponential and logarithmic functions and their derivatives. Some definitions of exponential and logarithmic functions involve no calculus at all. Some depend on learning about integrals before working with these functions. All of these approaches are ultimately equivalent. We'll use one approach here, but there may be a different presentation in your calculus text.

An exponential function is a function of the type $f(x) = b^x$ where $b > 0$ and is called the "base of the function." We call x the exponent of an exponential function. While we can easily define b^x for an integer x or even a rational number x, we need a way to define b^x for any real number x. Before we can do this, we need to do a lot of setup work.

Previously, we defined the irrational number e as $e = \lim_{n\to\infty}\left(1 + \frac{1}{n}\right)^n \approx 2.7182818..$

Now we'll define the exponential function $f(x) = e^x$ as $e^x = \lim_{n\to\infty}\left(1 + \frac{x}{n}\right)^n$.

You've already learned how to find e^x for specific values of x. The exponential function is a generalization of this idea. Sometimes we write $\exp(x)$ for e^x to distinguish it from exponential functions with other bases.

THE DERIVATIVE OF e^x

We can use the definition of the derivative and the definition of e to find the derivative of e^x:

$$(e^x)' = \lim_{h\to 0}\frac{e^{x+h} - e^x}{h} = \lim_{h\to 0}\frac{e^x(e^h - 1)}{h} = e^x \lim_{h\to 0}\frac{(e^h - 1)}{h}$$

Consider $\lim_{h\to 0}\frac{(e^h - 1)}{h}$.

If we can show that this limit is equal to 1, then we will conclude that the derivative of e^x is itself, i.e., $(e^x)' = e^x$. This point is not so straightforward, but the basic idea here is that because $e = \lim_{h\to 0}(1 + h)^{\frac{1}{h}}$ then, for small h, $e^h \approx 1 + h$, so $e^h - 1 \approx h$. There is, of course, sophisticated mathematics that lies between this hand-waving and actually proving that the limit is actually equal to one.

Theorem: The Derivative of e^x

$$\frac{d}{dx}(e^x) = e^x$$

Using the chain rule, we have $\frac{d}{dx}(e^{f(x)}) = f'(x)e^x$.

Inverse Functions

Recall from pre-calculus that if f is a 1-1 function, then f has an inverse function called f^{-1} with the property that $f(f^{-1}(x)) = f^{-1}(f(x)) = x$. The domain of the inverse function is the range of the original function.

The Natural Logarithm ln(x)

We define a function ln(x) to be the inverse of the exponential function. We call this function the natural logarithm function, i.e., $\ln(e^x) = e^{\ln x} = x$. The first equality holds for $x \in \mathbf{R}$, but the second one only for $x > 0$. The term *natural* in its name comes from several particularly nice properties possessed by this function and its inverse (the exponential function). Some of these properties will become clear shortly and some we'll see in later material on integrals. We can define the exponential function for all values of x, but e^x is positive for all values of x. That is, the range of e^x is the set of positive real numbers. Therefore, the domain of $\ln x$ is all positive numbers.

The Derivative of ln(x)

Using the chain rule, we can find the derivative of $\ln x$. Because $x = e^{\ln x}$, we have that $1 = (x)' = (e^{\ln x})'$. We apply the chain rule with the functions e^x and $\ln x$:

$$1 = (e^{\ln x})' = \underbrace{e^{\ln x}}_{\substack{\text{the derivative of} \\ e^x \text{evaluated at } \ln x; \\ \text{remember that} \\ (e^x)' = e^x}} \cdot (\ln x)'$$

We now solve the resulting equation for $(\ln x)'$: $\frac{1}{e^{\ln x}} = (\ln x)'$. Because e^x and $\ln x$ are inverse functions, $e^{\ln x} = x$, we have another useful theorem.

Theorem: The Derivative of ln x

$$\frac{d}{dx}(\ln x) = \frac{1}{x}$$

THE DERIVATIVE OF x^r

Earlier we saw that for an integer n, $\frac{d}{dx}(x^n) = nx^{n-1}$. Using the inverse relationship of e^x and $\ln x$ and properties of exponents, we can now show that this formula holds not only for integers n, but also for any real number r.

$$x^r = e^{\ln x^r} = e^{r \ln x} \Rightarrow$$

$$\frac{d}{dx}(x^r) = \frac{d}{dx}(e^{r \ln x}) = \frac{r}{x}e^{r \ln x} = \frac{r}{x}x^r = rx^{r-1}$$

Theorem: Generalized Power Rule

For any real number r, $\frac{d}{dx}(x^r) = rx^{r-1}$.

THE DERIVATIVE OF INVERSE FUNCTIONS

In general, we can find the derivative of any inverse function using the chain rule. It is useful to see this first for the exponential and logarithmic functions, because they are the motivating example for using inverse functions in calculus.

Just as with the natural logarithm function, we find the derivative of f^{-1} by taking the derivative of both sides of the equation $x = f(f^{-1}(x))$. We use the chain rule to take the derivative of the right-hand side and find that:

$$1 = f'(f^{-1}(x)) \cdot [f^{-1}(x)]'$$

$$[f^{-1}(x)]' = \frac{1}{\underbrace{f\,[f^{-1}(x)]}_{\substack{\text{the derivate of} \\ f \text{ evaluated at } f^{-1}(x)}}}$$

The biggest obstacle to understanding the Inverse Function Theorem is understanding the notation used to state it. Because the notation $f^{-1}(x)$ is often confused with reciprocals, let's state the theorem another way and illustrate its use with an example:

Let f and g be inverse functions and let (x_0, y_0) belong to f and let (y_0, x_0) belong to g. Then $f'(x_0) = \frac{1}{g'(y_0)}$, or equivalently, $g'(y_0) = \frac{1}{f'(x_0)}$.

This theorem allows us to find the derivative of a function's inverse at a given point without having to find the inverse function first.

Example:

Let $f(x) = x^2$, $x \geq 0$, and let $g(x)$ be the inverse of f. Find $g'(4)$.

Solution:

Because 4 is function g's input, it must be function f's output. Thus, we may write: (2, 4) belongs to f and (4, 2) belongs to g.

By the Inverse Function Theorem, $g'(4) = \frac{1}{f'(2)}$. Because $f'(x) = 2x$ and $f'(2) = 4$, then $g'(4) = \frac{1}{4}$. If we look at the graph below of $f(x) = x^2$, $x \geq 0$ and $g(x) = \sqrt{x}$ (Note: We didn't have to find this to answer the question), we see that the reciprocal relationship, 4 vs $\frac{1}{4}$ of the slope of the tangent line to f at $x = 2$ and the slope of the tangent line to g at $x = 4$ is reasonable.

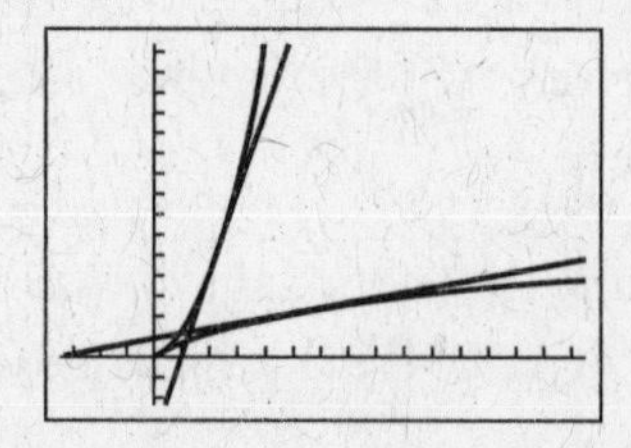

THE DERIVATIVE OF b^x AND LOG$_b x$

We are now—finally—ready to find the derivatives of the functions $f(x) = b^x$ and $f(x) = \log_b x$, for any positive base b. Using rules for manipulating logarithms and exponents (a quick review may be in order here!), we can rewrite b^x as $e^{\ln(b^x)} = e^{x \ln b}$. Now apply the rule for finding the derivative of $e^{f(x)}: (b^x)' = (e^{x \ln b})' = (x \ln b)' \cdot e^{x \ln b} = \ln b \cdot b^x$. Because $\log_b x$ is the inverse of b^x, we can compute its derivative:

$$(\log_b x)' = \frac{1}{(b^{\log_b x})'} = \frac{1}{\ln b \cdot b^{\log_b x}} = \frac{1}{\ln b \cdot x}.$$

Theorem: The Derivatives of b^x and $\log_b x$

For $b > 0$,

$$\frac{d}{dx}(b^x) = \ln b \cdot b^x$$

$$\frac{d}{dx}(\log_b x) = \frac{1}{\ln b \cdot x}$$

DERIVATIVES OF INVERSE TRIGONOMETRIC FUNCTIONS

We can use the formula for finding the derivative of an inverse function to find the derivatives of the inverse trigonometric functions. A quick review of inverse trigonometry from a pre-calculus text is probably a good idea before reading this material.

The arctangent function arctan x (or $\tan^{-1}(x)$) is the inverse of tan x on the interval $-\frac{\pi}{2} < x < \frac{\pi}{2}$. The derivative of arctan x is:

$$(\arctan x)' = \frac{1}{\sec^2(\arctan x)} = \cos^2(\arctan x) = [\cos(\arctan x)]^2$$

We can simplify this expression using relationships in right-triangle trigonometry. We are looking for the cosine of an angle θ whose tangent is x. Using right-triangle trigonometry, we don't actually need to know what θ is to determine its cosine. Using the triangle diagram below, we can determine that the cosine is $\frac{1}{\sqrt{1+x^2}}$.

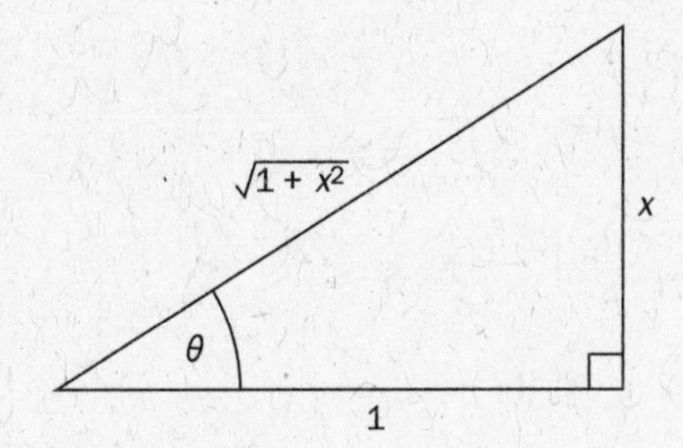

Therefore, $\frac{d}{dx}(\arctan x) = \frac{1}{1+x^2}$.

We can apply a similar procedure to find the derivatives of the other inverse trig functions.

$\frac{d}{dx}(\arctan x) = \frac{1}{1+x^2}$	$\frac{d}{dx}(\text{arccot } x) = \frac{-1}{1+x^2}$
$\frac{d}{dx}(\arcsin x) = \frac{1}{\sqrt{1-x^2}}$	$\frac{d}{dx}(\arccos x) = \frac{-1}{\sqrt{1-x^2}}$
$\frac{d}{dx}(\text{arcsec } x) = \frac{1}{\lvert x\rvert\sqrt{x^2-1}}$	$\frac{d}{dx}(\text{arccsc } x) = \frac{-1}{\lvert x\rvert\sqrt{x^2-1}}$

AP EXPERT TIP

For the AB exam, learning the formulas for the derivatives of arcsin x, arccos x, and arctan x is sufficient.

IMPLICIT DIFFERENTIATION

We can define mathematical relationships explicitly, meaning that we say what one variable or value of a function is equal to in terms of other variables. For example, $y = 2x + 3$ and $f(x) = \sin^2 x$ are explicitly defined functions.

A mathematical relationship can also be defined *implicitly*, meaning that we are given an equation relating the variables, but the equation is *not* solved for one of the variables. The equation $x^2 + y^2 = 2$ is an example of an implicitly defined relationship. We say "relationship" and not "function" there because the relationship may not be a function. In our example, for a given value of y there are up to two values of x that will make the equation true. We can also see that this relationship is not a function if we notice that the graph of the circle fails the vertical line test; it's revealed algebraically, as well, by solving the equation for y. The solution has two pieces:

$$y = +\sqrt{1 - x^2} \text{ and } y = -\sqrt{1 - x^2}.$$

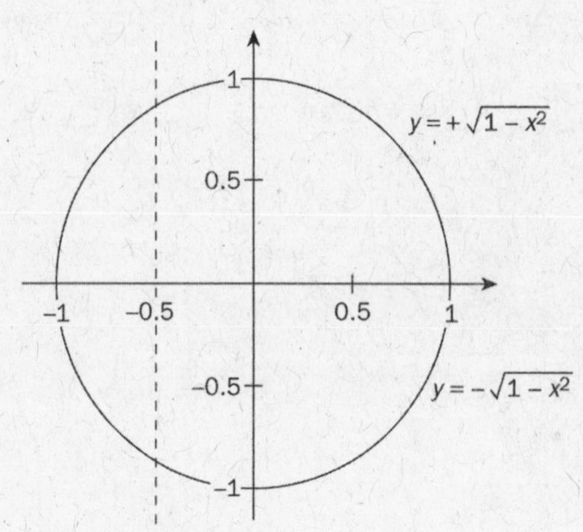

Even though this relationship is not a function, we can still find the line tangent to the curve at each point and we can ask what the slope of the tangent line is.

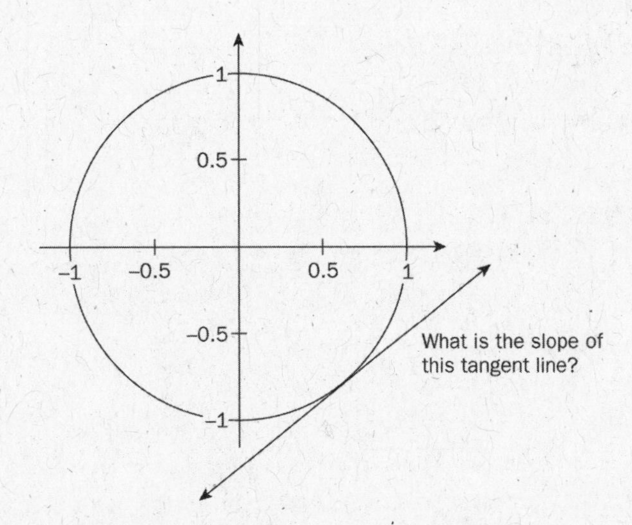

Using the chain rule we can differentiate equations that relate y and x. We think of x as the independent variable and y as a function of x, keeping in mind that we know that the derivative of y is $\frac{dy}{dx}$, even though we don't have an explicit expression for y in terms of x. We differentiate both sides of the equation that relates y and x and we find an implicit relationship between x, y, and $\frac{dy}{dx}$. In our example $x^2 + y^2 = 2$, we take the derivative of both sides of the equation with respect to x, keeping in mind that the derivative of y with respect to x is $\frac{dy}{dx}$.

$$\frac{d}{dx}(x^2 + y^2) = \frac{d}{dx}(2)$$

$$\frac{d}{dx}(x^2) + \frac{d}{dx}(y^2) = \underbrace{0}_{\text{the derivative of a constant is zero}}$$

$$2x + \underbrace{2y \cdot \frac{dy}{dx}}_{\text{chain rule}} = 0$$

The derivative $\frac{dy}{dx}$ is given in an equation relating x, y, and $\frac{dy}{dx}$. We can solve this equation for $\frac{dy}{dx}$, or leave it as is.

In this example, we could have started by solving the equation explicitly for y and computing the derivative directly; the answer would be the same.

Sometimes we can't find an explicit relationship between the variables. In these cases we must use implicit differentiation to find $\frac{dy}{dx}$. For example, if we are given the equation $\sin xy^2 = \cos x$, we can find an equation for $\frac{dy}{dx}$: $\frac{d}{dx}(\sin xy^2) = \frac{d}{dx}(\cos x)$.

Now we compute the derivative of each side.

The derivative of the left-hand side with respect to x is.

$$\frac{d}{dx}(\sin xy^2) = \underbrace{\cos xy^2 \frac{d}{dx}(xy^2)}_{\text{chain rule}}$$

$$= \cos xy^2 \underbrace{\left[y^2 + \underbrace{2xy \cdot \frac{dy}{dx}}_{\substack{\text{chain rule; } y \text{ is} \\ \text{a function of } x}} \right]}_{\text{product rule}}$$

The derivative of the right-hand side with respect to x is $\frac{d}{dx}(\cos x) = -\sin x$.

Because the relationship between x and y is given implicitly, the relationship between $\frac{dy}{dx}$, x, and y is also given implicitly.

$$\cos xy^2 \cdot \left(y^2 + 2xy \cdot \frac{dy}{dx} \right) = -\sin x$$

We can also use implicit differentiation to find the second derivative. For example, if x and y are related by the equation $2x \sin y = x + y$, we can find the second derivative with respect to x: $\frac{d^2y}{dx^2}$ by using implicit differentiation twice. We find an equation involving the first derivative:

$$\frac{d}{dx}(2x \sin y) = \frac{d}{dx}(x + y)$$

$$\underbrace{2 \sin y + 2x \cdot \frac{d}{dx}(\sin y)}_{\text{product rule}} = \underbrace{1 + \frac{dy}{dx}}_{\text{derivative of a sum}}$$

$$2 \sin y + 2x \cdot \underbrace{\cos y \cdot \frac{dy}{dx}}_{\text{chain rule}} = 1 + \frac{dy}{dx}$$

Now we differentiate this equation implicitly to find an equation involving the second derivative:

$$\frac{d}{dx}2\sin y + 2x \cdot \cos y \cdot \frac{dy}{dx} = \frac{d}{dx}1 + \frac{dy}{dx}$$

$$\underbrace{\frac{d}{dx}(2\sin y) + \frac{d}{dx}2x \cdot \cos y \cdot \frac{dy}{dx}}_{\text{derivative of a sum}} = \underbrace{\frac{d^2y}{dx^2}}_{\text{derivative of a sum}}$$

$$2\cos y \cdot \frac{dy}{dx} + \underbrace{2 \cdot \cos y \cdot \frac{dy}{dx} + 2x \cdot \frac{d}{dx}(\cos y) \cdot \frac{dy}{dx} + 2x \cdot \cos y \cdot \frac{d^2y}{dx^2}}_{\text{product rule applied two times}} = \frac{d^2y}{dx^2}$$

$$2\cos y \cdot \frac{dy}{dx} + 2\cos y \cdot \frac{dy}{dx} + \underbrace{2x(-\sin y)\frac{dy}{dx}}_{\text{chain rule}} \cdot \frac{dy}{dx} + 2x \cdot \cos y \cdot \frac{d^2y}{dx^2} = \frac{d^2y}{dx^2}$$

DERIVATIVES OF PARAMETRIC, POLAR, AND VECTOR FUNCTIONS*

In Chapter 10, we discussed functions defined in parametric and polar coordinates. We also discussed functions written in vector notation. Now we will use the chain rule to help calculate the derivative (and thus the tangent lines) of these types of functions.

Suppose we have a function defined with parametric equations $x = f(t)$ and $y = g(t)$. To calculate the tangent line of y (which is also a function of x), we must use the chain rule. We have $\frac{dy}{dt} = \frac{dy}{dx} \cdot \frac{dx}{dt}$. Provided $\frac{dx}{dt}$ is not equal to zero, we can rearrange this equality to get:

$$\frac{dy}{dx} = \frac{\frac{dy}{dt}}{\frac{dx}{dt}}, \text{ with } \left(\frac{dx}{dt}\right) \neq 0$$

This is the equation we will use to calculate the slope of tangent lines to parametric functions.

For example, suppose we want to find the equation of a tangent line through the parametric curve given by $x = \cos 5t$ and $y = 2\sin 3t$.

Using the chain rule and the formula above we can easily calculate:

$$\frac{dy}{dx} = \frac{\frac{dy}{dt}}{\frac{dx}{dt}} = \frac{2\cos 3t \cdot 3}{-\sin 5t \cdot 5} = \frac{-6\cos 3t}{5\sin 5t}.$$

When we fix a value of t, this will determine the slope of the line tangent to the curve at $(\cos 5t, 2\sin 3t)$.

*BC content

For example, let's suppose we fix the point $t = \frac{\pi}{10}$. This will determine the point $(x, y) = \left(\cos\frac{\pi}{2}, 2\sin\frac{3\pi}{10}\right) = \left(0, 2\sin\frac{3\pi}{10}\right)$. Furthermore, we can calculate the slope of the line tangent to our curve at $\left(0, 2\sin\frac{3\pi}{10}\right)$. We simply plug in the value $t = \frac{\pi}{10}$ into our equation for $\frac{dy}{dx}$ to get $\frac{dy}{dx} = \frac{-6\cos\frac{3\pi}{10}}{5\sin\frac{\pi}{2}} = \frac{-6}{5}\cos\frac{3\pi}{10}$. Thus we can now determine the equation for the line tangent to this planar curve at the point $\left(0, 2\sin\frac{3\pi}{10}\right)$.

Similarly, for polar curves we can compute $\frac{dy}{dx}$ using the chain rule. If the polar curve is described by the equation $r = f(\theta)$, then θ plays the role of t in a parametric equation and we have $x = r\cos\theta = f(\theta)\cos\theta$ and $y = r\sin\theta = f(\theta)\sin\theta$. Therefore, we can calculate (don't forget the product rule) $\frac{dy}{dx} = \frac{\frac{dy}{d\theta}}{\frac{dx}{d\theta}} = \frac{\frac{dr}{d\theta}\sin\theta + r\cos\theta}{\frac{dr}{d\theta}\cos\theta - r\sin\theta}$.

To find the slope of a line tangent to the curve, we follow the same method as with parametric equations.

To compute the derivative of a vector function, we simply differentiate each component in turn. For example, a vector function is usually written in the form $v(t) = (v_1(t), v_2(t), v_3(t))$ and thus the derivative of this vector function would be $v'(t) = (v'_1(t), v'_2(t), v'_3(t))$. If we need more than one derivative then we can repeat the process to find as many derivatives as we need. Pretty easy, huh? (Similarly, we can anti-differentiate a vector function term-by-term as well.)

TABLE OF DERIVATIVE FORMULAS

We covered a tremendous amount of ground in this chapter—more than we could cover in a few weeks of school. Here are all of the derivative formulas we discussed, organized in a table.

Name	Formula
Derivative of a constant	$\frac{d}{dx}(f(x)) = \frac{d}{dx}(c) = 0$
Power rule	$\frac{d}{dx}(x^n) = nx^{n-1}$, n an integer
Derivative of $\sqrt{x}$	$\frac{d}{dx}\left(\sqrt{x}\right) = \frac{1}{2\sqrt{x}}$

(*continued on next page*)

Name	Formula
Constant multiple rule	$\frac{d}{dx}[cf(x)] = c \cdot \frac{d}{dx}(f(x))$, c a real number
Sum and difference rule	$\frac{d}{dx}(f(x) \pm g(x)) = \frac{d}{dx}(f(x)) \pm \frac{d}{dx}(g(x))$ for f and g differentiable at x
Product rule	$\frac{d}{dx}(f(x) \cdot g(x)) = \frac{d}{dx}(f(x)) \cdot g(x) + f(x) \cdot \frac{d}{dx}(g(x))$ for f and g differentiable at x
Quotient rule	$\frac{d}{dx}\left(\frac{f(x)}{g(x)}\right) = \frac{\frac{d}{dx}(f(x)) \cdot g(x) - f(x) \cdot \frac{d}{dx}(g(x))}{(g(x))^2}$ for f and g differentiable at x, $g(x) \neq 0$
Chain rule	$\frac{d}{dx}h(g(x)) = \frac{dh}{dx}(g(x)) \cdot \frac{d}{dx}g(x)$ for g differentiable at x and h differentiable at $g(x)$
Derivatives of trigonometric functions	$\frac{d}{dx}(\sin x) = \cos x$; $\frac{d}{dx}(\cos x) = -\sin x$ $\frac{d}{dx}(\tan x) = \sec^2 x$; $\frac{d}{dx}(\cot) = -\csc^2 x$ $\frac{d}{dx}(\sec x) = \sec x \tan x$; $\frac{d}{dx}(\csc x) = -\csc x \cot x$
Derivative of e^x	$\frac{d}{dx}(e^x) = e^x$
Derivative of e^u	$\frac{d}{dx}(e^u) = e^u \frac{du}{dx}$
Generalized power rule	$\frac{d}{dx}(x^r) = rx^{r-1}$ for r a real number
Derivative of $\ln x$	$\frac{d}{dx}(\ln x) = \frac{1}{x}$
Derivative of $\ln u$	$\frac{d}{dx}(\ln u) = \left(\frac{1}{u}\right)\frac{du}{dx}$

(*continued on next page*)

Name	Formula
Derivative of f^{-1}	$\frac{d}{dx}(f^{-1}(x)) = \frac{1}{\frac{d}{dx}f(f^{-1}(x))}$ for all x where f is differentiable and $f'(f^{-1}(x)) \neq 0$
Derivative of b^x	$\frac{d}{dx}(b^x) = \ln b \cdot b^x$ for $b > 0$
Derivative of $\log_b x$	$\frac{d}{dx}(\log_b x) = \frac{1}{\ln b \cdot x}$, $b > 0$
Derivatives of inverse trigonometric functions	$\frac{d}{dx}(\arctan x) = \frac{1}{1+x^2}$ $\frac{d}{dx}(\text{arc}\cot x) = \frac{-1}{1+x^2}$ $\frac{d}{dx}(\arcsin x) = \frac{1}{\sqrt{1-x^2}}$ $\frac{d}{dx}(\arccos x) = \frac{-1}{\sqrt{1-x^2}}$ $\frac{d}{dx}(\text{arc}\sec x) = \frac{1}{\lvert x\rvert\sqrt{x^2-1}}$ $\frac{d}{dx}(\text{arc}\csc x) = \frac{-1}{\lvert x\rvert\sqrt{x^2-1}}$

REVIEW QUESTIONS

1. If $f(x) = x^5 - 2x^2 + \frac{3}{x}$, then $f'(-1) =$
   (A) -11
   (B) -2
   (C) 2
   (D) 6
   (E) 11

2. A particle is traveling along the x-axis. Its position is given by $x(t) = \frac{1-t^2}{t+3}$ at time $t \geq 0$. Find the instantaneous rate of change of x with respect to t when $t = 1$.
   (A) -2
   (B) $-\frac{1}{2}$
   (C) 0
   (D) $\frac{1}{2}$
   (E) 2

3. If $f(x) = \cos(\ln x)$ for $x > 0$, then $f'(x) =$
   (A) $-\sin(\ln x)$
   (B) $\sin(\ln x)$
   (C) $-\frac{\sin(\ln x)}{x}$
   (D) $\frac{\sin(\ln x)}{x}$
   (E) $\sin\left(\frac{\ln x}{x}\right)$

4. If $f(x) = x \cdot 2^x$, then $f'(x) =$
   (A) $2^x(x + \ln 2)$
   (B) $2^x(1 + \ln 2)$
   (C) $x \cdot 2^x \cdot \ln 2$
   (D) $2^x(1 + x \cdot \ln 2)$
   (E) $x \cdot 2^x(1 + \ln 2)$

5. Let $f(x) = x^3 - x + 2$. If h is the inverse of f, then $h'(2)$ could be
   (A) $\frac{1}{26}$
   (B) $\frac{1}{4}$
   (C) $\frac{1}{2}$
   (D) 2
   (E) 26

6. Let $f(x) = x \cdot g(h(x))$, where $g(4) = 2, g'(4) = 3, h(3) = 4$, and $h'(3) = -2$. Find $f'(3)$.
   (A) -18
   (B) -16
   (C) -7
   (D) 7
   (E) 11

7. If $f(x) = x^2 \ln x$, then $f'(x)$ is
   (A) $2x \cdot \ln x$
   (B) 2
   (C) x
   (D) $2x \cdot \ln x + x$
   (E) $2x \cdot \ln x + x^2e^x$

8. If $f(x) = \frac{\sin\sqrt{x}}{\sqrt{x}}$, then $f'(x)$ is
   (A) $\frac{\cos\sqrt{x}}{2x} - \frac{\sin\sqrt{x}}{2\sqrt{x^3}}$
   (B) $\frac{\cos\sqrt{x} - \sin\sqrt{x}}{2x}$
   (C) $\frac{\sqrt{x}\cos\sqrt{x} - \frac{\sin\sqrt{x}}{2\sqrt{x}}}{x}$
   (D) $\cos\sqrt{x}$
   (E) $\frac{\frac{\cos\sqrt{x}}{2} + \frac{\sin\sqrt{x}}{2\sqrt{x}}}{x}$

9. If $f(x) = \sin^3(x)$, then $f'\left(\frac{\pi}{3}\right)$ is

(A) $\frac{1}{8}$

(B) $\frac{3}{2}$

(C) $\frac{9}{4}$

(D) $\frac{3\sqrt{3}}{8}$

(E) $\frac{9}{8}$

10. A particle moves on the x-axis with position defined by: $x(t) = t^3 - 6t^2 + 2t + 1$; $t \geq 0$. What is the velocity of the particle when its acceleration is zero?

(A) −11

(B) −10

(C) 1

(D) 2

(E) 50

11. If $f(x) = x^3 + 3x - 1$ and $g(x) = f^{-1}(x)$, find $g'(3)$.

(A) $\frac{1}{30}$

(B) $\frac{1}{6}$

(C) 1

(D) 6

(E) 30

12. If $x^2y + xy^3 = 8 - 2y$ then the value of $\frac{dy}{dx}$ at (2,1) is

(A) $-\frac{5}{6}$

(B) $-\frac{5}{12}$

(C) 0

(D) $\frac{5}{12}$

(E) $\frac{5}{6}$

13. Which of the following is the equation of the tangent to the function $y = \tan^{-1} 2x$ at $\left(\frac{1}{2}, \frac{\pi}{4}\right)$?

(A) $y = x + \frac{\pi - 1}{4}$

(B) $y = \frac{x}{2} + \frac{\pi - 1}{4}$

(C) $y = x + \frac{\pi - 2}{4}$

(D) $y = \frac{x}{2} + \frac{\pi - 2}{4}$

(E) $y = x + \frac{\pi + 1}{4}$

14. Find $\frac{d^2y}{dx^2}$ if $xy + \pi = 2y^2$.

(A) $\frac{y}{4y - x}$

(B) $\frac{y^2}{(4y - x)^2}$

(C) $\frac{1}{4\left(\frac{y}{4y - x}\right) - x}$

(D) $\frac{-\frac{4y^2}{4y - 4}}{(4y - x)^2}$

(E) $\frac{2y - \frac{4y^2}{4y - 4}}{(4y - x)^2}$

15. *Which of the following is the equation of the tangent to the curve defined by $x(t) = t^3 + t + 1$ and $y(t) = \tan t$ at $t = 0$?

(A) $y = x - 1$

(B) $y = x$

(C) $y = x + 1$

(D) $y = 1$

(E) None of the above

*BC content

Free-Response Questions

16. Let $f(x) = \sqrt{1 - \sin x}$

    (a) Find $f'(x)$.

    (b) Write an equation for the line tangent to the graph of f at $x = 0$.

17. We are given the following information about two differentiable functions f and g.

x	$f(x)$	$f'(x)$	$g(x)$	$g'(x)$
1	4	−3	3	2
3	6	2	−2	3

    (a) $(f + g)'(3) =$

    (b) $\left(\frac{f}{g}\right)'(1) =$

    (c) If $B = f \cdot g$, then $B'(3) =$

    (d) $G(x) = \sqrt{f(x)}$, then $G'(1) =$

18. Given the curve: $x^2 + 4y = xy$

    (a) Find an expression for $\frac{dy}{dx}$.

    (b) Find the equation of a vertical tangent line (if one exists).

    (c) Find the coordinates of a point, with non-zero x-coordinate where the function has a horizontal tangent line (if such a point exists).

    (d) Find an expression for $\frac{d^2y}{dx^2}$ and use this to determine whether the curve has local maximum or minimum at the point found in part (c). Justify your answer.

19. Consider the function $f(x) = x \cdot \ln x$.

    (a) Find the instantaneous rate of change of f at $x = e$.

    (b) Find the average rate of change of f over the interval [2.5, 3].

20. The velocity of a particle moving along the x-axis is given by the equation $v(t) = \frac{\pi}{4} - \tan^{-1}(t^2 - 4t + 4)$, for $t \geq 0$. The following table gives information about $v(t)$ and $v'(t)$.

t	$0<t<1$	1	$1<t<2$	2	$2<t<3$	3	$t>3$
$v(t)$	Negative	0	Positive	Positive	Positive	0	Negative
$v'(t)$	Positive	Positive	Positive	0	Negative	Negative	Negative

    (a) Find an equation for the acceleration of the particle as a function of t. You do not need to simplify the equation.

    (b) Suppose that on the interval (0, 2), the particle lies in the positive ray of the x-axis. At what times in the interval (0, 2) is the particle moving away from the origin?

    (c) During what time intervals is the speed of the particle increasing?

ANSWERS AND EXPLANATIONS

1. D

We first rewrite f with negative exponents instead of denominators and use the power rule, the sum and difference rule, and the constant multiple rule to differentiate f:

$$f(x) = x^5 - 2x^2 + \frac{3}{x} = x^5 - 2x^2 + 3x^{-1}$$

$$f'(x) = 5x^4 - 2 \cdot 2x + 3(-1)x^{-2} = 5x^4 - 4x - \frac{3}{x^2}$$

We can now substitute $x = -1$ into $f'(x)$, to get $f'(-1) = 5(-1)^4 - 4(-1) - \frac{3}{(-1)^2} = 6$.

2. B

The instantaneous rate of change of a function at a point is its derivative at that point. In our case, we use the quotient rule to differentiate $x(t) = \frac{1 - t^2}{t + 3}$:

$$x(t) = \frac{1 - t^2}{t + 3}$$

$$x'(t) = \frac{(1 - t^2)'(t + 3) - (1 - t^2)(t + 3)'}{(t + 3)^2}$$

$$= \frac{(-2t)(t + 3) - (1 - t^2)(1)}{(t + 3)^2}$$

To find the instantaneous rate of change at $t = 1$, we substitute $t = 1$ into our expression for $x'(t)$:

$$x'(1) = \frac{(1 + 3)(-2 \cdot 1) - (1 - 1^2)}{(t + 3)^2} = \frac{4(-2) - 0}{4^2} = \frac{-8}{16} = -\frac{1}{2}$$

3. C

We use the chain rule and the derivatives $(\cos x)' = -\sin x$ and $(\ln x)' = \frac{1}{x}$ to differentiate f:

$$f'(x) = \underbrace{-\sin(\ln x)}_{\text{the derivative of } \cos x \text{ evaluated at } \ln x} \cdot \underbrace{\frac{1}{x}}_{\text{the derivative of } \ln x} = -\frac{\sin(\ln x)}{x}$$

4. D

We use the product rule and the derivative $(b^x)' = \ln b \cdot b^x$ to differentiate f:

$$\begin{aligned} f'(x) &= (x)' \cdot 2^x + x \cdot \left(2^x\right)' \\ &= 1 \cdot 2^x + x \cdot \left(\ln 2 \cdot 2^x\right) \\ &= 2^x\left(1 + x \cdot \ln 2\right) \end{aligned}$$

5. C

If h is the inverse of $f(x) = x^3 - x + 2$, then, according to the formula for the derivative of an inverse: $h'(2) = \dfrac{1}{f'(h(2))}$. To compute $h'(2)$ we must evaluate $f'(h(2))$, so we need find $f'(2)$ and $h(2)$. The function $f(x) = x^3 - x + 2$ is a polynomial; we compute its derivative using the power rule, the constant multiple rule, and the derivative of a constant: $f'(x) = 3x^2 - 1$.

Because h is the inverse of f, to find $h(2)$, we solve the equation

$2 = x^3 - x + 2 \Rightarrow x^3 - x = 0 \Rightarrow x(x + 1)(x - 1) = 0 \Rightarrow x = 0, -1, 1$.

Therefore, $h(2) = 0, -1$ or 1. We plug each of these options into the formula for $h'(2)$ above:

$h(2)$	0	-1	1
$f'\left(h(2)\right)$	-1	2	2
$h'(2) = \dfrac{1}{f'\left(h(2)\right)}$	-1	$\dfrac{1}{2}$	$\dfrac{1}{2}$

Because -1 is not one of the answer choices, the correct answer is $h'(2) = \dfrac{1}{2}$.

6. B

$$\begin{aligned} f(x) &= x \cdot g(h(x)) \\ f'(x) &= g(h(x)) + x\left(g'(h(x))h'(x)\right) \\ f'(3) &= g(h(3)) + 3\left(g'(h(3))h'(3)\right) \\ &= g(4) + 3\left(g'(4)\right)(-2) \\ &= 2 - 6(3) \\ &= -16 \end{aligned}$$

7. D

Product rule:

$$f' = u'v + uv'$$

$$u = x^2 \rightarrow u' = 2x$$

$$v = \ln x \rightarrow v' = \frac{1}{x}$$

$$f'(x) = 2x \cdot \ln x + x^2 \cdot \frac{1}{x}$$

$$= 2x \cdot \ln x + x$$

8. A

Quotient rule:

$$f' = \frac{u'v - uv'}{v^2}$$

$$u = \sin\sqrt{x} \rightarrow u' = \cos\sqrt{x} \cdot \frac{1}{2\sqrt{x}}$$

$$v = \sqrt{x} \rightarrow v' = \frac{1}{2\sqrt{x}}$$

$$f'(x) = \frac{\cos\sqrt{x} \cdot \frac{1}{2\sqrt{x}} \cdot \sqrt{x} - \sin\sqrt{x} \cdot \frac{1}{2\sqrt{x}}}{\left(\sqrt{x}\right)^2}$$

$$= \frac{\frac{\cos\sqrt{x}}{2} - \frac{\sin\sqrt{x}}{2\sqrt{x}}}{x}$$

$$= \frac{\cos\sqrt{x}}{2x} - \frac{\sin\sqrt{x}}{2\sqrt{x^3}}$$

9. E

Chain rule:

$$f = u^3 \Rightarrow f' = 3u^2 \frac{du}{dx}$$

$$u = \sin x \rightarrow \frac{du}{dx} = \cos x$$

$$f'(x) = 3(\sin x)^2 \cos(x)$$

$$\begin{aligned} f'\left(\frac{\pi}{3}\right) &= 3\left(\sin\left(\frac{\pi}{3}\right)\right)^2 \cos\left(\frac{\pi}{3}\right) \\ &= 3\left(\frac{\sqrt{3}}{2}\right)^2\left(\frac{1}{2}\right) \\ &= 3 \cdot \frac{3}{4} \cdot \frac{1}{2} \\ &= \frac{9}{8} \end{aligned}$$

10. B

$$x(t) = t^3 - 6t^2 + 2t + 1$$

$$v(t) = \frac{dx}{dt} = 3t^2 - 12t + 2$$

$$a(t) = \frac{dv}{dt} = 6t - 12$$

$$a = 0 \Rightarrow 6t - 12 = 0$$

$$t = 2$$

$$\begin{aligned} v(2) &= 3(2)^2 - 12(2) + 2 \\ &= -10 \end{aligned}$$

11. B

$$\begin{aligned} f(1) &= 3 \Rightarrow f^{-1}(3) = 1 \\ g'(x) &= \left.\frac{d}{dx} f^{-1}(x)\right|_{x=3} \\ &= \frac{1}{f'(1)} \\ f'(x) &= 3x^2 + 3 \end{aligned}$$

$$\therefore g'(x)\Big|_{x=3} = \frac{1}{3(1)^2 + 3} = \frac{1}{6}$$

12. B

$$x^2y + xy^3 = 8 - 2y$$
$$2xy + x^2y' + y^3 + x \cdot 3y^2y' = 0 - 2y'$$
$$x^2y' + 3xy^2y' + 2y' = -2xy - y^3$$
$$y'\left(x^2 + 3xy^2 + 2\right) = -2xy - y^3$$
$$y' = \frac{-2xy - y^3}{x^2 + 3xy^2 + 2}$$

$$y'(2,1) = \frac{-2(2)(1) - (1)^3}{(2)^2 + 3(2)(1)^2 + 2} = \frac{-4-1}{4+6+2} = -\frac{5}{12}$$

13. C

$$y = \tan^{-1} 2x$$
$$\frac{dy}{dx} = \frac{1}{1 + (2x)^2} \cdot (2)$$
$$\left.\frac{dy}{dx}\right|_{x=\frac{1}{2}} = \frac{2}{1 + \left(2\left(\frac{1}{2}\right)\right)^2} = \frac{2}{2} = 1$$

Equation of tangent line:

$$\frac{y - \frac{\pi}{4}}{x - \frac{1}{2}} = 1 \rightarrow y - \frac{\pi}{4} = \left(x - \frac{1}{2}\right)$$
$$y = x + \frac{\pi - 2}{4}$$

14. E

$$xy + \pi = 2y^2 \quad \text{(differentiate implicitly)}$$

$$\underbrace{y + xy'}_{\text{product rule}} = \underbrace{4yy'}_{\text{chain rule}}$$

$$y = 4yy' - xy'$$

$$y'(4y - x) = y$$

$$y' = \frac{y}{4y - x} \quad \text{(differentiate again - use quotient rule)}$$

$$y'' = \frac{y'(4y - x) - y(4y' - 1)}{(4y - x)^2} \quad \text{(substitute for } y \text{)}$$

$$y'' = \frac{\left(\frac{y}{\cancel{4y - x}}\right)\left(\cancel{4y - x}\right) - y\left(4\left(\frac{y}{4y - x}\right) - 1\right)}{(4y - x)^2}$$

$$= \frac{y - \frac{4y^2}{4y - x} + y}{(4y - x)^2}$$

$$= \frac{2y - \frac{4y^2}{4y - x}}{(4y - x)^2}$$

15. A

$$(x, y) = \left(t^3 + t + 1, \tan t\right)$$

$$(x, y)\big|_{t=0} = \left(0^3 + 0 + 1, \tan 0\right) = (1, 0)$$

$$\frac{dy}{dx} = \frac{\frac{dy}{dt}}{\frac{dx}{dt}} = \frac{\sec^2 t}{3t^2 + 1}$$

$$\left.\frac{dy}{dx}\right|_{t=0} = \frac{\sec^2 0}{3(0)^2 + 1} = \frac{1^2}{1} = 1$$

Equation of the tangent line:

$$\frac{y - 0}{x - 1} = 1 \rightarrow y = x - 1$$

Free-Response Answers

16. (a) We use the chain rule and the derivatives $\left(\sqrt{x}\right)' = \frac{1}{2\sqrt{x}}$ and $(\sin x)' = \cos x$ to compute

$$f'(x) = \underbrace{\frac{1}{2\sqrt{1-\sin x}}}_{\text{the derivative of } \sqrt{x} \text{ evaluated at } 1-\sin x} \cdot \underbrace{-\cos x}_{\text{the derivative of } 1-\sin x}$$

$$= -\frac{\cos x}{2\sqrt{1-\sin x}}$$

(b) To find the equation of a tangent line in point-slope form, $y - y_0 = m(x - x_0)$, we need to find a point on the line and the slope of the line.

To find the point we evaluate f at the point $x = 0$: $f(0) = \sqrt{1-\sin 0} = 1 \Rightarrow$ Point: $(0,1)$. The slope of the line is $f'(0)$. We substitute $x = 0$ into the formula for the derivative $f'(0) = \frac{-\cos 0}{2\sqrt{1-\sin 0}} = -\frac{1}{2}$ Slope: $m = -\frac{1}{2}$. We plug this information into the point-slope form of a line to find the tangent line: $y - 1 = -\frac{1}{2}(x - 0)$ or $y = -\frac{1}{2}x + 1$.

You generally won't be penalized for leaving equations in an unsimplified algebraic form, as long as the problem setup does not specifically ask you to simplify a result. The AP Calculus exam tests your knowledge of calculus, not algebra. If simplifying demonstrates a greater knowledge of calculus, then do it. If it only demonstrates skills in algebra, it isn't necessary. However, if you simplify and make an error in the process, you will lose points.

17. (a) $(f + g)'(3) = f'(3) + g'(3) = 2 + 3 = 5$

(b) Using the quotient rule $\left(\frac{f}{g}\right)(1) = \frac{f'(1) \cdot g(1) - f(1) \cdot g'(1)}{[g(1)]^2} = \frac{(-3) \cdot 3 - 4 \cdot 2}{3^2} = -\frac{17}{9}$

(c) If $B = f \cdot g$, then using the product rule

$B'(3) = (f \cdot g)'(3) = f'(3) \cdot g(3) + f(3) \cdot g'(3)$

$= 2 \cdot (-2) + 6 \cdot 3 = 14$

(d) $G(x) = \sqrt{f(x)}$, then using the chain rule

$$G\,(1) = \left(\sqrt{f}\right)'(1) = \underbrace{\frac{1}{2\sqrt{f(1)}}}_{\text{the derivative of } \sqrt{x} \text{ evaluated at } f(1)} \cdot f'(1) = \frac{-3}{2\sqrt{4}} = \frac{-3}{4}$$

18. (a) We differentiate $x^2 + 4y = xy$ implicitly and solve for $\frac{dy}{dx}$:

$$2x + 4\frac{dy}{dx} = x \cdot \frac{dy}{dx} + 1 \cdot y \Rightarrow \frac{dy}{dx} = \frac{2x - y}{x - 4}$$

(b) A vertical tangent may occur when $\frac{dy}{dx} = \infty$ or does not exist. In our case, this happens when the denominator of the expression for $\frac{dy}{dx}$ equals zero, so we set $x - 4 = 0 \Rightarrow x = 4$.

But when we let $x = 4$ in the original equation, we find:

$(4)^2 + 4y = 4y$
$16 = 0$

This is nonsense, and it means the function is not defined at $x = 4$ and so cannot have a vertical tangent there. The function does not have a vertical tangent anywhere. ($x = 4$ is actually the location of a vertical asymptote.)

(c) A horizontal tangent occurs when $\frac{dy}{dx} = 0$; in our case, when the numerator equals zero. We set $2x - y = 0 \quad \Rightarrow \quad y = 2x$.

Next, we substitute $y = 2x$ back into the original equation:

$$x^2 + 4(2x) = x(2x) \Rightarrow x^2 - 8x = 0 \Rightarrow x(x-8) = 0 \Rightarrow x = 0,8.$$

We see that the numerator equals zero at $x = 0$ and $x = 8$. We are asked to find a point the x-coordinate of which does not equal zero, so we restrict our attention to $x = 8$. At this point $y = 2x$; the y-coordinate of this point must be $y = 2 \cdot 8 = 16$. The coordinates of a point with non-zero x-coordinate where the curve has a horizontal tangent are $x = 8, y = 16$.

(d) To find the second derivative, $\frac{d^2y}{dx^2}$, we differentiate the first derivative $\frac{dy}{dx}$ with respect to x:

$$\frac{d^2y}{dx^2} = \frac{(x-4)\left(2 - \frac{dy}{dx}\right) - (2x - y)(1)}{(x-4)^2}$$

To determine whether the point (8, 16) is a local maximum or a local minimum, we can use the second derivative test. We plug the values $x = 8$, $y = 16$, and $\frac{dy}{dx} = 0$ into our expression for the second derivative, finding:

$$\frac{d^2y}{dx^2} = \frac{(8-4)(2-0) - (2 \cdot 8 - 16)}{(8-4)^2} = \frac{8}{16} = \frac{1}{2} \geq 0.$$

By the second derivative test, the point (8, 16) is a local minimum.

19. (a) The instantaneous rate of change of a function at a point is its derivative at that point. Using the product rule and the derivative $(\ln x)' = \frac{1}{x}$, we compute:

$$\begin{aligned} f'(x) &= (x)' \ln x + x(\ln x)' \\ &= 1 \cdot \ln x + x \cdot \frac{1}{x} = \ln x + 1 \end{aligned} \qquad \Rightarrow \qquad f'(e) = \ln e + 1 = 1 + 1 = 2$$

Therefore, the instantaneous rate of change at $x = e$ is $f'(e) = 2$.

(b) The average rate of change of a function over an interval is $\frac{\Delta f}{\Delta x}$. In our case, over the interval [2.5, 3], the average rate of change is $\frac{f(3) - f(2.5)}{3 - 2.5} = \frac{3\ln 3 - 2.5\ln 2.5}{0.5} \approx$ 2.010.

20. (a) Solving: $v(t) = \frac{\pi}{4} - \tan^{-1}(t^2 - 4t + 4)$

By the chain rule, $v'(t) = 0 - \frac{1}{1 + (t^2 - 4t + 4)^2} \cdot (2t - 4)$. Since acceleration is the derivative of velocity, this gives an equation for acceleration.

(b) The particle is moving away from the origin whenever both position and velocity are positive, or both position and velocity are negative. Because we are told that the position is positive during the interval $(0,2)$, the particle is moving away from the origin whenever $v(t)$ is also positive, i.e., on the interval $(1,2)$.

(c) Since the speed is the absolute value of velocity, the speed of the particle is increasing whenever $v(t)$ and $v'(t)$ are either both positive or both negative. This happens on the intervals $(1,2)$ and $9(3,\infty)$.

For part (a), be very careful when taking a derivative using a chain rule. If necessary, write down the formula and identifications to solve it. For example,

$$v(t) = h(g(t)) \text{ where } h(t) = \frac{\pi}{4} - \tan^{-1}(t) \text{ and } g(t) = t^2 - 4t + 4.$$

Then by the chain rule:

$$v'(t) = h'(g(t)) \cdot g'(t).$$

Compute $h'(t) = 0 - \frac{1}{1 + t^2}$ (don't forget the sign!) and $g(t) = 2t - 4$ to get the answer. Plug them into the chain rule to get the answer above. For free-response questions, you should never simplify unless asked to do so. You are too likely to make an algebraic mistake and lose points. This is true especially if all you need is the derivative at a single point. Then evaluate as soon as possible, again to avoid algebraic mistakes. For parts (b) and (c), think about the statements in the solution until they make sense and you are able to conjure them yourself. It may seem silly, but statements like these are at the heart of calculus. The calculations themselves are not as useful as the theorems they provide.

CHAPTER 13: THE DERIVATIVE AT A POINT

IF YOU LEARN ONLY THREE THINGS IN THIS CHAPTER . . .

1. The derivative of any function f can be computed at any point where the limit of the difference quotient is defined.
2. The Leibniz notation $\frac{df}{dx} = (x_0)$ is used to describe the rate of change, or the derivative, of f at point x_0.
3. The line tangent to a curve at a point (if it exists) is the line that best approximates the curve near that point.

THE DERIVATIVE FUNCTION

We can compute the derivative of any function f at any point where the limit of the difference quotient is defined. Therefore, if f is a function of x, the derivative of f is also a function of x. We use the notation $\frac{df}{dx}$ to describe the derivative function, i.e., the limit of the difference quotient $\frac{df}{dx} = \lim_{\Delta x \to 0} \frac{\Delta f}{\Delta x}$.

To describe the rate of change, or derivative, of f at a particular point x_0, we can use the notation $\frac{df}{dx}(x_0)$. This is called Leibniz notation; it's named after Wilhelm Gottfried Leibniz, who—along with Isaac Newton—is credited with inventing calculus. We also use prime notation to describe the derivative, in which we express the derivative as $f'(x)$. If we define a variable y as a function of x, $y = f(x)$, then we may also write the derivative of y using prime notation, $y' = f'(x)$. All of these notations, $\frac{df}{dx}$, $f'(x)$, and y' express the same concept: the derivative of the function $f(x)$.

One of the benefits of learning all of the derivative rules in Chapter 12 is that it is no longer necessary to constantly recompute the limit difference in order to calculate the derivative of a function. If we want to find the derivative of a function at a specific point, we can simply use our rules (power rule, product rule, quotient rule, etc.) to get a function for the derivative and then plug a our point into the resultant function.

For example, given $f(x) = x^2$, we can easily use the power rule to compute the derivative and realize that $f'(x) = 2x$. Now, if we are asked to find the derivative at the point $x = -1$, we simply plug that value into our derivative, such that $f'(-1) = -2$.

We can use Leibniz notation or prime notation to express the derivative of a function, and the AP exam expects you to be comfortable with both. Thus, if $f(x) = x^2$

$$
\begin{aligned}
(x^2)' &= \lim_{\Delta x \to 0} \frac{f(x + \Delta x) - f(x)}{\Delta x} = \lim_{x \to 0} \frac{(x + \Delta x)^2 - (x)^2}{\Delta x} \\
&= \lim_{\Delta x \to 0} \frac{[x + 2x\Delta x + (\Delta x)^2] - x^2}{\Delta x} \\
&= \lim_{\Delta x \to 0} \frac{2x\Delta x + (\Delta x)^2}{\Delta x} \\
&= \lim_{\Delta x \to 0} \frac{\cancel{\Delta x}(2x + \Delta x)}{\cancel{\Delta x}} = 2x
\end{aligned}
$$

then $f'(x) = 2x$ or $\frac{df}{dx} = 2x$. Both forms of notation are completely interchangeable, though prime notation is slightly more useful for working at specific points.

To summarize: don't reinvent the wheel when you're finding the derivative of a function at a given point. Use your rules to come up with a general formula for the derivative and then plug your point into the resultant function.

The Second Derivative

Because $f'(x)$ (the derivative of $f(x)$) is also a function, we can compute *its* derivative. The derivative of $f'(x)$ is called the second derivative of f and we write $f''(x)$, y'', or $\frac{d^2 f}{dx^2}$.

For example, we saw that the derivative of a linear function $y = mx + b$ is constant. That is, $y' = m$. The derivative of a constant function is 0, since the graph of a constant function is a horizontal line whose slope is, therefore, always 0. That is, $y'' = 0$.

In the example above, we saw that the derivative of $f(x) = x^2$ is $f'(x) = 2x$. The derivative is a linear function, so *its* derivative is equal to its slope. To compute the second derivative, simply apply the power rule to the first derivative. i.e., $f''(x) = 2$.

Points at Which the Derivative Is Undefined

The derivative $f'(x)$ may not be defined at every point, that is $f(x)$ may not be differentiable at every point x_0. The domain of $f'(x)$ is a subset of the domain of $f(x)$, because a function must be defined and continuous at x_0 to be differentiable there. If $f'(x_0)$ exists for every point in the domain of f, we say that f is differentiable; for clarity, we sometimes say that f is differentiable everywhere (on its domain). A function that has sharp corners is not differentiable everywhere, so we sometimes say that a differentiable function is a smooth function.

Remember, the derivative of $y = |x|$ does not exist at $x = 0$. Looking at the graph of $y = |x|$, we see that there is a sharp corner at $x = 0$. There is no line that is tangent to the curve at this point:

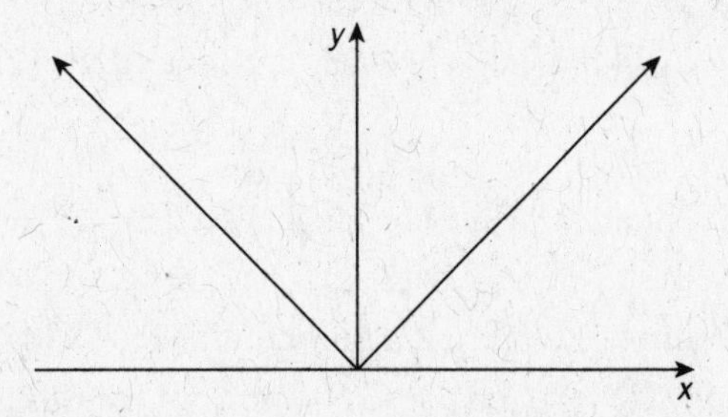

If the graph of f has a very sharp corner at x_0 where $\lim f'(x) = +\infty$ as $x \to x_0$ from one side and $\lim f'(x) = -\infty$ as $x \to x_0$ from the other side, we say that f has a cusp at x_0. The function $y = x^{\frac{2}{3}}$ is an example of a function with a cusp at $x = 0$. At a cusp, a function has a vertical tangent line:

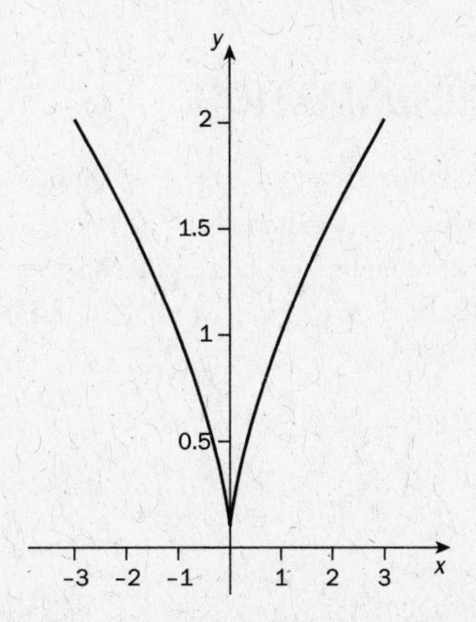

A function may have a vertical tangent at a point that is not a cusp. For example, the function $y = \sqrt[3]{x}$ has a vertical tangent at $x = 0$, but the slopes of the tangent lines of points approaching $x = 0$ from either side are both $+\infty$.

THE SLOPE OF A CURVE AT A POINT

We previously learned that m_{tan}—the slope of the line tangent to the graph of $f(x)$ at x_0—is the limit of the slopes of the secant lines between x_0 and points close to x_0. In other words, $x_0 + \Delta x$, for Δx close to zero. This slope is given by the limit of the difference quotient, which is the derivative at x_0. Using derivative notation, we write:

$$\underset{(\text{at } x_0)}{m_{\tan}} = f'(x_0) = \lim_{h \to 0} \frac{f(x_0 + h) - f(x_0)}{h}$$

We can use point-slope form to find the equation of any line given the slope of the line and a point on the line. Because $(x_0, f(x_0))$ is a point on the line tangent to $f(x)$ at x_0 and the slope of the tangent line is $f'(x_0)$, the equation of the line tangent to the curve $f(x)$ at x_0 is $y - f(x_0) = f'(x_0)(x - x_0)$.

For example, we can find the equation of the line tangent to $f(x) = x^2$ at $x = 3$ as follows. The slope of the tangent line is the derivative. We saw above that $f'(x) = 2x$, so the slope of the tangent line at $x = 3$ is $f'(3) = 2 \cdot 3 = 6$. We also need to know a point on the line. This is the point that is on the curve, so its coordinates are $(3, f(3)) = (3, 3^2) = (3,9)$. Now we use this information to write the equation of the tangent line:

$$\begin{aligned} y - f(x_0) &= f'(x_0)(x - x_0) \\ y - 9 &= 6(x - 3) \end{aligned}$$

LOCAL LINEAR APPROXIMATION

The line tangent to a curve at a point, if it exists, is the line that best approximates the curve near that point. This means that if a function is differentiable at a point, then locally it looks like the tangent line. That is to say, if you zoom in far enough with your graphing calculator, the function looks like a straight line—and the line that it looks like is the tangent line.

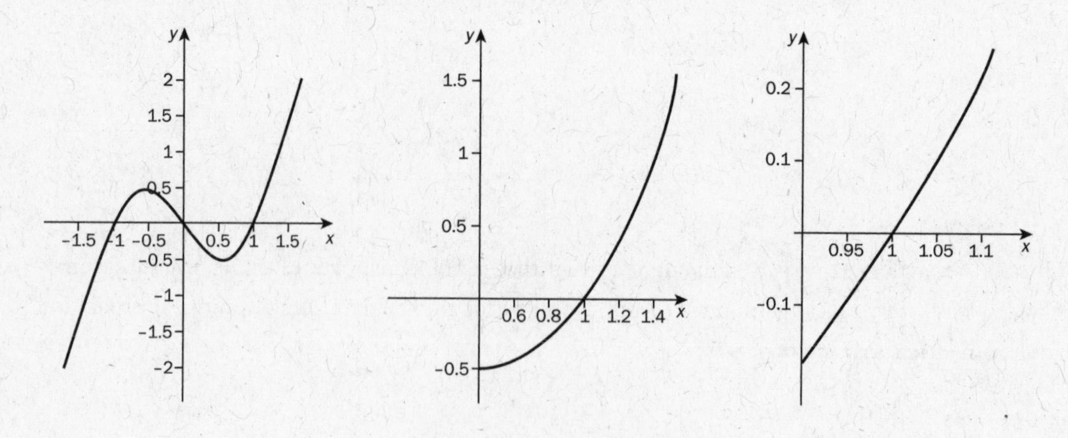

A function does not look like a line locally at a point where it is not differentiable—for instance, if the function has a sharp corner like $y = |x|$.

At points where a function is differentiable, the tangent line can be used to approximate the function. This is similar to the linear interpolation of data, in which we extend observed results by assuming that other data would lie on a straight line between observed data points. Using the tangent line to approximate a function requires a single data point and the direction of the data which is given by the slope of the tangent line.

We saw previously that if $f(x)$ is differentiable at x_0, the equation of the line tangent to $f(x)$ at x_0 is $y - f(x_0) = f'(x_0)(x - x_0)$. If we know $f(x_0)$, then, if x_1 is close to x_0, that is $x_1 - x_0$ is a number close to zero, then we can approximate $f(x_1)$ by the point (x_1, y_1) on the tangent line given by $f(x_1) \approx f(x_0) + f'(x_0)(x_1 - x_0)$:

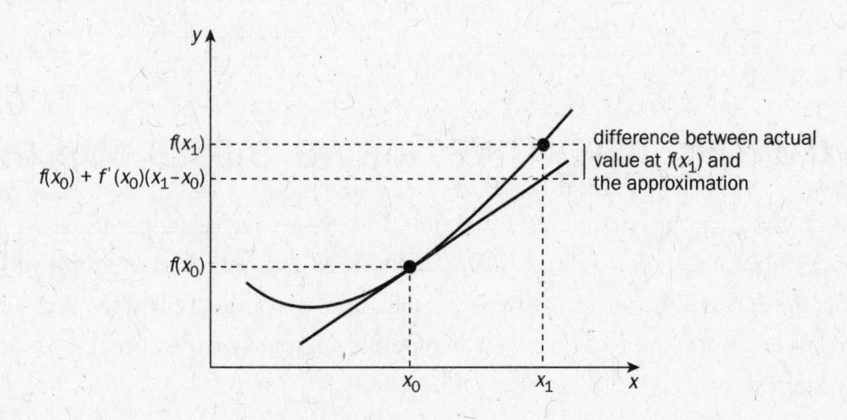

Before calculators existed, local approximations were used to approximate functions. For example, we can approximate $\sqrt{17}$ using the function $f(x) = \sqrt{x}$, and $x_0 = 16$. We start by computing the derivative $f'(16)$:

$$\begin{aligned} f'(16) &= \lim_{h \to 0} \frac{\sqrt{16+h} - \sqrt{16}}{h} \\ &= \lim_{h \to 0} \frac{(\sqrt{16+h} - 4)}{h} \cdot \frac{(\sqrt{16+h} + 4)}{(\sqrt{16+h} + 4)} \\ &= \lim_{h \to 0} \frac{\not{h}}{\not{h}\sqrt{16+h} + 4} = \frac{1}{8} \end{aligned}$$

The line tangent to the curve $f(x) = \sqrt{x}$ at $x_0 = 16$ is:

$$\begin{aligned} y - 4 &= \frac{1}{8}(x - 16) \\ y &= 4 + \frac{1}{8}(x - 16) \end{aligned}$$

Therefore, $\sqrt{17} \approx 4 + \frac{1}{8}(17 - 16) = 4.125$.

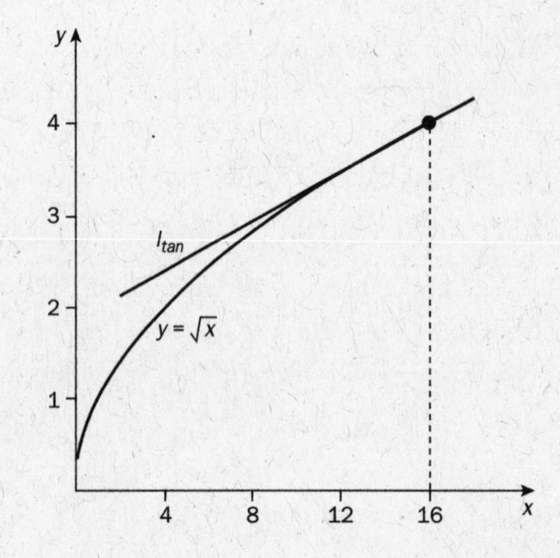

APPROXIMATING DERIVATIVES USING TABLES AND GRAPHS

We often don't have a formula describing a function. Instead, we have data points, as we do when we collect experimental data, for example. We can plot these data points on a graph and sketch a curve that fits the data points. We can also use the data points to approximate the derivative of the curve at a point, by finding the slope between two points near the specified point. We can use the graph to qualitatively analyze the derivative.

Example:

The distance a car has traveled on a straight road at time t is given by $p(t)$, where $t = 0$ is the starting time. A table of values for $p(t)$ in five-second intervals for $0 \leq t \leq 40$ is shown in the table. Find one approximation for the velocity of the car in feet per second at time $t = 25$. Show the computations you use to arrive at your answer.

t(seconds)	$p(t)$(feet)
0	0
5	30
10	80
15	155
20	265
25	440
30	640
35	840
40	1,020

Solution:

We are not given a formula for $p(t)$; rather we have only data points. To approximate the velocity of the car at $t = 25$, we recall that velocity is the rate of change, or derivative, of the position function. The velocity at $t = 25$ is, therefore, the slope of the line tangent to the position curve at this point. To approximate this slope we find the slope of the secant line between two points near $t = 25$. Because $t = 25$ *is* one of the data points, we have three options for approximating the slope at $t = 25$.

The slope between $t = 20$ and $t = 25$, given by $\dfrac{440 - 265}{5} = 35$ feet/second

The slope between $t = 25$ and $t = 30$, given by $\dfrac{640 - 440}{5} = 40$ feet/second

The slope between $t = 20$ and $t = 30$, given by $\dfrac{640 - 265}{10} = 37.5$ feet/second

Because $t = 25$ is midway between $t = 20$ and $t = 30$, the third slope approximation above is called the Symmetric Derivative. The symmetric derivative is the average of the derivative approximations using equally spaced points from the left and from the right of $t = 25$. $\left(37.5 = \dfrac{35 + 40}{2}\right)$.

In general, an approximation of the derivative of a function f at $x = x_0$ is given by the symmetric derivative denoted by: $f'(x_0) \approx \dfrac{f(x_0 + h) - f(x_0 - h)}{2h}$. In the above example,

$f'(25) \approx \dfrac{f(25 + 5) - f(25 - 5)}{2(5)} = \dfrac{f(30) - f(20)}{10} = \dfrac{640 - 265}{10} = 37.5$ feet per second. It is this method that the TI-83/84 calculator uses (with $h = 0.001$) to evaluate a derivative. If we plot these points on a curve and connect the data points we can sketch the graph of $p(t)$:

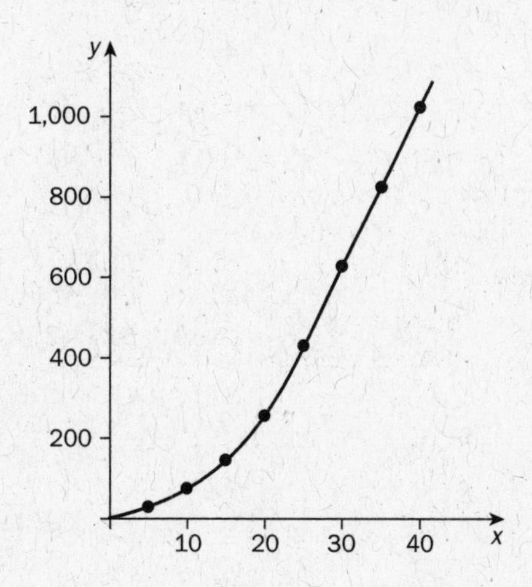

REVIEW QUESTIONS

1. Let $f(x) = \sin x$. If the derivative of f at $x = \frac{\pi}{3}$ is one-half the derivative of f at θ then $\theta =$

   (A) 0

   (B) $\frac{\pi}{6}$

   (C) $\frac{\pi}{4}$

   (D) $\frac{\pi}{3}$

   (E) $\frac{\pi}{2}$

2. The graph of $f(x)$ is shown below with the points A, B, C, D, and E labeled on the graph.

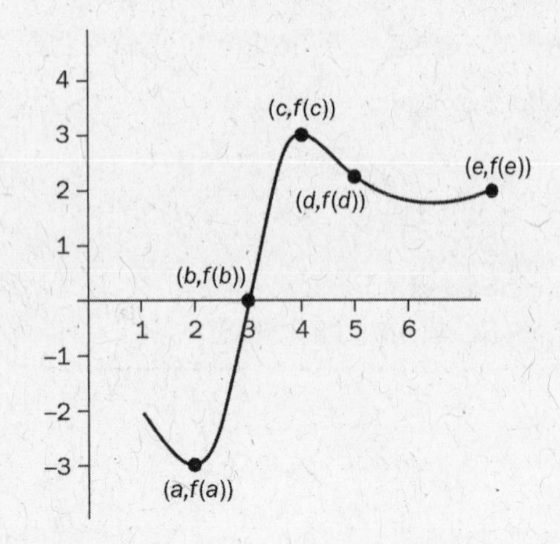

   Which of the lists below is presented in ascending order?

   (A) $f(b), f'(a), f'(d), f'(b)$

   (B) $f(a), f'(c), f'(e), f'(c)$

   (C) $f(a), f'(c), f'(b), f'(d)$

   (D) $f(a), f'(b), f'(c), f(d)$

   (E) $f'(a), f'(c), f(b), f'(d)$

3. The graph of $f(x)$ is shown below.

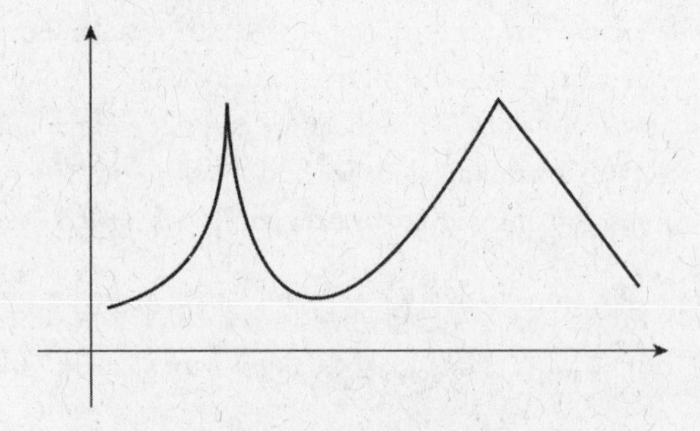

   Which of the following statements is true?

   (A) f is differentiable everywhere.

   (B) f has two cusps.

   (C) f has one cusp.

   (D) The graph of f has two vertical tangent lines.

   (E) $f'(x)$ is defined everywhere.

4. Suppose $f(x)$ is a differentiable function with $f(1) = 2, f(2) = -2, f'(2) = 5, f'(1) = 3$, and $f(5) = 1$. An equation of a line tangent to the graph of f is

   (A) $y - 3 = 2(x - 1)$

   (B) $y - 2 = (x - 1)$

   (C) $y - 3 = 5(x - 1)$

   (D) $y - 2 = 3(x - 1)$

   (E) $y - 1 = 5(x - 2)$

5. Which of the following expressions gives the derivative of $f(x) = \cos 2x$?

   (A) $f'(x) = \lim\limits_{x\to 0} \frac{\cos 2x}{2x}$

   (B) $f'(x) = \lim\limits_{x\to 0} \frac{\cos 2(x+h) - \cos 2x}{2x}$

   (C) $f'(x) = \lim\limits_{h\to 0} \frac{\cos 2(x+h) - \cos 2x}{h}$

   (D) $f'(x) = \lim\limits_{h\to 0} \frac{\cos (2x+h) - \cos (2x-h)}{h}$

   (E) $f'(x) = \lim\limits_{h\to 0} \frac{\cos 2x - \cos h}{h}$

6. If $f(x) = 4x + 1$, then $f'(-2) =$
   (A) −7
   (B) −2
   (C) 1
   (D) 4
   (E) None of these

7. If $f(x) = |x + 1|$, then $f'(-5) =$
   (A) −5
   (B) −1
   (C) 0
   (D) 1
   (E) 5

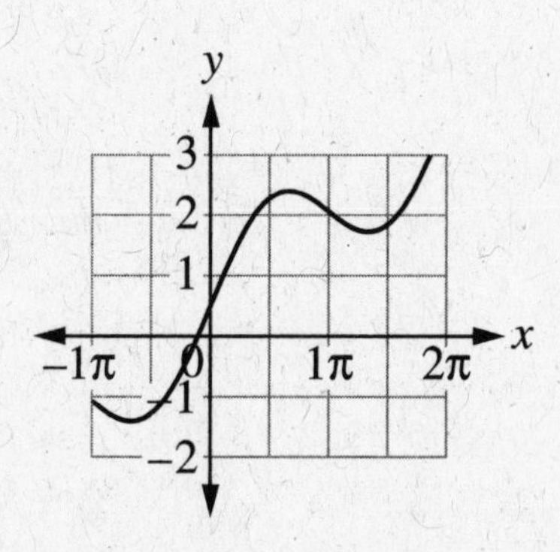

8. Consider the graph of $g(x)$ shown above. Which of the following statements is true?

   (A) $g'(0) < g'(\frac{\pi}{2}) < g'(\pi)$
   (B) $g'(\frac{\pi}{2}) < g'(\pi) < g'(0)$
   (C) $g'(\pi) < g'(\frac{\pi}{2}) < g'(0)$
   (D) $g'(\frac{\pi}{2}) < g'(0) < g'(\pi)$
   (E) $g'(0) < g'(\pi) < g'(\frac{\pi}{2})$

9. The function f is given by

$$f(x) = \begin{cases} x \text{ if } -3 < x < 3 \\ \dfrac{|x|}{x} \text{ if } |x| \geq 3 \end{cases}$$

Which of the following expressions is largest?

   (A) $f(-4)$
   (B) $f(0)$
   (C) $f'(0)$
   (D) $f'(4)$
   (E) $f'(5)$

This graph is for problems 10 and 11.

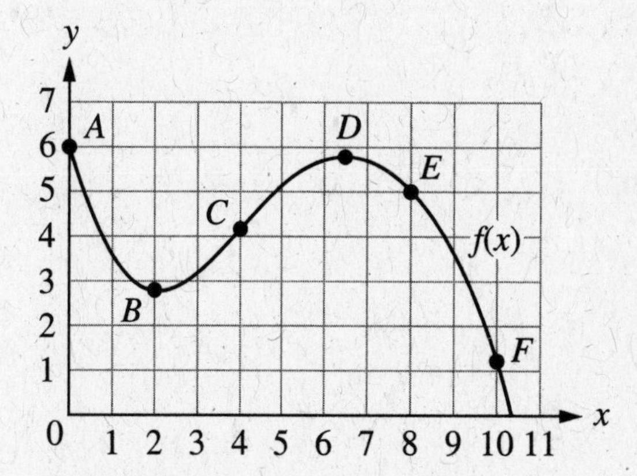

The points labeled A through F lie on the curve $f(x)$.

10. At which point(s) is $f'(x) = 0$?
   (A) B only
   (B) B and D
   (C) C and D
   (D) E and F
   (E) A, E, and F

11. At which point(s) is $f(x)f'(x) < 0$?

(A) A only
(B) B only
(C) C and D
(D) E and F only
(E) A, E, and F

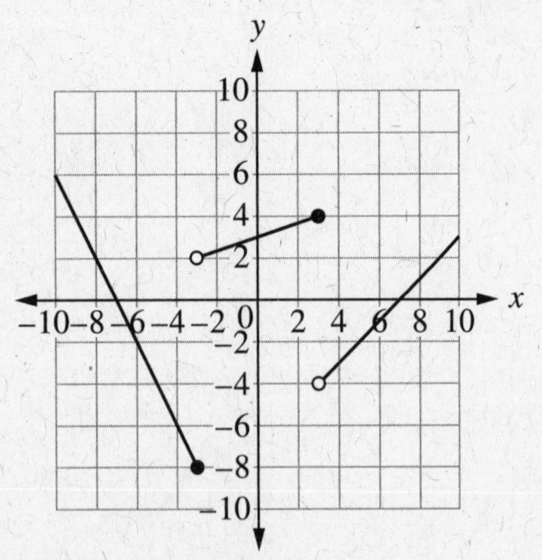

The graph of f(x)

12. What is the value of $f'(-5) + 2f'(-1) + 3f'(5)$?

(A) $-\frac{2}{3}$

(B) $\frac{1}{3}$

(C) $\frac{2}{3}$

(D) 1

(E) $\frac{5}{3}$

13. The line $2x - y = 9$ is tangent to the curve $f(x)$ at the point $(4, -1)$. What is the value of $f'(4)$?

(A) −2

(B) $\frac{1}{2}$

(C) 2

(D) 4

(E) 9

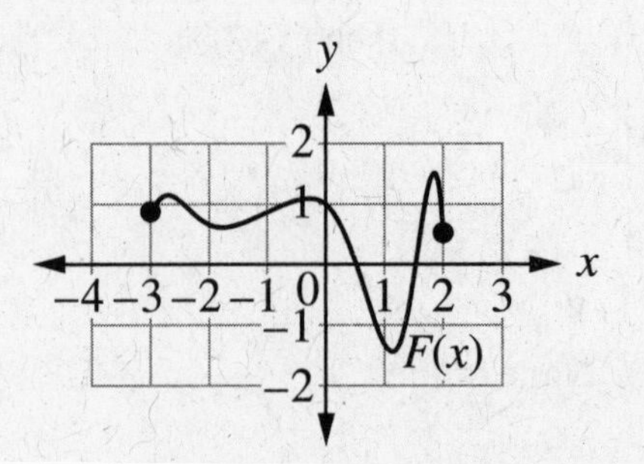

The graph of F(x)

14. At how many distinct points on the curve is $F'(x) = 0$?

(A) 2
(B) 3
(C) 4
(D) 5
(E) 6

15. If $f(x) = e^x$, which of the following is equal to $f'(e)$?

(A) $\lim_{h\to 0} \frac{e^{x+h}}{h}$

(B) $\lim_{h\to 0} \frac{e^{x+h} - e^e}{h}$

(C) $\lim_{h\to 0} \frac{e^{e+h} - e}{h}$

(D) $\lim_{h\to 0} \frac{e^{x+h} - 1}{h}$

(E) $\lim_{h\to 0} \frac{e^{e+h} - e^e}{h}$

FREE-RESPONSE QUESTIONS

16. The graph of a function f, shown below, consists of three line segments. Suppose $g(x)$ is a function whose derivative is f.

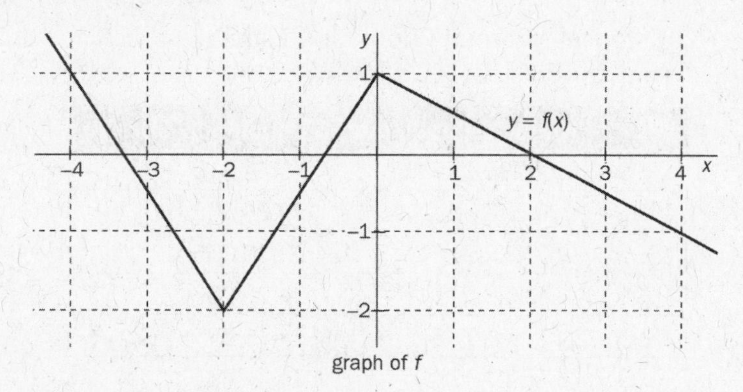

graph of f

(a) Find $g'(-1)$, $g'(1)$, $g''(-1)$, and $g''(1)$.

(b) Suppose $g(1) = 4$. Give an equation for the tangent line to the graph of $g(x)$ at $x = 1$.

(c) Describe the shape of the graph of $g(x)$ near $x = 2$.

(d) Give a piecewise defined equation for $g''(x)$.

17. You may use a calculator for this question.

During the course of a 15-hour storm, the water levels of a reservoir are measured. In addition, some data about the rate of change of the water level is collected. The data is summarized in the table below. Assume that h and h' are both continuous and differentiable functions of t for $0 \le t \le 15$.

Time, t (hours)	2	3	8	12	15
Water level $h(t)$ (feet)	428	432	457	477	483
Rate of change $h'(t)$ (feet/hour)	**	$3.5\,\frac{\text{ft}}{\text{h}}$	$4.3\,\frac{\text{ft}}{\text{h}}$	**	$6.4\,\frac{\text{ft}}{\text{h}}$

(a) Write the equation for the tangent line to the graph of h at $t = 3$.

(b) Compute the average rate of change over the interval [2,15]. Using the Intermediate Value Theorem and the data for $h'(t)$ determine in which time intervals there must be a time where the instantaneous rate of change is equal to the average rate of change.

(c) Is the data collected about the rate of change of the water level, $h'(t)$, consistent with the statement that $h''(t) > 0$ on the interval $2 < t < 15$? Explain your answer.

(d) Is the data collected about the water level, $h(t)$, consistent with the statement that $h''(t) > 0$ on the interval $0 \le t \le 15$?Explain your answer.

18. A table of historical national population estimates in the United States from 1990 to 1999 is shown below.

Historical National Population Estimates: 1990–1999*

Year	National Population	Population Change
1999	272,690,813	2,442,810
1998	270,248,003	2,464,396
1997	267,783,607	2,555,035
1996	265,228,572	2,425,296
1995	262,803,276	2,476,255
1994	260,327,021	2,544,413
1993	257,782,608	2,752,909
1992	255,029,699	2,876,607
1991	252,153,092	2,688,696
1990	249,464,396	2,645,166

* Source: Population Estimates Program, Population Division, U.S. Census

Bureau Internet Release Date: April 11, 2000
Revised date: June 28, 2000 http://www.census.gov/popest/data/intercensal/national/index.html

(a) Use the data on the table to approximate the rate of change of the U.S. population in 1999. Show the computations that lead to your answer. Indicate the units of measure.

(b) Use your answer from part (a) to estimate the U.S. population in the year 2000. Show the computations that lead to your answer.

ANSWERS AND EXPLANATIONS

1. A

To find the solution, remember that the derivative of $\sin x = \cos x$. Therefore, $f'\left(\frac{\pi}{3}\right) = \cos\frac{\pi}{3} = \frac{1}{2}$. Then,

$$\frac{1}{2} = \frac{1}{2}\cos\theta$$
$$1 = \cos\theta$$
$$0 = \theta$$

There are an infinite number of angles with cosine 1, but of the answer choices, only the cosine of (A)—$\theta = 0$—satisfies $\cos\theta = 1$.

2. B

To find the correct answer choice, we can use the graph to eliminate answer choices.

Choice (A) can be eliminated because the graph shows us that f has a horizontal tangent at $x = a$, so $f'(a) = 0$. The slope of the curve at d is negative, so $f'(d) < 0$ and the list in (A) is therefore not in ascending order.

Choice (B) is the correct choice. On the graph, we see that $f(a) < 0$. The graph has a horizontal tangent at $x = c$, so $f'(c) = 0$. The slope of the curve is slightly positive at $x = e$. If we visualize the line tangent to the curve at this point, we can determine that the slope of the line is less than 1, so $0 < f'(e) < 1$. From the graph we can see that $f(c) = 3$. Therefore, this list is in ascending order, and (B) is the correct answer choice.

Choice (C) can be eliminated. From the graph we see that the graph has a horizontal tangent at $x = a$; therefore $f'(a) = 0$. The slope of the graph is negative at $x = d$, so $f'(d) < 0$. Therefore, this list is not in ascending order.

Choice (D) can be eliminated. From the graph we see that the slope of the curve is positive at $x = b$, so $f'(b) > 0$. The graph has a horizontal tangent at $x = c$, so $f'(c) = 0$. Therefore, this list is not in ascending order.

Choice (E) can be eliminated. From the graph we see that $f(b) = 0$. The slope of the graph is negative at $x = d$, so $f'(d) < 0$. Therefore, this list is not in ascending order.

3. C

We can see from the graph that there are two points at which f is not differentiable, but that only one of them is a cusp according the definition of a cusp (i.e., a point where f has a vertical tangent line). The answer is therefore (C).

Even without knowing the definition of a cusp, we can arrive at the answer by eliminating choices. The graph of f is defined everywhere, but f has sharp corners. Therefore, f is not differentiable everywhere on its domain, because a function is not differentiable at a point where its graph is not smooth. We can thus eliminate choice (A). Choice (E) is equivalent to choice (A), so we can eliminate this choice as well.

Because a cusp is defined as a point at which the graph has a vertical tangent, choices (B) and (D) are equivalent. If one is true, then both are true, so we can eliminate these choices. We are left with only one possibility—(C).

4. D

The equation of the line tangent to the graph of $f(x)$ at the point $x = a$ is given by $y - f(a) = f'(a)(x - a)$.

We are given information about three values of $x: x = 1,\ x = 2$, and $x = 5$. It is difficult to try to retrofit the answer choices to the equation of a tangent line, so we use the given data to write the equations of tangent lines. To find the equation of a tangent line at *a*, we must know both the value of the *function* at *a* and the value of the *derivative* at *a*. This is clear if we organize the given information in a table:

a	$f(a)$	$f'(a)$	$y - f(a) = f'(a)(x - a)$ equation of the tangent line at $x=a$
1	2	3	$y - 2 = 3(x - 1)$
2	−2	5	$y + 2 = 5(x - 2)$
5	1	?	not enough information

Now we can easily match the information on the table with the answer choices. Only one of the answer choices matches the equation of a tangent line on the table—(D).

5. C

The easiest way to approach this problem is to write the definition of the derivative $(\cos 2x)'$ as the limit of a difference quotient and see if the expression matches one of the answer choices or is equivalent to one of the answer choices. We write the definition:

$$f'(x) = \lim_{h \to 0} \frac{\cos 2(x + h) - \cos 2x}{h}$$

This matches choice (C).

6. D

The graph of f is linear with a slope equal to 4 at all x. Therefore, $f'(-2) = 4$.

7. B

The graph of $f(x) = |x + 1|$ is v-shaped, consisting two rays. To the left of $x = -1$, the ray has a slope equal to −1 so it follows that $f'(-5) = -1$.

8. C

We need to imagine tangents drawn at $x = 0$, $x = \frac{\pi}{2}$, and $x = \pi$ and consider the slopes of these Tangents, which equal $g'(0)$, $g'\left(\frac{\pi}{2}\right)$, and $g'(\pi)$ respectively. We approximate from the graph and get $g'(0) \approx 2.5$, $g'\left(\frac{\pi}{2}\right) \approx 1$, and $g'(\pi) \approx \frac{-1}{2}$ so it follows that $g'(\pi) < g'\left(\frac{\pi}{2}\right) < g'(0)$.

9. C

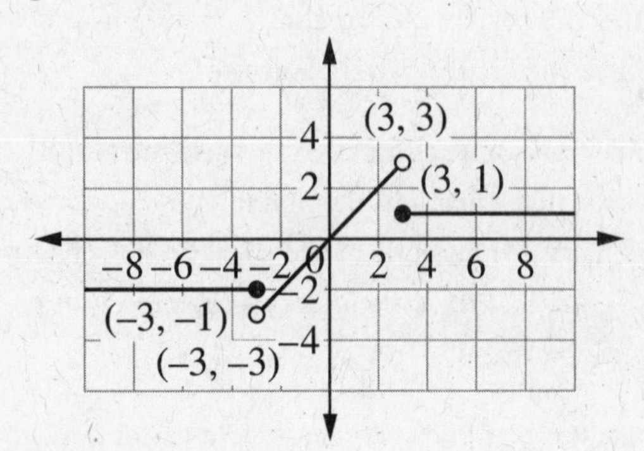

Let's consider the graph of f. From the graph, we evaluate each and get

$f(-4) = -1$
$f(0) = 0$
$f'(0) = 1$
$f'(4) = 0$
$f'(5) = 0$

10. B

Imagine tangents drawn at the labeled points. The tangents at B and D are horizontal with slopes of 0. Hence $f'(x) = 0$ at points B and D.

11. E

Remember, a negative times a positive is a negative. All of the labeled points are above the x-axis, which makes $f(x) > 0$. Now imagine tangents drawn at the labeled points. Only the tangents at A, E, and F have negative slopes making $f(x) \cdot f'(x) < 0$ at those points.

12. E

From the graph, it follows that

$f'(-5) = -2$, $f'(-1) = \frac{1}{3}$, and $f'(5) = 1$. With this, it follows that

$$f'(-5) + 2f'(-1) + 3f'(5) = -2 + 2 \cdot \frac{1}{3} + 3 \cdot 1 = \frac{5}{3}$$

13. C

The line $2x - y = 9$ in slope-intercept form is $y = 2x - 9$ making its slope 2. Since the line is tangent to f at $(4, -1)$, it follows that $f'(4) = 2$

14. D

Imagine tangents to the curve that are horizontal. From the graph below, there are five horizontal tangents making $f'(x) = 0$.

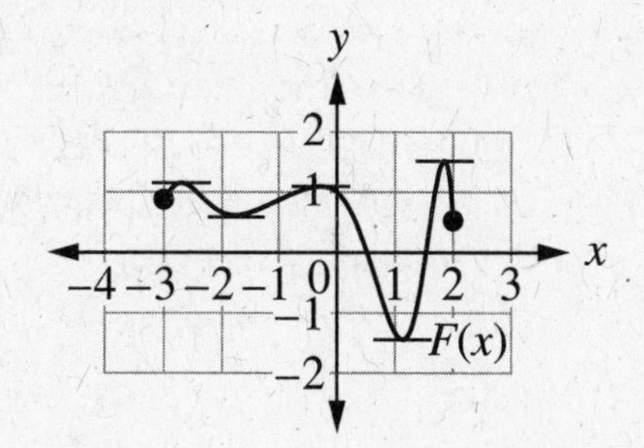

15. E

By the definition of the derivative at a point, $f'(a) = \lim_{h \to 0} \frac{f(a+h) - f(a)}{h}$, we substitute and get

$$f'(e) = \lim_{h \to 0} \frac{e^{e+h} - e^e}{h}.$$

FREE-RESPONSE ANSWERS

16. (a) $g'(-1) = -\frac{1}{2}$, $g'(1) = \frac{1}{2}$, $g''(-1) = \frac{3}{2}$, and $g''(1) = -\frac{1}{2}$

(b) The equation of the tangent line is given by the equation $y - g(x_0) = g'(x_0)(x - x_0)$.

At $x = 1$ we are given that $g(1) = 4$, so the line is $y - 4 = \frac{1}{2}(x - 1)$.

(c) Before $x = 2$, the derivative of g is positive, so the function is increasing. At $x = 2$ the derivative is 0, so the graph has leveled off. After $x = 2$, the derivative is negative, so the graph is decreasing. The graph looks something like:

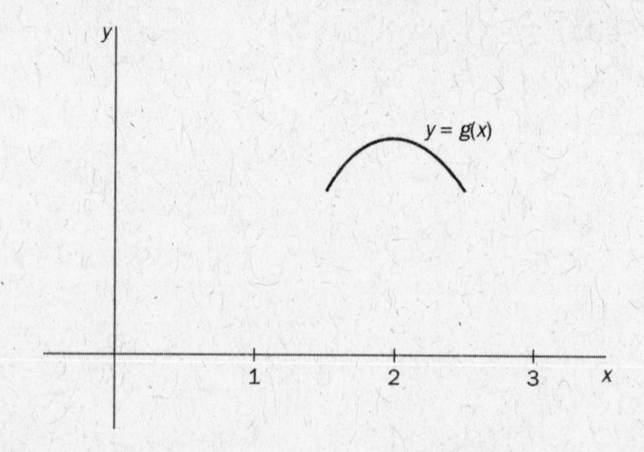

(d) f is the derivative of g, so the second derivative of g is f' and

$$g''(x) = f'(x) = \begin{cases} -\frac{3}{2} & \text{if } -4 \le x < -2 \\ \frac{3}{2} & \text{if } -2 < x < 0 \\ -\frac{1}{2} & \text{if } 0 < x \le 4 \end{cases}$$

Remember that f is giving you the derivative of g, so $g'(-1) = f(-1) = -\frac{1}{2}$ and so forth. For part (b) you do not need to (and should not) simplify the equation for the tangent line. A good choice is to use the point-slope formula, but use whatever formula you remember and leave it in whatever form you start with. For part (d) we are told that f consists of three line segments so all we need in order to write an equation of f' is the slopes of those lines.

17. (a) The equation for the tangent line is given by $y - h(t_0) = h'(t_0)(t - t_0)$. When $t_0 = 3$, the equation is $y - 438 = 3.5(t - 3)$.

(b) The average rate of change over the interval [2,15] is

$\dfrac{h(15) - h(2)}{15 - 2} = \dfrac{483 - 428}{13} = 4.231$ feet/hour. We assume h' is continuous on $0 \leq t \leq 15$ and because $h'(3) = 3.5$ feet/hour and $h'(8) = 4.3$ feet/hour, the Intermediate Value Theorem guarantees that $h'(t_0) = 4.231$ feet/hour for some t_0 in the interval [3,8]. There are no other intervals where IVT guarantees that the instantaneous rate of change is equal to the average rate of change.

(c)

Time, t (hours)	2	3	8	12	15
Rate of change $h'(t)$(feet/hour)	**	$3.5\,\frac{\text{ft}}{\text{h}}$	$4.3\,\frac{\text{ft}}{\text{h}}$	**	$6.4\,\frac{\text{ft}}{\text{h}}$

$\Delta h'(t)$		$0.8\,\frac{\text{ft}}{\text{h}}$	$2.1\,\frac{\text{ft}}{\text{h}}$	
$\Delta h'(t)$/hour		$0.16\,\frac{\text{ft}}{\text{h}}$	$0.3\,\frac{\text{ft}}{\text{h}}$	

Our approximation of h'' by the slope of the secant line between two values of h' gives two values, which are increasing. This is consistent with the statement that $h''(t) > 0$ on the interval $2 < t < 15$.

(d)

Time, t (hours)	2	3	8	12	15
Water level $h(t)$(feet)	428	432	457	477	483

$\Delta h(t)$		4	25	20	6	
$\Delta h(t)$/hour		4	5	5	2	

Our approximation of h' by the slope of the secant line between two values of h gives four values, not all of which are increasing. The approximation of h' [12,15] is less than the approximation of h' on the interval [8,12]. This is not consistent with the statement that $h''(t) > 0$ on the interval $2 < t < 15$.

18. (a) We can approximate the population's rate of change in 1999 by taking the slope between population data points near 1999. We use the points $x = 1999$ and x and the data on the chart. The slope between those two points is given by

$$\frac{\Delta\text{population (in people)}}{\Delta\text{time (in years)}} = \frac{2,442,810}{1} = 2,442,810 \text{ people/year}$$

(b) We assume that the function giving the population p at time t is differentiable. The answer in part (a) gives an approximation to the slope of the line tangent to the population function $p(t)$ at $t = 1999$. We use this slope and the data in the table to approximate the tangent line to population curve at $t = 1999$. The equation of the tangent line is:

$$y - 272,690,813 = 2,442,810(t - 1999)$$
$$y = 272,690,813 + 2,442,810(t - 1999)$$

The population in the year 2000 can be approximated by the point on the tangent line with the t-coordinate equal to 2000. We evaluate the tangent line at $t = 2000$ to obtain:

$$p(2000) \approx 272,690,813 + 2,442,810(2000 - 1999)$$
$$p(2000) \approx 275,133,623$$

From our calculation, we estimate that the population of the United States in the year 2000 was 275,133,623.

CHAPTER 14: THE DERIVATIVE AS A FUNCTION

IF YOU LEARN ONLY THREE THINGS IN THIS CHAPTER . . .

1. If f is defined and continuous on a given interval (a, b), then

   f is increasing on intervals where $f'(x) > 0$.

   f is decreasing on intervals where $f'(x) < 0$.

   f is constant on intervals where $f'(x) = 0$.

2. If f is continuous on the closed interval $[a, b]$, then f has both an absolute maximum and an absolute minimum on $[a, b]$.

3. **The First Derivative Test:** Suppose $f(x)$ has a critical point at x_0.

   If $f'(x) > 0$ on some open interval to the left of x_0 and $f'(x) < 0$ on some open interval to the right of x_0, then $f(x_0)$ is a relative maximum.

   If $f'(x) < 0$ on some open interval to the left of x_0 and $f'(x) > 0$ on some open interval to the right of x_0, then $f(x_0)$ is a relative minimum.

   If the sign of $f'(x)$ does not change on some interval containing x_0, then f does not have a relative extremum at x_0.

In Chapter 13, we saw that the derivative of a function $f(x)$ is also a function, called $f'(x)$ or $\frac{df}{dx}$. Its value at each point x in the domain of f is the slope of the line tangent to f at x (where there is a tangent line and it is not vertical). This slope is also the instantaneous rate of change of f at x.

Because f' is a function *derived* from f, there is a relationship between the two functions. That is, we can learn about one of the pair by analyzing the other.

INCREASING AND DECREASING BEHAVIOR OF *f* AND THE SIGN OF *f′*

If we examine the graph of a function f, in some places it looks like it is "going up" and in some places it looks like it is "going down." We can define these concepts precisely:

- We say that f is increasing or strictly increasing on an interval (a,b) if for all $x, y \in (a,b)$ where $x > y$, then $f(x) > f(y)$.
- We say that f is decreasing or strictly decreasing on an interval (a,b) if for all $x, y \in (a,b)$ where $x > y$, then $f(x) < f(y)$.
- We say that f is constant on an interval (a,b) if for all $x, y \in (a,b)$ where $x > y$, then $f(x) = f(y)$.

A function f is non-decreasing on an interval if for all $x, y \in (a,b)$ where $x > y, f(x) \geq f(y)$. A function f is nonincreasing on an interval if for all $x, y \in (a,b)$ where $x > y, f(x) \leq f(y)$. We make a distinction between functions that are strictly increasing (or decreasing) and those that are non-decreasing (or non-increasing). For example, the constant function $f(x) = 3$ is not strictly increasing, but it is non-decreasing:

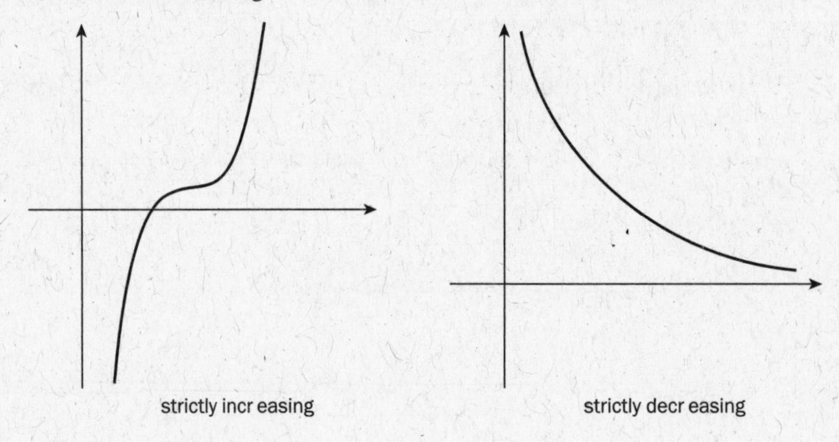

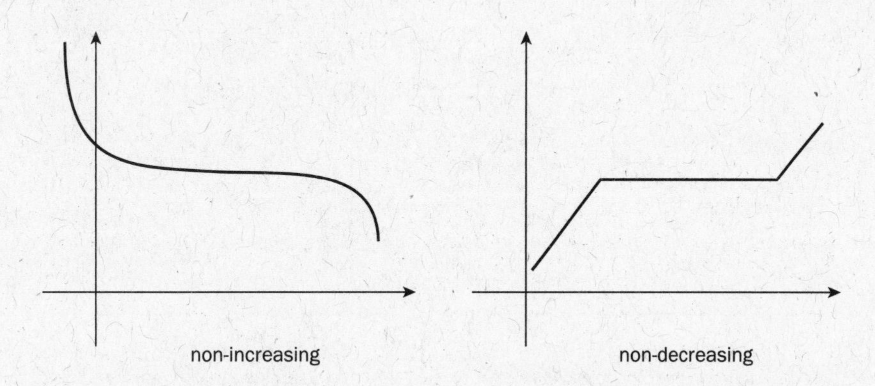

Now we'll relate these concepts to the derivative f'. We'll assume that the function f is continuous and differentiable (at least on the intervals we are talking about).

If f is increasing on an interval (a,b), then for all $x \in (a,b)$ and Δx small enough, the difference quotient is positive, i.e., $\frac{f(x+\Delta x) - f(x)}{\Delta x} > 0$.

Therefore, the limit of the difference quotient (which, remember, is $f'(x)$) is greater than or equal to zero.

Theorem:

If f is defined and continuous (at least on the intervals we are talking about), then

- f is increasing on intervals where $f'(x) > 0$.
- f is decreasing on intervals where $f'(x) < 0$.
- f is constant on intervals where $f'(x) = 0$.

To apply this theorem to f on an interval (a,b), then, f must be defined on $[a,b]$ and continuous on (a,b).

The theorem above says that if the derivative is strictly greater than zero on an interval, then we are certain that f is increasing on that interval. But a function can be increasing on an interval even if the derivative occasionally equals zero in that interval. For example, the function $f(x) = x^3$ is strictly increasing everywhere, even though $f'(0) = 0$.

CRITICAL POINTS AND LOCAL EXTREMA

We just defined and analyzed the increasing and decreasing behavior of a function. We are also interested in studying where a function f changes direction from increasing to decreasing, or vice versa.

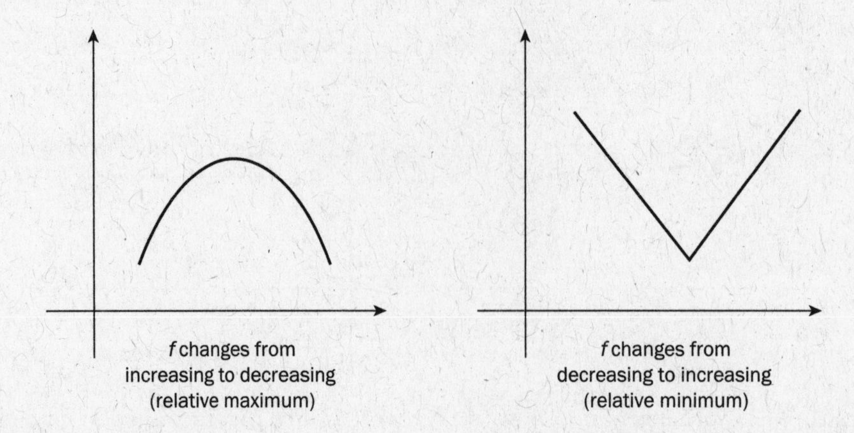

When we examine the graph of a function f, we may see points on the graph of f that are higher than all other nearby points and points that are lower than all other nearby points. These points are called *local extrema,* because locally—meaning if we stay close enough to the point—it is the highest or lowest point on the graph. Local extrema are also called *relative extrema,* because they are the highest or lowest relative to nearby points. Here are some definitions that make these ideas more precise:

- A function f has a local maximum or relative maximum at x_0 if there is an open interval containing x_0 on which $f(x_0) \geq f(x)$ for all x in the interval. The local maximum is the value of the function at x_0, $f(x_0)$.
- A function f is said to have a local minimum or relative minimum at x_0 if there is an open interval containing x_0 on which $f(x_0) \leq f(x)$ for all x in the interval. The local minimum is the value of the function at x_0, $f(x_0)$.
- The points at which f has either a local maximum or a local minimum are called local extrema or relative extrema (sing. *extremum*) and the values of the function at these points are the relative extrema.

A function may not have any relative extrema. For example, linear functions such as $y = 2x + 3$ have no relative extrema. The function $y = x^3$ also has no relative extrema. It is strictly increasing.

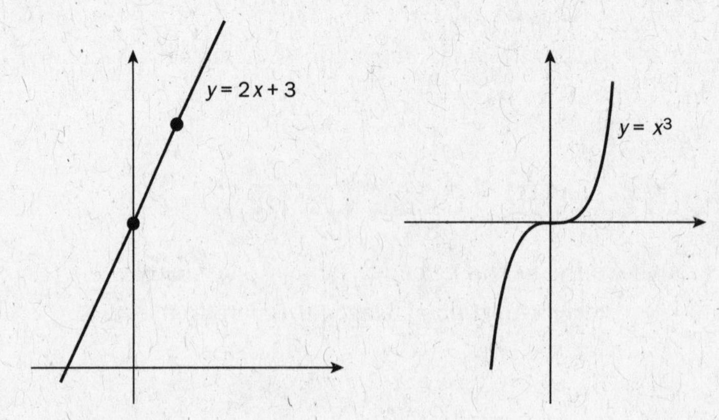

If a function's derivative does not change sign, then it has no relative extrema. For example, if we examine the graph of $f(x) = \sqrt{x}$, we can see that the function has no local extrema. The derivative of this function is $f'(x) = \frac{1}{2\sqrt{x}}$, which is always greater than zero (on its domain).

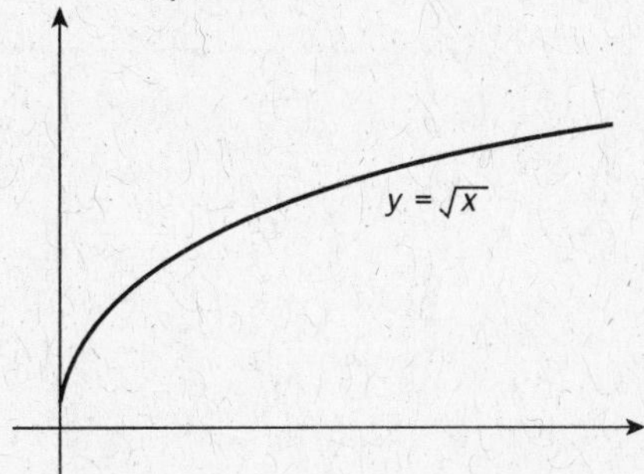

To understand a function, we need to identify its local extrema. The graph of a function provides an approximate idea of where the relative extrema occur, but we need a computational tool to identify the extreme values precisely.

The following theorem is a consequence of the Intermediate Value Theorem from Chapter 9. It shows the connection between the local extrema of f and the derivative of f.

Theorem:

Suppose f has a relative extremum at x_0. If f is defined on an open interval containing x_0, then either $f(x_0) = 0$ or $f(x_0)$ is undefined, i.e., f is not differentiable at x_0.

Because the relative extrema of a continuous function f occur only where the derivative is zero or undefined, the only points of f that could possibly be relative extrema are the points at which f is zero or undefined. This set of points plays an important role in understanding the behavior of f; we call them *critical points* or *critical numbers*. The set of points at which f has relative extrema is a subset of the set of critical points.

Not all critical points are points at which f has relative extrema. For example, consider the function $f(x) = x^3$. Because $f'(x) = 3x^2$, we can see that $f'(0) = 0$. Therefore $f(x) = x^3$ has a critical point at $x = 0$. The function $f(x)$ is always increasing, so $f(x) = x^3$ has no local extrema. We can also see this by looking at the graph of $f(x) = x^3$.

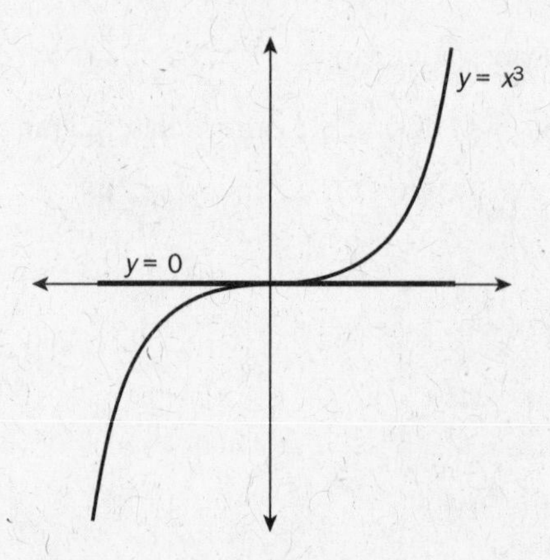

AP EXPERT TIP

The presence of a critical number does not mean that function has a relative extrema. This is one of the most common mistakes students make! Make sure you investigate the function's behavior to the left and right of your critical number to make a decision.

Even though f does not have a relative extremum at $x = 0$, it is still an important point on the graph because the graph of $f(x) = x^3$ has a horizontal tangent there.

The graphs below show several possibilities for the behavior of a graph at a relative extrema.

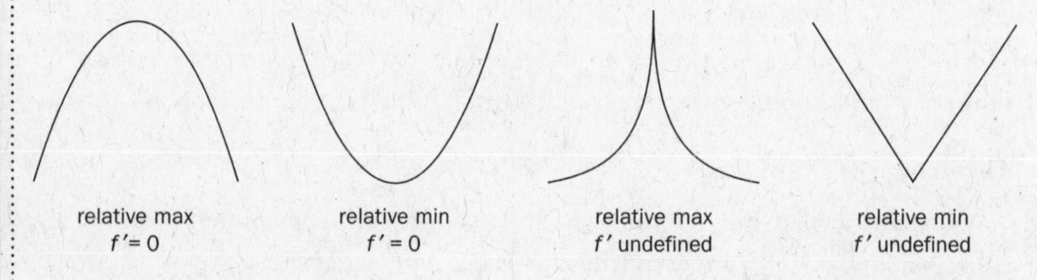

ABSOLUTE EXTREMA

Sometimes (but not always) a function will have a single biggest value on an interval. We say that the function f has an absolute maximum on an interval (a,b) if for all points in the interval, $f(x_0) \geq f(x)$. We say that f has an absolute minimum on (a,b), if for all points in the interval $f(x_0) \leq f(x)$. The absolute maximum or minimum is the value of the function at the point x_0. We call an absolute maximum or an absolute minimum an *absolute extremum*. Some calculus texts use the term global instead of absolute, but their meanings are interchangeable.

While an arbitrary function may or may not have an absolute extremum on an arbitrary interval, the following theorem guarantees the existence of an absolute extremum for a *continuous* function on a *closed* interval.

Theorem:

If f is continuous on the closed interval $[a,b]$, then f has both an absolute maximum and an absolute minimum on $[a,b]$. The absolute extrema occur either at critical points or at the endpoints of the interval.

Example:

> Let $f(x) = e^x \sin x$. Find the absolute maximum and minimum value of f on the interval $[0, \pi]$.

Solution:

Because $f(x) = e^x \sin x$ is a continuous function and the interval $[0, \pi]$ is a closed interval, we are guaranteed that f has an absolute maximum and minimum on $[0, \pi]$. The absolute extrema occur either at $[0, \pi]$ or at a critical point of f that lies in the interval. To find the critical points of f we compute its derivative (using the product rule from Chapter 12):

$$f(x) = e^x \sin x + e^x \cos x$$

This derivative is defined and continuous everywhere, so the only critical points occur where $f = 0$. We solve this equation to find:

$$f(x) = e^x \sin x + e^x \cos x = 0$$
$$e^x(\sin x + \cos x) = 0$$

$$e^x \neq 0 \text{ (never)} \quad \text{or} \quad \sin x = -\cos x$$

On the interval $[0, \pi]$, $\sin x = -\cos x$ only at the point $x = \frac{3\pi}{4}$, thus the only critical point of f on $[0, \pi]$ is $x = \frac{3\pi}{4}$.

To determine which of the three candidates, 0, π, or $\frac{3\pi}{4}$ are the absolute extrema on the interval, we plug each of them into the function; the extrema *must* occur at one of these three points.

x	0	π	$\frac{3\pi}{4}$
$f(x) = e^x \sin x$	$e^0 \sin 0 = 0$	$e^{\pi} \sin \pi = 0$	$e^{\frac{3\pi}{4}} \sin \frac{3\pi}{4} \approx 7.460$

From the table of values, we conclude that the absolute minimum of f on $[0,\pi]$ is 0, and this minimum occurs at the points $x = 0$ and $x = \pi$; the absolute maximum of f on $[0, \pi]$ is approximately 7.460 and occurs at the point $x = \frac{3\pi}{4}$.

To find the absolute maximum and minimum values of a function on a closed interval, make a table of "important points"—the endpoints of the interval and any

AP EXPERT TIP

Sometimes you will be asked to find an absolute extremum on an open interval (a, b). Use this theorem to justify under a specific condition:

Suppose f is continuous on an open interval (a, b), and f has exactly one relative extremum on (a, b) at $x = x_0$.

If $x = x_0$ is a relative minimum, then it is the absolute minimum on (a, b).

If $x = x_0$ is a relative maximum, then it is the absolute maximum on (a, b).

critical points that occur in the interval. Evaluate the function at each of these important points. The greatest value is the maximum; the least is the minimum.

THE FIRST DERIVATIVE TEST

If we cannot examine the graph of a function, we still need a method for determining whether a critical point is a local maximum, local minimum, or neither. If f' changes sign at a critical point, then f has a relative extremum at that point. We use the following theorem, called the First Derivative Test, to identify whether a critical point is a relative maximum, minimum, or neither. In the theorem, we assume that the function f is continuous in the neighborhood of a critical point x_0.

THEOREM: THE FIRST DERIVATIVE TEST

Suppose $f(x)$ has a critical point at x_0.

- If $f'(x) > 0$ on some open interval to the left of x_0 and $f'(x) < 0$ on some open interval to the right of x_0, then $f(x_0)$ is a relative maximum.
- If $f'(x) < 0$ on some open interval to the left of x_0 and $f'(x) > 0$ on some open interval to the right of x_0, then $f(x_0)$ is a relative minimum.
- If the sign of $f'(x)$ does not change on some interval containing x_0, then f does not have a relative extremum at x_0.

To apply the First Derivative Test, it is useful to make a chart. This chart is used to do a sign analysis of $f'(x)$ to determine where the derivative is positive, negative, and zero. We can do the analysis by checking the sign of $f'(x)$ on some test point in each of the intervals described in the test. It follows from the Intermediate Value Theorem that the sign of the derivative will be constant on each interval. This chart is also used to determine where $f(x)$ changes from increasing to decreasing, or vice versa, because these changes can only occur at a critical point for a continuous function.

Example:

Consider the function $y = x^2(x - 3)$.

(a) Find the critical points of the function.

(b) Do a sign analysis and make a sign graph for y' and indicate where the original function is increasing and decreasing. Label each line.

(c) Find the coordinates of all relative maxima.

(d) Find the coordinates of all relative minima.

Solution:

Hint: Always expand polynomials when possible. It's easier than applying the product rule.

(a) Expand the function: $y = x^2(x - 3) = x^3 - 3x^2$
Take the derivative of y and set it equal to zero: $y' = 3x^2 - 6x = 0$
Find the solutions by factoring: $y' = 3x(x - 2) = 0$
Critical Points: $x = 0$ or $x = 2$

(b) Now we do a sign analysis to answer this question:

Interval	$(-\infty, 0)$	$(0, 2)$	$(2, \infty)$
Test Point	$x = -1$	$x = 1$	$x = 4$
Sign of f'	$f'(-1) = (-3)(-3) = 9$ Positive	$f'(1) = (3)(-1) = -3$ Negative	$f'(4) = (12)(2) = 24$ Positive
Conclusion	f increasing	f decreasing	f increasing

Caution: **Sign graphs and charts must be clearly and accurately labeled. If they are not labeled or are labeled in an ambiguous manner, they will not be evaluated and you will not receive credit for your work.**

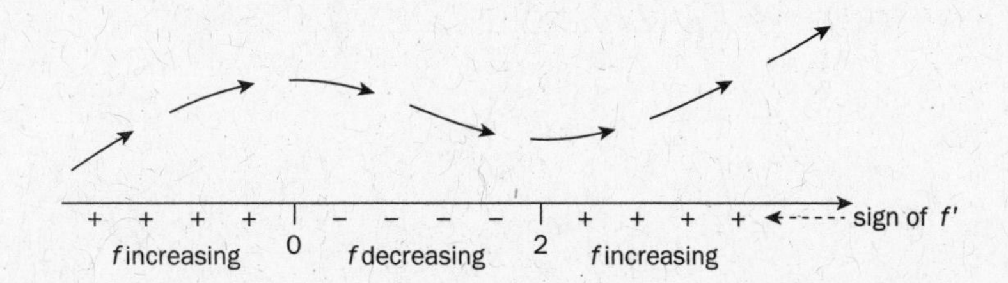

Your written explanation for this type of question must explicitly state that the function is increasing (or decreasing) on an interval because its derivative is positive (or negative) on that interval. The AP readers *will not* accept the sign chart, by itself, as justification. This is true for ALL questions involving sign charts. Don't depend on the sign chart to provide your answer or your reasoning.

(c) A relative maximum occurs at $x = 0$, because the first derivative changes sign from positive to negative.

(d) A relative minimum occurs at $x = 2$, because the first derivative changes sign from negative to positive.

CORRESPONDING CHARACTERISTICS OF THE GRAPHS OF f AND f'

We can use the relationship between f and f' to determine relationships between the graph of f and the graph of f'.

- Because f is increasing when $f' > 0$, the graph of f is going up only if the graph of f' is above the x-axis.
- Because f is decreasing when $f' > 0$, the graph of f is going down only if the graph of f' is below the x-axis.
- The graph of f has a horizontal tangent if the graph of f' crosses the x-axis.
- The graph of f is not smooth (i.e., it has no tangent line or a vertical tangent line) if f' is not continuous, i.e., where the graph of f' has a hole, a jump, or an asymptote.
- The graph of f has a vertical tangent if the graph of f' has an asymptote.

Example:

The graph of $f'(x)$ is shown below.

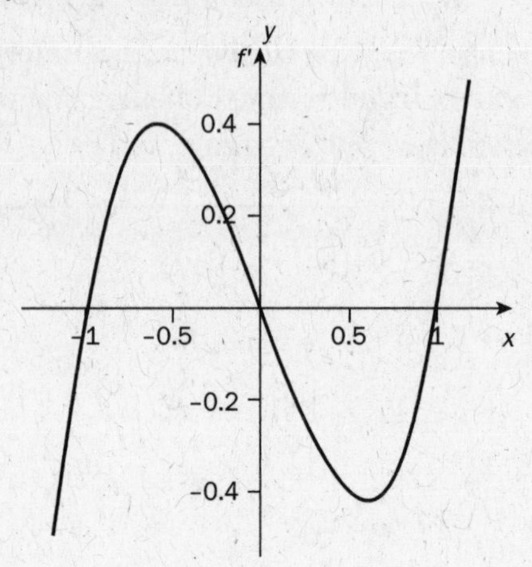

Which of the following graphs could be the graph of $f(x)$?

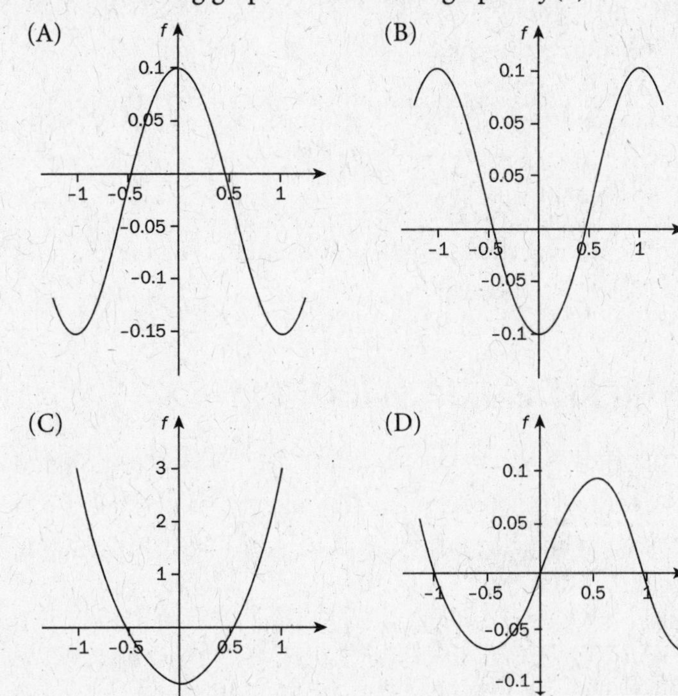

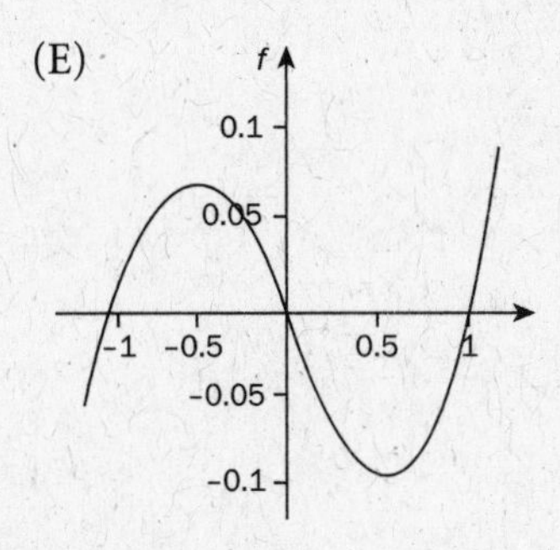

Solution: A

We can approach this problem by eliminating answer choices. Interpreting the graph of f', we see that, by the First Derivative Test,

- f has a relative minimum at $x = -1$; the function changes from decreasing to increasing at this point, because the sign of the first derivative changes from negative to positive.
- f has a relative maximum at $x = 0$; f changes from increasing to decreasing here, because the sign of the first derivative changes from positive to negative.
- f has another relative minimum at $x = 1$; the function changes from decreasing to increasing here, because the sign of the first derivative changes from negative to positive.

Using this information, we can eliminate

- (B) because this graph has the maxima and minima reversed
- (C) because this graph has only one relative extremum
- (D) and (E) because the relative extrema of these graphs are not located at the correct points

The only remaining choice is (A). We can confirm on the graph that the graph of (A) has relative minima at $x = -1, 1$ and a relative maximum at $x = 0$.

THE MEAN VALUE THEOREM

We've already seen that the slope of the tangent line at a point, or the derivative, is the limit of the slope of secant lines through the point. There is a further relationship between secant lines and the tangent line, which is described by an important theorem in calculus.

If a function f is defined and continuous on an interval $[a,b]$ and differentiable on (a,b), then there is at least one point $c \in (a,b)$ such that:

$$f'(c) = \frac{f(b) - f(a)}{b - a}$$

This tells us that the slope of the secant line between a and b is equal to the slope of the tangent line at c. We can see this on a graph.

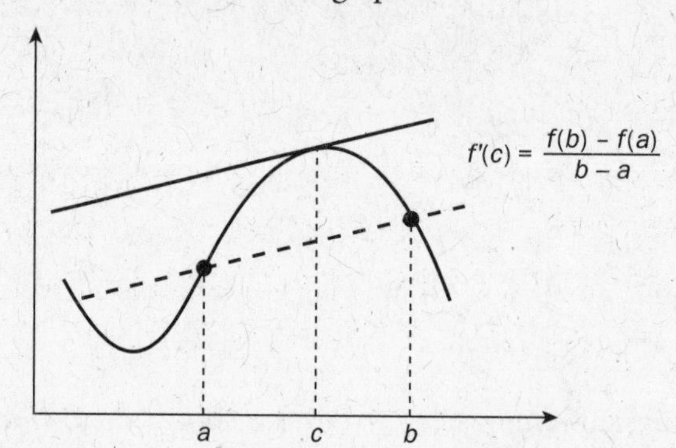

AP EXPERT TIP

Make sure ensure that f(a) = f(b) before applying Rolle's theorem. Don't just assume it can apply to any interval!

ROLLE'S THEOREM

Rolle's Theorem is a special case of the Mean Value Theorem. It deals with the special case in which $f(a) = f(b)$. In this case, $f(b) - f(a) = 0$, so $f'(c) = 0$ and there exists a point in (a,b) where f has a horizontal tangent.

We can apply the Mean Value Theorem and Rolle's Theorem to draw conclusions about the equations involving derivatives.

EQUATIONS INVOLVING DERIVATIVES: TRANSLATING BETWEEN VERBAL DESCRIPTIONS AND EQUATIONS INVOLVING DERIVATIVES

Functions are used to model the real world: We can translate verbal descriptions into equations and solve them. Derivatives give us more tools to use to answer real-life problems. Not only can we answer questions about functions, we can also model and answer questions about their rates of change.

For instance, the rate of change of a population is proportional to the current size of the population. This makes sense, because the number of births depends on the size of the population. We say that the rate of change of the population t is proportional to current population. Using derivatives, we can write this as $y' = ky$, where k is the proportionality constant.

In fact, this description characterizes all exponential growth and decay. That is, it models things such as the amount of medication present in a patient's bloodstream, the half-life of a radioactive substance, and the growth of an investment earning a fixed interest rate.

We can also use the derivative to answer specific questions about the behavior of a function.

Example:

A bug is sitting on the rim of a bicycle wheel that is 24 inches in diameter. The bug's vertical distance from the ground is given by the equation $y = 12(1 + \cos t)$. At time $t = 0$, the bug is at the top of the wheel. The vertical distance from the ground decreases and increases periodically. (We expect this, both from the equation governing the bug's distance and from our understanding of how a bicycle wheel travels.) What is the earliest time at which the bug's distance from the ground changes from decreasing to increasing?

Solution:

This question asks about the increasing-decreasing behavior of the bug's height. We use the relationship between the function and its derivative to rephrase the question in terms of the derivative. "What is the earliest time at which the derivative of the function describing the bug's height changes sign?" From the Intermediate Value Theorem, we know that the derivative of a function can only change sign at a point where the derivative is zero or undefined, i.e., at a critical point.

We need to find the derivative of $y = 12(1 + \cos t)$. We can use the definition of the derivative as a difference quotient to compute this derivative. We find that $y' = -12\sin t$. Because this function is continuous, the derivative can only change sign at a point where it is zero. That is, where

$$0 = -12 \sin t$$

$$0 = \sin t$$

$$t = 0, \pi\ 2\pi \ldots n\,\pi (n \geq 0)$$

We need to check at which, if any, of these points the derivative *actually* changes sign, so we do a sign analysis. We do not include the value $t = 0$, because the bug is beginning its journey at this time and therefore its direction is not changing there.

Interval	$(0, \pi)$	$(\pi, 2\pi)$
Test Point	$x = \frac{\pi}{2}$	$x = \frac{3\pi}{2}$
Sign of y'	$y'\left(\frac{\pi}{2}\right) = -12\sin\frac{\pi}{2} = -12(1) = -12 < 0$ negative	$y'\left(\frac{3\pi}{2}\right) = -12\sin\frac{3\pi}{2} = -12(-1) = 12 > 0$ positive
Conclusion	y decreasing	y increasing

We see that the derivative changes sign at the critical point, $t = \pi$. Therefore $t = \pi$ is the earliest time at which the bug's distance from the ground changes from decreasing to increasing.

REVIEW QUESTIONS

1. The graph of the differentiable function $y = f(x)$ is shown below. Which of the following is true?

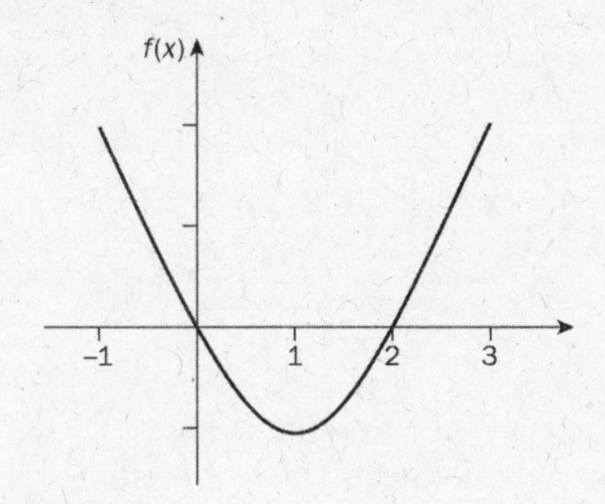

(A) $f'(0) > f(0)$
(B) $f'(1) < f(1)$
(C) $f'(2) < f(2)$
(D) $f'(1) = f(0)$
(E) $f'(2) = f(2)$

2. The graph of $y = f(x)$ is shown below. Which of the following graphs could be the derivative?

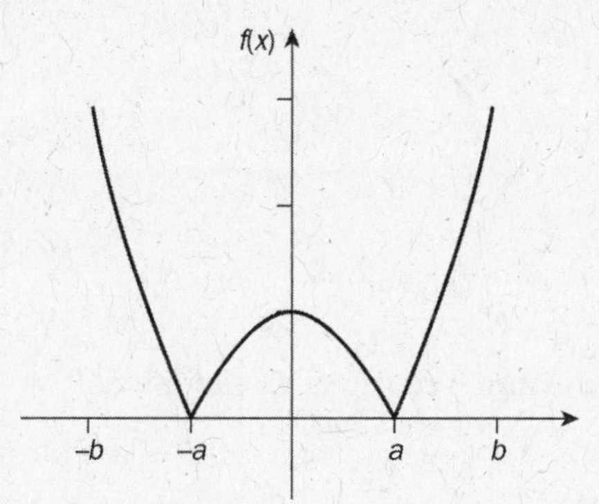

(A)

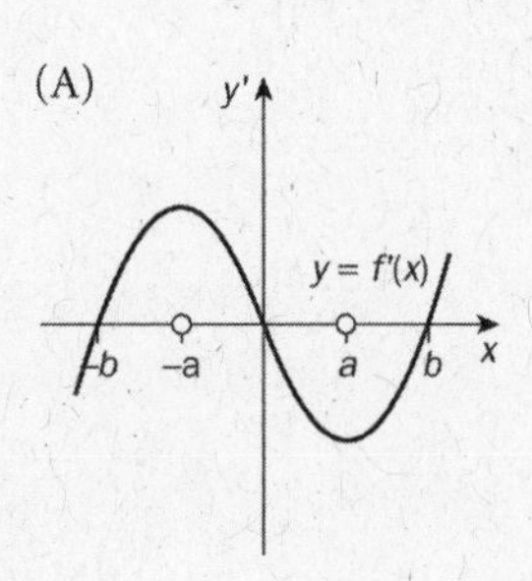

(B)

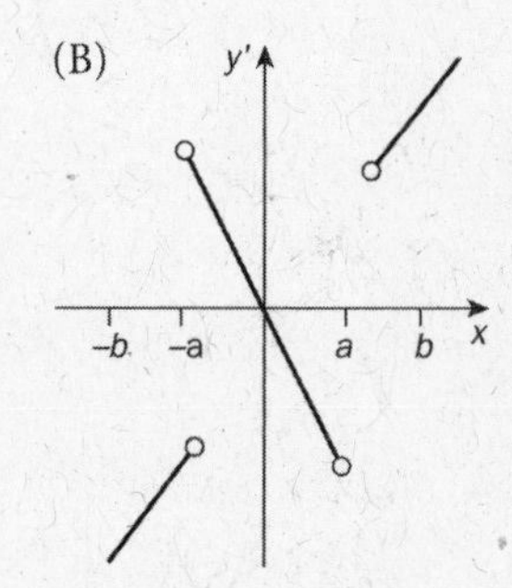

(C)

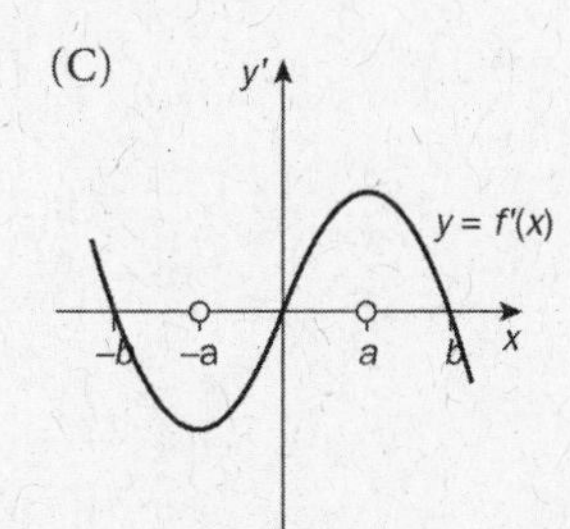

(D)

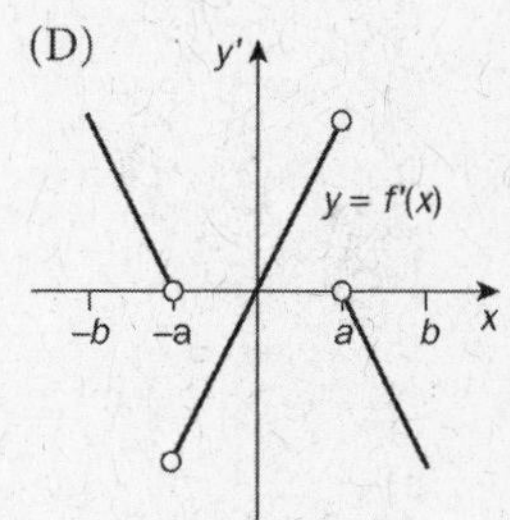

(E)

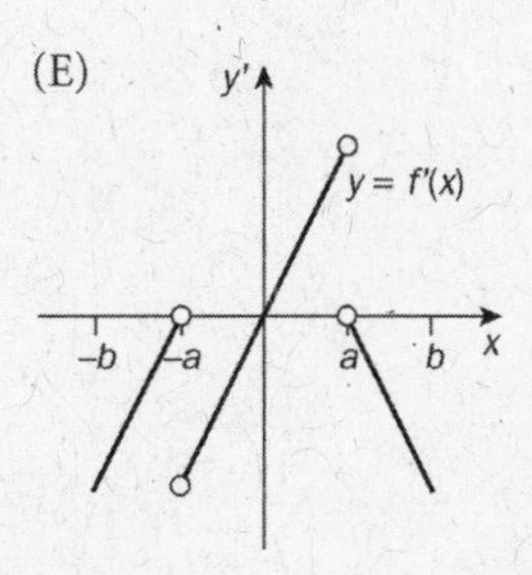

3. Find all values of c that satisfy the Mean Value Theorem for $f(x) = x^3 + 1$ on $[-2, 4]$.

(A) $c = 2$

(B) $c = \pm 2$

(C) $c = -2$

(D) $c = 0$

(E) No such value of c exists.

Question 4 requires a calculator.

4. Find the value of c that satisfies the Mean Value Theorem for $f(x) = x \cdot \sin x$ on $[1, 4]$.

(A) 1.239

(B) 1.290

(C) 2.029

(D) 2.463

(E) 3.027

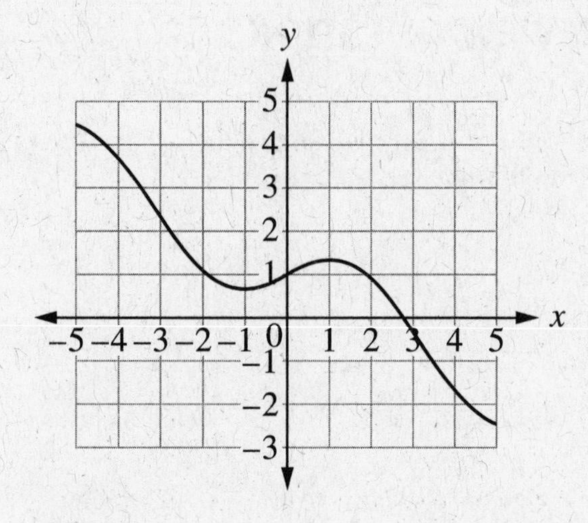

5. The graph of $F(x)$ is shown above. Which of the following is the graph of $F'(x)$?

(A)

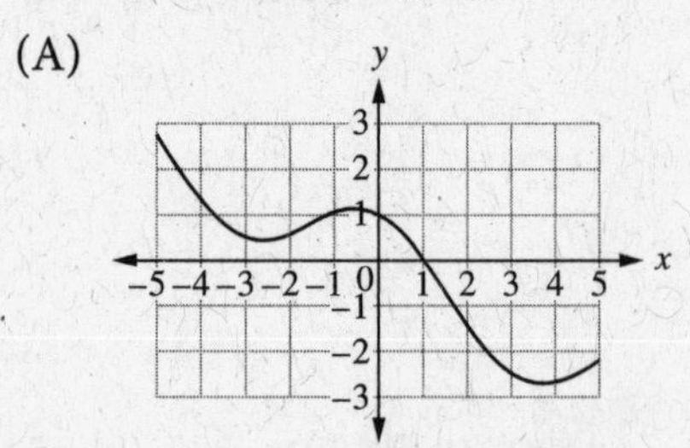

(B)

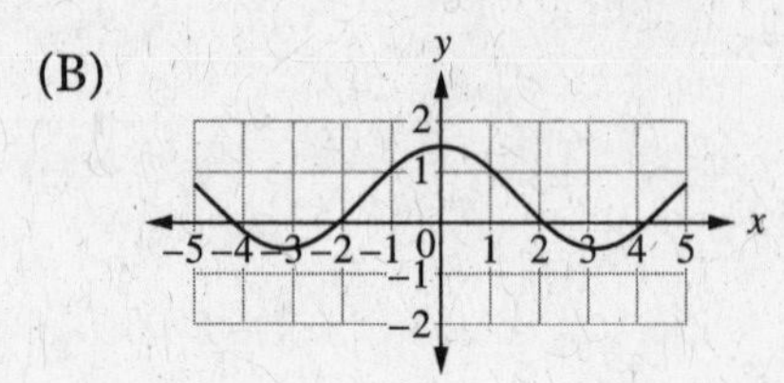

(C)

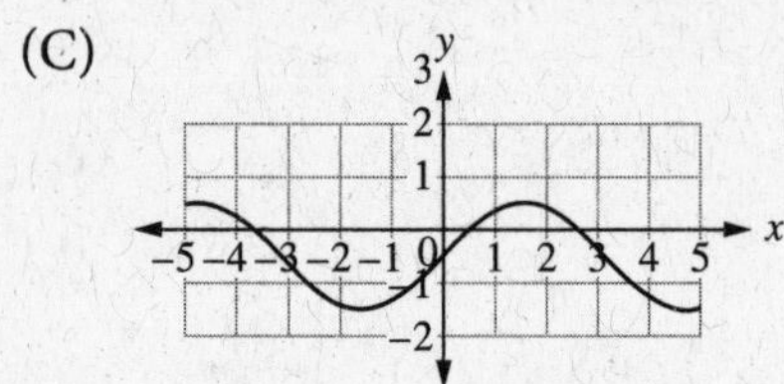

(D)

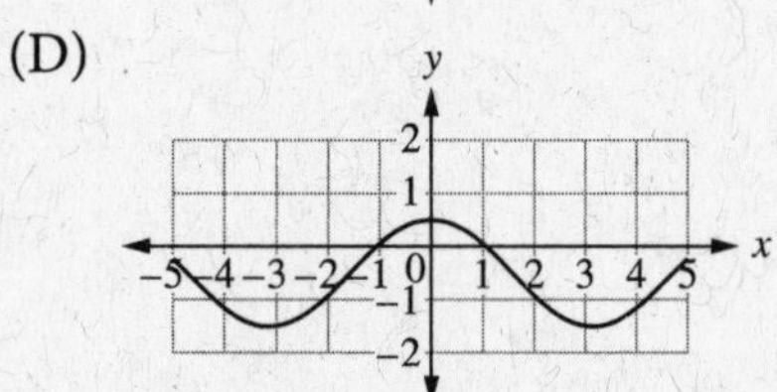

(E) None of these

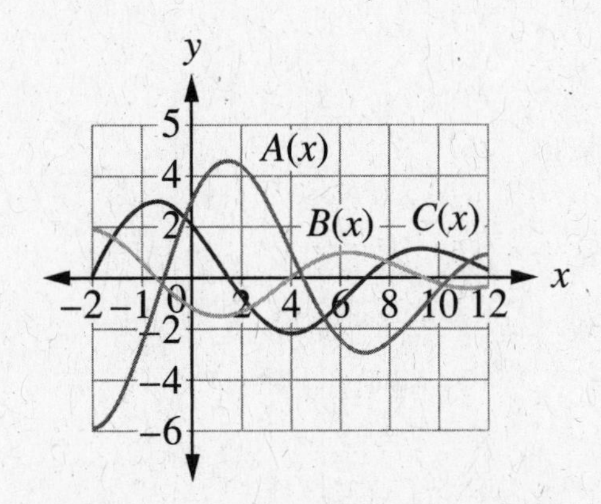

6. The graphs above include a function and its first and second derivatives. Which of the following describes the relationship?

(A) $A''(x) = B'(x) = C(x)$

(B) $A''(x) = C'(x) = B(x)$

(C) $B''(x) = A'(x) = C(x)$

(D) $B''(x) = C'(x) = A(x)$

(E) $C''(x) = B'(x) = A(x)$

7. Let $x(t) = t^{\frac{2}{3}}$ give the distance of a moving particle from its starting point as a function of time t. For what value of t is the instantaneous velocity of the particle equal to its average velocity over the interval [0,8]?

(A) $\frac{8}{27}$

(B) $\frac{27}{64}$

(C) $\frac{64}{27}$

(D) $\frac{27}{8}$

(E) $\frac{64}{9}$

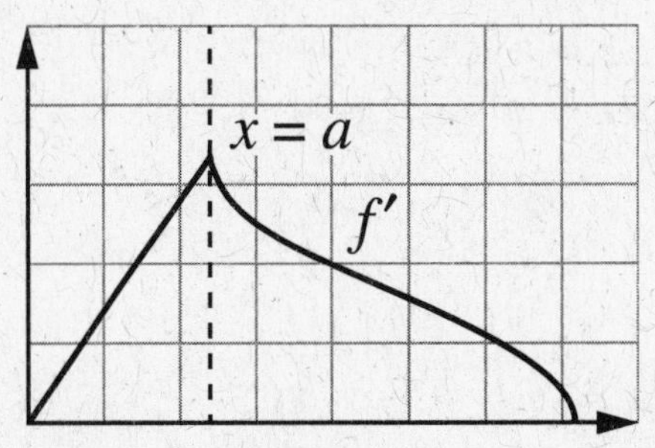

8. It follows from the graph of f', shown at the right, that

(A) f is not continuous at $x = a$.

(B) f is continuous but not differentiable at $x = a$.

(C) f has a relative maximum at $x = a$.

(D) f has an inflection point at $x = a$.

(E) None of these

Questions 9–11 refer to the following graph.

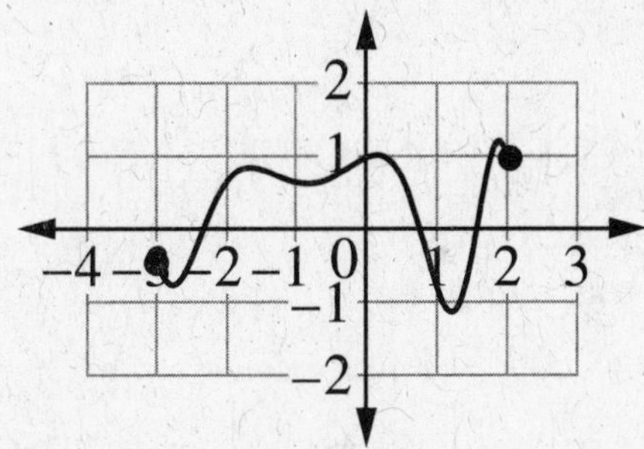

The graph of F

9. Which of the following statements is false?

(A) $F(-3) + F(2) > 0$

(B) $F(-1) + F'(-1) > 0$

(C) $F'(-1) \cdot F''(-1) < 0$

(D) $F(1) \cdot F'(1) < 0$

(E) $F(0) \cdot F'(0) > 0$

10. The function F has exactly this many critical numbers.

(A) 4

(B) 5

(C) 6

(D) 7

(E) 8

11. If the function G is defined by $G(x) = F(x^2 + x)$, then $G'(-1) \approx$

(A) -2

(B) $\frac{-1}{2}$

(C) 0

(D) $\frac{1}{2}$

(E) 2

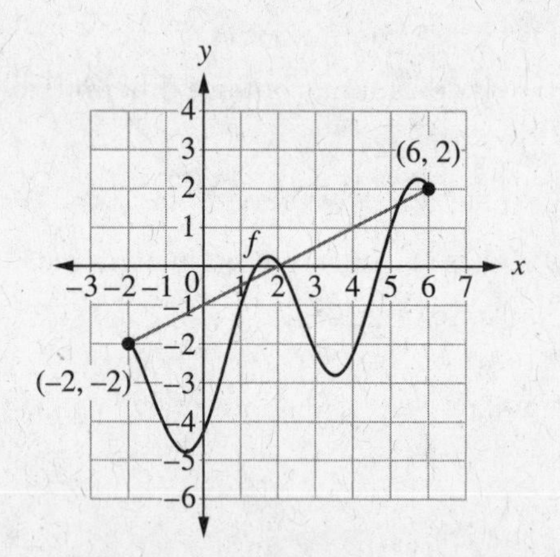

12. Pictured above is the graph of a function f defined for all x in the closed interval [−2, 6] and differentiable for all x in the open interval (−2, 6). The secant, shown in red, has a slope equal to $\frac{1}{2}$. With the graph as an aid, how many different values of $c \in (-2, 6)$ would make $f'(c) = \frac{1}{2}$?

(A) 3

(B) 4

(C) 5

(D) 6

(E) 7

13. What is the range of the function $f(x) = \frac{\ln x}{x}$ on the closed interval $[1, e^2]$?

(A) $f(1) \le f(x) \le f(e)$

(B) $f(1) \le f(x) \le f(e^2)$

(C) $f(2) \le f(x) \le f(e)$

(D) $f(e) \le f(x) \le f(e^2)$

(E) None of these

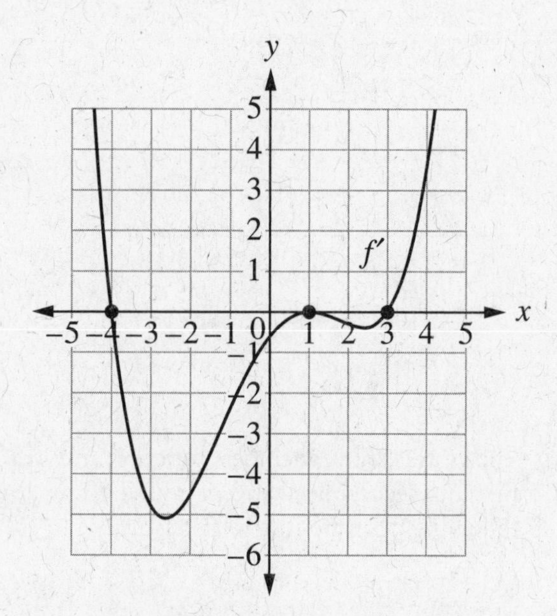

The graph of $f'(x)$, not the graph of $f(x)$

14. The graph of $f'(x)$ is shown. Which of the following statements is true?

(A) $f(-4)$ is a local minimum value.

(B) $f(-4)$ is a local maximum value.

(C) $f(1)$ is a local minimum value.

(D) $f(1)$ is a local maximum value.

(E) $f(3)$ is a local maximum value.

15. Let f be a function defined by

$$f(x) = \begin{cases} 2x - x^2 & \text{for } x \le 1 \\ x^2 + kx + p & \text{for } x > 1 \end{cases}$$

What values of k and p will make f both continuous and differentiable at $x = 1$?

(A) $k = 0$ and $p = 2$

(B) $k = 1$ and $p = 3$

(C) $k = 2$ and $p = 2$

(D) $k = -2$ and $p = 3$

(E) $k = -2$ and $p = 2$

Free-Response Questions

16. This is a graph of the derivative of f:

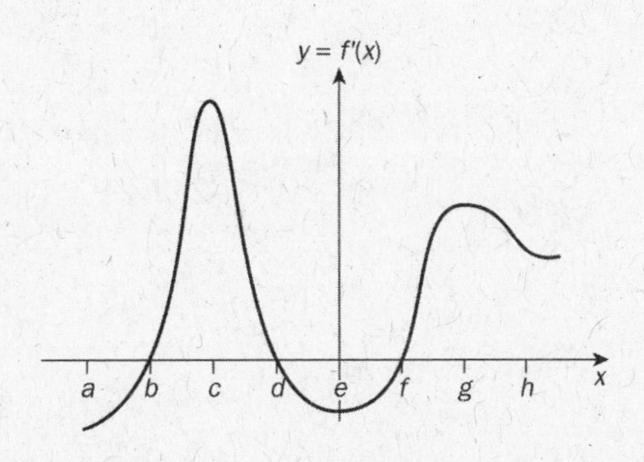

The function $y = f(x)$ is defined on $[a, h]$. Its derivative $y = f'(x)$ is shown above. Find the point(s) at which the graph of f

(a) has a relative maximum.

(b) has a relative minimum.

(c) attains its maximum value.

Justify all answers.

17. Let $y = f(x)$ be a differentiable function on $[-10,3]$ such that $f(-10) = 5, f(-1) = -2$, and $f(3) = 5$.

(a) What is the minimum number of zeros that this function could have?

(b) Is there a value of x for which $f(x) = 2$?

(c) Does f have any horizontal tangents?

(d) If f has only one extremum on $[-10,3]$, will it be a maximum or a minimum?

Justify all answers.

ANSWERS AND EXPLANATIONS

1. D

(A) False: $f'(0) < 0$ and $f(0) = 0$

(B) False: $f'(1) = 0$ and $f(1) < 0$

(C) False: $f'(2) > 0$ and $f(2) = 0$

(D) True: $f'(1) = 0$ and $f(0) = 0$

(E) False: $f'(2) > 0$ and $f(2) = 0$

2. B

Let's describe the graph in terms of its derivative, because this will allow us to eliminate answer choices.

On $x < -a$, the slope of the curve is negative, so the graph of y' lies below the x-axis. Thus choices (A) and (D) are eliminated. On $-a < x < 0$, the slope of the curve is positive, so the graph of y' lies above the x-axis. Thus the choices (C), (D), and (E) are not possible. We've eliminated all of the answer choices except one, so the correct answer is (B).

The graph described above is $y = |1 - x^2|$, so $a = 1$ and $b = 2$. There are corners (not true cusps) at $x = \pm 1$, so the derivative does not exist at these points. These points appear as jump discontinuities or tiny holes at the end of each segment of the graph of the derivative. Notice that the derivative consists of straight line segments. This is because the original function is the absolute value of a parabola, $y = x^2$, and its derivative is $y' = 2x$, which is a linear function.

3. A

The Mean Value Theorem (MVT) says that if f is continuous on $[-2, 4]$ and differentiable on $(-2, 4)$, there is a c on $(-2, 4)$ such that $f'(c) = \dfrac{f(4) - f(-2)}{4 - (-2)}$. We'll apply the theorem to our problem by first computing the quantities in the problem, $f(4), f(-2), f'(c)$:

$$f(4) = 4^3 + 1 = 65,\ f(-2) = (-2)^3 + 1 = -7$$
$$f'(x) = 3x^2 \Rightarrow f'(c) = 3c^2$$

We substitute this information into the equation in the MVT:

$$3c^2 = \frac{65 - (-7)}{4 - (-2)} \text{ or } 3c^2 = 12 \text{ or } c = \pm 2$$

Notice that $c = -2$ is an endpoint of the interval and the MVT says that c must lie in the open interval $(-2, 4)$. This eliminates $c = -2$, so $c = 2$ is the only correct answer.

4. D

The Mean Value Theorem (MVT) says that if f is continuous on $[1, 4]$ and differentiable on $(1, 4)$, there is a c on $(1, 4)$ such that $f'(c) = \dfrac{f(4) - f(1)}{4 - 1}$. We'll apply the theorem to our problem by first computing the quantities in the problem $f(1), f(4), f'(c)$:

$$f(1) = 1 \cdot \sin 1,\ f(4) = 4 \cdot \sin 4$$
$$f'(x) \underset{\text{product rule}}{=} \sin x + x\cos x \Rightarrow f'(c)$$
$$= \sin c + c\cos c$$

We substitute the given information into the MVT:

$$\sin c + c\cos c = \frac{4\sin 4 - 1\sin 1}{4 - 1}$$

Here we need to use a calculator (set in radian mode!) to simplify:

$$4\sin 4 \approx -3.0272$$
$$\sin 1 \approx 0.8414$$
$$\Rightarrow \sin c + c\cos c = \frac{4\sin 4 - 1\sin 1}{4 - 1} = -1.2896$$

Here, we once again need a graphing calculator to solve the above equation. One way to do this is to find the zero on the graph of $\sin x + x\cos x - (-1.2896)$.

TI-83: Press Y= and enter
$Y_1 = \sin(x) + x\cos(x) - (-1.2896)$.

Enter a window of X: [1,4] and Y: [−4,4] Press GRAPH. The graph should cross the x-axis only once. The value of x can be obtained by pressing 2nd TRACE 2 for the zero of the function.

Move the cursor to the left of the zero. Press ENTER. Move the cursor to the right of the zero. Press ENTER. Press ENTER again. The answer $x = 2.4629739$ will appear on the screen:

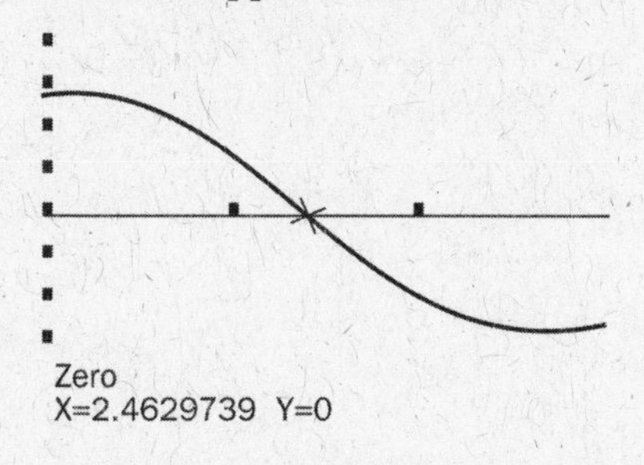

If you have a TI-89, you can have the calculator do most of the computational work for you:

TI-89: On the Y= screen, enter $y_1 = x^* \sin x$.

On the home screen, type F2: solve ($d(y,(x), x) = (y_1(4) - y_1(1))/(4-1), x)/1 < x < 4$. Press ENTER and the answer appears as $x = 2.46297$.

AP answers should be rounded to the nearest thousandth. On the free-response portion of the examination, any number that is not an exact answer must be rounded or truncated to the nearest thousandth. If it is rounded to the nearest hundredth, this will be counted as an incorrect answer.

5. D

The graph of $F(x)$ has horizontal tangents at the points where $x = \pm 1$. Of the choices for the graph $F'(x)$, only option (D) has zeros at $x = \pm 1$.

6. B

The function C is the derivative of A. The function B is the second derivative of A. On those intervals where the function A is increasing, C is positive and on those intervals where the function A is decreasing, C is negative. Similarly, on those intervals where the function A is concave up, B is positive and on those intervals where the function A is concave down, B is negative. It follows that $A''(x) = C'(x) = B(x)$.

7. C

$x(t) = t^{\frac{2}{3}}$ is the distance of the particle from its starting point as a function of time t. Its average velocity over the [0,8] is given by $v_{aver} = \frac{x(8) - x(0)}{8-0} = \frac{4-0}{8} = \frac{1}{2}$. Its instantaneous velocity is given by $x'(t) = \frac{2}{3}t^{\frac{-1}{3}} = \frac{2}{3\sqrt[3]{t}}$.
We set $x'(c) = \frac{2}{3\sqrt[3]{c}} = \frac{1}{2}$ and solve so
$3\sqrt[3]{c} = 4 \quad \sqrt[3]{c} = \frac{4}{3} \quad c = \frac{64}{27}$ which is in (0,8)

8. D

From the graph of f', we can see that $f'(a)$ exists so f is continuous and differentiable at $x = a$. At $x = a$, f is increasing from the left and to the right so f **does not** have a relative maximum at $x = a$. Remember, we're looking at a graph of the derivative, NOT the original function!

Finally, f has an inflection point at $x = a$. Inflection points occur where the second derivative is equal to zero or where the first derivative has a relative maximum or minimum.

9. D

Answer (D) is false since $F(1) \approx -0.8$, $F'(1) \approx -3$, and $F(1) \cdot F'(1) \approx 2.4 > 0$.

10. C

The function F has exactly six critical numbers because there are six places on the curve for $x \in (-3,2)$ where $F'(x) = 0$. We can visualize this by looking at horizontal tangents at the six locations shown below.

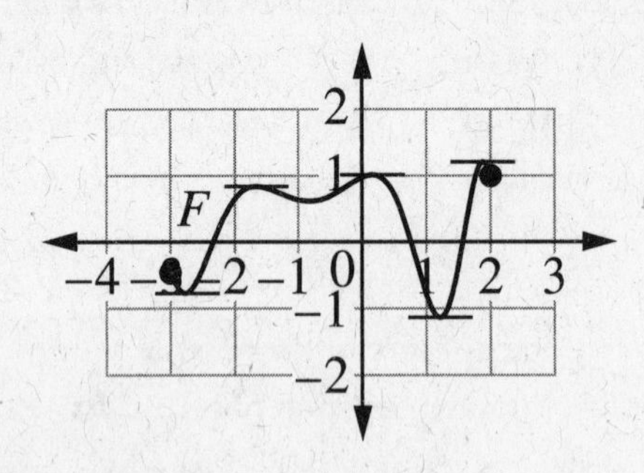

11. B

Given $G(x) = F(x^2 + x)$, by the chain rule $G'(x) = F'(x^2 + x)\cdot(2x + 1)$, which means that $G'(-1) = F'(1-1)\cdot(-2+1) = F'(0)\cdot(-1) \approx -\frac{1}{2}$

12. B

There are exactly four places where $f'(c) = \frac{1}{2}$ shown below.

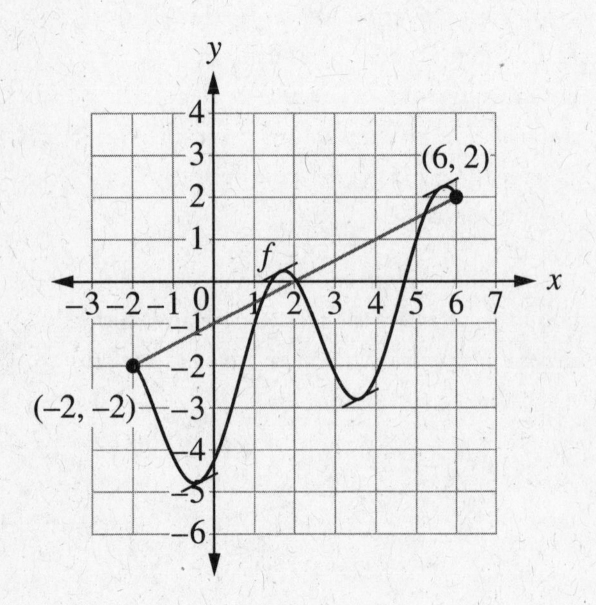

13. A

To find the range we will need to find the absolute extrema of our function. The candidates for extreme values are endpoint values, $f(1) = 0$ and $f(e^2) = \frac{\ln e^2}{e^2} = \frac{2}{e^2} \approx 0.271$ and function values at critical numbers c where $c \in (1,e^2)$. Differentiate by the quotient rule so $f'(x) = \frac{\frac{1}{x}\cdot x - \ln x}{x^2} = \frac{1-\ln x}{x^2}$ and solve $f'(c) = \frac{1-\ln c}{c^2} = 0$ which means $c = e$. The final candidate is $f(e) = \frac{1}{e} \approx 0.368$. $f(1) = 0$ is the absolute minimum and $f(e) = \frac{1}{e} \approx 0.368$is the absolute maximum so $f(1) \le f(x) \le f(e)$.

14. B

$f(-4)$ is a local maximum value by the first derivative test. To the left of $x = -4$, f is increasing and to the right of $x = -4$, f is decreasing. This means $f(-4)$ is a local maximum value. All of the other statements are false.

15. E

To start, for f to be continuous at $x = 1$, then

$$f(1) = \begin{cases} 2\cdot 1 - 1^2 & \text{for } x \le 1 \\ 1 + k\cdot 1 + p & \text{for } x > 1 \end{cases} = \begin{cases} 1 & \text{for } x \le 1 \\ 1 + k + p & \text{for } x > 1 \end{cases}$$

$$1 + k + p = 1 \Rightarrow k + p = 0$$

Secondly, for f to be differentiable at $x = 1$, then

$$f'(x) = \begin{cases} 2 - 2x & \text{for } x \le 1 \\ 2x + k & \text{for } x > 1 \end{cases} \Rightarrow f'(1) =$$

$$2 + k = 0 \Rightarrow k = -2$$

$$\begin{cases} 2 - 2\cdot 1 & \text{for } x \le 1 \\ 2\cdot 1 + k & \text{for } x > 1 \end{cases} \Rightarrow f'(1) = \begin{cases} 0 & \text{for } x \le 1 \\ 2 + k & \text{for } x > 1 \end{cases}$$

Finally we solve a system and get

$$\begin{cases} k + p = 0 \\ k = -2 \end{cases} \Rightarrow \begin{cases} p = 2 \\ k = -2 \end{cases}$$

FREE-RESPONSE ANSWERS

16. (a) f has a relative maximum at $x = d$, because the first derivative f' goes from positive to negative as x increases in an open interval around $x = d$.

(b) f has a relative minimum at $x = b$ and $x = f$, because the first derivative f' goes from negative to positive as x increases in open intervals around these points.

(c) f attains its maximum value at $x = h$, because the first derivative f' is mostly positive, indicating that the function is increasing most of the time as x increases. The function f is decreasing very little, as indicated by the fact that the graph of f' lies below the x-axis only on (a, b) and (d, f).

17. It's a good idea to start this problem by making a possible sketch of $y = f(x)$, using the information given in the problem.

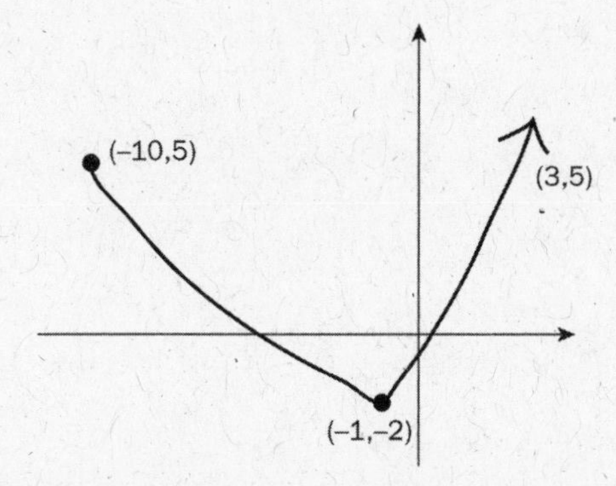

(a) Two. Because y is differentiable, it must be continuous (differentiability implies continuity). Furthermore, the Intermediate Value Theorem and the fact that $y > 0$ at both endpoints and $y < 0$ at a point in between imply that the graph must cross the x-axis at least two times, i.e., there must be at least two zeros, one in the interval $(-10,-1)$ and the other in the interval $(-1,3)$.

(b) Yes. We do not know exactly what that value of x is, but because 2 lies between 5 and -2, the Intermediate Value Theorem guarantees that there is at least one point on $(-10,-1)$ and another on $(-1,3)$ where $f(x) = 2$.

(c) Yes. Because the function is continuous and differentiable and $f(-10) = f(3) = 5$ (both endpoints have the same y-value), then by Rolle's Theorem f has at least one horizontal tangent.

(d) Minimum. If there is only one extremum, it must be a minimum because $f(-1) = -2$ lies below $f(-10) = 5$ and $f(-3) = 5$.

CHAPTER 15: SECOND DERIVATIVES

IF YOU LEARN ONLY THREE THINGS IN THIS CHAPTER . . .

1. **Second Derivative Test:** Suppose $f'(c) = 0$. Then:

   If $f''(c) > 0$, then f has a relative minimum at $x = c$.

   If $f''(c) < 0$, then f has a relative maximum at $x = c$.
2. Because f is concave up where the second derivative is positive (i.e., the first derivative is increasing), the graph of f is concave up where the graph of f' is increasing or where the graph of f'' is above the x-axis.
3. Because f is concave down where the second derivative is negative (i.e., the first derivative is decreasing), the graph of f is concave down where the graph of f' is decreasing or where the graph of f'' is below the x-axis.

In Chapter 13, we saw that because the derivative of a function $f(x)$ is also a function called $f'(x)$, we can find the derivative of f'. This derivative is called the *second derivative* of f and we write $f''(x)$ or $\frac{d^2f}{dx^2}$. In this chapter we will study the information that the second derivative $f''(x)$ provides about the function $f(x)$.

CONCAVITY AND THE SIGN OF f''

The sign of the first derivative of a function $f(x)$ tells us where f is increasing or decreasing. It does not tell us *how fast* the function is increasing or decreasing.

For instance, each of the following graphs shows the distance that two students, call them Sue and Sam, have traveled along a straight road on their way to school from their respective homes. Both graphs show that the students leave home at 8:00 a.m. and arrive at school at 8:15 a.m. Both graphs are increasing, indicating that the derivative of each position function, which is the velocity, is positive. That is, both Sam and Sue are continually heading *toward* school. The sign of the derivative, however, does not describe the difference in the way the graphs curve.

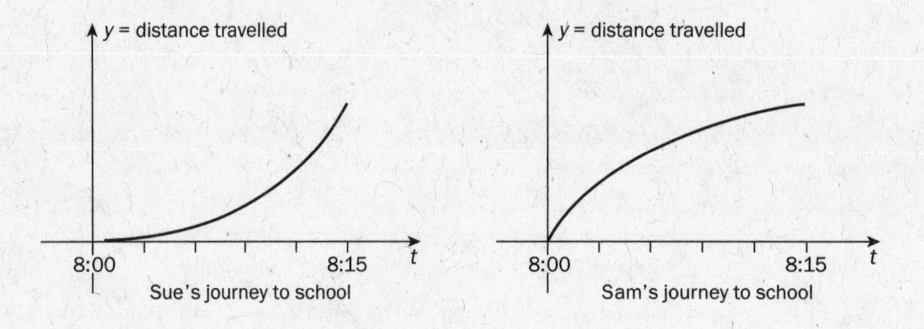

The first graph, showing Sue's position, is "cupping upward." We say that this graph is concave up. If we draw tangent lines at several points on the graph, we see that the slopes are increasing. The slopes of the tangent lines are Sue's velocity at each point on the graph. Because the slopes are increasing, the rate of change of her velocity, which is her acceleration, is increasing. Sue is traveling faster as she heads to school. Sue doesn't realize she is late for school until she checks her watch—and she speeds up to make it on time.

The second graph, showing Sam's position, is "cupping downward." We say that this graph is concave down. Here, even though the slopes of the tangent lines are all positive—indicating that Sam's velocity is positive as he heads to school—the slopes are getting less and less positive. The rate of change of Sam's velocity, or his acceleration, is negative. Sam is slowing down as he heads to school. Sam knows he's getting a late start; he sets off at a fast pace, but can't maintain his velocity. He gets tired and slows down as he approaches the school.

We'll make these notions precise with the following definitions:

- We say that f is concave up on intervals (a,b) where it is differentiable and f' is increasing.
- We say that f is concave down on intervals (a,b) where it is differentiable and f' is decreasing.

Both increasing and decreasing functions can be concave up or concave down. Four possibilities are shown below.

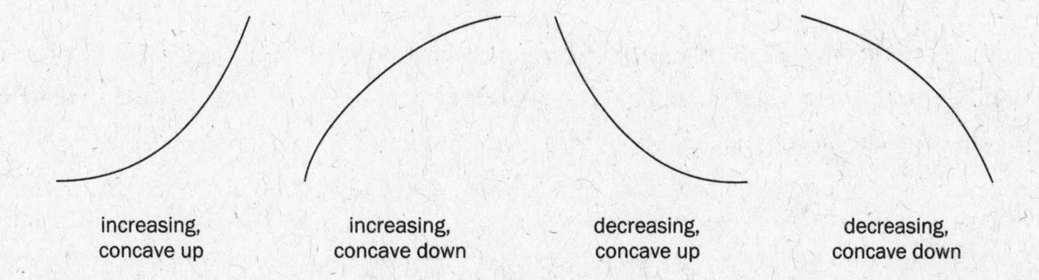

Notice that when f is concave up, the tangent lines lie below the curve. When f is concave down, the tangent lines lie above the curve.

Concavity tells us where the function's rate of change is increasing and where it is decreasing. For example, when acceleration is positive, the graph of the position function is concave up. Earlier, Sue was accelerating and the graph of her trip to school was concave up; Sam was decelerating, i.e., his acceleration was negative and his graph was concave down.

Because concavity tells us about the rate of change of the derivative, we can describe concavity in terms of the second derivative. We'll assume that f is twice differentiable (meaning both f' and f'' exist), at least on the intervals we are talking about. In this case we have the following relationship between the concavity of f and the sign of f'':

- If $f''(x) > 0$ on an interval, then $f(x)$ is concave up on that interval.
- If $f''(x) < 0$ on an interval, then $f(x)$ is concave down on that interval.

The Second Derivative Test

The second derivative can also be used to provide information about the critical points of a function $f(x)$.

Suppose c is a critical point of f where $f'(c) = 0$ (as opposed to the other kind of critical point for which $f'(c)$ does not exist). If $f''(c) > 0$, then $f'(c)$ is increasing. Because the derivative is equal to zero at c, we know that $f'(c)$ must be negative to the left of c and positive to the right of c. That is, the derivative changes from negative to positive at c. Therefore, $f(c)$ is a relative minimum. Similarly, if $f''(c) < 0$, $f(c)$ is a relative maximum.

We summarize these results in a theorem, assuming f is twice-differentiable on the intervals we are discussing.

THEOREM: SECOND DERIVATIVE TEST

Suppose $f'(c) = 0$. Then:

- If $f''(c) > 0$, then f has a relative minimum at $x = c$.
- If $f''(c) < 0$, then f has a relative maximum at $x = c$.

When we want to determine the behavior of a function at a critical point, it is often easier to use the second derivative test, when it applies (i.e., in places where the function is twice differentiable), than the first derivative test.

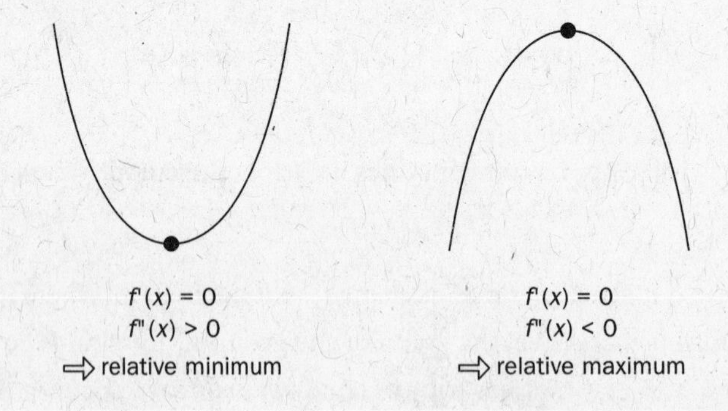

Example:

If $y = f(x) = x + 2\sin x$, find the critical points of f on the interval $0 \le x \le 2\pi$ and use the second derivative test to determine whether each point is a relative maximum or a relative minimum.

Note: **The College Board expects you to know the values of the trigonometric functions for $0, \frac{\pi}{6}, \frac{\pi}{4}, \frac{\pi}{3}, \frac{\pi}{2}$, and their multiples without using a calculator.**

Solution:

The critical points of f occur where $f' = 0$ or f' do not exist. We first compute the derivative of f: $f'(x) = 1 + 2\cos x$. The derivative is defined everywhere, so the only critical points occur where $f'(x) = 1 + 2\cos x = 0$. We solve this equation $1 + 2\cos x = 0 \Rightarrow \cos x = -\frac{1}{2}$. On the interval $[0, 2\pi]$, this occurs where $x = \frac{2\pi}{3}, \frac{4\pi}{3}$.

We now use the second derivative test to determine the nature of each of these critical points. To apply the test, we compute the second derivative of f and evaluate the second derivative at each of the critical points: $f''(x) = -2\sin x$.

For $x = \dfrac{2\pi}{3}$: $f''\left(\dfrac{2\pi}{3}\right) = -2\sin\left(\dfrac{2\pi}{3}\right) = (-2)\dfrac{\sqrt{3}}{2} = -\sqrt{3} < 0$. Because $f''\left(\dfrac{2\pi}{3}\right) < 0$, the second derivative test tells us that f has a relative maximum at $x = \dfrac{2\pi}{3}$.

For $x = \dfrac{4\pi}{3}$: $f''\left(\dfrac{4\pi}{3}\right) = (-2)\left(-\dfrac{\sqrt{3}}{2}\right) = \sqrt{3} > 0$. In this case the second derivative test tells us that f has a relative minimum at $x = \dfrac{4\pi}{3}$.

If $f'(c) = 0$ and $f''(c) = 0$, then the second derivative test does not tell us whether f has a relative extremum at $x = c$. For example, both the first and second derivatives of the functions $f(x) = x^4$ and $g(x) = x^3$ equal zero at zero, i.e., $f'(0) = 0, f''(0) = 0, g'(0) = 0, g''(0) = 0$. We can see from the graphs of f and g that $f(x) = x^4$ has a minimum at $x = 0$, but $g(x) = x^3$ does not have a relative extremum at $x = 0$.

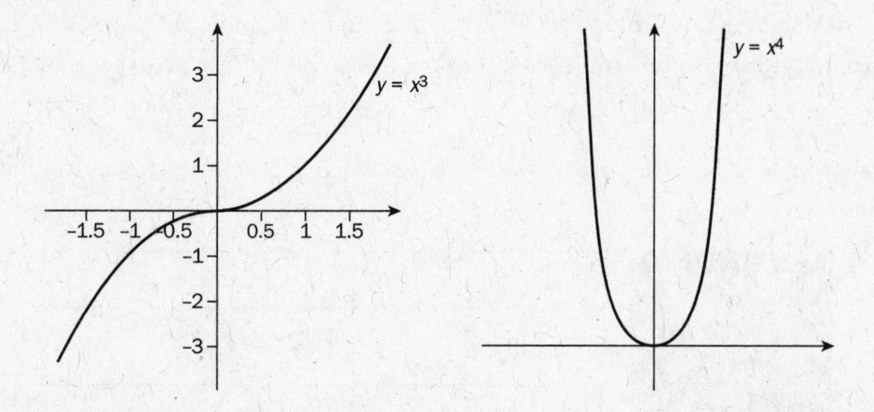

INFLECTION POINTS

The points at which a function changes concavity are of special interest. The graphs below show different ways in which a function might change concavity.

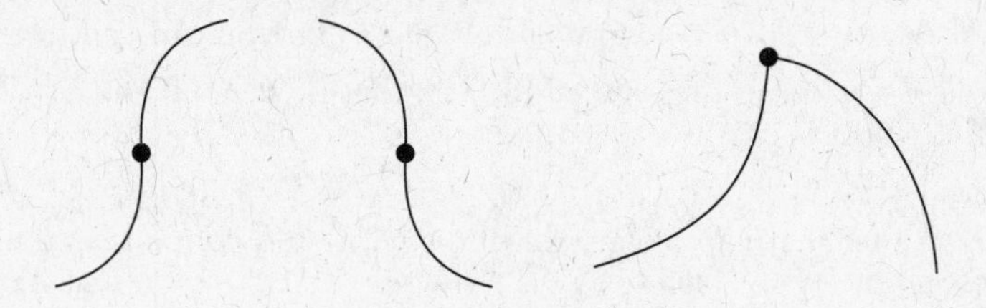

These points are called *inflection points*. Let's make this definition more precise: If f is continuous on an open interval containing a and f, *and* changes concavity at $(a, f(a))$, we say that the point is an inflection point or that f has an inflection point at $x = a$.

We do a sign analysis, similar to the first derivative sign analysis used in Chapter 14, to determine whether a point at which the second derivative is zero or undefined is an inflection point. We also use sign analysis to determine the intervals on which a function is concave up and the intervals on which it is concave down.

Example:

Using the function from the previous example, $f(x) = x + 2\sin x$:

(a) Do a sign analysis for the second derivative to determine the points of inflection on the interval $[0, 2\pi]$.

(b) Indicate the intervals on which f is concave down and on which it is concave up.

Solution:

(a) To find the inflection points of f on $[0, 2\pi]$, we first identify the points on $[0, 2\pi]$ where f'' is zero or undefined. We computed $f''(x) = -2\sin x$ in the example above. This function is defined everywhere, so inflection points can only occur at points where the second derivative is zero. We solve the equation $f'' = 0$ to find these points: $f''(x) = -2\sin x = 0 \Rightarrow \sin x = 0$.

On the interval $[0, 2\pi]$, this occurs where $x = n\pi$, where n is an integer. We do a sign analysis:

Interval	$0 < x < \pi$	$\pi < x < 2\pi$
Test Point	$x = \frac{\pi}{2}$	$x = \frac{3\pi}{2}$
Sign of f''	$f''\left(\frac{\pi}{2}\right) = -2\sin\frac{\pi}{2} = -2.$ negative	$f''\left(\frac{3\pi}{2}\right) = -2\sin\frac{3\pi}{2} = 2$ positive
Concavity of a	down	up

Because the sign of f'' changes at $x = \pi$, f has an inflection point at $x = \pi$. If you perform a sign analysis for the intervals $-\pi < x < 0$ and $2\pi < x < 3\pi$, you can confirm that f has inflection points at all $x = n\pi$ over the interval $[0, 2\pi]$; that is, at $x = 0$ and $x = 2\pi$ as well as at $x = \pi$.

(b) From the chart, we see that f is concave down on $(0, \pi)$ and concave up on $(\pi, 2\pi)$.

CORRESPONDING CHARACTERISTICS OF THE GRAPHS OF f, f', AND f''

We have now determined relationships between f, f', and f''. In Chapter 14, we discussed the relationships between the graph of f and the graph of f'; we can now add the following relationships:

- Because f is concave up where the second derivative is positive (i.e., the first derivative is increasing), the graph of f is concave up where the graph of f' is increasing or where the graph of f'' is above the x-axis.
- Because f is concave down where the second derivative is negative (i.e., the first derivative is decreasing), the graph of f is concave down where the graph of f' is decreasing or where the graph of f'' is below the x-axis.

Example:

The graph of the derivative of $y = f(x)$ on the interval $x_0 \leq x \leq x_4$ is shown below.

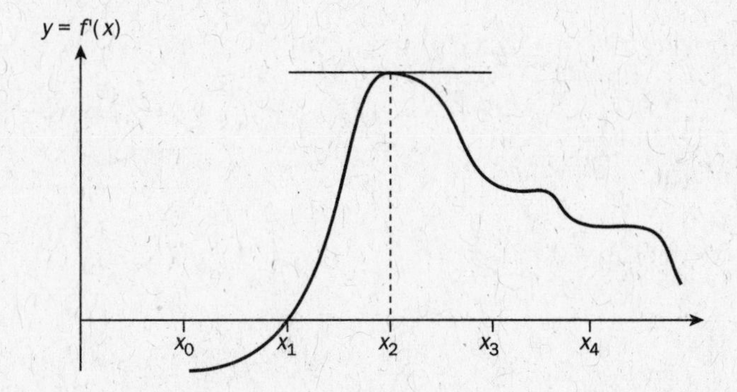

Which of the following lists all the points of inflection?

(A) x_1
(B) x_2
(C) x_3
(D) x_1, x_3
(E) x_2, x_3

Solution: B

The function f has inflection points at the points where f'' changes sign. The second derivative of f, that is f'', is the derivative of f'. The derivative of f' at a point is the slope of f' at that point. We can see on the graph of f', that the sign of the slope of f' changes from positive to negative at x_2—where the graph changes from increasing to decreasing—and at no other point, so the correct answer is (B).

REVIEW QUESTIONS

1. Find the point of inflection of $g(x) = x^2 - \frac{8}{x}$, $x > 0$.

   (A) 1
   (B) 2
   (C) 4
   (D) 8
   (E) 16

2. If $s''(t) = (t + 1)(t - 3)\sin^2 t$, then the inflection point(s) of the function $s(t)$ is/are at $t =$

   (A) $-1, 3$
   (A) -1 only
   (B) 3 only
   (C) $n\pi$, where n is an integer
   (D) $-1, 3, n\pi$, where n is an integer

Question 3 requires a calculator.

3. If $y = \frac{1}{3}x^3 - 4x^2 + x - 3\sin x$ then a point of inflection occurs at $x =$

   (A) 11.693
   (B) 9.128
   (C) 7.867
   (D) 6.597
   (E) 5.272

Question 4 requires a calculator.

4. If $f(x) = x^3 - \cos(2x)$, then $f''(x) = -1.5$ at $x =$

   (A) -1.338
   (B) -0.816
   (C) -0.551
   (D) -0.432
   (E) -0.300

5. The graph of the function f is shown in the figure below.

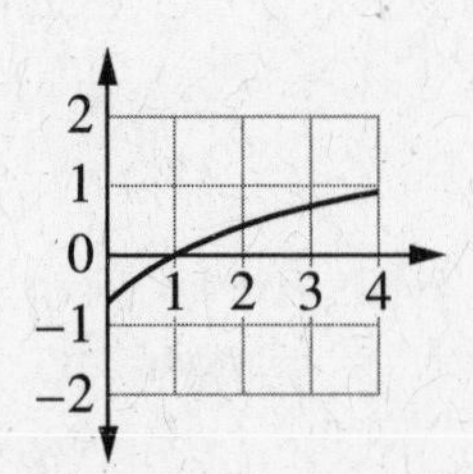

   Which of the following is true?

   (A) $f(1) < f'(1) < f''(1)$
   (B) $f(1) < f''(1) < f'(1)$
   (C) $f'(1) < f(1) < f''(1)$
   (D) $f''(1) < f(1) < f'(1)$
   (E) $f''(1) < f'(1) < f(1)$

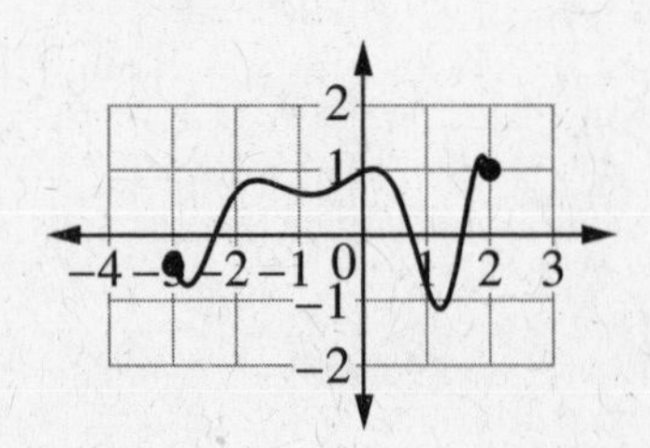

6. Which of the following statements is false?

   (A) $F(-2) + F''(-1) > 0$
   (B) $F'(-1) \cdot F''(-1) < 0$
   (C) $F'(0) \cdot F''(0) > 0$
   (D) $F'(0.5) \cdot F''(0.5) > 0$
   (E) $F'(1) - F''(1) < 0$

7. The graph of $y = x^4 + x^3 - 18x^2 + 4$ is concave down for

   (A) $x < 0$
   (B) $x > 0$
   (C) $x < -2$ or $x > \frac{3}{2}$
   (D) $x < \frac{3}{2}$ or $x > 2$
   (E) $-2 < x < \frac{3}{2}$

8. The domain of the function f is $x > 0$. If $f'(x) = x \ln x$, then $f(x)$ is concave down for all

(A) $0 < x < 1$

(B) $0 < x < e$

(C) $0 < x < \frac{1}{e}$

(D) $x > \frac{1}{e}$

(E) $x > e$

9. The x-coordinate of the inflection point of the curve $f(x) = e^{\tan-1x}$ is

(A) −1

(B) $-\frac{1}{2}$

(C) 0

(D) $\frac{1}{2}$

(E) 1

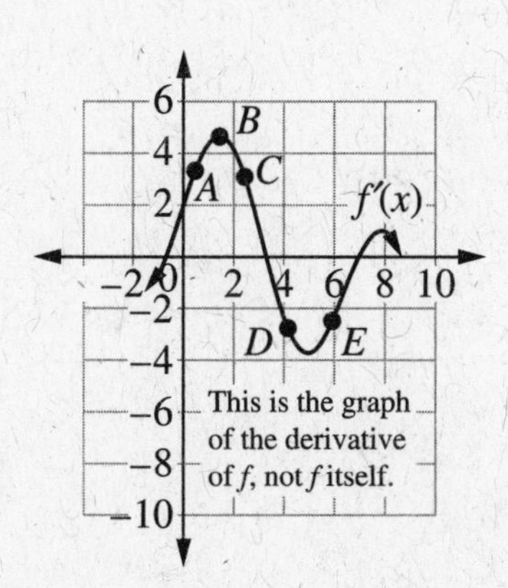

10. A graph of $y = f'(x)$, the derivative of a function f, is shown above. At which point(s) on the graph is $f'(x) \cdot f''(x) < 0$?

(A) A only

(B) C only

(C) D only

(D) D and E

(E) C and E

11. Consider a function f whose first derivative is given by $f'(x) = \frac{1-\ln x}{x^2}$. It is clear that $f'(e) = 0$, so e is a critical number. The value $f''(e)$ is

(A) negative, making $f(e)$ a local minimum

(B) positive, making $f(e)$ a local minimum

(C) negative, making $f(e)$ a local maximum

(D) positive, making $f(e)$ a local maximum

(E) none of the above

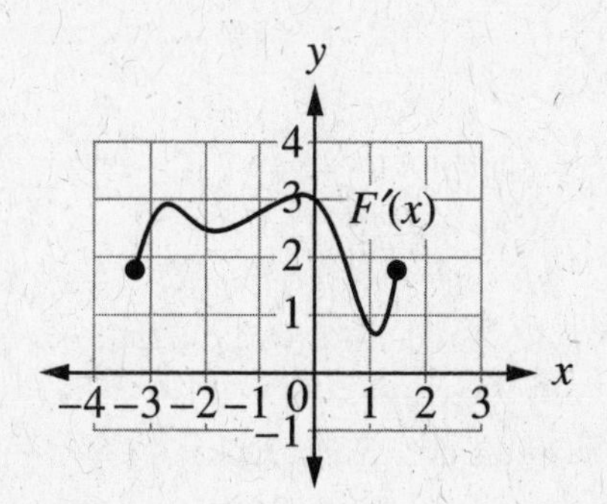

12. Consider the continuous function F whose first derivative $F'(x)$ is shown above. How many points of inflection does the function F have?

(A) None

(B) Two

(C) Three

(D) Four

(E) Five

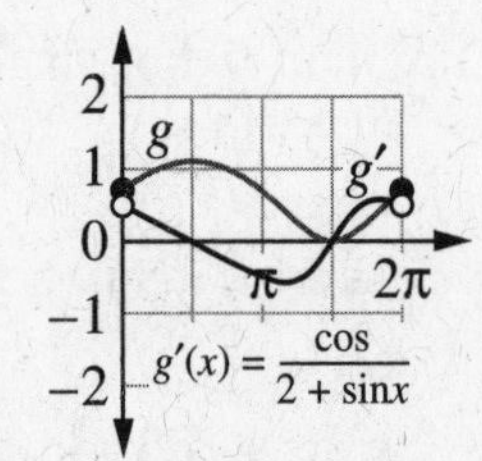

13. The function $g(x)$ is defined on the closed interval $[0, 2\pi]$ only. The graph of $g(x)$ and its derivative $g'(x)=\dfrac{\cos x}{2+\sin x}$ are shown above. Which of the following statements is false?

(A) $g'' = 0$ when $x=\dfrac{3\pi}{2}$.
(B) $g'(x)$ is increasing for $x \in \left(\frac{7\pi}{6}, \frac{11\pi}{6}\right)$ only.
(C) $g''(x) < 0$ when $x=\dfrac{\pi}{6}$.
(D) $g''(x) = 0$ at for two distinct x arguments on the closed interval $[0, 2\pi]$.
(E) $g(x)$ is concave up for $x \in \left(\frac{7\pi}{6}, \frac{11\pi}{6}\right)$ only.

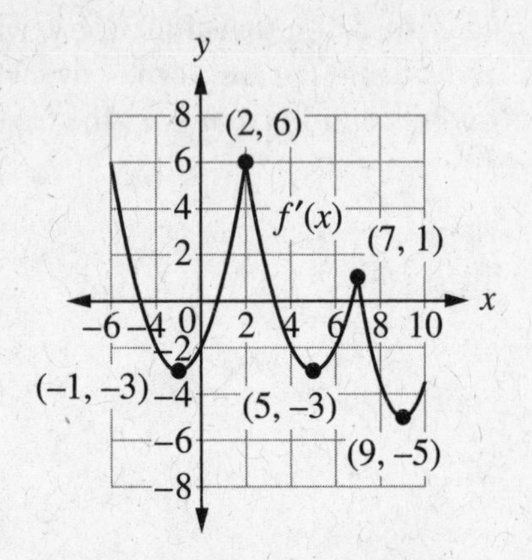

14. Consider the continuous function f whose first derivative $f'(x)$ is shown above. How many points of inflection does the function f have?

(A) None
(B) Two
(C) Three
(D) Four
(E) Five

15. A function g whose first derivative is given by $g'(x) = e^{-Kx}(2x - Kx^2)$ for $K > 0$. It is clear that $g'\left(\dfrac{2}{K}\right)=0$,, so $\dfrac{2}{K}$ is a critical number.

The value $g''\left(\dfrac{2}{K}\right)$ is

(A) negative, making $g\left(\dfrac{2}{K}\right)$ a relative maximum
(B) positive, making $g\left(\dfrac{2}{K}\right)$ a relative maximum
(C) negative, making $g\left(\dfrac{2}{K}\right)$ a relative minimum
(D) positive, making $g\left(\dfrac{2}{K}\right)$ a relative minimum
(E) none of the above

FREE-RESPONSE QUESTIONS

16. You are flying from Chicago to Atlanta. A little while after takeoff, the captain turns off the "fasten seatbelt" sign and announces that the plane has now reached a cruising speed of 635 miles per hour. Given that velocity is the derivative of position and that acceleration is the derivative of velocity, what is the acceleration of the plane at that moment?

17. If $f(x) = x^4 - 2x^3 + 4x - 3$, find the
    (a) point(s) of inflection
    (b) intervals on which the graph is concave up

18. Let $p(x)$ be a continuous function on the interval $[a, g]$ and differentiable on $(a, f) \cup (f, g)$.

x	a	b	c	d	e	f	g
$p'(x)$	negative	zero	positive	zero	negative	infinite	positive
$p''(x)$	positive	positive	zero	negative	negative	infinite	negative

Given the chart above for the first and second derivatives and assuming that the set $\{a,b,c,d,e,f,g\}$ contains all of the critical points and all the points at which inflection points occur, answer the following questions.

(a) What are the critical points of $p(x)$ on $[a, g]$?
(b) At which value(s) of x does a relative minimum occur? Justify your answer.
(c) At which point (or points) does a point of inflection occur? Justify your answer.
(d) If $p(c) = 0$, draw a reasonable sketch of the curve.

19. Let g be a function differentiable on the interval $[-2, 4]$, with $g(0) = -1$.

The graph of g', the derivative of g, is shown below. Answer the following questions.

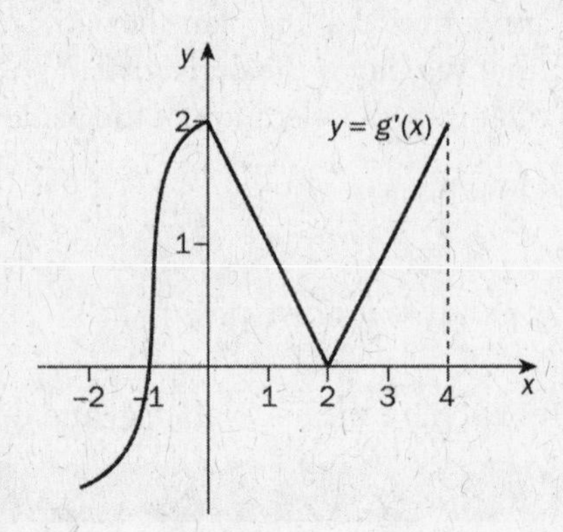

(a) Find all critical points.
(b) Find all the points at which relative minima occur.
(c) Find all the inflection points.
(d) On which interval(s) is/are the graph concave up?
(e) Find the equation of the tangent line at $(0,-1)$.
(f) Draw a reasonable sketch of the curve.

Note: **It helps to organize the relevant information with a sign chart or a sign analysis.**

20. A particle moves along the y-axis with velocity $v(t) = -\frac{2}{\pi}\sin\left(\frac{\pi}{2}t\right)$ cm/sec for time $t \geq 0$ in seconds.

(a) In what direction is the particle moving at $t = \frac{1}{3}$? What are the velocity and speed at this time?
(b) Find the earliest time, $t_1 > 0$, when the particle changes direction.
(c) What is the particle's average acceleration over the interval $[0, t_1]$?
(d) Does the concavity of the position function, $s(t)$, change sign over the interval $[0, t_1]$?

ANSWERS AND EXPLANATIONS

1. B

A point of inflection occurs when g'' changes sign. We compute:

$$g'(x) = 2x + \frac{8}{x^2}$$

$$g''(x) = 2 - \frac{16}{x^3}$$

We set $g''(x) = 0$ and solve the resulting equation:

$$2 - \frac{16}{X^3} = 0 \Rightarrow X^3 = 8 \Rightarrow X = 2$$

We use the point $x = 2$ to set up a sign analysis for the second derivative.

Interval	$0 < x < 2$	$x > 2$
Test Point	$x = 1$	$x = 3$
Sign of g''	$g''(1) = 2 - \frac{16}{1} = -14$ negative	$g''(3) = 2 - \frac{16}{27} \approx 1\frac{1}{2}$ positive
Concavity of g	down	up

Because g'' changes from negative to positive at $x = 2$, i.e., because g changes concavity, g has an inflection point at $x = 2$.

2. A

The only possible points at which the second derivative could change sign are places where it equals zero. We solve the equation $s''(t) = (t+1)(t-3)\sin^2 t = 0$ to find $s'' = 0$ at $t = -1, 3, n\pi$, n is an integer.

Notice that because the factor $\sin^2 t$ in the equation of s'' appears to the second power, it is always positive, i.e., $\sin^2(t) > 0$. Therefore the sign of s'' does not depend on the sign of this factor and the second derivative will not change sign at $n\pi$ (the roots of $\sin^2 t$); we only need to include $t = -1, 3$ in our sign analysis of the second derivative.

Interval	$t < -1$	$-1 < t < 3$	$t > 3$
Test Point	$t = -2$	$t = 0$	$t = 4$
Sign of s''	$s''(-2) = (-2+1)(-2-3)(+)$ $= (-)(-)(+) = (+)$ positive	$s''(0) = (0+1)(0-3)(+)$ $= (+)(-)(+) = (-)$ negative	$s''(4) = (4+1)(4-3)(+)$ $= (+)(+)(+) = (+)$ positive
Concavity	up	down	up

Because s'' changes sign at both $t = 3$ and $t = -1$, both of these are inflection points. This can also be phrased in terms of concavity. There is a point of inflection at $t = -1, 3$, because s'' changes concavity at both $t = 3$ and $t = -1$; both of these points are inflection points.

3. E

First compute $y' = x^2 - 8x + 1 - 3\cos x$ and $y'' = 2x - 8 + 3\sin x$ without using a calculator and then use a graphing calculator to view the graph of y''. Zoom in on the places where the graph appears to cross the x-axis; these are the points where the second derivative changes sign, hence they are the inflection points.

TI-83: We enter the equation for y'' in the Y= screen. The values of x in the window should contain the answer choices for x. To find the zero of y'', the values of y should contain zero. Set the window: $4 \le x \le 12$, $-2 \le y \le 2$. We now want to find the zero of the second derivative. Press GRAPH, 2nd TRACE, 2 for zero of the function. Toggle to the left of zero, press ENTER. Toggle to the right of zero and press ENTER twice. The answer appears at the bottom.

4. C

To find the solution we first need to find f' and f''. We compute:

$$f'(x) = 3x^2 + \underbrace{2\sin(2x)}_{\text{chain rule}}$$

$$f''(x) = 6x + \underbrace{4\cos(2x)}_{\text{chain rule; constant multiple rule}}$$

To solve the equation $f''(x) = -1.5$, we can use a graphing calculator to solve $f''(x) + 1.5 = 0$: $f''(x) + 1.5 = 6x + 4\cos(2x) + 1.5 = 0$.

TI-83: Enter $6x + 4\cos(2x) + 1.5$ in your Y= screen. To find the solution, the values of x shown in the graph should contain the answer choices for x; the range $-2 \le x \le 1$ makes sense. To find the zero of y'', the graph should show where $y = 0$; the range $-2 \le y \le 2$ makes sense here. Set the window: , $-2 \le x \le 1$, $-2 \le y \le 2$. To find the zero of the equation, press GRAPH, 2nd TRACE, 2 for zero of the function. Toggle to the left of zero, press ENTER. Toggle to the right of zero and press ENTER twice. The answer appears at the bottom.

5. D

We need to compare the values of $f(1), f'(1)$, and $f''(1)$. From the graph, it is clear that 1 is an x-intercept so $f(1) = 0$, that f is increasing at $x = 1$ so $f'(1) > 0$, and finally that f is concave down at $x = 1$ so $f''(1) < 0$. This $f''(1) < f(1) < f'(1)$.

6. C

We need to compare lots of values. To start it is true that $F(-2) + F''(-1) > 0$ because $F(-2) > 0$ and F is concave up at $x = -1$ so $F''(-1) > 0$. Now we focus our attention at $x = -1$ where F is decreasing and concave up making $F'(-1) < 0$ and $F''(-1) > 0$. This means $F'(-1).\ F''(-1) < 0$ is also true. Now we turn our attention at $x = 0$ where F is increasing and concave down making $F'(0) > 0$ and $F''(0) < 0$ so $F'(0).\ F''(0) < 0$, not $F'(0).\ F''(0) > 0$. So the statement (C) is false. The final two statements are true for similar reasons.

7. E

To find where $y = x^4 + x^3 - 18x^2 + 4$ is concave down, we need to find the second derivative and determine x values that make $y'' < 0$.

$$y = x^4 + x^3 - 18x^2 + 4 \qquad 6(2x^2 + x - 6) < 0$$
$$y' = 4x^3 + 3x^2 - 36x \qquad 6(2x - 3)(x + 2) < 0$$
$$y'' = 12x^2 + 6x - 36 \qquad \therefore -2 < x < \frac{3}{2}$$

8. C

The function $f(x)$ is concave down for all x in the domain where $f''(x) < 0$. To start, we find the second derivative, its zero, and look to its graph (see below). The graph reveals that $f''(x) < 0$ for $0 < x < \frac{1}{e}$.

$$f'(x) = x \ln x$$
$$f''(x) = \ln x + x \cdot \frac{1}{x}$$
$$f''(x) = \ln x + 1$$
$$\ln x + 1 = 0$$
$$\ln x = -1$$
$$x = e^{-1} = \frac{1}{e}$$

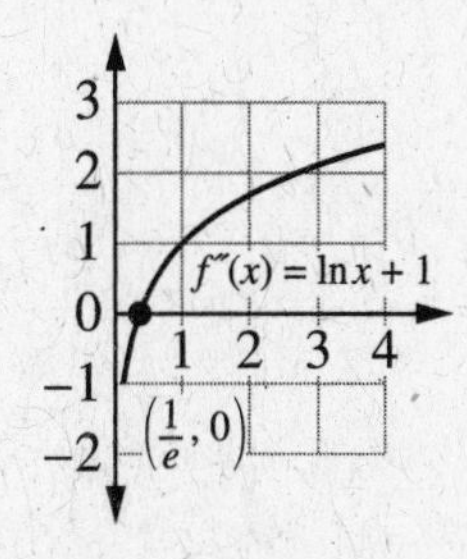

9. D

To find an *x*-coordinate of the inflection point of the curve $f(x) = e^{\tan^{-1}x}$ we need to find the second derivative.

$$f(x) = e^{\tan^{-1}x}$$

$$f'(x) = e^{\tan^{-1}x} \cdot \frac{1}{1+x^2}$$

$$f'(x) = \frac{e^{\tan^{-1}x}}{1+x^2}$$

$$f''(x) = \frac{\frac{e^{\tan^{-1}x}}{1+x^2} \cdot (1+x^2) - 2x \cdot e^{\tan^{-1}x}}{\left(1+x^2\right)^2}$$

$$f''(x) = \frac{e^{\tan^{-1}x} - 2x \cdot e^{\tan^{-1}x}}{\left(1+x^2\right)^2}$$

$$f''(x) = \frac{e^{\tan^{-1}x}(1-2x)}{\left(1+x^2\right)^2}$$

$$\frac{e^{\tan^{-1}x}(1-2x)}{\left(1+x^2\right)^2} = 0$$

$$x = \frac{1}{2}$$

$f''(x)$ changes signs at $x = \frac{1}{2}$

10. D

To find points where $f'(x) \cdot f''(x) < 0$, we need to start by recognizing that $f'(x) > 0$ at points *A*, *B*, and *C* and $f'(x) < 0$ at points *D* and *E*. Now we turn our attention to the second derivative. $f''(x) < 0$ at those points where $f'(x)$ is increasing, namely at points *A* and *E*. Similarly, $f'(x) < 0$ at those points where $f'(x)$ is decreasing, namely at points *C* and *D*. So we need to find points where the first and second derivative values are signed differently. At points *A* and *D*, we will have $f'(x) \cdot f'(x) < 0$.

11. C

The Second Derivative Test will help answer this question.

$$f''(x)=\frac{-\frac{1}{x}\cdot x^2-2x(1-\ln x)}{x^4}$$

$$f''(x)=\frac{-x-2x+2x\ln x}{x^4}$$

$$f''(x)=\frac{2x\ln x-3x}{x^4}$$

$$f''(x)=\frac{2\ln x-3}{x^3}$$

$$f''(e)=\frac{2\ln e-3}{e^3}=-\frac{1}{e^3}$$

Since $f'(e) = 0$ and $f''(e) < 0$, $f(e)$ is a local maximum.

12. D

The function $F(x)$ will have a point of inflection at every x in the domain where its derivative $F'(x)$ changes from increasing to decreasing or decreasing to increasing. There are four points on the graph where this happens.

13. A

To start, it appears that g changes concavity twice so we find the zeros of the second derivative to identify the x-coordinates of the inflection points.

$$g'(x)=\frac{\cos x}{2+\sin x}$$

$$g''(x)=\frac{-\sin x(2+\sin x)-\cos x\cdot\cos x}{(2+\sin x)^2}$$

$$g''(x)=\frac{-2\sin x-\sin^2 x-\cos^2 x}{(2+\sin x)^2}$$

$$g''(x)=\frac{-2\sin x-(\sin^2 x+\cos^2 x)}{(2+\sin x)^2}$$

$$g''(x)=\frac{-2\sin x-1}{(2+\sin x)^2}$$

$g''(x)$ will equal zero when

$$-2\sin x-1=0$$

$$\sin x=-\frac{1}{2}$$

$\therefore x=\frac{7\pi}{6},\frac{11\pi}{6}$ are the x-coordinates of inflection points

With this information, (B), (C), (D), and (E) are true as g is concave up when its derivative is increasing. (A) is false.

14. E

The function $f(x)$ will have a point of inflection at every x in the domain where its derivative $f'(x)$ changes from increasing to decreasing or decreasing to increasing. There are five points on the graph where this happens, at $x = -1, 2, 5, 7$, and 9.

15. A

To start, we confirm that $\frac{2}{K}$ is a critical number.

$$g'(x) = e^{-Kx}(2x - Kx^2) = 0$$

$$e^{-Kx} \cdot x(2 - Kx) = 0$$

Since $e^{-Kx} > 0$ for all x, then critical numbers are the solutions to

$$x(2 - Kx) = 0$$

$$\therefore x = 0, \frac{2}{k} \text{ are critical numbers.}$$

Since $\frac{2}{K}$ is a critical number then $g\left(\frac{2}{K}\right)$ is a candidate for being a relative extrema. We use the Second Derivative Test and test $g''\ \frac{2}{K}$.

$$\begin{aligned} g'(x) &= e^{-Kx}(2x - Kx^2) = 0 \\ g''(x) &= -Ke^{-Kx}(2x - Kx^2) + e^{-Kx}(2 - 2Kx) \\ g''(x) &= e^{-Kx}(-2Kx + K^2x^2 + 2 - 2Kx) \\ g''(x) &= e^{-Kx}(K^2x^2 - 4Kx + 2) \end{aligned}$$

The Second Derivative Test

$$g\left(\frac{2}{K}\right) = e^{-K \cdot 2/K}\left(K^2\left(\frac{2}{K}\right)^2 - 4K \cdot \frac{2}{K} + 2\right)$$

$$g\left(\frac{2}{K}\right) = e^{-2}(4 - 8 + 2) = \frac{-2}{e^2} < 0$$

Since $g'\left(\frac{2}{K}\right) = 0$, and $g''\left(\frac{2}{K}\right)$ then $g\left(\frac{2}{K}\right)$ is a relative maximum.

FREE-RESPONSE ANSWERS

16. Because speed is the absolute value of velocity, when the velocity is constant, the rate of change of the velocity—which is the acceleration—is zero.

17. (a) and (b) The inflection points occur where the second derivative changes sign. Therefore we start by computing f' and f'':

$$f'(x) = 4x^3 - 6x^2 + 4$$

$$f''(x) = 12x^2 - 12x$$

The points of inflection can only occur where the second derivative is zero or undefined. We see above that f'' is a polynomial; therefore it is defined everywhere and an inflection point can only occur where $f''(x) = 0$. We solve this equation:

$$f''(x) = 0 \Rightarrow 12x^2 - 12x = 0 \Rightarrow 12x(x-1) = 0 \Rightarrow x = 0 \quad or \quad x = 1$$

To determine which of these points are inflection points, we do a sign analysis of f'':

Interval	$x < 0$	$0 < x < 1$	$x > 1$
Test Point	$x = -1$	$x = \frac{1}{2}$	$x = 2$
Sign of f''	$f''(-1) = 12(-1)(-1-1)$ $= 24$ positive	$f''\left(\frac{1}{2}\right) = 12\left(\frac{1}{2}\right)\left(\frac{1}{2}-1\right)$ $= -3$ negative	$f''(2) = 12(2)(2-1)$ $= 24$ positive
Concavity	up	down	up

We see on the chart that at both $x = 0$ and $x = 1$, the second derivative changes sign (and so the graph changes concavity). Therefore, inflection points occur at both $x = 0$ and $x = 1$.

18. (a) b, d, f

Critical points occur when $p' = 0$ *or* p' does not exist.

$p'(b) = p'(d) = 0$ and $p'(f)$ does not exist.

(b) Sign graph for p':

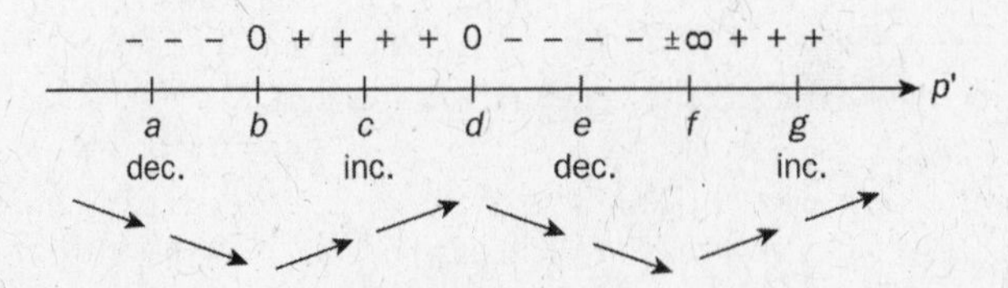

We see from the sign chart that a relative minimum occurs at $x = b, f$, because the slope of p changes sign from negative to positive at these points.

(c) Sign graph for p'':

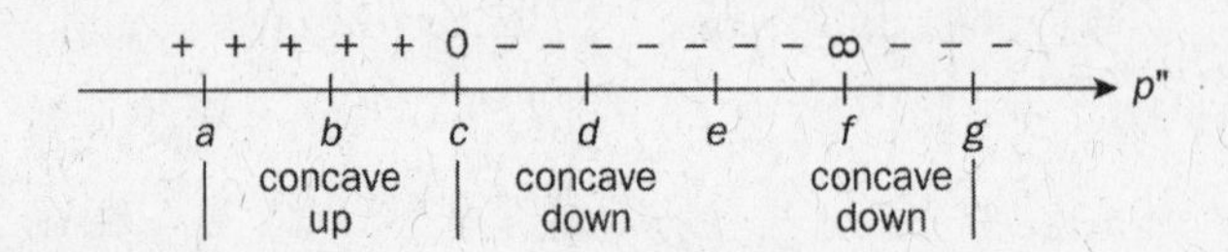

We see from the sign chart that an inflection point occurs at $x = c$, because the sign of p'' changes from positive to negative at this point.

(d)

19. *Note:* It helps to organize the relevant information with a sign chart or a sign analysis.

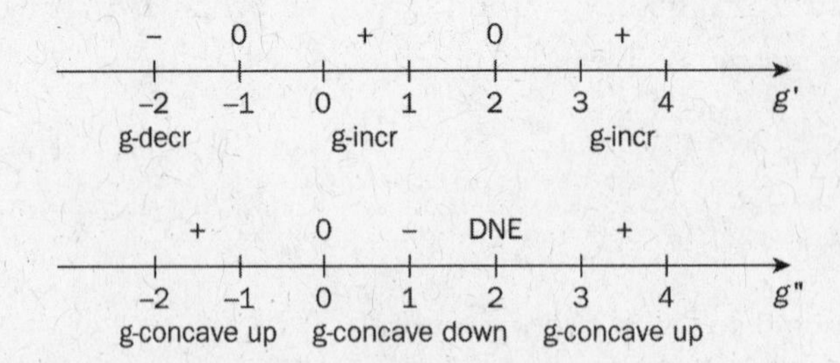

(a) Critical points occur when $g' = 0$ or g' DNE. We can see from the graph in the problem, or from the sign chart above, that g' exists everywhere on $[-2,4]$ and $g'(x) = 0$ for $x = -1, 2$. Therefore $x = -1, 2$ are the only critical points.

(b) We can see from the graph or from the sign chart that by the first derivative test, a relative minimum occurs at $x = -1$, because the sign of g' changes from negative to positive at this point.

(c) We can see from the graph or from the sign chart for g'' that inflection points occur at $x = 0$ and $x = 2$ because the sign of g'' changes at these points.

(d) We can see from the graph or from the sign chart for g'' that the graph of g is concave up on $(-2,0) \cup (2,4)$ when $g'' > 0$.

(e) The slope of the tangent line at $x = 0$ can be read off the graph: $m = 2$.

We use the point-slope form of the equation of a line:

$$y - y_0 = m(x - x_0),\ (x_0, y_0) = (0, -1)$$

$$y - (-1) = 2(x - 0)$$

$$y = 2x - 1$$

(f) Answers may vary, but one possible graph is as follows:

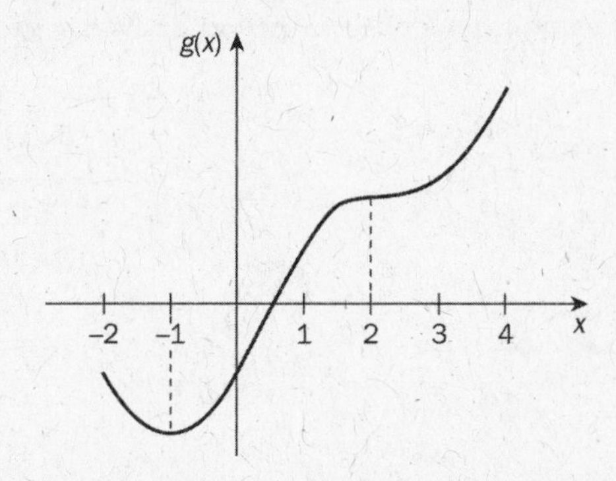

20. (a) We can determine the particle's direction by checking the sign of v; if v is positive, the height of the particle is increasing, so it is moving up. If v is negative, the height of the particle is decreasing, so it is moving down. We compute:

$$v\left(\frac{1}{3}\right) = -\frac{2}{\pi}\sin\left(\frac{\pi}{2}\cdot\frac{1}{3}\right) = -\frac{2}{\pi}\sin\left(\frac{\pi}{6}\right) = -\frac{2}{\pi}\cdot\frac{1}{2} = -\frac{1}{\pi} < 0$$

Because the velocity is negative, the particle is moving down the y-axis.

Because speed is the absolute value of velocity, the particle's speed at $t = \frac{1}{3}$ is $\frac{1}{\pi}$ cm/sec.

Velocity, $v(t)$, is the derivative of position $s(t)$. Acceleration is the derivative of velocity, $a(t) = v'(t)$.

***Note:* According to the AP Calculus Course Description booklet, you are expected to know the trigonometric functions of $0, \frac{\pi}{6}, \frac{\pi}{4}, \frac{\pi}{3}, \frac{\pi}{2}$ and their multiples without a calculator.**

(b) The particle changes direction when the sign of the velocity changes. Because the velocity function is defined everywhere, this can only happen when $v(t) = 0$. We set $v(t) = 0$ and solve this equation:

$$-\frac{2}{\pi}\sin\left(\frac{\pi}{2}t\right) = 0 \Rightarrow \sin\left(\frac{\pi}{2}t\right) = 0 \Rightarrow \frac{\pi}{2}t = n\pi \Rightarrow t = 2n,\ n \text{ an integer.}$$

According to the question, we need $t > 0$, therefore, the earliest possibility is when $n = 1$, i.e., $t = 2$ seconds. We now need to confirm that the velocity actually changes sign at $t = 2$, i.e., that this is a relative extremum and not merely a critical point.

To do this we can apply the second derivative test to $t = 2$. We compute v', the derivative of the velocity function (which is the second derivative of the position function) and evaluate $v'(2)$:

$$v(t) = -\frac{2}{\pi}\sin\left(\frac{\pi}{2}t\right)$$

$$v'(t)(= a(t)) = \underbrace{-\frac{2}{\pi}\cdot\frac{\pi}{2}\cos\frac{\pi}{2}t}_{\text{chain rule}} = -\cos\frac{\pi}{2}t$$

$$v'(2) = -\cos\pi = 1 > 0$$

It follows from the second derivative test that the point changes direction at time $t_1 = 2$.

(c) The average acceleration over the interval is the change in velocity over the change in time, $\frac{\Delta v}{\Delta t}$. We compute:

$$\frac{\Delta v}{\Delta t} = \frac{v(2) - v(0)}{2 - 0} = \frac{-\frac{2}{\pi}\sin\left(\frac{\pi}{2}\cdot 2\right) - \left(-\frac{2}{\pi}\sin\left(\frac{\pi}{2}\cdot 0\right)\right)}{2} = 0$$

The average acceleration over the interval $[0, t_1]$ is zero.

(d) Because the velocity function is continuous, the only points at which the concavity of the position function can change are points where its second derivative $s''(t) = a(t)$ is zero or undefined. In part (b) we computed $s''(t) = a(t) = -\cos\frac{\pi}{2}t$. This function is defined everywhere, so the concavity of s can only change at points where $a = 0$. We solve this equation $a(t) = -\cos\frac{\pi}{2}t = 0 \Rightarrow \cos\frac{\pi}{2}t = 0 \Rightarrow \frac{\pi}{2}t = \frac{\pi}{2} + n\pi \Rightarrow t = 1 + 2n$, where n is an integer, to find the value of t. When $n = 0$, $t = 1$ lies in the interval $[0,2]$. This is the only possible point in the interval $[0,2]$ where the concavity of s could change. To determine whether the concavity does, in fact, change at this point, we do a sign analysis of $s'' = a$.

Interval	$0 < t < 1$	$1 < t < 2$
Test Point	$x = \frac{1}{2}$	$x = \frac{3}{2}$
Sign of a	$a\left(\frac{1}{2}\right) = -\cos\left(\frac{\pi}{2}\cdot\frac{1}{2}\right) = -\frac{\sqrt{2}}{2}$ negative	$a(2) = -\cos\left(\frac{\pi}{2}\cdot\frac{3}{2}\right) = +\frac{\sqrt{2}}{2}$ positive
Concavity of a	down	up

The chart shows that $s(t)$ is concave down on $(0, 1)$ and concave up on $(1, 2)$. Therefore the position graph has an inflection point at $t = 1$, and the graph of s changes concavity at this point.

CHAPTER 16: APPLICATIONS OF DERIVATIVES

IF YOU LEARN ONLY TWO THINGS IN THIS CHAPTER . . .

1. The typical steps for solving an optimization problem are as follows:
   a) Draw a picture.
   b) Write down the quantity to be maximized or minimized.
   c) Define variables; one variable should be the quantity you are trying to optimize.
   d) Write a function for the quantity to be optimized in terms of the other variables in the problem.
   e) Write the constraint equation(s) (if there are any) in terms of your variables.
   f) If the function you are trying to optimize is a function of more than one variable, use the constraint equation to eliminate all but one variable.
   g) Optimize the function by finding the critical points, determining the nature of each critical point, and evaluating the function at the critical points and at the endpoints of the domain (if applicable).
   h) Make sure to answer the question that is asked, paying attention to units and rounding.
2. L'Hôpital's Rule: If we have a limit that is in indeterminate form of type $\frac{0}{0}$ *or* $\frac{\infty}{\infty}$, then the limit of the quotient of the functions is equal to the limit of the quotient of their derivatives.

GRAPHING USING DERIVATIVES

In Chapter 8, we saw how to graph rational functions and other functions with asymptotes without using the derivative. We can now use the first and second derivatives of *f* to add detail to our graphs.

INCREASING/DECREASING BEHAVIOR OF *f*

To determine the intervals on which *f* is increasing and the intervals on which it is decreasing, we do a sign analysis using the first derivative, as in Chapter 14.

CONCAVITY AND INFLECTION POINTS

To analyze the concavity of *f* and to find its inflection points, we compute f'' and determine the points at which the second derivative is zero or undefined. We do a sign analysis as in Chapter 15, to determine whether or not these points are inflection points and to determine the intervals on which the function is concave up and the intervals on which it is concave down. Remember that the AP readers expect more from your answers than just a sign chart, as mentioned in Chapter 14.

SKETCHING THE GRAPH OF *f*

We can make a table combining all the information about the function *f* and use that table to draw an accurate sketch. The table should include the following:

- The *y*-intercept of *f* (if 0 is in the domain of *f*)
- The *x*-intercepts of *f*, if there are any
- The vertical asympotes of *f*, where applicable
- The end behavior of *f* (using $\lim\limits_{x\to\infty} f(x)$ and $\lim\limits_{x\to-\infty} f(x)$), which will include the horizontal, oblique, or curvilinear asympotes of *f*, where applicable
- The critical points of *f* and then a sign analysis to determine whether these points are maxima, minima, or neither, and to analyze the increasing or decreasing behavior of *f*; remember that a sign chart alone will not be enough to get full credit on a free-response question that asks for this sort of information
- The points at which $f''(x) = 0$ and a sign analysis to determine whether each of these points is an inflection point, and to analyze the concavity of *f*

Example:

Which of the following could be the graph of $y = \dfrac{x^2 - 1}{x^2 - 3}$?

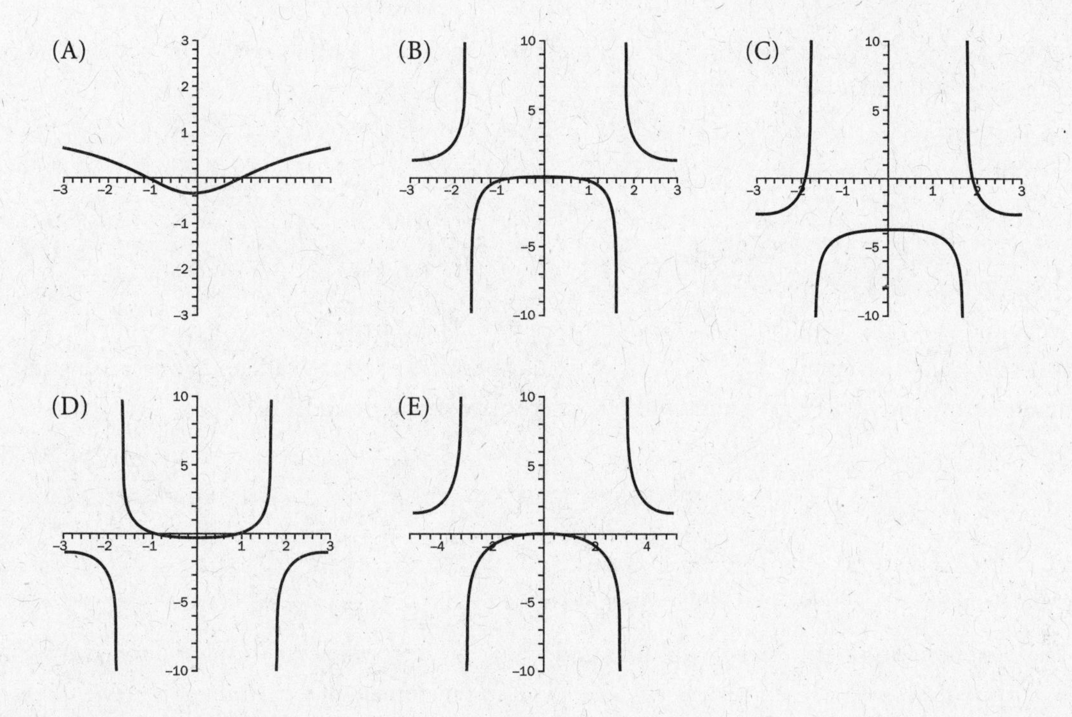

Solution: B

We can approach this problem two ways. Because it is a multiple-choice problem, one approach we can use is the process of elimination. The function $y = \dfrac{x^2 - 1}{x^2 - 3}$ has vertical asymptotes at $x = \pm\sqrt{3}$, since these values make the denominator zero and $\lim\limits_{x \to \sqrt{3}} \dfrac{x^2 - 1}{x^2 - 3}$ and $\lim\limits_{x \to -\sqrt{3}} \dfrac{x^2 - 1}{x^2 - 3}$ do not exist. Therefore, we can eliminate (A)—which does not have any vertical asymptotes—and (E), whose vertical asymptotes occur at about $x = \pm 3$. Our function has a horizontal asymptote at $y = 1$ because $\lim\limits_{x \to \infty} \dfrac{x^2 - 1}{x^2 - 3}$ and $\lim\limits_{x \to -\infty} \dfrac{x^2 - 1}{x^2 - 3}$ both equal 1. Using this information, we can eliminate (C) and (D). The only remaining choice is (B) and this is correct.

We saved ourselves a lot of work on this problem by taking the "eliminate answer choices" approach. Use the multiple-choice answers to your advantage!

However, we could also work this problem the traditional way; let's do it now to get some practice.

Following the list of steps on the previous page we find the *x*- and *y*-intercepts of the function by evaluating $y(0)$ and solving $y = 0$:

$$y(0) = \frac{0^2 - 1}{0^2 - 3} = \frac{1}{3}$$

$$0 = \frac{x^2 - 1}{x^2 + 3} \Rightarrow x^2 - 1 = 0 \Rightarrow x = \pm 1$$

We found the vertical and horizontal asymptotes above. There are vertical asymptotes at $x = \pm\sqrt{3}$ and a horizontal asymptote at $y = 1$. Next, we find the critical points by differentiating y using the quotient rule and finding the points where the derivative is zero or undefined:

$$y = \frac{x^2 - 1}{x^2 - 3}$$

$$y' = \frac{2x(x^2 - 3) - (x^2 - 1)2x}{(x^2 - 3)^2} = \frac{-4x}{(x^2 - 3)^2}$$

The critical points occur where y' is zero or undefined; y' is zero when $x = 0$ and undefined at $x = \pm\sqrt{3}$. Even though the points $x = \pm\sqrt{3}$ are not in the domain of y, the function may change from increasing to decreasing at these points, so we still include them in our sign analysis. We do a sign analysis to determine the increasing/decreasing behavior of y, to determine if a relative extremum occurs at the critical point $x = 0$.

Interval	Test Point	Sign of y'	Increasing/ Decreasing
$x < -\sqrt{3}$	$x = -2$	$y'(-2) = \frac{2(-2)((-2)^2 - 3) - ((-2)^2 - 1)2(-2)}{((-2)^2 - 3)^2}$ $= 8 > 0$ positive	increasing
$-\sqrt{3} < x < 0$	$x = -1$	$y'(-1) = \frac{2(-1)((-1)^2 - 3) - ((-1)^2 - 1)2(-1)}{((-1)^2 - 3)^2}$ $= 1 > 0$ positive	increasing
$0 < x < \sqrt{3}$	$x = 1$	$y'(1) = \frac{2 \cdot 1(1^2 - 3) - (1^2 - 1)2 \cdot 1}{(1^2 - 3)^2} =$ $= -1 < 0$ negative	decreasing
$x > \sqrt{3}$	$x = 2$	$y'(2) = \frac{2 \cdot 2(2^2 - 3) - (2^2 - 1)2 \cdot 2}{(2^2 - 3)^2}$ $= -8 < 0$ negative	decreasing

From the sign analysis, we see that the function is increasing on $x < -\sqrt{3}$ and $-\sqrt{3} < x < 0$, decreasing on $0 < x < \sqrt{3}$ and $x > \sqrt{3}$, and has a relative maximum at $x = 0$. The value of the maximum is $y(0)$; earlier we computed this to be $\frac{1}{3}$.

We continue with our analysis by determining the concavity of y and the location(s) of any inflection points. We compute:

$$y'' = \frac{-4(x^2 - 3)^2 + 4x \cdot 2x \cdot 2(x^2 - 3)}{(x^2 - 3)^4} = \frac{12(x^2 + 1)}{(x^2 - 3)^3}$$

The function can only change concavity where the second derivative is zero or undefined. We see from the equation of the second derivative that it is never equal to zero. It is undefined where the denominator equals zero, that is, where $x = \pm\sqrt{3}$. Even though these points are not in the domain of y, and, hence, we do not talk about inflection points occurring at $x = \pm\sqrt{3}$, the function could still change concavity at these points. We do a sign analysis of the second derivative to determine the concavity of the graph.

Interval	Test Point	Sign of y''	Concavity
$x < -\sqrt{3}$	$x = -2$	$y''(-2) = \frac{12((-2)^2 + 1)}{((-2)^2 - 3)^3} = 60 > 0$ positive	concave up
$-\sqrt{3} < x < \sqrt{3}$	$x = -1$	$y''(-1) = \frac{12((-1)^2 + 1)}{((-1)^2 - 3)^3} = -3 < 0$ negative	concave down
$x > \sqrt{3}$	$x = 2$	$y''(2) = \frac{12(2^2 + 1)}{(2^2 - 3)^3} = 60 > 0$ positive	concave up

We use this information to draw the graph of y and our graph looks like (B).

Wasn't it much easier the other way?

AP EXPERT TIP

Look for clues that can help you reduce your workload when graphing a function. For example, an even function will be symmetrical about the x-axis. Thus, when you find one inflection point, simply reflect across the axis to find its mate!

OPTIMIZATION

In math, optimization refers to finding the maximum or minimum value of a quantity, usually subject to specific conditions. For example, companies try to maximize profits and minimize costs; this optimization may be subject to constraints such as factory capacity, office space, or other limitations.

Optimization problems are sometimes called applied max/min problems because we are trying to find the maximum or minimum value of a function. This is where calculus comes in. If the quantity we are trying to optimize is described by a continuous function, we can use the information that calculus provides to find the maximum or minimum value of the function. In Chapter 14, we saw that the maximum or minimum of a continuous function can only occur at a critical point, or, if the function is continuous and we are looking at a closed interval, at one of the endpoints of the interval.

Example:

> Farmer Fred wants to build a 150-square-meter rectangular pigpen for his pigs. One wall of the pen will be the existing barn wall. The other three walls will be built from mesh fencing that costs $20/meter.
>
> (a) Find the dimensions of the pigpen that will minimize Farmer Fred's fencing cost.
>
> (b) Find the cost of the fencing materials for that pigpen, to the nearest dollar.
>
> (c) To the nearest decimeter, how much fencing material will Farmer Fred need to buy to construct the pigpen?

Solution:

The best way to start any modeling, real-life, or word problem is to draw a picture, if possible.

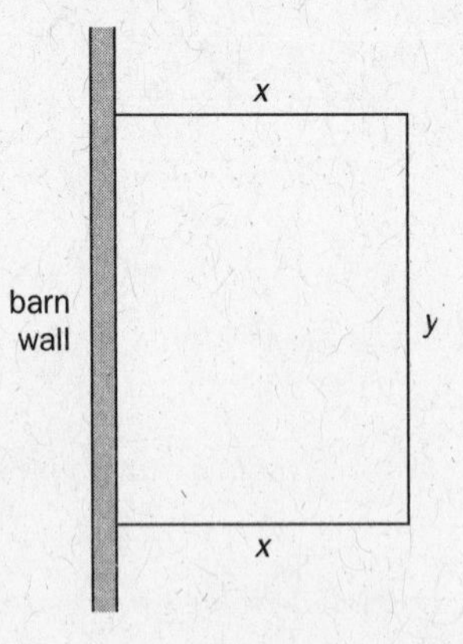

We are trying to minimize the cost of the fencing materials for the pigpen. Let's call this quantity C. Looking at our figure, the total amount of fencing is the sum of the length of the three free-standing sides of the fence. We'll let x be the length of the sides adjacent to the barn wall and y be the length of the side opposite the barn wall.

Now, we'll write an equation that expresses the cost in terms of these lengths. The cost is \$20 times the total length of fencing that needs to be purchased. That is, $C = 20(x + y + x) = 20(2x + y)$.

We call this the *equation to be optimized.* From Chapter 14, we know how to find the minimum value of a function of one variable. If we can express C in terms of just one variable, i.e., if we can eliminate x or y, then we will be able to use this knowledge to minimize the cost.

How do we eliminate x or y? Well, we have some additional information about Farmer Fred's pigpen. He needs the pigpen to have an area of 150 square meters. We can express this constraint in terms of x and y: $xy = 150$.

This is called a *constraint equation.* We solve the constraint equation for either one of the variables, let's say y: $y = \frac{150}{x}$. Now we rewrite the equation to be optimized by replacing y with the expression above: $C = 20(2x + y) = 20\left(2x + \frac{150}{x}\right)$. Because C is defined in terms of the length x of one of the sides of the pigpen, C is defined for $x > 0$.

Up to this point, all we've been doing is algebra. Now we are ready to use some calculus. From Chapter 14, we know that the minimum value of this function can only occur at a critical point or at an endpoint of its interval of definition, if the endpoint is included. Because the domain of C is $x > 0$, the minimum value of C can only occur at a critical point, since the endpoint 0 is not in the domain of C. To find the critical points of C we first find its derivative: $C' = 20\left(2 - \frac{150}{x^2}\right)$. The critical points are the points where $C' = 0$ or C' is undefined. C' is undefined at $x = 0$, but $x = 0$ is not in the domain of C. Hence, the only critical points of C occur where $C' = 0$. That is:

$$\begin{aligned}
20\left(2 - \frac{150}{x^2}\right) &= 0 \\
2 - \frac{150}{x^2} &= 0 \\
2x^2 - 150 &= 0 \\
x^2 - 75 &= 0 \\
x^2 &= 75 \\
x &= \pm 5\sqrt{3}
\end{aligned}$$

AP EXPERT TIP

Since we are working on an open interval and we only have ONE relative minimum, we know it must also be our absolute minimum, since we can't check endpoints and there are no other candidates!

AP EXPERT TIP

Since we are working on an open interval and we only have ONE relative minimum, we know it must also be our absolute minimum, since we can't check endpoints and there are no other candidates!

The point $x = -5\sqrt{3}$ is not in the domain of *C*, hence the only critical point of *C* is $x = 5\sqrt{3}$. We must now verify that this critical point is in fact a minimum. We can use the second derivative test. We compute the second derivative of *C* and evaluate C'' at the critical point.

Because $C''(5\sqrt{3}) > 0$, it follows from the second derivative test that *C* has a relative minimum at $x = 5\sqrt{3}$. Note that it really wasn't necessary to compute $C''(5\sqrt{3})$; it would have been sufficient to determine its sign.

Now we have determined that the cost of the fencing is minimized when $x = 5\sqrt{3}$, and we can answer the questions that were asked about the fence.

(a) We found that $x = 5\sqrt{3}$. Therefore $y = \dfrac{150}{5\sqrt{3}} = 10\sqrt{3}$. The dimensions of the pigpen that minimize the cost of the fencing materials are $5\sqrt{3}\text{ m} \times 10\sqrt{3}\text{ m}$.

(b) We find the cost of the pigpen by plugging the dimensions of the pigpen into the cost equation: $C = 20(2x + y) \Rightarrow C = 20(2 \cdot 5\sqrt{3} + 10\sqrt{3}) = \$693$.

(c) To build the pigpen, Farmer Fred needs two sides of length *x* and one side of length *y*. That is, he needs $2 \cdot 5\sqrt{3} + 10\sqrt{3} = 34.6\text{ m}$ of fencing material.

It is important to pay attention to the questions that are asked. That is, if you are asked about the cost, your answer should be about dollars, not about dimensions. It is also important to pay attention to rounding. In part (b) we are asked to find the cost to the nearest dollar (so, nothing should appear in our answer after the decimal point). In part (c), we are asked to find the length to the nearest decimeter, so we must round our answer to the nearest tenth.

The steps we followed in this example are typical of the procedure for solving an optimization problem:

1. Draw a picture.
2. Write down the quantity to be maximized or minimized.
3. Define variables; one variable should be the quantity you are trying to optimize.
4. Write a function for the quantity to be optimized in terms of the other variables in the problem.

5. Write the constraint equation(s) (if there are any) in terms of your variables.
6. If the function you are trying to optimize is a function of more than one variable, use the constraint equation to eliminate all but one variable.
7. Optimize the function by finding the critical points, determining the nature of each critical point, and evaluating the function at the critical points and at the endpoints of the domain (if applicable).
8. Make sure to answer the question that is asked, paying attention to units and rounding.

AP EXPERT TIP

Make sure the function you are optimizing is quadratic or higher! If you end up with a linear function, you've probably made a mistake as the derivative of a linear function is a constant. Thus, you can't easily set the derivative equal to zero and find a solution.

RELATED RATES

We are often given information about the rate at which one quantity changes and asked to use that information to provide data about changes in another, related quantity. We refer to problems of this type as related-rates problems.

Example:

Claire is rollerblading due east toward the mall at a speed of 10km/hr. John is biking due south away from the mall at a speed of 15 km/hr. Let x be the distance between Claire and the mall at time t and let y be the distance between John and the mall at time t.

(a) Find the distance, in kilometers, between Claire and John when $x = 5$ km and $y = 12$ km.

(b) Find the rate of change, in km/hr, of the distance between Claire and John when $x = 5$ km and $y = 12$ km.

(c) Let θ be the angle formed by the mall, Claire, and John (Claire is the vertex of the angle). Find the rate of change of θ, in radians per hour, when $x = 5$ km and $y = 12$ km.

Solution:

(a) We start by making a diagram to help us understand the information in the problem:

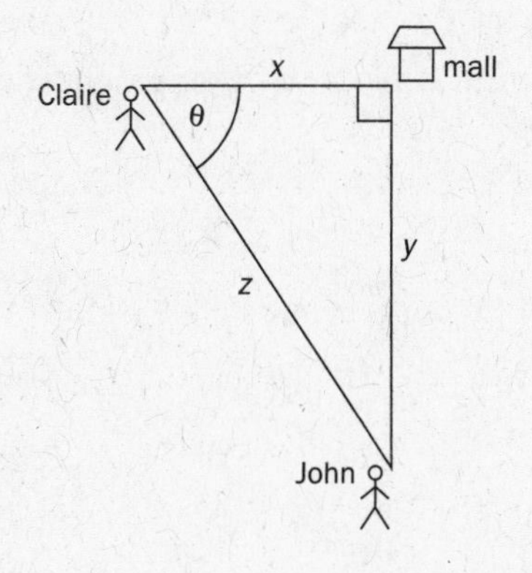

We let Claire's position be x, because Claire is traveling east. We let John's position be y, because John is traveling south. We call the distance between Claire and John z. The diagram is a right triangle; therefore, we can relate x, y, and z with the Pythagorean theorem: $x^2 + y^2 = z^2$.

We are given that $x = 5$, $y = 12$. Using the Pythagorean theorem, we find that $5^2 + 12^2 = z^2 \Rightarrow z = 13$, i.e., the distance between John and Claire is 13 km.

(b) In part (a), we found an equation relating x, y, and z. To find an equation relating the velocities, we use implicit differentiation to differentiate the equation from part (a): $2x\frac{dx}{dt} + 2y\frac{dy}{dt} = 2z\frac{dz}{dt}$. We are also given the information that Claire is traveling toward the mall at 10 km per hour; her distance from the mall is decreasing, so $\frac{dx}{dt} = -10$. We are told that John is traveling away from the mall at 15 km/hr. His distance from the mall is increasing, so $\frac{dy}{dt} = 15$. We are asked to find the rate of change of the distance between John and Claire when $x = 5$ and $y = 12$. We can cancel the 2's in the equation above and fill in all of the required information:

$$x\frac{dx}{dt} + y\frac{dy}{dt} = z\frac{dz}{dt}.$$

$$(5)(-10) + (12)(15) = 13\frac{dz}{dt} \Rightarrow 130 = 13\frac{dz}{dt} \Rightarrow \frac{dz}{dt} = 10 \text{ km/hr}$$

(c) Let θ be the angle formed by the Claire's path and the hypotenuse of the right triangle. The rate of change of θ with respect to time is $\frac{d\theta}{dt}$. Using right triangle trigonometry, we know $\sin\theta = \frac{y}{z}$.

We differentiate this equation with respect to time. Note both y and z are variables with respect to time:

$$\cos\theta \cdot \frac{d\theta}{dt} = \frac{z \cdot \frac{dy}{dt} - y \cdot \frac{dz}{dt}}{z^2}$$

We substitute in the values that were used in part (b). Using right triangle trigonometry we have $\cos\theta = \frac{x}{z} = \frac{5}{13}$. We find

$$\frac{5}{13} \cdot \frac{d\theta}{dt} = \frac{13 \cdot 15 - 12 \cdot \frac{130}{13}}{13^2} \Rightarrow \frac{d\theta}{dt} = \frac{13}{5} \cdot \frac{13 \cdot 15 - 12 \cdot \frac{130}{13}}{13^2} = \frac{15}{13}$$

$$\frac{d\theta}{dt} \approx 1.154 \text{ radians/hr}$$

The steps we followed in this example typify the method for solving related-rates problems:

1. Draw a diagram.
2. Identify the quantities that are changing and assign variables to represent them. Add labels to the diagram.
3. Write down an equation that relates the quantities.
4. Considering each variable as a function of t, use implicit differentiation to differentiate both sides of the equation with respect to t to find the rate of change (derivative) of each quantity.
5. Restate the given problem(s) in terms of your variables and their rates of change (derivatives).
6. Use Step 4 to solve the problem(s) in Step 5.

INTERPRETING THE DERIVATIVE AS A RATE OF CHANGE

Lots of real-life applications of the derivative involve modeling the rates of change of functions. Let's look at another example.

Example:

Suppose milk is leaking out a milk carton, the base of which is a square of length 7 cm and the sides of which are rectangles. Let h be the depth of the milk in the carton, measured in centimeters, where h is a function of time measured in minutes. The volume V of milk in the carton is changing at the rate of $-\frac{3}{h}$ ml (a milliliter is a cubic centimeter) per minute. (The volume of a solid whose base is a square with side length x and whose sides are rectangles of height h is $V = x^2h$.) What is the rate of change of the height of the milk in the carton?

Solution:

As usual, we start by drawing a diagram:

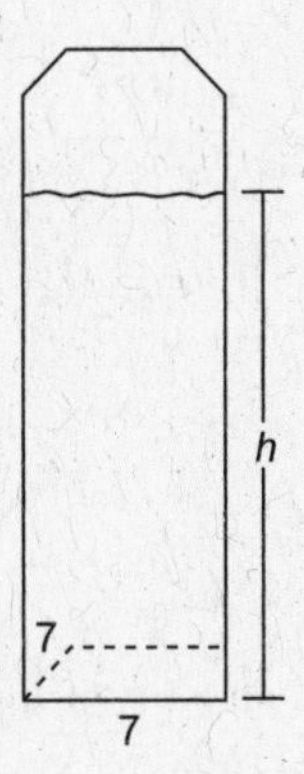

We can find the rate of change of the height of the milk in the carton by considering the volume of the milk and the height of the milk to be functions of *t*. (The length of the base of the carton, *x*, is constant, so it is not changing over time.) We can find the rate of change of the height of the milk in the carton by differentiating both sides of the equation $V = x^2 h$ with respect to *t*, $\frac{dV}{dt} = x^2 \frac{dh}{dt}$.

We are told that the rate of change of the volume of milk in the carton, $\frac{dV}{dt}$, is $-\frac{3}{h}$ ml/minute and that the length of the base of the carton is $x = 7$ cm. We can substitute this information into the equation above and solve for $\frac{dh}{dt}$:

$$\frac{-3}{h} = 7^2 \frac{dh}{dt} \Rightarrow \frac{dh}{dt} = -\frac{3}{49h} \text{ cm/minute}$$

Velocity, Speed, and Acceleration

Rates of change (derivatives) commonly appear on the AP exam in problems involving motion, where the ideas of velocity, speed, and acceleration come into play. Recall from Chapter 11 that velocity is the rate of change of position. Position and velocity are both signed quantities. For instance, if north is considered the positive direction, a distance of 10 km due north of a starting point is +10 km, while a distance of 10 km due south of a starting point is –10 km. Similarly, velocity is a signed quantity. Traveling 60 km/hr due north on a highway is +60 km/hr and traveling 60 km/hr due south on the highway is –60 km/hr. Speed is the absolute value of velocity. When we discuss speed we don't care about the direction—we are only concerned with how fast we will get where we are going.

Example:

A particle moves along a straight line. Its velocity at time $t \geq 0$ is given by $v(t) = 0.5 + \sin(\ln t^2)$. At time $t = 0$, the particle is at the zero position on the line.

(a) Find the acceleration of the particle at time $t = 4$.

(b) Is the speed of the particle increasing or decreasing at time $t = 4$? Justify your answer.

Solution:

(a) The acceleration of the particle is the derivative of velocity: We are given the velocity

$$v(t) = 0.5 + \sin(\ln t^2) \underset{\substack{\text{rules for}\\\text{manipulating}\\\text{logarithms}}}{=} 0.5 + \sin(2 \ln t)$$

We differentiate to find $a(t) = v'(t) = 2\cos(2\ln t) \cdot \frac{1}{t} = \frac{2\cos(\ln t^2)}{t}$. To find the acceleration at time $t = 4$, we substitute this value into the equation for acceleration $a(4) = \frac{2\cos(\ln 16)}{4} \approx -0.466 \Rightarrow a(4) \approx -0.466$.

(b) This part of the question is not as straightforward as it seems! The question asks whether the particle is speeding up or slowing down, which depends on the sign of the velocity. If the velocity is positive and getting more positive, i.e., $v > 0$ and $a > 0$, or if $v < 0$ and $a < 0$, then the speed, which is the absolute value of the velocity, is increasing.

In our case, $v(4) = 0.5 + \sin(\ln 16) \approx 0.861 > 0$, but (and from part (a)) $a(4) < 0$. Therefore the speed of the particle is decreasing.

Hint: If this logic is difficult to follow, think of the speed as decreasing when velocity as a function is going toward zero.

Velocity and Acceleration Vectors of Planar Curves*

Think back to our discussion of planar curves in Chapter 8. When we let the variable t denote time, we have an important mathematical and physical interpretation of these functions and their derivatives. We can think of parametric, polar, and vector functions as position functions of a particle moving through space. When we calculate the first derivative we are finding a formula for the velocity vector of this particle. Similarly, the second derivative of this position function gives us the acceleration vector of the particle.

Suppose we have a planar curve that is expressed by means of the parametric equation $y(t) = (f(t), g(t))$. We have already seen that at a fixed time t_0, $y(t_0)$ gives a point on the curve. If the curve represents the path of a particle on the plane, then $y(t_0)$ is the position of the particle. Just as we saw with vector functions, we can differentiate this parametric function to find the velocity of the curve at a point. We have $y'(t) = (f'(t), g'(t))$ and so at the time t_0, $y'(t_0) = (f'(t_0), g'(t_0))$. The ordered pair $(f'(t_0), g'(t_0))$ is a vector pointing in the direction of the velocity of the curve. The picture you should have in mind is an arrow (vector) at each point of the curve denoting the direction of motion of the particle moving along the curve.

Similarly, the ordered pair $(f''(t_0), g''(t_0))$ gives the acceleration vector of the curve at each point. The same ideas apply to polar and vector equations.

To compute the derivative of a vector function, we simply differentiate each component in turn. For example, a vector function is usually written in the form $v(t) = (v_1(t), v_2(t), v_3(t))$ and thus the derivative of this vector functions would be $v'(t) = (v'_1(t), v'_2(t), v'_3(t))$. If we need more than one derivative, then we can repeat the process to find as many derivatives as we need. Pretty easy, huh? Similarly, we can anti-differentiate a vector function term-by-term as well.

**BC content*

It is important to note the mathematical and physical interpretation of these derivative functions. We can think of parametric, polar, and vector functions as position functions of a particle moving through space. When we calculate the first derivative, we are finding a formula for the velocity vector of this particle. Similarly, the second derivative of this position function gives us the acceleration vector of the particle.

The derivative of a real-valued function expresses the rate of change of the function. When we evaluate the derivative at a point $x = a$, we are calculating the slope of the line tangent to the function at the point $x = a$.

AP EXPERT TIP

The function $F'(a) = (f'(a), g'(a))$ can also be used to find the slope of the line tangent to the curve at $t = a$ by taking the ratio of $g'(a)$ to $f'(a)$. This is because a velocity vector to a curve at a given point lies on the tangent line to the curve at that point.

The interpretation for parametric functions is slightly different. If we let $t = a$, then $F(a) = (f(a), g(a))$ is a point on the curve in the plane. What about the derivative? By definition, $F'(a) = (f'(a), g'(a))$, but this is an ordered pair. This is quite peculiar. How should we interpret this?

What happens here is that the derivative of a parametric function becomes a vector-valued function and the value of $F'(a) = (f'(a), g'(a))$ tells us the direction and velocity of the parametric function $F(t) = (f(t), g(t))$.

For example, suppose we have the parametric function $F(t) = (t^2 - 3t, \ln t - t)$, then $F\ (t) = \left(2t - 3, \frac{1}{t} - 1\right)$. Therefore, when $t = 1$, we have $F(1) = (1^2 - 3, \ln - 1) =$ $(-2, 0)$ and $F\ (1) = \left(2 - 3, \frac{1}{1} - 1\right) = (-1, 0)$.

This means that at the point $(-2, 0)$ on the planar curve generated by $F(t) = (t^2 - 3t, \ln t - t)$, the velocity vector points in the direction $(-1, 0)$ with respect to the point $(-2, 0)$.

So at the point $(-2, 0)$ the slope of the tangent to the curve is

$$\frac{g'(1)}{f'(1)} = \frac{\frac{1}{1} - 1}{2(1) - 3} = \frac{0}{-1} = 0.$$

AP EXPERT TIP

Questions related to motion along a curve in the *xy*-plane appear frequently on the free-response section of the BC exam.

Clearly, we can extend this idea to calculate the acceleration vector of a parametric curve as well. With $F(t)$ as above, we get $F\ (t) = \left(2, \frac{-1}{t^2}\right)$. When this function is evaluated at a point on the curve we get the acceleration vector of the curve at this point.

DIFFERENTIAL EQUATIONS

When we discussed related rates, we saw that if two quantities are related in an equation, we can differentiate both sides of the equation and find an equation that relates the rates of change of those quantities. Sometimes, we initially know that an equation relates the rate of change of one quantity to the quantity itself, or to a different quantity or quantities. Many things that we observe in the world are described by statements that talk about the rate at which things happen, such as "The economy is expanding at a rate of 3.5 percent," or "The rate at which medication is absorbed into the bloodstream is proportional to the amount of medication that is already present in the bloodstream." Using the language of calculus, we say that some equations relate both variables *and* their derivatives. Equations containing derivatives are called *differential equations.*

To solve most differential equations, we need the tools from later chapters, on integration. Using our knowledge of derivatives, however, we can work with differential equations of the type $\frac{dy}{dx} = f(x, y)$, where f is an expression that can involve both x and y (although f *could* include just one variable).

We can visualize solutions of differential equations of the type $\frac{dy}{dx} = f(x, y)$ by drawing a short line segment with slope $f(x, y)$ through each point (x, y) in the plane. (We can't actually do this for *every* point, but we can take a sampling of many points.) Following this procedure we see a pattern of lines. This pattern is called a *slope field* (or *direction field* or *vector field*) of the differential equation $\frac{dy}{dx} = f(x, y)$.

For example, here is the slope field of the differential equation $\frac{dy}{dx} = xy$:

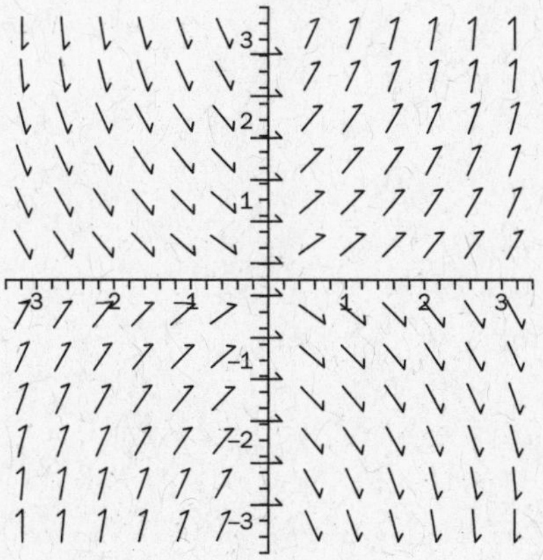

A solution of the differential equation is a function whose graph follows the slope field. We can sketch a solution by starting at one point on the graph and connecting line segments that flow into each other—a kind of calculus "connect-the-dots."

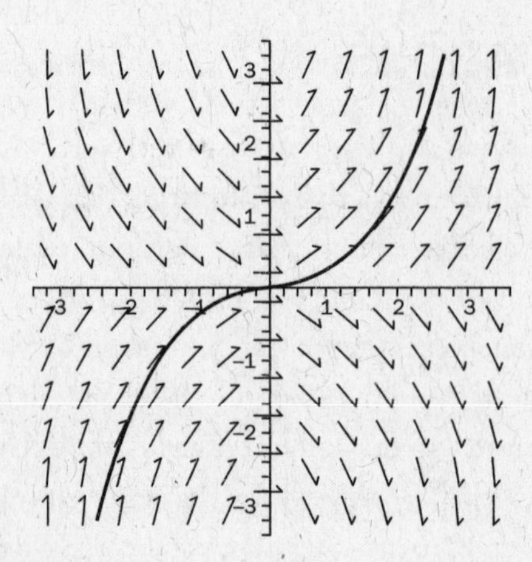

The slope field suggests many curves that can be solutions to the differential equation. That is, if we pick different starting points, we will get different solutions. Is there a "correct" solution among these? The answer is no—many different curves are solutions to differential equations.

For example, let's draw the slope field of the differential equation $\frac{dy}{dx} = 2x$.

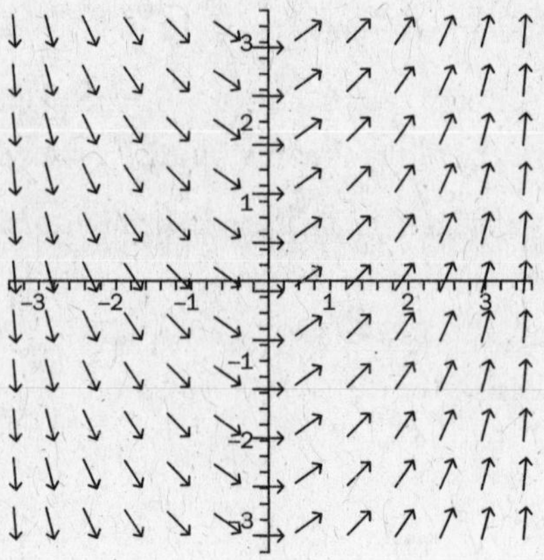

We can sketch a few solution curves.

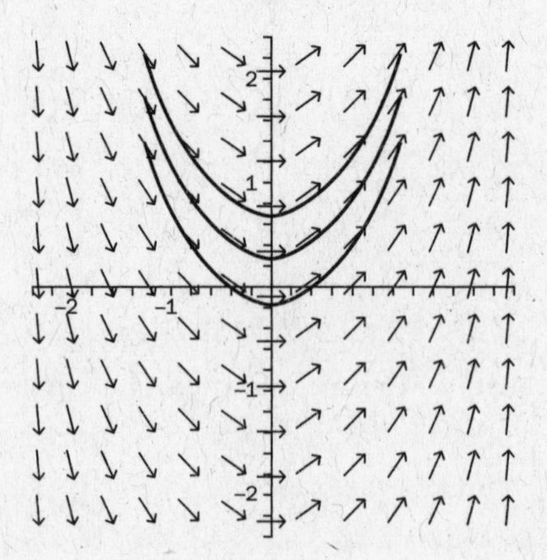

In this example, we can actually determine the equations of these solutions. The differential equation $\frac{dy}{dx} = 2x$ makes the statement: "The derivative of the function y is $2x$." The function $y = x^2$ is one such function; so are $x^2 + 2$, $x^2 - 9$, and, in general, $x^2 + c$, where c is any real number. The solution curves we sketched above are all parabolas $x^2 + c$—with different c's.

In general, we can't find equations for solutions to differential equations without using the tools for integration, found later in the text—but using slope fields, we can visualize solutions to equations of the type discussed here.

Slope fields appear on the AP exam in a few different ways. You may be asked to:

1. Sketch a slope field for a given differential equation.
2. Sketch a solution curve through a given point given a slope field.
3. Match a slope field to a differential equation.
4. Match a slope field to a solution of a differential equation.

Examples of some of these types of problems are given in the questions at the end of the chapter.

DIFFERENTIAL EQUATIONS VIA EULER'S METHOD*

There is a hard fact of life that you may not know about yet. Sometimes it is impossible to obtain an explicit formula for the solution of a differential equation. When we find ourselves in this situation, though, we must not give up because we can still obtain some important information about the shape of the solution. Euler's method is a very useful technique in this regard. We begin with an initial value problem and using not much more than the definition of derivatives, we can numerically plot points approximating the solution of our given differential equation to get a good idea of what the solution looks like.

Let's look at the very simple differential equation given by:

$$\frac{dy}{dt} = y - 3x$$
$$y(1) = 0$$

We would like to get an idea of what the solution function $y(t)$ looks like.

Let's start with the basics. What exactly is a derivative? Does this formula ring any bells?

$$\frac{dy}{dt}(a) = y'(a) = \lim_{h \to 0} \frac{y(a+h) - y(a)}{h}$$

A limit really only approximates a function, so if we let ≈ mean "approximately," we get

*BC content

$$\frac{dy}{dt}(a) \approx \frac{y(a+h)-y(a)}{h}$$, which we can rearrange to get

$$y(a+h) \approx y(a) + h\frac{dy}{dt}(a).$$

Thus, for a general first-order differential equation, $\frac{dy}{dt} = F(x, y),\ y(x_0) = y_0$, we can let $a = x_0$ and write the approximation function as:

$$\begin{aligned} y(x_0 + h) &\approx y(x_0) + h\frac{dy}{dt}(x_0) \\ &\approx y_0 + h \cdot F(x_0, y_0) \end{aligned}$$

To use Euler's method, we will generalize this process and develop a large collection of points that approximate actual values of $y(t)$. The x-values will all differ by a prescribed step increment, h. This step increment will determine the accuracy of our approximation (relative to the actual solution to the differential equation). The smaller h is, the more accurate our graph will be. Unfortunately, it's also true that the smaller h is, the more complicated the computations become. We have to pick an h that's small enough to give us useful information but that's still large enough to be manageable.

AP EXPERT TIP

On the Calc BC exam you may be asked to determine whether an approximation obtained by Euler's method is an overestimate or an underestimate of the exact value. To do this, evaluate the second derivative at the indicated point to determine the curve's concavity. If the solution curve is concave up (the second derivative is positive), Euler's approximation is an underestimate. If the solution curve is concave down, Euler's approximation is an overestimate.

To calculate more values of $y(t)$ we get:

$$\begin{aligned} y_1 &= y(x_0 + h) \approx y_0 + h \cdot F(x_0, y_0) \\ y_2 &= y(x_0 + 2h) \approx y_1 + h \cdot F(x_1, y_1) \qquad \text{where } x_k = x_{k-1} + h \\ y_3 &= y(x_0 + 3h) \approx y_2 + h \cdot F(x_2, y_2) \end{aligned}$$

These are the points we will plot that will approximate the actual solution, $y(t)$. Now let's use this method to solve the differential equation above.

It is given in the question that $x_0 = 1$, $y(x_0) = y_0 = 0$, $F(x, y) = y - 3x$ and we will take $h = 0.1$. Using the equations above we get:

$$\begin{aligned} y_1 &\approx y_0 + h \cdot F(x_0, y_0) = 0 + 0.1 \cdot F(1,0) \\ &= 0 + 0.1(0 - 3 \cdot 1) = 0 - 0.3 = -0.3 \\ y_2 &\approx y_1 + h \cdot F(x_1, y_1) = -0.3 + 0.1 \cdot F(1 + 0.1, -0.3) \\ &= -0.3 + 0.1(-0.3 - 3(1.1)) = -0.66 \\ y_3 &\approx y_2 + h \cdot F(x_2, y_2) = -0.66 + 0.1 \cdot F(1.1 + 0.1, -0.66) \\ &= -0.66 + 0.1(-0.66 - 3(1.2)) = -1.086 \\ y_4 &\approx y_3 + h \cdot F(x_3, y_3) = -1.086 + 0.1 \cdot F(1.2 + 0.1, -1.086) \\ &= -1.086 + 0.1(-1.086 - 3(1.3)) = -1.5846 \end{aligned}$$

...

Clearly, these computations get more complicated with each iteration, but it's all worth it! If $y(t)$ is the actual solution of this differential equation, then we know:

$$y(1.1) \approx -0.03$$
$$y(1.2) \approx -0.66$$
$$y(1.3) \approx -1.086$$

If we cannot obtain the solution by other means, we can always use Euler's method to get an approximation of the solution to any differential equation.

L'HÔPITAL'S RULE*

In Chapter 7, we learned about limits as a way to analyze the behavior of functions near a point. If the function in question was continuous, then the limit was found by simply plugging in the limiting x value. If, however, the function wasn't continuous at the limit point, we had to employ special techniques to determine the value of the limit. In this section we will learn another very useful technique that can be applied to a wide variety of limit questions.

AP EXPERT TIP

Even though L'Hôpital's rule is a BC-only topic, AB students take heed! You can use it to easily simplify some otherwise tricky limits. Even though this concept isn't explicitly tested on your exam, you can certainly use it to save time and gain an advantage.

DETERMINING LIMITS

Suppose we are asked to find $\lim\limits_{x \to 0} \frac{e^x - 1}{x}$. This function is clearly not continuous at $x = 0$ and thus we cannot simply plug in the value $x = 0$. Furthermore, we can't simplify the term $\frac{e^x - 1}{x}$ at all, so we cannot apply any of the techniques we learned in previous chapters. It looks as though we might as well give up now. Of course, giving up is never the answer in mathematics. We simply need more tricks, and the trick we need here is L'Hôpital's Rule.

Observe that both the numerator and the denominator of this limit approach 0 as x approaches 0. If we did naively try to plug in the value $x = 0$ into the function, we would get $\frac{0}{0}$, which is an undefined quantity. In general, when this happens the limit is said to be in indeterminate form of type $\frac{0}{0}$ and it is not known whether the limit exists or not.

Suppose we are trying to find the horizontal asymptote of our function $\frac{e^x - 1}{x}$. To do so, we want to find $\lim\limits_{x \to \infty} \frac{e^x - 1}{x}$. However, we are presented with another problem. Both the numerator and the denominator approach infinity as x approaches infinity. In general, when this happens the limit is said to be in indeterminate form of type $\frac{\infty}{\infty}$ and it is not known whether the limit exists or not.

**BC content*

Luckily all is not lost; L'Hôpital's Rule can be used to solve *both* types of indeterminate forms. This truly valuable rule is stated on the following page.

L'Hôpital's Rule

Suppose f and g are differentiable functions and $g'(x) \neq 0$ near $x = a$. Suppose also that:

$$\lim_{x \to a} f(x) = 0 \quad \text{and} \quad \lim_{x \to a} g(x) = 0$$

or that,

$$\lim_{x \to a} f(x) = \pm\infty \quad \text{and} \quad \lim_{x \to a} g(x) = \pm\infty$$

Then

$$\lim_{x \to a} \frac{f(x)}{g(x)} = \lim_{x \to a} \frac{f'(x)}{g'(x)}$$

In short, L'Hôpital's Rule says that if we have a limit that is in indeterminate form of type $\frac{0}{0}$ *or* $\frac{\infty}{\infty}$, then the limit of the quotient of the functions is equal to the limit of the quotient of their derivatives. Note, however, that we must first begin with a limit that is in *indeterminate form.*

Let's use L'Hôpital's Rule to solve the limit that began this discussion. We wanted to find $\lim_{x \to 0} \frac{e^x - 1}{x}$. This limit is clearly in indeterminate form so we compute:

$$\lim_{x \to 0} \frac{e^x - 1}{x} \overset{L'Hôpital}{=} \lim_{x \to 0} \frac{(e^x - 1)'}{(x)'} = \lim_{x \to 0} \frac{e^x}{1} = \lim_{x \to 0} e^x = 1$$

Similarly, to find $\lim_{x \to \infty} \frac{e^x - 1}{x}$, we get:

$$\lim_{x \to \infty} \frac{e^x - 1}{x} \overset{L'Hôpital}{=} \lim_{x \to \infty} e^x = \infty$$

Thus, $\frac{e^x - 1}{x}$ has no horizontal asymptotes. More importantly we have a valuable tool that we can use to evaluate any limit in indeterminate form.

Suppose we are asked to find $\lim_{x \to 1} \frac{x^2 - 1}{x - 1}$. We could use techniques we learned in previous chapters (such as factoring) to solve this limit. However, because the limit is in indeterminate form, let's use our new toy, L'Hôpital's Rule. The solution becomes one line:

$$\lim_{x\to 1}\frac{x^2-1}{x-1} \overset{L'Hôpital}{=} \lim_{x\to 1}\frac{(x^2-1)'}{(x-1)'} = \lim_{x\to 1}\frac{2x}{1} = \lim_{x\to 1} 2x = 2$$

Clearly, L'Hôpital's Rule is a great technique! However, the real power of L'Hôpital's Rule emerges when we calculate limits that we previously wouldn't have been able to solve, such as $\lim_{x\to\infty}\frac{\ln(\ln x)}{x}$. First, we must always check that the limit is in indeterminate form. We know that as x approaches infinity so does $\ln x$, so both the numerator and the denominator approach infinity as x approaches infinity. Now, justified in our use of L'Hôpital's Rule, we get:

$$\lim_{x\to\infty}\frac{\ln(\ln x)}{x} \overset{L'Hôpital}{=} \lim_{x\to\infty}\frac{(\ln(\ln x))'}{(x)'}$$

$$\overset{Chain\ Rule}{=} \lim_{x\to\infty}\frac{\frac{1}{\ln x}\cdot\frac{1}{x}}{1}$$

$$= \lim_{x\to\infty}\frac{1}{x\ln x} = 0$$

It is also important to recognize when to use L'Hôpital's Rule. Suppose we want to calculate $\lim_{x\to a} f(x)\cdot g(x)$ and we know $\lim_{x\to a} f(x) = 0$ and $\lim_{x\to a} g(x) = \infty$. What will the limit of the product be? The function f is pulling to 0 and the function g is pulling to infinity. This limit is said to be in indeterminate form of type $0\cdot\infty$. However, it is important to notice that by rewriting this limit we can employ L'Hôpital's Rule. Because $f\cdot g = \frac{g}{\frac{1}{f}}$, we get $\lim_{x\to a} f(x)\cdot g(x) = \lim_{x\to a}\frac{g(x)}{\frac{1}{f}(x)}$ which is now in indeterminate form of type $\frac{\infty}{\infty}$ and so we can use L'Hôpital's Rule.

For example, to calculate $\lim_{x\to\infty} x^2e^{-x}$, first notice that $\lim_{x\to\infty} x^2 = \infty$ and $\lim_{x\to\infty} e^{-x} = \lim_{x\to\infty}\frac{1}{e^x} = 0$.

Therefore, we have $\lim_{x\to\infty} x^2e^{-x} = \lim_{x\to\infty}\frac{x^2}{\frac{1}{e^{-x}}} = \lim_{x\to\infty}\frac{x}{e^x} \overset{L'Hôpital}{=} \lim_{x\to\infty}\frac{(x)}{(e^x)} = \lim_{x\to\infty}\frac{1}{e^x} = 0.$

By now you're starting to appreciate how incredibly useful L'Hôpital's Rule can be. And if you haven't already committed it to memory, you should. However, for the few remaining skeptics out there we'll bring out one more application of L'Hôpital's Rule, for determining indeterminate powers.

Indeterminate powers arise in limits of the form $\lim\limits_{x \to a} [f(x)]^{g(x)}$ such that when:

1. $\lim\limits_{x \to a} f(x) = 0$ and $\lim\limits_{x \to a} g(x) = 0$, the limit has type 0^0
2. $\lim\limits_{x \to a} f(x) = \infty$ and $\lim\limits_{x \to a} g(x) = 0$, the limit has type ∞^0
3. $\lim\limits_{x \to a} f(x) = 1$ and $\lim\limits_{x \to a} g(x) = \pm\infty$, the limit has type 1^∞

To solve these limits via L'Hôpital's Rule, we use the identity $[f(x)]^{g(x)} = e^{g(x)\ln f(x)}$.

Thus, solving $\lim\limits_{x \to a} [f(x)]^{g(x)}$ reduces to solving $\lim\limits_{x \to a} g(x)\ln f(x)$ and taking $\lim\limits_{ex \to a} g(x)\ln f(x)$.

For example, use L'Hôpital's Rule to find $\lim\limits_{x \to 0^+} x^{\sin x}$. This limit is clearly indeterminate of form 0^0, so to use L'Hôpital's Rule we must first calculate $\lim\limits_{x \to 0^+} x\ln(\sin x)$. However, this limit is also indeterminate (of form $0 \cdot \infty$). No worries, we simply use L'Hôpital's Rule again to evaluate this limit to get:

$$\lim_{x \to 0^+} x\ln(\sin x) = \lim_{x \to 0^+} \frac{\ln(\sin x)}{\frac{1}{x}} \overset{L'\,Hôpital}{=} \lim_{x \to 0^+} \frac{\frac{\cos x}{\sin x}}{\frac{-1}{x^2}} = \lim_{x \to 0^+} \frac{-x^2 \cos x}{\sin x}.$$

However the limit on the right needs L'Hôpital's Rule as well. We get:

$$\lim_{x \to 0^+} \frac{-x^2 \cos x}{\sin x} \overset{L'\,Hôpital}{=} \lim_{x \to 0^+} \frac{x^2 \sin x - 2x\cos x}{\cos x} = \frac{0}{1} = 0$$

Finally, we are finished (with L'Hôpital to the rescue). Combining all these steps into one, we get:

$$\lim_{x \to 0^+} x^{\sin x} = e^{\lim\limits_{x \to 0^+} x\ln(\sin x)} = e^{\lim\limits_{x \to 0^+} \frac{-x^2 \cos x}{\sin x}} = e^0 = 1$$

When working with indeterminate forms such as $0 \cdot \infty$, 0^0, or 1^∞, it is tempting to make erroneous conclusions, yet still get a lucky right answer (kind of like reducing the fraction $\frac{19}{95}$ by canceling the 9's. After all, $\frac{1}{5}$ is right, isn't it?). One might argue that $0 \cdot \infty = 0$ (anything times zero is zero?), or that $0^0 = 1$ (anything to the zero power is one?). The examples in this section would support that argument: See $\lim\limits_{x \to \infty} x^2 e^{-x} = 0$ and $\lim\limits_{x \to 0^+} x^{\sin x} = 1$ above. However, it is possible to construct limit problems with these indeterminate forms that do not have answers of 0 or 1. That is because we are not really performing operations with these quantities. Rather, we are performing operations with quantities that are *approaching* 0 or 1. A tug-of-war ensues and the end result could be anything. So, using the techniques of this section, you should confirm that $\lim_{x\to0^+}\left(\frac{1}{x}\right)\sin 2x = 2$ $(0\cdot\infty \neq 0)$, and that $\lim_{x\to0^+} x^{\frac{3}{\ln x}} = e^3$ $(0^0 \neq 1)$. When working with limits involving 0 and ∞, not all is as it seems.

L'Hôpital's Rule is very important and useful. We will see more applications of this technique in upcoming chapters on improper integrals and series.

REVIEW QUESTIONS

1. A particle is moving along the x-axis with a velocity, $v(t) = \sin t + \cos t$, for $t \geq 0$. What is its maximum acceleration over the interval $[0, 2\pi]$?

(A) $\cos\frac{3\pi}{4} + \sin\frac{3\pi}{4}$

(B) $\cos 2\pi + \sin 2\pi$

(C) $\cos\frac{7\pi}{4} + \sin\frac{7\pi}{4}$

(D) $\cos\frac{7\pi}{4} - \sin\frac{7\pi}{4}$

(E) $\cos 0 + \sin 0$

2. The lengths of the sides of a square are decreasing at a constant rate of 4 ft/min. In terms of the perimeter, P, what is the rate of change of the area of the square in square feet per minute?

(A) $-2P$

(B) $2P$

(C) $-4P$

(D) $4P$

(E) $8P$

3. The base of a triangle is decreasing at a constant rate of 0.2 cm/sec and the height is increasing at 0.1 cm/sec. If the area is increasing, which answer best describes the constraints on the height h at the instant when the base is 3 centimeters?

(A) $h > 3$

(B) $h < 1$

(C) $h > 1.5$

(D) $h < 1.5$

(E) $h > 2$

4. Dashing Dan is driving south on Sunset Boulevard, where the speed limit is 40 miles per hour. He is traveling at a constant speed. He crosses an intersection. A police officer is sitting 33 feet due east of the intersection with her radar tuned to Dan's car. One second after Dan crosses the intersection, Dan is 55 feet away from the police officer and she records Dan going away from her at a speed of 37.5 miles per hour. She issues Dan a ticket because Dan's speed on Sunset Boulevard is

(A) between 35 and 40 miles per hour

(B) between 40 and 45 miles per hour

(C) between 45 and 50 miles per hour

(D) between 50 and 55 miles per hour

(E) more than 55 miles per hour

5. If the rate of change of $f(x) = 2^x$ at $x = 5$ is four times the rate of change at $x = a$, then $a =$

(A) 2

(B) 3

(C) 4

(D) 5

(E) 6

6. Calculate $\lim_{x\to\infty} e^{-2x} \ln x$.

(A) 1

(B) e^2

(C) 0

(D) e

(E) $\frac{\pi}{2}$

7. Use Euler's method with step increment $h = 0.1$ to approximate $y(0.\ 3)$, where $y(x)$ is the solution to the initial value problem $y' = y + xy, y(0) = 1$.

(A) 1.221

(B) 0.1

(C) 1.37

(D) 1.1

(E) 1.54

8. What is the y-intercept of the line that is tangent to the curve $f(x)=\sqrt{2x-3}$ at the point on the curve where $x = 6$?

(A) 0

(B) $\frac{1}{3}$

(C) $\frac{2}{3}$

(D) 1

(E) $\frac{4}{3}$

9. At what point x on the curve $f(x) = \ln(2x-1)$ does the tangent line have a slope of $\frac{2}{5}$?

(A) 1

(B) e

(C) 3

(D) 5

(E) 8

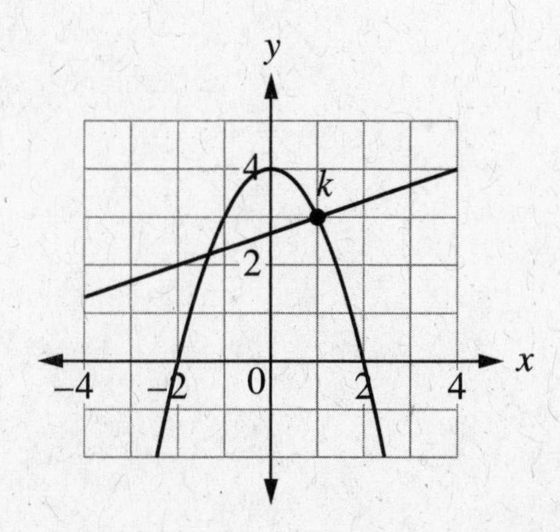

10. Consider the graph of $f(x) = 4 - x^2$ and a normal line to the graph at the point $x = k$, as shown. In terms of k, what is the y-intercept of the normal line at the point on the curve where $x = k$? Note: In the graph above, $k = 1$.

(A) $3.5 - k^2$

(B) $1.5 + k^2$

(C) $2 + 0.5k^2$

(D) $1.5 + k$

(E) $4 - 1.5k^2$

This graph is for problems 11–12.

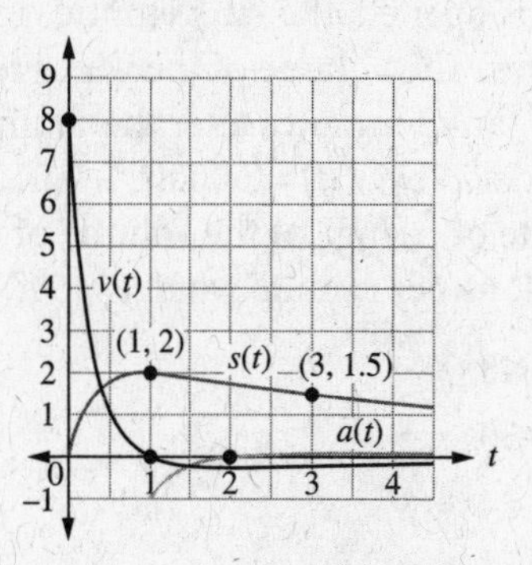

11. A particle moves along a number line (not shown) for $t \geq 0$. Its position function is $s(t)$, its velocity function is $v(t)$, and its acceleration function is $a(t)$. All are graphed with respect to time t in seconds. For the time interval when $1 < t < 2$, the particle is

(A) Not moving

(B) Moving to the left and slowing down

(C) Moving to the left and speeding up

(D) Moving to the right and slowing down

(E) Moving to the right and speeding up

12. The total distance traveled by the particle for $0 \leq t \leq 3$ is

(A) 0.5

(B) 1.5

(C) 2

(D) 2.5

(E) 3.5

13. The position of a particle moving along a number line is given by $s(t)=\frac{2}{3}t^3-6t^2+10t$ where t is time in seconds. When is the particle at rest?

(A) 1 second only

(B) 2 seconds only

(C) 5 seconds only

(D) 1 and 2 seconds

(E) 1 and 5 seconds

14. A water tank holds 6000 gallons. When the drain is opened, the tank empties in 100 minutes. The volume of the water remaining in the tank t minutes after the drain is opened is modeled by $V(t) = 6000(1 - 0.01t)^2$. What is the rate of change of the volume of the water in gallons per minute when $t = 60$?

(A) −48

(B) −30

(C) −24

(D) 24

(E) 48

15. After an injection, the concentration in mg/L (milligrams per liter) of a drug in the blood-stream is modeled by the function $C(t) = K(e^{-at} - e^{-bt})$ where a, b, and K are positive constants with $b > a$ and t is time measured in hours after the injection is given.

What is the rate of change of the drug's concentration in mg/L per hour exactly one hour after an injection?

(A) $K(e^{-a} - e^{-b})$

(B) $K(e^{-b} - e^{-a})$

(C) $K(be^{-b} - ae^{-a})$

(D) $K(ae^{-a} - be^{-b})$

(E) None of the above

Free-Response Questions

Question 16 requires a calculator.

16. The rate of change of the temperature in March is modeled by $T\,(t) = t \cdot \cos\left(\frac{t}{7}\right)$.

    (a) Is the temperature increasing or decreasing at $t = 7$? Explain.

    (b) Is the rate of change of the temperature at $t = 7$ increasing or decreasing? Justify your answer.

17. Water is flowing into a tank at the following rate: $E(t) = \dfrac{t+2}{t^2+1}$. The water is leaving the tank at the rate: $D(t) = \sin t \cdot \dfrac{10t}{\sqrt{t^4+5}}$.

    (a) Let $A(t)$ be the amount of water in the tank at time t. What is the rate of change of $A(t)$?

    (b) At $t = 2$ is the amount of water in the tank increasing or decreasing?

    (c) At $t = 2$ is the rate of change of the amount of water in the tank increasing or decreasing?

18. Consider the differential equation: $\dfrac{dy}{dx} = 2x - y$.

    (a) On the grid below, sketch a slope field for the given differential equation at the 12 points.

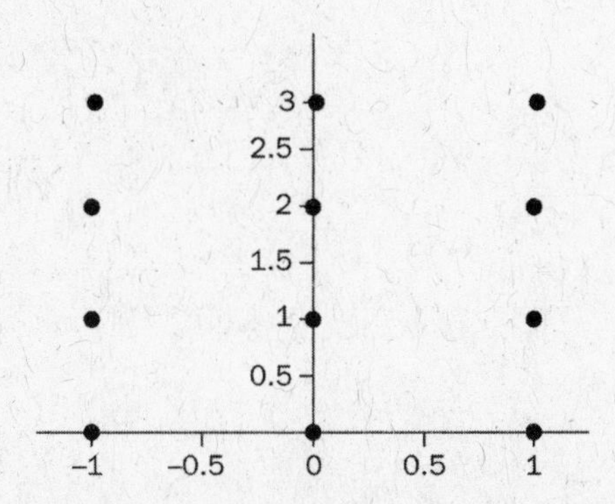

    (b) Although the slope field in part (a) is drawn for a finite number of points, it is really defined at every point on the xy-plane. Describe all the points where the slopes are negative.

19. If $y = 18 - 3x$, what is the maximum value of xy?

20. Consider the function $y = 3x^{\frac{2}{3}} - 2x$.

    (a) Find the critical points. Indicate the nature of each critical point (i.e., $y' = 0$ or y' does not exist, or DNE).

    (b) Make a sign graph for y' and indicate where the original function is increasing and where it is decreasing. Label each graph.

    (c) Make a sign graph for y'' and indicate where the original function is concave up and where it is concave down. Label each graph.

    (d) Find the coordinates of all extrema. Indicate the cusp(s) and the horizontal tangent(s).

    (e) Sketch the curve. Check your work with a calculator.

ANSWERS AND EXPLANATIONS

1. D

Acceleration is the derivative of velocity, so to find an equation for the acceleration, we differentiate the velocity function $v(t) = \sin t + \cos t \Rightarrow v'(t) = a(t) = \cos t - \sin t$. Because the acceleration function is continuous, the maximum acceleration can only occur at the critical points (where $a'(t)$ is zero or undefined) or at one of the endpoints of the interval $[0, 2\pi]$.

$$a(t) = \cos t - \sin t \Rightarrow a'(t) = -\sin t - \cos t$$

$$a'(t) = -\sin t - \cos t = 0 \Rightarrow \sin t = -\cos t \Rightarrow \tan t = -1$$

$\tan t = -1$ in the second and fourth quadrants at $t = \frac{3\pi}{4}, \frac{7\pi}{4}$.

Notice that the derivative $a'(t)$ is defined everywhere, so the only critical points are the points at which $a'(t) = 0$. We can make a table of the critical points and endpoints and the value of the acceleration $a(t)$ at each of these points.

t	$\frac{3\pi}{4}$	$\frac{7\pi}{4}$	0	2π
$a(t)$	$\cos\frac{3\pi}{4} - \sin\frac{3\pi}{4} = -\frac{\sqrt{2}}{2} - \frac{\sqrt{2}}{2} = -\sqrt{2} \approx -1.414$	$\cos\frac{7\pi}{4} - \sin\frac{7\pi}{4} = \frac{\sqrt{2}}{2} - \left(-\frac{\sqrt{2}}{2}\right) \approx 1.414$	$\cos 0 + \sin 0 = 1 + 0 = 1$	$\cos 2\pi + \sin 2\pi = 1 + 0 = 1$

Of these four choices, the maximum acceleration is $\sqrt{2}$; we are guaranteed that the maximum acceleration is one of these four values, so the correct answer is (D).

In this problem, you really have to pay attention to just what you are maximizing. Usually, position is $x(t)$ and for this, you need to set $x'(t) = v(t) = 0$ and look for $x''(t) = a(t) < 0$. This problem is changed around, and we are asked to maximize acceleration, so we need to look at $a'(t)$ and $a''(t)$. This concept can be difficult at first, but becomes easier with practice. They like to ask problems like this on the AP exam.

2. A

This is a related-rates question, with a twist. We are given information about the rate of change of the side of the square and we need to answer a question about the rate of change of the area. After we do this we are done with the related-rates part of the problem and we have to express our answer in terms of the perimeter.

Following the procedure for related-rates problems, we first draw a diagram and identify the quantities in the problem.

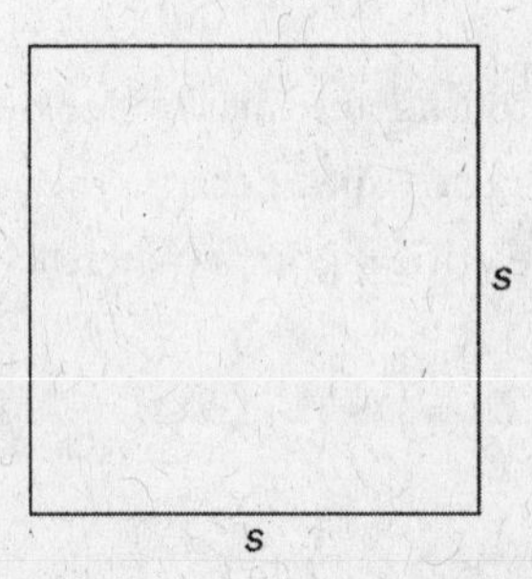

We find the area of the square in terms of our variables $A = s^2$. We differentiate both sides of this equation with respect to t to find an equation relating $\frac{dA}{dt}$ and $\frac{ds}{dt}$: $\frac{dA}{dt} = \underbrace{2s\frac{ds}{dt}}_{\text{chain rule}}$. We are given that the lengths of the sides of the square are decreasing at a rate of 4 ft/min, i.e., $\frac{ds}{dt} = -4$. We substitute this information into the equation for $\frac{dA}{dt}$ above and find $\frac{dA}{dt} = 2s(-4) = -8s$.

To finish the problem we need to express $\frac{dA}{dt}$ in terms of the perimeter. The perimeter of a square is $P = 4s$. Therefore $s = \frac{P}{4}$. Substituting this expression for s into the expression above, we find:

$$\frac{dA}{dt} = -8 \cdot \frac{P}{4} = -2P$$

3. D

Following the procedure for related-rates problems, we first draw a diagram and label the quantities in the problem:

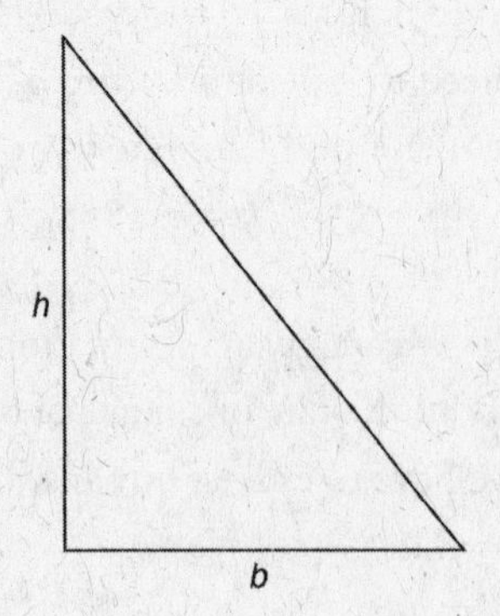

Next we find an equation relating the quantities in the problem, in this case, the formula for the area of a triangle: $A = \frac{1}{2}bh$. We differentiate both sides of the equation with respect to t using the product rule:

$A'(t) = \frac{dA}{dt} = \frac{1}{2}\left(b\frac{dh}{dt} + h\frac{db}{dt}\right)$. We express the given information in terms of our variables:
$\frac{db}{dt} = -0.2 \text{ cm/sec}$, $\frac{dh}{dt} = 0.1 \text{ cm/sec}$, $b = 3$ cm and substitute the information into the expression above,

giving $A' = \frac{1}{2}(3(+0.1) + h(-0.2))$.

We are given that the area is increasing, i.e., $A'(t) > 0$; therefore we set the expression for A' greater than zero and solve for h:

$$\frac{1}{2}(3(+0.1) + h(-0.2)) > 0 \quad \Rightarrow \quad h(-0.2) > -0.3 \Rightarrow h < 1.5$$

4. C

Following the procedure for related-rates problems, we first draw a diagram and identify the quantities in our problem.

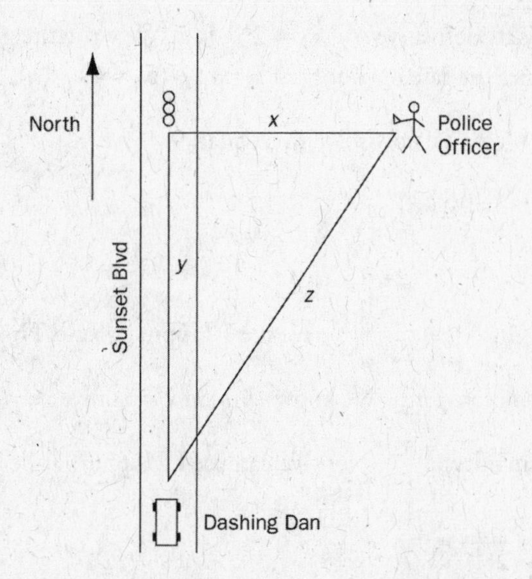

We use the Pythagorean theorem to find an equation relating our variables: $x^2 + y^2 = z^2$. We differentiate both sides of the equation with respect to time: $2x\frac{dx}{dt} + 2y\frac{dy}{dt} = 2z\frac{dz}{dt}$. The police officer's distance from the intersection is given in feet; Dan's speed is given in miles per hour, and we are told this speed at a time t given in seconds. We need to express everything in terms of the same units. We'll use feet, seconds, and feet per second. Dan is traveling away from the police officer at 37.5 mph, which is
$\frac{37.5 \text{ miles}}{\text{hour}} \cdot \frac{5280 \text{ feet}}{\text{mile}} \cdot \frac{1 \text{ hour}}{3600 \cdot \text{seconds}} = 55$ feet/second, i.e., $\frac{dz}{dt} = 55$ ft/sec.

We are told that one second after crossing the intersection, Dan is 55 feet away from the police officer.

We use the Pythagorean theorem to find Dan's distance due south from the intersection:

$33^2 + y^2 = 55^2 \Rightarrow y = 44$ ft. We fill this information into the equation relating the derivatives of the quantities and solve to $\frac{dy}{dt}$, which is the speed at which Dan is traveling on Sunset Boulevard:

$$2 \cdot 33 \cdot 0 + 2 \cdot 44 \cdot \frac{dy}{dt} = 2 \cdot 55 \cdot 55 \Rightarrow \frac{dy}{dt} = 68.75 \text{ ft/sec.}$$

We computed our answer in feet/second, but the answer choices are given in miles per hour. Thus we must express the answer in miles per hour:

$$68.75 \frac{\text{feet}}{\text{second}} \cdot \frac{1 \text{ mile}}{5280 \text{ feet}} \cdot \frac{3600 \text{ seconds}}{1 \text{ hour}} = 46.875 \text{ mph.}$$

It is important to pay attention to units. In this problem we are given information in miles and feet, and in seconds and hours. To work the problem, we need to make our units uniform. There are many ways to do this.

5. B

The rate of change of $f(x) = 2^x$ is its derivative $f'(x) = 2^x \cdot \ln 2$. We write the given information that the rate of change of f at $x = 5$ is four times the rate of change at $x = a : f'(5) = 4 \cdot f'(a)$.

That is, $f'(5) = 2^5 \cdot \ln 2 = 4 \cdot 2^a \cdot \ln 2$. We now solve this equation for a:

$2^5 \cdot \ln 2 = 2^2 \cdot 2^a \cdot \ln 2 \Rightarrow 2^5 = 2^{2+a} \Rightarrow a = 3.$

6. C

To calculate the limit $\lim_{x \to \infty} e^{-2x} \ln x = \lim_{x \to \infty} \frac{\ln x}{e^{2x}}$ we use L'Hôpital's Rule. First we have to make sure the limit is in indeterminate form. We know $\lim_{x \to \infty} \ln x = \lim_{x \to \infty} e^{2x} = \infty$, so indeed the limit is in indeterminate form of type $\frac{\infty}{\infty}$. Now we can use L'Hôpital's Rule to compute:

$$\lim_{x \to \infty} e^{-2x} \ln x = \lim_{x \to \infty} \frac{\ln x}{e^{2x}} \overset{L'Hôpital}{=} \lim_{x \to \infty} \frac{\frac{1}{x}}{2e^{2x}} = \lim_{x \to \infty} \frac{1}{2xe^{2x}} = 0.$$

7. C

In this question we are given that $h = 0.1$, $x_0 = 0$, $y_0 = 1$, and $F(x, y) = y + xy$. Using Euler's method for approximating solutions, we have:

$y_0 = y(x_0) = y(0) = 1$

$y_1 = y_0 + hF(x_0, y_0) = 1 + (0.1)F(0,1) = 1 + 0.1 = 1.1$

$y_2 = y_1 + hF(x_1, y_1) = 1.1 + (0.1)F(0.1,1.1) = 1.1 + (0.1)1.21 = 1.221$

$y_3 = y_2 + hF(x_2, y_2) = 1.221 + (0.1)F(0.2, 1.221) = 1.37$

Therefore, since $y_3 = y(x_0 + 0.3) = y(0.3)$, we have the approximation $y(0.3) \approx 1.37$. The answer is (C).

8. D

We first need to find the derivative of $f(x) = \sqrt{2x-3}$. By the chain rule, we get $f'(x) = \dfrac{2}{2\sqrt{2x-3}} = \dfrac{1}{\sqrt{2x-3}}$. The slope of the tangent line when $x = 6$ is $f'(6) = \dfrac{1}{\sqrt{2 \cdot 6 - 3}} = \dfrac{1}{3}$. We use the point-slope form of a line $y - y_1 = m(x - x_1)$ where $m = \dfrac{1}{3}$ is the slope and (6,3) is the point. Then the equation of the tangent line is $y - 3 = \dfrac{1}{3}(x-6)$. To determine the y-intercept, we need to determine y when $x = 0$. So $y - 3 = \dfrac{1}{3}(0-6)$, $y - 3 = -2$, and finally $y = 1$. Therefore the y-intercept is 1.

9. C

We first need to find the derivative of $f(x) = \ln(2x - 1)$. By the chain rule, $f'(x) = \dfrac{1}{2x-1} \cdot 2 = \dfrac{2}{2x-1}$. Now we set $\dfrac{2}{2x-1} = \dfrac{2}{5}$ which means $x = 3$.

10. A

To start, we need to find the derivative so $f'(x) = -2x$. Next we need to determine the slope of the normal line to f at the point where $x = k$. The slope of the tangent at $x = k$ is $f'(k) = -2k$. It follows that the slope m of the normal line $m = \dfrac{1}{2k}$, as the normal line is perpendicular to the tangent line (remember geometry?). The point of normalcy is $(k, f(k))$. Now we use point-slope form to get the equation of the normal line, $y - f(k) = \dfrac{1}{2k}(x-k)$. Since $f(k) = 4 - k^2$, then the normal line can be written $y - (4 - k^2) = \dfrac{1}{2k}(x-k)$. Finally, the y-intercept is when $x = 0$ which means $y - (4-k^2) = \dfrac{1}{2k}(0-k) \Rightarrow y - 4 + k^2 = -\dfrac{1}{2} \Rightarrow y = 3.5 - k^2$.

11. C

For $1 < t < 2$, both the velocity and the acceleration functions are negative which means the particle is moving to the left (i.e., $v(t) < 0$) and the velocity is becoming more negative (i.e., $a(t) < 0$) so we choose (C).

12. D

The particle moves to the right for $0 \le t \le 1$ advancing from 0 to 2, a distance of 2 units. Then the particle moves to the left for $1 \le t \le 3$ retreating from 2 to 1.5, a distance of 0.5 units. The total distance traveled is $2 + 0.5 = 2.5$ units.

13. E

A particle is at rest when its velocity is zero. In order to get a velocity equation, we take the first derivative of the position function. Correct application of the power rule yields the following: $v(t) = 2t^2 - 12t + 10 = 0 \Rightarrow 2(t^2 - 6t + 5) = 0 \Rightarrow 2(t-1)(t-5) = 0$. So $t = 1$ and 5 seconds.

14. A

The rate of change is the derivative so $V'(t) = 12000(1 - 0.01t)\cdot(-0.01) = -120(1 - 0.01t)$. To find the rate of change at $t = 60$, evaluate $V'(60) = -120(1 - 0.01\cdot 60) = -120(1 - 0.6) = -48$.

15. C

The rate of change is the derivative of $C(t) = K(e^{-at} - e^{-bt})$ so it follows that $C'(t) = K(-ae^{-at} + be^{-bt})$ by the chain rule. Now we need to evaluate at $t = 1$ so $C'(1) = K(-ae^{-a} + be^{-b}) = K(be^{-b} - ae^{-a})$.

FREE-RESPONSE ANSWERS

16. (a) The temperature, T, is increasing when $T'(t) > 0$. We use a calculator (set in radians mode!) to compute: $T'(7) = 7 \cdot \cos\left(\frac{7}{7}\right) = 3.782 > 0$. Therefore, the temperature is increasing at $t = 7$.

(b) The rate of change of temperature, $T'(t) > 0$, is increasing when $T''(t) > 0$. We use a calculator (set in radian mode!) to compute: $T''(7) = -.301 < 0 \Rightarrow T'(t)$ is decreasing, i.e., the rate of change of the temperature is decreasing.

17. (a) We let $A(t)$ be the amount of water in the tank at time t. The rate of change of the amount of water in the tank is $A'(t)$ = rate in – rate out, i.e.,

$$A'(t) = E(t) - D(t) = \frac{t+2}{t^2+1} - \sin t \cdot \frac{10t}{\sqrt{t^4+5}}$$

(b) The amount of water in the tank is increasing when $A'(t) > 0$ and decreasing when $A'(t) < 0$. To determine the situation at $t = 2$, we compute $A'(2)$:

$$A'(2) = E(2) - D(2) \approx -3.188 < 0$$

Therefore, the amount of water in the tank is decreasing at $t = 2$.

(c) The rate of change of the amount of water is given by $A'(t)$. It is increasing when *its* derivative is positive, i.e., $A''(t) > 0$, and decreasing when $A''(t) < 0$. To determine the situation at $t = 2$, we compute $A''(t)$ and then evaluate $A''(2)$ using a calculator (set in radian mode!):

$$A'(t) = E(t) - D(t) = \frac{t+2}{t^2+1} - \sin t \cdot \frac{10t}{\sqrt{t^4+5}}$$

$$A''(t) = \frac{t^2+1-(t+2)\cdot 2t}{(t^2+1)^2} - \cos t \cdot \frac{10t}{\sqrt{t^4+5}} - \sin t \cdot \left(\frac{10t}{\sqrt{t^4+5}}\right)'$$

$$= \frac{-t^2-4t+1}{(t^2+1)^2} - \cos t \cdot \frac{10t}{\sqrt{t^4+5}} - \sin t \cdot \frac{10 \cdot \sqrt{t^4+5} - 10t \cdot \frac{4t^3}{2\sqrt{t^4+5}}}{t^4+5}$$

$A''(2) \approx 2.41 > 0$. Because $A''(2) > 0$, the rate of change of the amount of water is increasing.

18. (a)

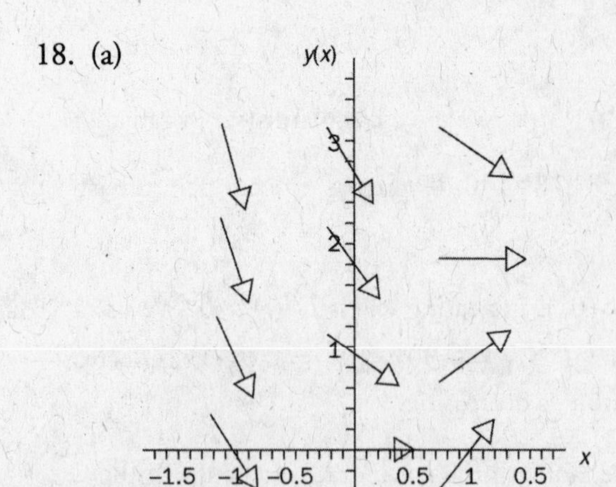

(b) The slopes are negative where the derivative is less than zero, that is, where $2x - y < 0 \Rightarrow 2x < y$ or $y > 2x$ for $x \neq 0 \neq y$.

19. We let $A = xy = x(18 - 3x) = 18x - 3x^2$. If this function has a maximum value, it will occur where $\frac{dA}{dx}$ is equal to zero or is undefined. We compute $\frac{dA}{dx} = 18 - 6x$. Because the derivative of the function is defined everywhere, if the function has a maximum value it must occur at a point where $\frac{dA}{dx} = 0$. We solve this equation $\frac{dA}{dx} = 0 \Rightarrow 18 - 6x = 0 \Rightarrow x = 3$.

We apply the second derivative test to determine whether the function has a *relative* maximum at $x = 3$. We compute $A''(x) = -6 < 0$ for all x; therefore, the function has a relative maximum at $x = 3$. Because A is a quadratic function, its graph is a parabola. The vertex of a parabola is the only relative extremum on the graph and it is a global extremum. Therefore, A has a global maximum at $x = 3$. To compute the maximum value we now substitute $x = 3$ back into the original equation for A: $18 \cdot 3 - 3 \cdot 3^2 = 27$.

20. (a) $y = 3x^{\frac{2}{3}} - 2x$. We start by computing the derivative y':

$$y' = 2x^{-\frac{1}{3}} - 2 = 2\left(\frac{1}{\sqrt[3]{x}} - 1\right)$$

We find the points where the derivative is zero or undefined:

$y' = 0$ or $2\left(\frac{1}{\sqrt[3]{x}} - 1\right) = 0$ when $x = 1$, and

y' is undefined when $x = 0$. Therefore, we find the critical points: $x = 0$ and $x = 1$.

(b and c) $y' = 2x^{-\frac{1}{3}} - 2$. Take the second derivative, $y'' = \frac{-2}{3}x^{\frac{-4}{3}} = -\frac{2}{3^3\sqrt{x^4}}$.

Note that $x^4 > 0$ for all $x \neq 0 \Rightarrow y'' < 0$ for all $x \neq 0$.

Remember: $\sqrt[3]{8} = 2$.

Interval	$(-\infty,0)$	$(0,1)$	$(1, \infty)$
Test Point	$x = -1$	$x = \frac{1}{8}$	$x = 8$
Sign of f'	$f'(-1) = (-2)(2) = -4$ negative	$f'\left(\frac{1}{8}\right) = (2)(2-1) = 2$ positive	$f'(8) = (2)\left(\frac{1}{2}-1\right) = -1$ negative
Conclusion	f decreasing	f increasing	f decreasing
Sign of f''	negative	negative	negative

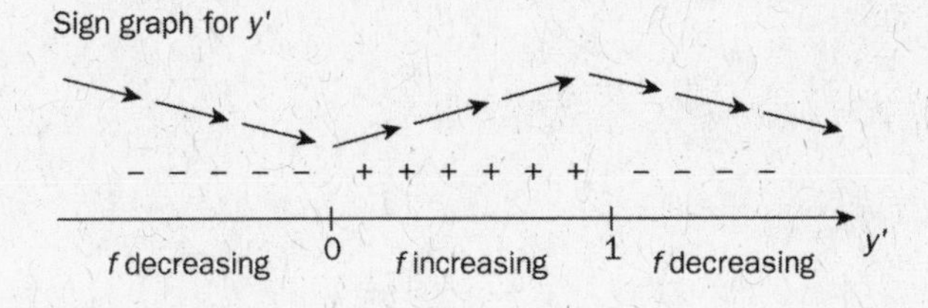

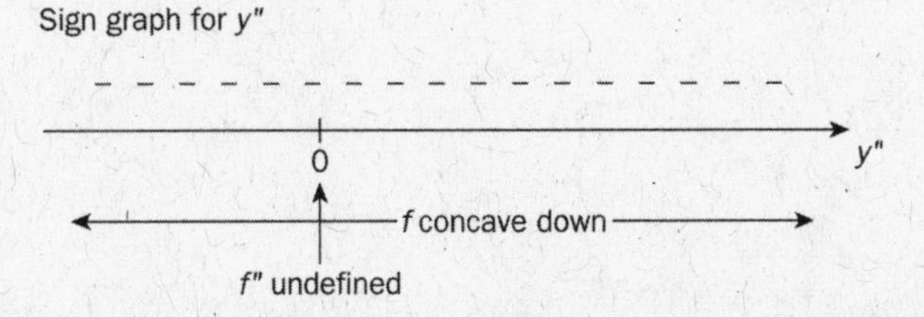

(d) A relative maximum occurs at $x = 1$, when the first derivative changes from positive to negative here. $f(1) = 3 \cdot \sqrt[2]{1^3} - 2 \cdot 1 = 1$. The relative maximum point is $(1,1)$. We can also use the second derivative test to evaluate the critical point $x = 1$. Because $f'(1) = 0$ and $f''(1) < 0$, therefore, by the second derivative test, f has a maximum at this point.

A relative minimum occurs at $x = 0$, because the first derivative changes from negative to positive here. It can be seen from the sketch that there is a cusp at

$x = 0 \cdot f(0) = 3 \cdot \sqrt[2]{0^3} - 2 \cdot 0 = 0$. The relative minimum point is $(0,0)$. We cannot use the second derivative test to evaluate this critical point; the second derivative test can only be applied to critical points where the first derivative equals zero.

(e) We use window: $[-1,4] \times [-1.1,2.1]$:

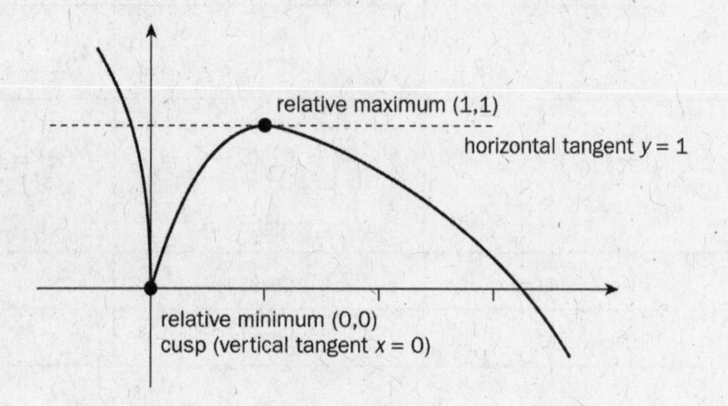

CHAPTER 17: INTRODUCTION TO INTEGRALS

IF YOU LEARN ONLY THREE THINGS IN THIS CHAPTER . . .

1. The Riemann sum, defined as $R_n \approx \sum_{i=0}^{n} f(c_i)\Delta x$, where c_i is a point in the ith subdivision, can provide a close approximation of the area under a curve.
2. The change in the value of a function over an interval $[a, b]$ is the area under the derivative curve, or the rate of change of the function from a to b.
3. Properties of the Definite Integral:

$$\int_a^b f(x) \pm g(x)\ dx = \int_a^b f(x)\ dx \pm \int_a^b g(x)\ dx$$

$$\int_a^b c \cdot f(x)\, dx = c\int_a^b f(x)\, dx$$

In the last few chapters we learned about derivatives, which are one of the two legs of calculus. In this chapter we'll learn about the other leg—integrals. You recall that we have a key phrase that sums up derivatives: "The derivative is the rate of change." Similarly, we have a key phrase for definite integrals: "A definite integral is the area under the curve."

These two phrases are so important that it's not a bad idea to say each of them to yourself a few times each night before going to bed.

In the last chapter we built up to what exactly we mean by saying the derivative is a rate of change. We started by considering the rate of change of a linear function, that is, a line; the rate of change of a line is its slope. We then took the idea of the slope of a line and generalized it to other functions—*voila*, the derivative was born.

Let's do something similar now to build up the theory of definite integrals.

COMPUTING DISTANCE TRAVELED

Suppose Sam starts out at 3 p.m. walking at a steady rate of 5 km/hr. If he walks until 5 p.m., how far does he travel? He walks for 5 – 3 = 2 hours; $5\frac{\text{km}}{\text{hr}} \cdot 2\text{ hr} = 10\text{ km}$. We can think of the distance traveled as the velocity multiplied by the change in time. Let's look at this on a graph.

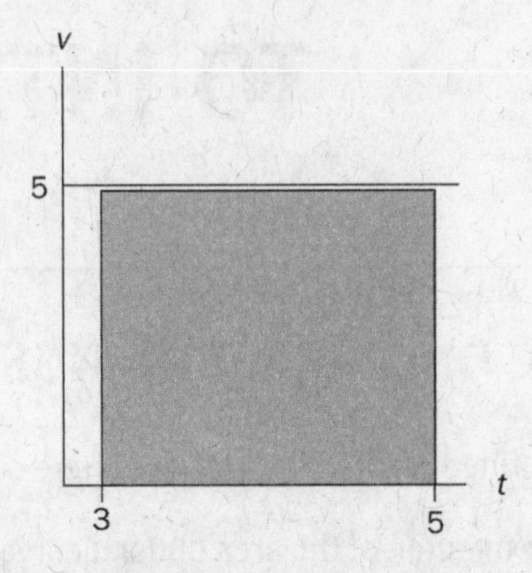

Velocity is given on the vertical axis and time is given on the horizontal axis. Distance is velocity multiplied by time, so we can also think of the distance traveled as the area—velocity given on the vertical axis times time given on the horizontal axis. The distance Sam travels is given by the area under the line $v = 5$ from time $t = 3$ to the time $t = 5$.

What if the velocity function isn't constant? Suppose velocity is given by some arbitrary curvy function, $v(t)$. How can we compute the total distance traveled from time $t = a$ to time $t = b$? Suppose the graph of $v(t)$ is shown below.

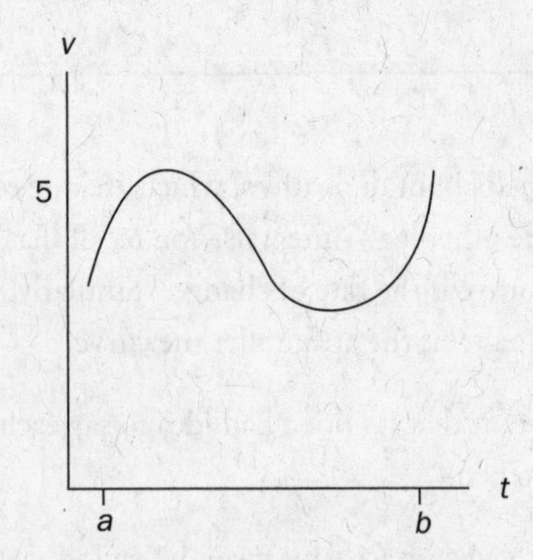

Using our example with Sam, we can think of the distance traveled from time $t = a$ to time $t = b$ as the velocity multiplied by the time, or the area under the velocity curve from a to b.

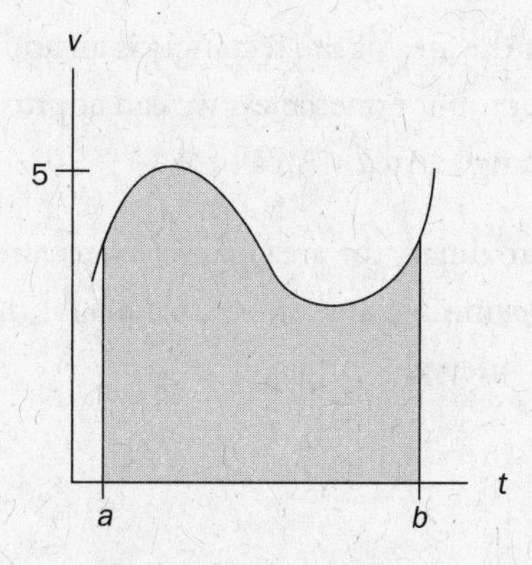

Great. So now we know that the distance traveled from time *a* to time *b* is the area shown above.

But, what *is* this area? Is it 10 km? Is it 20 km? We can't compute it (yet!), but we can approximate it.

FINDING THE AREA UNDER A CURVE

Suppose we want to find the area under the curve $f(x)$ from $x = a$ to $x = b$.

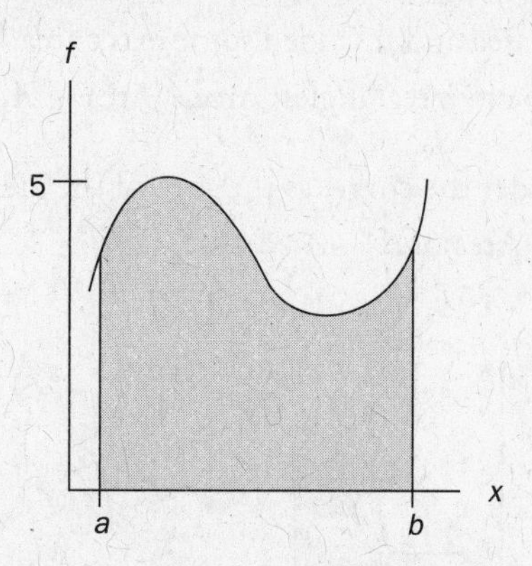

Well, first, let's simplify matters by assuming that f is constant, that is, that f is equal to its starting value, at $x = a$; i.e., $f(a)$ on the entire interval.

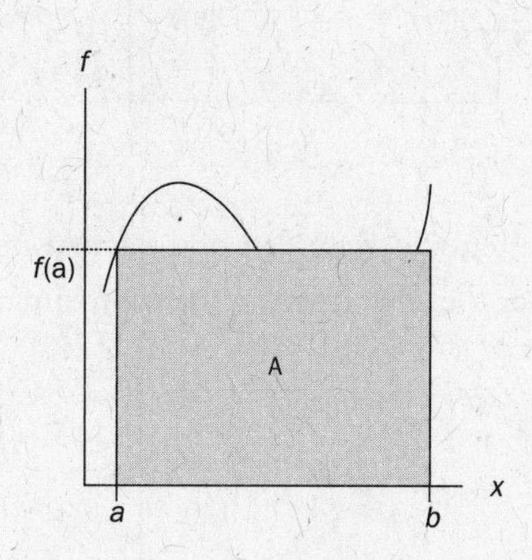

Looking at the graph, we see that the area of the rectangle is an approximation. We included some extra area and we left some area out, but nonetheless we can approximate the area under the curve from a to b as the area of the rectangle: Area ≈ Area of A.

Let's refine this idea. We can approximate the area using two rectangles—one with a height that is the value of the function at the beginning, and one with a height that is the value of the function at a point $x = c$ in the middle of the interval.

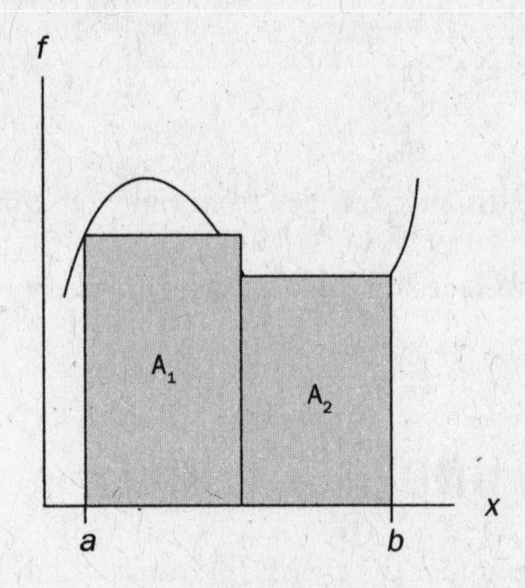

Once again, we've left out some area and included some extra area. This time we approximate the area by the sum of the areas of the two rectangles: Area ≈ Area of A_1 + Area of A_2.

We can approximate the area under the curve, as the sum of the areas of the four rectangles: Area ≈ Area of A_1 + Area of A_2 + Area of A_3 + Area of A_4.

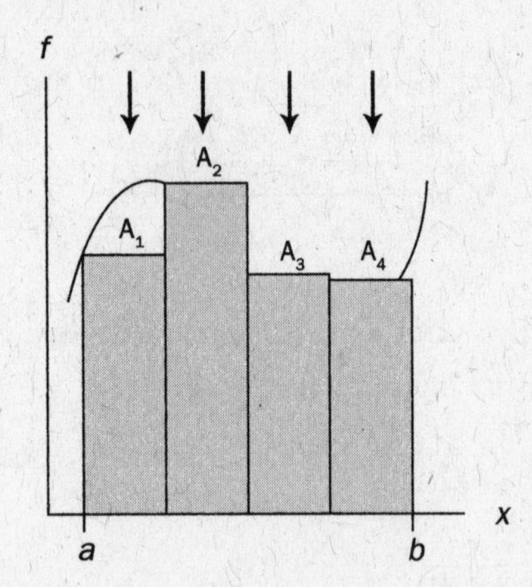

Our approximation is getting better; we left out less of the actual area this time and we included less superfluous area. The junk, or the error between the approximation and the actual area under the curve, is decreasing.

We can rewrite this using "sigma" notation: Area $\approx \sum_{i=1}^{4}$ Area of A_i. (If you don't remember sigma notation, this would be a good time to pause and review.)

The advantage of using sigma notation is that now we can generalize this approximation to any number of rectangles, say n rectangles, and write Area $\approx \sum_{i=1}^{n}$ Area of A_i. Let's observe a couple of things here. With more rectangles, the error—the extra junk we include and the stuff we leave out—is smaller. Also, as the number of rectangles increases, the length of each rectangle decreases.

It makes sense that if we could use an infinite number of rectangles, the error would disappear; that is, the actual area would be equal to the sum of the areas of the infinite number of rectangles. We can't actually compute the sum of the areas of infinitely many rectangles, but we *can* take the limit of the sum of the areas of n rectangles: Area $= \lim_{n \to \infty} \sum_{i=1}^{n}$ Area of A_i.

So far so good, but now we need a uniform way to determine the rectangles and compute the area of each one. To do this, we divide the interval $[a, b]$ into n equal pieces $\frac{b-a}{n}$. We'll call this distance Δx and it is the length of the base of each rectangle. For the height of each rectangle, we use the value of the function at some point in the rectangle, say, the left endpoint. Here are pictures of this division with two rectangles and with n rectangles, where the height of each rectangle is the value of the function at the left endpoint of the rectangle.

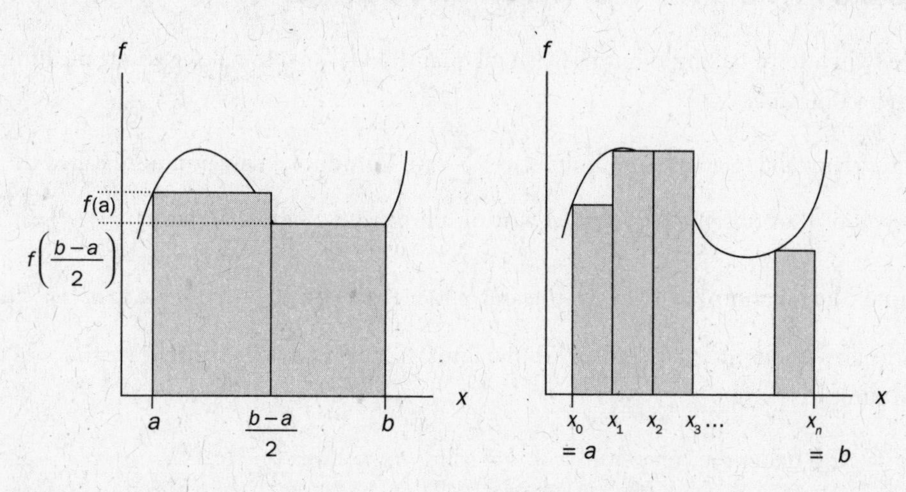

In the picture on the right, the area under the curve is approximated by
Area $\approx f(x_0)\Delta x + f(x_1)\Delta x + f(x_2)\Delta x + f(x_3)\Delta x + \ldots f(x_{n-1})\Delta x$, or, in sigma notation,

Area $\approx \sum_{i=0}^{n-1} f(x_i)\Delta x$.

Notice that the height of each rectangle is $f(x_{i-1})$ because we are using the left endpoint, which is x_{i-1} for the height of each rectangle.

RIEMANN SUMS

This approximation is an example of a Riemann sum. We define the nth Riemann sum, using the notation R_n, to be $R_n \approx \sum_{i=1}^{n} f(c_i)\Delta x$, where c_i is a point in the ith subdivision.

Above we observed that using more rectangles gives a better approximation and the greater the number of rectangles, the smaller the length of each one. It makes sense that as the number of rectangles gets bigger and bigger—as it approaches infinity—the error term gets very, very, small, and, in the limit, it disappears. That is, while the nth Riemann sum *approximates* the area under the curve, the "infinity-th" Riemann sum is *equal* to the area. We can't actually compute the "infinity-th" Riemann sum, but we can take the *limit* as n approaches infinity of the nth Riemann sum. Whether we use the left endpoint, right endpoint, midpoint, or some other point c_i in each rectangle to evaluate the height of the rectangle, it's still the case that

$$\text{Area} = \lim_{n \to \infty} R_n = \lim_{n \to \infty} \sum_{i=1}^{n} f(c_i)\Delta x.$$

If the limit exists, we say that the function f is integrable.

INTRODUCING THE DEFINITE INTEGRAL

When we switch from talking about really small quantities—Δ's—to talking about the limits of Δ's, the Δ becomes a d.

Also, rather than talking about an infinite sum, $\sum_{i=1}^{\infty}$, we introduce a new notation, called the integral symbol $\int$, or *integrand*. Instead of writing a limit and a sigma, we write $\int_a^b f(x)\,dx$.

We interpret the notation $\int_a^b f(x)\,dx$ as "the area under the curve $f(x)$ from $x = a$ to $x = b$," and we say that the area under the curve is equal to the limit, if it exists, as n approaches infinity of the nth Riemann sum.

$$\int_a^b f(x)dx = \text{Area under the curve from } a \text{ to } b = \lim_{n \to \infty} R_n = \lim_{n \to \infty} \sum_{i=1}^{n} f(c_i)\Delta x$$

We call this notation the *definite integral.* If f is integrable, we say that $\int_a^b f(x)\,dx$ is the definite integral of the function $f(x)$ from a to b. Putting the notation and the interpretation together gives us a definition: If f is integrable on the interval $[a, b]$, then the definite integral $\int_a^b f(x)\,dx$ of $f(x)$

from a to b is the area under the curve $f(x)$ from a to b. This definition is what lies behind our phrase, "the definite integral is the area under the curve."

Previously, we defined the derivative as the limit of the slopes of the secant lines; then, we developed techniques that allowed us to find the derivative without actually computing the limit. We'll do the same thing in this chapter with definite integrals.

In Chapter 22, we'll develop techniques that allow us to find the area under a curve (i.e., the definite integral) without actually computing the areas of rectangles and finding the limit of the sum areas of rectangles. In Chapter 23, we'll use the idea of a Riemann sum to do some actual approximations of areas under specific curves.

RATES OF CHANGE AND THE DEFINITE INTEGRAL

We started this chapter by computing a distance traveled given information about the velocity. That is, we were given information about the rate of change of a quantity and we were asked to determine the change in the quantity. We used this idea to build up the notion of the area under a curve.

We can extend this idea to any function. If we are given the rate of change of a function, $f'(x)$, then the area under the curve $f'(x)$ from $x = a$ to $x = b$ is the change in the value of the function f from a to b – $f(b)$ – $f(a)$. The area under the curve is the definite integral (remember our mantra), so

$$\int_a^b f'(x)\,dx = f(b) - f(a).$$

The change in the value of a function over an interval $[a,b]$ is the area under the derivative curve, or the rate of change of the function from a to b. We do not need to know how much *stuff* we started out with to do this computation. The definite integral provides information about the *change* in quantity, which does not depend on knowing the quantity at the outset.

For example, suppose air is being pumped into a balloon at the rate $r(t) = 10(1.5 + \sin(t))$ ml/second. If the balloon is completely deflated at time $t = 0$, then the amount of air in the balloon after, say, 10 seconds is the area under the curve $r(t) = 10(1.5 + \sin(t))$ from $t = 0$ to $t = 10$:

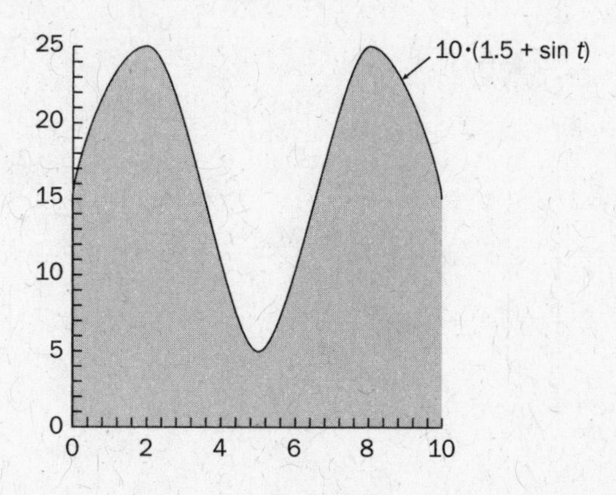

We can also write the volume of the definite integral of the rate of air filling the balloon:

$$V(10) = \int_0^{10} 10(1.5 + \sin t)dt$$

PROPERTIES OF THE DEFINITE INTEGRAL

Because the definite integral is defined in terms of a limit of Riemann sums, the rules for manipulation of limits come into play when we want to evaluate definite integrals. We'll give a list of rules here. The proofs all follow from the fact that the definite integral is defined as a limit; they can be found in any standard calculus text. For these theorems, we'll assume that the functions *f* and *g* are basically continuous, meaning that they are continuous except for a few point or jump discontinuities.

Theorem: Properties of the Definite Integral

$$\int_a^b f(x) \pm g(x)\, dx = \int_a^b f(x)\, dx \pm \int_a^b g(x)\, dx$$

$$\int_a^b c \cdot f(x)\, dx = c\int_a^b f(x)\, dx$$

AREA UNDER THE *X*-AXIS IS NEGATIVE

Because the definite integral is defined as the area of a sum of rectangles whose heights are the value of the function at various points, the definite integral considers any area that lies *under* the *x*-axis to be negative.

If we take a basic function, say, $f(x) = -x$, and compute the definite integral $\int_0^{10} -x\, dx$ (i.e., the area under the curve $f(x) = -x$, from $x = 0$ to $x = 10$), this area lies under the *x*-axis so the height of each rectangle in a Riemann sum is *negative*. If we look at this definite integral on a graph, we see that it is a triangle.

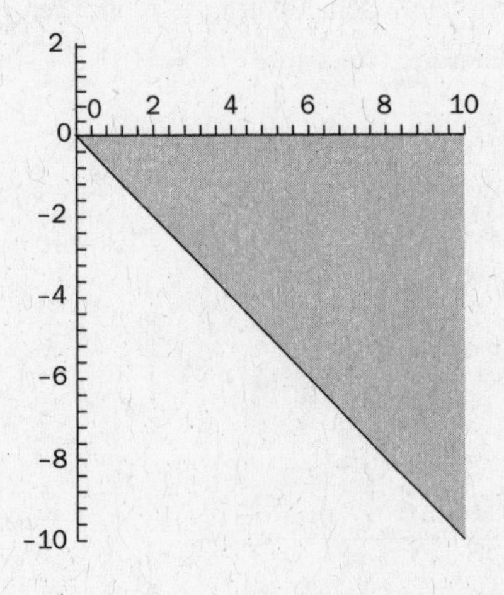

Its height, given by $f(10)$, is negative. We can compute its area $\frac{1}{2}bh = \frac{1}{2}(10 - 0)(-10) = -50$, that is, $\int_0^{10} -x\,dx = -50$.

If part of the curve is above the x-axis and part of the curve is below the x-axis, then the positive and negative areas will partially or totally cancel each other out. For example, the curve $y = x^3$ is symmetric around the y-axis, so $\int_{-3}^{3} x^3 dx = 0$. We can see this by looking at the graph.

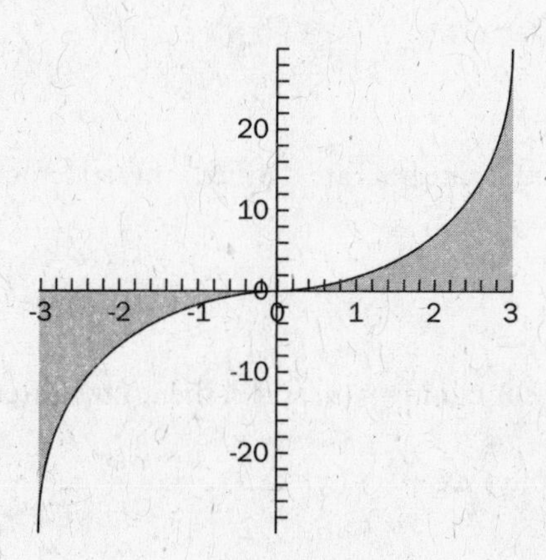

Note that x^3 is an odd function. Since all odd functions have the same rotational symmetry, we can generalize this finding into a time-saving tip! The integral of any odd function over an interval that is symmetric about the y-axis (−3 to 3, −1 to 1, etc.) will be 0. Thus, the same holds true for $y = x^5$, $y = x^7$ and even $y = x$.

COMPARISON: $\int_a^b f(x)\,dx$ VERSUS $\int_b^a f(x)\,dx$

We just saw that the area under the curve—the definite integral—can be negative. When we set up the definition of the definite integral as the limit of a Riemann sum, we defined the length of each rectangle to be $\frac{b-a}{n}$. In the diagrams we drew to motivate this definition, we assumed that $a < b$, so that $\frac{b-a}{n} > 0$. If $a > b$, then we have $\frac{b-a}{n} < 0$. It follows that:

Theorem:

$$\int_a^b f(x)\,dx = -\int_b^a f(x)\,dx$$

It also follows from the definition of the definite integral as the area under the curve that if we compute the definite integral from a point to itself—i.e., we don't really have an *area* to compute, the definite integral is zero.

Theorem:

$$\int_a^a f(x)\,dx = 0$$

Adding Areas

Because the definite integral is defined in terms of the area under the curve, it makes sense that you can add areas. If f is integrable on the interval $[a, b]$ and c is some point between a and b, then

$$\int_a^b f(x)\,dx = \int_a^c f(x)\,dx + \int_c^b f(x)\,dx.$$

In fact, because of the way definite integrals are defined, this statement is true even if c is not between a and b.

Theorem:

If f is integrable on an interval containing a, b, and c, then, no matter how a, b, and c are ordered, it is true that $\int_a^b f(x)\,dx = \int_a^c f(x)\,dx + \int_c^b f(x)\,dx$.

REVIEW QUESTIONS

1. Let f be a continuous function and let

$$A = \int_a^b -f(x)dx \quad B = \int_b^a f(x)dx$$

$$C = \int_b^a -f(x)dx \quad D = \int_a^b f(x)dx$$

Which of following statements is true?

(A) $A = -B$
(B) $C = D$
(C) $A = C$
(D) All three statements are true.
(E) None of the statements is true.

2. Using the definition of the definite integral as the area under the curve, compute $\int_0^6 2x-2\ dx$.

(A) 36
(B) 18
(C) −18
(D) 24
(E) −24

3. A particle moves along the x-axis so that its velocity at time t is given by the function whose graph is shown below. At which time is the particle farthest from the origin?

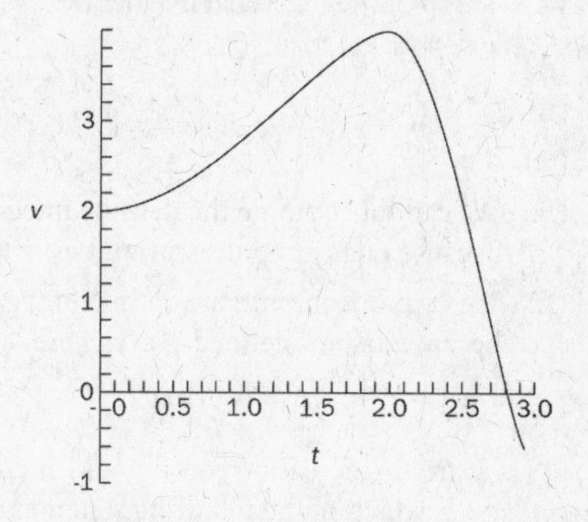

(A) $t = 0.5$
(B) $t = 1.5$
(C) $t = 2$
(D) $t = 2.8$
(E) $t = 3$

4. Let $f(x)$ be a continuous function and suppose $a < b$. Which of the following statements is always true?

I. $\int_a^b f(x)dx \le \int_a^b |f(x)|\,dx$

II. $\int_a^b f(x)dx \le \left|\int_a^b f(x)dx\right|$

III. $\int_a^b |f(x)|\,dx \le \left|\int_a^b f(x)dx\right|$

(A) Only I
(B) Only II
(C) Only III
(D) I and II
(E) II and III

5. Let $f(x) = [x]$ be the greatest integer. That is, $f(x)$ is equal to the next smallest integer less than or equal to x. For instance, $[4.5] = 4$, $[-3.6] = -4$. What is $\int_1^4 f(x)dx$?
   (A) 4
   (B) 5
   (C) 6
   (D) We cannot compute the definite integral because f has jump discontinuities.
   (E) We cannot compute the definite integral because f is not defined everywhere on [0,4].

6. If f is a strictly increasing continuous function and $a < b$, which of the following statements is always true?

   I. $R_{n \text{ right endpoint}} \geq \int_a^b f(x)dx$

   II. $R_{n \text{ midpoint}} = \int_a^b f(x)dx$

   III. $R_{n \text{ left endpoint}} \leq \int_a^b f(x)dx$

   (A) Only I
   (B) Only II
   (C) I and III
   (D) I, II, and III
   (E) None of the statements is true

7. Suppose $\int_1^2 f(x)\,dx = 2$ and $\int_0^3 f(x)\,dx = 1$, and $\int_1^3 3f(x) = -6$. What is $\int_0^1 f(x)\,dx$?
   (A) −2
   (B) 2
   (C) 3
   (D) 5
   (E) Cannot be determined from the information given.

8. $\int_{-1}^{4} |2x-4|\,dx =$
   (A) −5
   (B) 5
   (C) $\frac{19}{2}$
   (D) 13
   (E) $\frac{45}{2}$

9. On what interval $[a, b]$ will the maximum value of $\int_a^b (4+3x-x^2)dx$ be achieved?
   (A) [−1, 4]
   (B) [1, 4]
   (C) [−1, ∞]
   (D) [1, ∞]
   (E) [4, ∞]

10. If $\int_2^7 f(x)\,dx = 11$ and $\int_7^4 f(x)dx = 5$, then $\int_2^4 f(x)dx =$
    (A) −16
    (B) −6
    (C) 6
    (D) 16
    (E) None of these

11. If $\int_{-1}^{1} e^{-x^2}\,dx = k$, then $\int_{-1}^{0} e^{-x^2}\,dx =$
    (A) $-2k$
    (B) $-k$
    (C) $-\frac{k}{2}$
    (D) $\frac{k}{2}$
    (E) $2k$

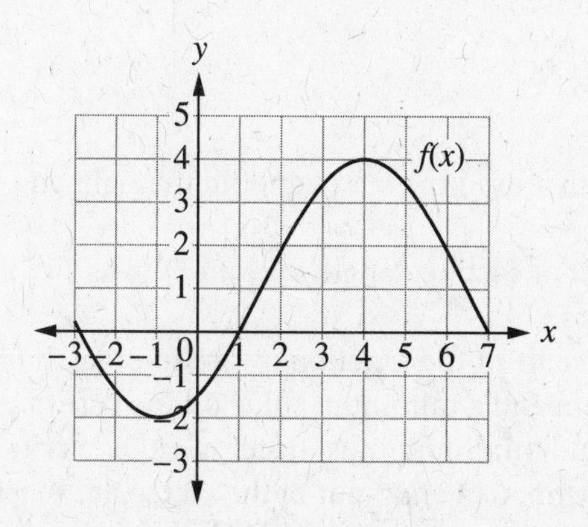

12. Given the graph of f shown above, which of the following statements are true?

   I. $\int_{-1}^{5} f(x)dx < \int_{0}^{5} f(x)dx$

   II. $\int_{-2}^{2} f(x)dx < \int_{2}^{-2} f(x)dx$

   III. $2\int_{1}^{2} f(x)dx < \int_{2}^{3} f(x)dx$

   (A) I only
   (B) I and II only
   (C) I and III only
   (D) II and III only
   (E) I, II, and III

13. Which of the following integrals have the same value?

   I. $\int_{0}^{\frac{\pi}{4}} \sin x \, dx$

   II. $\int_{0}^{\frac{-\pi}{4}} \sin x \, dx$

   III. $\int_{\frac{\pi}{4}}^{\frac{\pi}{2}} \cos x \, dx$

   (A) I and II only
   (B) I and III only
   (C) II and III only
   (D) I, II, and III
   (E) No two necessarily have the same value

14. The function f, continuous for all real numbers x, has the following properties:

   I. $\int_{1}^{3} f(x)\,dx = 7$

   II. $\int_{1}^{5} f(x)\,dx = 10$

   What is the value of k if $\int_{3}^{5} k\,f(x)\,dx = 33$?

   (A) −11
   (B) −3
   (C) 0
   (D) 3
   (E) 11

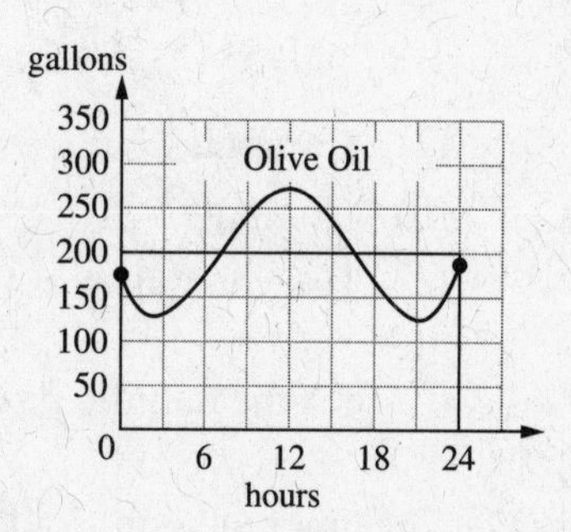

15. The rate of production of olive oil, in gallons per hour, at a processing facility on a given day is given by the graph shown to the right. Which value best approximates the total number of gallons of olive oil produced in the 24-hour period?

   (A) 1200
   (B) 1800
   (C) 2400
   (D) 3600
   (E) 4800

FREE-RESPONSE QUESTION

16. Question 16 requires a calculator.

    In this chapter, we discussed Riemann sums that used the left endpoint, the right endpoint, and the midpoint to evaluate the height of each rectangle to approximate $\int_a^b f(x)\,dx$.

    We can also use the maximum or minimum value of the function on the interval. If f is continuous, we know that f attains a maximum and a minimum value on the base of each rectangle, and we know that the maximum and minimum must occur at either a critical point that lies within the base of the rectangle or at an endpoint of the rectangle. When we approximate the area using the maximum value to evaluate the height of the rectangle, we call this an upper sum, and we use the notation U_n to compute the upper sum using n rectangles of equal length. The upper sum is always greater than or equal to the actual area under the curve. When we approximate the area using the minimum value, we call this a lower sum and use the notation L_n. The lower sum is always less than the actual area.

    Let $f(x) = x^3 - 3x^2 + 2x + 1$. Answer the following questions about $\int_0^3 f(x)\,dx$.

    (a) Compute U_3.

    (b) Compute L_3.

    (c) Use your answers from parts (a) and (b) to find an upper bound and a lower bound for the definite integral $\int_0^3 f(x)\,dx$.

17. The graph of a sinusoidal function $f(x) = 4 \sin \Phi x$ for some unknown value Φ is shown. The area under one peak of this sine wave, shaded below, is exactly 15.25.

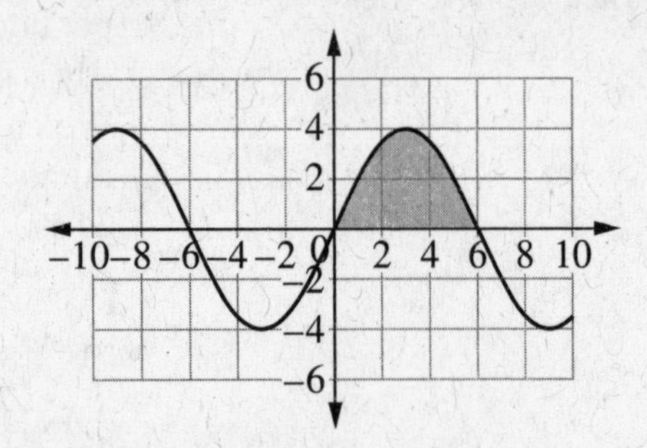

    Use the graph to aid in your inquiry to answer the questions. Justify your reasoning.

    (a) Evaluate: $\int_0^{12} f(x)dx$.

    (b) Evaluate: $\int_0^{-3} 3f(x)dx$.

    (c) Solve for c: $\int_8^{c} f(x+1)dx = 0$.

    (d) If $\int_{12}^{15} k\,dx = \int_{-9}^{-6} f(x)dx$, what is k?

ANSWERS AND EXPLANATIONS

1. B

Using the properties of the definite integral, we can express each of A, B, C in terms of D.

$$A = \int_a^b -f(x)dx = -\int_a^b f(x)dx = -D$$

$$B = \int_b^a f(x)dx = -\int_a^b f(x)dx = -D$$

$$C = \int_b^a -f(x)\,dx = -\int_b^a f(x)\,dx =$$

$$-\left(-\int_a^b f(x)\,dx\right) = \int_a^b f(x)\,dx = D$$

It follows that (B) is correct.

2. D

The definite integral is the area under the curve, where the area under the x-axis is negative. If we graph this function we can see that there are two regions of area; a small triangle below the x-axis and a large triangle above the axis.

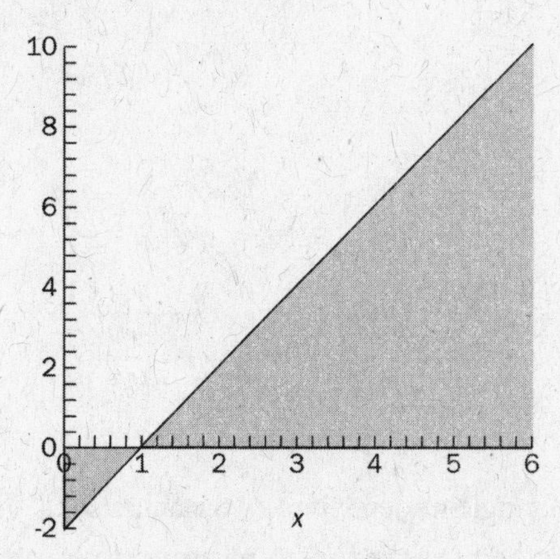

The area of the small triangle is $\frac{1}{2} \cdot 1 \cdot 2 = 1$; the area of the big triangle is $\frac{1}{2} \cdot 5 \cdot 10 = 25$. We count the area of the small triangle as negative, so:

$$\int_0^6 2x - 2\ dx = -(\text{Area of the small triangle}) +$$

$$(\text{Area of the big triangle}) = -1 + 25 = 24$$

3. D

Velocity is the rate of change of distance, so the area under the velocity curve from $t = 0$ to a time t on the graph is the change in distance over the time interval (counting the area under the axis as negative). The area that is above the t-axis represents distance traveled *away* from the origin. The area under the t-axis represents distance traveled *toward* the origin. The distance from the origin is greatest when the definite integral is the greatest, i.e., when the total area (counting the area beneath the axis as negative) is greatest. This occurs at the point $t = 2.8$.

4. D

The definite integral of the absolute value of a function counts all the area as positive, because the area that lies beneath the x-axis in the original function is flipped above the x-axis. Because the definite integral counts the area under the x-axis as negative, and the absolute value of a function is always positive, it follows that $\int_a^b f(x)\,dx \leq \int_a^b |f(x)|\,dx$; statement I is true.

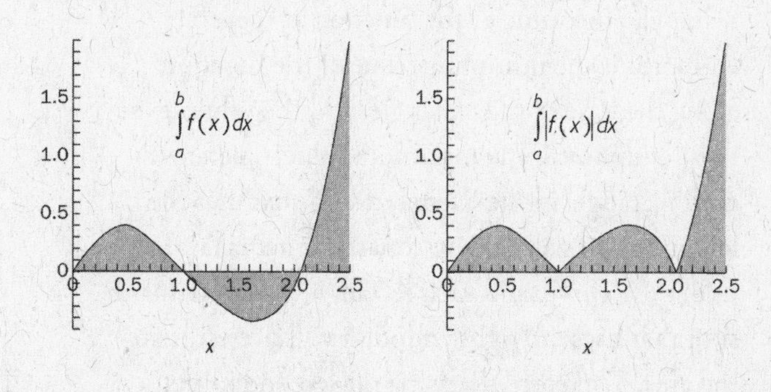

The absolute value of an expression is always greater than or equal to the expression, so statement II is true.

Using the same reasoning as for statement I above, it follows that $-\int_a^b f(x)\,dx \le \int_a^b |f(x)|\,dx$. Because the definite integral of the absolute value is greater than both the definite integral and "minus" the definite integral, it follows that:

$$\left|\int_a^b f(x)\,dx\right| \le \int_a^b |f(x)|\,dx$$

Therefore statement III is not true and the correct answer is (D).

5. C

The definite integral is the area under the curve. We graph the function $f(x) = [x]$ on the interval $[0,4]$ and see that the area is the sum of three rectangles, the areas of which are 1, 2, and 3. Therefore, the area under the curve is the sum of the areas of the three rectangles, that is $\int_1^4 f(x)\,dx = 1 + 2 + 3 = 6$.

Note that it is not necessary for a function to be continuous for the definite integral to exist. This follows from the definition of the definite integral and (D) is therefore not correct. The function *is* defined for all x, so (E) is also incorrect.

6. C

If f is a strictly increasing function, then for each rectangle, the value of the function at the left endpoint is the minimum value of the function along the base of the rectangle, and the value at the right endpoint is the maximum value. Therefore, each rectangle in the Riemann sum that uses the left endpoint will underestimate the actual area under the curve, and each rectangle in the Riemann sum that uses the right endpoint will overestimate the area. Therefore, R_n using the left endpoints is less than the actual area, and R_n using the right endpoints is greater than the actual area. We conclude that statements I and III are true.

(This would be true even for a function that was not always greater than zero, say $f(x) = x^3$ on $[-3,3]$.)

Let's take a look at pictures of the left endpoint and the right endpoint Riemann sums for a strictly increasing function, say $f(x) = x^3$ on the interval $[0,3]$.

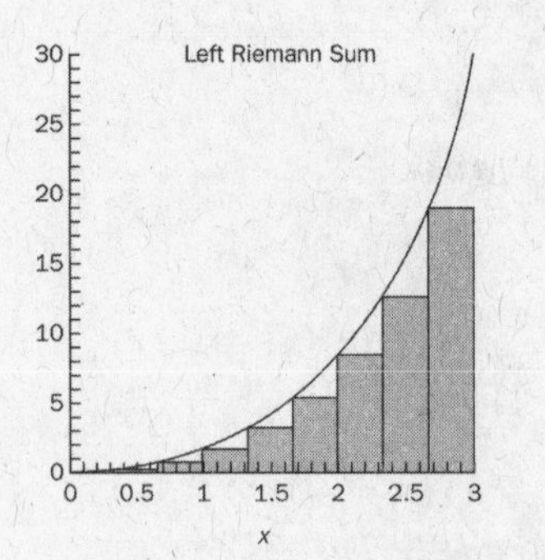

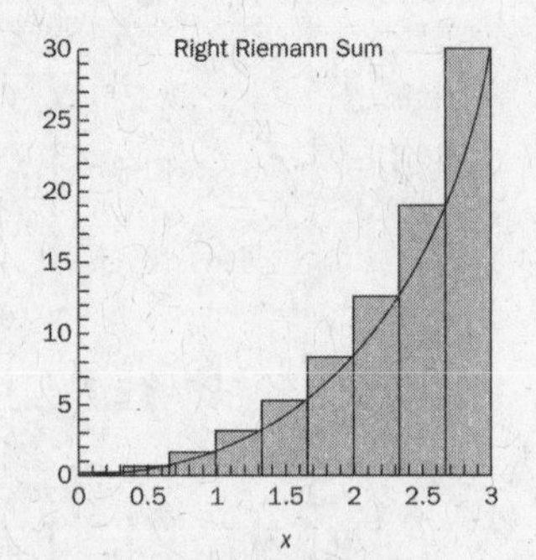

Let's pause for a little test strategy and notice that at this point we can cross off (A), (B), and (E). The correct answer must be (C) or (D).

Now we'll tackle statement II. By taking the first Riemann sum, where we approximate the area under the curve using just one rectangle, we can see that this statement is false.

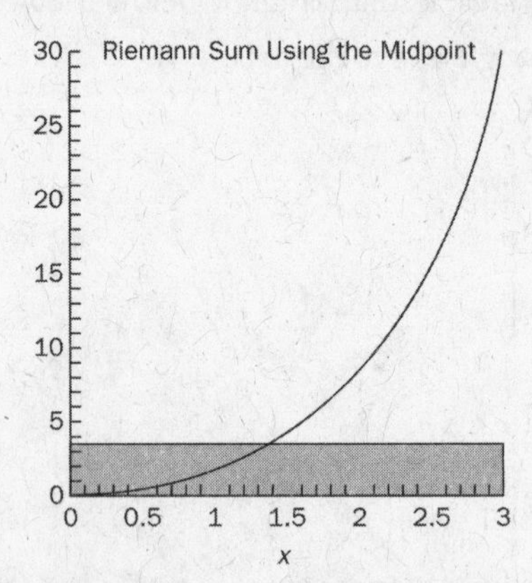

The rectangle underestimates the actual area. (This happens because the function is concave up on the interval [0, 3]—can you say why this causes the midpoint Riemann sum to underestimate the area?)

In this example, the midpoint Riemann sum underestimates the area under the curve, but it can also equal the area under the curve, such as when $f(x) = x$, or overestimate the area, such as when $f(x) = x^{\frac{1}{3}}$. Thus, statement II is not *always* true, and the question asks which of the statements is *always* true. Because statement II is false, the correct answer must be (C).

7. C

Using properties of the definite integral,

$$\int_1^3 3f(x)\,dx = -6 \Rightarrow 3\int_1^3 f(x)\,dx = -6 \Rightarrow$$

$$\int_1^3 f(x)\,dx = -2.$$

Using the additivity of the definite integral,

$$\int_0^3 f(x)\,dx = \int_0^1 f(x)\,dx + \int_1^3 f(x)\,dx.$$

We are given that $\int_0^3 f(x)\,dx = 1$ and we showed above that $\int_1^3 f(x)\,dx = -2$. We substitute these values into the equation above and solve $1 = \int_0^1 f(x)\,dx + (-2) \Rightarrow \int_0^1 f(x)\,dx = 3$—which is (C).

8. D

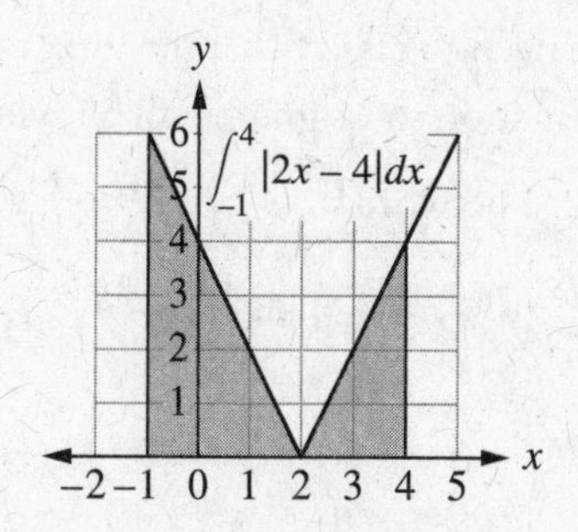

This integral can be interpreted as the area shown in the figure. We need to find the sum of the areas of the two triangles.

$$\int_{-1}^{4} |2x-4|\,dx = \int_{-1}^{2} |2x-4|\,dx + \int_{2}^{4} |2x-4|\,dx$$

$$= \frac{1}{2}(3)(6) + \frac{1}{2}(2)(4) = 13$$

9. A

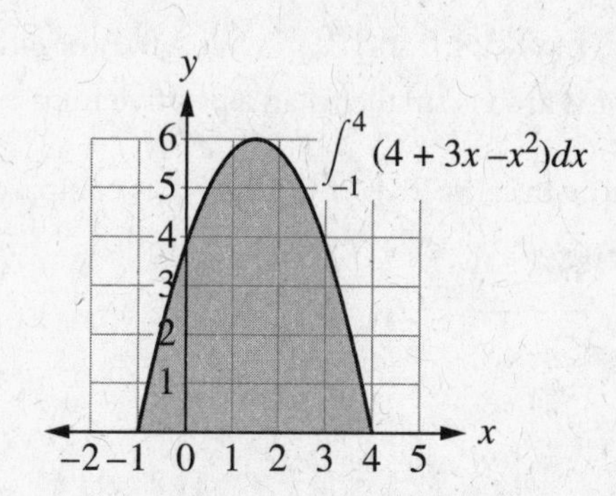

The integral $\int_a^b (4+3x-x^2)\,dx$ will take on its largest value on the interval $[a, b] = [-1, 4]$ as it will capture the largest region above the x-axis as shown in the figure.

10. D

Since $\int_2^7 f(x)\,dx = 11$, it follows that $\int_2^4 f(x)\,dx + \int_4^7 f(x)\,dx = 1$. Given $\int_7^4 f(x)\,dx = 5$, it follows that $\int_4^7 f(x)\,dx = -5$. This means that $\int_2^4 f(x)\,dx - 5 = 11$. So $\int_2^4 f(x)\,dx = 16$.

11. D

To start, we need to realize that the integrand e^{-x^2} is an even function so its graph will be symmetric to the y-axis. Therefore $\int_{-1}^{0} e^{-x^2}\,dx = \int_{0}^{1} e^{-x^2}\,dx$.

Since $\int_{-1}^{1} e^{-x^2}\,dx = k$ then $\int_{-1}^{0} e^{-x^2}\,dx = \frac{k}{2}$.

12. E

This problem requires that we compare integral values from the graph.

I. is true because $\int_{-1}^{5} f(x)\,dx$ is approximately 1.8 units less than $\int_{0}^{5} f(x)\,dx$ because the region $\int_{-1}^{0} f(x)\,dx \approx -1.8$.

II. is true. To start, we recognize that $\int_{-2}^{2} f(x)\,dx$ and $\int_{2}^{-2} f(x)\,dx$ are opposites. From the graph, it is clear that $\int_{-2}^{2} f(x)\,dx < 0$ so $\int_{2}^{-2} f(x)\,dx > 0$. A negative number is always smaller than a positive number.

III. is true because $2\int_{1}^{2} f(x)\,dx \approx 2$ as compared to $\int_{2}^{3} f(x)\,dx \approx 2.8$

13. D

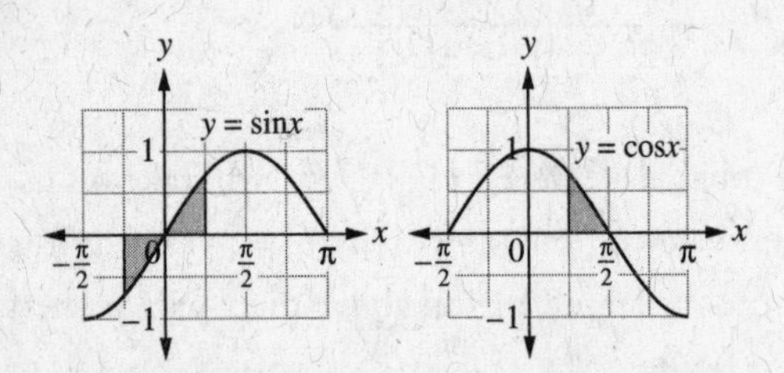

A quick sketch of $y = \sin x$ and $y = \cos x$ will make comparing these integrals an easy task. It is clear all three shaded regions are congruent and since

$$\int_{0}^{\frac{-\pi}{4}} \sin x\,dx = -\int_{\frac{-\pi}{4}}^{0} \sin x\,dx > 0$$ we know that all the integrals are equal in value.

14. E

Since f is continuous and $\int_{1}^{3} f(x)\,dx = 7$ and $\int_{1}^{5} f(x)\,dx = 10$, it follows that

$$\int_{1}^{5} f(x)\,dx = \int_{1}^{3} f(x)\,dx + \int_{3}^{5} f(x)\,dx = 10.$$

By substitution, $7 + \int_{3}^{5} f(x)\,dx = 10$ so $\int_{3}^{5} f(x)\,dx = 3$. Now we consider $\int_{3}^{5} k\,f(x)\,dx = 33$, which means that $k\int_{3}^{5} f(x)\,dx = 33$ and by substitution $3k = 33$ and finally $k = 11$.

15. E

Our solution is the area under the rate curve between 0 and 24 hours. To approximate this area, we can draw a rectangle as shown that is close in value to the area under the curve. The area of the rectangle is 4800.

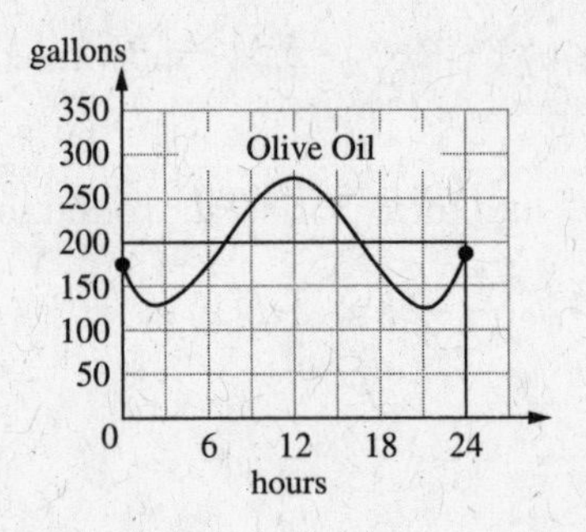

FREE-RESPONSE ANSWER

16. Let's start by graphing this curve on the interval $0 \leq x \leq 3$. The window $[0,3] \times [0,8]$ makes sense.

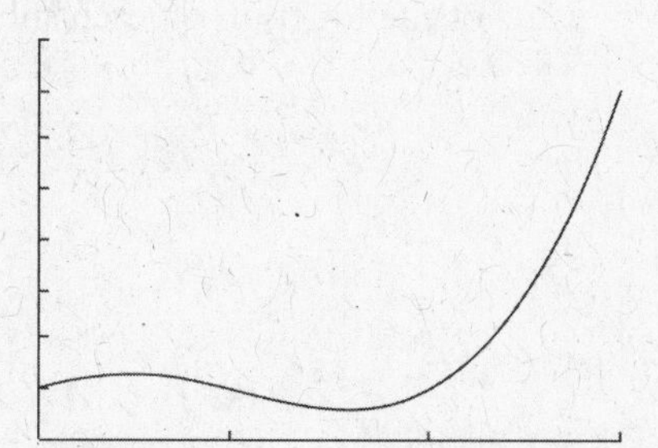

(a) Because we are computing the third upper sum, the interval $0 \leq x \leq 3$ is divided into three equal pieces—each with length 1. Thus the length of each rectangle is 1. By looking at the graph, we can see that the height of the first rectangle is obtained at the local maximum that occurs on the interior of the interval $[0,1]$. The height of the second rectangle is obtained from one of the endpoints—it is not clear which one from looking at this particular graph. The height of the third rectangle is obtained from the right endpoint.

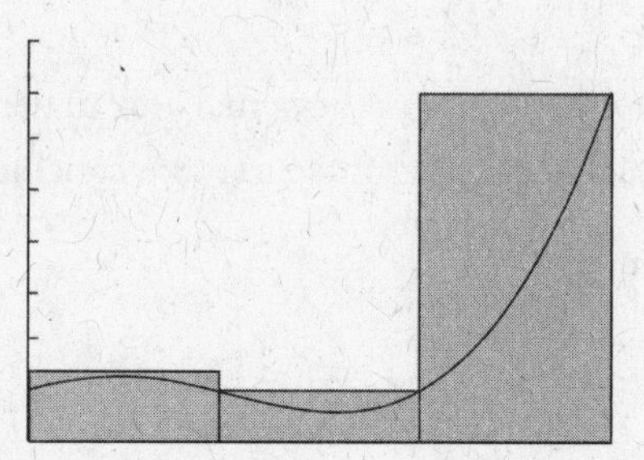

We can use the calculator to determine the maximum value of the function on each subinterval. We find that on the subinterval $0 \leq x \leq 1$, the maximum value occurs at the point $x = 0.423$, $y = 1.385$. Therefore, the height of the first rectangle is 1.385 and the area is $1 \cdot 1.385 = 1.385$.

On the subinterval $1 \leq x \leq 2$, the value of the function is $y = 1$ at the endpoints (the right endpoint and the left endpoint, because the value of the function is the same at these two points), so the height of the second rectangle is 1 and its area is $1 \cdot 1 = 1$.

On the subinterval $2 \leq x \leq 3$, the maximum value of the function occurs at the right endpoint $x = 3$. The value of the function at this point is $y = 7$, so the height of the third rectangle is 7 and its area is $1 \cdot 7 = 7$.

The value of the third upper sum, U_3, is the sum of the three areas:
$U_3 = 1.385 + 1 + 7 = 9.385.$

(b) We repeat the procedure from part (a) to find the lower sum, this time using the calculator to find the minimum value of the function on each subinterval.

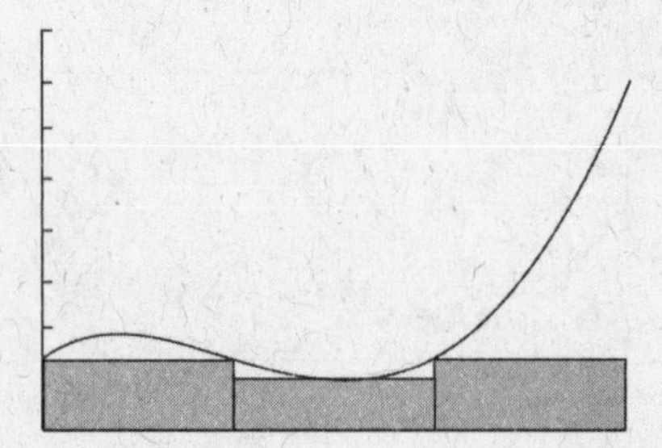

We find that on the first subinterval, the minimum value of the function is 1 and the area of the first rectangle is 1. On the second subinterval, the minimum value of the function is 0.615 so the area of the second rectangle is 0.615. On the third subinterval, the minimum value of the function is 1, so the area of the third rectangle is 1.

The value of the third lower sum, L_3, is the sum of the three areas:
$L_3 = 1 + 0.615 + 1 = 2.615.$

(c) Because the upper sum overestimates the actual area under the curve and the lower sum underestimates the actual area under the curve, we conclude that

$$2.615 \le \int_0^3 f(x)\,dx \le 9.385.$$

17. (a) To start, $\int_0^6 f(x)\,dx = 15.25$. Since this transformation of this sine wave has no vertical displacement, we know: $\int_6^{12} f(x)\,dx = -\int_0^6 f(x)\,dx = -15.25$. So it follows that $\int_0^{12} f(x)\,dx = \int_0^6 f(x)\,dx + \int_6^{12} f(x)\,dx = 15.25 + (-15.25) = 0$

(b) Again, $\int_0^6 f(x)\,dx = 15.25$. It follows that $\frac{1}{2}\int_0^6 f(x)\,dx = \int_0^3 f(x)\,dx = 7.625$ because of the symmetry of the sine wave. The value of the integral indicated by $\int_{-3}^0 f(x)\,dx$ will be the opposite of $\int_0^3 f(x)\,dx$ because the function is below the x-axis for the interval [-3, 0]. Therefore $\int_{-3}^0 f(x)\,dx = -7.625$. By the properties of the definite integral, $\int_0^{-3} 3f(x)\,dx = 3\int_0^{-3} f(x)\,dx = -3\int_{-3}^0 f(x)\,dx$. So it follows that $\int_0^{-3} 3f(x)\,dx = -3\int_{-3}^0 f(x)\,dx = -3(-7.625) = 22.875.$

(c) To start, let $u = x + 1$ so $du = dx$. Now let's find solutions c where $\int_9^c f(u)du = 0$. Notice that $\int_9^{15} f(u)\,du = 0$ and $\int_9^3 f(u)du = -\int_3^9 f(u)du = 0$. Because of the periodic nature of this sine wave, we know that $\int_9^{9+6m} f(u)du = 0$ for all integers m. It follows that $\int_9^c f(u)du = 0$ for $c = 9 + 6m$ for all integers m. Now we need to make a slight adjustment to get the values of c where $\int_9^c f(x+1)du = 0$. Since $X = u - 1$ we choose $c = 8 + 6m$ for all integers m.

(d) Because of symmetry, we know that $\int_{12}^{15} k\,dx = 7.625$. We want to find k so that $\int_{-9}^{-6} k\,dx = 7.625$. We need a geometric meaning to $\int_{-9}^{-6} k\,dx$. If we let $y = k$ and recognize that this is the graph of a horizontal line, then $\int_{-9}^{-6} k\,dx$ equals the area of a rectangle with width 3 and height k. So $\int_{-9}^{-6} k\,dx = 3k = 7.625$ and finally $k = \frac{7.625}{3} = \frac{61}{24} \approx 2.542$.

CHAPTER 18: APPLICATIONS OF INTEGRALS

IF YOU LEARN ONLY FOUR THINGS IN THIS CHAPTER . . .

1. The area between two curves is the integral of the length of the typical cross section.
2. The area between a curve and the x-axis is given by the integral $\int_a^b |f(x)|dx$.
3. If we can express the area of a cross section of S in a plane perpendicular to the x-axis as $A(x)$, where A is a function we can integrate, and if S is bounded by $x = a$ and $x = b$, then the volume of the solid S is $\int_a^b A(x)\,dx$.
4. If we can express the area of a cross section of S in a plane perpendicular to the y-axis as $A(y)$, where A is a function we can integrate, and if S is bounded by $y = a$ and $y = b$, then the volume of the solid S is $\int_a^b A(y)\,dy$.

Definite integrals are used to model many real-world phenomena. In recent years the focus of the AP exam has shifted away from testing computational skills and toward testing your conceptual grasp of calculus. The folks that write the exam want to make sure that you know what a definite integral is and what it represents, how to set up various kinds of integrals, and how to use definite integrals to model things in the real world.

In this chapter we'll go over the applications of the definite integral that usually show up on the exam. On any particular exam, though, the exam writers may decide to include a new or unusual application. That's fine. If we understand the basic information that a definite integral provides, we can set up definite integrals to model any applicable situation.

On almost every recent AP exam, there has been at least one multipart free-response question based on the topics covered in this chapter. There is often more than one question like this.

AREA BETWEEN TWO CURVES

A definite integral is the area under a curve. We can also use the definite integral to find the area *between* two curves, either between specified limits of integration or between points where the curves intersect.

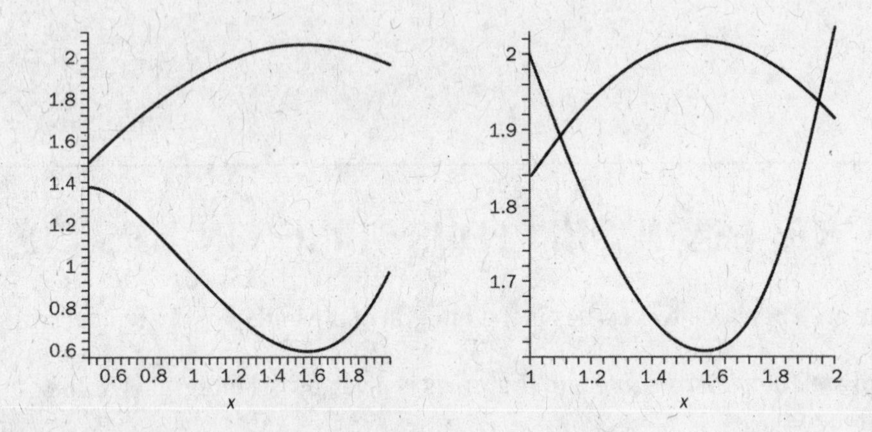

It is helpful to think of the area between two curves as a line segment (really an infinitely narrow rectangle) sweeping out an area. We are integrating the *length of the typical cross section* and moving in the direction that sweeps out the full area.

The area between two curves is the integral of the length of the typical cross section.

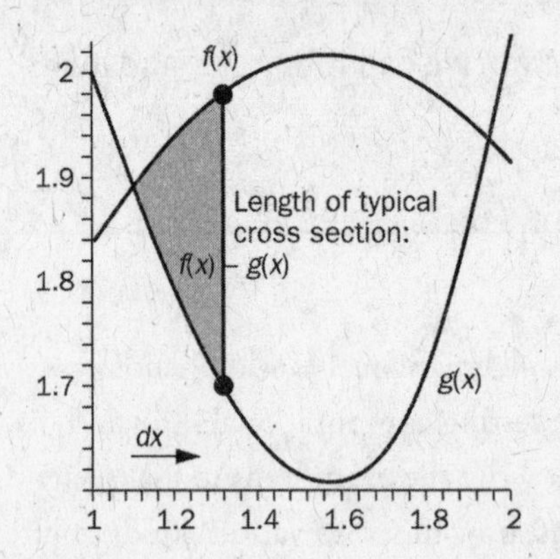

To compute the area between two curves between $x = a$ and $x = b$, we evaluate the integral $\int_a^b$ function on top – function on bottom $\cdot\, dx$. The proof for this can be found in any standard calculus text.

For example, to find the area between the curves $f(x) = \ln(x + 1)$ and $g(x) = -x^2 + 5$ between $x = 0$ and $x = 2$, we first need to determine which curve is on top. We can do this with a quick sketch in which we shade the area we are finding.

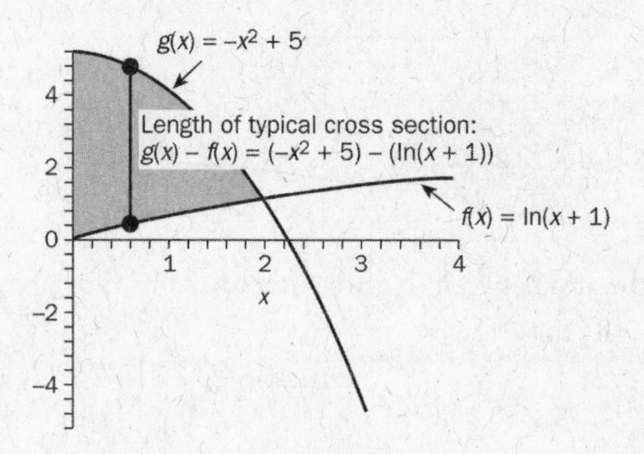

We see from the graph that the curve $g(x) = -x^2 + 5$ lies on top, so the area between the curves is given by:

$$\int_0^2 g(x) - f(x)\,dx$$

$$= \int_0^2 (-x^2 + 5) - (\ln(x + 1))\,dx$$

We can then evaluate this integral using a technique of integration or by using a graphing calculator. (Note that evaluating this integral without a calculator requires using integration by parts, which is not part of the Calculus AB syllabus but *is* part of the BC syllabus. Integration by parts is covered in Chapter 21.)

Using our calculator's utility to compute a definite integral, we find that

$$= \int_0^2 (-x^2 + 5) - (\ln(x + 1))\,dx \approx 6.037$$

What If One (or Both) of the Curves Lies Beneath the x-axis?

It doesn't matter if one or both of the curves lies under the x-axis; the computation remains the same. Let's look at the previous example, but shift each of the curves down by 5. That is, let $f(x) = \ln(x + 1) - 5$ and let $g(x) = -x^2$. If we graph these two functions, the area *between* the curve looks the same. Only the vertical placement of the curves has changed; their relative positions remain the same.

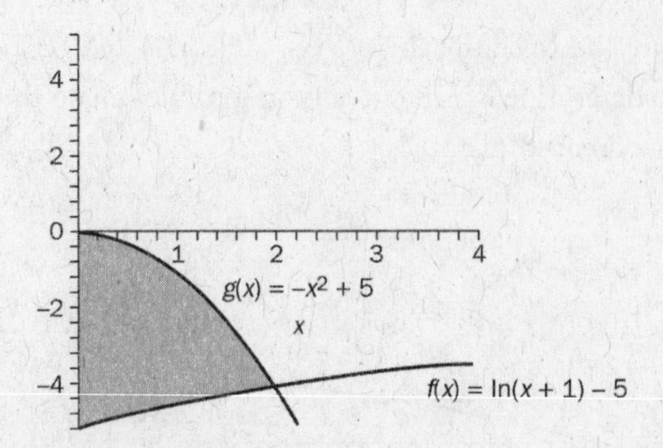

We can see visually that the area between the two curves remains the same and we can compute the new definite integral to verify this:

$$\int_0^2 g(x) - f(x)\,dx$$

$$= \int_0^2 (-x^2) - (\ln(x+1) - 5)\,dx \underset{\text{properties of definite integrals}}{=} \int_0^2 (-x^2 + 5) - (\ln(x+1))\,dx \approx 6.037$$

Setting up the Integral: Which Curve Is on Top?

Sometimes the curves switch places on the interval of integration; that is, sometimes one of the curves starts out on top, but somewhere in the middle the curves intersect and the curve that used to be on the bottom moves to the top position. When this happens we must break the integral into pieces.

Example:

Find the area between the curves $f(x) = \sin x$ and $g(x) = \cos x$ on the interval $[0, 2\pi]$.

Solution:

First, we graph these two functions on the same set of axes to determine which function is on top and where the curves intersect. We can shade the area we are trying to find.

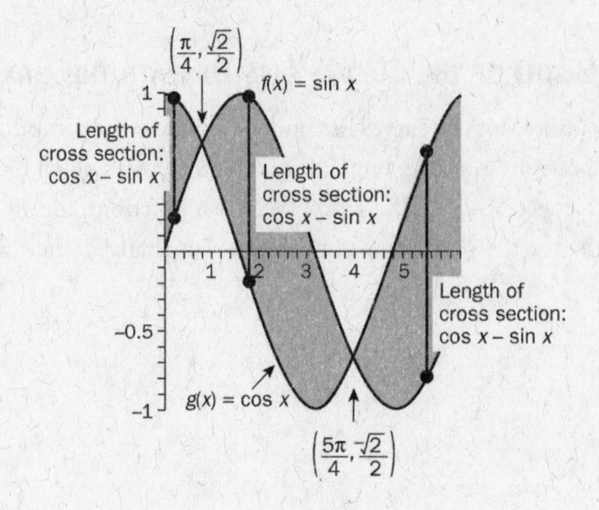

As excellent calculus students who have memorized the trigonometric functions of the angles $0, \frac{\pi}{6}, \frac{\pi}{4}, \ldots$etc., as required by the AP folks, we know that on the interval $[0, 2\pi]$, $\sin x$ and $\cos x$ intersect at $x = \frac{\pi}{4}$ and $x = \frac{5\pi}{4}$. We see from looking at the graph that on the interval $\left[0, \frac{\pi}{4}\right]$, the cosine curve lies on top; on $\left[\frac{\pi}{4}, \frac{5\pi}{4}\right]$, the sine curve is on top, and on $\left[\frac{5\pi}{4}, 2\pi\right]$, $\cos x$ is back on top. The area between the two curves is therefore expressed by the sum of three definite integrals:

$$\int_0^{\frac{\pi}{4}} \cos x - \sin x \; dx + \int_{\frac{\pi}{4}}^{\frac{5\pi}{4}} \sin x - \cos x \; dx + \int_{\frac{5\pi}{4}}^{2\pi} \cos x - \sin x \; dx$$

The technique for computing these integrals appears in Chapter 20; we can also evaluate these definite integrals using a calculator. Doing so, we find:

$$\int_0^{\frac{\pi}{4}} \cos x - \sin x \, dx = \sqrt{2} - 1$$

$$\int_{\frac{\pi}{4}}^{\frac{5\pi}{4}} \sin x - \cos x \, dx = 2\sqrt{2}$$

$$\int_{\frac{5\pi}{4}}^{2\pi} \cos x - \sin x \, dx = 1 + \sqrt{2}$$

The sum of these three definite integrals is $4\sqrt{2} \approx 5.657$.

Area Between a Curve and the *x*-axis

Asking "What is the area between the curve *f*(*x*) and the *x*-axis?" is not (necessarily) the same as asking "What is $\int_a^b f(x)\,dx$?" The area between two curves is always positive, while the definite integral is signed area. The area between a curve and the *x*-axis is given by the integral $\int_a^b |f(x)|\,dx$.

AP EXPERT TIP

If you calculate definite integrals using a calculator like the TI-83, you won't get exact solutions with square roots - instead, you'll get non-repeating decimal approximations. Make sure your answers are accurate to three decimal places in the FRQ section, and be ready to convert multiple choice options to decimals in order to find a choice that matches your answer.

AREA BOUNDED BY CURVES

A typical question on the AP exam asks you to find the area of the region bounded by given curves; often a graph of the curves is provided as part of the problem setup.

This is a twist on finding the area between curves. The twist is that we need to examine the graph of the curves to determine what the limits of integration are and to decide which curves lie above and which lie below.

Example:

The region R shown below is bounded by the curves $f(x) = x$, $g(x) = -(x-1)^2 + 1$ and the x-axis. Find the area of the region R.

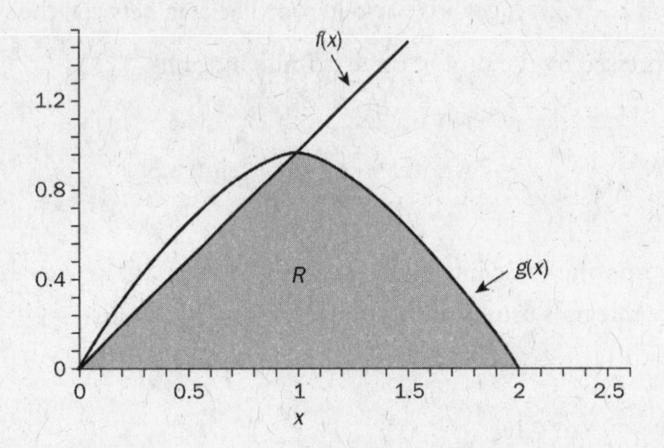

Solution:

Looking at the graph we see that there are two distinct regions. The left region R_1 starts at the point $x = 0$ and runs until the two curves intersect at $x = 1$. The second region R_2 runs from $x = 1$ to the point where $g(x) = -(x - 1)^2 + 1$ intersects the x-axis, $x = 2$.

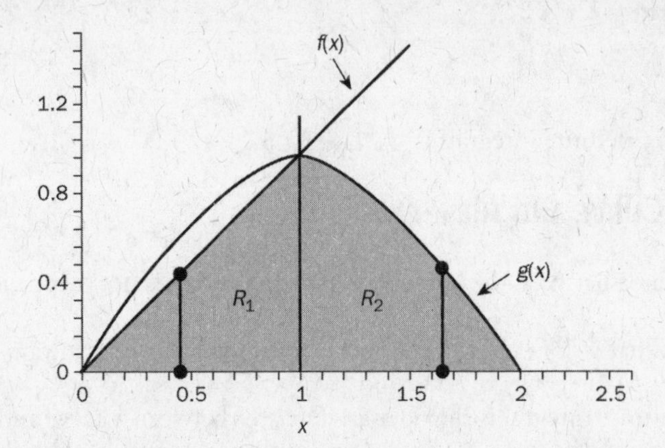

In region R_1, the cross section lies between the curve $f(x) = x$ and the x-axis. Therefore, the length of this cross section is x, and the area of region R_1 is computed by:

$$\text{Area of } R_1 = \int_0^1 x\,dx = \frac{1}{2}$$

In region R_2, the cross section lies between the curve $g(x) = -(x - 1)^2 + 1$ and the x-axis. Therefore, the length of this cross section is $-(x - 1)^2 + 1$, and the area of region R_2 is computed by:

$$\text{Area of } R_2 = \int_1^2 -(x-1)^2 + 1\, dx = \frac{2}{3}$$

The area of R is equal to the sums of the areas of R_1 and R_2:

$$\begin{aligned}\text{Area of } R &= \text{Area of } R_1 + \text{Area of } R_2 \\ &= \int_0^1 x\, dx + \int_1^2 -(x-1)^2 + 1\, dx \\ &= \frac{1}{2} + \frac{2}{3} \approx 1.167\end{aligned}$$

The area of this region can also be found with a single integral expression by using *horizontal* cross-sectional segments that sweep through the region from bottom to top. If we use the lengths of horizontal segments, we must rewrite our original curves from $y = f(x)$ and $y = g(x)$ to $x = f(y)$ and $x = g(y)$, and travel through the region from the horizontal boundaries $y = c$ to $y = d$. The basic setup is as follows:

$$\text{Area} = \int_c^d (\text{function on right} - \text{function on left}) \cdot dy$$

where the function on the right $= f(y)$ and the function on the left $= g(y)$. (Notice that we are integrating with respect to y this time.)

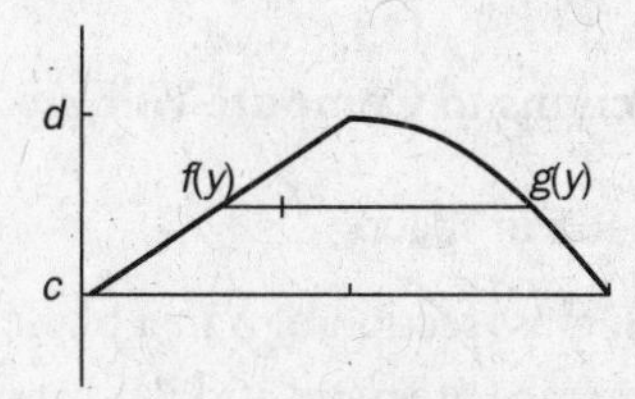

Now, if we return to the last example, we rewrite $f(x) = x$ as $f(y) = y$, and we rewrite $g(x) = -(x - 1)^2 + 1$ as $g(y) = \sqrt{1-y} + 1$ (Note: only the top half of the parabola applies here.)]

Because the parabola is to the right of the line, and the point of intersection of the two curves is (1, 1), we set up an integral representing the area of the region in terms of y as y goes from 0 to 1:

$$\text{Area} = \int_0^1 [g(y) - f(y)]\, dy$$

$$\text{Area} = \int_0^1 \left[\sqrt{1-y} + 1 - y\right] dy = 1.167.$$

It should be noted here that sometimes we don't have a choice regarding the horizontal or vertical setup, so it's important to know them both. A multiple-choice question may force the issue, giving you possible setups in one direction only. Or, in a free-response question, rewriting the given functions in terms of the other variable may not be possible or convenient.

VOLUMES OF SOLIDS

We can extend the idea of thinking of the area between two curves as being swept out by the typical cross section to thinking of volumes the same way. We can think of a volume of a solid as being swept out by the typical cross section, i.e., an area. Just as the area between two curves is the integral of (the length of) the typical cross section, the volume of a solid is the integral of the (area) of the typical cross section.

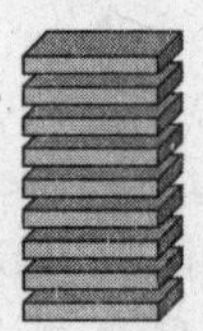

The solids above show cross sections that move along the x-axis, and cross sections that move along the y-axis. If the cross sections are moving along the x-axis, that is, if the cross sections are perpendicular to the x-axis, then we are integrating in the x-direction and we need the area of the typical cross section expressed as a function of x. If the cross sections are moving along the y-axis, we are integrating in the y-direction and we need the area of the typical cross section expressed as a function of y.

Theorem: Using Parallel Cross Sections to Compute Volume

- Let S be a solid, bounded region in space.
- If we can express the area of a cross section of S in a plane perpendicular to the x-axis as $A(x)$, where A is a function we can integrate, and if S is bounded by $x = a$ and $x = b$, then the volume of the solid S is $\int_a^b A(x)dx$.
- If we can express the area of a cross section of S in a plane perpendicular to the y-axis as $A(y)$, where A is a function we can integrate, and if S is bounded by $y = a$ and $y = b$, then the volume of the solid S is $\int_a^b A(y)dy$.

The most straightforward example—which often shows up the exam—is when we are given that the base of a solid is a region in the plane, and we are given information about the cross section in a plane perpendicular to the x- or y-axis.

Example:

Suppose R is the region in the following plane, bounded by the curves $f(x) = e^x - 5$, $g(x) = (x - 1)^3$, and the y-axis, and R is the base of a solid.

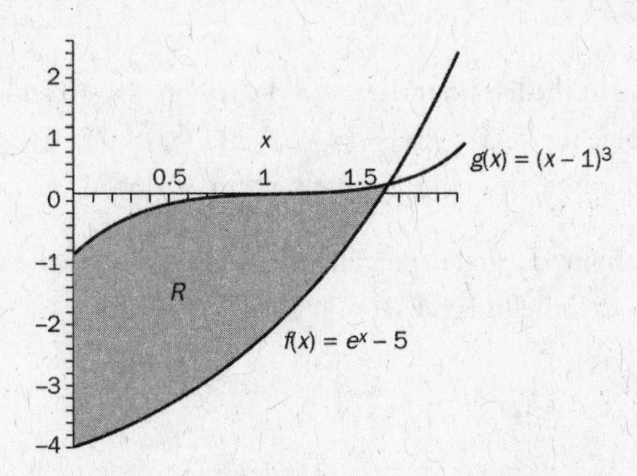

If each cross section of the solid perpendicular to the x-axis is an isosceles right triangle with one leg in the base, what is the volume of the solid?

Solution:

We are told that the typical cross section perpendicular to the x-axis is an isosceles right triangle with one leg in the base. Therefore the area of the typical cross section is $\frac{1}{2}l^2$ where l is the length of the leg in the base. If we express the area of this cross section as a function of x, $A(x)$, and find the boundaries a and b of the region R, then the volume of the solid can be expressed as $\int_a^b A(x)\,dx$.

We start by drawing the typical cross section.

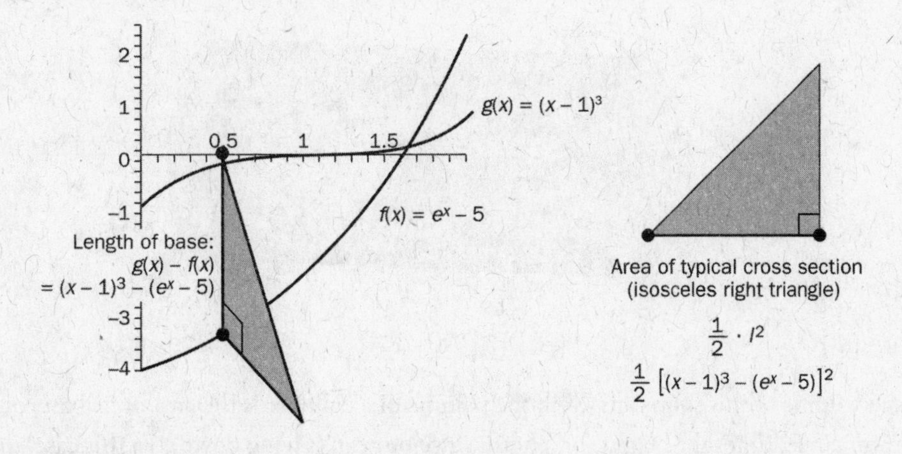

Looking at the graph of the region R, we see that on the entire region, cross sections of R perpendicular to the x-axis are bounded by the curve $g(x) = (x - 1)^3$ above and $f(x) = e^x - 5$ below. These are the legs of the isosceles right triangles that form the cross sections of the solid.

Therefore, the length of the leg in terms of x is $l = (x - 1)^3 - (e^x - 5)$, so the area of the typical cross section is given by:

$$A(x) = \frac{1}{2}[(x-1)^3 - (e^x - 5)]^2$$

We also see on the graph that the left boundary a of the region is $x = 0$ and the right boundary of the region is the point where the curves $g(x) = (x - 1)^3$ and $f(x) = e^x - 5$ intersect. We can use a graphing calculator to determine this point of intersection: ≈ 1.6671.

We can now set up the definite integral expressing the volume of the solid and use a graphing calculator to compute the definite integral:

$$\int_0^{1.6671} \frac{1}{2}[(x-1)^3 - (e^x - 5)]^2\,dx = 5.271$$

SOLIDS OF REVOLUTION

Volumes of solids of revolution are a big favorite of the AP folks. A solid of revolution is created when a region is rotated around a line, usually the x-axis, the y-axis, or a line parallel to one of these axes. Solids of revolution are a special case of the solids we saw in the previous section. What's so special about them? The cross sections are always circles or washers (i.e., a circle with a hole in the middle).

Close your eyes and imagine for a minute that you are revolving the line $y = 1$ around the x-axis. Actually, let's use just a piece of the line—say the segment from $x = 0$ to $x = 3$. What figure is created? Did you say a cylinder or, in technical math lingo, a can of soup? That's right.

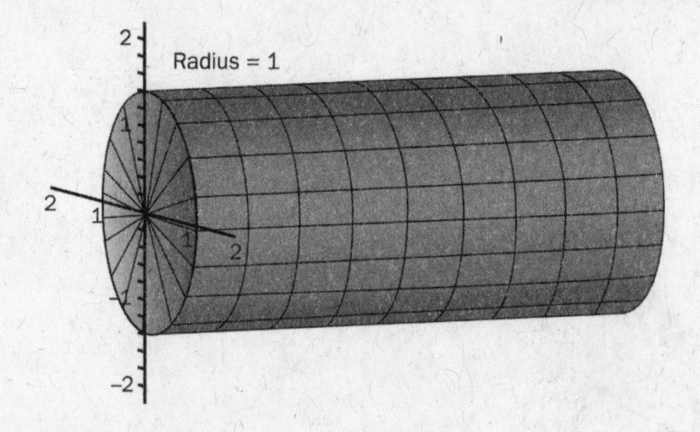

What's the volume of this soup can? Well, the volume of a cylinder is the area of its base, or any cross section circle, times the height (or length, since our can is lying down). In this case, the base is a circle of radius 1, so its area is $\pi \cdot 1^2 = \pi$. The length of the can is 3, because we took the line segment from $x = 0$ to $x = 3$, so the volume of the can is:

$$\pi \cdot 3 = 3\pi$$

We can also attack this problem from a calculus perspective and integrate the area of the typical cross section, which in this case is a circle, with constant area $A(x) = \pi \cdot 1^2 = \pi$. The boundaries of the tin can are $x = 0$ and $x = 3$. The volume of the can is, therefore, $\int_0^3 A(x)\,dx = \int_0^3 \pi\,dx = 3\pi$, just as we found on the previous page.

Thinking about a soup can as a solid of revolution—obtained by rotating a horizontal line around the x-axis—gives us a method to find the volumes of other solids of revolution. As with the solids in the previous subsection, we integrate the area of the typical cross section. However, because the cross section of a solid of revolution is always a circle or a circle with a hole in the middle, we only need to find the radius of the circle to find the area of the typical cross section.

Example:

The region bounded by $y = x^2$, $x = 1$, $x = 2$, and the x-axis is revolved around the x-axis. What is the volume of the resulting solid?

Solution:

We start by graphing the solid and trying to get a sense of what the solid of revolution looks like. We can draw the region and a two-dimensional slice of the solid of revolution. Trying to draw a picture of the three-dimensional solid can be tricky. It isn't necessary to draw the three-dimensional solid, but if your artistic abilities can handle it, it can't hurt.

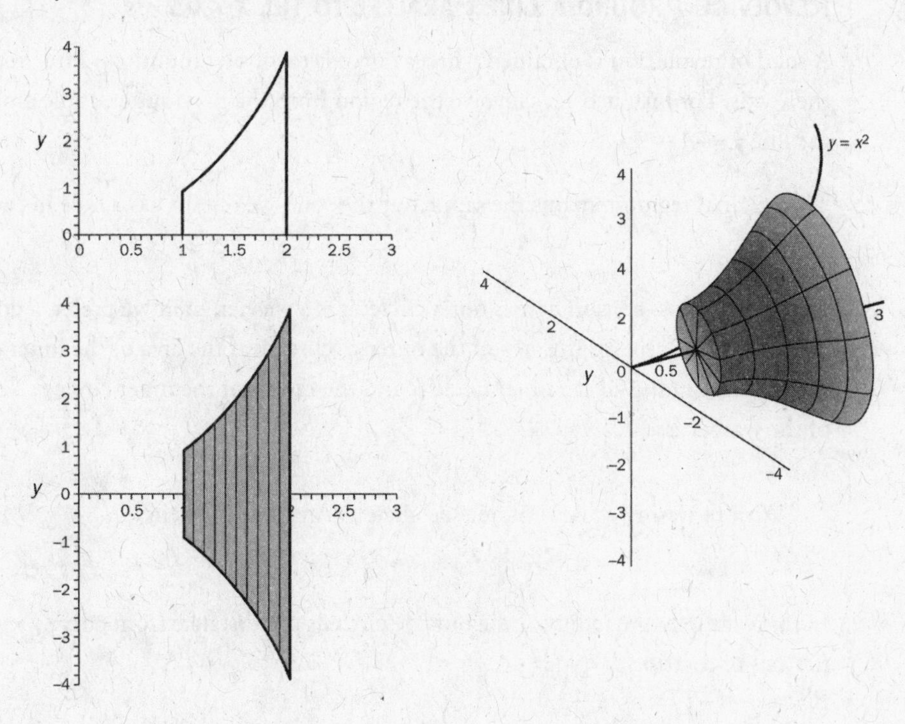

AP EXPERT TIP

It's always a good idea to write down your radius length next to your diagram. Then you can substitute your radius length into $V = \int_a^b \pi r^2\, dx$ as shown in the diagram.

Because this is a solid of revolution and there is no hole in the middle, we see that the typical cross section is a circle. The radius of the typical cross section is x^2:

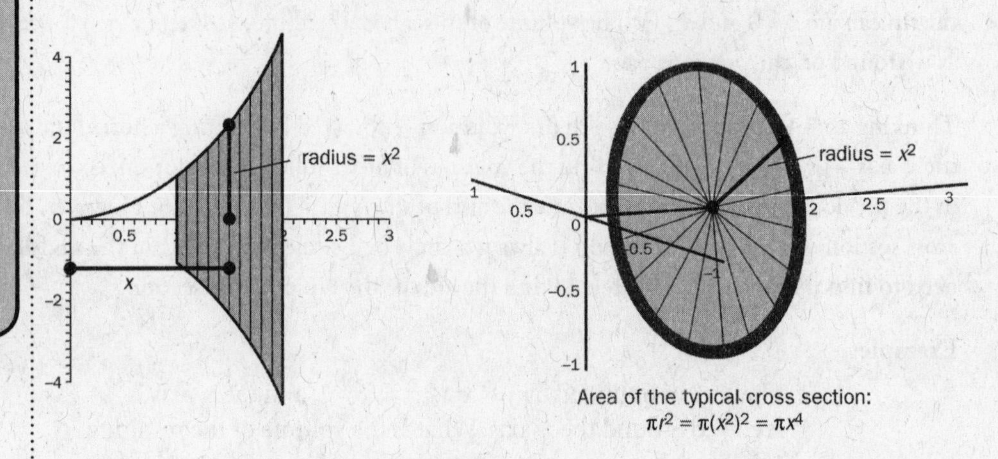

Area of the typical cross section:
$\pi r^2 = \pi(x^2)^2 = \pi x^4$

The boundaries of the solid are $x = 1$ and $x = 2$, so the volume of the solid is:

$$\int_1^2 \pi x^4\, dx = \frac{31\pi}{5} \approx 19.478$$

Revolving Around a Line Parallel to the x-axis

A solid of revolution is obtained when a curve is revolved around *any* line, not just the x-axis. For instance, let's revolve the region from the previous example around the line $y = -1$.

The original region remains the same, but the solid we obtain has a hole in the middle.

The typical cross section here is not a circle; it is a washer, or, if we prefer, a donut. The area of a washer is the area of the outer circle minus the area of the inner circle. Let's call the radius of the outer circle R and the radius of the inner circle r. The area of the washer is:

$$\begin{aligned}\text{Area of washer} &= \text{Area of outside circle} - \text{Area of inside circle}\\ &= \pi R^2 - \pi r^2\end{aligned}$$

In this example, the radius of the outside circle is the distance from curve $y = x^2$ to the axis of rotation, $y = -1$:
$R = x^2 - (-1) = x^2 + 1$.

The radius of the inner circle is the distance from the x-axis to the axis of rotation $y = -1$: $r = 0 - (-1) = 1$.

The area of the typical cross section, which is a washer, is:

$$\begin{aligned} A(x) &= \pi R^2 - \pi r^2 = \pi(x^2+1)^2 - \pi \cdot 1^2 \\ &= \pi(x^2+1)^2 - \pi \\ &= \pi[(x^2+1)^2 - 1] \end{aligned}$$

The volume of the solid is the integral of the area of the typical cross section. Because the boundaries of the region are $x = 1$ and $x = 2$, the volume of the solid is:

$$\int_1^2 \pi[(x^2+1)^2 - 1]\,dx = \frac{163\pi}{15} \approx 34.139$$

The method for finding the volume of a solid of revolution (i.e., for a region revolved around the x-axis or a line parallel to the x-axis) is as follows:

- Draw a picture of the region R that is being rotated.
- To the best of your artistic ability, draw a picture of the solid of revolution.
- Draw the typical cross section.
- Determine its shape (circle or washer).
- Determine the radius of the circle (or, for a washer, the radii of the outer and inner circle).
- Determine a formula for $A(x)$, the area of the typical cross section in terms of x.
- Determine the boundaries of the region R, $x = a$, and $x = b$.
- Compute the definite integral of $A(x)$ with limits a and b.

It is also important to be able to find the volume of a solid that is generated when a plane region is revolved about a vertical line (the y-axis, or any line parallel to the y-axis).

Example:

The region bounded by $x = \sqrt{y} + 1$, $x = 1$, and $y = 4$ is revolved about the y-axis. What is the volume of the resulting solid?

Solution:

Adjusting the procedure listed previously for a vertical axis of rotation, we begin with a diagram showing the given region (in Quadrant I) followed by another diagram showing a cross-sectional washer.

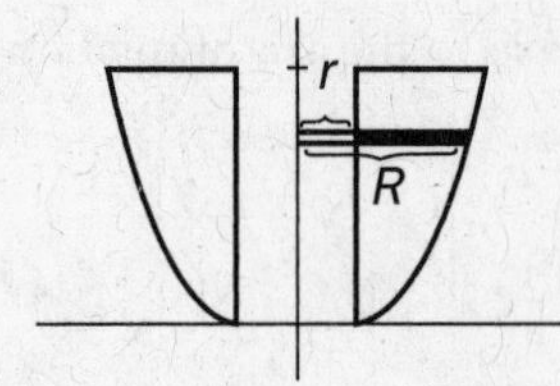

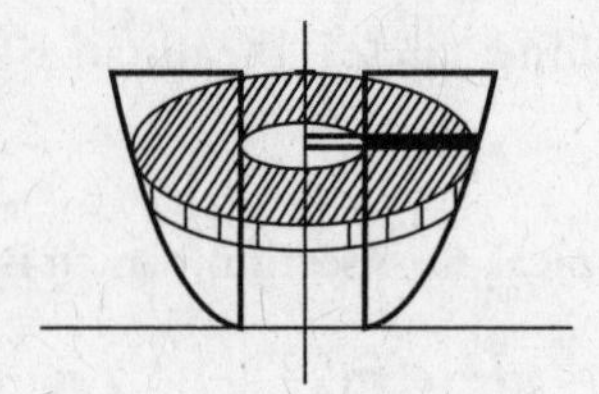

The radius of the outer circle is $R = \sqrt{y} + 1 - 0 = \sqrt{y} + 1$. The radius of the inner circle is $r = 1 - 0 = 1$.

To help represent R and r correctly, always remember that both R and r are measured from a point on the axis of rotation (in this case, $x = 0$). This point is the center of the washer.

Now we can substitute these radius lengths into $V = \int_a^b \pi(R^2 - r^2)\, dy$, where a and b are the horizontal boundaries of the region, $y = 0$ and $y = 4$.

Thus, $V = \int_0^4 \pi[(\sqrt{y} + 1)^2 - (1)^2]\, dy = 58.643$.

There is another method used to compute volumes of solids of revolution—the method of cylindrical shells. This method is typically taught in calculus courses, but it is not needed for the AP exam. It is generally easier to use the shell method to compute the volume of a solid obtained by revolving a region around the y-axis, but these types of solids do not typically show up on the AP exam.

ACCUMULATED CHANGE

We saw in Chapter 17 that we can express the total change in a quantity over a time period as the definite integral of its rate of change over that time period. We can also use the definite integral to think about change thus far or to evaluate how much of a total quantity has accumulated at a certain point in time.

Suppose f is a function that gives the rate of change of some quantity. The total change in the quantity over the time period $t = a$ to $t = b$ is the definite integral $\int_a^b f(t)\, dt$.

We can tweak this notion a little bit to use the definite integral to answer questions about how much stuff has accumulated *so far*. If f is a function describing the rate of change of some quantity, then we can define a function $G(t)$:

$$G(t) = \int_0^t f(t)\, dt$$

The function $G(t)$ is the area under the curve $f(t)$ from some specified starting time a to time t. It tells us how much stuff accumulates from the starting time to time t. Functions like these are called *accumulation functions.*

Over intervals where f is negative, nothing will be accumulating—the quantity will be decreasing.

Example:

Let $G(t) = \int_a^t f(t)\, dt$ where the graph of $f(t)$ is shown below.

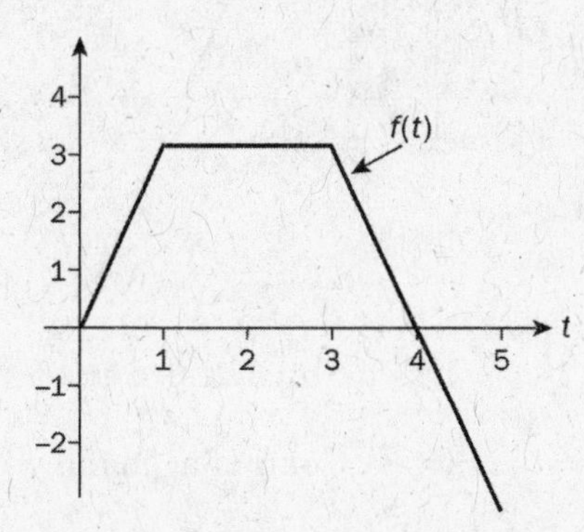

(a) Compute $G(0), G(1), G(2), G(3), G(4)$, and $G(5)$.

(b) Suppose $f(t)$ gives the rate of change of the water level in Lake Esmerelda where the height of the water is described in meters above sea level and the time t is given in months. If the level of the lake is 110 m above sea level at the starting time, what is the level of the lake at time $t = 4$ months?

Solution:

(a) Because $G(t)$ is defined in terms of a definite integral, to compute $G(0), G(1)$, etc., we need to compute the area under the curve f from $t = 0$ to 0, 1, 2, 3, 4, and 5. $G(0) = 0$, because no area has accumulated. $G(1)$ is the area shown:

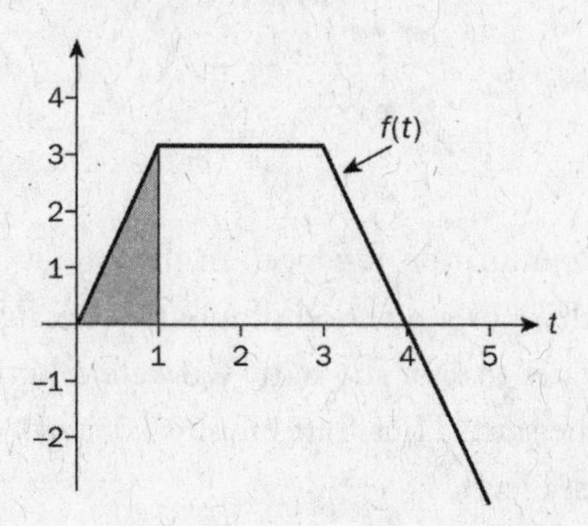

We can see from the graph that this is a triangle with area $\frac{1}{2}bh = \frac{1}{2} \cdot 1 \cdot 3 = \frac{3}{2}$.

To compute $G(2)$ we add the area of the adjacent rectangle:

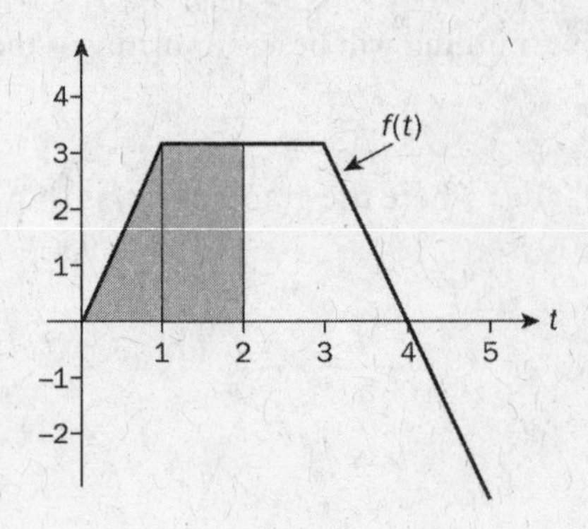

The area of this rectangle is $1 \cdot 3 = 3$, so the total shaded area is the area of the triangle plus the area of the rectangle: $G(2) = \frac{3}{2} + 3 = \frac{9}{2}$. Continuing in this way, we compute $G(3) = \frac{9}{2} + 3 = \frac{15}{2}$; $G(4) = \frac{15}{2} + \frac{3}{2} = 9$.

When we compute $G(5)$ we subtract the area that lies below the x-axis.

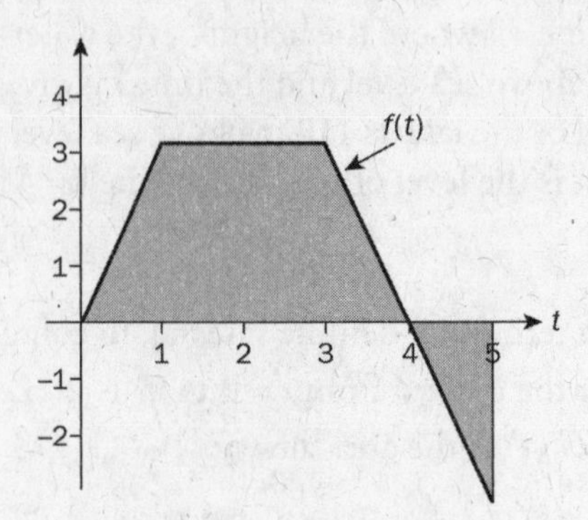

$G(5) = 9 - \frac{3}{2} = \frac{15}{2}$.

(b) The function $G(t)$ is an accumulation function. In this case it describes the accumulated change in the lake's water level over a period of months. Because $G(4) = 9$ at time $t = 4$, the level of the lake is 9 m greater than at the start. We were told that the level of the lake was 110 m above sea level at the start. Thus four months later, at $t = 4$, the level of the lake is 110 + 9, or 119 m above sea level.

DISTANCE TRAVELED BY A PARTICLE ALONG A LINE

Questions about a particle traveling along a straight line are another big favorite with the AP folks. Typically, we are given information about the particle's velocity and asked questions about the particle's position and its acceleration. Earlier, we saw how we can use derivatives to go from a particle's position function, $x(t)$, to its velocity and acceleration functions, $v(t)$ and $a(t)$, respectively. Succinctly:

$$v(t) = x'(t)$$

$$a(t) = v'(t) = x''(t)$$

We can use integration to work in reverse! If we are given a velocity or acceleration function, we need only take the integrals to work backwards to a position function. Specifically, the definite integral of the velocity function can give us either the total distance a particle travels or its displacement.

These types of problems are another application of the definite integral. They can be presented numerically with data tables, graphically, or algebraically with functions. Let's take a look at an algebraically presented problem here and look at the other presentations in the questions at the end of the chapter.

Example:

A particle moves along the x-axis with initial position $x(0) = 1$. The velocity of the particle at time $t \geq 0$ is given by $v(t) = \sin\sqrt{t}$.

(a) What is the change in position of the particle over the time period $0 \leq t \leq 15$?

(b) What is the position of the particle at time $t = 15$?

Solution:

(a) Velocity is the rate of change of position, so by thinking of the change in position as the accumulated change in the velocity, we can find the change in the particle's position using a calculator to compute the definite integral:

$$\int_0^{15} \sin\sqrt{t}\, dt \approx 4.429$$

(b) The position of the particle at time $t = 15$ is the particle's position at the start plus the change in position over the time period. We were given that the particle's position at time $t = 0$ is $x(0) = 1$. Therefore the position of the particle at time $t = 15$ is its initial position plus its change in position over the time period $0 \leq t \leq 15$:

$$x(15) = 1 + 4.429 = 5.429$$

AP EXPERT TIP

Questions relating to the average value of a function appear regularly on the AP exam.

AVERAGE VALUE OF A FUNCTION

When we want to find the average of a list of numbers, we add the numbers to get their total and divide that total by the number of numbers in the list:

$$Average = \frac{a_1 + a_2 + \cdots a_n}{n}$$

We have a similar concept for the average value of a function. To find the average value of a function on an interval $[a, b]$, we find the area under the graph of the function (that's the "total" in the average) and divide by the length of the interval $b - a$. That is, if f is continuous on the closed interval $[a, b]$, then the average value of f is given by:

$$C = \frac{\int_a^b f(x)\,dx}{b - a}$$

We can interpret this definition as saying, there is some point C such that the area of the rectangle with base $b - a$ is the same as the area under $f(x)$ on $[a, b]$.

REVIEW QUESTIONS

The numbers in the graph are the areas of the enclosed regions. Questions 1 and 2 refer to the graph. Note: Figure not drawn to scale!

The graph below shows the velocity in cm/sec of a particle moving along the y-axis.

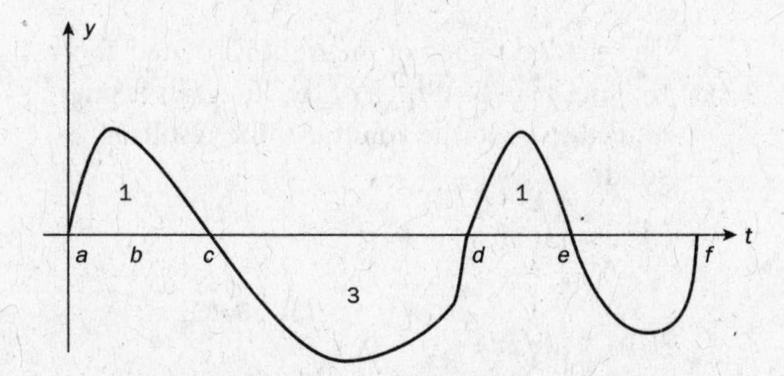

1. The particle is at $y = 1$ at time $t = a$. What is the particle's distance from the origin at time $t = e$?

   (A) $y = 1$
   (B) $y = 0$
   (C) $y = 3$
   (D) $y = 4$
   (E) $y = 5$

2. If the particle travels a total distance of 8 units from time $t = a$ to time $t = f$, what is the total distance traveled by the particle from time $t = e$ to time $t = f$?

   (A) 2
   (B) −7
   (C) 3
   (D) −3
   (E) Cannot be determined from the information given

3. Find the area between the curves $f(x) = x^2 - 3x + 2 = (x - 1)(x - 2)$ and $g(x) = -(x^2 - 3x + 2) = -(x - 1)(x - 2)$ on the interval from $x = 0$ to $x = 2$.

   (A) 0
   (B) $\frac{2}{3}$
   (C) 1
   (D) $\frac{4}{3}$
   (E) 2

4. Let $f(x)$ be an integrable function on the interval $a \le x \le b$. Which of the following statements is true?

   I. The area between $f(x)$ and the x-axis on $a \le x \le b$ is equal to $\int_a^b f(x)\,dx$.

   II. If $f(x) \le 0$ on $a \le x \le b$, then the area between $f(x)$ and the x-axis on $a \le x \le b$ is equal to $-\int_a^b f(x)\,dx$.

   III. The area between $f(x)$ and the x-axis on $a \le x \le b$ could be less than zero.

   (A) Only I
   (B) Only II
   (C) Only III
   (D) I and II
   (E) I and III

Question 5 requires a calculator.

5. If the area under the curve $f(x) = \sqrt{x}$ over the interval $[k, 30]$ is 50, find k.

   (A) 19.609
   (B) 19.981
   (C) 20.013
   (D) 20.683
   (E) 20.892

6. What is the average value of $f(x) = e^{kx}$ on the closed interval $[0,k]$?

(A) $\frac{1}{k^2}e^{k^2}$

(B) $\frac{1}{k}e^{k^2}$

(C) $\frac{1}{k}e^{k^2} - \frac{1}{k}$

(D) $\frac{1}{k^2}e^{k^2} - 1$

(E) $\frac{1}{k^2}e^{k^2} - \frac{1}{k^2}$

Question 7 requires a calculator.

7. There are two regions enclosed by the curves $y = \ln x$ and $y = -(x - 1)(x - 2)(x - 4)$. What is the sum of the areas of the two regions?

(A) 1.21
(B) 1.47
(C) 1.53
(D) 1.77
(E) 2.17

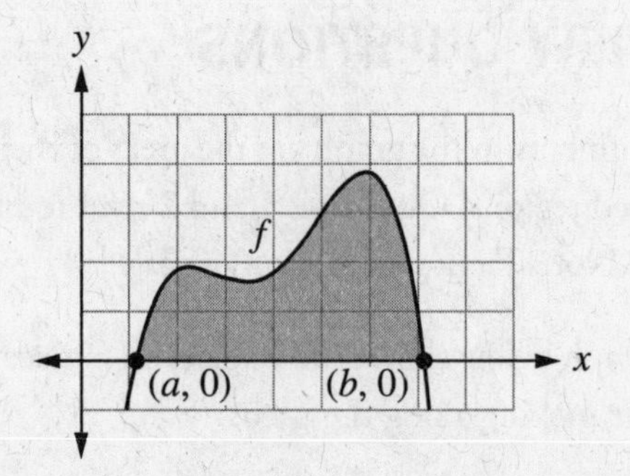

8. The shaded region at the right is rotated about the line $y = -3$. Which of the integrals is guaranteed to yield the volume of the resulting solid?

(A) $\pi\int_a^b ((f(x))^2 + 9)\,dx$

(B) $\pi\int_a^b (f(x) - 3))^2\,dx$

(C) $\pi\int_a^b (9 - f(x))^2\,dx$

(D) $\pi\int_a^b ((f(x) + 3)^2 + 9)\,dx$

(E) $\pi\int_a^b ((f(x) + 3)^2 - 9)\,dx$

9. The base of a solid is the region bounded by the parabola $x^2 = 8y$ and $y = 4$, and each plane section perpendicular to the base and the y-axis is an equilateral triangle. The volume of the solid is

(A) $\frac{64\sqrt{3}}{3}$
(B) $64\sqrt{3}$
(C) $32\sqrt{3}$
(D) 32
(E) None of these

10. What is the average value of $y = \tan^2 x$ over the interval from $x = 0$ to $x = \frac{\pi}{4}$?

(A) $1-\frac{\pi}{4}$

(B) $1+\frac{\pi}{4}$

(C) $\frac{4}{\pi}-1$

(D) $1-\frac{4}{\pi}$

(E) $\sqrt{2}$

11. What is the area enclosed by the curve $y = e^{2x}$ and the lines $x = 1$ and $y = 1$?

(A) $\frac{2-e^2}{2}$

(B) $\frac{e^2-3}{2}$

(C) $\frac{3-e^2}{2}$

(D) $\frac{e^2-2}{2}$

(E) $\frac{e^2-1}{2}$

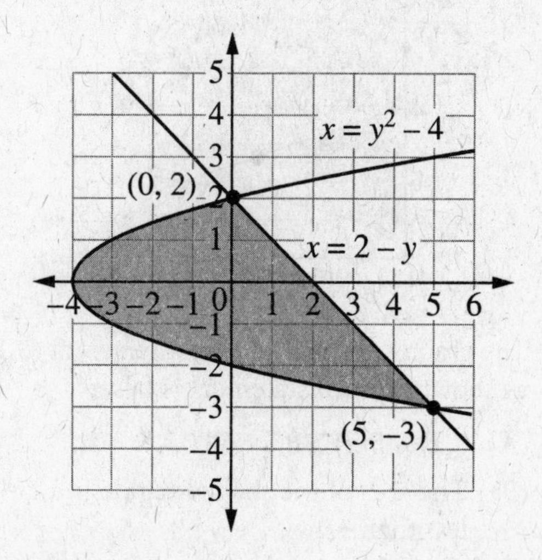

12. Which of the following gives the area of the shaded region above?

(A) $\int_{-4}^{5}(y^2+y-6)\,dy$

(B) $\int_{-3}^{2}(y^2+y-6)\,dy$

(C) $\int_{-4}^{5}(6-y-y^2)\,dy$

(D) $\int_{-3}^{2}(6-y-y^2)\,dy$

(E) None of these

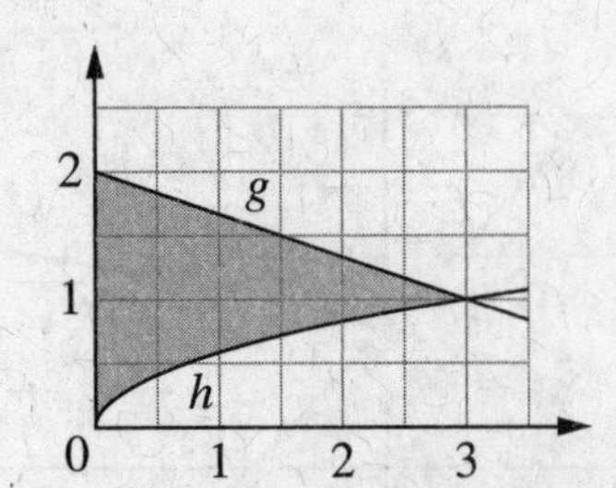

13. The graphs of $g(x)$ and $h(x)$ intersect at (3,1) as shown. Which statement is false?

(A) $g(x) - h(x) > 0$ for all $0 \le x < 3$.

(B) The area of the shaded region is a little less than 3.

(C) $0 \le \frac{1}{3}\int_0^3 (g(x)-h(x))\,dx \le 2$

(D) The area of the shaded region is given by: $\int_0^3 (g(x)-h(x))\,dx$.

(E) If the shaded region is rotated about the x-axis, the volume is $\pi\int_0^3 (g(x)-h(x))^2\,dx$.

14. A particle moves along the x-axis so that its acceleration at any time t is given by $a(t) = 6t - 18$. At time $t = 0$, the velocity v of the particle is 24 units/sec and at the time $t = 1$, the position x of the particle is 20. Which statement is false?

(A) $u(t) = 3t^2 - 18t + 24$ for all $t \ge 0$.

(B) The particle is moving to the left when $2 < t < 4$.

(C) The starting position of the particle is $x(0) = 4$.

(D) The total distance traveled by the particle for $0 \le t \le 4$ is $\int_0^4 (3t^2 - 18t + 24)\,dt$.

(E) $x(3) - x(0) = \int_0^3 (3t^2 - 18t + 24)\,dt$.

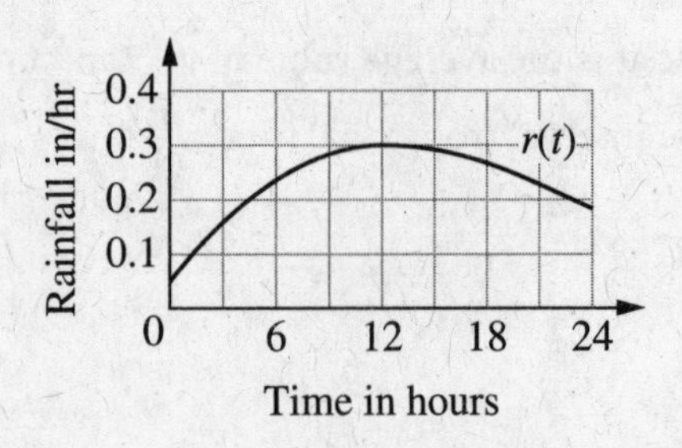

15. For a 24-hour period, the graph of the rate of rainfall $r(t)$, in inches per hour, is shown. Which statement is false?

(A) The rain fell hardest in the middle hours of the 24-hour period.

(B) During the 24-hour period, the total rainfall in inches is $\int_0^{24} r(t)\,dt$.

(C) During the 24-hour period, the total rainfall is between 1 and 7 inches.

(D) During the 24-hour period, the average rate of rainfall is $\frac{r(24)-r(0)}{24}$ in/hr.

(E) If $k < 12$, then $\int_0^k r(t)\,dt < \int_k^{24} r(t)\,dt$.

16. If the average value of $f(x) = \sqrt{x-3}$ on the interval $[3, k]$ is 8, what is k?

(A) 8

(B) 24

(C) 51

(D) 147

(E) 151

FREE-RESPONSE QUESTIONS

17. The graph below shows the velocity of two bicyclists 1 and 2 in a race from time $t = 0$ to time $t = 16$. The racers start side by side and travel along the same road. The race begins at time $t = 0$. At time $t = 16$ one of the racers completes the course.

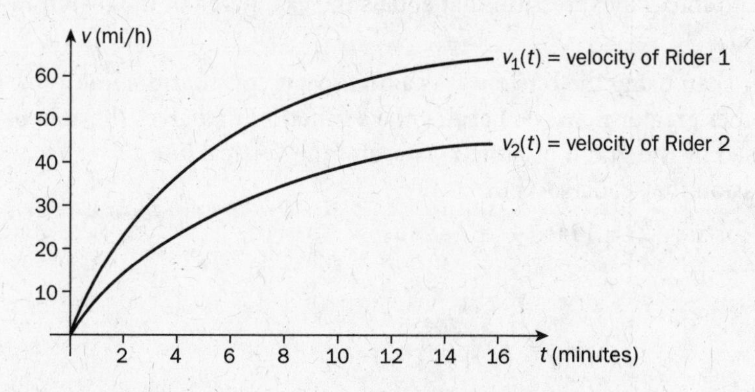

(a) What does the area between the two curves represent?

(b) Which bicyclist finishes the course at time $t = 16$?

(c) Assuming that the velocity of the other racer continues as an integrable function beyond the portion shown on the graph, set up (but do not solve) an equation to find the time at which the second racer finishes the course.

18. Consider the region R enclosed by the graphs of $f(x) = \cos(x - 0.4)$ and $g(x) = x^2$, shown below.

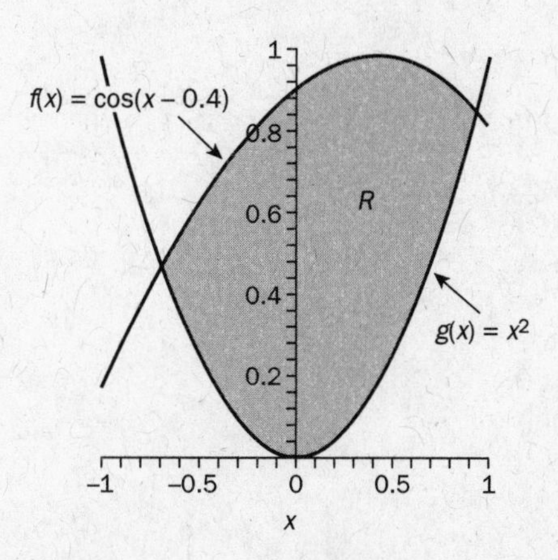

(a) Find the points of intersection of the graphs of f and g.

(b) Find the area of the region R.

(c) Find the volume generated by revolving the region R about the line $y = 0$.

(d) The region R is the base of a solid; each cross section of the solid, in a plane perpendicular to the x-axis, is a rectangle, the shorter side of which lies in the base and the longer side of which is three times its width. Find the volume of the solid.

19. Let f be the function given by $f(x)=\sqrt{x+3}$ and R be the region enclosed by the graph of f, the x-axis, and the vertical line $x = 6$.

    (a) Sketch a graph of the enclosed region R.

    (b) Write an integral expression that equals the exact area of the region R. Evaluate the integral.

    (c) Rather than using the line $x = 6$ as a boundary, consider the line $x = k$ where k is a number greater than -3. Let $A(k)$ be a function that gives the area of the region enclosed by the graph of f, the x-axis, and the vertical line $x = k$. Write an integral expression that equals $A(k)$.

    (d) Solve for k: $A(k) = 144$.

ANSWERS AND EXPLANATIONS

1. B

From time $t = a$ to time $t = c$, the particle is traveling up the y-axis and travels a total distance of 1 unit in the upward direction. From time $t = c$ to time $t = d$, the particle is traveling down the y-axis and travels a distance of 3. From time $t = d$ to $t = e$, the particle travels 1 unit upward. Therefore over the time period $t = a$ to $t = e$, the net change in the particle's position is 1 unit downward. Because the particle's initial position was $y = 1$, the ending position of the particle is $y = 0$.

2. C

The total distance traveled on an interval is the area between the curve and the x-axis on that interval. From the graph we see that the total distance traveled from time $t = a$ to time $t = e$ is 5. Because the total distance traveled on the entire interval—from $t = a$ to time $t = f$—is 8, by the additivity of the definite integral, the distance traveled from time $t = e$ to time $t = f$ must be 3.

Notice that we can eliminate (B) and (D) immediately because total distance traveled cannot be negative.

3. E

We start by sketching the graphs of these curves on the same set of axes. They are each a parabola that intersects the x-axis at $x = 1$ and $x = 2$. The parabola $f(x)$ opens up and the parabola $g(x)$ opens down.

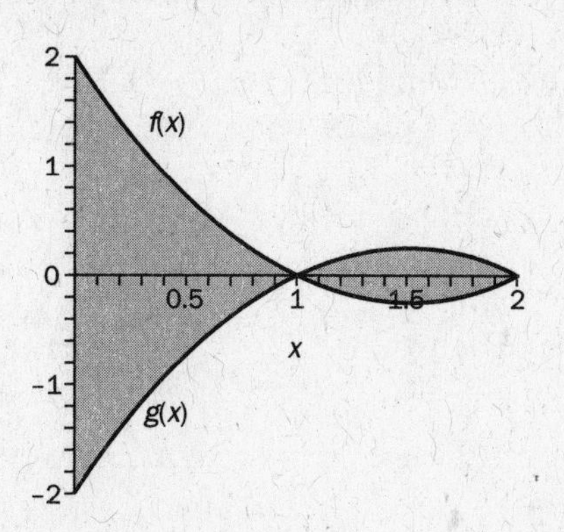

The area between the curves is the integral of the typical cross section, i.e., the top curve minus the bottom curve. From $x = 0$ to $x = 1$, the curve $f(x)$ is on top; from $x = 1$ to $x = 2$, $g(x)$ is on top. The area is therefore given by two integrals:

$$\text{Area} = \int_0^1 f(x) - g(x)\,dx + \int_1^2 g(x) - f(x)\,dx.$$

Notice that $g(x) = -f(x)$. We can therefore simplify the calculation:

$$\begin{aligned}\text{Area} &= \int_0^1 f(x) - (-f(x))\,dx + \int_1^2 (-f(x)) - f(x)\,dx \\ &= \int_0^1 2f(x)\,dx + \int_1^2 -2f(x)\,dx \\ &= 2\int_0^1 f(x)\,dx - 2\int_1^2 f(x)\,dx \\ &= 2\int_0^1 x^2 - 3x + 2\,dx - 2\int_1^2 x^2 - 3x + 2\,dx\end{aligned}$$

Using the techniques of integration from Chapter 21 we compute

$$\begin{aligned}\text{Area} &= 2\int_0^1 x^2 - 3x + 2\,dx - 2\int_1^2 x^2 - 3x + 2\,dx \\ &= 2\left(\frac{x^3}{3} - \frac{3x^2}{2} + 2x\right)\Bigg|_0^1 - 2\left(\frac{x^3}{3} - \frac{3x^2}{2} + 2x\right)\Bigg|_1^2\end{aligned}$$

4. B

Statement II is correct. If $f(x) \le 0$ on $a \le x \le b$, then on this interval $|f(x)| = -f(x)$. Therefore $\int_a^b |f(x)|\,dx = \int_a^b -f(x)\,dx = -\int_a^b f(x)\,dx$.

Statement I is incorrect. The area between $f(x)$ and the x-axis on $a \le x \le b$ is equal to the definite integral of the *absolute value* of the function $\int_a^b |f(x)|\,dx$.

Statement III is incorrect; the area between two curves is always positive. Therefore only statement II is correct and (B) is the correct answer.

5. B

We use the calculator's solver utility to solve the equation $\int_k^{30} \sqrt{x} = 50$ for k. We find $k = 19.981$.

EQUATION SOLVER eqn:0=fnInt(√(X) ,X,K,30)−50	fnInt(√(X),X,...=0 X=0 K=19.98121979 ▮... bound={⁻1ᴇ99,1...	F1 Tools F2 Algebra F3 Calc F4 Other F5 PrgmIO F6 Clean up ■ solve($\int_k^{30} \sqrt{x}\,dx$ = 50, k) k = 19.9812198 solve(∫(√(x),x,k,30)=50, k... MAIN RAD APPROX FUNC 1/30

We can also solve this problem by working backward. Plug each of the answer choices in for k in the integral $\int_k^{30} \sqrt{x} = 50$ and see which one gives a correct answer.

6. E

To find the average value we evaluate

$$\frac{1}{k-0}\int_0^k e^{kx}\,dx = \frac{1}{k^2}e^{kx}\Big|_0^k = \frac{1}{k^2}(e^{k^2}-1).$$

7. D

This is a calculator problem so let $Y_1 = \ln x$ and $Y_2 = -(x-1)(x-2)(x-4)$. The graphs intersect at points where $x = 1, 2.389, 3.719$. The area of the region to the left is $\int_1^{2.389}(Y_1 - Y_2)\,dx$ and to the right $\int_{2.389}^{3.719}(Y_2 - Y_1)\,dx$. The sum of the areas is $\int_1^{2.389}(Y_1 - Y_2)\,dx + \int_{2.389}^{3.719}(Y_2 - Y_1)\,dx = 1.77$ or with the absolute value function you get $\int_1^{3.719}|Y_2 - Y_1|\,dx = 1.77$

8. E

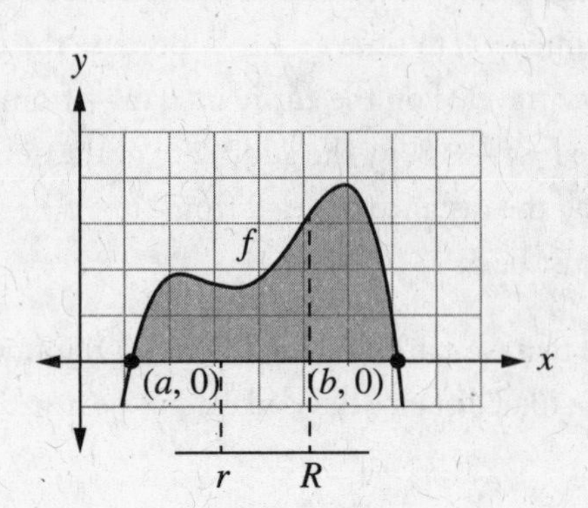

Use the washer method given by $\pi\int_a^b (R^2 - r^2)\,dx$ where R is the outer radius and r is the inner radius. Since $y = -3$ is the axis of rotation, $R = f(x) + 3$ and $r = 3$. This gives $\pi\int_a^b ((f(x)+3)^2 - 3^2)\,dx$.

9. B

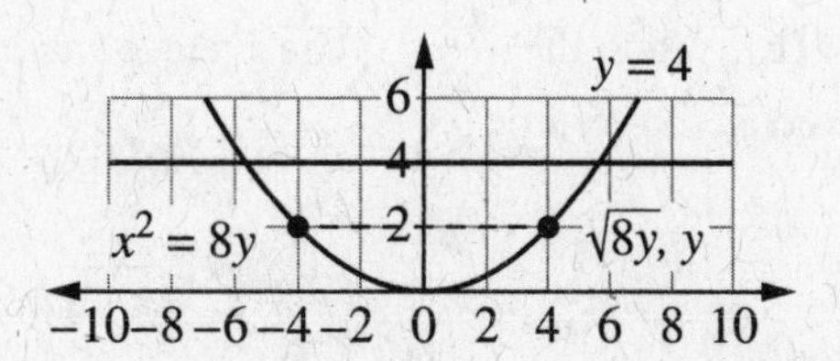

The region bounded by $x^2 = 8y$ and $y = 4$ is shown. Since the equilateral cross-sections are perpendicular to the y-axis, then we need function of y not x, so $x^2 = 8y \Rightarrow x = \pm\sqrt{8y}$. Now we label a point on the curve $\left(\sqrt{8y}, y\right)$. Since the area of an equilateral triangle is given by $A = \frac{s^2\sqrt{3}}{4}$, then it follows that cross sectional area $A = \frac{\left(2\sqrt{8y}\right)^2\sqrt{3}}{4} = 8\sqrt{3}\,y$. With this, the integral that gives the volume of the solid is cubic units.

10. C

To find the average value, we evaluate $\frac{1}{\frac{\pi}{4}-0}\int_0^{\frac{\pi}{4}} \tan^2 x\,dx = \frac{4}{\pi}\int_0^{\frac{\pi}{4}} \tan^2 x\,dx$. We recall that $\tan^2 x + 1 = \sec^2 \Rightarrow \tan^2 x = \sec^2 x - 1$.

By substitution and FTC, $\frac{4}{\pi}\int_0^{\frac{\pi}{4}} \tan^2 x\,dx = \frac{4}{\pi}\int_0^{\frac{\pi}{4}} (\sec^2 x - 1)\,dx = \frac{4}{\pi}(\tan x - x)\Big|_0^{\frac{\pi}{4}} = \frac{4}{\pi}\left(1 - \frac{\pi}{4}\right) = \frac{4}{\pi} - 1$.

11. B

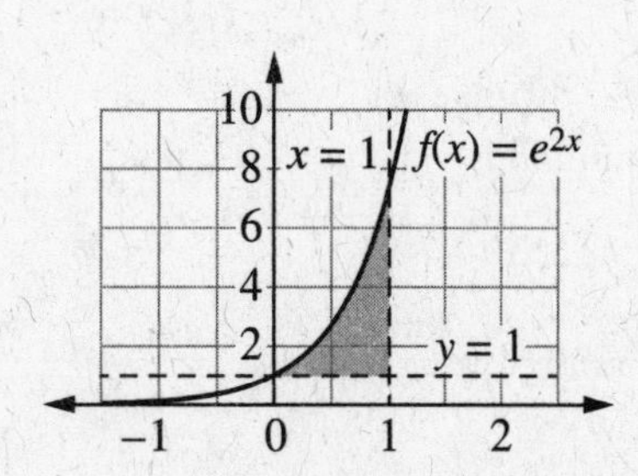

The graph of the region is shown above. The shaded region is bounded above by $y = e^{2x}$ and below by the line $y = 1$. The area of the region is given by $A = \int_0^1\left[e^{2x} - 1\right]dx = \frac{1}{2}e^{2x} - x\Big|_0^1 = \frac{1}{2}e^2 - 1 - \frac{1}{2} = \frac{1}{2}e^2 - \frac{3}{2} = \frac{e^2 - 3}{2}$

12. D

The area of the shaded region is bounded to the right by $x = 2 - y$ and to the left by $x = y^2 - 4$.
For $-3 \le y \le 2$, we know that $2 - y \ge y^2 - 4$ so it follows that the area of the shaded region is given by $\int_{-3}^{2}(2 - y - (y^2 - 4))\,dy = \int_{-3}^{2}(6 - y - y^2)\,dy$.

13. E

All of the statements are true except (E). When the shaded region is rotated about the x-axis, the volume is $\pi\int_0^3\left((g(x))^2-(h(x))^2\right)dx$;

it is not $\pi\int_0^3(g(x)-h(x))^2\,dx$.

14. D

All of the statements are true except (D). The total distance traveled by the particle for $0 \le t \le 4$ is $\int_0^4 \left|3t^2-18t+24\right| dt$ and not $\int_0^4(3t^2-18t+24)\,dt$ which represents the displacement for $0 \le t \le 4$. Note: the particle changes direction at $t = 2$.

15. D

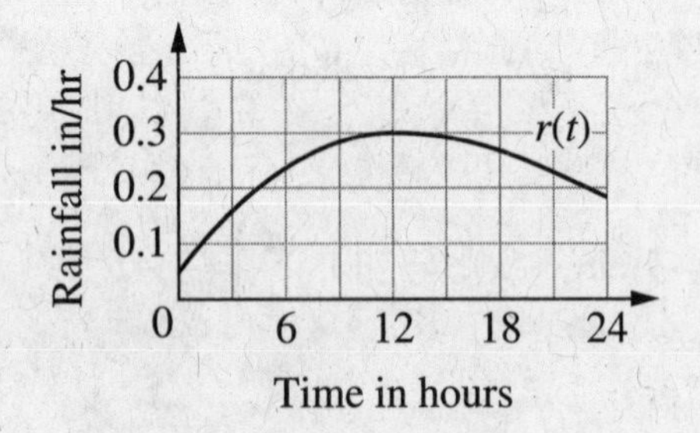

All of the statements are true except (D). During the 24-hour period, the average rate of rainfall is $\frac{1}{24}\int_0^{24} r(t)\,dt$ in/hr. The quantity $\frac{r(24)-r(0)}{24}$ is the average rate of change in the rate of rainfall, a different idea.

16. D

The average value of $f(x)=\sqrt{x-3}$ over $[3, k]$ is given by $\frac{1}{k-3}\int_3^k\sqrt{x-3}\,dx$. We solve

$$\frac{1}{k-3}\int_3^k\sqrt{x-3}\,dx=8 \Rightarrow \frac{2}{3(k-3)}\cdot(x-3)^{\frac{3}{2}}\Big|_3^k = 8 \Rightarrow \frac{2}{3(k-3)}(k-3)^{\frac{3}{2}}=8 \text{ and}$$

$$\frac{2}{3}(k-3)^{\frac{1}{2}}=8 \Rightarrow (k-3)^{\frac{1}{2}}=12 \Rightarrow k-3=144 \Rightarrow k=147$$

FREE-RESPONSE ANSWERS

17. (a) Because the area under each curve represents the total distance traveled by the respective racers from $t = 0$ to $t = 16$, the area between the curves represents the distance between the riders at time $t = 16$.

(b) Because the area under $v_1(t)$ is greater than the area under $v_2(t)$, Rider 1 has traveled farther at time $t = 16$. Therefore, this rider has completed the course.

(c) Rider 1 completes the course at time $t = 16$. Therefore the total length of the course is given by the definite integral $\int_0^{16} v_1(t)\,dt$. We set A equal to the value of this integral, i.e., A is the total distance of the course. To find the time at which Rider 2 finishes the course we solve the equation $\int_0^{t} v_2(t)\,dt = A$ for t.

18. (a) Part (a) of this question appeared in the questions at the end of Chapter 4. The two intersection points are $x = -0.6839$ and $x = 0.9291$.

(b) The area between the curves is the definite integral of the upper curve minus the lower curve, with the limits of integration equal the left and right intersection points of the curve:

$$\text{Area} = \int_{-0.6839}^{0.9291} (\cos(x - 0.4) - x^2)\,dx$$

We use the graphing calculator's utility to compute the definite integral:

$$\int_{-0.6839}^{0.9291} (\cos(x - 0.4) - x^2)\,dx = 1.015$$

(c) To find the volume of the solid of revolution, we need to integrate the area of the typical cross section. First, we draw the solid of revolution and the typical cross section.

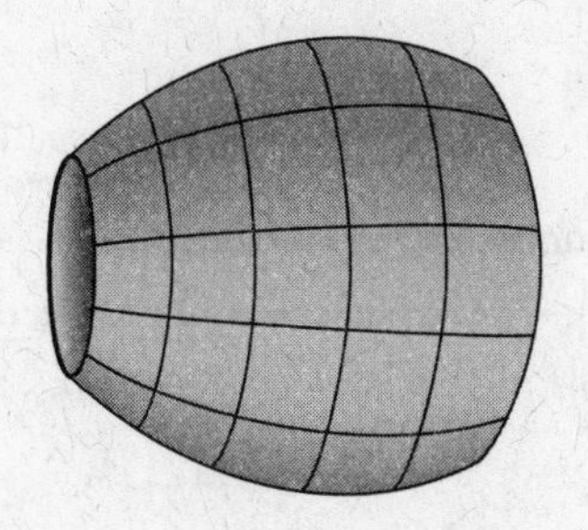

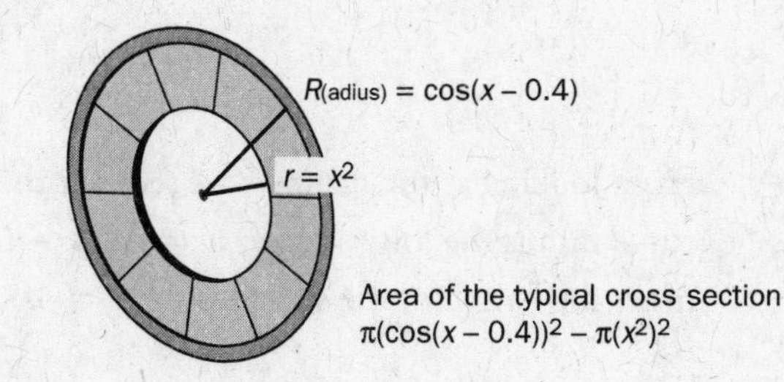

To compute the volume of the solid revolution of the region R about the line $y = 0$, we need to integrate the area of the typical cross section. In this case, the typical cross section is a washer. The boundaries of the definite integral are the intersection points of the curves; $x = -0.6839\ldots$ and $x = 0.9291\ldots$, thus:

$$\int_{-0.6839}^{0.9291} \pi\left(\cos(x - 0.4)\right)^2 - \pi\left(x^2\right)^2 \, dx$$

We can use the utility of the graphing calculator to compute the value of the definite integral: 3.339.

(d) To compute the volume of a solid, we need to compute the definite integral of the area of the typical cross section. The limits of integration are the boundaries of the base region—in this case the intersection points of the two curves, $x = -0.6839$ and $x = 0.9291$.

We are told that the typical cross section is a rectangle, the shorter side of which, w, lies in the base of the rectangle and the longer side of which, l, is three times its width.

First we draw a picture of the typical cross section.

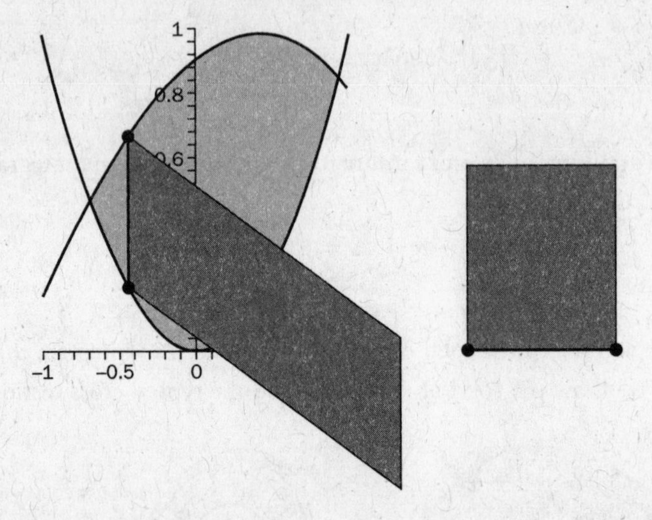

From looking at the diagram we see that the length of the shorter side, w, is the upper curve minus the lower curve: $w = \cos(x - 0.4) - x^2$. Because the longer side is three times the length of the shorter side, $l = 3w$, so $l = 3(\cos(x - 0.4) - x^2)$.

The typical cross section is a rectangle, whose area is lw, so the area of the typical cross section is:

$$A(x) = [\cos(x - 0.4) - x^2][3(\cos(x - 0.4) - x^2)]$$
$$= 3(\cos(x - 0.4) - x^2)^2$$

The volume of the solid is the integral of $A(x)$ with limits of integration is the intersection points of the two curves that form the boundary of the base:

$$V = \int_{-0.6839}^{0.9291} 3(\cos(x - 0.4) - x^2)^2\, dx$$

We can compute this definite integral with a graphing calculator to find that the volume is 2.302.

19. (a) We sketch the region and shade the region R.

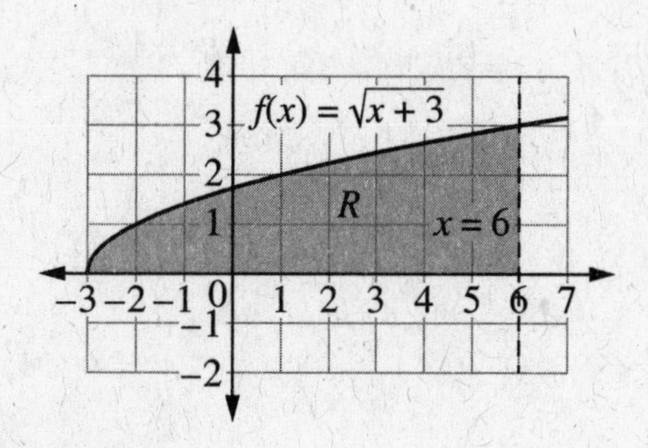

(b) The area of R is given by

$$A_R = \int_{-3}^{6} \sqrt{x+3}\, dx = \int_{-3}^{6} (x+3)^{\frac{1}{2}}\, dx = \frac{2}{3}(x+3)^{\frac{3}{2}}\Big|_{-3}^{6}$$

$$A_R = \frac{2}{3}(6+3)^{\frac{3}{2}} - \frac{2}{3}(-3+3)^{\frac{3}{2}} = \frac{2}{3}9^{\frac{3}{2}} - 0 = \frac{2}{3}\cdot 27$$

$$\therefore A_R = 18$$

(c) The area of the region is given by

$$A(k)=\int_{-3}^{k}\sqrt{x+3}\,dx$$

(d) The area of the region is given by

$$\int_{-3}^{k}\sqrt{x+3}\,dx=144$$

$$\int_{-3}^{k}\sqrt{x+3}\,dx=144$$

$$\int_{-3}^{k}(x+3)^{\frac{1}{2}}\,dx=144$$

$$\frac{2}{3}(x+3)^{\frac{3}{2}}\Big|_{-3}^{k}=144$$

$$\frac{2}{3}(k+3)^{\frac{3}{2}}-\frac{2}{3}(-3+3)^{\frac{3}{2}}=144$$

$$\frac{2}{3}(k+3)^{\frac{3}{2}}=144$$

$$(k+3)^{\frac{3}{2}}=216$$

$$\left((k+3)^{\frac{3}{2}}\right)^{\frac{2}{3}}=216^{\frac{2}{3}}$$

$$k+3=36$$

$$\therefore k=33$$

CHAPTER 19: ANTIDERIVATIVES—THE INDEFINITE INTEGRAL

IF YOU LEARN ONLY TWO THINGS IN THIS CHAPTER . . .

1. If $F(x)$ and $G(x)$ are both antiderivatives of $f(x)$, then $F(x) - G(x) = C$, for some real number C.
2. Properties of the Indefinite Integral:

   $\int f(x)\,dx \pm \int g(x)\,dx = \int f(x) \pm g(x)\,dx$

   $\int c \cdot f(x)\,dx = c\int f(x)\,dx$

In the previous two chapters we talked about definite integrals. One of the ways we thought about definite integrals was as the accumulated change in a function over an interval. If $f'(x)$ is the rate of change of a function, then $\int_a^b f'(x)\,dx$ is the accumulated change in the function over the interval $[a,b]$. This way of thinking about definite integrals establishes a relationship between derivatives and integrals. In this chapter and the next, we will develop this connection.

WHAT IS AN ANTIDERIVATIVE?

When you learned about square roots, the topic was probably introduced by asking a question like "What number squared equals 4?" One answer is 2 and this question motivated the idea of the square root.

A similar idea comes into play in calculus. We begin by asking the question, "What function's derivative is x?" If we think for a while, we come up with an answer: $\frac{1}{2}x^2$. Why is this the correct answer? Because $\frac{d}{dx}\left(\frac{1}{2}x^2\right) = x$. This is not the only possible answer; just as the question "What number squared equals 4?" has more than one answer, so does this question. Other answers are $\frac{1}{2}x^2 + 1$, $\frac{1}{2}x^2 - 6$, and, in fact, $\frac{1}{2}x^2 + C$, where C is any real number. Because the derivative of a constant is zero, we have $\frac{d}{dx}\left(\frac{1}{2}x^2 + C\right) = \frac{d}{dx}\left(\frac{1}{2}x^2\right) + \frac{d}{dx}(C) = x + 0 = x$.

This example motivates the following definition of the antiderivative: If $F(x)$ is a function whose derivative is $f(x)$, that is, if $\frac{d}{dx}(F(x)) = f(x)$, we say that $F(x)$ is an antiderivative of $f(x)$.

We say *an* antiderivative and not *the* antiderivative because, as we saw above, an antiderivative is not uniquely associated with one function. It follows from the definition of the antiderivative that if $F(x)$ is an antiderivative of $f(x)$, then for all real numbers C, $F(x) + C$ is also an antiderivative of $f(x)$.

We get a similar statement in reverse, as well.

Theorem:

If $F(x)$ and $G(x)$ are both antiderivatives of $f(x)$, then $F(x) - G(x) = C$, for some real number C.

This is true because if $F(x)$ and $G(x)$ are both antiderivatives of $f(x)$, then

$$\frac{d}{dx}(F(x)) - \frac{d}{dx}(G(x)) = f(x) - f(x) = 0$$

Using the properties of the derivatives we can rewrite the left hand side as:

$$\frac{d}{dx}(F(x)) - \frac{d}{dx}(G(x)) = \frac{d}{dx}(F(x) - G(x))$$

So:

$$\frac{d}{dx}(F(x) - G(x)) = 0$$

We learned earlier that if the derivative of a function is zero, the function must be constant, that is, $F(x) - G(x) = C$.

THE INDEFINITE INTEGRAL

The notation for the antiderivative is the symbol $\int dx$.

If $F(x)$ is an antiderivative of $f(x)$, i.e., if $\frac{d}{dx}(F(x)) = f(x)$, we write $\int f(x)\ dx = F(x) + C$.

We call the symbol $\int dx$ the *indefinite integral* and we can read the expression $\int f(x)dx$ as "the indefinite integral of $f(x)$."

It's not an accident that the symbol for the antiderivative and the symbol for the definite integral are similar. We'll see their connection in our upcoming discussions.

THE INDEFINITE INTEGRAL IS UNIQUE

We include the constant + C in the solution to an indefinite integral to indicate that an antiderivative is not unique. This also allows us to uniquely define the indefinite integral, because if two functions are antiderivatives, they differ by a constant. An antiderivative is not unique, but the indefinite integral is unique because its definition includes the constant.

For example, $\int 2\sin 2x\,dx = -\cos 2x + C$, because $\frac{d}{dx}(\cos 2x) = -2\sin 2x$. Another possible answer here is $\int 2\sin 2x\,dx = -2\cos^2 x$, because $2\sin 2x\,dx = \frac{d}{dx}(-2\cos^2 x) \underset{\text{chain rule}}{=} 4\cos x \sin x$

$$= 2 \cdot 2\sin x \cdot \cos x \underset{\text{trig identity}}{=} 2\sin 2x$$

Which antiderivative is correct? The answer is *both*, because via a series of trig identities

$-\cos 2x = -\cos^2 x + \sin^2 x = -\cos^2 x + (1 - \cos^2 x) = 1 - 2\cos^2 x.$

The two antiderivatives differ by a constant, so the answers are equivalent.

PROPERTIES OF THE ANTIDERIVATIVE

Because antiderivatives are defined in terms of derivatives, the antiderivative has properties similar to the derivative.

Theorem: Properties of the Indefinite Integral

- $\int f(x)\,dx \pm \int g(x)\,dx = \int f(x) \pm g(x)\,dx$
- $\int c \cdot f(x)\,dx = c\int f(x)\,dx$

In Chapter 22, we'll see how to compute antiderivatives.

REVIEW QUESTIONS

Note: While the following questions all test your understanding of the concept of antiderivatives, they require knowledge of some specific techniques that you should have covered in your calculus class. If you're a little rusty, feel free to read Chapter 21 and return to these problems afterwards.

1. $\int \frac{d}{dx}(F(x))\,dx =$

   (A) $F(x)$
   (B) $F'(x)$
   (C) $F(x) + C$
   (D) $F(x) = G(x) + C$
   (E) Cannot be determined from the information given

2. Suppose $G(x)$ is an antiderivative of $g(x)$ and $F(x)$ is an antiderivative of $f(x)$. If $f(x) = g(x)$, which of the following statements must be true?

   (A) $F(x) = G(x)$
   (B) $F(x) = -G(x)$
   (C) $F(x) = G(x) + 1$
   (D) $F(x) = G(x) + C$
   (E) There is no relationship between F and G.

3. The graph $f(x)$ is shown below.

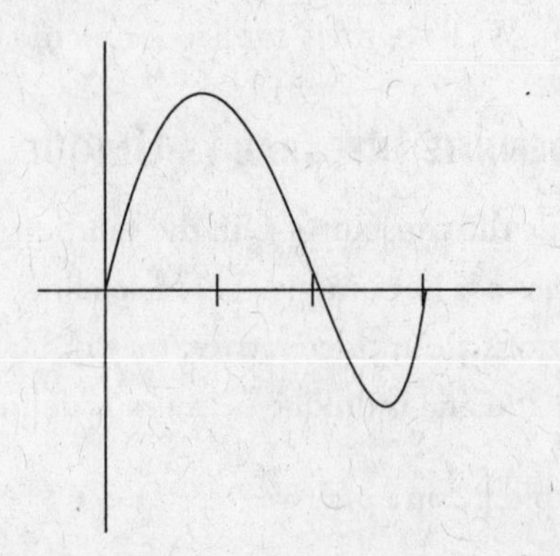

If $F(x)$ is an antiderivative of $f(x)$, which of the following graphs could be the graph of F?

(A)

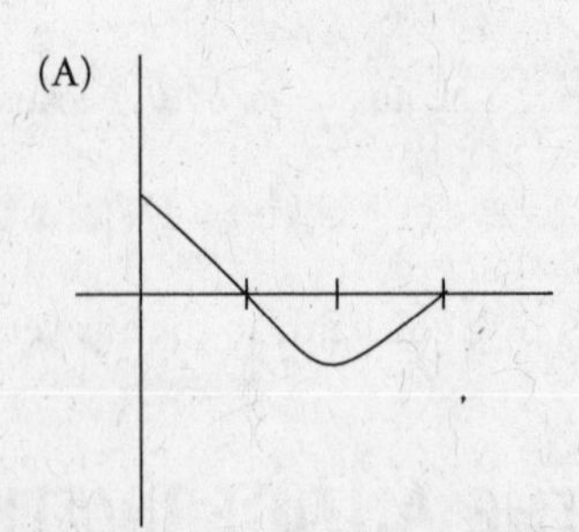

(B)

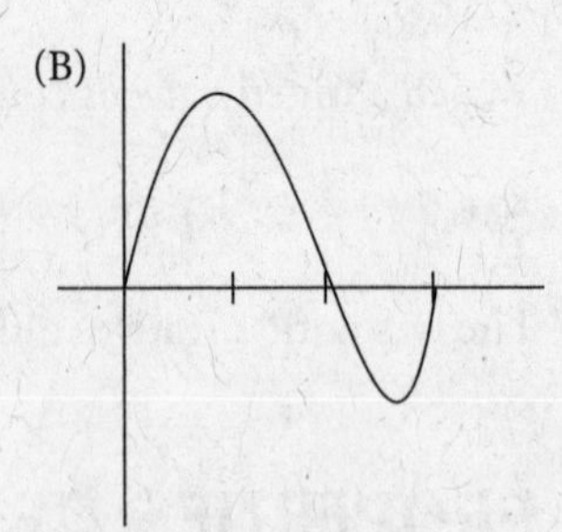

(C)

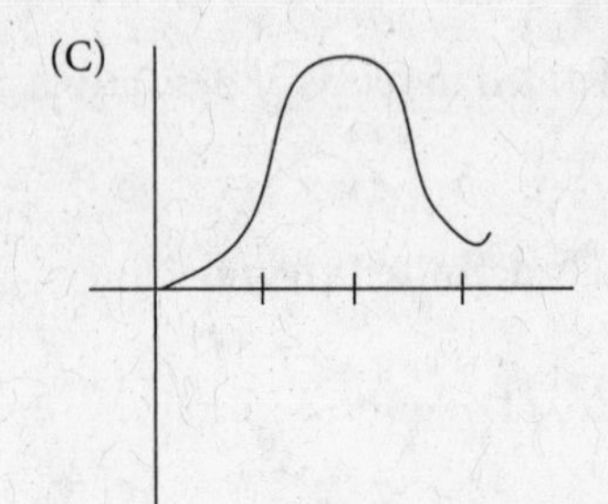

(D)

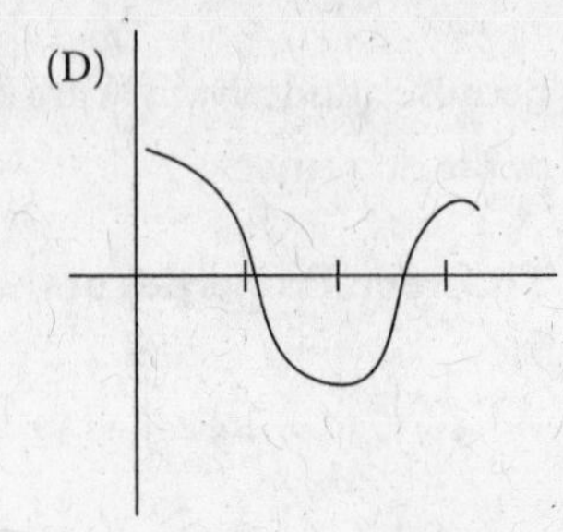

(E)

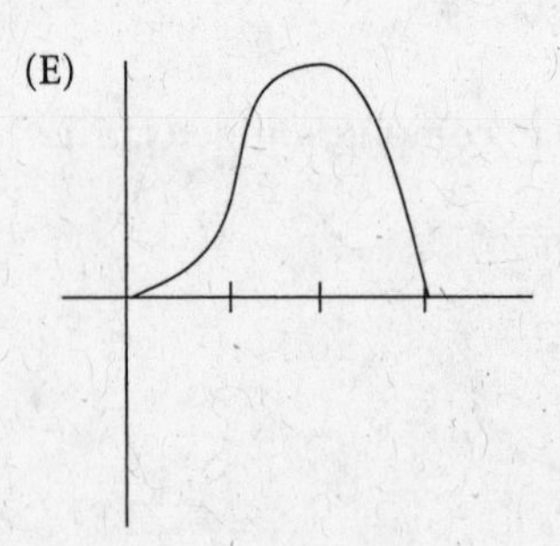

4. If $\int f(x)+g(x)\,dx = R(x)+C$ and
$\int f(x)-h(x)\,dx = S(x)+D$ and
$\int 2g(x)\,dx = T(x)+E$, then $\int h(x)\,dx =$

(A) $R(x) + S(x) + F$

(B) $R(x) - S(x) - \frac{1}{2}T(x) + F$

(C) $R(x) + S(x) - \frac{1}{2}T(x) + F$

(D) $T(x) - S(x) + F$

(E) Cannot be determined from the information given

5. If $\tan^2 x$ is an antiderivative of $f(x)$, which of the following *cannot* be an antiderivative of $f(x)$?

(A) $\sec^2 x$

(B) $\tan^2 x + 3$

(C) $\dfrac{\sin^2 x}{\cos^2 x}$

(D) $\dfrac{\sin^2 x}{1-\cos^2 x}$

(E) $\dfrac{1}{\cos^2 x}$

6. Which of the following is an antiderivative of $f(x) = \ln x$?

(A) $F(x) = \frac{1}{x}$

(B) $F(x) = x\ln x$

(C) $F(x) = x\ln x - x$

(D) $F(x) = \ln x - \frac{1}{x}$

(E) Cannot be determined from the information given.

7. If $F' = f$ and $f(x) = 18x^5 - 12x^3 + 2x$, which of the following could be $F(x)$?

(A) $3x^6 - 3x^4 + x^2 + 1$

(B) $3x^6 - 4x^4 + x^2 + 5$

(C) $18x^6 - 12x^4 + 2x^2$

(D) $90x^4 - 36x^2 + 2$

(E) None of these

8. Let $G'(x) = g(x)$ and $g(x) = \sin x + \cos x$. Which of the following could be $G(x) + 1$?

(A) $\sin x + \cos x + 1$

(B) $\sin x - \cos x + 2$

(C) $\cos x - \sin x - 1$

(D) $-\sin x - \cos x - 2$

(E) None of these

9. Let $H(x)$ be an antiderivative of $h(x)$ and $h(x) = e^x + x$. It follows that $H(x) + x =$

(A) $xe^{x-1} + x^2 + C$

(B) $e^x + x^2 + C$

(C) $e^x + \frac{1}{2}x^2 + C$

(D) $e^x + \frac{1}{2}x^2 + x + C$

(E) None of these

10. If $f(t) = 13t^{\frac{8}{5}}$ and $f(t)$ is the derivative of a function $F(t)$, than $\dfrac{F(t)}{t}$ could be

(A) $5t^{\frac{8}{5}} + 1$

(B) $5t^{\frac{8}{5}} + \frac{1}{t}$

(C) $5t^{\frac{13}{5}} + 2$

(D) $5t^{\frac{13}{5}} + \frac{5}{t}$

(E) None of these

11. $\int(\beta+\sec^2\beta)d\beta=$

(A) $\frac{1}{2}\beta^2+\frac{1}{3}\sec^3\beta+C$

(B) $\beta^2+\frac{1}{3}\sec^3\beta+C$

(C) $\frac{1}{2}\beta^2+\frac{1}{3}\tan\beta+C$

(D) $\frac{1}{2}\beta^2+\tan\beta+C$

(E) None of these

12. $\int\sin(2x+1)dx=$

(A) $\cos(2x + 1) + C$
(B) $-\cos(2x + 1) + C$
(C) $-2\cos(2x + 1) + C$
(D) $-\frac{1}{2}\cos(2x+1)+C$
(E) None of these

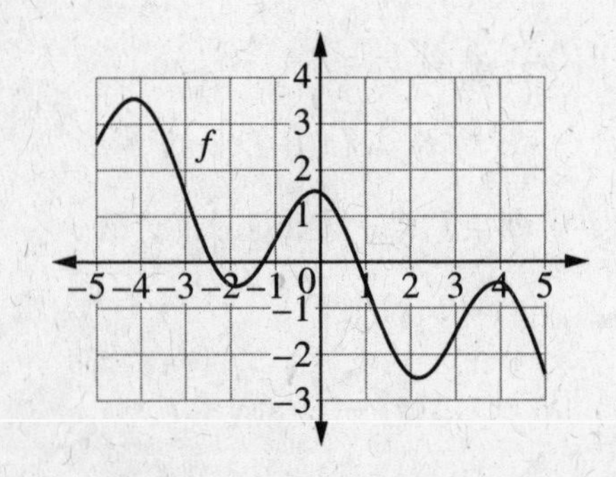

13. Given $F(x)$ is an antiderivative of $f(x)$ and the graph of f is shown above. Which of the following could be the graph of $F(x)$?

(A)

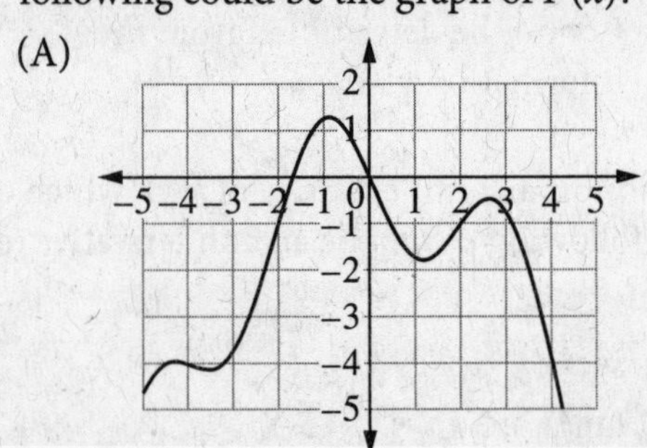

(B)

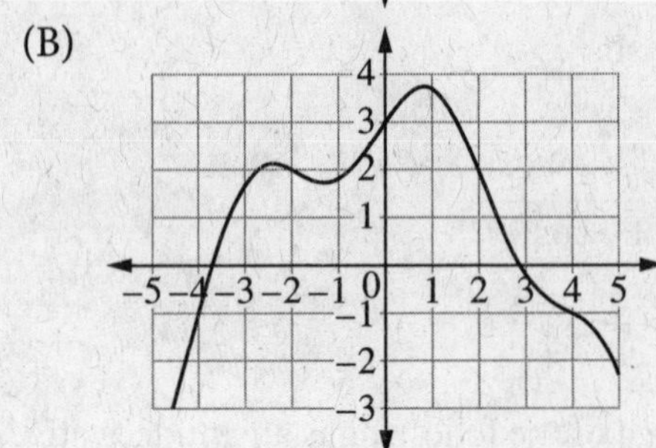

(C)

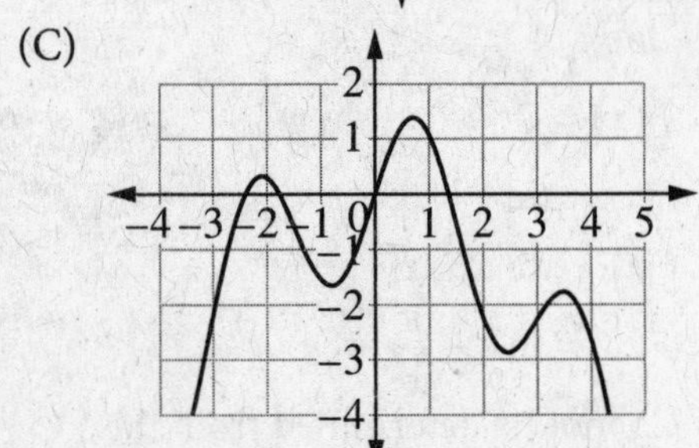

(D)

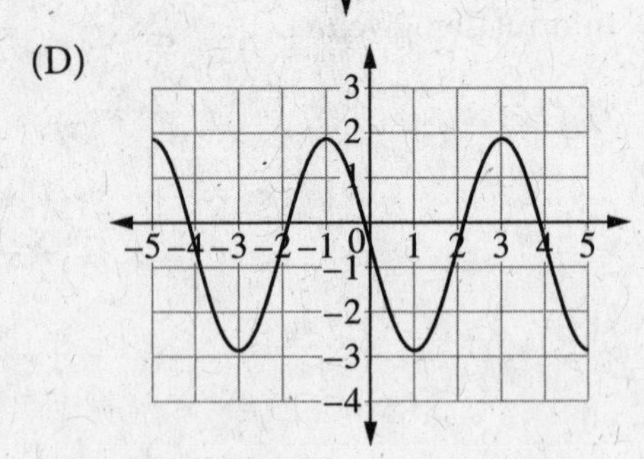

(E) None of these

14. Consider the indefinite integrals shown below. Which of these statements are true?

   I. $\int \sin^2 x\,dx = \sin^2 x + C$

   II. $\int \sin^2 x\,dx = \cos^2 x + C$

   III. $\int (ax+b)^n\,dx = \frac{1}{a(n+1)}(ax+b)^{n+1} + C$

   (A) I only
   (B) II only
   (C) III only
   (D) I and III
   (E) I, II, and III

15. $F'(t) = f(t)$ and $G'(t) = g(t)$ for all real numbers t. Which of the following could be false?

   (A) If $F(t) = G(t)$, then $F'(t) + G'(t) = 2f(t)$.
   (B) If $f(t) = g(t)$, then $F(t) - G(t) = k$ where k is a real constant.
   (C) If $F(t) = G(t) + 5$, then $f(t) = g(t)$.
   (D) If $f(t) = g(t)$, then $F(t) = G(t)$.
   (E) If $F(t) = 3G(t)$, then $F'(t) = 3G'(t)$.

16. f and g are continuous for all real numbers. If $f'(t) > g'(t)$ for all t, then which of the following is necessarily true?

   (A) The graphs of f and g never intersect.
   (B) The graphs of f and g intersect exactly once.
   (C) The graphs of f and g intersect at most once.
   (D) The graphs of f and g intersect more than once.
   (E) There is too little information to answer the question.

ANSWERS AND EXPLANATIONS

1. C

The indefinite integral of a function $\int f(x)\,dx$ is any antiderivative of f plus a constant. The function $F(x)$ is an antiderivative of $\frac{d}{dx}(F(x))$; therefore $\int \frac{d}{dx}(F(x))\,dx = F(x) + C$.

2. D

Two antiderivatives of the same function must differ by a constant. Because $f(x) = g(x)$, $F(x)$ and $G(x)$ must differ by a constant—but we don't know what the constant is. Therefore, the answer is (D).

3. C

Because $F(x)$ is an antiderivative of $f(x)$, the graph shown in the statement of the problem is the graph of the derivative of $f(x)$. Using what we already know from previous chapters about the connection between functions and their derivatives, we know that where $f(x)$ is increasing, $F(x)$ is concave up. The graph of $f(x)$ is increasing from the origin to the first tick-mark, so f must be concave up on this region. Only choices (C) and (E) meet this criteria, so we can eliminate the other options.

The function $f(x)$ has a minimum—the graph changes from decreasing to increasing in between the second and third tick-marks. Therefore, $F(x)$ has an inflection point in this region and the graph must change from concave down to concave up in this region. The graph shown in (C) is the only one that satisfies this criterion.

4. B

We view these antiderivatives as a system of three linear equations in three unknowns. We set $a = \int f(x)\,dx$, $b = \int g(x)\,dx$, $c = \int h(x)\,dx$. Using the properties of the indefinite integral, we rewrite the three statements we are given as:

$$\int f(x) + g(x)\,dx = R(x) + C \Rightarrow \quad \int f(x)\,dx + \int g(x)\,dx = R(x) + C$$

$$\int f(x) - h(x)\,dx = S(x) + D \Rightarrow \quad \int f(x)\,dx - \int h(x)\,dx = S(x) + D$$

$$\int 2g(x)\,dx = T(x) + E \Rightarrow \quad 2\int g(x)\,dx = T(x) + E$$

We can interpret this as the system of equations:

$$a + b = R$$
$$a - c = S$$
$$2b = T$$

We solve this system of equations for a, b, and c:

$$a + b = R \qquad \Rightarrow a + \frac{T}{2} = R \Rightarrow a = R - \frac{T}{2}$$

$$a - c = S \qquad \Rightarrow R - \frac{T}{2} - c = S \Rightarrow c = R - S - \frac{1}{2}T$$

$$2b = T \qquad \Rightarrow b = \frac{T}{2}$$

Substituting the integral $\int h(x)\, dx$ back for c and adding the constant, we see that the correct answer is (B).

5. D

Two antiderivatives of the same function must differ by a constant. We can use trig identities to assess each answer option.

(A) $\sec^2 x = \tan^2 x + 1$ is an antiderivative of $f(x)$.
(B) $\tan^2 x + 3$ is an antiderivative of $f(x)$.
(C) $\frac{\sin^2 x}{\cos^2 x} = \tan^2 x$ is an antiderivative of $f(x)$.
(E) $\frac{1}{\cos^2 x} = \sec^2 x = \tan^2 x + 1$ is an antiderivative of $f(x)$.

By process of elimination, (D) is correct; we can also see this directly using trig identities:

$$\frac{\sin^2 x}{1 - \cos^2 x} = \frac{\sin^2 x}{\sin^2 x} = 1.$$

6. C

We do not need to compute the antiderivative of $f(x)$ to answer this question. By definition, $F(x)$ is an antiderivative of $f(x)$ if $\frac{d}{dx}(F(x)) = f(x)$. To determine the correct answer, we differentiate the answer choices.

(A) $\frac{d}{dx}\left(\frac{1}{x}\right) = -\frac{1}{x^2}$

(B) $\frac{d}{dx}(x \ln x) = \ln x + 1$

(C) $\frac{d}{dx}(x \ln x - x) = \ln x$

(D) $\frac{d}{dx}\left(\ln x - \frac{1}{x}\right) = \frac{1}{x} + \frac{1}{x^2}$

The correct answer choice is therefore (C).

There is a technique for computing antiderivatives, called *integration by parts*, which can be used to determine the antiderivative of $f(x) = \ln x$. It is not part of the Calculus AB syllabus, but it is part of the BC syllabus. Integration by parts is covered in Chapter 21.

7. A

Since $F' = f$ and $f(x) = 18x^5 - 12x^3 + 2x$, then $F(x) = \int (18x^5 - 12x^3 + 2x)\,dx$

$$\begin{aligned} F(x) &= \int (18x^5 - 12x^3 + 2x)\,dx \\ F(x) &= \frac{18}{6}x^6 - \frac{12}{4}x^4 + \frac{2}{2}x^2 + C \\ F(x) &= 3x^6 - 3x^4 + x^2 + C \end{aligned}$$

The only choice that agrees is option (A) $3x^6 - 3x^4 + x^2 + 1$.

8. B

Since $G'(x) = g(x)$ and $g(x) = \sin x + \cos x$, $G(x) = \int g(x)\,dx = \int (\sin x + \cos x)\,dx$.

$$\begin{aligned} G(x) &= \int (\sin x + \cos x)\,dx \\ G(x) &= \int \sin x\,dx + \int \cos x\,dx \\ G(x) &= -\cos x + \sin x + C \\ G(x) &= \sin x - \cos x + C \end{aligned}$$

It follows that $G(x) + 1 = \sin x - \cos x + C + 1$. If $C = 1$, then option (B) would agree $G(x) + 1 = \sin x - \cos x + 2$

9. D

Since $H(x)$ is an antiderivative of $h(x)$, it follows that $H(x) = \int h(x)\,dx = \int (e^x + x)\,dx$.

$$\begin{aligned} H(x) &= \int (e^x + x)\,dx \\ H(x) &= e^x + \frac{1}{2}x^2 + C \end{aligned}$$

It follows that $H(x) + x = e^x + \frac{1}{2}x^2 + x + C$

10. B

Since $f(t) = 13t^{\frac{8}{5}}$ and $f(t)$ is the derivative of $F(t)$, then

$$F(t) = \int f(t)\,dt = \int 13t^{\frac{8}{5}}\,dt$$

$$F(t) = \int f(t)\,dt = \int 13t^{\frac{8}{5}}\,dt$$

$$F(t) = \int 13t^{\frac{8}{5}}\,dt = 13\int t^{\frac{8}{5}}\,dt = 13 \cdot \frac{5}{13}t^{\frac{13}{5}} + C$$

$$\therefore F(t) = 5t^{\frac{13}{5}} + C$$

This means $\dfrac{F(t)}{t} = \dfrac{5t^{\frac{13}{5}}}{t} + \dfrac{C}{t} = 5t^{\frac{8}{5}} + \dfrac{C}{t}$ so only option (B) $5t^{\frac{8}{5}} + \dfrac{1}{t}$ agrees.

11. D

$$\int (\beta + \sec^2 \beta)\, d\beta = \frac{1}{2}\beta^2 + \tan\beta + C$$

12. D

We know that $\int \sin x\, dx = -\cos x + C$ which makes (A) implausible so we focus on the (B)-(D). We need to imagine the derivatives of options (B), (C), and (D) producing sin $(2x + 1)$. Because of the chain rule, we differentiate option (D) and get $\dfrac{d}{dx}\left(-\dfrac{1}{2}\cos(2x+1)+C\right) = -\dfrac{1}{2}(-\sin(2x+1))\cdot 2 = \sin(2x+1)$.

13. B

We want to find $F(x)$ so that $F'(x) = f(x)$. Option (B) increases for $x \in (-\infty - 2.4) \cup (-1.3, 0.8)$; on the same intervals the graph of f is positive. Similarly, option (B) decreases for $x \in (-2.4, -1.3) \cup (0.8, \infty)$; on the same intervals the graph of f is negative. This is true for option (B) only.

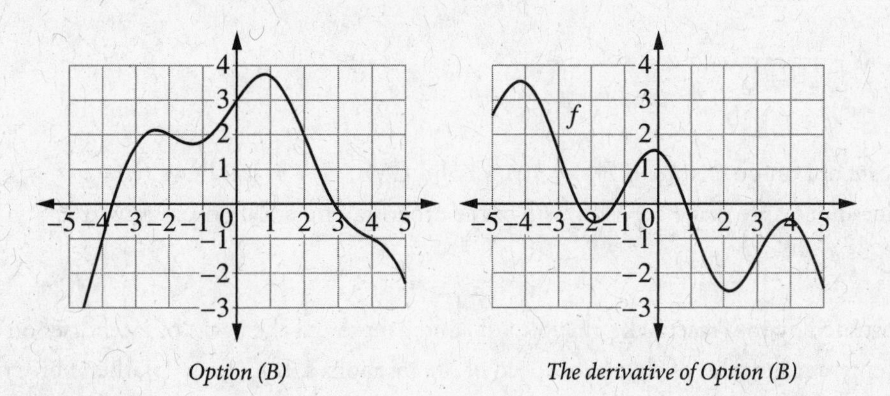

Option (B) *The derivative of Option (B)*

14. D

We need to consider each statement.

I. Consider $\int \sin 2x\,dx = \sin^2 x + C$ Differentiate $\sin^2 x + C$ using the chain rule.

$$\frac{d}{dx}(\sin^2 x + C) = \frac{d}{dx}\left((\sin x)^2 + C\right) = 2\sin x \cos x + 0 = \sin 2x$$

This means $\sin 2x$ is an antiderivative of $\sin^2 x + C$. $\int \sin 2x\,dx = \sin^2 x + C$ is a true statement.

II. Consider $\int \sin 2x\,dx = \cos^2 x + C$ Differentiate $\cos^2 x + C$ using the chain rule.

$$\frac{d}{dx}(\cos^2 x + C) = \frac{d}{dx}\left((\cos x)^2 + C\right) = 2\cos x(-\sin x) + 0 = -\sin 2x$$

This means $-\sin 2x$ is an antiderivative of $\cos^2 x + C$. $\int \sin 2x\,dx = \cos^2 x + C$ is a false statement.

III. Consider $\int (ax+b)^n\,dx = \frac{1}{a(n+1)}(ax+b)^{n+1} + C$ Differentiate $\frac{1}{a(n+1)}(ax+b)^{n+1} + C$ using the chain rule.

$$\frac{d}{dx}\left(\frac{1}{a(n+1)}(ax+b)^{n+1} + C\right) = \frac{1}{\cancel{a}\,\cancel{(n+1)}}\cancel{(n+1)}(ax+b)^n \cdot \cancel{a} + 0 = (ax+b)^n$$

This means $(ax+b)^n$ is an antiderivative of $\frac{1}{a(n+1)}(ax+b)^{n+1} + C$.

$$\int (ax+b)^n\,dx = \frac{1}{a(n+1)}(ax+b)^{n+1} + C$$

15. D

Antiderivatives are not unique. Note: If $F(t) = 3(t) + 1$ and $G(t) = 3t + 5$, it follows $f(t) = g(t) = 3$. However $f(t) = g(t) = 3$ does not imply that $F(t) = G(t)$. All of the other statements are necessarily true.

16. C

This can be illustrated by the "racetrack principle." If f and g represent the paths of racehorses on a track, then $f'(t)$ and $g'(t)$ represent the rates of change or speed of the racehorses. If $f'(t) > g'(t)$, then horse f is running faster than horse g at all times. Now, if horse f starts behind horse g,then horse f will run past horse g eventually (the horses intersect). However, if horse f starts ahead of horse g,then horse f will run further ahead leaving horse g in the dust and the horses will never intersect. Given the conditions stated in the problem, horses f and g intersect at most once.

CHAPTER 20: THE FUNDAMENTAL THEOREM OF CALCULUS

IF YOU LEARN ONLY TWO THINGS IN THIS CHAPTER . . .

1. **The Fundamental Theorem of Calculus:** If f is continuous on $[a,b]$ and F is any antiderivative of f then $\int_a^b f(x)\,dx = F(b) - F(a)$.
2. **Second Fundamental Theorem of Calculus:** The function $G(x) = \int_a^x f(t)\,dt = F(x) - F(a)$ is differentiable and $\frac{d}{dx}(G(x)) = f(x)$, that is, $\frac{d}{dx}\int_a^x f(t)\,dt = f(x)$.

We'll make two connections explicit in this chapter. First, we'll see the connection between the definite integral and the indefinite integral. Then we'll see an explicit connection between derivatives and integrals—the bridge between the two legs of calculus.

THE FUNDAMENTAL THEOREM OF CALCULUS (FTC)

In Chapter 18 we discussed the idea of the total change in a function over a period of time.

If $g'(x)$ is the rate of change of a function, then the area under the curve $g'(x)$ from a to b, i.e., $\int_a^b g'(x)\,dx$, is the total change in the quantity of the function from $x = a$ to $x = b$. We can thus think of the accumulated change as $\int_a^b g'(x)\,dx = g(b) - g(a)$.

We can think about this idea in terms of antiderivatives. If f is an integrable function and F is an antiderivative of f, that is $F'(x) = f(x)$, then the above integral becomes:

$$\int_a^b f(x)\,dx = F(b) - F(a)$$

OK...except that an antiderivative isn't unique. Suppose we choose another antiderivative. We also saw in Chapter 18 that if F and G are both antiderivatives of a function f, then they differ by a constant, i.e., $G(x) = F(x) + C$. Let's see how changing the antiderivative changes the definite integral on the previous page. In that equation, we chose the antiderivative F of f. If instead we choose the antiderivative G, we obtain:

$$\int_a^b f(x)\,dx = G(b) - G(a) = (F(b) + C) - (F(a) + C) = F(b) - F(a)$$

...the same thing! The value of the definite integral doesn't depend on which antiderivative we choose.

This is it...our *big* theorem...

Theorem: The Fundamental Theorem of Calculus

If f is continuous on $[a,b]$ and F is any antiderivative of f, then $\int_a^b f(x)\,dx = F(b) - F(a)$.

We often use the shorthand "FTC" instead of writing "Fundamental Theorem of Calculus." We sometimes use the notation $F(x)|_a^b$ or $F(x)]_a^b$ instead of writing $F(b) - F(a)$, i.e., $F(x)|_a^b \underset{\text{definition}}{=} F(b) - F(a)$.

Recall our discussion in Chapter 17 about the area under the curve $f(x)$ on the interval $a \le x \le b$.

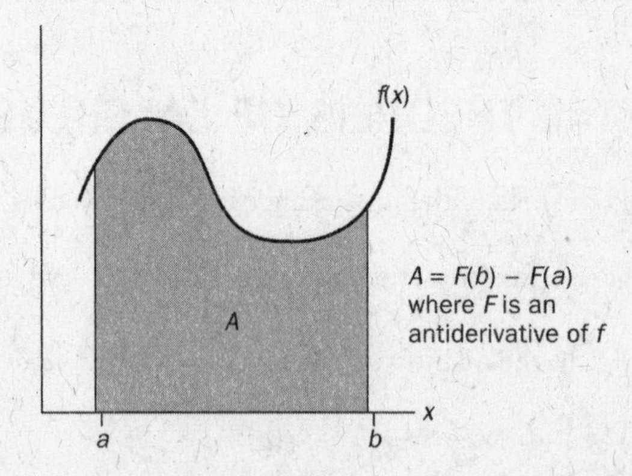

We spent lots of time approximating this area using rectangles, more rectangles, and finally an infinite number of rectangles. Using the FTC, we now have a way to compute this area *without* finding areas of rectangles. We only need to find a function F whose derivative is f, that is, an antiderivative of f, and compute $F(b) - F(a)$.

For example, suppose we want to compute the area under the curve $f(x) = \cos x$ on the interval $0 \le x \le \frac{\pi}{4}$. We can rewrite this as a definite integral, $\int_0^{\frac{\pi}{4}} \cos x\, dx$. According to the FTC, we can compute this area by finding an antiderivative of $\cos x$ and evaluating it at $x = \frac{\pi}{4}$ and $x = 0$. That is, we need to find a function the derivative of which is $\cos x$. The derivative of $\sin x$ is $\cos x$; $\frac{d}{dx}(\sin x) = \cos x$, so $\sin x$ is an antiderivative of $\cos x$. By the FTC:

$$\int_0^{\frac{\pi}{4}} \cos x\, dx = \sin x\Big|_0^{\frac{\pi}{4}} = \sin\frac{\pi}{4} - \sin 0 = \frac{\sqrt{2}}{2} - 0 = \frac{\sqrt{2}}{2}.$$

That is, the area under the cosine curve on the interval $0 \le x \le \frac{\pi}{4}$ is $\frac{\sqrt{2}}{2} \approx .707$.

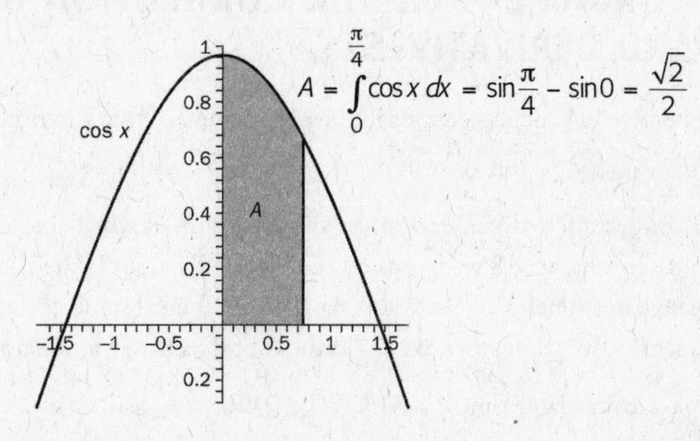

In Chapter 21 we'll develop some techniques for computing antiderivatives that will allow us to work with definite integrals for a wide variety of functions.

The Connection Between Definite and Indefinite Integrals

We can express the FTC in terms of indefinite integrals instead of antiderivatives. Remember that the indefinite integral of a function is unique, because it includes the constant. An antiderivative, on the other hand, is not unique. Using the indefinite integral we can rewrite the FTC as:

$$\int_a^b f(x)\, dx = \left(\int f(x)\, dx\right)\Big|_a^b$$

VARIABLES ARE DUMB, OR DUMMY VARIABLES

Variables are place holders. They have no meaning other than the meaning we assign to them. When we say "let f be a function of t-time," we are using t for convenience. There is no reason we can't say "let f be a function of W-time." When we compute a definite integral, it doesn't matter what we call the variable in the function because no variables appear in the solution:

$$\int_a^b f(t)\,dt = \int_a^b f(x)\,dx = \int_a^b f(\Theta)\,d\Theta = F(b) - F(a).$$

Sometimes it's convenient to change the variable of integration—we'll see this in the next few pages and also in Chapter 22.

WHAT'S SO FUNDAMENTAL? THE CONNECTION BETWEEN INTEGRALS AND DERIVATIVES

The Fundamental Theorem of Calculus sounds like a serious name. What's so fundamental about the FTC?

The FTC provides a connection between derivatives and integrals—it works like a bridge between the two legs of calculus. In Chapter 18 we discussed the idea of the accumulated change of a function, i.e., the change in a function from a starting point $t = a$ until some point t. If f was the rate of change of a quantity over time, then we could think of the change in the quantity from some set starting time a to a variable time t as $G(t) = \int_a^t f(t)\,dt$.

We can define the function above for any continuous function. Now that we have the FTC at our disposal, let's work with this idea a little more. From the FTC we know that:

$$\int_a^b f(t)\,dt = F(b) - F(a).$$

From properties of the definite integral, we know that a and b don't have to be in any particular order. We can use the idea of the accumulation function from Chapter 18 and define a function that measures the accumulated area under a curve. For convenience, we'll use the variable t for the function we're integrating and the variable x for the accumulation function we're defining:

$$G(x) = \int_a^x f(t)\,dt = F(x) - F(a)$$

The difference between the function we're defining here and the accumulation function is that now we have a concrete way of computing $G(x)$: $G(x) = F(x) - F(a)$ where $F(x)$ is an antiderivative of $f(x)$. Let's see what happens when we take the derivative of G.

$$\frac{d}{dx}(G(x)) = \frac{d}{dx}\int_a^x f(t)\,dt = \frac{d}{dx}(F(x) - F(a))$$

But F is an antiderivative of f, so $\frac{d}{dx}(F(x)) = f(x)$. $F(a)$ is a constant, so $\frac{d}{dx}(F(a)) = 0$. Therefore, $\frac{d}{dx}(G(x)) = f(x)$.

THE SECOND FUNDAMENTAL THEOREM OF CALCULUS

We can write this conclusion as a theorem. We'll assume that $f(x)$ is continuous, at least on any interval we're thinking about.

Theorem: Second Fundamental Theorem of Calculus

The function $G(x) = \int_a^x f(t)\,dt = F(x) - F(a)$ is differentiable and $\frac{d}{dx}(G(x)) = f(x)$, that is, $\frac{d}{dx}\int_a^x f(t)\,dt = f(x)$.

Some books refer to the First and Second Fundamental Theorems of Calculus as one big Fundamental Theorem. Other books reverse the order in which the theorems are presented here. It doesn't matter what names these theorems have, and the order in which they appear is not important. What's important is what they tell us: the connections between definite integrals, indefinite integrals, and derivatives.

Let's think of derivatives and integrals in terms of a staircase: think of computing derivatives as going down a stair and computing integrals as going up a stair. It follows from the definition of the antiderivative that computing the antiderivative of a derivative gives you back the original function: $\int \frac{d}{dx} F(x)\,dx = F(x) + (C)$.

In short, going down a stair and then up a stair leaves you where you started. The Second Fundamental Theorem tells us that taking the derivative of an integral leaves you where you started. If you go up and then down, you don't go anywhere. For example, $\frac{d}{dx}\int_1^x e^{t^2}\,dt = e^{x^2}$.

We can also combine the Second Fundamental Theorem and the chain rule. Suppose $G(x) = \int_2^{u(x)} f(t)\,dt$. What is $G'(x)$ in this case? We aren't computing the definite integral from a constant to a variable, but from a constant to a *function.*

First we put ourselves back in the realm of the Second Fundamental Theorem. We set up an integral from a constant to a variable, i.e., we set $H(u) = \int_a^u f(t)\,dt$ and rewrite $G(x) = H(u(x))$. By the chain rule, $G'(x) = H'(u(x)) \cdot u'(x)$. According to the Second Fundamental Theorem, $H'(u) = f(u)$, so $G'(x) = f(u) \cdot u'(x)$.

For example, $\frac{d}{dx}\int_0^{x^2} \ln t\,dt = (\ln x^2) \cdot (x^2)' = (\ln x^2) \cdot 2x$.

Using the properties of the definite integral, we can also differentiate functions that are defined in terms of the definite integral evaluated between *two* functions.

$$\begin{aligned}
\frac{d}{dx}\int_{u(x)}^{v(x)} f(t)\,dt &= \frac{d}{dx}\int_{u(x)}^{c} f(t)\,dt + \frac{d}{dx}\int_{c}^{v(x)} f(t)\,dt \\
&= \frac{d}{dx}\left(-\int_{c}^{u(x)} f(t)\,dt\right) + \frac{d}{dx}\int_{c}^{v(x)} f(t)\,dt \\
&= -\frac{d}{dx}\int_{c}^{u(x)} f(t)\,dt + \frac{d}{dx}\int_{c}^{v(x)} f(t)\,dt \\
&= -f(u(x))\cdot u'(x) + f(v(x))\cdot v'(x) \\
&= f(v(x))\cdot v'(x) - f(u(x))\cdot u'(x)
\end{aligned}$$

For example:

$$\begin{aligned}
\frac{d}{dx}\int_{x}^{\sin x^2} \sqrt{t}\,dt &= \sqrt{\sin x^2}\cdot\frac{d}{dx}\left(\sin x^2\right) - \sqrt{x}\cdot 1 \\
&= \sqrt{\sin x^2}\cdot 2x\cos x^2 - \sqrt{x}
\end{aligned}$$

INTERPRETING THE FUNDAMENTAL THEOREMS GRAPHICALLY

Using the connection between definite integrals and area under a curve, we can use the FTC and the Second Fundamental Theorem to analyze functions presented graphically.

Example:

The graph of the function $f(x)$ is shown below. The semicircular region of the function is the top half of the circle $(x - 2)^2 + (y - 2)^2 = 1$.

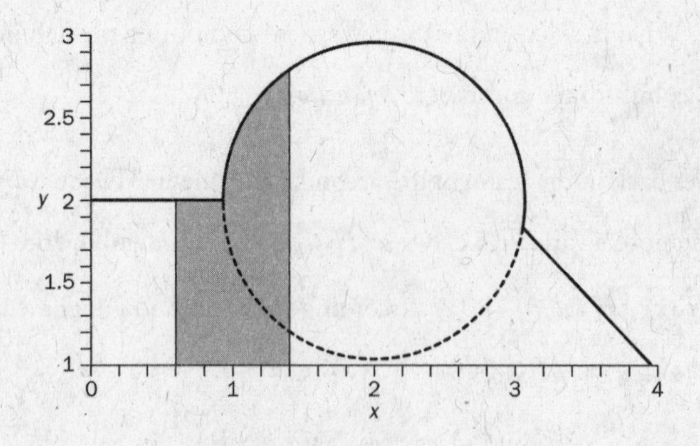

The function $F(x)$ is defined by $F(x) = \int_{0.6}^{x} f(t)\,dt$ on the interval $0 \le x \le 4$.

(a) Compute $F(2)$.

(b) Compute $F'(2)$.

Solution:

(a) The value of $F(2)$ is $F(2) = \int_{0.6}^{2} f(t)\,dt$, so we need to compute the area under the curve f on the interval $0.6 \le x \le 2$. Let's shade this area on the graph.

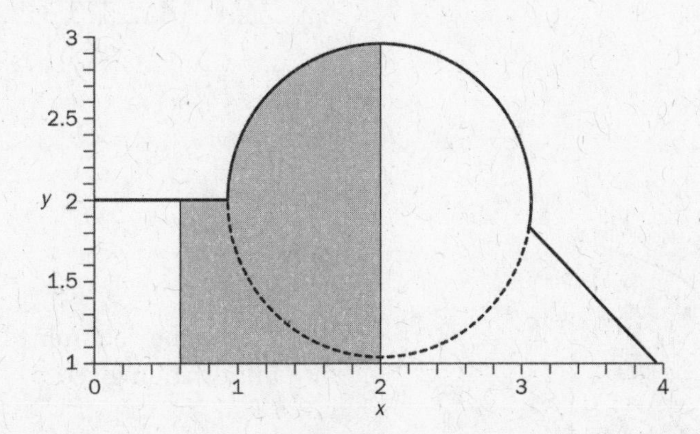

We can compute this area by dividing it into two pieces.

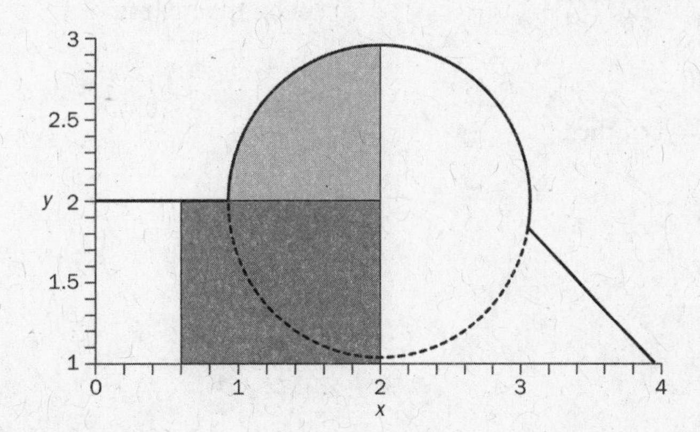

The more lightly shaded area is one-quarter of the area of the circle $(x - 2)^2 + (y - 2)^2 = 1$. Because the radius of this circle is 1, its area is $\pi \cdot 1^2 = \pi$. Therefore, the area of the lightly shaded region is $\frac{\pi}{4}$. The darkly shaded region is a rectangle with height 2 and length $2 - 0.6 = 1.4$. The area of the darkly shaded rectangle is, therefore, $2 \cdot 1.4 = 2.8$. Therefore the area of the entire shaded region is $\frac{\pi}{4} + 2.8 \approx 3.585$, i.e., $F(2) = \int_{0.6}^{2} f(t)\,dt \approx 3.585$.

(b) By the Second Fundamental Theorem, $F'(x) = f(x)$. Therefore, $F'(2) = f(2) = 3$.

REVIEW QUESTIONS

1. If $\int_a^b f(x)\,dx = 2a - 3b$, then $\int_a^b f(x) + 3\,dx =$

    (A) $2a - 3b + 3$

    (B) $3b - 3a$

    (C) $-a$

    (D) $5a - 6b$

    (E) $a - 6b$

2. The graph of f is shown below.

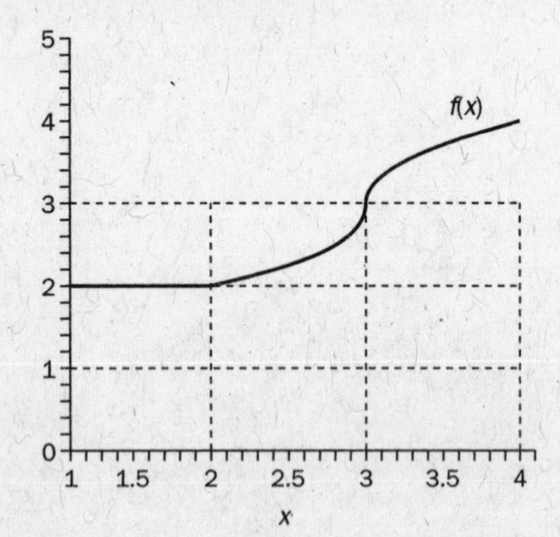

If $\int_2^4 f(x)\,dx = 6$ and $F'(x) = f(x)$, then $F(4) - F(1) =$

    (A) 4

    (B) 5

    (C) 6

    (D) 7

    (E) 8

The following information is used in questions 3 and 4:

Suppose $f(x)$ is a continuous function and f is the derivative of the function $F(x)$.

Below is a table of values for $f(x)$ and $F(x)$:

x	0	1	2	3
$f(x)$	-1	0	1	-2
$F(x)$	4	3	A	8

3. What is $\int_1^3 f(x)\,dx$?

    (A) 5

    (B) 8

    (C) -2

    (D) 19

    (E) Cannot be determined from the information given

4. If the area under the curve $f(x)$ on the interval $0 \le x \le 2$ is equal to the area under the curve $f(x)$ on the interval $2 \le x \le 3$, then $A =$

    (A) 4

    (B) 5

    (C) 5.5

    (D) 6

    (E) 7

5. For $x > 0$, compute $\frac{d}{dx}\left(\int_{x}^{x^2} \sqrt{t}\, dt\right)$.

(A) $x - \sqrt{x}$

(B) $2x^2 - \sqrt{x}$

(C) $\sqrt{x}$

(D) $-\frac{1}{2\sqrt{x}}$

(E) $-\frac{1}{2x} + \frac{1}{2\sqrt{x}}$

6. Let $f(x) = \int_{\frac{2\pi}{3}}^{x^2} \cos t\, dt$. Which of the following is a critical point of f?

(A) π

(B) $\frac{\pi}{2}$

(C) $\sqrt{\pi}$

(D) $\sqrt{\frac{\pi}{2}}$

(E) None of the above

7. $\frac{d}{dx}\int_0^x \frac{1}{1+t^4}\,dt =$

(A) $\frac{1}{1+x^4}$

(B) $\frac{4x^3}{1+x^4}$

(C) $\frac{1}{1+x^4} - 1$

(D) $\frac{1}{4x^3}$

(E) None of the above

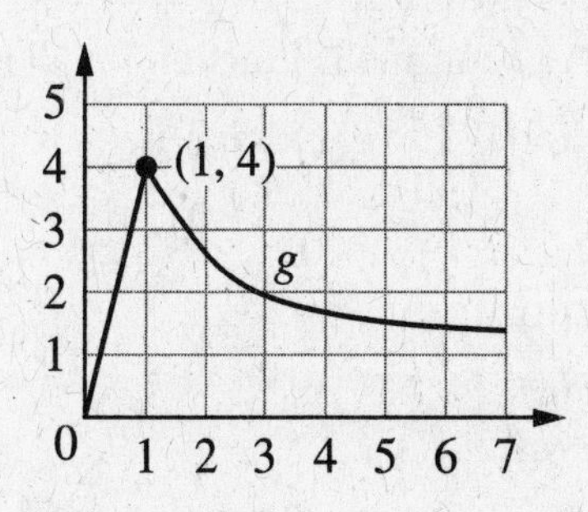

8. The graph of g is shown above. Let $G'(x) = g(x)$ for $x > 0$ and $\int_1^5 g(x)\,dx = 8.9$. What is the exact value of $G(5) - G(0)$?

(A) 8.9

(B) 9.9

(C) 10.9

(D) 11.9

(E) 12.9

9. If $x^2 - 3x = \int_2^x f(t)\,dt$, what is the value of $f(4)$?

(A) 2

(B) 4

(C) 5

(D) 6

(E) 8

10. Let g be the function given by $g(x) = \int_0^x |t - 1|\,dt$. An equation of the line tangent to g at $x = 3$ is

(A) $5x - 2y = -4$

(B) $2x - 2y = -11$

(C) $2x - y = -8$

(D) $5x - 2y = 15$

(E) $4x - 2y = 7$

11. If F' is a continuous function for all real x, then $\lim_{h \to 0}\left(\frac{1}{h}\int_a^{a+h} F'(x)\,dx\right)$ is

(A) 0
(B) $F(0)$
(C) $F(a)$
(D) $F'(0)$
(E) $F'(a)$

12. If $f(x)=\int_2^x \frac{1}{1+t^3}\,dt$, then the value of $f'(2)$ is

(A) $-\frac{7}{18}$
(B) 0
(C) $\frac{1}{9}$
(D) $\frac{1}{3}$
(E) None of the above

13. If $f(x)=\sqrt{2x+7}$, where $F'(x) = f(x)$ and $F(1) = 7$, then $F(-1.5) =$

(A) −2
(B) $-\frac{3}{2}$
(C) $-\frac{2}{3}$
(D) $\frac{2}{3}$
(E) $\frac{3}{2}$

14. If $F'(x)=(3x+1)^{\frac{3}{2}}$ and $F(1) = 4$, then $F(x) =$

(A) $\frac{6\sqrt{3x+1}}{3}$
(B) $\frac{4\sqrt{3x+1}}{9}$
(C) $\frac{2(3x+1)^{\frac{5}{2}}-4}{15}$
(D) $\frac{2(3x+1)^{\frac{5}{2}}-44}{5}$
(E) $(3x+1)^{\frac{5}{2}}+4$

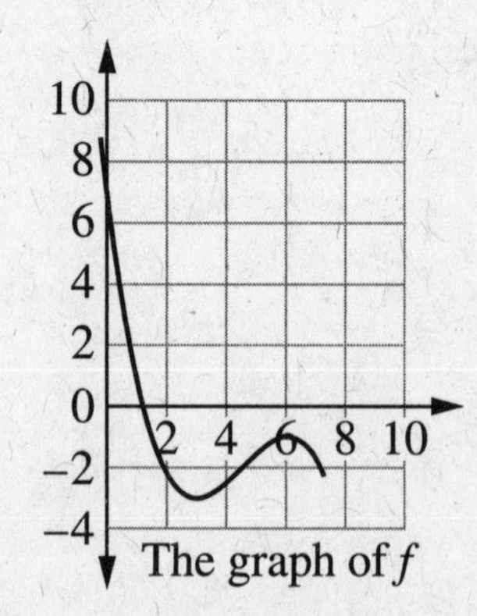

The graph of f

15. Consider the graph of above. If $g(x)=\int_0^{2x} f(t)\,dt$, what is the value of $g'(3)$?

(A) 0
(B) −1
(C) −2
(D) −3
(E) −6

Free-Response Question

16. The graph of the function f shown below consists of three line segments and a semicircle.

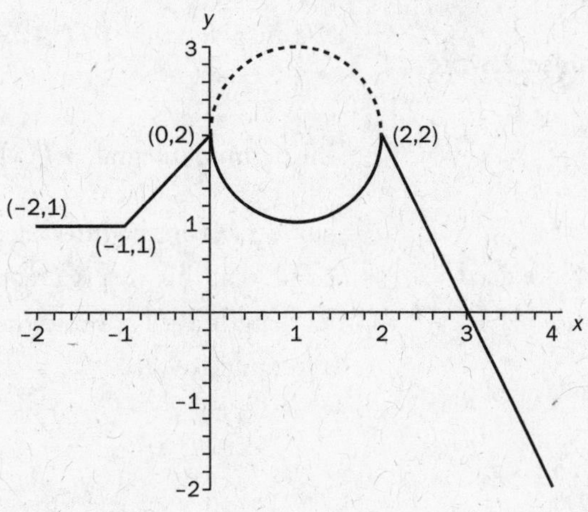

Let g be the function given by $g(x) = \int_{-2}^{x} f(x)\, dx$

(a) Find $g(1)$ and $g'(1)$.

(b) Find all the values of x on the open interval $-2 < x < 4$ at which g attains a relative maximum. Justify your answer.

17. A graph of the function f, defined for $\in [-6, 3]$, is shown below. The graph consists of three segments and two quarter circles.

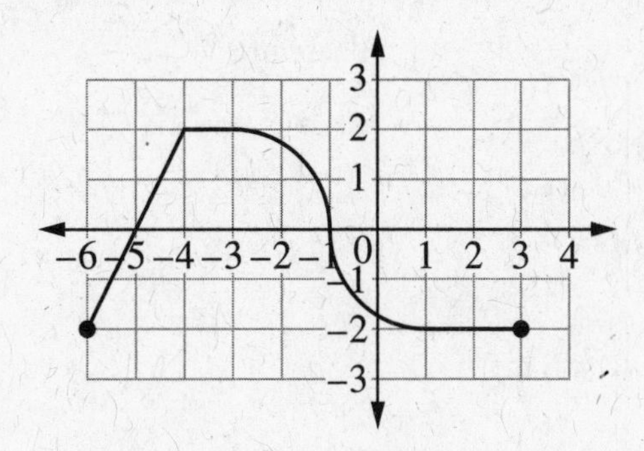

The function A is defined by setting $A(x) = \int_{-4}^{x} f(t)\, dt$

(a) Evaluate $A'(1)$

(b) Find all the solutions of the equation $A(x) = 0$

(c) The graph of $y = A(x)$ has one relative minimum point. Find the exact coordinates of this point.

(d) Determine an equation of the line tangent to the graph of A at the point on the graph of A where $x = 1$.

ANSWERS AND EXPLANATIONS

1. C

Using the properties of definite integrals we rewrite $\int_a^b f(x) + 3\, dx = \int_a^b f(x)\, dx + \int_a^b 3\, dx.$

We are given that $\int_a^b f(x)\, dx = 2a - 3b$; we need to compute $\int_a^b 3\, dx$. To apply the FTC, we need to find an antiderivative of 3; the function $3x$ is an antiderivative of 3, because $\frac{d}{dx}(3x) = 3$. Therefore, applying the FTC gives us $\int_a^b 3\, dx = 3x\Big|_a^b = 3b - 3a.$

Adding the two definite integrals we obtain:

$$\int_a^b f(x) + 3\, dx = \int_a^b f(x)\, dx + \int_a^b 3\, dx$$
$$= (2a - 3b) + (3b - 3a) = -a.$$

2. E

Because $F'(x) = f(x)$, F is an antiderivative of f. Therefore, by the FTC, $F(4) - F(1) = \int_1^4 f(x)\, dx.$

By definition of the definite integral, $\int_1^4 f(x)\, dx$ is the area under the curve $f(x)$ on the interval $1 \le x \le 4$. By properties of the definite integral:

$$\int_1^4 f(x)\, dx = \int_1^2 f(x)\, dx + \int_2^4 f(x)\, dx.$$

We are given that $\int_2^4 f(x)\, dx = 6$. By definition of the definite integral, $\int_1^2 f(x)\, dx$ is the area under the curve $f(x)$ on the interval $1 \le x \le 2$. We see on the graph that this area is a rectangle, the base of which has length $2 - 1 = 1$ and the height of which is 2; it is a rectangle of area $1 \cdot 2 = 2$.

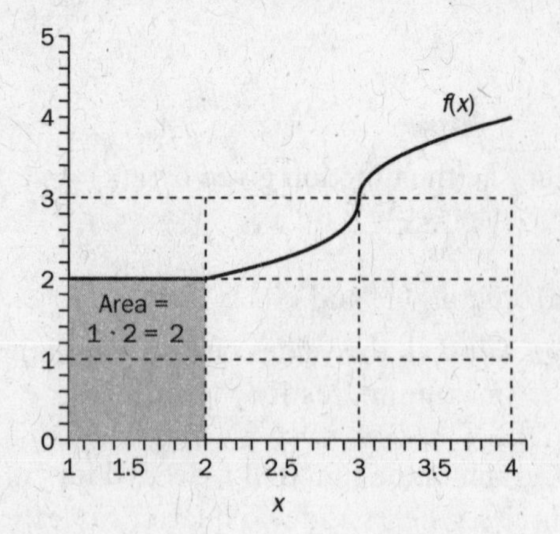

Therefore,

$$\int_1^4 f(x)\, dx = \int_1^2 f(x)\, dx + \int_2^4 f(x)\, dx$$
$$= 6 + 2 = 8.$$

3. A

Because we are given that F is antiderivative of f, by the FTC, $\int_1^3 f(x)\, dx = F(3) - F(1).$

Filling in the values for $F(3)$ and $F(1)$ from the chart, we obtain $\int_1^3 f(x)\, dx = F(3) - F(1) = 8 - 3 = 5.$

4. D

Because the definite integral is the (signed) area under the curve, we can express the information in the question as $\int_0^2 f(x)\,dx = \int_2^3 f(x)\,dx$. Because F is an antiderivative of f, by applying the FTC, we can rewrite the above as $F(2) - F(0) = F(3) - F(2)$. We fill in the values for $F(0)$, $F(2)$, and $F(3)$ from the chart, and solve for A:

$$A - 4 = 8 - A$$
$$2A = 12$$
$$A = 6$$

5. B

It follows from the Second Fundamental Theorem that:

$$\frac{d}{dx}\left(\int_x^{x^2} \sqrt{t}\,dt\right) = \frac{d}{dx}\left(\int_c^{x^2} \sqrt{t}\,dt\right) - \frac{d}{dx}\left(\int_c^{x} \sqrt{t}\,dt\right)$$
$$= \sqrt{(x^2)} \cdot \frac{d}{dx}(x^2) - \sqrt{x}$$
$$= x \cdot 2x - \sqrt{x}$$
$$= 2x^2 - \sqrt{x}$$

6. D

To find the critical points of f we need to compute the derivative $\frac{df}{dx}$, and determine

$f'(x) = \frac{d}{dx}\left(\int_{\frac{2\pi}{3}}^{x^2} \cos t\,dt\right)$. It follows from the Second Fundamental Theorem and the chain rule that

$$\frac{d}{dx}\left(\int_{\frac{2\pi}{3}}^{x^2} \cos t\,dt\right) = (\cos(x^2)) \cdot \frac{d}{dx}(x^2) = 2x \cdot \cos x^2.$$

This derivative is defined everywhere, so the only critical points occur where this expression is equal to zero. We solve:

$$f'(x) = 2x \cdot \cos x^2 = 0$$
$$x = 0 \quad \text{or} \quad \cos x^2 = 0$$
$$x^2 = \pm\frac{\pi}{2}, \pm\frac{3\pi}{2}, \pm\frac{5\pi}{2}, \ldots \Rightarrow$$
$$x = \pm\sqrt{\frac{\pi}{2}}, \pm\sqrt{\frac{3\pi}{2}}, \pm\sqrt{\frac{5\pi}{2}}, \ldots$$

Choice (D) matches one of the solutions.

A little test strategy: Instead of solving the equation $f'(x) = 2x \cdot \cos x^2 = 0$, we can solve backward. We can plug each of the answer choices into the equation and see which one gives a true statement.

7. A

By the FTC,

$$\frac{d}{dx}\int_0^x \frac{1}{1+t^4}\,dt = \frac{1}{1+x^4}$$

8. C

Since $G'(x) = g(x)$ for $x > 0$, then by FTC $G(5) - G(0) = \int_0^5 g(x)\,dx$. We know that $G(5) - G(0) = \int_0^5 g(x)\,dx = \int_0^1 g(x)\,dx + \int_1^5 g(x)\,dx$.
From the graph, $\int_0^1 g(x)\,dx = 2$ and we are given $\int_1^5 g(x)\,dx = 8.9$. So $G(5) - G(0) = \int_0^1 g(x)\,dx + \int_1^5 g(x)\,dx = 2 + 8.9 = 10.9$.

9. C

We are given $x^2 - 3x = \int_2^x f(t)\,dt$. It follows that

$$\frac{d}{dx}(x^2 - 3x) = \frac{d}{dx}\int_2^x f(t)\,dt$$

and by the FTC $2x - 3 = f(x)$; therefore $f(4) = 5$.

10. E

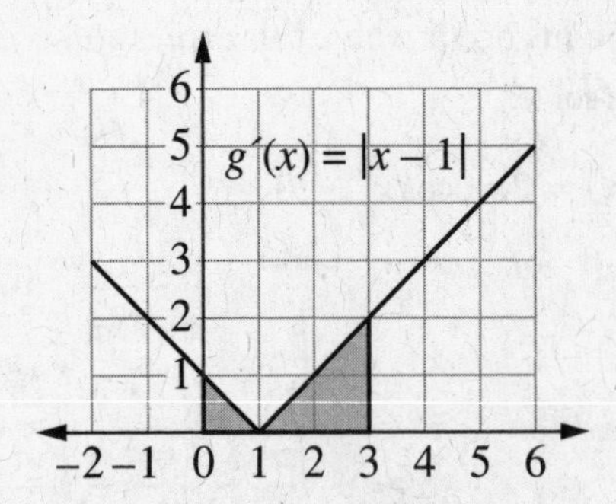

We are given $g(x) = \int_0^x |t-1|\,dt$. It follows that $g'(x) = |x - 1|$ by the FTC. To find the equation of the tangent line to $g(x)$ at x = 3, we need to find $g(3)$ and $g'(3)$. From the graph, $g(3) = \int_0^3 |t-1|\,dt = 2.5$ and $g'(3) = |3 - 1| = 2$. By point-slope, $y - 2.5 = 2(x - 3) \Rightarrow y - 2.5 = 2x - 6 \Rightarrow$ $2y - 5 = 4x - 12 \Rightarrow 4x - 2y = 7$.

11. E

We are given F' is a continuous function for all real x and we want to evaluate

$$\lim_{h\to 0}\left(\frac{1}{h}\int_a^{a+h} F'(x)\,dx\right) = \lim_{h\to 0}\frac{\int_a^{a+h} F'(x)\,dx}{h}.$$

By the FTC, it follows that

$$\lim_{h\to 0}\frac{\int_a^{a+h} F'(x)\,dx}{h} = \lim_{h\to 0}\frac{F(a+h)-F(a)}{h}$$

and by the definition of the derivative at a point

$$\lim_{h\to 0}\frac{F(a+h)-F(a)}{h} = F'(a).$$

12. C

We are given $f(x) = \int_2^x \frac{1}{1+t^3}\,dt$ and it follows that $f'(x) = \frac{d}{dx}\int_2^x \frac{1}{1+t^3}\,dt$, which means that $f'(x) = \frac{1}{1+x^3}$. It follows that $f'(2) = \frac{1}{9}$.

13. D

Given $f(x) = \sqrt{2x+7}$ and $F'(x) = f(x)$ so $F(x) = \int_a^x \sqrt{2t+7}\,dt$. By the FTC, $\int_{-1.5}^{1} \sqrt{2t+7}\,dt = F(1) - F(-1.5)$. We know that $F(1) = 7$ so it follows that

$$F(-1.5) = F(1) - \int_{-1.5}^{1}\sqrt{2t+7}\,dt = 7 - \int_{-1.5}^{1}\sqrt{2t+7}\,dt.$$

We need to evaluate the

$$\int_{-1.5}^{1}\sqrt{2t+7}\,dt = \frac{1}{2}\int_{-1.5}^{1}(2t+7)^{\frac{1}{2}}\,2\,dt = \frac{1}{2}\cdot\frac{2}{3}$$

$$(2t+7)^{\frac{3}{2}}\Big|_{-1.5}^{1} = \frac{1}{3}\left(9^{\frac{3}{2}} - 4^{\frac{3}{2}}\right) = \frac{1}{3}(27-8) = \frac{19}{3}$$

This means the answer is $7 - \frac{19}{3} = \frac{2}{3}$

14. C

By the FTC, $F(x) = \int_a^x F'(t)\,dt$ and since $F'(x) = (3x+1)^{\frac{3}{2}}$ it follows that $F(x) = \int_a^x (3t+1)^{\frac{3}{2}}\,dt$. We integrate so

$$F(x) = \frac{1}{3}\int_a^x (3t+1)^{\frac{3}{2}}\,3\cdot dt = \frac{1}{3}\cdot\frac{2}{5}(3t+1)^{\frac{5}{2}}\Big|_a^x$$

$$= \frac{2}{15}(3x+1)^{\frac{5}{2}} + C.$$

Since $F(1) = 4$ then

$$F(1) = \frac{2}{15}(3\cdot 1+1)^{\frac{5}{2}} + C = 4 \Rightarrow C =$$

$$4 - \frac{2}{15}\cdot 32 = \frac{60}{15} - \frac{64}{15} = -\frac{4}{15}$$

Finally, $F(x) = \dfrac{2(3x+1)^{\frac{5}{2}} - 4}{15}$.

15. C

We want to find $g(x) = \int_a^{2x} f(t)\,dt$. To start, let $h(x) = \int_a^x f(t)\,dt \Rightarrow h'(x) = f(x)$. Now it follows that $g(x) = h(2x) \Rightarrow g'(x) = 2h'(2x)$ by the chain rule. Therefore $g'(3) = 2f(6) = -2$.

FREE-RESPONSE ANSWER

16. (a) By the definition of g and by the definition of the definite integral, $g(1)$ is the area under the curve $f(x)$ on the interval $-2 \le x \le 1$. This area consists of the sums of the three areas shown below:

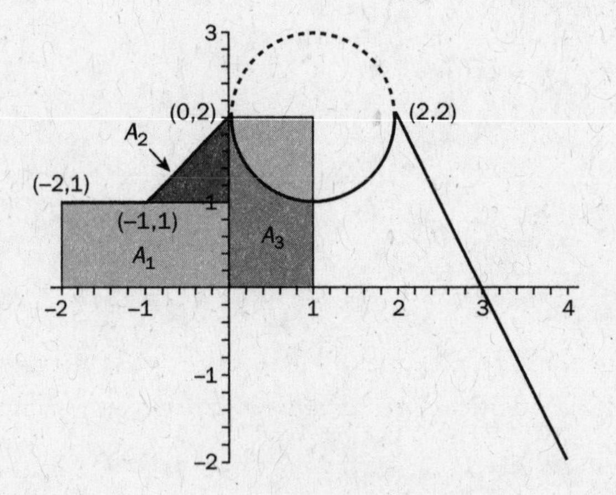

On this graph, region A_1 is a rectangle of length 2 and height 1; its area is $2 \cdot 1 = 2$. Region A_2 is a triangle with base 1 and height 1; its area is $\frac{1}{2} \cdot 1 \cdot 1 = \frac{1}{2}$. Region A_3 is the light gray rectangle minus one-fourth of the circle. The gray rectangle has base 1 and height 2; its area is $1 \cdot 2 = 2$. We see on the graph that the radius of the circle is 1. Therefore the area of the entire circle is $\pi \cdot 1^2 = \pi$, so the area of the shaded quarter circle is $\frac{\pi}{4}$. The area of A_3 is $2 - \frac{\pi}{4}$.

Therefore:

$$g(1) = \int_{-2}^{1} f(x)\,dx = \text{Area of } A_1 + \text{Area of } A_2 + \text{Area of } A_3$$

$$= 2 + \frac{1}{2} + \left(2 - \frac{\pi}{4}\right) = \frac{18 - \pi}{4}$$

To find $g'(1)$, we apply the Second Fundamental Theorem, which tells us that:

$$g'(x) = \frac{d}{dx}\left(\int_{-2}^{x} f(x)\,dx\right) = f(x).$$

Therefore, $g'(1) = f(1)$. We see on the graph that $f(1) = 1$, therefore $g'(1) = 1$.

(b) The function g attains a relative maximum at the points where it changes from increasing to decreasing, which correspond to the points at which the sign of the derivative of g changes from positive to negative. We saw above that by the Second Fundamental Theorem, $g'(x) = f(x)$. We see on the graph of f that the sign of f changes from positive to negative only at $x = 3$ on the interval $-2 < x < 4$. Therefore, g has a relative maximum at $x = 3$ and at no other point on this interval.

17. (a) By the FTC, $A'(x) = f(x)$ so $A'(1) = f(1) = -2$.

(b) To start, we need to recognize that -4 is a zero because

$$A(-4) = \int_{-4}^{-4} f(t)\,dt = 0$$

$\therefore x = -4$ is a solution

To find other solutions of the equation $A(x) = 0$, we need to carefully look to the graph and recognize that congruent regions above and below the x-axis will produce a zero. To that end,

$$\int_{-4}^{-1} f(t)\,dt = 2 + \pi \text{ and}$$

$$\int_{-1}^{2} f(t)\,dt = -\pi - 2 \text{ so}$$

$$A(2) = \int_{-4}^{2} f(t)\,dt = \int_{-4}^{-1} f(t)\,dt + \int_{-1}^{2} f(t)\,dt = 2 + \pi + (-\pi - 2) = 0$$

$\therefore x = 2$ is a solution

Similarly, we consider the triangle regions to find a third zero.

$$\int_{-6}^{-5} f(t)\,dt = -\frac{1}{2}\,base \cdot height = -\frac{1}{2}(1)(2) = -1 \text{ and}$$

$$\int_{-5}^{-4} f(t)\,dt = \frac{1}{2}\,base \cdot height = \frac{1}{2}(1)(2) = 1 \text{ so}$$

$$\int_{-6}^{-4} f(t)\,dt = \int_{-6}^{-5} f(t)\,dt + \int_{-5}^{-4} f(t)\,dt = -1 + 1 = 0$$

$$A(-6) = \int_{-4}^{-6} f(t)\,dt = -\int_{-6}^{-4} f(t)\,dt = -0 = 0$$

$\therefore x = -6$ is a solution

The solutions are $x = -6, -4, 2$.

(c) To find the exact coordinates of the relative minimum point of $y = A(x)$ we start with the First Derivative Test. Create a table of $A'(x) = f(x)$.

x	$(-6, -5)$	-5	$(-5, -1)$	-1	$(-1, 3)$
$A'(x)$	negative	0	positive	0	negative
$A(x)$	decreases	rel. min.	increases	rel. max.	decreases

The relative minimum is the point $(-5, A(-5))$. We need to evaluate $A(-5)$.

$A(-5) = \int_{-4}^{-5} f(t)\,dt = -\int_{-5}^{-4} f(t)\,dt = -1.$

Therefore the relative minimum point is $(-5, -1)$.

(d) To determine an equation of the tangent line to the graph of A where $x = 1$, we need $(1, A(1))$ and $A'(1)$ $A(1) = \int_{-4}^{1} f(t)\,dt = \int_{-4}^{-3} f(t)\,dt + \int_{-3}^{1} f(t)\,dt = 2 + 0 = 2$ and $A'(1) = f(1) = -2$

The tangent line is $y - 2 = -2(x - 1)$.

CHAPTER 21: TECHNIQUES OF ANTIDIFFERENTIATION

IF YOU LEARN ONLY FOUR THINGS IN THIS CHAPTER . . .

1. For $r \neq 1$, $\int x^r \, dx = \frac{x^{r+1}}{r+1} + C$.
2. The rules of the substitution game say that any x disappears, and that there must be a du.
3. The integration by parts formula: $\int u \, dv = uv - \int v \, du$.
4. Graphically, an improper integral has the same interpretation as a definite integral: An improper integral measures the area under the graph of the function where one endpoint is at infinity.

Now that we can use the FTC (Fundamental Theorem of Calculus) as a powerful tool for computing the area under the curve, we should put it to work for us. To apply the FTC to a definite integral $\int_a^b f(x) \, dx$, we need to know an antiderivative of f. In this chapter we'll build up a library of antiderivatives of basic functions and we'll develop one more advanced technique for computing indefinite integrals (which, remember, is just an antiderivative plus a constant).

FINDING ANTIDERIVATIVES

Let's start simple. What is $\int 2x \, dx$? The indefinite integral asks a question: "What function has the derivative $2x$?" Well, the only way to get a polynomial as the derivative is to start with a polynomial. If the derivative has degree 1, then the original polynomial had degree 2, so the answer has something to do with x^2. This is, in fact, the answer, because $\frac{d}{dx}(x^2) = 2x$; x^2 is an antiderivative of $2x$, so $\int 2x\ dx = x^2 + C$.

What about $\int x\ dx$? Going through the same thought process we used above, we know that the answer has something to do with x^2. But we saw above that the derivative of x^2 isn't x—it's $2x$. The function x^2 isn't quite the antiderivative we're looking for; it's off by a multiplicative constant of 2. Well, that's OK. Using the properties of derivatives and integrals, we can always fudge constants. If x^2 is too big by a factor of 2, then the correct antiderivative should be $\frac{x^2}{2}$. We also always have a way of checking whether we've found the correct antiderivative—we differentiate. That is, we check $\frac{d}{dx}\left(\frac{x^2}{2}\right) = \frac{1}{2} \cdot 2x = x$. Thus we know that $\frac{x^2}{2}$ is the antiderivative we're seeking, i.e.,

$$\int x\ dx = \frac{x^2}{2} + C.$$

We could keep on working backward from derivatives, but it makes sense to streamline this process. Let's develop a procedure.

AP EXPERT TIP

Never forget the + C when calculating an indefinite integral. Doing so is a great way to lose easy points on the FRQ section and fall for trap answers in the multiple choice questions. The best way to avoid forgetting is to make it second-nature. Don't cut corners when you practice. Always write the + C.

CONSIDER $\int x^r\ dx$

Using the same reasoning we used to find $\int x\ dx$, we can compute $\int x^r\ dx$, as long as $r \neq -1$. The answer must have something to do with x^{r+1}. When we differentiate x^{r+1} we get $(r + 1)x^r$, which is too big by a multiplicative factor of $(r + 1)$. Therefore the antiderivative of x^r is $\frac{x^{r+1}}{r+1}$.

Theorem:

For $r \neq -1$, $\int x^r\ dx = \frac{x^{r+1}}{r+1} + C$.

We can verify that this is, in fact, correct, by computing the derivative $\frac{d}{dx}\left(\frac{x^{r+1}}{r+1}\right) = \frac{1}{r+1} \cdot (r+1)x^r = x^r$, confirming that $\frac{x^{r+1}}{r+1}$ *is* an antiderivative of x^r.

Take time to check! We have a way to make sure that our antiderivative is correct. If $\int f(x)\ dx = F(x) + C$, then $F'(x) = f(x)$. Take the derivative of the answer to make sure you've computed correctly.

This rule doesn't work for $r = -1$ because the rule for derivatives $\frac{d}{dx}(x^r) = rx^{r-1}$ was valid for all r *except* $r = 0$. The formula $\int 1\, dx$ tells us that $\int 1\, dx = x + C$, which makes sense because the derivative of x is 1.

From the properties of derivatives or integrals, it follows that $\int m\, dx = mx + C$. This also makes sense; if we think of the integral as the accumulated change in a function, we can interpret this result as saying that if the rate of change of a function is constant, equal to m, then the function is a line. The constant of integration, C, is the y-intercept of the line—in a real-life situation, it's the quantity present when $x = 0$.

We saw above that we cannot apply the rule $\int x^r\, dx = \frac{x^{r+1}}{r+1} + C$ to $r = -1$, i.e., to compute the indefinite integral $\int \frac{1}{x}\, dx$. So what is $\int \frac{1}{x}\, dx$? The indefinite integral asks the question, "What function has the derivative $\frac{1}{x}$?" Can you think of a function the derivative of which is $\frac{1}{x}$? Did you say $\ln x$? Good for you—almost. The natural logarithm function is only defined for $x > 0$ and we want an antiderivative for $\frac{1}{x}$ that is defined for all values of x in the domain of $\frac{1}{x}$ (i.e., for all $x \neq 0$), so we use the antiderivative $\ln|x|$ for $\frac{1}{x}$: $\int \frac{1}{x}\, dx = \ln|x| + C$.

Why is this the case? For $x > 0$, $\ln|x| = \ln x$, so $\frac{d}{dx}(\ln|x|) = \frac{d}{dx}(\ln x) = \frac{1}{x}$. For $x < 0$, $\ln|x| = \ln(-x)$. We differentiate this function using the chain rule: $\frac{d}{dx}(\ln(-x)) = \frac{1}{-x}(-1) = \frac{1}{x}$.

COMPUTING ANTIDERIVATIVES DIRECTLY

When we need to find an antiderivative of a function f—say, to compute an indefinite or a definite integral, $\int f(x)\, dx$ or $\int_a^b f(x)\, dx$—the first thing we ask is "What function has $f(x)$ as its derivative?" We want to build up a library of basic functions for which we know the answer to this question. For other, more complicated functions, we will need to develop more advanced techniques.

For example, $\int \cos x\, dx = \sin x + C$ because $\frac{d}{dx}(\sin x) = \cos x$; $\int \sin x\, dx = -\cos x + C$, because $\frac{d}{dx}(\cos x) = -\sin x \Rightarrow \frac{d}{dx}(-\cos x) = \sin x$.

We could go on like this, randomly picking functions and figuring out their antiderivatives, but it makes more sense to organize the information in a table.

Because the derivative is...	The indefinite integral is...
$\frac{d}{dx}(x) = 1$	$\int 1\,dx = x + C$
$\frac{d}{dx}\left(\frac{x^{r+1}}{r+1}\right) = x^r,\ r \neq -1$	$\int x^r\,dx = \frac{x^{r+1}}{r+1} + C$
$\frac{d}{dx}(\sin x) = \cos x$; $\frac{d}{dx}(-\cos x) = \sin x$ $\frac{d}{dx}(\tan x) = \sec^2 x$; $\frac{d}{dx}(-\cot x) = \csc^2 x$ $\frac{d}{dx}(\sec x) = \sec x \tan x$; $\frac{d}{dx}(-\csc x) = \csc x \cot x$	$\int \cos x\,dx = \sin x + C$; $\int \sin x\,dx = -\cos x + C$ $\int \sec^2 x\,dx = \tan x + C$; $\int \csc^2 x\,dx = -\cot x + C$ $\int \sec x \tan x\,dx = \sec x + C$; $\int \csc x \cot x\,dx = -\csc x + C$
$\frac{d}{dx}(e^x) = e^x$	$\int e^x\,dx = e^x + C$
$\frac{d}{dx}(\ln x) = \frac{1}{x}$	$\int \frac{1}{x}\,dx = \ln\lvert x\rvert + C$
$\frac{d}{dx}\left(\frac{b^x}{\ln b}\right) = b^x$ for $b > 0$	$\int b^x\,dx = \frac{b^x}{\ln b} + C$
$\frac{d}{dx}(\arctan x) = \frac{1}{1+x^2}$ $\frac{d}{dx}(\arcsin x) = \frac{1}{\sqrt{1-x^2}}$ $\frac{d}{dx}(\text{arcsec}\, x) = \frac{1}{\lvert x\rvert\sqrt{x^2-1}}$	$\int \frac{1}{1+x^2}\,dx = \arctan x + C$ $\int \frac{1}{\sqrt{1-x^2}}\,dx = \arcsin x + C$ $\int \frac{1}{x\sqrt{x^2-1}}\,dx = \text{arcsec}\,\lvert x\rvert + C$

We can use this table, combined with the properties of definite and indefinite integrals, to answer many questions involving definite and indefinite integrals.

For example:

$$\int_{\frac{\pi}{4}}^{\frac{\pi}{3}} 2\sec^2 x\,dx \underset{\substack{\text{properties of}\\\text{the definite}\\\text{integral}}}{=} 2\int_{\frac{\pi}{4}}^{\frac{\pi}{3}} \sec^2 x\,dx \underset{\substack{\tan x\text{ is an}\\\text{antiderivative}\\\text{of }\sec^2 x}}{=} 2\cdot \tan x\Big|_{\frac{\pi}{4}}^{\frac{\pi}{3}} = 2\tan\frac{\pi}{3} - 2\tan\frac{\pi}{4} = 2\sqrt{3} - 2.$$

WHEN *F* IS COMPLICATED: THE SUBSTITUTION GAME

Suppose we want to compute the indefinite integral $\int 2x\cos x^2 dx$. Someone who is *really* good at recognizing antiderivatives could compute this directly. By the chain rule,

$$\frac{d}{dx}(\sin x^2) = \underbrace{(\cos x^2)}_{\substack{\text{derivative of sin} \\ \text{evaluated at } x^2}} \cdot \underbrace{2x}_{\substack{\text{derivative} \\ \text{of } x^2}} = 2x\cos x^2, \text{ so } \int 2x\cos x^2 dx = \sin x^2 + C$$

What if we're *not* really good at computing antiderivatives directly?

There is a method, called *substitution*, which works like the chain rule in reverse. It allows us to simplify complicated antiderivatives and transform them into something we can compute directly.

The motivation behind integration by substitution is the derivative

$$\frac{d}{dx}(F(u(x))) \underset{\text{chain rule}}{=} F'(u(x)) \cdot u'(x), \text{ which tells us that } \int F'(u(x)) \cdot u'(x)\ dx = F(u(x)) + C.$$

If we can recognize that an integrand—that is a function whose antiderivative we are seeking—is of this form, then we can use substitution to compute the antiderivative.

We mentioned earlier that variables are just place holders. We can use variables in creative ways to make an integral easier to work with. In the indefinite integral $\int 2x\cos x^2 dx$, we can set

$u = x^2$. Then $\frac{du}{dx} = 2x$, or we can write this in the form $du = 2x\ dx$. We can then rearrange the integral above as $\int 2x\cos x^2 dx = \int \cos \underbrace{x^2}_{u} \cdot \underbrace{2x\,dx}_{du} = \int \cos u\ du.$

The integral written in terms of u is something we can integrate directly.

$$\int \cos u\ du = \sin u + C$$

To recover the x we substitute back

$$\int \cos u\ du = \sin u + C = \sin x^2 + C$$

The substitution method gives us a way to recognize the pieces of the chain rule.

Here's another example: Compute $\int \frac{(\ln x)^{\frac{3}{2}}}{x} dx$. You recall that the derivative of $\ln x$ is $\frac{1}{x}dx$, which is in the integrand. So this integral is begging for a substitution! Set $u = \ln x$. Then $du = \frac{1}{x}dx$, and we rearrange the integral as $\int u^{\frac{3}{2}}\ du = \frac{2}{5}u^{\frac{5}{2}} + C = \frac{2}{5}(\ln x)^{\frac{5}{2}} + C.$

MANIPULATING CONSTANTS

The example above was a straight application of the chain rule in reverse. Sometimes things are not presented quite so cleanly. For example, computing $\int xe^{x^2}\ dx$ directly is tricky. We can try to make the u-substitution $u = x^2$; then $du = 2x\,dx$, which is too bad for us, because we don't have a $2x$ under the integrand. We only have an x. On second thought, maybe it isn't too bad. Using the properties of the indefinite integral, we can rewrite $\int xe^{x^2}\,dx$ as:

$$\int \underbrace{\frac{1}{2}\cdot 2}_{\text{we've multiplied by one!}}\ xe^{x^2}\,dx$$

$$\underset{\text{properties of integrals}}{=} \frac{1}{2}\int 2xe^{x^2}\,dx$$

We can always fudge with *constants* this way.

Now we *can* use substitution. We rearrange terms and make the u-substitution:

$$\frac{1}{2}\int 2xe^{x^2}\,dx = \frac{1}{2}\int e^{\overbrace{x^2}^{u}}\cdot \underbrace{2x\,dx}_{du} = \frac{1}{2}\int e^u\,du = \frac{1}{2}e^u + C$$

Substituting back we find:

$$\frac{1}{2}e^u + C = \frac{1}{2}e^{x^2} + C$$

Think of solving indefinite integrals by this method as the "Substitution Game." The only rules of the game are that all of the x's must disappear and there must be a du in the new integral. The goal of the game is to wind up with an integral that you can compute directly.

WHICH SUBSTITUTION SHOULD I MAKE?

Sometimes there is more than one choice for a substitution, and sometimes you may want to substitute more than once—the super-zoom-whammy-chain-rule-in-reverse.

For example, to integrate $\int \sin 3\theta \cdot \cos^4 3\theta\ d\theta$ we can substitute in stages. (If the θ in the problem makes you queasy, go ahead and replace it with an x-variable.)

We start with the substitution:

$$u = 3\theta \quad \text{so}$$
$$du = 3d\theta$$

We rewrite

$$\int \underbrace{\frac{1}{3}\cdot 3}_{\text{multiply by 1}}\ \sin 3\theta\cdot\cos^4 3\theta\,d\theta \underset{\text{properties of integral}}{=} \frac{1}{3}\int \sin \underbrace{3\theta}_{u}\cdot\cos^4 \underbrace{3\theta}_{u}\cdot \underbrace{3d\theta}_{du} = \frac{1}{3}\int \sin u\cdot\cos^4 u\,du$$

This looks better. Can we integrate it directly? Maybe—it depends on how good we are at recognizing antiderivatives. If we can't recognize this as any particular antiderivative, we can play the substitution game again.

$$\frac{1}{3}\int \sin u \cdot \cos^4 u \, du \qquad \text{we set} \qquad \begin{aligned} w &= \cos u \\ dw &= -\sin u \, du \end{aligned}$$

And rewrite:

$$\begin{aligned}
\frac{1}{3}\int \sin u \cdot \cos^4 u \, du &= -1 \cdot \frac{1}{3}\int -1 \cdot \sin u \cdot \cos^4 u \, du \\
&= -\frac{1}{3}\int \cos^4 u(-1 \cdot \sin u \, du) \\
&= -\frac{1}{3}\int \Big(\underbrace{\cos u}_{w}\Big)^4 \left(\frac{-\sin u \, du}{dw}\right) \\
&= -\frac{1}{3}\int w^4 \, dw = -\frac{1}{3} \cdot \frac{w^5}{5} + C = -\frac{1}{15}w^5 + C
\end{aligned}$$

Now we substitute back twice.

$$-\frac{1}{15}w^5 + C = -\frac{1}{15}\cos^5 u + C = -\frac{1}{15}\cos^5 3\theta + C$$

It would also have been possible to make one substitution right at the beginning—$u = \cos 3\theta$—but substituting twice works just fine.

Here's another example which we'll revisit in the next chapter: Compute $\int \sin^3 x \, dx$.

First, break up $\sin^3 x$ to $\sin^2 x \cdot \sin x$. Next, substitute $\sin^2 x = 1 - \cos^2 x$. We now have $\int \sin^3 x = \int (1 - \cos^2 x)\sin x \, dx = \int \sin x \, dx + \int \cos^2 x(-\sin x) \, dx$. The first integral is $-\cos x$. We can use substitution to figure out the second integral.

Set $u = \cos x$ so $du = -\sin x \, dx$. This integral then becomes $\int u^2 \, du = \frac{1}{3}u^3 + C = \frac{1}{3}\cos^3 x + C$. Therefore the final answer is

$$\int \sin^3 x = \frac{1}{3}\cos^3 x - \cos x + C.$$

Getting good at substitution takes some practice. So practice! Open your textbook to the chapter on integration by substitution and do a bunch of problems and then do some more. Keep doing problems until you start to get the hang of which substitution will help in a given situation.

MANIPULATING WITH CONSTANTS IS GOOD; MANIPULATING ANYTHING ELSE IS BAD

Just about everyone who has ever sat through calculus has tried to do something like the following, at least once—but you need to make sure that you don't do it on the AP exam.

Imagine that you're asked to compute $\int e^{x^2} dx$. Mr. or Ms. Calculus will say, "Gee, let's make a u-substitution; $u = x^2$ so $du = 2x\,dx$. OK, now I need to multiply by 1, in the form of $\frac{2x}{2x}$."

$$\int e^{x^2} dx = \int 2x \cdot \frac{1}{2x} e^{x^2} dx$$

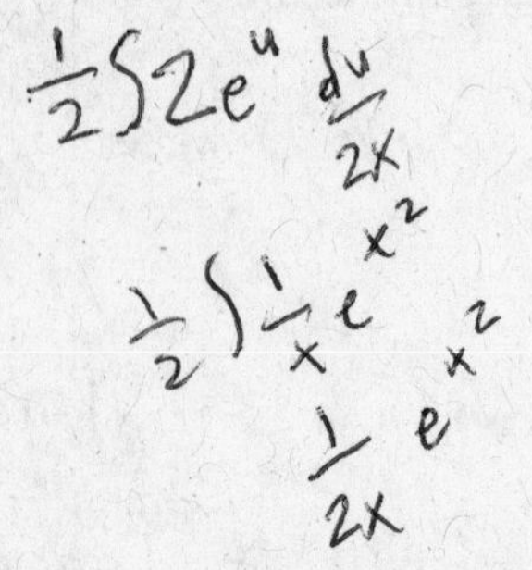

"Now I'll pull the $2x$ out of the integrand...using..."

$$\int e^{x^2} dx = 2x \int \frac{1}{2x} e^{x^2} dx$$

WAIT—*STOP!* Just what property of the indefinite integral were you planning to use to pull the $2x$ outside of the integrand? There isn't any such property. We can't do it. We can only pull constants out of the integrand, not functions . *Not ever.* Don't do it.

Backwards Substitutions

The rules of the substitution game say the x's have to disappear and that there must be a du. That's it. We can use the limited rules to our advantage.

For example, how do we integrate $\int x^2 \sqrt{x-1}\, dx$?

First, we see if we can find an antiderivative directly. Nope—unless you're a genius at this. We could try the substitution $u = x^2$, but that would blow the rules of the substitution game. What if we try $u = x - 1$?

$$\int x^2 \sqrt{\underbrace{x-1}_{u}}\, \underbrace{dx}_{du}$$

Well, we've taken care of the $\sqrt{x-1}$ and the dx, but what about the x^2? This is where the rules (or lack of rules) can help us. If $u = x - 1$, then $u + 1 = x$. So we can replace x^2 by $(u + 1)^2$. That is:

$$\int \underbrace{x^2}_{(u+1)^2} \sqrt{\underbrace{x-1}_{u}}\, \underbrace{dx}_{du} = \int (u+1)^2 \cdot \sqrt{u}\, du$$

Have we won the substitution game? Well we followed the rules—x is gone and we have a du. But, can we integrate this expression? We can! First we rewrite

$$\int (u-1)^2 \cdot \sqrt{u}\, du = \int (u^2 + 2u + 1) \cdot u^{\frac{1}{2}}\, du$$
$$= \int u^{\frac{5}{2}} + 2u^{\frac{3}{2}} + u^{\frac{1}{2}}\, du$$

We can compute this integral directly.

$$= \int u^{\frac{5}{2}} + 2u^{\frac{3}{2}} + u^{\frac{1}{2}}\, du = \frac{2}{7}u^{\frac{7}{2}} + 2 \cdot \frac{2}{5}u^{\frac{5}{2}} + \frac{2}{3}u^{\frac{3}{2}} + C$$

$$= \frac{2}{7}u^{\frac{7}{2}} + \frac{4}{5}u^{\frac{5}{2}} + \frac{2}{3}u^{\frac{3}{2}} + C$$

Now we substitute back.

$$= \frac{2}{7}(x-1)^{\frac{7}{2}} + \frac{4}{5}(x-1)^{\frac{5}{2}} + \frac{2}{3}(x-1)^{\frac{3}{2}} + C$$

Substitution in Definite Integrals*

When we compute a definite integral using substitution, we have two options. Under the first option, we can use substitution to find the antiderivative of the original function, substitute back, and plug in the original limits of integration.

For example, to compute $\int_0^{\frac{\pi^2}{4}} \frac{\cos\sqrt{x}}{\sqrt{x}}\, dx$ we can make the u-substitution $\begin{aligned} u &= \sqrt{x} \\ du &= \frac{1}{2\sqrt{x}}dx \end{aligned}$.

We then rewrite and rearrange the original integral.

$$\int_0^{\frac{\pi^2}{4}} \underbrace{2 \cdot \frac{1}{2}}_{\substack{\text{multiply} \\ \text{by } 1}} \frac{\cos\sqrt{x}}{\sqrt{x}}\, dx = 2\int_0^{\frac{\pi^2}{4}} \underbrace{\cos\sqrt{x}}_{u} \cdot \underbrace{\frac{1}{2\sqrt{x}}dx}_{du}$$

$$= 2\int_{x=0}^{x=\frac{\pi^2}{4}} \cos u\, du$$

We write the limits of integration as $x = 0$ and $x = \frac{\pi^2}{4}$ to remind ourselves that these values refer to x and not to u. We can compute the antiderivative of $\cos u$ (it is $\sin u$) and then substitute back:

$$= 2\int_{x=0}^{x=\frac{\pi^2}{4}} \cos u\, du = 2\left[\sin u\Big|_{x=0}^{x=\frac{\pi^2}{4}}\right] = 2\left[\sin\sqrt{x}\Big|_{x=0}^{x=\frac{\pi^2}{4}}\right] = 2\left(\sin\sqrt{\frac{\pi^2}{4}} - \sin\sqrt{0}\right)$$

$$= 2\sin\frac{\pi}{2} - 2\sin 0 = 2$$

Change of Variables*

With definite integrals, we also have the option of putting *everything*, including the limits of integration, in terms of u and getting rid of the x's completely. We call this method change of variables.

* *BC content*

In the example above we substituted, then substituted back to compute the definite integral. Let's do this problem again, this time changing variables.

We start out the same way, with $\int_0^{\frac{\pi^2}{4}} \frac{\cos\sqrt{x}}{\sqrt{x}}\,dx$, and make the substitution $\begin{aligned} u &= \sqrt{x} \\ du &= \frac{1}{2\sqrt{x}}\,dx \end{aligned}$.

We use the equation relating *x* and *u* to change the limits of integration to be in terms of *u*. Because $u = \sqrt{x}$, the lower limit of integration $x = 0$ becomes $u = \sqrt{0} = 0$. The upper limit of integration $x = \frac{\pi^2}{4}$ becomes $u = \sqrt{\frac{\pi^2}{4}} = \frac{\pi}{2}$.

When we substitute *u* and *du* into the original integral, we also replace the limits of integration with their corresponding *u*-values. The integral becomes:

$$\int_{x=0}^{x=\frac{\pi^2}{4}} \frac{\cos\sqrt{x}}{\sqrt{x}}\,dx \Rightarrow 2\int_{u=0}^{u=\frac{\pi}{2}} \cos u\,du = 2\left[\sin u\Big|_{u=0}^{u=\frac{\pi}{2}}\right] = 2\left(\sin\frac{\pi}{2} - \sin 0\right) = 2$$

When we solve a definite integral by changing variables, we throw the *x*'s away completely.

INTEGRATION BY PARTS*

Clearly, the substitution game is a very useful trick when evaluating integrals. However, not all integrals can be solved using the substitution method. In this section you will learn some new tricks that will substantially increase the types of integrals that you can solve. The first of such methods is integration by parts.

Integration by parts can be thought of as the antidifferentiation of the product rule. The formula we will use is $\int u\,dv = uv - \int v\,du$. If we let $u = f(x)$ and $v = g(x)$ then this formula is written $\int f(x)g'(x)\,dx = f(x)g(x) - \int g(x)f'(x)\,dx$, which, when we differentiate both sides, becomes $f(x)g'(x) = \frac{d}{dx}[f(x)g(x)] - g(x)f'(x)$. After rearranging, this gives us back the product rule:

$$\frac{d}{dx}[fg] = gf' + fg'$$

Let's use the initial formula to solve $\int xe^x\,dx$. In this integrand we have two functions. To use the integration by parts formula, we need to decide which function is *u* and which function is *dv*. Much like the substitution method, this choice really just takes practice. If the formula doesn't work the way you think it should with your initial guess, then just guess again. In general, however, it helps to keep in mind that we would like *u* to be simpler after *differentiating* and we want *dv* to be simpler after *integrating*.

* *BC content*

With this motivation, let's take $u = x$ and $dv = e^x\,dx$. Therefore,

$$u = x \qquad dv = e^x dx$$

so, $$du = dx \qquad v = \int dv = \int e^x\,dx = e^x$$

Then using the integration by parts formula, we have:

$$\begin{aligned}\int xe^x\,dx &= xe^x - \int e^x\,dx \\ &= xe^x - e^x \\ &= e^x(x-1)\end{aligned}$$

and we're done.

If at First You Don't Succeed: A Useful Trick

Let's look at another exercise that contains a very helpful trick. Suppose we are asked to find $\int e^x \cos x\,dx$. Neither function becomes simpler after we differentiate or integrate, so it isn't clear how we should define u and dv. When this happens just make an arbitrary assignment and see what happens. Let $u = \cos x$ and $dv = e^x\,dx$. Thus

$$u = \cos x \qquad dv = e^x dx$$

so, $$du = -\sin x\,dx \qquad v = e^x$$

Using the integration by parts formula we get:

$$\int e^x \cos x\,dx = e^x \cos x - \int e^x(-\sin x)\,dx = e^x \cos x + \int e^x \sin x\,dx$$

However, the integral on the right still needs to be evaluated. We can try integration by parts on $\int e^x \sin x\,dx$.

Let:

$$u_1 = \sin x \qquad dv_1 = e^x dx$$

so, $$du_1 = \cos x \qquad v_1 = e^x$$

We get $\int e^x \sin x\,dx = e^x \sin x - \int e^x \cos x\,dx$; when we plug this into our equation for $\int e^x \cos x\,dx$, we get:

$$\begin{aligned}\int e^x \cos x\,dx &= e^x \cos x + \int e^x \sin x\,dx \\ &= e^x \cos x + e^x \sin x - \int e^x \cos x\,dx\end{aligned}$$

Initially, it doesn't look like we've made any progress—it looks more like this approach will just lead us in circles. However, if we add $\int e^x \cos x\,dx$ to both sides, we get:

$$2\int e^x \cos x\,dx = e^x \cos x + e^x \sin x$$

And thus, dividing by 2, $\int e^x \cos x\,dx = \frac{1}{2}(e^x \cos x + e^x \sin x) + C$ and we're done!

It All Depends on How You Express the Integral: Another Useful Trick

Let's look at another example that uses integration by parts. Suppose we want to find $\int \ln x\,dx$. At first glance it may not be clear that we should *or even can* use integration by parts to compute this integral. However, to help you see that there are two functions in the integrand let's write $\int \ln x \cdot 1\,dx$.

To solve by integration by parts we let:

$$u = \ln x \qquad dv = dx$$

and so, $$du = \frac{1}{x}dx \qquad dv = x$$

Therefore we get:

$$\int \ln x\,dx = x\ln x - \int x \cdot \tfrac{1}{x}\,dx$$
$$= x\ln x - \int dx$$
$$= x\ln x - x + C$$

This example, in particular, should illustrate the power of integration by parts.

Let's revisit an example we used in the substitution section: Compute $I = \int \sin^3 x\,dx$.

We set $u = \sin^2 x$; $dv = \sin x\,dx$. Then $du = 2\sin x\cos x$ and $v = -\cos x$. Therefore we get:

$$I = \int \sin^3 x\,dx = -\sin^2 x\cos x - \int -\cos x(2\sin x\cos x)\,dx = -\sin^2 x\cos x + 2\int \cos^2 x\sin x\,dx.$$

Now using $\cos^2 x = 1 - \sin^2 x$, we get:

$$I = -\sin^2 x\cos x + 2\int (1-\sin^2 x)\sin x\,dx = -\sin^2 x\cos x + 2\int \sin x\,dx - 2\int \sin^3 x\,dx \text{ or}$$
$$I = -\sin^2 x\cos x - 2\cos x - 2I$$

The rest is simple algebra:

$$3I = -\sin^2 x\cos x - 2\cos x$$
$$I = -\frac{1}{3}\sin^2 x\cos x - \frac{2}{3}\cos x + C = -\frac{1}{3}(1-\cos^2 x)\cos x - \frac{2}{3}\cos x + C$$
$$= -\frac{1}{3}\cos x + \frac{1}{3}\cos^3 x - \frac{2}{3}\cos x + C = \frac{1}{3}\cos^3 x - \cos x + C$$

Here's one final example: Compute $\int \tan^{-1} x\,dx$.

Let $u = \tan^{-1}x$; $dv = dx$. Then $du = \dfrac{dx}{1+x^2}$ and $v = x$. Therefore,

$$\int \tan^{-1} x\,dx = x\tan^{-1} x - \int \frac{x\,dx}{1+x^2}.$$

Now, we can solve the remaining integral by substitution, or you can immediately see the answer:

$$\int \tan^{-1} x\, dx = x\tan^{-1} x - \frac{1}{2}\ln(1+x^2) + C$$

ANTIDERIVATIVES BY SIMPLE PARTIAL FRACTIONS*

The last technique we will cover in this chapter is that of partial fractions. The idea is that we can take a fractional integrand that we don't know how to integrate and break it up into the sum of two simpler integrands that we *do* know how to integrate.

Consider the integral $\int \frac{7x+39}{x^2+9x+14}\,dx$. We surely do not know a formula to compute this integral; neither substitution nor integration by parts will be much help. Instead, let's work at splitting up the fractional integrand into two smaller fractions. Notice that by factoring the denominator we get $\frac{7x+39}{x^2+9x+14} = \frac{7x+39}{(x+2)(x+7)}$.

Now, the goal is to split this fraction into two smaller fractions. Suppose we have such a decomposition. Let A and B be two constants such that $\frac{7x+39}{(x+2)(x+7)} = \frac{A}{x+2} + \frac{B}{x+7}$.

This is exactly the form we would like because both of these fractions are easy to integrate. To determine the value of the constants, we simply cross multiply and create a system of equations that we can solve. We get $\frac{A}{x+2} + \frac{B}{x+7} = \frac{Ax+7A+Bx+2B}{(x+2)(x+7)} = \frac{(A+B)x+(7A+2B)}{(x+2)(x+7)}$.

Now, if $\frac{7x+39}{(x+2)(x+7)} = \frac{A}{x+2} + \frac{B}{x+7}$, then we must have $A + B = 7$ and $7A + 2B = 39$. Solving this system of two equations and two unknowns, we get that $A = 5$ and $B = 2$. Therefore, we can simplify our initial integral to get:

$$\begin{aligned}\int \frac{7x+39}{x^2+9x+14}\,dx &= \int \frac{5}{x+2} + \frac{2}{x+7}\,dx \\ &= 5\int \frac{1}{x+2}\,dx + 2\int \frac{1}{x+7}\,dx \\ &= 5\ln(x+2) + 2\ln(x+7) + C\end{aligned}$$

In this example the denominator was the product of two linear factors; as a result we had two partial fractions. If, instead, there were three factors in the denominator, we would follow the same procedure, only we would have a system of three equations and three unknowns. For example, we would set $\frac{x+3}{x(x+2)(x+7)} = \frac{A}{x} + \frac{B}{x+2} + \frac{C}{x+7}$, cross multiply, and solve. What if we had a denominator like $\frac{1}{x^2(x-2)^3}$? You might think of creating partial fractions with denominators x^2 and $(x-2)^3$. However, we need to consider all factors in establishing partial fractions. Let's say we

* *BC content*

wish to compute $\int \frac{(x^3 - 1)\,dx}{x^2(x-2)^3}$. We need to write the fraction as the sum of the following partial fractions:

$$\frac{(x^3 - 1)}{x^2(x-2)^3} = \frac{A}{x^2} + \frac{B}{x} + \frac{C}{(x-2)^3} + \frac{D}{(x-2)^2} + \frac{E}{(x-2)}.$$

To solve this, multiply both sides by $x^2(x - 2)^3$ to get:

$$x^3 - 1 = A(x-2)^3 + Bx(x-2)^3 + Cx^2 + Dx^2(x-2) + Ex^2(x-2)^2$$

We can easily compute C and A by substituting 2 and 0 for x, respectively. You should get $C = \frac{7}{4}$ and $A = \frac{1}{8}$. You can substitute these values and expand powers to determine the other values, if you care to try. You should get $B = \frac{3}{16}$, $D = \frac{5}{4}$, and $E = -\frac{3}{16}$.

ANTIDERIVATIVES BY IMPROPER INTEGRALS*

So far we have only considered indefinite integrals (i.e., $\int f(x)\,dx$) and definite integrals (i.e., $\int_a^b f(x)\,dx$). In this section we will examine how to solve definite improper integrals—integrals where one or both of the endpoints are at ± infinity, i.e., $\int_{-\infty}^{b} f(x)\,dx$, or $\int_a^{\infty} f(x)\,dx$, or $\int_{-\infty}^{\infty} f(x)\,dx$.

Before you get too worried about the sudden appearance of these infinity symbols, you should know that solving an improper integral is really just a small step beyond solving a definite integral. Intuitively, we solve it by finding a definite integral and then taking a limit and letting one of the endpoints tend to infinity, thus calculating the integral of the function with an endpoint at infinity. More specifically, we use one of the following formulas:

1) $\int_a^{\infty} f(x)\,dx = \lim_{t\to\infty} \int_a^t f(x)\,dx$, provided this limit exists and is finite.
2) $\int_{-\infty}^{b} f(x)\,dx = \lim_{t\to-\infty} \int_t^b f(x)\,dx$, provided this limit exists and is finite.
3) $\int_{-\infty}^{\infty} f(x)\,dx = \int_{-\infty}^{a} f(x)\,dx + \int_a^{\infty} f(x)\,dx$, provided both integrals on the right exist.

Graphically, an improper integral has the same interpretation as a definite integral. That is, an improper integral measures the area under the graph of the function where one endpoint is at infinity.

* *BC content*

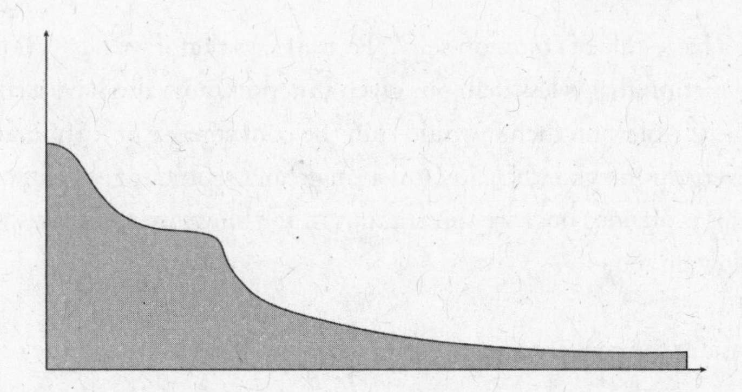

Clearly, the graph must decrease quickly enough or else the integral will have infinite value. If the corresponding limit of the improper integral is finite, we say the integral is *convergent*. If however, the limit that defines the improper integral approaches infinity, we say the integral is *divergent*. (We'll see these terms again in the next chapter.)

Let's take a look at some improper integrals. Suppose we are asked to evaluate $\int_2^\infty \frac{1}{(x+4)^{\frac{3}{2}}}\,dx$. According to the definition of an improper integral, we have:

$$\int_2^\infty \frac{1}{(x+4)^{\frac{3}{2}}}\,dx = \lim_{t\to\infty} \int_2^t \frac{1}{(x+4)^{\frac{3}{2}}}\,dx$$

$$= \lim_{t\to\infty}\left[\frac{-2}{\sqrt{x+4}}\right]_2^t = \lim_{t\to\infty}\left[\frac{-2}{\sqrt{t+4}} - \frac{-2}{\sqrt{6}}\right]$$

$$= \frac{2}{\sqrt{6}} + \lim_{t\to\infty}\frac{-2}{\sqrt{t+4}}$$

$$= \frac{\sqrt{6}}{3}, \text{ because the limit on the right has value 0.}$$

See? Improper integrals are nothing to be worried about. They are just a combination of integrals and limits.

It is very useful to know some basic rules for distinguishing between convergent and divergent improper integrals. In fact, knowing an integral is divergent can sometimes save you the headache of assuming it's convergent and trying to calculate its value. An important rule of this type is the comparison test for improper integrals.

If $f(x)$ and $g(x)$ are continuous functions and $f(x) \geq g(x) \geq 0$ for $x \geq a$, then:

(a) If $\int_a^\infty f(x)\,dx$ is convergent, then $\int_a^\infty g(x)\,dx$ is also convergent.

(b) If $\int_a^\infty g(x)\,dx$ is divergent, then $\int_a^\infty f(x)\,dx$ is also divergent.

This test illustrates a basic rule of common sense. Part (a) says that if we have a convergent (finite) integral, then any function that is less than our given function must also have a convergent integral. This is true. If *g* blew up then it would ruin the convergence of *f*. Similarly, part (b) says that if we have a divergent integral, then any function greater than our given function will also be divergent. Again, this is obvious once we think about it. If *g* blows up and *f* is always greater than *g*, then *f* had better blow up too.

As an example, consider the convergence of $\int_1^\infty \frac{1}{(3x+1)^2}\,dx$. Let's try to use the comparison test here. What shall we compare $\frac{1}{(3x+1)^2}$ to, though? How about $\frac{1}{x^2}$? This is a very good candidate because we know $\int_1^\infty \frac{1}{x^2}\,dx$ is convergent. Let's see if we have a bound.

$$\frac{1}{(3x+1)^2} \overset{?}{\leq} \frac{1}{x^2}$$

$$x^2 \overset{?}{\leq} (3x+1)^2$$

This relationship clearly holds for $x > 1$. Thus we can use the comparison test to show that, because $\int_1^\infty \frac{1}{x^2}\,dx$ is convergent and $\frac{1}{x^2} \geq \frac{1}{(3x+1)^2}$, we know that $\int_1^\infty \frac{1}{(3x+1)^2}\,dx$ is convergent as well.

When trying to use the comparison test for improper integrals, it is necessary to know a comparison function that has either a convergent or divergent improper integral. To this end, we state the following result (without proof): $\int_1^\infty \frac{1}{x^p}\,dx$ is convergent if $p > 1$ and divergent if $p \leq 1$.

We encourage you to prove this result on your own—it's not difficult; all you need is the limit definition of improper integrals.

REVIEW QUESTIONS

1. $\int 2e^x + \frac{5}{x^2}\,dx =$

   (A) $e^{2x} - \frac{5}{x} + C$

   (B) $e^{2x} - \frac{10}{x^3} + C$

   (C) $2e^x - \frac{5}{x} + C$

   (D) $e^{2x} - \frac{10}{x^3} + C$

   (E) $e^{2x} + \frac{10}{x^3} + C$

2. $\int_{\frac{5\pi}{6}}^{\frac{5\pi}{4}} \sec^2\theta\,d\theta =$

   (A) $1+\frac{\sqrt{3}}{3}$

   (B) $1+\sqrt{3}$

   (C) $1-\frac{\sqrt{3}}{3}$

   (D) $1-\sqrt{3}$

   (E) $-\frac{\sqrt{3}}{3}-1$

3. $\int_1^{\sqrt{3}} \frac{2}{x^2+1}\,dx =$

   (A) $\frac{\pi}{3}$

   (B) $-\frac{\sqrt{3}}{4}+1$

   (C) $\frac{\pi}{6}$

   (D) $\frac{\sqrt{3}}{4}+1$

   (E) $\ln 4$

4. $\int_0^{\sqrt{2}} (3x-\sqrt{2})^5 dx =$

   (A) 504

   (B) 168

   (C) 84

   (D) 28

   (E) –84

5. The u-substitution $u=\sqrt{x}$ transforms the indefinite integral $\int \sin\sqrt{x}\,dx$ into which of the following indefinite integrals?

   (A) $\int \sin u\,du$

   (B) $2\int \sin u\,du$

   (C) $\int u\ \sin u\,du$

   (D) $2\int u\ \sin u\,du$

   (E) None of the above

6. $\int_{-1}^{0} \frac{8x^3-12x^2+7x+2}{2x-2}\,dx =$

   (A) $\frac{23}{6}$

   (B) 8

   (C) $\frac{23}{6}-\frac{5}{2}\ln 2$

   (D) $-\frac{5}{2}\ln 2$

   (E) $8+\frac{5}{2}\ln 2$

7. Evaluate the integral $\int_0^1 \frac{x-1}{x^2+3x+2}\,dx$.

   (A) $3\ln 3$

   (B) $-5\ln 2 + 3\ln 3$

   (C) $-5\ln 3 + 2\ln 2$

   (D) $\ln 2 + 3\ln 3$

   (E) $\ln 2$

8. Evaluate the integral $\int e^{2\theta} \cos 3\theta \, d\theta$.

(A) $\frac{1}{13}(2e^{2\theta} \cos 3\theta + 3e^{2\theta} \sin 3\theta)$

(B) $(2e^{2\theta} \cos 3\theta)$

(C) $\frac{1}{13}(e^{2\theta} \cos 3\theta + e^{3\theta} \sin 3\theta)$

(D) $(3e^{3\theta} \cos 2\theta + 2e^{3\theta} \sin 2\theta)$

(E) $\frac{1}{13}(2e^{2\theta} \sin 3\theta)$

9. Evaluate the integral $\int_{-\infty}^{1} xe^{2x^2} \, dx$.

(A) $\frac{e^2}{3}$

(B) e

(C) $\frac{e}{4}$

(D) ∞

(E) $-\infty$

10. $\int x\sqrt{4-x^2} \, dx =$

(A) $\left(x^2-4\right)^{\frac{3}{2}}+C$

(B) $\left(4-x^2\right)^{\frac{3}{2}}+C$

(C) $-\frac{1}{3}\left(4-x^2\right)^{\frac{3}{2}}+C$

(D) $-\frac{1}{3}x\left(4-x^2\right)^{\frac{3}{2}}+C$

(E) $\frac{1}{3}x\left(4-x^2\right)^{\frac{3}{2}}+C$

11. $\int_{\frac{\pi}{3}}^{\frac{\pi}{2}} \frac{2\sin x}{\sqrt{1-\cos x}} \, dx =$

(A) $-4+2\sqrt{2}$

(B) 1

(C) $4-2\sqrt{2}$

(D) $4+2\sqrt{2}$

(E) None of the above

12. If the substitution $u=\sqrt{x+1}$ is used, then $\int_0^3 \frac{dx}{x\sqrt{x+1}}$ is equivalent to

(A) $\int_1^2 \frac{du}{u^2-1}$

(B) $\int_1^2 \frac{2\cdot du}{u^2-1}$

(C) $2\int_0^3 \frac{du}{u^2-1}$

(D) $2\int_1^2 \frac{du}{u(u^2-1)}$

(E) $2\int_0^3 \frac{du}{u(u-1)}$

13. $\int_0^1 \frac{e^x}{e^x+1} \, dx =$

(A) $\ln\frac{e+1}{2}$

(B) $\ln(e+1)$

(C) $\ln(2e+2)$

(D) $\ln 2$

(E) e

14. $\int_0^1 \frac{x^2}{x^2+1} \, dx =$

(A) $1-\frac{\pi}{4}$

(B) $1+\frac{\pi}{4}$

(C) $2+\frac{\pi}{4}$

(D) $\frac{1}{2}\ln 2$

(E) $\ln 2$

15. $\int sin^{-1} x\, dx =$

(A) $\dfrac{(\sin^{-1} x)^2}{2} + C$

(B) $\sin x - \int \dfrac{x}{\sqrt{1-x^2}}\, dx$

(C) $x \cos^{-1} x - \int \dfrac{x\, dx}{\sqrt{1-x^2}}$

(D) $\sin^{-1} x + \int \dfrac{dx}{\sqrt{1-x^2}}$

(E) $x \sin^{-1} x - \int \dfrac{x}{\sqrt{1-x^2}}\, dx$

16. $\int \dfrac{1}{x^2+x}\, dx =$

(A) $\dfrac{1}{2}\arctan\left(x+\dfrac{1}{2}\right)+C$

(B) In $|x^2 + x| + C$

(C) $\ln\left|\dfrac{x+1}{x}\right|+C$

(D) $\ln\left|\dfrac{x}{x+1}\right|+C$

(E) None of these

Free-Response Question

17. The volume of water in a water tank changes at a rate given by the equation
$V'(t) = \dfrac{4t}{\sqrt[3]{t^2+3}}$, for $t \geq 0$.
$V(t)$ has units in gallons and t has units in hours. At $t = \sqrt{5}$ the tank holds 9 gallons.
   (a) What is the volume of water in the tank after 4 hours?
   (b) Find the time that the tank contains 24 gallons of water.
   (c) Suppose that at the beginning of filling the tank a valve was opened so that the volume of water at time t is given by the formula $V(t) = (t^2+2) + 3(t^2+3)^{\frac{2}{3}} - 10$. Find a formula for the rate at which the water escapes through the valve. Indicate units of measure.

Question 18 requires a calculator.

18. Let $f(x) = \dfrac{x-3}{x+1}$.
   (a) The function f can be written in the form $f(x) = A - \dfrac{B}{x+1}$. Find A and B.
   (b) Find f_{ave} (the average value of f) on the interval [0, 2].
   (c) Approximate c (to the nearest 0.001) so that $f(c) = f_{ave}$.
   (d) Define $a(k)$ to be the average value of f on the interval $[0, k]$, where $k > 0$.
      (i) Clearly show that $a(k) = 1 - \ln(k+1)^{\frac{4}{k}}$
      (ii) Approximate k so that $a(k) = 0$.

ANSWERS AND EXPLANATIONS

1. C

We use the properties of indefinite integrals to rewrite the given integral as $\int 2e^x + \frac{5}{x^2}\,dx = 2\int e^x\,dx + 5\int x^{-2}\,dx$. The antiderivative of e^x is e^x, because $\frac{d}{dx}(e^x) = e^x$, so $2\int e^x\,dx = 2e^x + C$. The antiderivative of x^{-2} is $-x^{-1}$ because $\frac{d}{dx}(-x^{-1}) = x^{-2}$, so $5\int x^{-2}\,dx = 5(-x^{-1}) + D = -\frac{5}{x} + D$. Combining these two indefinite integrals, we find $\int 2e^x + \frac{5}{x^2}\,dx = 2\int e^x\,dx + 5\int x^{-2}\,dx = 2e^x - \frac{5}{x} + C$, corresponding to choice (C).

2. A

To apply the FTC (Fundamental Theorem of Calculus), we first need to find an antiderivative of $\sec^2\theta$. It's possible to proceed directly because the antiderivative of $\sec^2\theta$ is $\tan\theta$, since $\frac{d}{d\theta}(\tan\theta) = \sec^2\theta$. We apply the FTC to get $\int_{\frac{5\pi}{6}}^{\frac{5\pi}{4}} \sec^2\theta\,d\theta = \tan\theta\Big|_{\frac{5\pi}{6}}^{\frac{5\pi}{4}} = \tan\left(\frac{5\pi}{4}\right) - \tan\left(\frac{5\pi}{6}\right) = 1 - \left(-\frac{1}{\sqrt{3}}\right) = 1 + \frac{\sqrt{3}}{3}$.

3. C

To apply the FTC, we first need to find an antiderivative of $\frac{2}{x^2+1}$. As excellent calculus students, we recognize that the antiderivative of $\frac{1}{x^2+1}$ is arctan x, because $\frac{d}{dx}(\arctan x) = \frac{1}{x^2+1}$. Using properties of the definite integral, we rewrite the integral as

$$\int_1^{\sqrt{3}} \frac{2}{x^2+1}\,dx = 2\int_1^{\sqrt{3}} \frac{1}{x^2+1}\,dx$$

Now we can apply the FTC:

$$\begin{aligned} 2\int_1^{\sqrt{3}} \frac{1}{x^2+1}\,dx &= 2\left[\arctan x\Big|_1^{\sqrt{3}}\right] \\ &= 2(\arctan\sqrt{3} - \arctan 1) \\ &= 2\left(\frac{\pi}{3} - \frac{\pi}{4}\right) = 2\frac{\pi}{12} = \frac{\pi}{6} \end{aligned}$$

4. D

To apply the FTC, we first need to find an antiderivative of $(3x - \sqrt{2})^5$. Although it is possible to compute this antiderivative directly, in this case it makes sense to make a u-substitution. Following the rules of the substitution game, we let:

$u = (3x - \sqrt{2})$, so
$du = 3dx$.

Using the properties of definite integrals, we can rewrite the integral as:

$$\int_0^{\sqrt{2}} (3x-\sqrt{2})^5 dx = \int_{x=0}^{x=\sqrt{2}} \underbrace{\frac{1}{3}\cdot 3}_{\substack{\text{multiply}\\ \text{by 1}}} (3x-\sqrt{2})^5 dx = \frac{1}{3}\int_{x=0}^{x=\sqrt{2}} \left(\underbrace{3x-\sqrt{2}}_{u}\right)^5 \cdot \underbrace{3\,dx}_{du}$$

At this stage our limits of integration are given as *x*-values. We can transform them to *u*-values, changing variables completely and eliminating the need to substitute back at the end of the problem.

Because $u = (3x - \sqrt{2})$,

$$x = \sqrt{2} \Rightarrow u = 2\sqrt{2}$$
$$x = 0 \Rightarrow u = -\sqrt{2}$$

We rewrite the integral—changing the variable, including the limits of integration $\frac{1}{3}\int_{u=-\sqrt{2}}^{u=2\sqrt{2}} (u)^5\, du$.

The antiderivative of u^5 is $\frac{u^6}{6}$, because $\frac{d}{du}\left(\frac{u^6}{6}\right) = u^5$.

We apply the FTC:

$$\frac{1}{3}\int_{u=-\sqrt{2}}^{u=2\sqrt{2}} (u)^5\, du = \frac{1}{3}\left[\frac{u^6}{6}\Big|_{u=-\sqrt{2}}^{u=2\sqrt{2}}\right]$$
$$= \frac{1}{3}\left(\frac{(2\sqrt{2})^6}{6} - \frac{(-\sqrt{2})^6}{6}\right)$$
$$= \frac{1}{3}\left(\frac{512}{6} - \frac{8}{6}\right) = \frac{1}{18}(512-8)$$
$$= \frac{1}{18}(504) = 28$$

5. D

This is an example of a sneaky substitution.

If $u = \sqrt{x}$, then

$$\frac{du}{dx} = \frac{1}{2\sqrt{x}}$$
$$du = \frac{1}{2\sqrt{x}}\,dx$$

Remembering the rules of the substitution game, we can rewrite the original integral to squeeze in a *du*:

$$\int \sin\sqrt{x}\,dx = \int \underbrace{2\sqrt{x}\cdot\frac{1}{2\sqrt{x}}}_{\text{multiply by 1}}\sin\sqrt{x}\,dx$$

$$= \int 2\underbrace{\sqrt{x}}_{u}\cdot\sin\underbrace{\sqrt{x}}_{u}\,\underbrace{\frac{1}{2\sqrt{x}}\,dx}_{du}$$

We introduced an additional term of $\sqrt{x}$, but that's okay; we swallow it up with a u.

We transform the integral $\int 2u\sin u\ du = 2\int u\sin u\ du$. The correct answer is (D).

6. C

To apply the FTC, we first need to find an antiderivative of the expression under the integrand. Right now, it looks hopelessly complicated, but it isn't. If we carry out the division, we'll be left with "polynomial + $\frac{\text{constant}}{2x-2}$" and we *can* find an antiderivative for this expression.

We divide:

$$\begin{array}{r} 4x^2 - 2x + \frac{3}{2} \\ 2x-2\overline{)8x^3 - 12x^2 + 7x + 2} \\ \underline{-(8x^3 - 8x^2)} \\ -4x^2 + 7x \\ \underline{-(-4x^2 + 4x)} \\ 3x + 2 \\ \underline{-(3x - 3)} \\ 5 \end{array}$$

And we rewrite the integrand

$$\int_{-1}^{0} 4x^2 - 2x + \frac{3}{2} + \frac{5}{2x-2}\,dx$$

Using properties of the definite integral, we can rewrite this integral as

$$4\int_{-1}^{0} x^2\,dx - 2\int_{-1}^{0} x\,dx + \int_{-1}^{0}\frac{3}{2}\,dx + \frac{5}{2}\int_{-1}^{0}\frac{1}{x-1}\,dx$$

To find the antiderivatives in the first three integrals, we use the fact that the antiderivative of x^n is $\frac{x^{n+1}}{n+1}$, because $\frac{d}{dx}\left(\frac{x^{n+1}}{n+1}\right) = x^n$. To find the antiderivative in the fourth integral, we can use a u-substitution; $u = x - 1$ or we can compute directly that an antiderivative of $\frac{1}{x-1}$ is $\ln|x-1|$.

We now apply the FTC:

$$4\int_{-1}^{0} x^2\,dx - 2\int_{-1}^{0} x\,dx + \int_{-1}^{0} \frac{3}{2}\,dx + \frac{5}{2}\int_{-1}^{0} \frac{1}{x-1}\,dx$$

$$= 4\left[\frac{x^3}{3}\Big|_{-1}^{0}\right] - 2\left[\frac{x^2}{2}\Big|_{-1}^{0}\right] + \frac{3}{2}\left[x\Big|_{-1}^{0}\right] + \frac{5}{2}\left[\ln|x-1|\Big|_{-1}^{0}\right]$$

$$= 4\left(0 - \frac{(-1)^3}{3}\right) - 2\left(0 - \frac{(-1)^2}{2}\right) + \frac{3}{2}(0 - (-1)) + \frac{5}{2}(\ln|0-1| - \ln|(-1)-1|)$$

$$= \frac{4}{3} + 1 + \frac{3}{2} - \frac{5}{2}\ln 2 = \frac{23}{6} - \frac{5}{2}\ln 2$$

7. **B**

To evaluate the integral, we need to first use partial fractions to break up the integrand. We write $\frac{x-1}{x^2+3x+2} = \frac{x-1}{(x+1)(x+2)} = \frac{A}{x+1} + \frac{B}{x+2}$ and try to solve for A and B.

By cross multiplying and setting the numerators equal to one another we get $A(x+2) + B(x+1) = x(A+B) + (2A+B) = x - 1$.

Thus, to solve for A and B we have to solve the system of equations below:

$A + B = 1$

$2A + B = -1$

We get that $A = -2$, $B = 3$. Therefore, we can rewrite the integral to get:

$$\int_0^1 \frac{x-1}{x^2+3x+2}\,dx = -2\int_0^1 \frac{1}{x+1}\,dx + 3\int_0^1 \frac{1}{x+2}\,dx = -2\ln|x+1|\Big|_0^1 + 3\ln|x+2|\Big|_0^1$$

$$= -2(\ln 2 - \ln 1) + 3(\ln 3 - \ln 2) = -2\ln 2 + 3\ln 3 - 3\ln 2 = -5\ln 2 + 3\ln 3$$

and we are finished.

8. A

As you might have guessed we have to use integration by parts to solve this integral. This is a good question because it requires integration by parts twice and we have to use the trick that we learned in the section on integration by parts. So let's get started.

Given $\int e^{2\theta}\cos 3\theta\, d\theta,$ we get

$$u = \cos 3\theta, \qquad dv = e^{2\theta}d\theta$$
$$du = -3\sin 3\theta d\theta, \qquad v = \tfrac{1}{2}e^{2\theta}$$
$$\int e^{2\theta}\cos 3\theta\, d\theta = \tfrac{1}{2}e^{2\theta}\cos 3\theta - \int \tfrac{1}{2}e^{2\theta}(-3\sin 3\theta)\, d\theta = \tfrac{1}{2}e^{2\theta}\cos 3\theta + \tfrac{3}{2}\int e^{2\theta}\sin 3\theta\, d\theta$$

Now we have to use integration by parts again on $\int e^{2\theta}\sin 3\theta\, d\theta,$ we get

$$u = \sin 3\theta, \qquad dv = e^{2\theta}d\theta$$
$$du = 3\cos 3\theta d\theta, \qquad v = \tfrac{1}{2}e^{2\theta}$$
$$\int e^{2\theta}\sin 3\theta\, d\theta = \tfrac{1}{2}e^{2\theta}\sin 3\theta - \int \tfrac{1}{2}e^{2\theta}(3\cos 3\theta)\, d\theta = \tfrac{1}{2}e^{2\theta}\sin 3\theta - \tfrac{3}{2}\int e^{2\theta}\cos 3\theta\, d\theta$$

Now just as it seems that we are going in circles we rewrite our equations and make some grand simplifications. From the very beginning,

$$\begin{aligned}\int e^{2\theta}\cos 3\theta\, d\theta &= \tfrac{1}{2}e^{2\theta}\cos 3\theta + \tfrac{3}{2}\int e^{2\theta}\sin 3\theta\, d\theta\\ &= \tfrac{1}{2}e^{2\theta}\cos 3\theta + \tfrac{3}{2}\Big[\tfrac{1}{2}e^{2\theta}\sin 3\theta - \tfrac{3}{2}\int e^{2\theta}\cos 3\theta\, d\theta\Big]\\ &= \tfrac{1}{2}e^{2\theta}\cos 3\theta + \tfrac{3}{4}e^{2\theta}\sin 3\theta - \tfrac{9}{4}\int e^{2\theta}\cos 3\theta\, d\theta\end{aligned}$$

Now all we have to do is solve this equation for $\int e^{2\theta}\cos 3\theta\, d\theta$ and we are done. We get

$$\int e^{2\theta}\cos 3\theta\, d\theta = \tfrac{1}{2}e^{2\theta}\cos 3\theta + \tfrac{3}{4}e^{2\theta}\sin 3\theta - \tfrac{9}{4}\int e^{2\theta}\cos 3\theta\, d\theta$$
$$\tfrac{13}{4}\int e^{2\theta}\cos 3\theta\, d\theta = \tfrac{1}{2}e^{2\theta}\cos 3\theta + \tfrac{3}{4}e^{2\theta}\sin 3\theta$$
$$\int e^{2\theta}\cos 3\theta\, d\theta = \tfrac{4}{13}\left(\tfrac{1}{2}e^{2\theta}\cos 3\theta + \tfrac{3}{4}e^{2\theta}\sin 3\theta\right)$$
$$\int e^{2\theta}\cos 3\theta\, d\theta = \tfrac{1}{13}(2e^{2\theta}\cos 3\theta + 3e^{2\theta}\sin 3\theta)$$

The same result can be achieved by initial choosing $u = e^{2\theta}d\theta$ and $dv = \cos 3\theta d\theta$.

9. E

The integral $\int_{-\infty}^{1} xe^{2x^2}dx$ is an improper integral; therefore we must first calculate $\int_{-t}^{1} xe^{2x^2}dx$ and then take the limit as t approaches infinity. We begin by calculating the definite integral using the substitution $u = 2x^2$ and so $du = 4x\,dx$. We get:

$$\int_{-t}^{1} xe^{2x^2}\, dx = \int_{-t}^{1} e^{2x^2}x\, dx = \int_{-t}^{1} e^{u}\tfrac{1}{4}\, du = \frac{1}{4}\int_{2t^2}^{2} e^{u}\, du = \frac{1}{4}(e^2 - e^{2t^2})$$

Therefore, we get

$$\int_{-\infty}^{1} xe^{2x^2}\,dx = \lim_{t\to\infty}\int_{-t}^{1} xe^{2x^2}\,dx = \lim_{t\to\infty}\frac{1}{4}(e^2 - e^{2t^2}) = \lim_{t\to\infty} - e^{2t^2} = -\infty$$

Because there is a negative sign in front of e^{2t^2}, this number becomes increasingly negative as t approaches infinity. Thus, $\int_{-\infty}^{1} xe^{2x^2}\,dx = -\infty$.

10. C

$\int x\sqrt{4-x^2}\,dx = \int (4-x^2)^{\frac{1}{2}}\,x\,dx = -\frac{1}{2}\int (4-x^2)^{\frac{1}{2}}(-2x)\,dx =$ Let $u = 4-x^2, \frac{du}{dx} = -2x, du = -2x\,dx,$

and by substitution $-\frac{1}{2}\int (4-x^2)^{\frac{1}{2}}(-2x)\,dx = -\frac{1}{2}\int u^{\frac{1}{2}}du.$ Now we antidifferentiate and

$-\frac{1}{2}\int u^{\frac{1}{2}}du = -\frac{1}{2}\cdot\frac{2}{3}u^{\frac{3}{2}} + C = -\frac{1}{3}u^{\frac{3}{2}} + C$

Finally we substitute and $4 - x^2$ and we have $-\frac{1}{3}(4-x^2)^{\frac{3}{2}} + C$

11. C

$$\int_{\frac{\pi}{3}}^{\frac{\pi}{2}} \frac{2\sin x}{\sqrt{1-\cos x}}\,dx = 2\int_{\frac{\pi}{3}}^{\frac{\pi}{2}} \frac{\sin x}{\sqrt{1-\cos x}}\,dx$$

Let $u = 1-\cos x, \frac{du}{dx} = \sin x$, and rearrange so $2\int_{\frac{\pi}{3}}^{\frac{\pi}{2}} \frac{1}{\sqrt{1-\cos x}}\sin x\,dx = 2\int_{x=\frac{\pi}{3}}^{x=\frac{\pi}{2}} \frac{1}{\sqrt{u}}\frac{du}{dx}dx =$ We change the bounds of integration. When $x = \frac{\pi}{3}$, it follows that $u = 1-\cos\frac{\pi}{3} = \frac{1}{2}$. So $u = \frac{1}{2}$. When $x = \frac{\pi}{2}$, it follows that $u = 1-\cos\frac{\pi}{2} = 1$. So $u = 1$. $2\int_{u=\frac{1}{2}}^{u=1} \frac{1}{\sqrt{u}}\,du = 2\int_{\frac{1}{2}}^{1} u^{\frac{-1}{2}}\,du = 2\cdot\left.\frac{u^{\frac{1}{2}}}{\frac{1}{2}}\right|_{\frac{1}{2}}^{1} = \left.4u^{\frac{1}{2}}\right|_{\frac{1}{2}}^{1} = 4\left(\sqrt{1} - \sqrt{\frac{1}{2}}\right) = 4 - 2\sqrt{2}$

12. B

To start, with $u = \sqrt{x+1}$ $u^2 = x + 1 \Rightarrow x = u^2 - 1$

$$\therefore \frac{dx}{du} = 2u \Rightarrow dx = 2u\,du$$

The bounds of integration yields $\begin{array}{l} x = 0 \Rightarrow u = \sqrt{0+1} = 1 \\ x = 3 \Rightarrow u = \sqrt{3+1} = 2 \end{array}$

$$\int_0^3 \frac{dx}{x\sqrt{x+1}} = \int_1^2 \frac{2u\,du}{(u^2-1)\cdot u} = \int_1^2 \frac{2\,du}{u^2-1}$$

13. A

Let $u = e^x + 1 \Rightarrow \frac{du}{dx} = e^x \Rightarrow du = e^x dx$. We change bounds of integration so $x = 0 \Rightarrow u = e^0 + 1 = 2$ and $x = 1 \Rightarrow u = e^1 + 1 = e + 1$ so

$$\int_0^1 \frac{e^x}{e^x + 1} dx = \int_0^1 \frac{1}{e^x + 1} e^x \, dx = \int_2^{e+1} \frac{1}{u} du = \ln|u|\Big|_2^{e+1} = \ln(e+1) - \ln 2 = \ln\frac{e+1}{2}.$$

14. A

$$\int_0^1 \frac{x^2}{x^2+1} dx = \int_0^1 \frac{x^2+1-1}{x^2+1} dx = \int_0^1 \left(\frac{x^2+1}{x^2+1} - \frac{1}{x^2+1}\right) dx =$$

$$\int_0^1 1 \, dx - \int_0^1 \frac{1}{x^2+1} dx = (x - \tan^{-1} x)\Big|_0^1 =$$

$$(1 - \tan^{-1} 1) - (0 - 0) = 1 - \frac{\pi}{4}$$

15. E

Integration by parts

$$\int \sin^{-1} dx = \int u \, dv \text{ with}$$

$$u = \sin^{-1} x \qquad dv = dx$$

$$du = \frac{1}{\sqrt{1-x^2}} dx \qquad v = x$$

$$\int u \, dv = uv - \int v \, du \text{ means}$$

$$\int \sin^{-1} dx = \sin^{-1} x \cdot x - \int x \cdot \frac{1}{\sqrt{1-x^2}} dx$$

$$\int \sin^{-1} dx = x \sin^{-1} x - \int \frac{x}{\sqrt{1-x^2}} dx$$

16. D

Integration by partial fractions

$$\int \frac{1}{x^2+x} dx = \int \frac{1}{x(x+1)} dx = \int \frac{A}{x} + \frac{B}{x+1} dx$$

$$\frac{1}{x(x+1)} = \frac{A}{x} + \frac{B}{x+1} \Rightarrow 1 = A(x+1) + Bx$$

$$0 \cdot x + 1 = (A+B)x + A$$

$$\begin{cases} A + B = 0 \\ A = 1 \end{cases} \Rightarrow B = -1$$

$$\int \frac{A}{x} + \frac{B}{x+1} dx = \int \frac{1}{x} - \frac{1}{x+1} dx = \ln|x| - \ln|x+1| + C$$

$$\therefore \ln\left|\frac{x}{x+1}\right| + C$$

FREE-RESPONSE ANSWER

17. (a) The volume of water in the tank at time t is given by the expression $V(t) = \int_0^t V'(s)ds + V(0)$.

Solving, we find an antiderivative for $V'(t)$ using the u-substitution, $\begin{matrix} u = s^2 + 3 \\ du = 2sds \end{matrix}$:

$$\int V'(t) = \int \frac{2}{\sqrt[3]{(s^2+3)}}(2s)ds = \int \frac{2}{\sqrt[3]{u}}du = 2\int u^{-\frac{1}{3}}du = 2\left(\frac{3}{2}u^{\frac{2}{3}}\right) = 3(s^2+3)^{\frac{2}{3}} + C.$$

To determine the constant, C, we plug in the initial condition: at $t = \sqrt{5}$, $V(t) = 9$.

$$9 = 3((\sqrt{5})^2 + 3)^{\frac{2}{3}} + C$$

$$9 = 3(5+3)^{\frac{2}{3}} + C = 3(4) + C$$

$$C = -3$$

So our antiderivative is $V(t) = 3(t^2+3)^{\frac{2}{3}} - 3$. Computing $V(4) = 3(4^2+3)^{\frac{2}{3}} - 3 = 3 \cdot 19^{\frac{2}{3}} - 3$, we find that the volume of water in the tank at $t = 4$ is $3 \cdot 19^{\frac{2}{3}} - 3$ gallons.

(b) We computed $V(t) = 3(t^2+3)^{\frac{2}{3}} - 3$ in part (a). The tank holds 24 gallons of water when $V(t) = 24$. Solving:

$$24 = 3(t^2+3)^{\frac{2}{3}} - 3$$

$$27 = 3(t^2+3)^{\frac{2}{3}}$$

$$9 = (t^2+3)^{\frac{2}{3}}$$

$$9^{\frac{3}{2}} = (t^2+3)$$

$$24 = t^2$$

$$t = \sqrt{24}$$

This happens at $t = \sqrt{24}$ hours.

(c) The rate of change of the volume of water $(V'(t))$ is given by the rate of water flowing in, ($\frac{4t}{\sqrt[3]{t^2+3}}$ gallons per hour) minus the rate of water flowing out of the valve. Since

$V(t) = (t^2+2) + 3(t^2+3)^{\frac{2}{3}} - 10$, we compute:

$$V'(t) = (2t + 0) + 3\frac{2}{3}(t^2 + 3)^{-\frac{1}{3}}(2t + 0) - 0$$

$$V'(t) = 2t + 4t(t^2 + 3)^{-\frac{1}{3}}$$

Because we know the rate of water flowing in is $\frac{4t}{\sqrt[3]{t^2 + 3}}$ gallons per hour, this means the rate of the flow of water out is $2t$ gallons per hour.

For part (a), integrate using *u*-substitution and don't forget the units in the final answer. There are a lot of pitfalls in using *u*-substitution, so be careful: write down all the details, and don't skip steps. Most importantly, don't forget to re-substitute *u* before you evaluate a definite integral! For part (c), the chain rule used to compute the derivative is exactly the *u*-substitution we used earlier.

18. (a) $f(x) = \frac{x-3}{x+1} = \frac{x+1-4}{x+1} = 1 - \frac{4}{x+1} = A - \frac{B}{x+1}$

$\therefore A = 1, B = 4$

Or by long division,

$$\begin{array}{r} 1 \\ x+1\overline{)x-3} \\ \underline{x+1} \\ -4 \end{array}$$

$\therefore f(x) = 1 - \frac{4}{x+1} = A - \frac{B}{x+1}$

$\therefore A = 1, B = 4$

(b) $f_{\text{ave}} = \frac{1}{2-0}\int_0^2 \left(1 - \frac{4}{x+1}\right) dx = \frac{1}{2}(x - 4\ln(x+1))|_0^2 =$

$\frac{1}{2}(2 - 4\ln 3) - \frac{1}{2}(0 - 4\ln 1) = 1 - 2\ln 3 - 0$

$\therefore f_{\text{ave}} = 1 - 2\ln 3$

(c) This is a calculator problem!

$f(c) = f_{\text{ave}}$

$f(c) = \frac{c-3}{c+1} = 1 - 2\ln 3$

$\therefore c \approx 0.820$

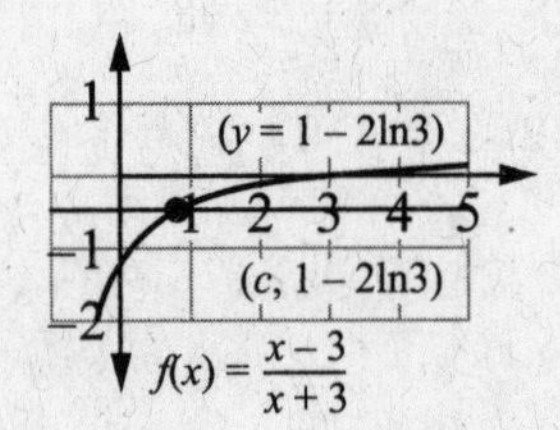

(d) Let $a(k)$ be the average value of $f(x) = \dfrac{x-3}{x+1}$ on $[0, k]$. This means

$$a(k) = \frac{1}{k-0}\int_0^k \frac{x-3}{x+1}\,dx = \frac{1}{k}\int_0^k \left(1 - \frac{4}{x+1}\right)dx =$$

$$a(k) = \frac{1}{k}(x - 4\ln(x+1))|_0^k = \frac{1}{k}(k - 4\ln(k+1)) = 1 - \frac{4}{k}\ln(k+1)$$

$$\therefore\ a(k) = 1 - \frac{4}{k}\ln(k+1) \text{ QED}$$

Now we solve $a(k) = 0$. This is a calculator problem!

$$a(k) = 1 - \frac{4}{k}\ln(k+1)$$

Enter in your calculator.

$$Y_1 = 1 - \left(\frac{4}{x}\right)\ln(x+1)$$

Find its zero.

$k \approx 9.347$

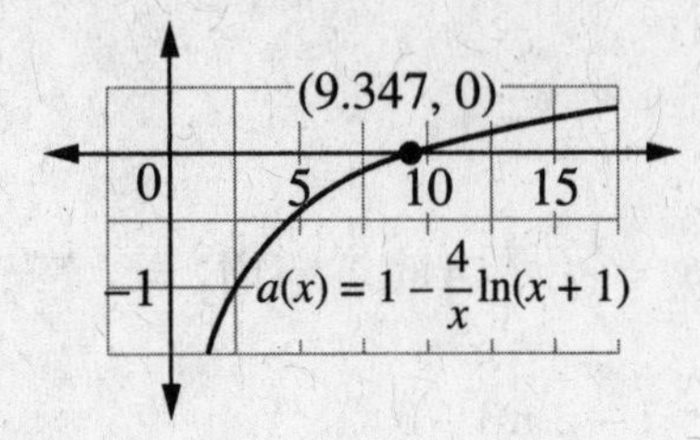

CHAPTER 22: APPLICATIONS OF ANTIDIFFERENTIATION

IF YOU LEARN ONLY THREE THINGS IN THIS CHAPTER . . .

1. There is a connection between accumulation functions and antiderivatives. If f is the rate of change of quantity g, then the change in g from some set starting point c to x is $\int_c^x f(x)\,dx$.
2. The most basic type of differential equation tells us that the rate of change of a function is $f(x)$, $\frac{dy}{dx} = f(x)$.
3. The logistic differential equation for population growth:

$$\frac{dP}{dt} = kP\left(1 - \frac{P}{K}\right),$$

P = population
K = carrying constant
k = constant

Antiderivatives or indefinite integrals give us lots of options. An antiderivative or an indefinite integral is only determined up to a constant—that pesky C tacked on to the end of every indefinite integral. In a modeling situation where we are given information about the rate of change of a function and we need to determine the function itself, we're playing the antiderivative game—that is to say, we're looking for an antiderivative of a given function. If we need to determine a specific antiderivative, we need enough information to determine the constant as well. Often this information is presented as the value of a function at a specific point.

INITIAL CONDITIONS—WHAT IS C?

Information that allows us to determine the C in the indefinite integral $\int f(x)\,dx = F(x) + C$ is called an initial condition. For example, suppose Carol is a first-year college student. She's saving money because she wants to buy a car when she graduates; in real life she'd probably put the money in some kind of bank account or investment where she'd earn interest, but to simplify our task here let's assume Carol is putting the money under her mattress. She saves at a rate of $f(t) = 300t^2$ dollars/year, where t is in years. Will she have enough money to buy a car at the end of four years of college?

To answer the question, we need to know how much money Carol starts with. If $f(t)$ is the rate of change of Carol's money, then the amount of money Carol has, $g(t)$, is an antiderivative of f. Which antiderivative? We don't know. That is, $g(t) = \int 300t^2\, dt = 100t^3 + C$. If we know C, then we know g. Suppose Carol starts college with \$5,000 saved. This is our initial condition; we can phrase it in terms of g: $g(0) = 5{,}000$, and now we can solve for C: $g(0) = 5{,}000 = 100 \cdot 0^3 + C \Rightarrow C = 5{,}000$.

Now we can determine an equation for the amount of money Carol has t years after starting college: $g(t) = 100t^3 + 5{,}000$. After four years of school Carol has $g(4) = 100 \cdot 4^3 + 5{,}000 =$ \$11,400—not enough to buy an SUV, but enough for a nice budget sedan.

There Is More Than One Way to Give an Initial Condition

In the previous example, we were provided with the starting value, i.e., $g(0)$. We would have reached the same conclusion if we'd been provided with the amount of money Carol had at any other time. For instance, suppose that instead of being told that Carol started college with \$5,000, we were told that after two years of college Carol has \$5,800. We can find the constant C from this information, as well.

$$g(2) = 5{,}800 = 100 \cdot 2^3 + C$$
$$5{,}800 = 800 + C$$
$$C = 5{,}000$$

ACCUMULATION FUNCTIONS AND ANTIDERIVATIVES

There is a connection between accumulation functions and antiderivatives. If f is the rate of change of quantity g, then the change in g from some set starting point c to x is $\int_c^x f(x)\, dx$. If we know the value of $g(c)$, then $g(x) = g(c) + \int_c^x f(x)\, dx$: the starting amount plus the change.

We can also think of this in terms of antiderivatives. Because f is the rate of change of g, g is an antiderivative of f. To determine the particular antiderivative we need to know the value of g at any single point—say c.

Initial Conditions and the Motion of a Particle

Initial conditions often pop up on the AP exam in questions that talk about the motion of the particle along a line—those AP folks love moving particles. In this type of problem, we are given a function describing the velocity of the particle and information about the initial position of the particle. We are asked to find the position of the particle at some other time.

Example:

A particle moves along the x-axis so that its velocity at time t is given by $v(t) = t^2 - 3t + 2$. At time $t = 0$ the particle is located at $x(0) = -2$. What is the position of the particle at time $t = 3$?

Solution:

Let's use two different approaches to solve this problem. An antiderivative-with-initial-condition approach and an accumulation function approach.

Antiderivative-with-Initial Condition Approach:

Because velocity is the derivative of position, the position of the particle at time t is given by:

$$x(t) = \int v(t)\,dt = \int t^2 - 3t + 2\,dt = \tfrac{1}{3}t^3 - \tfrac{3}{2}t^2 + 2t + C$$

To determine $x(t)$, we need to find C. We are told that the initial position of the particle is $x(0) = -2$. We solve $x(0) = -2 = \frac{1}{3}\cdot 0^3 - \frac{3}{2}\cdot 0^2 + 2\cdot 0 + C \Rightarrow C = -2$. Therefore the position of the particle at time t is given by $x(t) = \frac{1}{3}t^3 - \frac{3}{2}t^2 + 2t - 2$. The position of the particle at time $t = 3$ is $x(3) = \frac{1}{3}\cdot 3^3 - \frac{3}{2}\cdot 3^2 + 2\cdot 3 - 2 = -\frac{1}{2}$.

Accumulation Function Approach:

The change in the particle's position from time $t = 0$ to $t = 3$ is given by the definite integral

$$\int_0^3 t^2 - 3t + 2\,dt = \left[\tfrac{1}{3}t^3 - \tfrac{3}{2}t^2 + 2t\right]_0^3$$

$$= \left(\tfrac{1}{3}\cdot 3^3 - \tfrac{3}{2}\cdot 3^2 + 2\cdot 3\right) - \left(\tfrac{1}{3}\cdot 0^3 - \tfrac{3}{2}\cdot 0^2 + 2\cdot 0\right)$$

$$= \frac{3}{2}$$

The position of the particle at time $t = 3$ is its initial position $x(0) = -2$, plus the change in position over the interval $0 \le x \le 3$, which is $-2 + \frac{3}{2} = -\frac{1}{2}$.

DIFFERENTIAL EQUATIONS

In Chapter 16 we discussed equations containing derivatives—differential equations. In a differential equation we are given information about the rate of change of a function and asked to find the function. Does this sound familiar? Perhaps like an antiderivative?

Because solutions to differential equations involve taking antiderivatives, their solutions contain constants. When we are given initial conditions that allow us to determine the constants, the solution is called a *particular solution* to the differential equation. If we don't have initial conditions and express the solution with the unknown constants, we call the solution a *general solution.*

The most basic type of differential equation tells us that the rate of change of a function is $f(x)$. In the language of differential equations, we are given $\frac{dy}{dx} = f(x)$.

For example, imagine that we are told that the volume of water in a lake, $V(t)$ (in thousands of liters), is changing at a rate of $f(t) = \sin\left(\frac{\pi}{6}t\right)$ thousands of liters per month. We are also told that at time $t = 0$, the volume of water in the lake is 250 thousand liters. Because the rate of change, or derivative, of the volume is $f(t) = \sin\left(\frac{\pi}{6}t\right)$, the volume of water in the lake at time t is the indefinite integral

$$V(t) = \int f(t)\,dt = \int \sin\left(\frac{\pi}{6}t\right)dt = -\frac{6}{\pi}\cos\left(\frac{\pi}{6}t\right) + C.$$

To determine the volume of water in the lake at time t, we need an initial condition. We are told that $V(0) = 250$. We solve

$$V(0) = 250 = -\frac{6}{\pi}\cos\left(\frac{\pi}{6}\cdot 0\right) + C$$

$$250 = -\frac{6}{\pi} + C$$

$$C = 250 + \frac{6}{\pi} \approx 251.910$$

That is, $V(t)$ is given by the function $V(t) = -\frac{6}{\pi}\cos\left(\frac{\pi}{6}t\right) + 251.910$.

SEPARABLE DIFFERENTIAL EQUATIONS

Not all differential equations are of the form $\frac{dy}{dx} = f(x)$. Most are considerably more complicated and far beyond the scope of the AP exam. In Chapter 16, we saw how to draw qualitative solutions to differential equations of the form $\frac{dy}{dx} = f(x, y)$. Sometimes we can rewrite equations of this form as $g(y)\frac{dy}{dx} = h(x)$, i.e., we can separate the x-stuff from the y-stuff. Differential equations that can be written this way are called *separable differential equations*, and we can solve them with the tools at our disposal.

Let's go back to the example above. We can write this problem as a differential equation $\frac{dV}{dt} = \sin\left(\frac{\pi}{6}t\right)$. Because there are no V's in the problem, this equation is separable. To solve a differential equation of this type, we bring the dt over to the other side and integrate both sides of the equation, $dV = \sin\left(\frac{\pi}{6}t\right)dt$ and $\int dV = \int \sin\left(\frac{\pi}{6}t\right)dt$.

Remembering that variables are dumb, the integral of the left side is just V (plus a constant)—the integral doesn't care that V is a function of t.

We computed the indefinite integral on the right-hand side above. We find the general solution to the differential equation $V(t) = -\frac{6}{\pi}\cos\left(\frac{\pi}{6}t\right) + C$. We use the initial condition as we did above to find that $C = 251.910$. The particular solution is $V(t) = -\frac{6}{\pi}\cos\left(\frac{6}{\pi}t\right) + 251.910$.

The advantage to separating the variables is that we can now work with more complicated differential equations. For example, consider the differential equation $\frac{dy}{dx} = \frac{\sin x}{y}$, given the initial condition $y(0) = 1$ (i.e., when $x = 0$, $y = 1$). To solve it, we separate the variables, $y\,dy = \sin x\,dx$, and integrate both sides to find the general solution:

$$\int y\,dy = \int \sin x\,dx$$

$$\frac{y^2}{2} = -\cos x + C$$

We only need to put the constant on one side of the equation. If P + constant = Q + constant and we don't know what the constants are, we might as well write $P = Q$ + constant.

We now use the initial condition and solve for C:

$$\frac{1^2}{2} = -\cos 0 + C \Rightarrow \frac{1}{2} = -1 + C \Rightarrow C = \frac{3}{2}$$

We plug in for C and solve for y.

$$\frac{y^2}{2} = -\cos x + \frac{3}{2}$$

$$y^2 = 2\left(-\cos x + \frac{3}{2}\right)$$

$$y = \pm\sqrt{2\left(-\cos x + \frac{3}{2}\right)}$$

To receive full credit for the solution to a differential equation, we must solve for y. When we need to find a particular solution, the best way to find C correctly is to use the initial condition in the original solution (the one that isn't solved for y) and only then solve for y. Solving for y before using the initial conditions can lead to all sorts of messy complications with C.

EXPONENTIAL GROWTH: $y' = ky$

The Calculus AB syllabus specifically mentions equations of the type $y' = ky$, where the rate of change of a quantity is proportional to the amount present. Earlier in the text, we mentioned that this type of equation models phenomena such as radioactive growth and decay, absorption of medicine in the bloodstream, and investments.

This type of problem is an example of a separable differential equation. We write the equation as $\frac{dy}{dx} = ky$. We separate the variables and integrate both sides to find:

$$\frac{dy}{y} = k\,dx$$

$$\int \frac{dy}{y} = \int k\,dx$$

$$\ln|y| = kx + C$$

To solve this equation for y we have to be careful. Because $\ln|y| = kx + C$, $e^{\ln|y|} = e^{kx+C}$.

That is $|y| = e^{kx+C}$. Because y is always positive, we can drop the absolute value signs. We use the laws of exponents to write:

$$y = e^{kx+C}$$

$$y = e^{kx} \cdot e^{C}$$

We can rewrite e^{c} as some other constant A and find the solution $y = Ae^{kx}$.

When we solved for y in the general solution, the additive constant C became a multiplicative constant A. This constant represents the amount we start with; that is, if $y(0) = y_0$, we can use this initial condition to find A: $y_0 = Ae^0 = A \cdot 1 = A$. For this reason, we often call the constant A_0 and understand that it is the amount present at $x = 0$.

IDENTIFYING EXPONENTIAL-GROWTH TYPE PROBLEMS

Sometimes it's easy to identify an exponential-growth or decay problem. If we are given a differential equation of the form $\frac{dy}{dx} = ky$, it's clear, but Things aren't always this simple.

Any question that has the form $\frac{dy}{y} = f(x)dx$ after we separate variables is likely to have an exponential solution—i.e., the solution will be of the form $y = Ae^{g(x)}$.

Example:

A bandana falls off a tall building. At time $t \geq 0$, the velocity of the bandana satisfies the differential equation $\frac{dv}{dt} = -10v - 32$ with initial condition $v(0) = 0$. Use separation of variables to find an expression for v in terms of t when t is measured in seconds.

Solution:

We are directed to separate variables—so let's start by doing just that: $\frac{dv}{-10v - 32} = dt$. We integrate both sides of the equation, $\int \frac{dv}{-10v - 32} = \int dt$. Integrating the right side is no problem. Integrating the left side requires a u-substitution, $\begin{matrix} u = -10v - 32 \\ du = -10\,dv \end{matrix}$:

$$\int \underbrace{-\frac{1}{10}(-10)}_{\text{multiply by 1}} \frac{dv}{-10v - 32} = \int dt$$

$$-\frac{1}{10}\int \frac{\overbrace{-10dv}^{du}}{\underbrace{-10v - 32}_{u}} = \int dt$$

$$-\frac{1}{10}\int \frac{1}{u}\,du = \int dt$$

It's easier to bring the $-\frac{1}{10}$ to the other side of the equation before integrating, mostly to help keep track of the constant when we are done:

$$\begin{aligned} \int \frac{1}{u}\,du &= -10\int dt \\ \ln|u| &= -10t + C \\ u &= e^{-10t+C} \\ u &= Ae^{-10t} \end{aligned}$$

Now we substitute back: $-10v - 32 = Ae^{-10t}$.

Keeping in mind our rule to find the constant before solving for v we use the information:

$$\begin{aligned} v(0) &= 0: \\ -10 \cdot 0 - 32 &= Ae^{-10 \cdot 0} \\ -32 &= A \end{aligned}$$

We fill this information into the equation and solve for v:

$$-10v - 32 = -32e^{-10t}$$
$$-10v = -32e^{-10t} + 32$$
$$v = 3.2e^{-10t} - 3.2$$

BUILDING AN EXPONENTIAL GROWTH EQUATION

Sometimes we are not given the differential equation at all. Instead, we need to figure out from the context that we are looking at a problem of the type $\frac{dy}{dt} = ky$. Population growth is an example of this type of problem. If we are told that a population doubles every n years, this means that in differential-equation-speak we are being told "The rate of growth of a population is proportional to its size and $y(n) = 2\,y(0)$."

From here we can set up the differential equation $\frac{dy}{dt} = ky$. The information that $y(n) = 2\,y(0)$ allows us to find k. There is a problem of this type in the questions at the end of the chapter.

SOLVING LOGISTIC DIFFERENTIAL EQUATIONS AND USING THEM IN MODELING*

In this section we will discuss a mathematical equation that is used as a model for population growth. It is called the *logistic model* and it is slightly more accurate than the model we already have seen for exponential growth.

To better motivate the logistic model, let's think about population growth. When a population is small it grows quite steadily and the rate is almost constant. During this time, the exponential growth model is in effect. However, will the population continue to grow undeterred in this way for all time? Of course not! After a while, the population becomes too big for its own good and the relative rate of growth actually slows down. In fact, if the population exceeds its carrying capacity, then the population is likely to decrease and the growth rate will be negative. What we want is a new model that will better reflect this more complex behavior.

The equation below is called the *logistic differential equation* and it can do this more complicated job for us.

$$\frac{dP}{dt} = kP\left(1 - \frac{P}{K}\right),$$

P = population
K = carrying constant
k = constant

Note that if P is very close to the carrying constant, K, then $\frac{P}{K}$ is close to 1 and so $\frac{dP}{dt}$ is close to 0, as we would expect. If P is instead small compared to the carrying constant, K, then $\frac{P}{K}$ is close to 0 and so the population P is growing at a constant rate k.

* *BC content*

The logistic differential equation is a separable differential equation and thus can be solved using methods we covered in the last section. To save time however, we will simply state the result. The solution to the logistic differential equation is:

$$P(t) = \frac{K}{1 + Ae^{-kt}}, \quad \text{where } A = \frac{K - P_0}{P_0} \quad \text{and} \quad P_0 = \text{initial population}$$

For some practice using the logistic differential equation, write the solution to the initial value problem $\frac{dP}{dt} = 0.7\left(1 - \frac{P}{150}\right)$, $P(0) = 100$ and use the solution to calculate $P(30)$.

In this situation, we have $K = 150$, $k = 0.7$, and $P_0 = 100$; thus $A = \frac{150 - 100}{100} = \frac{1}{2}$ and so we get $P(t) = \frac{150}{1 + \frac{1}{2}e^{-0.7t}} = \frac{300}{2 + e^{-0.7t}}$. Furthermore, we get $P(30) = \frac{300}{2 + e^{-0.7 \cdot 30}} \approx 150$.

REVIEW QUESTIONS

1. A particle moves along the x-axis with a velocity given by $v(t) = 2 + \sin t$. When $t = 0$ the particle is at $x = -2$. Where is the particle when $t = \pi$?

   (A) π

   (B) 2π

   (C) $\pi - 1$

   (D) $\pi - 2$

   (E) $\pi + 1$

2. Water flows into a pond at a rate of $300\sqrt{t}$ gallons/hour and flows out at a rate of 400 gallons/hour. After 1 hour there are 10,000 gallons of water in the pond. How much water is in the pond after 9 hours?

   (A) 10,000

   (B) 11,000

   (C) 12,000

   (D) 14,000

   (E) 16,000

3. The graph of the derivative of f, f', is shown below.

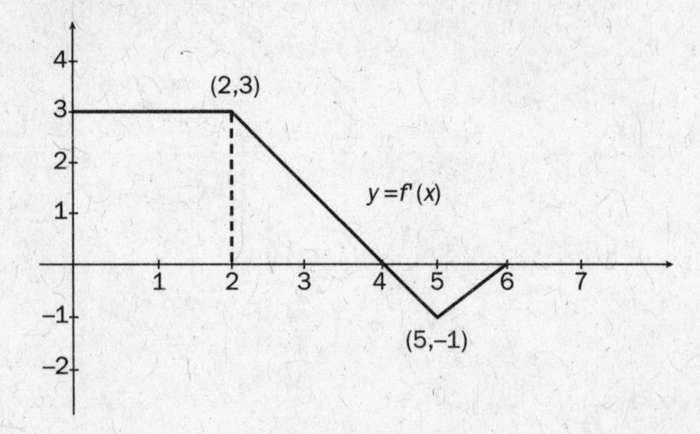

   If $f(0) = 7$, find $f(6)$.

   (A) 9

   (B) 11

   (C) 12

   (D) 14

   (E) 15

4. Consider the differential equation $\frac{dy}{dx} = \frac{e^x - 1}{2y}$. If $y = 4$ when $x = 0$, what is a value of y when $x = 1$?

   (A) $\sqrt{e+14}$

   (B) $\sqrt{e+15}$

   (C) $\sqrt{e^2+11}$

   (D) $\sqrt{e^2+15}$

   (E) $\sqrt{e^2-12}$

5. Consider the differential equation $\frac{dy}{dx} = (1-2x)\cdot y$. If $y = 10$ when $x = 1$, find an equation for y.

   (A) $y = e^{x-x2}$

   (B) $y = 10 + e^{x-x2}$

   (C) $y = e^{x-x2+10}$

   (D) $y = 10 \cdot e^{x-x2}$

   (E) $y = x - x^2 + 10$

6. If $\frac{dy}{dx} = \frac{1}{4}y$ and $y(0) = 5$, then $y(4) =$

   (A) $5e$

   (B) $5 + e$

   (C) $10e$

   (D) $10 + e$

   (E) $5e^2$

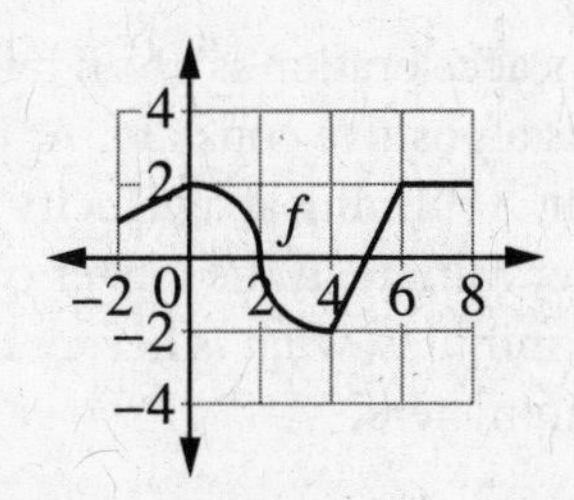

7. Given the graph of f above, if $F'(x) = f(x)$ and $F(0) = 3$, what is $F(7)$?

(A) 2
(B) 3
(C) 5
(D) $2\pi + 5$
(E) $2\pi + 7$

8. If $\frac{dy}{dx} = y\sec^2 x$ and $y = 8$ when $x = 0$, then $y =$

(A) $\tan x + 8$
(B) $8e^{\tan x}$
(C) $e^{\tan x} + 7$
(D) $e^x \tan x + 8$
(E) $\tan e^x + 7$

9. If $\frac{dy}{dx} = xy^2$ with the initial condition $y(0) = 1$, then $y =$

(A) $\frac{2}{2-x^2}$
(B) $\frac{1}{1-x^2}$
(C) $\frac{2}{2+x^2}$
(D) $e^{\frac{x^2}{2}}$
(E) $e^{\frac{-x^2}{2}}$

10. At every point (x, y) on a curve, the slope of the curve is $4x^3y$. If the curve contains the point $(0, 4)$, then its equation is

(A) $y = 4e^{x^4}$
(B) $y = e^{x^4} + 3$
(C) $y = \ln(x + 1) + 4$
(D) $y = x^4 + 4$
(E) $y^2 = x^3 + 16$

11. If the graph of $y = f(x)$ contains the point $(0,2)$, $\frac{dy}{dx} = \frac{-x}{ye^{\frac{x^2}{2}}}$, and $f(x) > 0$ for all x, then $f(x) =$

(A) $1+e^{\frac{-x^2}{2}}$
(B) $3-e^{\frac{-x^2}{2}}$
(C) $1 + e^{-x^2}$
(D) $\sqrt{2e^{\frac{-x^2}{2}}+2}$
(E) $\sqrt{e^{\frac{-x^2}{2}}+3}$

12. The graph of the function shown to the right is a solution to one of the differential equations below. Which one?

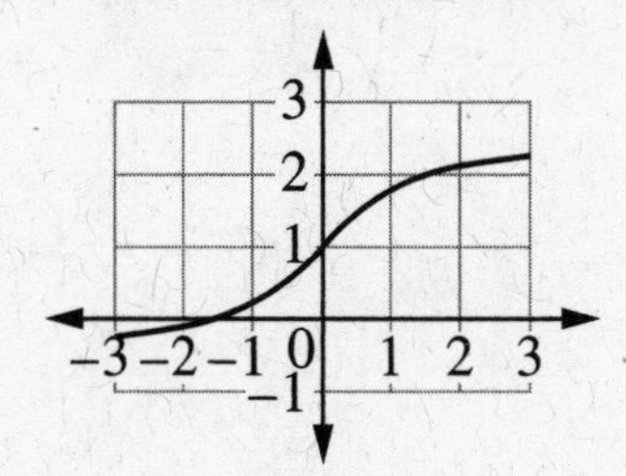

(A) $y' = 1 - x^2y^2$
(B) $y' = 1 + x^2y^2$
(C) $y' = \frac{1}{1+y^2}$
(D) $y' = \frac{1}{1-x^2}$
(E) $y' = \frac{1}{1+x^2}$

13. Which of the following could be a path through the slope field created by the differential equation $\frac{dy}{dx} = \sec^2(0.5x)$?

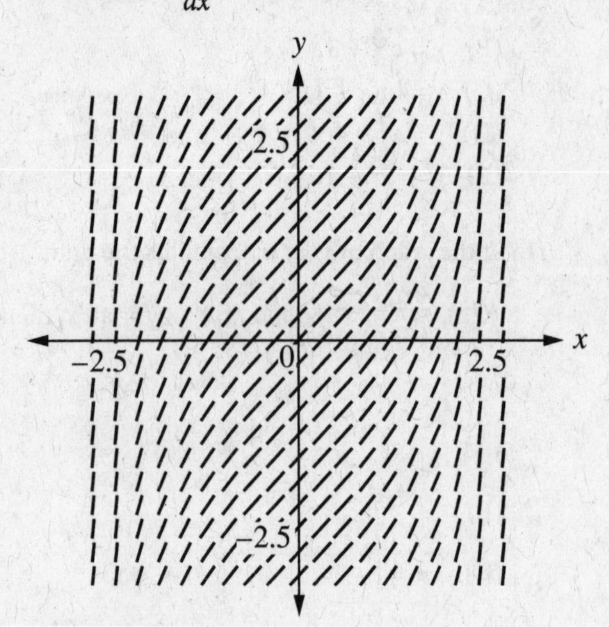

(A) $y = \tan x$
(B) $y = 2 \tan x + 1$
(C) $y = \tan(0.5x)$
(D) $y = 2\tan(0.5x)$
(E) $y = \tan(0.5x) + 2$

14. A jogger's acceleration is given by $a(t) = -kt$ where k is a positive constant. At time $t = 0$, the jogger is running at a velocity of 192 meters per minute. If the jogger comes to a stop in 8 minutes, what is her total distance covered in meters?

(A) 960
(B) 1024
(C) 1440
(D) 1600
(E) 1820

15. The population $P(t)$ of the island fox on Catalina Island satisfies the logistic differential equation $\frac{dP}{dt} = P\left(2 - \frac{P}{5000}\right)$. According to this model, the $\lim_{t \to \infty} P(t)$ is equal to

(A) 2,500
(B) 3,000
(C) 4,200
(D) 5,000
(E) 10,000

Free-Response Questions

16. The population of a country doubles every 20 years. If the population (in millions) of the country was 100 in 2002, what will the population be in 2014?

17. The acceleration of an airplane from the moment of liftoff ($t = 0$) to 20 minutes into the flight is shown below.

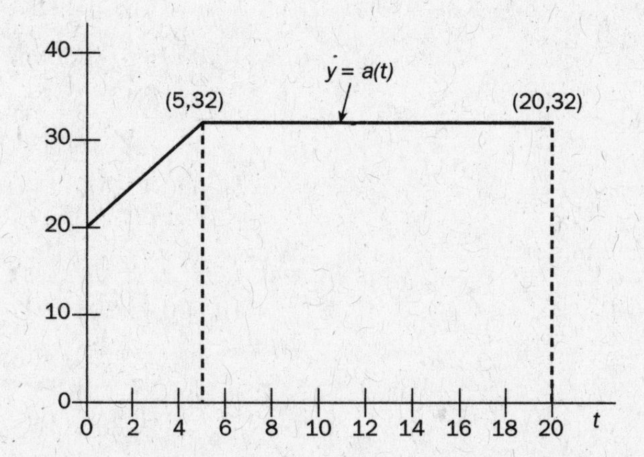

If the speed at liftoff is 900 ft/min, what is the speed of the plane after 20 minutes?

ANSWERS AND EXPLANATIONS

1. B

The position of the particle $x(t)$ is an antiderivative of the velocity function, so we start by computing the indefinite integral of the velocity:

$$v(t) = 2 + \sin t$$
$$x(t) = \int v(t)\,dt = 2t - \cos t + C \Rightarrow$$
$$x(t) = 2t - \cos t + C$$

We can use the initial condition to determine C:

$$x(0) = -2 \Rightarrow -2 = 2 \cdot 0 - \cos 0 + C \Rightarrow$$
$$-2 = -1 + C \Rightarrow C = -1$$
$$x(t) = 2t - \cos t - 1$$

Now we evaluate the position function at time $t = \pi$:

$$x(\pi) = 2\pi - \cos \pi - 1 \Rightarrow$$
$$x(\pi) = 2\pi - (-1) - 1 = 2\pi$$

2. C

The total rate of change of the volume of water in the pond is $r(t) =$ *Rate in* – *Rate out.*

We are given that the rate in is $300\sqrt{t}$ gallons/hr and the rate out is 400 gallons per hour. Therefore, $r(t) = 300\sqrt{t} - 400$.

If we let $A(t)$ be the amount of water in the pond at time t, then A is an antiderivative of r. Let's approach this problem as an accumulation function. Because we are given the volume of water at $t = 1$ and we are asked to find the volume of water at $t = 9$, we can set the problem up as $A(9) = A(1) +$ *change in volume on* $1 \leq t \leq 9$. The change in the volume of water is given by the definite integral $\int_1^9 r(t)\,dt = \int_1^9 300\sqrt{t} - 400\,dt$, so:

$$A(9) = 10{,}000 + \int_1^9 300\sqrt{t} - 400\,dt$$
$$= 10{,}000 + \int_1^9 300t^{\frac{1}{2}} - 400\,dt$$
$$= 10{,}000 + \left(\frac{2}{3} \cdot 300t^{\frac{3}{2}} - 400t\right)\Big]_1^9$$
$$= 10{,}000 + \left(200t^{\frac{3}{2}} - 400t\right)\Big]_1^9$$
$$= 10{,}000 + \left(200 \cdot 9^{\frac{3}{2}} - 400 \cdot 9\right)$$
$$- \left(200 \cdot 1^{\frac{3}{2}} - 400 \cdot 1\right) = 12{,}000$$

3. E

To find $f(6)$, we can compute $f(6) = f(0) +$ *total change in* f *on* $0 \leq x \leq 6$. The area under the graph of f' from $x = 0$ to $x = 6$ gives the total change in f on $0 \leq x \leq 6$. We can express this as a definite integral: $f(6) = f(0) + \int_0^6 f'(x)\,dx$. The definite integral $\int_0^6 f'(x)\,dx$ is the area under the graph of f'. To compute this area, we break it into three regions, shown below:

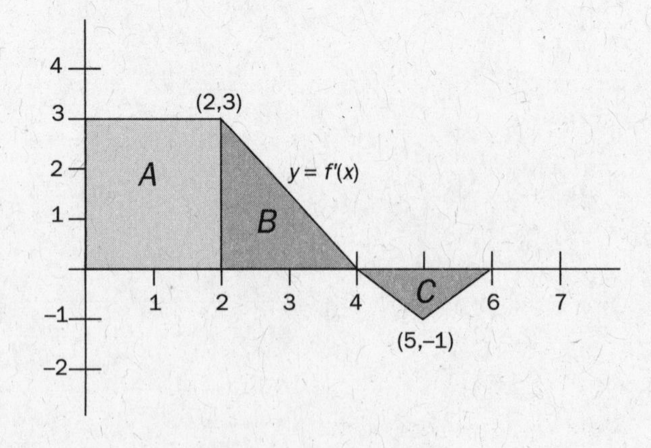

The area under the curve is therefore *Area of rectangle A* + *Area of triangle B* – *Area of triangle C*. Note that we *subtract* the area of triangle *C* because the definite integral considers the area beneath the *x*-axis to be negative. We find:

$$\int_0^6 f'(x)\,dx = 3\cdot 2 + \frac{1}{2}\cdot 2\cdot 3 - \frac{1}{2}\cdot 2\cdot 1 = 8$$

We are given that $f(0) = 7$, so

$$f(6) = f(0) + \int_0^6 f'(x)\,dx = 7 + 8 = 15.$$

4. A

To solve this differential equation we use the technique of separation of variables:

$$\frac{dy}{dx} = \frac{e^x - 1}{2y}$$

$$2y\,dy = (e^x - 1)\,dx$$

$$\int (2y)\,dy = \int (e^x - 1)\,dx$$

$$y^2 = e^x - x + C$$

We use the initial condition $x(0) = 4$ to find C by plugging the information into the solution above:

$$4^2 = e^0 - 0 + C = 1 + C \Rightarrow C = 15$$

$$y^2 = e^x - x + 15 \Rightarrow y = \pm\sqrt{e^x - x + 15}$$

The initial condition requires us to use the positive square root. So, $y = \sqrt{e^x - x + 15}$.

We now substitute $x = 1$ into this expression and solve for y:

$$y^2 = e^1 - 1 + 15 = e + 14 \Rightarrow y = \sqrt{e + 14}$$

5. D

To solve this differential equation we use the technique of separation of variables:

$$\frac{dy}{dx} = (1 - 2x)\cdot y$$

$$\frac{1}{y}dy = (1 - 2x)\,dx$$

Notice that after separating the variables, the equation is of the form $\frac{dy}{y} = f(x)\,dx$, which means the answer will be of the form $y = Ae^{g(x)}$, and we will need to pay special attention to finding the constant.

We integrate both sides of the equation to find:

$$\int \frac{1}{y}dy = \int (1 - 2x)\,dx$$

$$\ln|y| = x - x^2 + C$$

$$y = e^{x - x^2 + C} = e^{x - x^2}\cdot e^C$$

$$y = A\cdot e^{x - x^2}$$

We use the initial condition $y(1) = 10$ to find A:

$y = A\cdot e^{x-x^2} \Rightarrow 10 = A\cdot e^{1-1^2} = A \Rightarrow A = 10$

$y = 10\cdot e^{x-x^2}$

6. A

We use the technique of separation of variables to solve this differential equation:

$$\frac{dy}{dx} = \frac{1}{4}y$$

$$\frac{1}{y}dy = \frac{1}{4}dx$$

Notice that after separating the variables, the equation is of the form $\frac{1}{y}dy = f(x)\,dx$, so the solution will involve an exponential function and a multiplicative constant.

$$\int \frac{1}{y}dy = \int \frac{1}{4}dx$$

$$\ln|y| = \frac{1}{4}x + C$$

$$y = Ae^{\frac{1}{4}x}$$

We use the initial condition that when $x = 0$, $y = 5$ to find A: $5 = Ae^{\frac{1}{4}\cdot 0} = A \Rightarrow A = 5$. We now evaluate the function when $x = 4$: $y(4) = 5e^{\frac{1}{4}\cdot 4} = 5e$.

7. C

Since $F'(x) = f(x)$ and $F(0) = 3$, it follows that $\int F'(x)\,dx = \int f(x)\,dx \Rightarrow F(x) = \int_0^x f(t)\,dt + 3$. So $F(7) = \int_0^7 f(t)\,dt + 3 = 2 + 3 = 5$.

8. B

$\frac{dy}{dx} = y\sec^2 x \Rightarrow \frac{1}{y}dy = \sec^2 x\,dx \Rightarrow \int \frac{1}{y}dy = \int \sec^2 x\,dx$ and antidifferentiate so

$\ln|y| = \tan x + c \Rightarrow |y| = e^{\tan x + c} \Rightarrow |y| = Ce^{\tan x}$. Since $y = 8$ when $x = 0$, then $y = 8e^{\tan x}$

9. A

If $\frac{dy}{dx} = xy^2 \Rightarrow \frac{1}{y^2}dy = x\,dx \Rightarrow \int \frac{1}{y^2}dy = \int x\,dx$ and antidifferentiate so

$\frac{-1}{y} = \frac{1}{2}x^2 + C_1 \Rightarrow \frac{1}{y} = C_2 - \frac{1}{2}x^2 \Rightarrow y = \frac{1}{C_2 - \frac{1}{2}x^2}$. Since $y(0) = 1$, then $1 = \frac{1}{C_2} \Rightarrow C_2 = 1$ which means

$$y = \frac{1}{1 - \frac{1}{2}x^2} = \frac{2}{2 - x^2}$$

10. A

Since the slope of the curve is $4x^3y$, it follows that $\frac{dy}{dx} = 4x^3y$. It follows that

$\frac{1}{y}dy = 4x^3dx \Rightarrow \int \frac{1}{y}dy = \int 4x^3\,dx \Rightarrow \ln|y| = x^4 + c \Rightarrow |y| = e^{x^4 + c} \Rightarrow |y| = Ce^{x^4}$

Given the point (0, 4), then its equation is $y = 4e^{x^4}$

11. D

$\frac{dy}{dx} = \frac{-x}{ye^{\frac{x^2}{2}}} \Rightarrow y\,dy = -xe^{\frac{-x^2}{2}}\,dx \Rightarrow \int y\,dy = \int -xe^{\frac{-x^2}{2}}\,dx$. It follows that

$\frac{1}{2}y^2 = e^{-\frac{x^2}{2}} + C_1 \Rightarrow y^2 = 2e^{-\frac{x^2}{2}} + C_2 \Rightarrow y = \pm\sqrt{2e^{-\frac{x^2}{2}} + C_2}$. Given $f(x) > 0$ and the point (0, 2), substitute and

$2 = \sqrt{2 + C_2} \Rightarrow C_2 = 2$ which means $f(x) = \sqrt{2e^{-\frac{x^2}{2}} + 2}$.

12. E

The slope at the point (0, 1) appears to be 1, which means we can dismiss (C) as a possibility. Now focus on the point (2, 2) where the slope is a small positive number. Only the slope indicated by option (E) $y' = \frac{1}{1+x^2}$ makes sense, $y' = \frac{1}{1+2^2} = \frac{1}{5}$, while the slopes at the other options are clearly different.

Note: $\int y'\,dx = \int \frac{1}{1+x^2}\,dx \Rightarrow y = \tan^{-1} x + c$.

The graph looks like the transformation of the inverse tangent graph $y = \tan^{-1}x + 1$ where $y' = \frac{1}{1+x^2}$.

13. D

To start, $\frac{dy}{dx} = \sec^2(0.5x) \Rightarrow dy = \sec^2(0.5x)dx \Rightarrow \int dy = \int \sec^2(0.5x)\,dx$, now we antidifferentiate and $y = 2\tan(0.5x) + C$. Only option (D) agrees.

14. B

We need to antidifferentiate twice in this problem and resolve constants on route. $v(T) = \int_0^T -kt\,dt$, it follows that $v(T) = -k\frac{T^2}{2} + C$. We know that $v(0) = 192$, therefore $v(T) = -k\cdot\frac{T^2}{2} + 192$. Since $v(8) = 0$, Substitute $-k\frac{8^2}{2} + 192 = 0$, $-32k + 192 = 0$, so $k = 6$.

This means $v(T) = -6\cdot\frac{T^2}{2} + 192 - 3T^2 + 192$. Now
$s(8) = \int_0^8 (-3T^2 + 192)\,dT$ so $s(8) = (-T^3 + 192T)\Big|_0^8 = 1024$ meters.

15. E

The logistic differential equation $\frac{dP}{dt} = P\left(2 - \frac{P}{5000}\right)$ achieves population equilibrium for those values of P where $\frac{dP}{dt} = 0$. We solve $\frac{dP}{dt} = 0$ and get $P = 0$ or $P = 10000$. The carrying capacity is $\lim_{t\to\infty} P(t)$ which is equal to 10000.

FREE-RESPONSE ANSWERS

16. This question is an exponential growth question in disguise. We interpret the given information as "The rate of growth of the population is proportional to its size and $y(20) = 2y(0)$." That is $\frac{dy}{dt} = ky$. The information $y(20) = 2y(0)$ will enable us to find k—the growth constant.

We solve this differential equation using the technique of separation of variables:

$$\frac{1}{y}\,dy = k\,dt$$

$$\int \frac{1}{y}\,dy = \int k\,dt$$

$$\ln|y| = kt + C$$

$$e^{\ln|y|} = e^{kt+C} = e^{kt} \cdot e^{C} = Ae^{kt} \Rightarrow$$

$$|y| = Ae^{kt}$$

When $t = 0$, the population of the town is $y(0) = Ae^{k \cdot 0} = A$, so $y(20) = 2y(0) = 2A$.

We use this information to find k:

$$y(20) = Ae^{k \cdot 20} = 2A$$

$$e^{20k} = 2$$

$$20k = \ln 2$$

$$k = \frac{\ln 2}{20} \approx 0.034657359$$

That is,

$$y = Ae^{\frac{\ln 2}{20}t} \underset{\text{laws of exponents}}{=} A(e^{\ln 2})^{\frac{t}{20}} \underset{e^{\ln 2}=2}{=} A \cdot 2^{\frac{t}{20}}$$

We set $t = 0$ to be the year 2002, and use the given information $y(0) = 100$ to find A. $100 = A2^{\frac{0}{20}} = A \Rightarrow A = 100$. The population at time t is therefore given by $y(t) = 100 \cdot 2^{\frac{t}{20}}$ (where $t = 0$ is the year 2002).

We are asked to find the population in 2014, that is $y(12)$:

$$y(12) = 100 \cdot 2^{\frac{12}{20}} \approx 151.172.$$

The population in the year 2014 will be approximately 151.172 million people.

17. We can think of this as an accumulation problem. The speed of the airplane 20 minutes after liftoff is the initial speed plus the change in speed on $0 \le t \le 20$. The change in speed is given by the definite integral $\int_0^{20} a(t)\,dt$. Therefore:

$$s(20) = s(0) + \int_0^{20} a(t)\,dt = 900 + \int_0^{20} a(t)\,dt$$

The definite integral $\int_0^{20} a(t)\,dt$ is the area under the acceleration curve on $0 \le t \le 20$.

We break this area into two regions: trapezoid *A* and rectangle *B*:

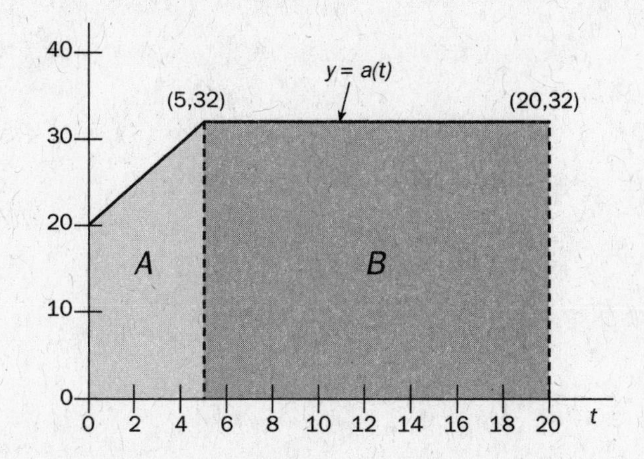

$$\int_0^{20} a(t)\,dt = Area\ of\ A + Area\ of\ B$$

$$= \frac{1}{2} \cdot 5(20 + 32) + (15 \times 32) = 610$$

Therefore the speed at $t = 20$ is $s(20) = 900 + 610 = 1510$ ft/min.

If the problem contains units, units must appear in the answer to receive full credit!

CHAPTER 23: NUMERICAL APPROXIMATIONS

IF YOU LEARN ONLY THREE THINGS IN THIS CHAPTER . . .

1. Left Riemann Sum:

$$\underset{\text{left endpoint}}{R_n} \approx \sum_{i=1}^{n} f(x_{i-1})\Delta x = (f(x_0)+f(x_1)+\cdots+f(x_{i-1}))\frac{b-a}{n}$$

2. Right Riemann Sum:

$$\underset{\text{right endpoint}}{R_n} = \sum_{i=1}^{n} f(x_i)\Delta x = (f(x_1)+f(x_2)+\cdots+f(x_n))\frac{b-a}{n}$$

3. Midpoint Riemann Sum:

$$\underset{\text{midpoint}}{R_n} \approx \sum_{i=1}^{n} f\left(x_{i-1}+\tfrac{b-a}{2n}\right)\Delta x = \left(f\left(x_0+\tfrac{b-a}{2n}\right)+f\left(x_1+\tfrac{b-a}{2n}\right)+\cdots+f\left(x_{i-1}+\tfrac{b-a}{2n}\right)\right)\frac{b-a}{n}$$

In Chapter 17, we defined a definite integral as the limit of a sum of areas of rectangles. Later we learned the Fundamental Theorem of Calculus (FTC), which allows us to compute definite integrals without thinking about rectangles at all. Great—now we never need to think about rectangles! Well—not exactly.

To apply the FTC, we need to be able to find an antiderivative, but we can't always do this. Some functions do not have an antiderivative that we can write as a tidy function. For example, if we try to find an antiderivative for $f(x) = e^{x^2}$, we'll be looking for a very long time because there is no elementary formula for the antiderivative of $f(x) = e^{x^2}$.

Another situation in which we can't apply the FTC is when we're given a function via a graph or a table of values. If we don't have a formula for f, we can't find a formula for its antiderivative.

Are we stuck? Nope. We can always go back to our definition of the definite integral as the area under the curve.

APPROXIMATING THE AREA UNDER THE CURVE USING RIEMANN SUMS

In Chapter 17, we saw that we could use rectangles to approximate the area under a curve $f(x)$ on the interval $a \le x \le b$, that is, to approximate $\int_a^b f(x)\,dx$.

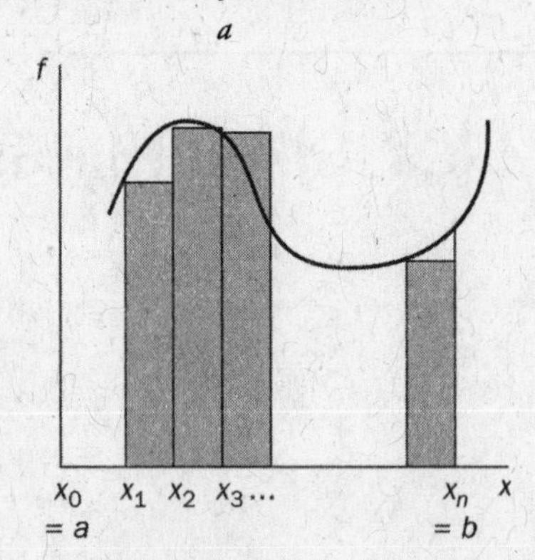

We can divide the interval $[a,b]$ into n equal pieces each of length $\frac{b-a}{n}$. We call the points dividing the subintervals *sampling points* and we denote them by $x_0, x_1, \ldots x_n$, where $x_0 = a$ and $x_n = b$. We can use a point c_i in the base of each rectangle, i.e., a point that lies in the ith subinterval $[x_{i-1}, x_i]$, to determine the height of that rectangle. We call the approximation of the area under the curve by the rectangles determined by these points a Riemann sum, as we defined in Chapter 17.

$$R_n \approx \sum_{i=1}^{n} f(c_i)\Delta x = (f(c_1) + f(c_2) + \ldots + f(c_n))\Delta x$$

RIEMANN SUM USING THE LEFT ENDPOINT

When we use the left endpoint of each rectangle to evaluate the height of the rectangle, we call this a left Riemann sum.

$$\underset{\text{left endpoint}}{R_n} \approx \sum_{i=1}^{n} f(x_{i-1})\Delta x = (f(x_0) + f(x_1) + \ldots + f(x_{i-1}))\frac{b-a}{n}$$

For example, as we said above, we can't use the FTC to compute the area under the curve $f(x) = e^{x^2}$, because the antiderivative of this function cannot be expressed by a simple mathematical formula. We can, however approximate the area under this curve using a Riemann sum. Let's approximate the area under f on the interval [1,2] using the fifth left Riemann sum, i.e., compute R_5 using the left endpoints as the sampling numbers to evaluate the height of each rectangle.

The area we are computing is shown on the graph below.

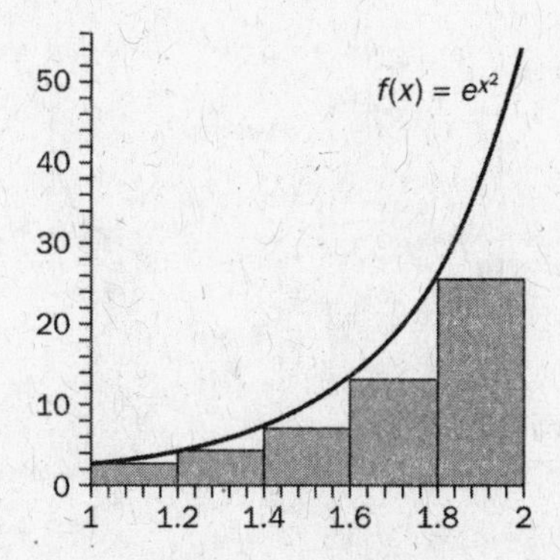

We divide the interval [1, 2] into five equal pieces to get the sampling points.

x_0	x_1	x_2	x_3	x_4	x_5
1	1.2	1.4	1.6	1.8	2

There are five rectangles, and the height of the ith rectangle is $f(x_{i-1})$. We can make a table of values:

Rectangle	1	2	3	4	5
Sampling point (left endpoint)	$x_0 = 1$	$x_1 = 1.2$	$x_2 = 1.4$	$x_3 = 1.6$	$x_4 = 1.8$
Height	e^{1^2}	$e^{1.2^2}$	$e^{1.4^2}$	$e^{1.6^2}$	$e^{1.8^2}$

Because there are five rectangles, Δx, the base of each rectangle is $\frac{2-1}{5} = 0.2$.

Therefore, we approximate

$$\int_1^2 e^{x^2}\,dx \text{ by } \underset{\text{left endpoint}}{R_5} = \sum_{i=1}^{n} f(x_{i-1})\Delta x = \left(e^{1^2} + e^{1.2^2} + e^{1.4^2} + e^{1.6^2} + e^{1.8^2}\right) \cdot 0.2$$

We use a calculator to compute this value and find: $\int_1^2 e^{x^2}\,dx \approx 10.502$

RIEMANN SUM USING THE RIGHT ENDPOINT

We can also use the right endpoint of each rectangle to approximate the area of the rectangle—a right Riemann sum. The right endpoint of each rectangle is *xi*, so the height of each rectangle is *f*(*xi*). In this case, the area is approximately:

$$\underset{\text{right endpoint}}{R_n} = \sum_{i=1}^{n} f(x_i)\Delta x = (f(x_1) + f(x_2) + \ldots + f(x_n))\Delta x$$

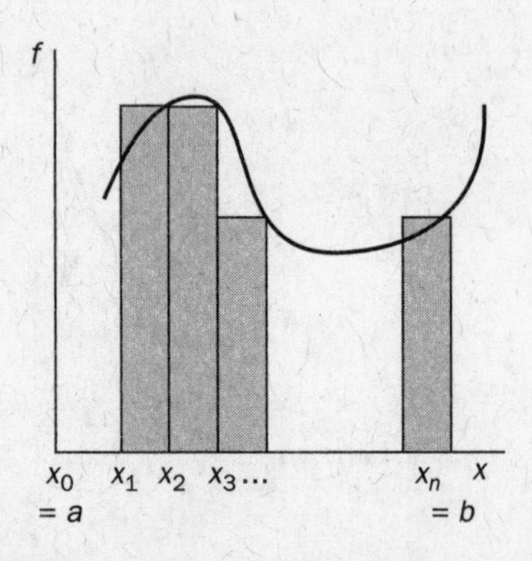

RIEMANN SUM USING THE MIDPOINT

If we use the midpoint of each rectangle to evaluate the height, then the height of each rectangle is half the distance from the left endpoint (x_{i-1}) to the right endpoint (x_i). The distance between the two endpoints is $\Delta x = \dfrac{b-a}{n}$, so half of this distance is $\dfrac{1}{2} \cdot \dfrac{b-a}{n} = \dfrac{b-a}{2n}$. The midpoint of the rectangle is, therefore, $x_{i-1} + \frac{b-a}{2}$, and the height of each rectangle is $f\left(x_{i-1} + \frac{b-a}{2}\right)$.

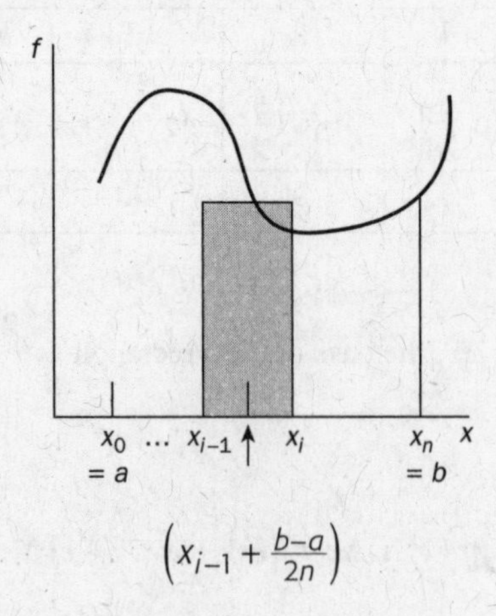

$$\underset{\text{midpoint}}{R_n} \approx \sum_{i=1}^{n} f\left(x_{i-1} + \tfrac{b-a}{2n}\right)\Delta x = \left(f\left(x_0 + \tfrac{b-a}{2n}\right) + f\left(x_1 + \tfrac{b-a}{2n}\right) + \ldots + f\left(x_{n-1} + \tfrac{b-a}{2n}\right)\right)\frac{b-a}{n}$$

TRAPEZOIDAL APPROXIMATION

Instead of approximating the area under the curve using a sum of rectangles, we can approximate the area under the curve using n trapezoidal regions, as shown below.

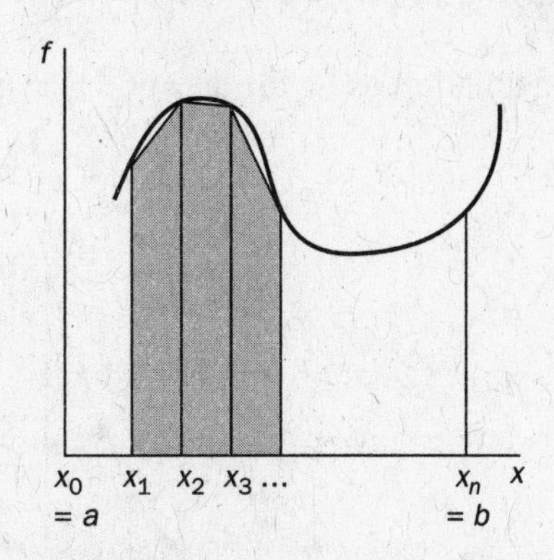

This type of approximation is called *trapezoidal approximation* and we denote the approximation using n trapezoids by T_n.

The area of a trapezoid is $\frac{1}{2}(b_1 + b_2)h$, where b_1 and b_2 are the lengths of the parallel sides and h is the perpendicular distance between them. The area of the ith trapezoid in a trapezoidal approximation is shown below:

$$b_1 = f(x_{i-1}),\ b_2 = f(x_i) \text{ and } h = x_i - x_{i-1} = \Delta x = \frac{b-a}{n}$$

That is, the area of each trapezoid is given by $\frac{1}{2}(f(x_{i-1}) + f(x_i)) \cdot \Delta x$.

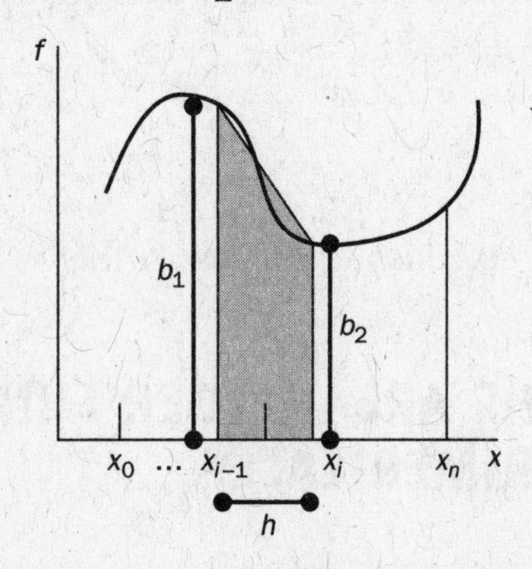

When we add up the sum of the areas of all the trapezoids we obtain:

$$\begin{aligned} T_n &= \left(\frac{1}{2}(f(x_0) + f(x_1)) \cdot \Delta x\right) + \left(\frac{1}{2}(f(x_1) + f(x_2)) \cdot \Delta x\right) + \cdots + \left(\frac{1}{2}(f(x_{n-1}) + f(x_n)) \cdot \Delta x\right) \\ &= \left(\frac{1}{2}f(x_0) + f(x_1) + f(x_2) + \cdots + f(x_{n-1}) + \frac{1}{2}f(x_n)\right)\Delta x \end{aligned}$$

Unlike approximation by rectangles, in which we choose the sampling point used to evaluate the height of each rectangle, there are no choices to be made with trapezoidal approximation beyond the number of trapezoids n we wish to use to approximate the area.

For comparison, let's now evaluate the area under the curve $f(x) = e^{x^2}$ on the interval $1 \leq x \leq 2$, i.e., $\int_1^2 e^{x^2}\,dx$, using T_5—an approximation using the area of five trapezoids.

The area is depicted below.

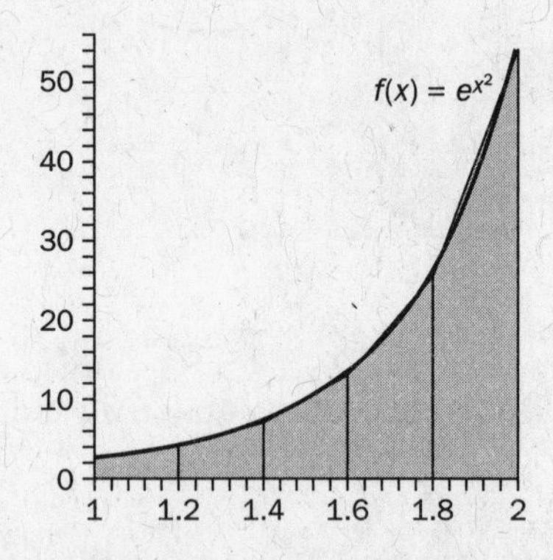

We use the same chart as earlier—now adding the data for $x_5 = 2$.

Sampling point x_i	$x_0 = 1$	$x_1 = 1.2$	$x_2 = 1.4$	$x_3 = 1.6$	$x_4 = 1.8$	$x_5 = 2$
$f(x_i)$	e^{1^2}	$e^{1.2^2}$	$e^{1.4^2}$	$e^{1.6^2}$	$e^{1.8^2}$	e^{2^2}

As earlier, $\Delta x = 0.2 = \frac{21-1}{5}$. We now use this data in the formula for T_5.

$$T_5 = \left(\frac{1}{2}e^{1^2} + e^{1.2^2} + e^{1.4^2} + e^{1.6^2} + e^{1.8^2} + \frac{1}{2}\cdot e^{2^2}\right)\cdot 0.2$$

We use a calculator to compute this value and find that $\int_1^2 e^{x^2}\,dx \approx T_5 = 15.690$.

APPROXIMATING THE AREA UNDER FUNCTIONS GIVEN GRAPHICALLY OR NUMERICALLY

The approximation methods in this section also work for functions given graphically or numerically (with a table of values), as long as we can determine $f(x_i)$ for each sampling point x_i in the interval $[a,b]$.

Example:

The speed (in km/hr) of a migrating bird is a differentiable function s of time t (in hours). The table below shows the velocity of the bird measured every hour from 7:00 a.m. to 5:00 p.m. on a single day of travel.

time	7:00 A.M.	8:00 A.M.	9:00 A.M.	10:00 A.M.	11:00 A.M.	12:00 P.M.	1:00 P.M.	2:00 P.M.	3:00 P.M.	4:00 P.M.	5:00 P.M.
velocity	35	36	40	42	40	42	41	37	35	33	33

Approximate the distance traveled by the bird over the 10-hour period using a trapezoidal approximation with subintervals of length 2 hours.

Solution:

The distance traveled by the bird is the area under the velocity curve from time $t = 0$ to time $t = 10$, where time $t = 0$ corresponds to 7:00 a.m. Because the length of each subinterval is 2 hours, we are approximating the area using five trapezoids. Our sampling points are therefore

t_0	t_1	t_2	t_3	t_4	t_5
0	2	4	6	8	10
7.00 A.M.	9.00 A.M.	11.00 A.M.	1.00 P.M.	3.00 P.M.	5.00 P.M.

We use the data given in the table and these sampling points to compute T_5; in this computation, the length of each trapezoid is 2:

$$T_5 = \left(\frac{1}{2} \cdot 35 + 40 + 40 + 41 + 35 + \frac{1}{2} \cdot 33\right) \cdot 2 = 380 \text{ km}$$

The distance traveled by the bird is thus 380 km.

REVIEW QUESTIONS

Note: All review questions require a calculator.

1. Let f be a continuous function on the closed interval [0, 8]. If the values of f are given below at five points, find the trapezoidal approximation of $\int_0^8 f(x)dx$ using four subintervals of equal length.

x	0	2	4	6	8
$f(x)$	7	4	11	5	5

(A) 26
(B) 36
(C) 42
(D) 48
(E) 52

2. Let g be a continuous function on the closed interval [1, 11]. If the values of f are given below at three points, use trapezoids to approximate $\int_1^{11} g(x)dx$ using two subintervals.

x	1	9	11
$f(x)$	23	14	10

(A) 165
(B) 172
(C) 190.5
(D) 40
(E) 80

3. The expression $\frac{1}{10}\left[\left(\frac{1}{10}\right)^2+\left(\frac{2}{10}\right)^2+\left(\frac{3}{10}\right)^2+\dots+\left(\frac{10}{10}\right)^2\right]$ is a Riemann sum approximation of which integral?

(A) $\int_1^{10} x^2 dx$

(B) $\int_1^{10} \left(\frac{1}{x}\right)^2 dx$

(C) $\int_0^{1} \left(\frac{1}{x}\right)^2 dx$

(D) $\int_0^{1} x^2 dx$

(E) $\frac{1}{10}\int_0^{1} x^2 dx$

4. The integral $\int_0^1 \sin(x)dx$ is approximated by several Riemann sum approximations below. Which of the following is NOT a Riemann sum approximation?

(A) $\frac{1}{16}\left[\sin\left(\frac{1}{16}\right)+\sin\left(\frac{2}{16}\right)+\sin\left(\frac{3}{16}\right)+\dots+\sin\left(\frac{16}{16}\right)\right]$

(B) $\frac{1}{8}\left[\sin\left(\frac{0}{16}\right)+\sin\left(\frac{2}{16}\right)+\sin\left(\frac{4}{16}\right)+\dots+\sin\left(\frac{14}{16}\right)\right]$

(C) $\frac{1}{8}\left[\sin\left(\frac{2}{16}\right)+\sin\left(\frac{4}{16}\right)+\sin\left(\frac{6}{16}\right)+\dots+\sin\left(\frac{16}{16}\right)\right]$

(D) $\frac{1}{4}\left[\sin\left(\frac{2}{16}\right)+\sin\left(\frac{6}{16}\right)+\sin\left(\frac{10}{16}\right)+\sin\left(\frac{14}{16}\right)\right]$

(E) $\frac{1}{8}\left[\sin\left(\frac{0}{16}\right)+\sin\left(\frac{2}{16}\right)+\sin\left(\frac{4}{16}\right)+\dots+\sin\left(\frac{16}{16}\right)\right]$

5. The rate at which gas is flowing through a large pipeline is given in thousands of gallons per month in the chart below.

t (months)	0	3	6	9	12
$R(t)$ (1000 gallons/month)	43	62	56	60	68

Use a midpoint Riemann sum with two equal subintervals ($\Delta t = 6$) to approximate the number of gallons that pass through the pipeline in a year.

(A) 594
(B) 672
(C) 732
(D) 744
(E) 1068

6. The graph of a differentiable function f over the interval $[-5,15]$ is shown below. Find the trapezoidal approximation of the area under the curve using four equal subintervals.

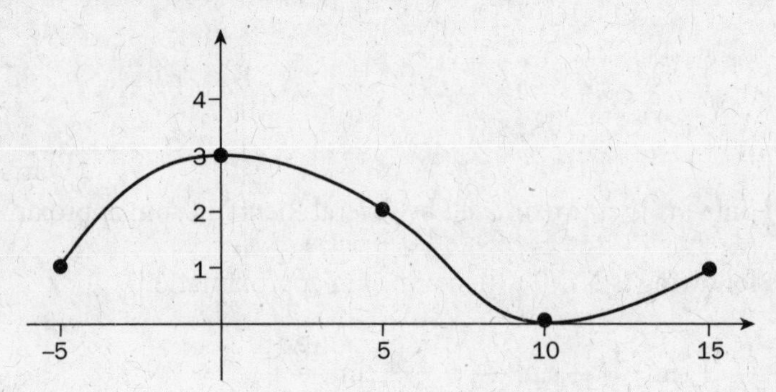

(A) 22
(B) 24
(C) 26
(D) 28
(E) 30

7. The temperature at the base of a mountain at a ski resort is measured in degrees Fahrenheit over a 12-day period in January. Approximate the average temperature over that period using a trapezoidal approximation with $\Delta t = 3$.

t (days)	0	3	6	9	12
F (°F)	20	25	22	28	15

(A) 21
(B) 23
(C) 24
(D) 25
(E) 26

8. Consider the function $y = f(x)$ whose graph is shown below, for which $f'(x) > 0$ and $f'(x) > 0 \forall x > 0$.

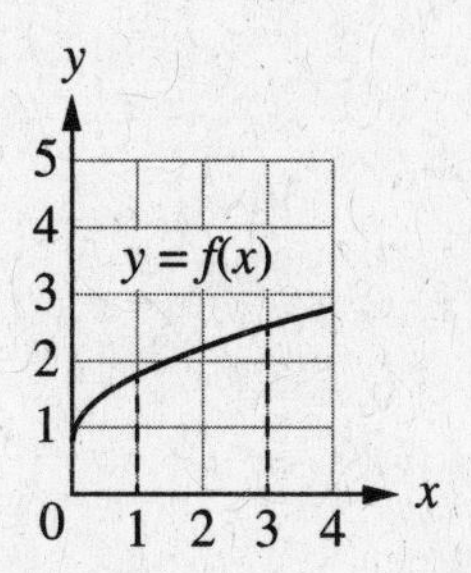

Which of the following will overestimate the value of $\int_1^3 f(x)dx$?

I. A left Riemann sum
II. A right Riemann sum
III. A trapezoidal approximation

(A) I only
(B) II only
(C) III only
(D) I and III only
(E) II and III only

Questions 9–11 refer to the table below, which gives values for the continuous function $f(x)$:

x	2	3	8	10	14
$f(x)$	3	7	9	5	1

9. Which of the following gives a ***left*** Riemann approximation using 4 subintervals of $\int_2^{14} f(x)dx$?

(A) 66
(B) 68
(C) 71
(D) 72
(E) 76

10. Which of the following gives a ***right*** Riemann approximation using 4 subintervals of $\int_{2}^{14} f(x)dx$?

(A) 66
(B) 68
(C) 71
(D) 72
(E) 76

11. Which of the following gives a ***trapezoidal*** approximation using 4 subintervals of $\int_{2}^{14} f(x)dx$?

(A) 66
(B) 68
(C) 71
(D) 72
(E) 76

12. Which of the following gives a ***midpoint*** Riemann approximation using 2 subintervals of $\int_{2}^{14} f(x)dx$ where *f* has values as given in the table below?

x	2	5	8	11	14
f(*x*)	3	7	9	5	1

(A) 66
(B) 68
(C) 71
(D) 72
(E) 76

13. If $\int_{-3}^{3} e^{\sin x}\, dx$ is approximated using a midpoint Riemann sum with 3 equal subintervals, the approximation is nearest to

(A) 2.885
(B) 3.885
(C) 4.537
(D) 5.771
(E) 7.771

14. The expression $\frac{1}{10}\sum_{x=1}^{10} e^{\frac{x}{10}}$ is an approximation of which of the following?

(A) $\frac{1}{10}\int_0^1 e^x\,dx$

(B) $\int_0^1 e^x\,dx$

(C) $\int_1^{10} e^{\frac{x}{10}}\,dx$

(D) $\int_0^{10} e^x\,dx$

(E) $\frac{1}{10}\int_0^{10} e^x\,dx$

15. A child rides a toboggan down a snow-covered hill, eventually stopping after 30 seconds when the hill levels out at the bottom. The speed of the toboggan (in meters per second) at various times is given in the chart below. Use a right Riemann sum to estimate the average speed of the toboggan during the 30-second ride.

time (seconds)	speed (m/s)
0	2
5	3
10	5
15	8
20	5
25	3
30	0

(A) 0.867 m/s
(B) 3.714 m/s
(C) 4 m/s
(D) 4.333 m/s
(E) 5 m/s

FREE-RESPONSE QUESTION

16. This problem requires a calculator.

    Let $f(x) = x + \sin x$ over the interval $[0,16]$. Evaluate the following for $f(x)$:

    (a) $\int_0^{16} (x + \sin x)\, dx$

    (b) Midpoint Reimann sum approximation over four equal subintervals

    (c) Trapezoidal approximation over four equal subintervals

17. A large water container develops a crack at its base and water begins to leak out. The rate at which the water was leaking out for the first 12 hours after the crack occurred is a strictly decreasing function, several values of which are given in the table below:

Time (hours)	0	2	5	7	12
Rate (liters/hour)	11	7.4	4.0	2.7	1.0

    (a) Use a left Riemann sum to estimate the amount of water that leaked out of the container during the first 12 hours.

    (b) Is this estimate greater or less than the actual amount of water that leaked out? Justify your answer.

    (c) Using the value obtained in part (a), estimate the average rate at which water was leaking out of the container over the 12-hour period.

    (d) A function that models the rate at which water flows out of the container is given by:

    $R(t) = 11e^{-\frac{t}{5}},\ 0 \le t \le 12$

    Use this model to determine the error in your solution to the Riemann sum in part (a).

ANSWERS AND EXPLANATIONS

1. E

The formula for a trapezoidal approximation of $\int_a^b f(x)dx$ with n subintervals is:

$$T_n = \Delta x\left[\frac{1}{2}f(x_0) + f(x_1) + f(x_2) + \cdots + f(x_{n-1}) + \frac{1}{2}f(x_n)\right]$$

where $\Delta x = \dfrac{b-a}{n}$. Therefore:

$$\int_0^8 f(x)dx \approx \frac{8-0}{4}\left[\frac{1}{2}\cdot 7 + 4 + 11 + 5 + \frac{1}{2}\cdot 5\right] = 52$$

2. B

Because the intervals are not equal in length, we cannot use the formula for T_n, which assumes that the intervals have equal length. Instead, we form two trapezoids, shown below, and add their areas.

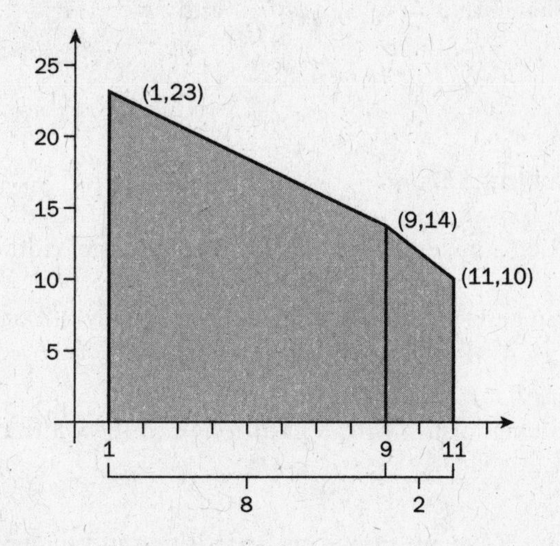

The area of a trapezoid is $A = \frac{1}{2}h(b_1 + b_2)$ where $h = \Delta x$ and b_1 and b_2 are y values.

$$\int_1^{11} f(x)dx \approx \frac{1}{2}\cdot 8(23+14) + \frac{1}{2}\cdot 2(14+10) = 172$$

3. D

The formula for R_n is $R_n = (f(c_1) + f(c_2) + \cdots f(c_n))\Delta x$. In the expression given, the number of summands is 10; we set the expression equal to R_{10}:

$$R_{10} = \Delta x(f(c_1) + f(c_2) + f(c_3) + \ldots + f(c_{10}))$$
$$= \frac{1}{10}\left[\left(\frac{1}{10}\right)^2 + \left(\frac{2}{10}\right)^2 + \left(\frac{3}{10}\right)^2 + \cdots + \left(\frac{10}{10}\right)^2\right]$$

Examining the equality, we see that $\Delta x = \frac{1}{10}$.

$$f(c_1) = \left(\frac{1}{10}\right)^2, f(c_2) = \left(\frac{2}{10}\right)^2, f(c_3) = \left(\frac{3}{10}\right)^2, \ldots, f(c_{10}) = 1^2$$

If $c_1 = \frac{1}{10}, c_2 = \frac{2}{10}, \ldots, c_n = 1$, then $f(c_i) = (c_i)^2 \Rightarrow f(x) = x^2$. This is just one possibility—let's see if it matches one of the answer choices.

Recall that $x_{i-1} \leq c_i \leq x_i$. Therefore $a = x_0 \leq c_1 = \frac{1}{10}$ and $b = x_n \geq c_n = 1$. Choice (D) matches these criteria.

4. E

The formula is given in problem 3 above.

(A) True. This is a right Riemann sum with 16 partitions and the width of each partition is $\frac{1}{16}$.

(B) True. This is a left Riemann sum with 8 partitions and the width of each partition is $\frac{2}{16} = \frac{1}{8}$.

(C) True. This is a right Riemann sum with 8 partitions and the width of each partition is $\frac{2}{16} = \frac{1}{8}$.

(D) True. This is a midpoint Riemann sum with 4 partitions and the width of each partition is $\frac{4}{16} = \frac{1}{4}$.

(E) False. This looks like a left Riemann sum with 8 partitions, but the last entry is wrong.

5. C

The formula for the second Riemann sum approximation using midpoints is:

$$\underset{\text{midpoint}}{R_2} = \left(f\left(x_0 + \tfrac{b-a}{4}\right) + f\left(x_1 + \tfrac{b-a}{4}\right)\right)\Delta x.$$

Because we are using two subdivisions of equal length, our sampling points are $a = t_0 = 0$, $t_1 = 6$, $b = t_2 = 12$, and $\Delta t = 6$. We plug this information into the formula for the midpoint sum and use the chart to evaluate the function.

$$\underset{\text{midpoint}}{R_2} = (f(0+3) + f(6+3)) \cdot 6 = (f(3) + f(9)) \cdot 6 = (62 + 60) \cdot 6 = 732 \text{ gallons}$$

6. E

We use the formula for the fourth trapezoidal approximation:

$$T_4 = \Delta x\left[\frac{1}{2}f(x_0) + f(x_1) + f(x_2) + f(x_3) + \frac{1}{2}f(x_4)\right]$$

where $\Delta x = \dfrac{b-a}{4}$. In our case $\Delta x = \dfrac{15-(-5)}{4} = 5$

and

Sampling point	x_0	x_1	x_2	x_3	x_4
x_i	–5	0	5	10	15
$f(x_i)$	1	3	2	0	1

where we read the values $f(xi)$ from the graph. We compute:

$$T_4 = 5\left[\frac{1}{2} \cdot 1 + 3 + 2 + 0 + \frac{1}{2} \cdot 1\right] = 5 \cdot 6 = 30$$

7. B

The average value of a continuous function over a closed interval is given by $\dfrac{\int_a^b f(x)\,dx}{b-a}$.

In this case, we want to compute $\dfrac{\int_0^{12} F(t)\,dt}{12-0}$. We are told to approximate $\int_0^{12} F(t)\,dt$ using trapezoidal approximation with $\Delta t = 3$, that is, $\dfrac{12-0}{n} = 3 \Rightarrow n = 4$. The formula for T_4 is

$$T_4 = \Delta x\left[\frac{1}{2}f(x_0) + f(x_1) + f(x_2) + f(x_3) + \frac{1}{2}f(x_4)\right].$$

In our case, $T_4 = 3\left[\frac{1}{2}F(0) + F(3) + F(6) + F(9) + \frac{1}{2}F(12)\right]$

$$= 3\left[\frac{1}{2} \cdot 20 + 25 + 22 + 28 + \frac{1}{2} \cdot 15\right] = 277.5$$

Thus we approximate $\int_{0}^{12} F(t)\,dt \approx T_4 = 277.5$, and we approximate the average temperature

by $\dfrac{\int_{0}^{12} F(t)\,dt}{12-0} \approx \dfrac{277.5}{12} = 23.125.$

8. B

$\int_{1}^{3} f(x)\,dx$ refers to the area between $y = f(x)$ and the x-axis on the interval (1,3).

We know that the function is increasing, since $f'(x) > 0$, so a left approximation will lie under the curve and therefore underestimate the area, while a right approximation will lie above the curve and overestimate as shown below:

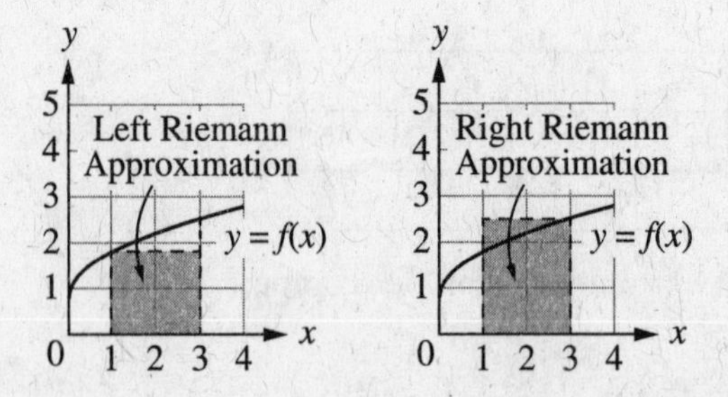

Therefore, statement I is false and statement II is true. This means we have only to consider the trapezoidal approximation. A trapezoidal approximation will always lie under the curve if the curve is concave down, and above the curve if it is concave up. In this case, we know that $f''(x) < 0$ and therefore the curve is concave down and the trapezoidal approximation will underestimate the area as shown below:

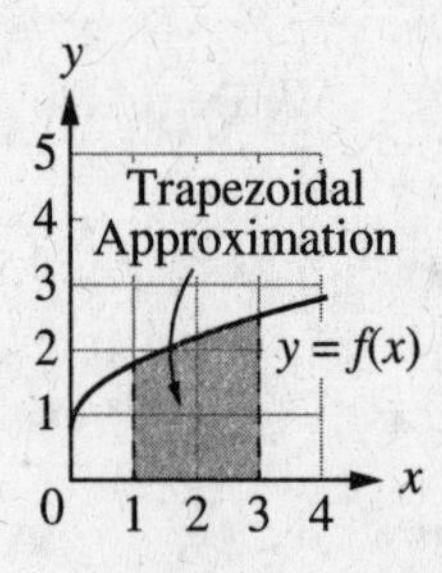

Therefore, statement III is false, and we choose option (B).

9. E

To calculate a left Riemann sum, we choose the left endpoint for each interval and multiply by the width of the interval. In this case, we have:

$$\begin{aligned}\int_2^{14} f(x)\,dx &\approx 3(3-2)+7(8-3)+9(10-8)+5(14-10)\\ &= 3(1)+7(5)+9(2)+5(4)\\ &= 3+35+18+20\\ &= 76\end{aligned}$$

10. A

To calculate a right Riemann sum, we choose the right endpoint for each interval and multiply by the width of the interval. In this case, we have:

$$\begin{aligned}\int_2^{14} f(x)\,dx &\approx 7(3-2)+9(8-3)+5(10-8)+1(14-10)\\ &= 7(1)+9(5)+5(2)+1(4)\\ &= 7+45+10+4\\ &= 66\end{aligned}$$

11. C

To calculate a trapezoidal approximation, we take the average of the heights at the left and right endpoints and multiply by the width of the interval. In this case, we have:

$$\begin{aligned}\int_2^{14} f(x)\,dx &\approx \left(\frac{3+7}{2}\right)(3-2)+\left(\frac{7+9}{2}\right)(8-3)+\left(\frac{9+5}{2}\right)(10-8)+\left(\frac{5+1}{2}\right)(14-10)\\ &= 5(1)+8(5)+7(2)+3(4)\\ &= 5+40+14+12\\ &= 71\end{aligned}$$

Another approach is to recognize that the trapezoidal approximation is always the average of the left and right Riemann sum approximations. Since we had already calculated these in questions 2 and 3, we could have calculated our trapezoidal approximation simply by averaging these two results:

$$\begin{aligned}Trap_4 &= \frac{LRAM_4 + RRAM_4}{2}\\ &= \frac{76+66}{2}\\ &= 71\end{aligned}$$

12. D

To calculate a midpoint Riemann sum, we choose the midpoint for each interval and multiply by the width of the interval. In this case, we have:

$$\begin{aligned}\int_2^{14} f(x)\,dx &\approx 7(8-2)+5(14-8)\\ &= 7(6)+5(6)\\ &= 42+30\\ &= 72\end{aligned}$$

13. E

The width of the interval from -3 to 3 is 6 units, therefore each of the three subintervals will have width 2, creating the intervals: (-3, -1), (-1, 1) and (1, 3), which have midpoints at x = -2, x = 0 and x = 2 respectively.

We can evaluate the function using our calculator at each of these three midpoints, leading to

$$\begin{aligned}&f(-2)=e^{\sin(-2)}\approx 0.403\\ &f(0)=e^{\sin(0)}=1\\ &f(2)=e^{\sin(2)}\approx 2.483\end{aligned}$$

We multiply each of these midpoint values by the width of the interval, which is 2, to arrive at the midpoint Riemann sum approximation:

$$\begin{aligned}\int_{-3}^{3} e^{\sin x}\,dx &\approx 2[f(-2)+f(0)+f(2)]\\ &\approx 2[0.403+1+2.483]\\ &\approx 7.771\end{aligned}$$

14. B

Perhaps a good way to begin this problem is by expanding the expression that is given in sigma notation so that we can make more sense of it:

$$\begin{aligned}\frac{1}{10}\sum_{x=1}^{10} e^{\frac{x}{10}} &= \left(\frac{1}{10}\right)\cdot\left(e^{\frac{1}{10}}+e^{\frac{2}{10}}+e^{\frac{3}{10}}+e^{\frac{4}{10}}+\ldots+e^{\frac{10}{10}}\right)\\ &= \underbrace{\left(\frac{1}{10}\right)}_{\text{rectangle width}}\cdot\underbrace{e^{\frac{1}{10}}}_{\text{rectangle height}}+\left(\frac{1}{10}\right)\cdot e^{\frac{2}{10}}+\left(\frac{1}{10}\right)\cdot e^{\frac{3}{10}}+\ldots+\left(\frac{1}{10}\right)\cdot e^{\frac{10}{10}}\end{aligned}$$

We can see that this is a series of 10 Riemannian rectangles, each of which has a width of $\frac{1}{10}$ and a height that is given by $e^{\frac{x}{10}}$ as x goes from 1 to 10. Since we have 10 rectangles, each of width $\frac{1}{10}$, we can determine that our interval is of length 1, going from 0 to 1 and ruling out choices (C), (D), and (E). If we graph the rectangles whose sum is given above, we have the following:

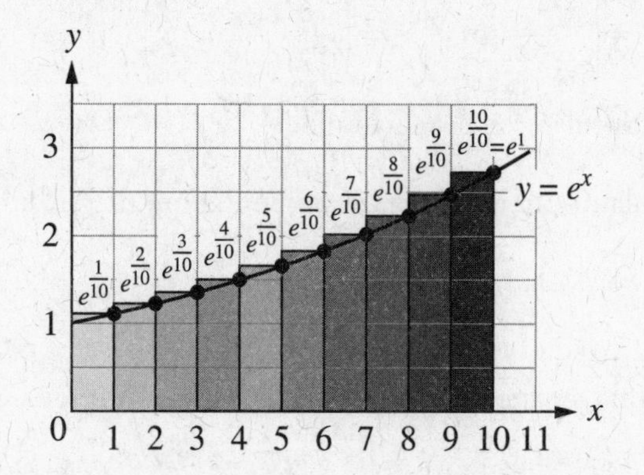

From this, we can see that the height of each rectangle is given by the function $y = e^x$, and that the sum of the areas of the rectangles yields an approximation for $\int_0^{10} e^x\,dx$, which is option (B).

15. C

If we let the speed of the toboggan be represented by $v(t)$, then the average speed over the 30-second interval would be given by $\frac{1}{30-0}\int_0^{30} v(t)\,dt$. Since we are not given a function for $v(t)$, we must estimate the value of $\int_0^{30} v(t)\,dt$, in this case using a right Riemann sum. We do this by taking the right endpoint for each rectangle and multiplying the width of the rectangle (5 for each in this case as the times are given at equal intervals of 5 seconds) by the function value at that point. Thus:

$$\begin{aligned}\int_0^{30} v(t)\,dt &\approx 5\cdot 3+5\cdot 5+5\cdot 8+5\cdot 5+5\cdot 3+5\cdot 0\\ &= 5\cdot(3+5+8+5+3)\\ &= 5\cdot 24\\ &= 120\end{aligned}$$

We use this estimate in our average value formula to determine the average speed:

$$\begin{aligned}\frac{1}{30-0}\int_0^{30} v(t)\,dt &= \frac{1}{30}(120)\\ &= 4\text{ m/s}\end{aligned}$$

Therefore, the average speed of the toboggan is approximately 4 meters per second.

FREE-RESPONSE ANSWER

16. (a) $\int_0^{16} (x + \sin x)\, dx = \left[\frac{x^2}{2} - \cos x\right]_0^{16} = \left(\frac{16^2}{2} - \cos 16\right) - \left(\frac{0^2}{2} - \cos 0\right) \underset{\substack{\text{evaluate with}\\ \text{a calculator}}}{\approx} 129.958$

Make sure your calculator is in radian mode!

A graphing calculator can do the entire computation; that is, it can compute $\int_0^{16} (x + \sin x)\, dx$. See Chapter 6 for instructions.

(b) We use the formula for the midpoint Riemann sum, with four subintervals:

$$\underset{\text{midpoint}}{R_4} = \left(f\left(x_0 + \tfrac{b-a}{2n}\right) + f\left(x_1 + \tfrac{b-a}{2n}\right) + f\left(x_2 + \tfrac{b-a}{2n}\right) + f\left(x_3 + \tfrac{b-a}{2n}\right)\right)\Delta x$$

In this question:

$$a = x_0 = 0,\ x_1 = 4,\ x_2 = 8,\ x_3 = 12,\ b = x_4 = 16$$

$$\Delta x = \frac{16-0}{4} = 4; \qquad \frac{b-a}{2n} = \frac{16-0}{2 \cdot 4} = \frac{16}{8} = 2$$

We substitute these values into the expression for the approximation:

$$\underset{\text{midpoint}}{R_4} = ((2 + \sin 2) + (6 + \sin 6) + (10 + \sin 10) + (14 + \sin 14)) \cdot 4$$

$$\underset{\substack{\text{evaluate with}\\ \text{a calculator}}}{\approx} 132.306$$

Make sure your calculator is in radians mode!

(c) We use the formula for a trapezoidal approximation with four subintervals:

$$T_4 = \Delta x\left[\frac{1}{2}f(x_0) + f(x_1) + f(x_2) + f(x_3) + \frac{1}{2}f(x_4)\right].$$

In this question:

$$a = x_0 = 0,\ x_1 = 4,\ x_2 = 8,\ x_3 = 12,\ b = x_4 = 16$$

$$\Delta x = \frac{16-0}{4} = 4$$

We substitute this information into the approximation formula

$$T_4 = 4\left[\frac{1}{2}(0+\sin 0)+(4+\sin 4)+(8+\sin 8)+(12+\sin 12)+\frac{1}{2}(16+\sin 16)\right] \underset{\substack{\text{evaluate using}\\ \text{a calculator}}}{\approx} 126.208.$$

Make sure your calculator is in radians mode!

17. (a) To calculate a left Riemann sum, we choose the left endpoint for each interval and multiply by the width of the interval. In this case, we have

$$11 \cdot (2-0) + (7.4) \cdot (5-2) + (4) \cdot (7-5) + (2.7) \cdot (12-7)$$
$$= 22 + 22.2 + 8 + 13.5$$
$$= 65.7$$

Therefore, approximately 65.7 L of water leaked out of the container in the first 12 hours.

(b) As this is a strictly decreasing function, choosing the left endpoint at each subinterval will cause the rectangle to be above the curve. Thus, the left Riemann sum will overestimate the actual amount of water that leaked out.

(c) The average value of a function f on the interval [a, b] is given by $\frac{1}{b-a}\int_a^b f(x)\,dx$. We use the estimated value from our Riemann sum in part (a) to get:

$$f_{av} = \frac{1}{12-0}\int_a^b f(x)\,dx \approx \frac{1}{12}(65.7) = 5.475\,\frac{\text{L}}{\text{hr}}$$

(d) The amount of water predicted to leak out of the container by the model is given by:

$$\int_0^{12} 11e^{-\frac{t}{5}}\,dt = 50.011\,\text{L}$$

As predicted in part (b), this is less than our estimated value from the right Riemann sum. The error in our estimate is the difference between the estimated value and the actual value predicted by the model 65.7 – 50.011 = 15.689 L. Therefore, our Riemann sum overestimated the actual amount by 15.689 liters.

CHAPTER 24: THE CONCEPT OF SERIES*

IF YOU LEARN ONLY TWO THINGS IN THIS CHAPTER . . .

1. A sequence $\{a_n\}$ has a limit L if the terms a_n become successively closer to L as n approaches infinity: $\lim_{n \to \infty} a_n = L$ or simply $a_n \to L$ as $n \to \infty$.
2. Notation for an infinite series: $\sum_{n=1}^{\infty} a_n = a_1 + a_2 + a_3 + \ldots + a_n + \ldots$

THE CONCEPT OF SERIES

Before we can begin working with series we need to develop the language of sequences. A sequence is just a list of numbers that are written down in a particular order:

$$a_1, a_2, a_3, a_4, \ldots, a_n, \ldots$$

To avoid having to write down infinitely many terms each time we want to refer to a given sequence, we use the notation $\{a_n\}$ or $\{a_n\}_{n=1}^{\infty}$ to represent the sequence above. Note that in this and many other sequences we'll use in this section, the n is indexed from 1 to infinity. In general, however, the index n can begin at any number.

For each positive integer n we have a corresponding element of the sequence a_n. Many times, a formula is used to define the nth element of a sequence.

For example, consider the sequence below.

$$1, \frac{1}{2}, \frac{1}{3}, \frac{1}{4}, \frac{1}{5}, \ldots$$

*BC content

Clearly, the nth element is given by the formula $a_n = \frac{1}{n}$.

What are the elements of the sequence determined by $\left\{a_n = \frac{n^2 - 3n}{n+1}\right\}$?

Let's see; the first element is a_1. When $n = 1$, we have $\frac{1^2 - 3(1)}{1+1} = \frac{-2}{2} = -1$. That isn't too bad.

Now for the second element, we let $n = 2$. We get $a_2 = \frac{2^2 - 3(2)}{2+1} = \frac{-2}{3}$. What is the third element? The fourth? Using this formula we can determine each element of this sequence. We get the infinite sequence

$$\left\{\frac{n^2 - 3n}{n+1}\right\}_{n=1}^{\infty} = \left\{-1, \frac{-2}{3}, 0, \frac{4}{5}, \ldots\right\}.$$

Consider the sequence given by $\left\{\frac{n-1}{2n}\right\}_{n=1}^{\infty}$. Starting at $n = 1$, we can write out the elements of this sequence to get

$$\left\{\frac{n-1}{2n}\right\}_{n=1}^{\infty} = \left\{0, \frac{1}{4}, \frac{1}{3}, \frac{3}{8}, \frac{2}{5}, \frac{5}{12}, \frac{3}{7}, \ldots\right\}$$

To get some idea of what this sequence looks like we can plot the points on a graph as the function $f(n) = \frac{n-1}{2n}$, where n only takes the values $n = 1, 2, 3, 4, 5, \ldots$

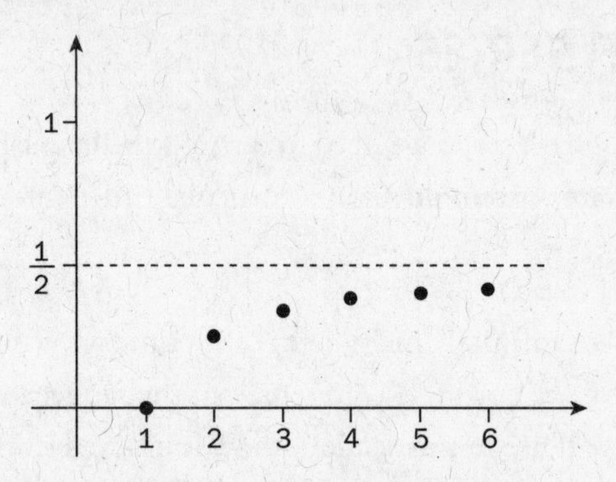

It is natural to ask if a sequence of numbers approaches a limit. For example, in the figure above, it is clear that the sequence approaches the number $\frac{1}{2}$. We call this number the limit of the sequence.

Definition:

A sequence $\{a_n\}$ has a limit L if the terms a_n become successively closer to L as n approaches infinity. We write $\lim_{n\to\infty} a_n = L$ or simply $a_n \to L$ as $n \to \infty$. Furthermore, if a sequence has a finite limit L, then we say the sequence is *convergent*. If, on the other hand, the sequence does not approach any limit L, then we say the sequence is *divergent*.

We have already seen an example of a convergent sequence. However, consider the sequence $\{-1^n\}_{n=1}^{\infty} = \{-1, 1, -1, 1, -1, 1, \ldots\}$. This sequence is divergent. If we plot the points on a graph this becomes immediately clear.

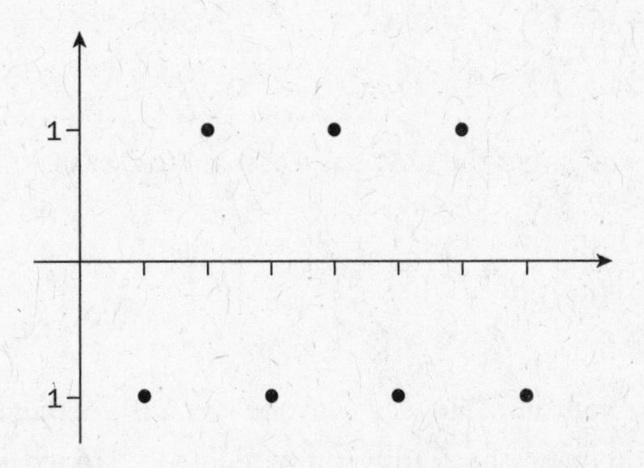

Clearly, $\lim_{n\to\infty} (-1)^n =$ does not exist. Another important type of divergence occurs when the terms a_n become progressively larger as n becomes larger, as in the sequence $\{2n\}$. When this occurs, we say the sequence diverges to infinity and we write $\lim_{n\to\infty} a_n = \infty$.

To conclude our discussion of sequences we mention some important classifications of sequences. We say a sequence is increasing if the terms of the sequence are increasing, meaning $a_n < a_{n+1}$ for each n. Similarly, a sequence is called decreasing if the terms of the sequence are decreasing, meaning $a_n > a_{n+1}$. Lastly, a sequence is said to be bounded if there exists a real number M such that $a_n \leq M$, for all n.

FROM SEQUENCES TO SERIES VIA PARTIAL SUMS

When we add all the terms of an infinite sequence we get an infinite series. This infinite series is denoted:

$$\sum_{n=1}^{\infty} a_n = a_1 + a_2 + a_3 + \ldots + a_n + \ldots$$

You may be thinking to yourself that if we add together infinitely many numbers then surely the result will also be infinite. After all, infinitely many numbers is a lot of numbers, right? Indeed, this can happen. The series:

$$\sum_{n=1}^{\infty} n = 1 + 2 + 3 + 4 + 5 + \ldots + n + \ldots$$

cannot have a finite sum. However, this is certainly not the end of the story. Sometimes the infinite sum will actually equal a finite number.

If each term of the series becomes progressively small enough then our infinite series will yield a finite sum. For example, if we try to evaluate the series:

$$\sum_{n=1}^{\infty} \frac{1}{2^n} = \frac{1}{2} + \frac{1}{4} + \frac{1}{8} + \frac{1}{16} + \ldots + \frac{1}{2^n} + \ldots$$

we see that as n gets larger and larger the sum gets closer and closer to 1. Thus we can say that

$$\sum_{n=1}^{\infty} \frac{1}{2^n} = \frac{1}{2} + \frac{1}{4} + \frac{1}{8} + \frac{1}{16} + \ldots + \frac{1}{2^n} + \ldots = 1$$

So now you must be asking yourself, "How can anyone ever tell whether a series is finite or infinite?" This is a very good question; to answer it, we look back to what we know about sequences.

TO CONVERGE OR NOT TO CONVERGE?

To determine whether a given series is finite or infinite, we examine the partial sums of the series. With these partial sums, we can create a new sequence $\{s_k\}$ which has terms defined by

$$\begin{aligned} s_1 &= a_1 \\ s_2 &= a_1 + a_2 \\ s_3 &= a_1 + a_2 + a_3 \\ &\ldots \\ s_k &= a_1 + a_2 + a_3 + \ldots + a_k \\ &\ldots \end{aligned}$$

If this new sequence $\{s_k\}$ has a limit then the series is finite and is equal to the limit of the sequence $\{s_k\}$. So, if $\lim_{k \to \infty} s_k = S$ (and S is finite) then we write:

$$\sum_{n=1}^{\infty} a_n = S$$

But what happens if our sequence of partial sums does not have a finite limit? What if $\lim_{k \to \infty} s_k = \infty$? Intuitively, this means that as we sum up more and more terms in our series, we are approaching infinity. Clearly, any series with this property cannot have a finite sum.

Now is a good time to introduce some terminology to label these two very different cases.

Definition:

Consider the series

$$\sum_{n=1}^{\infty} a_n = a_1 + a_2 + a_3 + \ldots + a_n + \ldots$$

with partial sums given by

$$s_k = \sum_{n=1}^{k} a_n = a_1 + a_2 + a_3 + \ldots + a_k$$

This series is said to converge (or is convergent) if $\lim_{k \to \infty} s_k = S$, where S is a real number less than infinity. This series is said to diverge (or is divergent) if $\lim_{k \to \infty} s_k = \infty$.

Great! Now using what we know about convergent and divergent sequences we can determine if a given series converges to a finite number or diverges to infinity by simply looking at the sequence of its partial sums.

Unfortunately, the entire story is not that simple. As we saw in some of the examples, writing down an explicit formula for the partial sums can be a challenge on its own. In addition, even if we know that a series converges, we still do not have enough tools to fully determine *exactly* what number the series converges to. In the coming chapters we hope to tie up these loose ends. We will see some tests that can be used to determine whether or not a series converges. We will also discover some important series that will help evaluate a given series.

REVIEW QUESTIONS

1. Find a formula for the general term a_n for the sequence below.

$$\left\{-2, \frac{3}{4}, -\frac{4}{9}, \frac{5}{16}, -\frac{6}{25}, \ldots\right\}$$

(A) $a_n = \dfrac{(n+1)}{n^2}$

(B) $a_n = \dfrac{(-1)^n(n+2)}{n^2}$

(C) $a_n = \dfrac{(-1)^n(n+1)}{n^2}$

(D) $a_n = \dfrac{n}{(n+1)^2}$

(E) $a_n = \dfrac{(-1)^{n+1}(n+1)}{n^2}$

2. Determine whether the series $\sum_{n=1}^{\infty} \dfrac{(2n+1)^2}{n(2n+4)}$ is convergent or divergent and evaluate the series.

(A) The series is convergent and equals 56.
(B) The series is convergent and equals 31π.
(C) The series is convergent and equals 42.
(D) The series is convergent and equals 16.
(E) The series is divergent.

3. Which sequence below follows the rule $a_n = \dfrac{(-1)^{n+1}2^{n-1}}{n^2}$?

(A) $\left\{1, \frac{-1}{2}, \frac{4}{9}, \frac{-1}{2}, \ldots\right\}$

(B) $\left\{1, \frac{-1}{2}, \frac{4}{7}, \frac{-1}{3}, \ldots\right\}$

(C) $\left\{1, \frac{1}{2}, \frac{4}{9}, \frac{1}{2}, \ldots\right\}$

(D) $\left\{1, \frac{-3}{2}, \frac{2}{9}, \frac{-2}{3}, \ldots\right\}$

(E) $\left\{\frac{-1}{3}, \frac{1}{2}, \frac{-4}{3}, \frac{1}{4}, \ldots\right\}$

4. Which choice below best describes a sequence with terms defined by $a_n = \dfrac{1}{2n+3}$?

(A) Increasing, bounded
(B) Decreasing, bounded
(C) Increasing, not bounded
(D) Decreasing, not bounded
(E) Neither increasing nor decreasing, bounded

5. What is the tenth term in the following sequence $\left\{\frac{-1}{2}, \frac{2}{4}, \frac{-3}{8}, \frac{4}{16}, \frac{-5}{32}, \ldots\right\}$?

(A) $\dfrac{10}{512}$

(B) $\dfrac{9}{1024}$

(C) $\dfrac{-6}{79}$

(D) $\dfrac{-9}{512}$

(E) $\dfrac{5}{512}$

6. Suppose $\sum_{n=1}^{\infty} a_n$ is a series with corresponding partial sums $\{s_k\}$. The graph below plots the values of $\{s_k\}$ for $k = 1, 2, 3\ldots$

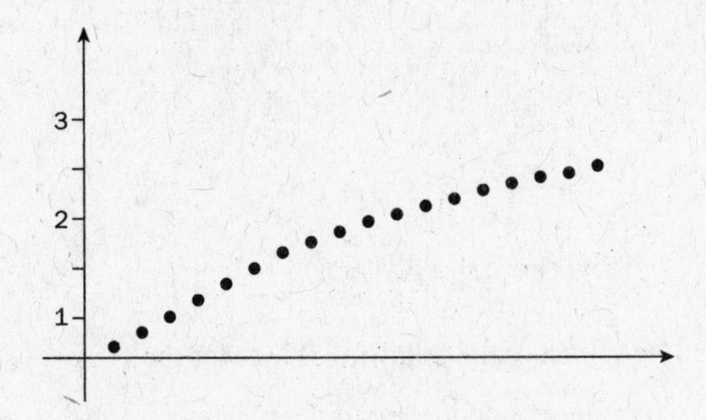

Which statement below is FALSE?

(A) $\lim_{n\to\infty} a_n = \frac{1}{2}$

(B) $\lim_{n\to\infty} a_n = 0$

(C) $\lim_{k\to\infty} s_k = 2\frac{1}{2}$

(D) This series $\sum_{n=1}^{\infty} a_n$ is convergent.

(E) The a_n are all positive.

7. For what values of r is the sequence $\{r^n\}$ convergent?

(A) $r > 1$

(B) $-1 < r \leq 1$

(C) $-1 \leq r \leq 1$

(D) $-1 < r < 1$

(E) The sequence is never convergent.

8. Which series below is divergent?

(A) $\sum_{n=1}^{\infty} \frac{n^3+1}{3e^n+1}$

(B) $\sum_{n=1}^{\infty} \frac{n^2+n}{3n^3-7}$

(C) $\sum_{n=1}^{\infty} \frac{n^3+n^2}{3n^3-7}$

(D) $\sum_{n=1}^{\infty} \frac{(-1)^n(n+1)^2}{3n^3}$

(E) $\sum_{n=1}^{\infty} \frac{7}{3n^3}$

9. Which of the following sequences is generated by the formula $\left\{\frac{n-1}{n^2}\right\}_{n=1}^{\infty}$?

(A) $\left\{2, \frac{3}{4}, \frac{4}{9}, \frac{5}{16}, \frac{6}{25} \ldots\right\}$

(B) $\left\{\frac{1}{4}, \frac{2}{9}, \frac{3}{64}, \frac{4}{125} \ldots\right\}$

(C) $\left\{0, \frac{1}{4}, \frac{2}{9}, \frac{3}{16}, \frac{4}{25} \ldots\right\}$

(D) $\left\{0, \frac{1}{2}, \frac{2}{3}, \frac{3}{4}, \frac{4}{5} \ldots\right\}$

(E) $\left\{0, \frac{1}{4}, \frac{2}{9}, \frac{3}{64}, \frac{4}{125} \ldots\right\}$

10. Which of the following formulas represent the sequence $-1, \frac{1}{2}, -\frac{1}{4}, \frac{1}{8} \ldots$?

(A) $\left\{\frac{(-1)^n}{2^n}\right\}_{n=0}^{\infty}$

(B) $\left\{\frac{(-1)^n}{2^n}\right\}_{n=1}^{\infty}$

(C) $\left\{\frac{(-1)^n}{2n}\right\}_{n=0}^{\infty}$

(D) $\left\{\frac{(-1)^{n+1}}{2^n}\right\}_{n=0}^{\infty}$

(E) $\left\{\frac{(-1)^{n+1}}{2^n}\right\}_{n=1}^{\infty}$

11. Which (if any) of the three series listed below are equivalent?

I. $\sum_{n=1}^{\infty} a_n$

II. $\sum_{n=0}^{\infty} a_{n-1}$

III. $\sum_{n=0}^{\infty} a_{n+1}$

(A) I and II only
(B) I and III only
(C) II and III only
(D) I, II, and III
(E) None of these are equivalent.

12. Determine the sum of the series generated by $\sum_{n=0}^{\infty} \frac{2n^3 + \cos(n)}{(2n+1)^3 + e^{-n}}$.

(A) 0
(B) $\frac{4}{3}$
(C) $\frac{1}{1+\frac{1}{e}}$
(D) 1
(E) The series diverges.

13. Which of the following is true for the sequence defined by $\sum_{n=1}^{\infty} 1 - n^2$?

(A) The series is increasing and unbounded.
(B) The series is decreasing and unbounded.
(C) The series is increasing and bounded.
(D) The series is decreasing and bounded.
(E) The series is neither increasing or decreasing, and it is unbounded.

14. Consider the sequence $a_1, a_2, a_3, a_4 \ldots a_n$ and the corresponding series $\sum_{n=1}^{\infty} a_n$ with partial sums given by $S_k = a_1 + a_2 + a_3 + a_4 + \ldots + a_k$. Which of the following statements MUST be true?

    I. If $\lim_{n\to\infty} a_n = 6$, the series converges.

    II. If $\lim_{k\to\infty} s_k = 6$, the series converges.

    III. If the sequence is divergent, then $\lim_{n\to\infty} a_n = \infty$.

    (A) I only
    (B) II only
    (C) III only
    (D) I and II only
    (E) II and III only

15. Consider the sequence $a_1, a_2, a_3, a_4 \ldots a_n$ and the corresponding series $\sum_{n=1}^{\infty} a_n$. Which of the following statements MUST be true?

    I. If $\sum_{n=1}^{\infty} a_n$ converges, then $\lim_{n\to\infty} a_n = 0$.

    II. If $\lim_{n\to\infty} a_n \neq 0$, then $\sum_{n=1}^{\infty} a_n$ diverges.

    III. If $\lim_{n\to\infty} a_n = 0$, then $\sum_{n=1}^{\infty} a_n$ converges.

    (A) I and II only
    (B) I and III only
    (C) II and III only
    (D) I, II, and III.
    (E) None of the above

FREE-RESPONSE QUESTION

16. Determine whether the sequence defined by $\left\{\dfrac{\sin^2 n}{2^n}\right\}$ is convergent or divergent.
17. Determine whether the series: $\sum_{n=1}^{\infty} n^{\frac{1}{n}}$ is convergent or divergent

ANSWERS AND EXPLANATIONS

1. C

We have the terms $a_1 = -2$, $a_2 = \frac{3}{4}$, $a_3 = -\frac{4}{9}$, $a_4 = \frac{5}{16}$, $a_5 = -\frac{6}{25}$, ... and we notice three things: First, the denominators change by n^2; second, the numerators obey $n + 1$; third, the series alternates by $(-1)^n$. Combining all this information we get $a_n = \frac{(-1)^n(n+1)}{n^2}$.

2. E

If the terms of a series don't go to zero as n approaches infinity then the series cannot possibly be finite; that is, it cannot possibly converge. Note that $\lim_{n\to\infty} \frac{(2n+1)^2}{n(2n+4)} = \lim_{n\to\infty} \frac{4n^2+4n+1}{2n^2+4n} = \frac{4}{2} = 2 \neq 0.$ Therefore the series is divergent.

3. A

To find the proper series we need to only evaluate a few terms of the series given by $\left\{a_n = \frac{(-1)^{n+1}2^{n-1}}{n^2}\right\}$.

We see that $a_1 = 1$, $a_2 = \frac{-2}{4} = \frac{-1}{2}$, $a_3 = \frac{2^2}{9} = \frac{4}{9}$, $a_4 = \frac{-2^3}{16} = \frac{-1}{2}$, ... so the series must be $\left\{1, \frac{-1}{2}, \frac{4}{9}, \frac{-1}{2}, \ldots\right\}$.

4. B

We are given $a_n = \frac{1}{2n+3}$ so $a_{n+1} = \frac{1}{2n+2+3} = \frac{1}{2n+5}$. Because $2n+5 > 2n+3$ for all $n = 1, 2, 3, \ldots$ we get that $a_n > a_{n+1}$, so the sequence must be decreasing. Furthermore, since the sequence is decreasing, a_1 is the largest term and so the series must also be bounded.

5. E

This question really has two parts. We need to determine the general formula for the nth term of the sequence and then substitute $n = 10$ to find the tenth term. When we look at the sequence we notice three things:

1. It is alternating, so we have a factor of $(-1)^n$.
2. The denominators are increasing in powers of 2^n.
3. The numerator is increasing with n.

Therefore, the general term for the sequence must be $a_n = \frac{(-1)^n n}{2^n}$ and if we check a couple of terms, this is definitely correct. Now to find the tenth term in the sequence, we let $n = 10$ to get $a_{10} = \frac{(-1)^{10}10}{2^{10}} = \frac{10}{1024} = \frac{5}{512}$ and we are done.

6. A

Clearly this series is convergent because the sequence of partial sums is approaching the value $2\frac{1}{2}$. Therefore, $\lim_{n\to\infty} a_n \neq \frac{1}{2}$. In fact, if a series is convergent then $\lim_{n\to\infty} a_n = 0$, and we were told that the series was convergent.

7. B

While discussing limits we learned how to compute the limit of an exponential function. Using this information, we have $\lim_{n\to\infty} r^n = \begin{cases} \infty & \text{if } r > 1 \\ 0 & \text{if } 0 < r < 1 \end{cases}$.
So we know the sequence converges for $0 < r < 1$ and diverges for $r > 1$. We still have more values of r to check.

When $r = 1$: $\lim_{n\to\infty} r^n = \lim_{n\to\infty} 1^n = 1$ so the sequence converges.

When $r = 0$: The limit is clearly 0, so the sequence converges.

When $-1 < r < 0$: Then $0 < |r| < 1$ and so $\lim_{n\to\infty} |r^n| = \lim_{n\to\infty} |r|^n = 0$ and this forces our original sequence to converge.

When $r = -1$: We can write out some terms of the sequence and see that it diverges; we get $\{-1, 1, -1, 1, -1, 1, \ldots\}$. Clearly the sequence does not converge to anything.

Compiling this information, our result is that the sequence $\{r^n\}_{n=1}^{\infty}$ converges for $-1 < r \leq 1$.

8. C

We want to find which series is divergent. Recall, the divergence theorem says that if $\lim_{n \to \infty} a_n \neq 0$, then the series is divergent. Given the series $\sum_{n=1}^{\infty} \frac{n^3 + n^2}{3n^3 - 7}$, we calculate $\lim_{n \to \infty} \frac{n^3 + n^2}{3n^3 - 7} = \frac{1}{3} \neq 0$. Thus, $\sum_{n=1}^{\infty} \frac{n^3 + n^2}{3n^3 - 7}$ is divergent.

9. C

$$\left\{ \underbrace{0}_{\frac{1-1}{1^2}}, \underbrace{\frac{1}{4}}_{\frac{2-1}{2^2}}, \underbrace{\frac{2}{9}}_{\frac{3-1}{3^2}}, \underbrace{\frac{3}{16}}_{\frac{4-1}{4^2}}, \underbrace{\frac{4}{25}}_{\frac{5-1}{5^2}} \ldots \right\}$$

10. D

We can try each formula until we find the one that works:

(A) $\left\{\frac{(-1)^n}{2^n}\right\}_{n=0}^{\infty} = \frac{(-1)^0}{2^0}$

$= \frac{1}{1}$ ... wrong!

(B) $\left\{\frac{(-1)^n}{2^n}\right\}_{n=1}^{\infty} = \frac{(-1)^1}{2^1}$

$= \frac{-1}{2}, \ldots$ wrong!

(C) $\left\{\frac{(-1)^n}{2n}\right\}_{n=0}^{\infty} = \frac{(-1)^0}{2(0)}, \ldots$ wrong!

(D) $\left\{\frac{(-1)^{n+1}}{2^n}\right\}_{n=0}^{\infty} = \frac{(-1)^1}{2^0}, \frac{(-1)^2}{2^1}, \frac{(-1)^3}{2^2}, \frac{(-1)^4}{2^3}$

$= -1, \frac{1}{2}, -\frac{1}{4}, \frac{1}{8} \ldots$ this one works!

(E) $\left\{\frac{(-1)^{n+1}}{2^n}\right\}_{n=1}^{\infty} = \frac{(-1)^2}{2^1}, \ldots$

$= \frac{1}{2}, \ldots$ wrong!

11. B

This is called an index shift, and is used to adjust indices of a given series. Whenever we increase (decrease) the initial value in the index, we decrease (increase) the value in each n in the series. In this case, in order to get from the first series, $\sum_{n=1}^{\infty} a_n$, to either of the other two, we decrease the initial value by one. Therefore, we need to increase the value of n in order to maintain equivalence. Thus:

$$\sum_{n=1}^{\infty} a_n = \sum_{n=1-\underline{1}}^{\infty} a_{n\underline{+1}} = \sum_{n=0}^{\infty} a_{n+1},$$

so we see that series I and III are equivalent, while series II is not.

12. E

We begin by considering the limit of the terms of the series as n goes to infinity:

$$\lim_{n \to \infty} \frac{2n^3 + \cos(n)}{(2n+1)^3 + e^{-n}} \quad \text{(Consider the dominant terms)}$$

$$= \lim_{n \to \infty} \frac{2n^3}{(2n+1)^3}$$

$$= \lim_{n \to \infty} \frac{2n^3}{8n^3 + \ldots}$$

$$= \frac{1}{4}$$

This means that the terms of the series get closer and closer to $\frac{1}{4}$ as n gets larger, which means we are, in effect, adding an infinite number of $\frac{1}{4}$'s.

Adding an infinite number of positive terms leads to infinity unless (possibly) the terms are going to zero. This is not the case here, and so the series diverges.

13. B

We begin by considering the terms a_n and a_{n+1}:

$$\begin{aligned} a_n &= 1-n^2 \\ a_{n+1} &= 1-(n+1)^2 \\ &= 1-\left(n^2+2n+1\right) \\ &= -n^2-2n \end{aligned}$$

Since $n \geq 1: 1-n^2 > -2n-n^2$

$$\therefore a_n > a_{n+1}$$

$\therefore$ the series is decreasing.

Since the series is decreasing and it begins at n = 1, $a_1 = 1 - (1)^2 = 0$, the series is bounded above by 0. It is not bounded below, however, as $\lim\limits_{n\to\infty} a_n = -\infty$. Since the series is not bounded **both** above and below, we say that it is unbounded.

14. B

We treat each of the three statements separately as true or false:

Statement I is false. If the limit of the sequence is 6, it means that as we go toward infinity in the series, we would be adding an infinite number of 6's—clearly, this will make the sum grow to infinity. In order for a series to converge, the corresponding *sequence* must converge to zero. In other words: if $\sum a_n$ converges, we know that $\lim\limits_{n\to\infty} a_n = 0$.

Statement II is true. A series is said to converge if the limit of its partial sums exists and is finite. In this case, since the limit of the partial sums is 6, we say that the series converges to 6.

Statement III is false. Or, perhaps a better way of saying this is that statement III might be false. If $\lim\limits_{n\to\infty} a_n = \infty$ then the sequence is divergent. However, the converse is not necessarily true. A sequence will also diverge if the limit does not exist. For example, consider the sequence: $a_n = (-1)^n = -1, 1, -1, 1, -1, 1\ldots$. Here we have a sequence that diverges and yet $\lim\limits_{n\to\infty} a_n \neq \infty$.

15. A

This question deals with the necessary and sufficient conditions for the convergence/divergence of series.

$\lim\limits_{n\to\infty} a_n = 0$ is *necessary* for convergence, but not *sufficient*. In other words, for a series to converge, $\lim\limits_{n\to\infty} a_n$ must be zero, but it is possible that $\lim\limits_{n\to\infty} a_n = 0$ and the series diverges. For example: the series $\sum\limits_{n=1}^{\infty} \frac{1}{n}$ diverges, even though $\lim\limits_{n\to\infty} \frac{1}{n} = 0$. Thus, statement I is true, and statement III may be false. Statement II is a *sufficient* condition for divergence. In other words, $\lim\limits_{n\to\infty} a_n \neq 0$ is sufficient to guarantee the divergence of $\sum\limits_{n=1}^{\infty} \frac{1}{n}$, so statement II is true.

FREE-RESPONSE ANSWERS

16. To determine if the sequence converges or diverges we must calculate $\lim\limits_{n\to\infty} \frac{\sin^2 n}{2^n}$; however, $\lim\limits_{n\to\infty} \sin^2 n$ is undefined because sine oscillates between −1 and 1. It may seem like we don't know where to go from here, but using exactly this information we can solve the question. That is, because $|\sin x| \le 1$, we get $\lim\limits_{n\to\infty}\left|\frac{\sin^2 n}{2^n}\right| = \lim\limits_{n\to\infty}\frac{|\sin^2 n|}{2^n} = \lim\limits_{n\to\infty}\frac{|\sin n|^2}{2^n} \le \lim\limits_{n\to\infty}\frac{1}{2^n} = 0$. Now since the absolute value converges to zero, the actual sequence must converge to zero as well. Thus, $\left\{\frac{\sin^2 n}{2^n}\right\}$ is convergent.

17. As is true in most cases when considering the convergence/divergence of series, we begin with the divergence test i.e., if $\lim\limits_{n\to\infty} a_n \neq 0$ then the series is necessarily divergent.

    In this case, we need to evaluate $\lim\limits_{n\to\infty} n^{\frac{1}{n}}$.

    This limit is of the indeterminate form ∞^0 and so we can apply L'Hospital's rule (also written as L'Hôpital's).

    Let $y = n^{\frac{1}{n}}$

    then: $\ln y = \ln n^{\frac{1}{n}} = \frac{1}{n}\ln n$

    $$\lim_{n\to\infty} \ln y = \lim_{n\to\infty}\left(\frac{1}{n}\ln n\right) = \lim_{n\to\infty}\left(\frac{\ln n}{n}\right)$$

    This is of the form: $\frac{\infty}{\infty}$, so we can differentiate numerator and denominator:

    $$\lim_{n\to\infty}\left(\frac{\ln n}{n}\right) = \lim_{n\to\infty}\left(\frac{\frac{1}{n}}{1}\right) = 0 = \lim_{n\to\infty}(\ln y)$$

    $$\text{So: } \lim_{n\to\infty} y = \lim_{n\to\infty} e^{\ln y} = e^{\lim\limits_{n\to\infty}\ln y} = e^0 = 1$$

    $$\therefore \lim_{n\to\infty} n^{\frac{1}{n}} = 1$$

    $$\therefore \lim_{n\to\infty} n^{\frac{1}{n}} \neq 0$$

    Therefore, the series is divergent.

CHAPTER 25: THE PROPERTIES OF SERIES*

IF YOU LEARN ONLY THREE THINGS IN THIS CHAPTER . . .

1. Notation for a geometric series: $1 + r + r^2 + r^3 + \cdots + r^{n-1} + \cdots = \sum_{n=1}^{\infty} r^{n-1}$
2. Notation for a harmonic series: $\sum_{n=1}^{\infty} \frac{1}{n} = 1 + \frac{1}{2} + \frac{1}{3} + \frac{1}{4} + \cdots + \frac{1}{n} + \cdots$
3. The integral test: If f is a continuous, positive, decreasing function on $[1, \infty]$ and we let $a_n = f(n)$, then
   a) If $\int_1^{\infty} f(x)dx < \infty$, then $\sum_{n=1}^{\infty} a_n$ is convergent.
   b) If $\int_1^{\infty} f(x)dx > \infty$, then $\sum_{n=1}^{\infty} a_n$ is divergent.

THE GEOMETRIC SERIES WITH APPLICATIONS

Some series arise very frequently and are important enough to be given a name. The first of these that we will encounter is the geometric series. Each term in this series is obtained by multiplying the preceding one by a common factor, r.

$$1 + r + r^2 + r^3 + \cdots + r^{n-1} + \cdots = \sum_{n=1}^{\infty} r^{n-1}$$

*BC content

Clearly if $r = 1$ then this series is divergent because then $s_k = \underbrace{1+1+1+\cdots+1}_{k\text{-times}} = k$ and as k approaches infinity so does s_k. Furthermore, the same must be true if r is any number bigger than 1 by the same reasoning.

However, if $r = \frac{1}{2}$ we get

$$1+\frac{1}{2}+\frac{1}{4}+\frac{1}{8}+\cdots+\frac{1}{2^n}+\cdots=\sum_{n=1}^{\infty}\left(\frac{1}{2}\right)^{n-1}$$

which we saw from the previous chapter is equal to 1 and thus convergent. In fact, the geometric series is very well behaved for any value of r between -1 and 1, and the exact value of the series is easily obtained. We state this result below.

Fact: The geometric series $\sum_{n=1}^{\infty} r^{n-1} = 1 + r + r^2 + r^3 + \cdots + r^n + \cdots$ converges when $|r| < 1$ and, in fact, $\sum_{n=1}^{\infty} r^{n-1} = 1 + r + r^2 + r^3 + \cdots + r^n + \cdots = \frac{1}{1-r}$, where $|r| < 1$.
If, however, $|r| \geq 1$, then the geometric series is divergent.

Now we are equipped with a very powerful tool that we can use to explicitly calculate a given series. We need only think of a clever way to write the series so that it looks like a geometric one.

Example:

Use the geometric series to calculate the value of $\sum_{n=1}^{\infty} 9^{-n+1} \cdot 8^{n+1}$.

Solution:

At first glance this may not seem like a very easy series to evaluate. However, the geometric series is here to save the day; all we have to do is try to make this series look as much like the geometric series as possible. Let's get started.

Using the basic rules of algebra we can simplify this expression. These simplifications may seem like they are from out of the blue but keep in mind that we are trying to end with a series that resembles the geometric series. We get:

$$\sum_{n=1}^{\infty} 9^{-n+1} \cdot 8^{n+1} = \sum_{n=1}^{\infty} \frac{8^{n+1}}{9^{n-1}} = \sum_{n=1}^{\infty} \frac{8^{n-1} \cdot 8^2}{9^{n-1}} = 8^2 \sum_{n=1}^{\infty} \frac{8^{n-1}}{9^{n-1}} = 8^2 \sum_{n=1}^{\infty} \left(\frac{8}{9}\right)^{n-1}$$

Now we're getting somewhere! The series on the right is exactly in the form of the geometric series. Even better, the series is convergent because $\left|\frac{8}{9}\right| < 1$. Even better still (and here is the real power of the geometric series), we can explicitly calculate the value of this geometric series. We get:

$$8^2\sum_{n=1}^{\infty}\left(\frac{8}{9}\right)^{n-1} = 8^2\frac{1}{1-\frac{8}{9}} = 8^2\frac{1}{\frac{1}{9}} = 8^2 \cdot 9 = 576$$

Thus, by rearranging terms in the original series to get them in the form of the geometric series, we have that $\sum_{n=1}^{\infty} 9^{-n+1} \cdot 8^{n+1} = 576$ and we are done.

Example:

Use the geometric series to calculate the value of $\sum_{n=1}^{\infty} 5^{-n}(-9)^{n-1}$.

Solution:

Again we use basic algebra to get this series into a more manageable form.

$$\sum_{n=1}^{\infty} 5^{-n}(-9)^{n-1} = \sum_{n=1}^{\infty}\frac{(-9)^{n-1}}{5^n} = \sum_{n=1}^{\infty}\frac{(-9)^{n-1}}{5^{n-1}\cdot 5^1} = \frac{1}{5}\sum_{n=1}^{\infty}\frac{(-9)^{n-1}}{5^{n-1}} = \frac{1}{5}\sum_{n=1}^{\infty}\left(-\frac{9}{5}\right)^{n-1}.$$

However, since $\left|-\frac{9}{5}\right| > 1$, the geometric series tells us that this series is actually divergent and so $\sum_{n=1}^{\infty} 5^{-n}(-9)^{n-1}$ has no finite value.

By now you can appreciate the importance and usefulness of the geometric series. Unfortunately, not all series can be written this way so there is still more to learn. Let's look at some more series that are important enough to be given a name.

THE HARMONIC SERIES

In the last section we saw a very helpful convergent series that we can possibly use to calculate the value of a given series. Almost of equal importance is the harmonic series. As we will see, this series diverges and thus if we can write a given series in this form, we can save ourselves the headache of trying to determine its value.

Definition: The harmonic series is given by $\sum_{n=1}^{\infty}\frac{1}{n} = 1 + \frac{1}{2} + \frac{1}{3} + \frac{1}{4} + \cdots + \frac{1}{n} + \cdots$

As we claimed earlier this series is divergent. Let's examine the partial sums to see why this is so. We expect that the limit of the partial sums approaches infinity. If we can show that some of the terms of this series approaches infinity then we will surely be done. With this in mind it is more convenient to consider partial sums of the form s_2k. We get:

$$s_{2^1} = s_2 = 1 + \tfrac{1}{2}$$

$$s_{2^2} = s_4 = 1 + \tfrac{1}{2} + \tfrac{1}{3} + \tfrac{1}{4} > 1 + \tfrac{1}{2} + \tfrac{1}{4} + \tfrac{1}{4} = 1 + \tfrac{2}{2}$$

$$s_{2^3} = s_8 = 1 + \tfrac{1}{2} + \tfrac{1}{3} + \tfrac{1}{4} + \tfrac{1}{5} + \tfrac{1}{6} + \tfrac{1}{7} + \tfrac{1}{8}$$

$$> 1 + \tfrac{1}{2} + \tfrac{1}{4} + \tfrac{1}{4} + \tfrac{1}{8} + \tfrac{1}{8} + \tfrac{1}{8} + \tfrac{1}{8} = 1 + 1 + \tfrac{1}{2} = 1 + \tfrac{3}{2}$$

$$s_{2^4} = s_{16} = 1 + \tfrac{1}{2} + \tfrac{1}{3} + \tfrac{1}{4} + \tfrac{1}{5} + \tfrac{1}{6} + \tfrac{1}{7} + \tfrac{1}{8} + \tfrac{1}{9} + \tfrac{1}{10} + \tfrac{1}{11} + \tfrac{1}{12} + \tfrac{1}{13} + \tfrac{1}{14} + \tfrac{1}{15} + \tfrac{1}{16}$$

$$> 1 + \tfrac{1}{2} + \tfrac{1}{4} + \tfrac{1}{4} + \tfrac{1}{8} + \tfrac{1}{8} + \tfrac{1}{8} + \tfrac{1}{8} + \tfrac{1}{16} + \tfrac{1}{16} + \tfrac{1}{16} + \tfrac{1}{16} + \tfrac{1}{16} + \tfrac{1}{16} + \tfrac{1}{16} + \tfrac{1}{16}$$

$$= 1 + \tfrac{1}{2} + \tfrac{1}{2} + \tfrac{1}{2} + \tfrac{1}{2} = 1 + \tfrac{4}{2}$$

If we continue this process we get $s_{2^k} > 1 + \frac{k}{2}$. Thus $\lim_{k\to\infty} s_{2^k} > \lim_{k\to\infty}\left(1 + \frac{k}{2}\right) = \infty$ and so the sequence $\{s_k\}$ is divergent. Furthermore, this means that the harmonic series is also divergent.

THE INTEGRAL TEST AND THE CONVERGENCE OF THE *p*-SERIES

We have come a long way from the definition of convergent and divergent series. We know that the geometric series is convergent and its value is easily determined, and we have seen that the harmonic series is divergent. However, getting a given series in one of these two forms can be a tricky process and, in general, determining the exact value of a series can be difficult.

In this section we will look at a test that will help us determine if a series is convergent or divergent. In short, we compare the area of an improper integral with the series itself. To do this, we consider each element a_n of the series as a box of width 1 and height a_n; that is, a box of area a_n. If each of these boxes can fit *below* a function whose improper integral is *finite,* then clearly the series itself must also be finite (recall the integral of a function measures the area under the curve). If, however, each of these boxes can fit *above* a function whose improper integral is *infinite,* then clearly the series itself must also be infinite. Let's look at some examples to help this become clear.

Example:

Consider the series $\sum_{n=1}^{\infty} \frac{1}{n^2} = \frac{1}{1} + \frac{1}{4} + \frac{1}{9} + \cdots \frac{1}{n^2} + \cdots$. To see if this series converges or diverges, we could try to write out the partial sums s_k and try to evaluate $\lim_{k\to\infty} s_k$. However, there is not a simple formula for these partial sums. Instead, let's compare the series with the function $f(x) = \frac{1}{x^2}$. The following figure shows the graph of $f(x) = \frac{1}{x^2}$ and the rectangles that lie completely below that graph.

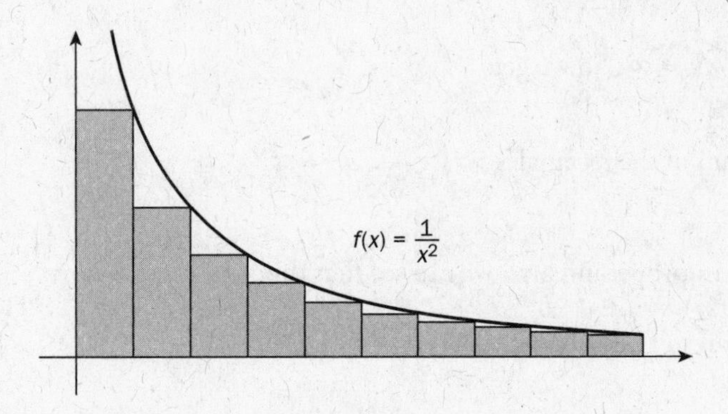

As we mentioned earlier, each rectangle has an area equal to one of the elements of the series. Thus:

$$\sum_{n=1}^{\infty} \frac{1}{n^2} = \sum \text{area of the rectangles}$$

And since $\int_1^{\infty} \frac{1}{x^2}\,dx$ determines the area under the graph from $x = 1$ to $x = \infty$, we get that the sum must be finite if this improper integral is finite; because then

$$\sum_{n=1}^{\infty} \frac{1}{n^2} = \sum \text{area of the rectangles} < 1 + \int_1^{\infty} \frac{1}{x^2}\,dx = 1 - \frac{1}{x}\Big|_1^{\infty} = 1 - (0 - 1) = 2 < \infty.$$

Therefore, using an improper integral, we can see that this sum must converge even though we do not have a formula for its partial sums.

Example:

Consider the series $\sum_{n=1}^{\infty} \frac{1}{\sqrt[4]{n}} = 1 + \frac{1}{\sqrt[4]{2}} + \frac{1}{\sqrt[4]{3}} + \cdots + \frac{1}{\sqrt[4]{n}} + \cdots$. To see if this series is convergent or divergent we use a similar technique; we compare this series to the graph of the function $f(x) = \frac{1}{\sqrt[4]{x}}$. The figure below shows this graph and the rectangles representing the elements of the series that lie completely above the graph.

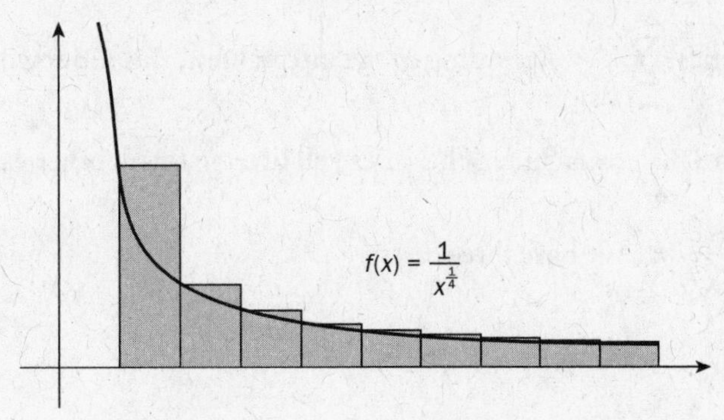

In this case, $\int_1^{\infty} \frac{1}{\sqrt[4]{x}}\,dx = \infty$. So we get:

$$\sum_{n=1}^{\infty} \frac{1}{\sqrt[4]{n}} = \sum \text{sum of the rectangles} > \int_1^{\infty} \frac{1}{\sqrt[4]{x}}\,dx = \infty$$

Therefore, using an improper integral, we can see that this sum must *diverge*.

Let's summarize our results as follows.

The Integral Test:

If f is a continuous, positive, decreasing function on $[1,\infty]$ and we let $a_n = f(n)$, then

1) If $\int_1^{\infty} f(x)dx$ is finite, then $\sum_{n=1}^{\infty} a_n$ is convergent.
2) If $\int_1^{\infty} f(x)dx$ is infinite, then $\sum_{n=1}^{\infty} a_n$ is divergent.

The integral test says that a series is convergent (or divergent) if and only if its corresponding integral is convergent (or divergent).

This method of using improper integrals to determine the convergence or divergence of series is called the integral test.

Closely related to the integral test is the remainder estimate of the integral test. Recall the remainder of a series $\sum a_n$ is given by $R_n = s - s_n$, where s is the full sum and s_n is the nth partial sum. To control the remainder term we use the following inequality:

$$\int_{n+1}^{\infty} f(x)dx \le R_n \le \int_n^{\infty} f(x)dx$$

These upper and lower bounds give a way of determining how close a remainder is to the actual sum of the series. Not surprisingly, we can also use this inequality to determine how many values of n are necessary to approximate the series to a given error.

We can also use this integral test to evaluate the convergence of another important series in mathematics, the p-series $\sum_{n=1}^{\infty} \frac{1}{n^p}$. According to the integral test, this series will converge exactly for those values of p when $\int_1^{\infty} \frac{1}{x^p}dx$ is finite. The series will diverge for all other values of p. Thus, we need only evaluate $\int_1^{\infty} \frac{1}{x^p}dx$. We have three cases:

1) If $p < 0$, then $\lim_{n\to\infty} \frac{1}{n^p} = \lim_{n\to\infty} n^{|p|} = \infty$, and so the partial sums cannot converge to a finite number if each of the terms goes to infinity.

2) If $p = 0$, then $\sum_{n=1}^{\infty} \frac{1}{n^p} = \sum_{n=1}^{\infty} 1 = \infty$; so the sum diverges.

3) If $p > 0$, the function $f(x) = \frac{1}{x^p}$ is continuous, positive, and decreasing on $[1, \infty]$; so we can apply the integral test. We have $\int_1^{\infty} \frac{1}{x^p}\,dx = \left(\lim_{x\to\infty} \frac{x^{1-p}}{1-p}\right) - \frac{1}{1-p}$ and thus is finite only if $1 - p < 0$, that is, if $p > 1$. So, $\int_1^{\infty} \frac{1}{x^p}\,dx$ is finite if $p > 1$ and $\int_1^{\infty} \frac{1}{x^p}\,dx$ is infinite if $p \le 1$.

Let's summarize what we have learned of the *p*-series.

Fact: The *p*-series $\sum_{n=1}^{\infty} \frac{1}{n^p}$ converges if $p > 1$ and diverges if $p \le 1$.

COMPARING SERIES TO TEST FOR CONVERGENCE OR DIVERGENCE

In the last section we compared a series with an improper integral to see if it converged or diverged. We stated that if the improper integral converged and the series was less than the integral, then the series too must converge. Similarly, we stated that if the integral diverged to infinity and the series was always greater than the integral, then the series too must diverge to infinity.

In this section, we will perform similar techniques to determine if a given series converges or diverges. Note that in this section, as in the last, we require the terms of our series to be positive. We will primarily use the comparison test, which is described below.

Comparison Test:

Suppose $\sum a_n$ and $\sum b_n$ are series with positive terms.

1) If $\sum b_n$ is convergent and $a_n \le b_n$ for each n, then $\sum a_n$ must also be convergent.

2) If $\sum b_n$ is divergent and $a_n \ge b_n$ for each n, then $\sum a_n$ must also be divergent.

Before we begin to look at examples, let's think about why the comparison test should be true. In case 1), $\sum a_n$ must be convergent because otherwise it would have some terms that go off to infinity. However, because $a_n \le b_n$, this would force some b_n terms to go off to infinity, which would ruin the convergence of $\sum b_n$. Similarly, in case 2), if $\sum b_n$ is divergent then it must have some terms that go off to infinity. Because $a_n \ge b_n$, this would force some of the a_n terms to also go to infinity and thus $\sum a_n$ would also be divergent.

Keep in mind that to use the comparison test we need to have two things: 1) a series $\sum b_n$ whose convergence/divergence we know about, and 2) a relation between the elements a_n and b_n.

Most of the time we can let $\sum b_n$ be the geometric series, the harmonic series, or the p-series, and use these known series to compare to the unknown series.

Example:

Does the series $\sum_{n=1}^{\infty} \frac{1}{n^2+1}$ converge or diverge?

Solution:

Now this series is very close to another series that we have seen before, namely $\sum_{n=1}^{\infty} \frac{1}{n^2}$. From previous examples, we know that $\sum_{n=1}^{\infty} \frac{1}{n^2}$ converges. Furthermore, we know that $\frac{1}{n^2+1} \leq \frac{1}{n^2}$ simply by cross-multiplying. Now because each of the terms in $\sum_{n=1}^{\infty} \frac{1}{n^2+1}$ is positive and bounded above by $\frac{1}{n^2}$, we can use the comparison test and conclude that $\sum_{n=1}^{\infty} \frac{1}{n^2+1}$ converges.

ALTERNATING SERIES WITH ERROR BOUND

So far all the series we have been looking at have all been positive series, meaning the terms of the series are positive. In general, of course, this need not be the case. For example, consider the series below

$$\frac{1}{2} - \frac{4}{3} + \frac{9}{4} - \frac{16}{5} + \frac{25}{6} - \frac{36}{7} + \cdots = \sum_{n=1}^{\infty} (-1)^{n+1} \frac{n^2}{n+1}$$

The terms in this series are alternatively positive and negative; thus, we call a series like this one an alternating series.

We have already seen that the harmonic series is divergent. However, should we expect the same to be true for an alternating harmonic series? In fact, the alternating harmonic series, i.e., $\sum_{n=1}^{\infty} (-1)^{n+1} \frac{1}{n} = 1 - \frac{1}{2} + \frac{1}{3} - \frac{1}{4} + \cdots$, *is* convergent. The idea is that the negative terms in the series are just enough to keep the series from diverging to infinity. To help determine the convergence and divergence of alternating series we will often use the following theorem (which we state without proof).

The Alternating Series Test:

Given an alternating series $\sum_{n=1}^{\infty} (-1)^{n+1} a_n = a_1 - a_2 + a_3 - a_4 + \cdots$ where each $a_n > 0$, if we know that

1) $a_{n+1} \le a_n$

2) $\lim_{n \to \infty} a_n = 0$

then the alternating series must be convergent.

Remark: Condition 1) in the alternating series test does not necessarily have to hold for every *n*. As long as we eventually get $a_{n+1} \le a_n$ for large enough *n*, we can employ this test.

In short, this test says that if the absolute value of the terms of the series is decreasing to zero, then the series is convergent.

Using this test, it is clear that the alternating harmonic series is convergent. Both condition 1) and 2) are satisfied. Let's consider another example that utilizes the alternating series test.

Suppose we are asked to determine the convergence or divergence of the series $\sum_{n=1}^{\infty} (-1)^n \frac{\sqrt{n}}{1 + 2\sqrt{n}}$.

This is clearly an alternating series and we can use the alternating series test to determine if it converges or not. For this series, we have $a_n = \frac{\sqrt{n}}{1 + 2\sqrt{n}}$; however, it is not immediately clear if these terms are decreasing. In order to check condition 1) of the test above, we create a related function $f(x) = \frac{\sqrt{x}}{1 + 2\sqrt{x}}$. Keep in mind we are only concerned with the positive values of *x* because $n \ge 1$.

Now we have to recall some things we learned about derivatives. Namely, a function is decreasing precisely for those values of *x* when $f'(x) < 0$. So, is $f(x)$ decreasing?

$$f'(x) = \frac{(1 + 2\sqrt{x})\frac{1}{2\sqrt{x}} - \sqrt{x}\frac{2}{2\sqrt{x}}}{(1 + 2\sqrt{x})^2} \overset{?}{<} 0$$

$$\frac{1 + 2\sqrt{x} - 2\sqrt{x}}{2\sqrt{x}(1 + 2\sqrt{x})^2} \overset{?}{<} 0$$

$$\frac{1}{2\sqrt{x}(1 + 2\sqrt{x})^2} \overset{?}{<} 0$$

Because $f'(x)$ is definitely not less than 0, $f(x)$ is not decreasing and so the *an* are not decreasing either. Furthermore, $\lim_{n\to\infty} a_n = \lim_{n\to\infty} \frac{\sqrt{n}}{1+2\sqrt{n}} = \frac{1}{2} \neq 0$. Therefore, both conditions of the alternating series test have failed. We can safely say that our series is divergent.

Now, a quick bit of philosophy. When we say that a series *equals* a finite number, say S, what this really means is that as we add more and more terms of the infinite series together, we get closer and closer to the number S. If we have a convergent positive series then we can picture the partial sums steadily getting closer and closer to S as we add on more terms. If we have an alternating series, however, the partial sums continually overshoot and undershoot the limit value S. This occurs precisely because of the alternating positive and negative terms.

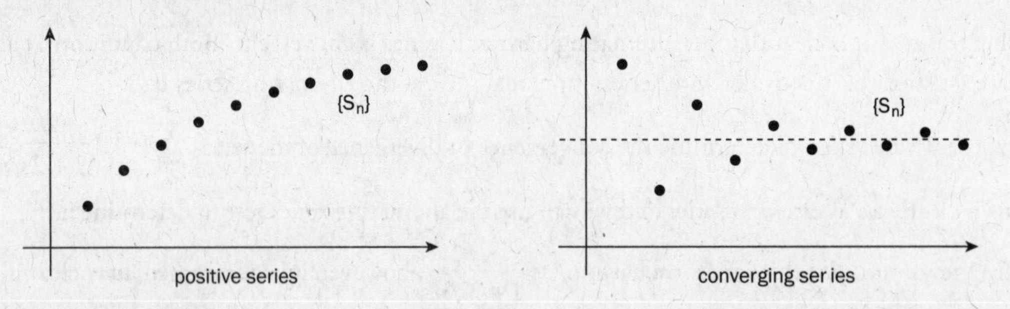

In both these figures the partial sums get closer and closer to the actual limit. This difference between the partial sums and the actual limit is called the *error term of the partial sums.* We define this error term, E_n, as the difference

$E_n = S - s_n$, where $\{s_n\}$ is the nth partial sum of the series.

For a convergent series, the error term must approach zero as n approaches infinity. In fact, we can get an even better bound on the error term of a convergent alternating series. We have the following theorem.

Alternating Series Estimation Theorem

Suppose $S = \sum_{n=1}^{\infty} (-1)^n a_n$ is the finite value of the alternating series. Suppose also that

1) $a_{n+1} \leq a_n$ and

2) $\lim_{n\to\infty} a_n = 0$

then $|E_n| = |S - s_n| \leq a_{n+1}$.

In short, this theorem says that given a convergent alternating series, the nth error term of the partial sums is always less than the $(n + 1)^{\text{th}}$ coefficient, a_{n+1}.

This theorem is a very valuable tool when calculating the approximate value of a given series. Using this theorem we can determine exactly how many terms of the series we need to approximate the sum to a given error.

ABSOLUTELY CONVERGENT SERIES

So far we have worked with positive series and alternating series. We have discussed conditions necessary for the convergence of positive series as well as seen tests that result in the convergence of alternating series. Given any series $\sum_{n=1}^{\infty} a_n$, we can consider the series $\sum_{n=1}^{\infty} |a_n| = |a_1| + |a_2| + |a_3| + \cdots + |a_n| + \cdots$ where each term is the absolute value of the corresponding term in the original series. In this way we create a positive series related to our original series.

Definition: We say that the original series $\sum_{n=1}^{\infty} a_n$ is absolutely convergent if the corresponding series of positive terms $\sum_{n=1}^{\infty} |a_n|$ is convergent.

If we think about it for a bit, it makes sense that absolute convergence is a stronger property than just convergence. For example, consider our good old friend the harmonic series and its close relative the alternating harmonic series. The alternating harmonic series is written as

$$\sum_{n=1}^{\infty} \frac{(-1)^{n+1}}{n} = 1 - \frac{1}{2} + \frac{1}{3} - \frac{1}{4} + \ldots$$

We saw in a previous section that the alternating harmonic series is convergent. However, this series is NOT absolutely convergent because this would imply that $\sum_{n=1}^{\infty} \left| \frac{(-1)^{n+1}}{n} \right| = \sum_{n=1}^{\infty} \frac{1}{n}$ is convergent and we know that the harmonic series is NOT convergent.

This example shows that a convergent series is not necessarily absolutely convergent. On the other hand, the reverse statement *is* true.

Fact: Suppose that the series $\sum_{n=1}^{\infty} a_n$ is absolutely convergent—then this series must be convergent.

So, if we can show that a series is absolutely convergent then we will automatically get that it is convergent.

THE RATIO TEST FOR CONVERGENCE AND DIVERGENCE

As we have seen, the comparison test is a very useful tool when determining the convergence or divergence of infinite series. However, many times it is not clear what we should compare our given series to. When this happens we are back at square one. What we would hope for is a simple test that we can apply to the terms of any given series to check convergence and divergence. Luckily we have such a test! It is called the ratio test and this test measures absolute convergence.

Intuitively, the ratio test checks to see how fast the terms in our series are growing. If the ratio of two consecutive terms very far in the sequence is small, then the terms are growing slowly and thus the series will converge. If the ratio of two consecutive terms is too large, then the terms are not decreasing fast enough and thus the series will diverge.

Let us now carefully state the conditions and result of the ratio test.

The Ratio Test

Let $\sum_{n=1}^{\infty} a_n$ be any given series.

(a) If $\lim_{n\to\infty}\left|\frac{a_{n+1}}{a_n}\right| < 1$, then the series is absolutely convergent.

(b) If $\lim_{n\to\infty}\left|\frac{a_{n+1}}{a_n}\right| > 1$ or $\lim_{n\to\infty}\left|\frac{a_{n+1}}{a_n}\right| = \infty$, then the series is divergent.

Clearly, the ratio test is a very valuable tool. Using only the terms of the series itself, we can apply the ratio test to check the series' convergence or divergence. It is important to note, however, a caveat of the ratio test. If $\lim_{n\to\infty}\left|\frac{a_{n+1}}{a_n}\right| = 1$, the ratio test is inconclusive and does not give us any information. When this happens we must use another means to evaluate the convergence of the series.

For example, let's use the ratio test to examine the convergence of the series $\sum_{n=1}^{\infty}\frac{3^n}{n!}$. In this series, $a_n = \frac{3^n}{n!}$. So, to use the ratio test we calculate:

$$\lim_{n\to\infty}\left|\frac{a_{n+1}}{a_n}\right| = \lim_{n\to\infty}\left|\frac{\frac{3^{n+1}}{(n+1)!}}{\frac{3^n}{n!}}\right| = \lim_{n\to\infty}\left|\frac{3^{n+1}}{(n+1)!}\cdot\frac{n!}{3^n}\right| = \lim_{n\to\infty}\frac{3}{n+1} = 0 < 1$$

Therefore, $\sum_{n=1}^{\infty}\frac{3^n}{n!}$ is a convergent series.

Now suppose we are asked to determine the convergence of $\sum_{n=1}^{\infty}\frac{(-3)^n}{n^3}$.

Again we employ our new best friend, the ratio test. For this series, we have $a_n = \frac{(-3)^n}{n^3}$; therefore, we have:

$$\left|\frac{a_{n+1}}{a_n}\right| = \left|\frac{\frac{(-3)^{n+1}}{(n+1)^3}}{\frac{(-3)^n}{n^3}}\right| = \left|\frac{(-3)^{n+1}}{(n+1)^3} \cdot \frac{n^3}{(-3)^n}\right| = 3\left(\frac{n}{n+1}\right)^3$$

To test the convergence of this series, we take the limit of this ratio as n approaches infinity. We get

$$\lim_{n\to\infty} 3\left(\frac{n}{n+1}\right)^3 = 3\left(\lim_{n\to\infty} \frac{n}{n+1}\right)^3 = 3 \cdot 1^3 = 3$$

Therefore, because $3 > 1$, the ratio test tells us that this series must be divergent.

REVIEW QUESTIONS

1. Evaluate the infinite series $\sum_{n=1}^{\infty} \frac{3^n + 2^n}{6^n}$.

(A) $\frac{\pi}{2}$

(B) $2\frac{3}{2}$

(C) $5\frac{1}{2}$

(D) $\frac{5}{3}$

(E) $1\frac{1}{2}$

2. Determine whether the series $\sum_{n=1}^{\infty} \frac{6n+1}{n^2}$ is convergent or divergent. If it is convergent, find its sum.

(A) The series is convergent and equals $6\frac{3}{4}$.

(B) The series is convergent and equals $5\frac{4}{5}$.

(C) The series is convergent and equals $\frac{2\pi}{3}$.

(D) The series is divergent and equals $\pi + \frac{1}{2}$.

(E) The series is divergent.

3. Calculate the value of $\sum_{n=1}^{\infty} 3^{-n+2}(-7)^{n-1}$.

(A) $\frac{9}{10}$

(B) $\frac{4}{9}$

(C) $2\frac{3}{10}$

(D) $\frac{\pi-2}{3}$

(E) It is divergent.

4. Which series below is NOT absolutely convergent?

(A) $\sum_{n=1}^{\infty} (-1)^n \frac{n^5}{5^n}$

(B) $\sum_{n=1}^{\infty} (-5)^n \frac{1}{\sqrt[3]{n}}$

(C) $\sum_{n=1}^{\infty} \frac{\cos 3n}{n^3}$

(D) $\sum_{n=1}^{\infty} \frac{(-1)^n}{n^2}$

(E) All the above series are absolutely convergent.

5. How many terms of the series do we need in order to show that $\sum_{n=1}^{\infty} \frac{3}{n^2}$ approximates its limit within an error of 0.001?

(A) 3,000

(B) 2,650

(C) 2,525

(D) 1,800

(E) 1,730

6. Suppose that $\sum_{n=1}^{\infty} b_n$ and $\sum_{n=1}^{\infty} c_n$ are two series with positive terms. Which statement below is definitely TRUE?

(A) If $\sum_{n=1}^{\infty} b_n$ converges and $c_n \geq b_n$ for all n, then $\sum_{n=1}^{\infty} c_n$ diverges.

(B) If $\sum_{n=1}^{\infty} b_n$ converges and $b_n \geq c_n$ for all n, then $\sum_{n=1}^{\infty} c_n$ diverges.

(C) If $\sum_{n=1}^{\infty} b_n$ diverges and $b_n \geq c_n$ for all n, then $\sum_{n=1}^{\infty} c_n$ converges.

(D) If $\sum_{n=1}^{\infty} b_n$ diverges and $c_n \geq \frac{1}{2} b_n$ for all n, then $\sum_{n=1}^{\infty} c_n$ also diverges.

(E) None of the above

7. Consider the series $\sum_{n=1}^{\infty} \frac{7}{5n^p}$. Which inequality below contains *all* the values of p where this series is divergent?

(A) $p > \frac{7}{5}$

(B) $p \geq 1$

(C) $p \leq 1$

(D) $p \leq \frac{7}{5}$

(E) $p \leq 0$

8. For which of the following series is the ratio test inconclusive?

(A) $\sum_{n=1}^{\infty} \frac{5^{n+1}}{n^3}$

(B) $\sum_{n=1}^{\infty} \frac{n}{2^n}$

(C) $\sum_{n=1}^{\infty} \frac{(-3)^{n-1}}{\sqrt{n}}$

(D) $\sum_{n=1}^{\infty} \frac{\sqrt{n}}{1+n^2}$

(E) $\sum_{n=1}^{\infty} \frac{n^n}{n!}$

9. Evaluate the series $\sum_{n=1}^{\infty} 2^{n+1} \cdot 3^{1-n}$.

(A) 0

(B) $\frac{2}{3}$

(C) 6

(D) 12

(E) The series diverges.

10. Which of the following series diverge?

I. $24 + 18 + \frac{27}{2} + \frac{81}{8} + \ldots$

II. $1 - 2 + 4 - 8 + 16 \ldots$

III. $6 - 6 + 6 - 6 + 6 \ldots$

(A) I only

(B) II only

(C) III only

(D) I and II only

(E) II and III only

11. A certain function $f(x)$ is known to be positive and decreasing on $(1, \infty)$, and $\int_1^{\infty} f(x)dx = 2$.
Which of the following statements must be true if $f(n) = a_n$?

I. $\sum_{n=1}^{\infty} a_n$ is convergent.

II. $\sum_{n=1}^{\infty} a_n = 2$

III. $\sum_{n=1} a_n < 2$

(A) I only
(B) II only
(C) III only
(D) I and II only
(E) I and III only

12. Which of the following statements are true for the two series below:

I. $\sum_{n=1}^{\infty} \frac{3n^2}{e^{n^3}}$.

II. $\sum_{n=1} \frac{1}{n}(\ln(n))^4$.

(A) Series I and II both converge.
(B) Series I converges and series II diverges.
(C) Series I diverges and series II converges.
(D) Series I and II both diverge.
(E) There is not enough information given to determine the convergence of these series.

13. Consider the two series $\sum_{n=1}^{\infty} u(n)$ and $\sum_{n=1}^{\infty} c(n)$.

It is known that $\sum_{n=1}^{\infty} c(n)$ converges, but the convergence/divergence of $\sum_{n=1}^{\infty} c(n)$ is unknown. A student correctly uses the comparison test to prove that $\sum_{n=1}^{\infty} u(n)$ converges.
Which of the following statements must be true?

I. $u_n \le c_n \forall n \in \mathbb{R}$.

II. Both $u_n > 0$ and $c_n > 0$ for all $n \in \mathbb{R}$.

III. $\lim_{n \to \infty} c_n = 0$.

(A) I and II only
(B) I and III only
(C) II and III only
(D) I, II, and III
(E) None of the above

14. Which (if any) of the series listed below converge?

I. $\sum_{n=1}^{\infty} \frac{(-1)^{n+1}}{n+10}$.

II. $\sum_{n=1}^{\infty} \frac{(-1)^{n+1} n^3}{n^3+10}$.

III. $\sum_{n=1}^{\infty} \frac{(-1)^{n+1} n^2}{n^3+10}$.

(A) I and II only
(B) I and III only
(C) II and III only
(D) I, II, and III only
(E) None of the above

Question 15 requires a calculator.

15. Consider the alternating series $\sum_{n=1}^{\infty}(-1)^{n-1}\frac{2n-3}{2n^3+7}$. If the exact sum is approximated using the finite sum $\sum_{n=1}^{10}(-1)^{n-1}\frac{2n-3}{2n^3+7}$, the alternating series estimation theorem will predict a maximum error of:

(A) 0.007119

(B) 0.008470

(C) 0.629630

(D) 1

(E) None of the above

FREE-RESPONSE QUESTION

16. Using the integral test, determine whether the series $\sum_{n=1}^{\infty} \frac{\ln n}{n^2}$ is divergent or convergent and justify your answer in detail. Show *all* of your work.

17. Determine whether the series $\sum_{x=1}^{\infty} \frac{\cos^2 x}{x^2+2x+1}$ converges or diverges. Show your work.

ANSWERS AND EXPLANATIONS

1. E

This question relies on the convergence of the geometric series. We know that $\sum_{n=1}^{\infty} r^{n-1} = \frac{1}{1-r}$ provided $|r| < 1$. To make this clear, let's simplify the equation and use the geometric series to compute:

$$\sum_{n=1}^{\infty} \frac{3^n + 2^n}{6^n} = \sum_{n=1}^{\infty} \frac{3^n}{6^n} + \frac{2^n}{6^n} = \sum_{n=1}^{\infty} \frac{3^n}{6^n} + \sum_{n=1}^{\infty} \frac{2^n}{6^n}$$

$$= \sum_{n=1}^{\infty} \left(\tfrac{1}{2}\right)^n + \sum_{n=1}^{\infty} \left(\tfrac{1}{3}\right)^n$$

$$= \frac{1}{2}\sum_{n=1}^{\infty} \left(\tfrac{1}{2}\right)^{n-1} + \frac{1}{3}\sum_{n=1}^{\infty} \left(\tfrac{1}{3}\right)^{n-1}$$

$$= \frac{1}{2}\left(\frac{1}{1-\frac{1}{2}}\right) + \frac{1}{3}\left(\frac{1}{1-\frac{1}{3}}\right)$$

$$= 1 + \frac{1}{2} = 1\frac{1}{2}$$

2. E

Our first check for convergence should always be to look at the limit $\lim_{n\to\infty} \frac{6n+1}{n^2}$. If this limit is not zero, then, by the divergence test, the series MUST be divergent. If it is zero, then the series *could* be convergent. We get $\lim_{n\to\infty} \frac{6n+1}{n^2} = 0$. So it could be convergent; we need to take a closer look. Let's simplify the series to get:

$$\sum_{n=1}^{\infty} \frac{6n+1}{n^2} = \sum_{n=1}^{\infty} \frac{6}{n} + \frac{1}{n^2} = 6\sum_{n=1}^{\infty} \frac{1}{n} + \sum_{n=1}^{\infty} \frac{1}{n^2}$$

What do we know about these series? The series $\sum_{n=1}^{\infty} \frac{1}{n^2}$ is convergent; however, the series $\sum_{n=1}^{\infty} \frac{1}{n}$ is the harmonic series and we know that this diverges. Thus, the entire series $\sum_{n=1}^{\infty} \frac{6n+1}{n^2}$ must diverge as well.

3. E

The series $\sum_{n=1}^{\infty} 3^{-n+2}(-7)^{n-1}$ doesn't make much sense to us as it is. However, a discerning eye may have picked out the geometric series hidden in there somewhere. Keep in mind that we want to get a series of the form $\sum_{n=1}^{\infty} r^{n-1}$ where $|r| < 1$. We have:

$$\sum_{n=1}^{\infty} 3^{-n+2}(-7)^{n-1} = \sum_{n=1}^{\infty} \frac{(-7)^{n-1}}{3^{n-2}}$$

$$= \sum_{n=1}^{\infty} \frac{(-7)^{n-1}}{3^{n-1} \cdot 3^{-1}}$$

$$= \sum_{n=1}^{\infty} 3\left(\frac{-7}{3}\right)^{n-1}$$

$$= 3\sum_{n=1}^{\infty} \left(\frac{-7}{3}\right)^{n-1}$$

Because $\left|\frac{-7}{3}\right| > 1$, the series diverges.

4. B

We will show that (B) is not absolutely convergent. The readers should show, using any method they prefer, that (A), (C), (D), and (E) are, in fact, absolutely convergent.

To show that $\sum_{n=1}^{\infty} (-5)^n \frac{1}{\sqrt[3]{n}}$ is NOT absolutely convergent, we will use the ratio test. This means we must show that $\lim_{n\to\infty} \left|\frac{a_{n+1}}{a_n}\right| > 1$.

We get:

$$\lim_{n\to\infty} \left|\frac{a_{n+1}}{a_n}\right| = \lim_{n\to\infty} \left|\frac{(-5)^{n+1} \frac{1}{(n+1)^{\frac{1}{3}}}}{(-5)^n \frac{1}{n^{\frac{1}{3}}}}\right|$$

$$= \lim_{n\to\infty} \left|\frac{(-5)n^{\frac{1}{3}}}{(n+1)^{\frac{1}{3}}}\right|$$

$$= 5 \lim_{n\to\infty} \left|\frac{n^{\frac{1}{3}}}{(n+1)^{\frac{1}{3}}}\right|$$

$$= 5 \cdot 1 = 5 > 1$$

5. A

An accuracy of 0.001 means that we must determine our value of n such that the remainder $R^n \leq 0.001$. From the remainder estimate for the integral test, we have the following inequality:

$$R_n \leq \int_n^\infty \frac{3}{x^2}\,dx = \lim_{t\to\infty}\int_n^t \frac{3}{x^2}\,dx$$
$$= \lim_{t\to\infty}\left[\frac{-3}{x}\right]_n^t$$
$$= \lim_{t\to\infty}\frac{-3}{t}+\frac{3}{n}=\frac{3}{n}.$$

Thus, we want $\frac{3}{n} < 0.001$, and solving this inequality for n, we get $n > \frac{3}{0.001} = 3{,}000$. So to get an accuracy of 0.001, we must sum up 3,000 terms of the series.

6. D

If you guessed (A), (B), or (C), then go back and read the rules for the comparison test. Each of these answers is *similar* to the statement, but none of them is the same. Alas, the devil is in the details. In fact, you can even find examples of series that contradict the statements in (A), (B), and (C).

Now, suppose we know that $\sum_{n=1}^{\infty} b_n$ is divergent. Then clearly, $\sum_{n=1}^{\infty}\frac{1}{2}b_n = \frac{1}{2}\sum_{n=1}^{\infty} b_n$ must also be divergent simply because $\sum_{n=1}^{\infty} b_n$ is. If we also know that $c_n \geq \frac{1}{2}b_n$ for all n, then by the comparison test, $\sum_{n=1}^{\infty} c_n$ must also be divergent.

7. C

The important point here is that we need to find the inequality that contains ALL (not just some of) the values of p where the series is divergent.

Perhaps this series looks familiar. The material that we covered on the p-series, $\sum_{n=1}^{\infty}\frac{1}{n^p}$ should ring a bell for you. If not, go back and read that section again.

We know that $\sum_{n=1}^{\infty}\frac{1}{n^p}$ converges for $p > 1$ and diverges for $p \leq 1$.

Clearly, $\sum_{n=1}^{\infty}\frac{7}{5n^p}$ diverges precisely when $\sum_{n=1}^{\infty}\frac{1}{n^p}$ diverges; the factor $\frac{7}{5}$ doesn't change anything.

Thus, $\sum_{n=1}^{\infty}\frac{7}{5n^p}$ diverges if $p \leq 1$; this is the largest such interval for p.

8. D

Recall that the ratio test is inconclusive if $\lim_{x\to a}\left|\frac{a_{n+1}}{a_n}\right| = 1$. We will show that this occurs for $\sum_{n=1}^{\infty}\frac{\sqrt{n}}{1+n^2}$. We get:

$$\lim_{n\to\infty}\left|\frac{a_{n+1}}{a_n}\right| = \lim_{n\to\infty}\left|\frac{\dfrac{\sqrt{n+1}}{1+(n+1)^2}}{\dfrac{\sqrt{n}}{1+n^2}}\right|$$
$$= \lim_{n\to\infty}\left|\frac{\sqrt{n+1}}{\sqrt{n}}\cdot\frac{1+n^2}{1+(n+1)^2}\right|$$
$$= \lim_{n\to\infty}\frac{\sqrt{n+1}}{\sqrt{n}}\cdot\lim_{n\to\infty}\frac{1+n^2}{1+(n+1)^2}$$
$$= 1\cdot 1 = 1$$

So, the ratio test is inconclusive.

The reader is encouraged to calculate the limits of (A), (B), (C), and (E) to show that the ratio test does indeed determine absolute convergence (or divergence) in these cases.

9. D

With some rearranging, we can recognize this as a geometric series:

$$\sum_{n=1}^{\infty} 2^{n+1}\cdot 3^{1-n} = \sum_{n=1}^{\infty}\frac{2^{n+1}}{3^{n-1}} = \sum_{n=1}^{\infty}\frac{2^2\cdot 2^{n-1}}{3^{n-1}} = 4\sum_{n=1}^{\infty}\frac{2^{n-1}}{3^{n-1}} = 4\sum_{n=1}^{\infty}\left(\frac{2}{3}\right)^{n-1}$$

Therefore, the sum to infinity is given by: $S_\infty = \frac{a}{1-r} = \frac{4}{1-\frac{2}{3}} = \frac{4}{\frac{1}{3}} = 12$

10. E

Each of these is a geometric series. The convergence/divergence of a geometric series is based on the value of the common ratio.

I. $24+18+\frac{27}{2}+\frac{81}{8}+\ldots \quad r=\frac{3}{4}$ series converges.

II. $1-2+4-8+16\ldots \quad r=-2 \Rightarrow$ series diverges.

III. $6-6+6-6+6\ldots \quad r=-1 \Rightarrow$ series diverges.

11. E

The integral test tells us that if $\int_1^{\infty} f(x)\,dx = 2$ then $\sum_{n=1}^{\infty} a_n$ is convergent. It does not tell us what the series converges to, but gives us an upper bound. In this case, $\int_1^{\infty} f(x)\,dx = 2$ implies that $\sum_{n=1}^{\infty} a_n < 2$. Therefore, statements I and III are true.

12. B

We use the integral test to determine the convergence of both series.

Series I

$$\int_1^{\infty}\frac{3x^2}{e^{x^3}}\,dx = \lim_{t\to\infty}\int_1^{t}\frac{3x^2}{e^{x^3}}\,dx$$

$$\begin{aligned} \textit{Let } u &= x^3 \\ du &= 3x^2dx \end{aligned}$$

$$\int\frac{3x^2}{e^{x^3}}\,dx = \int e^{-u}\,du = -e^{-u}+C = -e^{-x^3}+C$$

$$\begin{aligned} \lim_{t\to\infty}\int_1^{t}\frac{3x^2}{e^{x^3}}\,dx &= \lim_{t\to\infty} -e^{-x^3}\Big|_1^t \\ &= \lim_{t\to\infty}\left(-e^{-t^3}-\,-\frac{1}{e}\right) = \frac{1}{e} \end{aligned}$$

$\therefore \sum_{n=1}^{\infty}\frac{3n^2}{e^{n^3}}$ converges by the integral test.

Series II

$$\int_1^{\infty}\frac{1}{x}(\ln(x))^4\,dx = \lim_{t\to\infty}\int_1^{t}\frac{1}{x}(\ln(x))^4\,dx$$

$$\begin{aligned} \textit{Let } u &= \ln(x) \\ du &= \frac{1}{x}dx \end{aligned}$$

$$\int\frac{1}{x}(\ln(x))^4\,dx = \int u^4\,du = \frac{1}{5}u^5+C = \frac{1}{5}(\ln(x))^5+C$$

$$\begin{aligned} \lim_{t\to\infty}\int_1^{t}\frac{1}{x}(\ln(x))^4\,dx &= \lim_{t\to\infty}\frac{1}{5}(\ln(x))^5\Big|_1^t \\ &= \frac{1}{5}\lim_{t\to\infty}\left((\ln(t))^5-0\right) = \infty \end{aligned}$$

$\therefore \sum_{n=1}^{\infty}\frac{1}{n}(\ln(n))^4$ diverges by the integral test.

13. D

In order to use the comparison test to prove that a series converges, the unknown series must be smaller than a known convergent series for all values of n. Therefore, if $\sum_{n=1}^{\infty} c(n)$ is to be used to prove that $\sum_{n=1}^{\infty} u(n)$ converges, then $u(n)$ must be less than $c(n)$ for all values of n, so statement I is true. Furthermore, the convergence test requires that both series are positive for all values of n and so statement II must be true. Statement III is a necessary condition for the convergence of $c(n)$ by the divergence test: if the terms of a series do not go to zero as n approaches infinity, the series cannot be finite and therefore must diverge. Thus, all three statements must be true.

14. B

We recognize all three of these as alternating series (because of the presence of the $(-1)^{n+1}$ in all three series) and can try the alternating series test to test for the convergence in all three cases. In order to do so, we need to know two things: for $\sum_{n=1}^{\infty} (-1)^{n+1} \cdot a^n$

- is $\lim_{n \to \infty} a_n = 0$?
- is $a_{n+1} \leq a_n$ beyond some value of n?

We examine each of the three series individually using these criteria.

In series I, we see that the terms are: $\frac{1}{11}, -\frac{1}{12}, \frac{1}{13}, -\frac{1}{14} \ldots$

Clearly, as n approaches infinity, these terms are approaching zero and so the first condition is met. Also, if we consider the magnitudes of the terms: $\frac{1}{11}, \frac{1}{12}, \frac{1}{13}, \frac{1}{14} \ldots$

we see that each term is smaller than the previous one, and so these are decreasing. Thus, the alternating series test tells us that this series converges.

In series II, we see that $\lim_{n \to \infty} \frac{n^3}{n^3 + 10} = 1 \neq 0$. Therefore, the second condition is not met and the alternating series test fails. The divergence test shows us that: $\lim_{n \to \infty} \frac{(-1)^{n+1} n^3}{n^3 + 10}$ does not exist (as n grows to infinity, the terms bounce back and forth between values approaching +1 and −1, so the limit does not exist) and therefore the series diverges.

In series III, we see that $\lim_{n \to \infty} \frac{n^2}{n^3 + 10} = 0$ and so the first condition is met. In order to determine if the values of $\frac{n^2}{n^3 + 10}$ are decreasing, we can apply differential calculus:

$$\begin{aligned}
\frac{d}{dn}\left(\frac{n^2}{n^3+10}\right) &= \frac{2n\left(n^3+10\right)-n^2\left(3n^2\right)}{\left(n^3+10\right)^2} \\
&= \frac{2n^4+20n-3n^4}{\left(n^3+10\right)^2} \\
&= \frac{20n-n^4}{\left(n^3+10\right)^2} \\
&= \frac{n\left(20-n^3\right)}{\left(n^3+10\right)^2} \\
\frac{d}{dn}\left(\frac{n^2}{n^3+10}\right) &= 0 \Rightarrow n=0,\ n=\sqrt[3]{20}
\end{aligned}$$

$\frac{d}{dn}\left(\frac{n^2}{n^3+10}\right)$ does not exist: $n=\sqrt[3]{-10}$

$\therefore$ we have critical values at $n=0,\ n=\sqrt[3]{20}$ and $n=\sqrt[3]{-10}$.

$\left(n^3+10\right)^2 \geq 0$ for all real numbers, $\therefore$ the sign of the derivative will not change at $n=\sqrt[3]{-10}$.

Checking the signs of the derivative around the other two critical values, $n = 0$, $n = \sqrt[3]{20}$, we see that:

$$\frac{d}{dn}\left(\frac{n^2}{n^3+10}\right) > 0 \ on\ \left(0,\sqrt[3]{20}\right)$$

$$\frac{d}{dn}\left(\frac{n^2}{n^3+10}\right) < 0 \ on\ \left(\sqrt[3]{20},\infty\right)$$

Thus, the series will decrease once n is greater than $\sqrt[3]{20}$. Since the alternating series test does not require the terms to be decreasing for all values of n, but only past some finite value of n, the condition is met and the alternating series test tells us that series III is convergent.

15. A

The alternating series estimation theorem states that $\left|\sum_{n=1}^{10}(-1)^{n-1}\frac{2n-3}{2n^3+7}\right|$ will differ from $\sum_{n=1}^{\infty}(-1)^{n-1}\frac{2n-3}{2n^3+7}$ by less then the value of $\frac{2(11)-3}{2(11)^3+7} = 0.007119$.

FREE-RESPONSE ANSWER

(Note: The calculation is slightly involved and we use techniques such as integration by parts and L'Hôpital's rule. Try the computation on your own first and only look below if you get stuck.)

16. Using the integral test, we will show that $\sum_{n=1}^{\infty} \frac{\ln n}{n^2}$ is convergent. Recall, the integral test states that if we define $f(x) = \frac{\ln x}{x^2}$ and f is continuous, positive, and decreasing on $[1,\infty]$, then $\sum_{n=1}^{\infty} \frac{\ln n}{n^2}$ is convergent if and only if $\int_1^{\infty} \frac{\ln x}{x^2}\,dx$ is convergent. Clearly, $f(x)$ is continuous and positive on $[1,\infty]$. We do, however, need to check that $\frac{\ln x}{x^2}$ is decreasing as well. To do this, we will show that the first derivative is not positive; then, using what we know about the nature of f' and f, we can say that $\frac{\ln x}{x^2}$ is decreasing. We calculate $f'(x) = \frac{x^2 \frac{1}{x} - \ln x \cdot 2x}{x^4} = \frac{x - 2x\ln x}{x^4} = \frac{1 - 2\ln x}{x^3}$. It follows that f' is negative when $1 - 2\ln x < 0$, i.e., when $x > e^{\frac{1}{2}}$. Thus, f is decreasing when $x > e^{\frac{1}{2}}$ and so we can use the integral test. Now we can employ the integral test. What follows is the computation of $\int_1^{\infty} \frac{\ln x}{x^2}\,dx$. We compute:

$$\int_1^{\infty} \frac{\ln x}{x^2}\,dx = \lim_{t\to\infty} \int_1^{t} \frac{\ln x}{x^2}\,dx$$

$$u = \ln x \quad dv = \frac{dx}{x^2}$$

$$du = \frac{dx}{x} \quad v = \frac{-1}{x}$$

Using integration by parts:

$$= \lim_{t\to\infty} \left(\frac{-\ln x}{x}\bigg|_1^t - \int_1^t \frac{-1}{x^2}\,dx \right)$$

$$= \lim_{t\to\infty} \left(\frac{-\ln t}{t} - \frac{1}{x}\bigg|_1^t \right) = \lim_{t\to\infty} \left(\frac{-\ln t}{t} - \left[\frac{1}{t} - 1\right] \right)$$

$$= \lim_{t\to\infty} \left(\frac{-\ln t}{t} - \frac{1}{t} + 1 \right) = -\lim_{t\to\infty} \frac{\ln t}{t} - 0 + 1$$

$$\overset{L'H\hat{o}pital}{=} -\lim_{t\to\infty} \frac{\frac{1}{t}}{1} + 1 = 1 - \lim_{t\to\infty} \frac{1}{t} = 1 - 0 = 1$$

Because this integral is finite (convergent), by the integral test, $\sum_{n=1}^{\infty} \frac{\ln n}{n^2}$ is also convergent.

17. We start by using the divergence test:

$$\lim_{x\to\infty}\frac{\cos^2 x}{x^2+2x+1}=0$$

This tells us that the series *may* converge, but does not guarantee that it does, so we need to look farther.

We observe that antidifferentiating $\dfrac{\cos^2 x}{x^2+2x+1}$ will be too difficult, which rules out the integral test, and this series is not alternating, which rules out the alternating series test.

We also observe that the terms of this series are always positive, as:

$$\frac{\cos^2 x}{x^2+2x+1}=\frac{\cos^2 x}{(x+1)^2}=\left(\frac{\cos x}{x+1}\right)^2\geq 0 \text{ for all } x\in\mathbb{R}.$$

Further: $\dfrac{\cos^2 x}{x^2+2x+1}=\dfrac{\cos^2 x}{(x+1)^2}\leq\dfrac{1}{(x+1)^2}<\dfrac{1}{x^2}$ and $\displaystyle\sum_{x=1}^{\infty}\frac{1}{x^2}$ is known to converge as this is a p-series with $p = 2 > 1$.

Therefore, by the comparison test: $\displaystyle\sum_{x=1}^{\infty}\frac{\cos^2 x}{x^2+2x+1}$ converges.

That is, $\dfrac{\cos^2 x}{x^2+2x+1}<\underbrace{\dfrac{1}{x^2}}_{\substack{\text{converges}\\ p-\text{series}\\ p>1}}$ $\quad\displaystyle\sum_{x=1}^{\infty}\frac{\cos^2 x}{x^2+2x+1}$ converges.

CHAPTER 26: TAYLOR SERIES*

IF YOU LEARN ONLY FOUR THINGS IN THIS CHAPTER . . .

1. A power series is a series of the form
$\sum_{n=0}^{\infty} a_n x^n = a_0 + a_1 x^1 + a_2 x^2 + a_3 x^3 + \ldots + a_n x^n + \ldots$ where x is a variable and the a terms are called the coefficients of the power series.
2. The Taylor series of the function $f(x)$ at $x = c$ is given by
$$f(x) = \sum_{n=0}^{\infty} \frac{f^{(n)}(c)}{n!}(x-c)^n = f(c) + \frac{f'(c)}{1!}(x-c) + \frac{f''(c)}{2!}(x-c)^2 + \frac{f'''(c)}{3!}(x-c)^3 + \ldots$$
3. The Maclaurin series of the function $f(x)$ is given by
$$f(x) = \sum_{n=0}^{\infty} \frac{f^{(n)}(0)}{n!}x^n = f(0) + \frac{f'(0)}{1!}x + \frac{f''(0)}{2!}x^2 + \frac{f'''(0)}{3!}x^3 + \ldots$$
4. The binomial series: If k is any real number and $|x|<1$, then
$$(1+x)^k = \sum_{n=0}^{\infty} \binom{k}{n} x^n = 1 + kx + \frac{k(k-1)}{2!}x^2 + \frac{k(k-1)(k-2)}{3!}x^3 + \ldots$$
where $\binom{k}{n} = \frac{k!}{n!(k-n)!}$.

FROM SERIES TO POWER SERIES

Previously we explored infinite series of the form $\sum_{n=1}^{\infty} a_n$, where the elements a_n are real numbers. We saw that some series, like the geometric series $\sum_{n=1}^{\infty} r^{n-1}$ (with $|r|<1$), converge to

*BC content

a finite number while others, like the harmonic series $\sum_{n=1}^{\infty} \frac{1}{n}$, diverge to infinity. We then learned some very useful techniques that help us determine when a given series converges and when it diverges. In this chapter we will apply much of the same techniques to study *power series.*

A power series is a series in which the elements of the series are not just real numbers but can actually be functions of x. In general, a power series is a series of the form:

$$\sum_{n=0}^{\infty} a_n x^n = a_0 + a_1 x^1 + a_2 x^2 + a_3 x^3 + \cdots + a_n x^n + \cdots$$

where x is a **variable** and the a_n's are called the *coefficients of the power series.*

Even more generally, a power series can be written in the form:

$$\sum_{n=0}^{\infty} a_n (x-c)^n = a_0 + a_1(x-c)^1 + a_2(x-c)^2 + \cdots + a_n(x-c)^n + \cdots$$

A power series in this form is called a *power series centered at c.*

Much like infinite series, we are interested in the convergence of a power series. As one might expect, however, a single power series may converge for one value of x and diverge for a different value of x. To study convergence of a power series, therefore, we ask "For what values of x is a given power series convergent or divergent?"

Let's look at a specific power series. We ask, for what values of x does the series $\sum_{n=1}^{\infty} \frac{(x-2)^n}{n}$ converge?

To answer this question we employ the ratio test, keeping in mind that x can vary. To apply the ratio test we look at the terms $A_n = \frac{(x-2)^n}{n}$. Thus, we are concerned with the ratio

$$\left|\frac{A_{n+1}}{A_n}\right| = \left|\frac{\frac{(x-2)^{n+1}}{n+1}}{\frac{(x-2)^n}{n}}\right| = \left|\frac{(x-2)^{n+1}}{n+1} \cdot \frac{n}{(x-2)^n}\right| = |x-2| \frac{n}{n+1}.$$

We know that $\lim_{n\to\infty} \frac{n}{n+1} = 1$, so $\lim_{n\to\infty} |x-2| \frac{n}{n+1} = |x-2|$. Therefore, according to the ratio test, our series above will converge precisely when $|x - 2| < 1$, and it will diverge when $|x - 2| > 1$. Furthermore, we know that

$$\begin{aligned} |x-2| &< 1 \\ -1 < x-2 &< 1 \\ 1 < x &< 3 \end{aligned}$$

Therefore, $\sum_{n=1}^{\infty} \frac{(x-2)^n}{n}$ converges when $1 < x < 3$.

Recall that the ratio test is inconclusive when $\lim_{n\to\infty}\left|\frac{A_{n+1}}{A_n}\right| = 1$. In our example, this occurs when $|x - 2| = 1$. This occurs precisely when $x = 3$ or 1. To examine the convergence of the power series at these points, we must plug each one into the equation separately and examine the convergence of the resulting infinite series.

If $x = 3$, we get $\sum_{n=1}^{\infty}\frac{(3-2)^n}{n} = \sum_{n=1}^{\infty}\frac{1^n}{n} = \sum_{n=1}^{\infty}\frac{1}{n}$, which is the harmonic series. We know this is divergent.

If $x = 1$, we get $\sum_{n=1}^{\infty}\frac{(1-2)^n}{n} = \sum_{n=1}^{\infty}\frac{(-1)^n}{n}$, which is convergent by the alternating series test from the last chapter.

Therefore, compiling all this information we see that our original series converges when $1 < x < 3$ and when $x = 1$. We write $\sum_{n=1}^{\infty}\frac{(x-2)^n}{n}$ converges for $1 \le x < 3$, and we can graphically represent this information on a number line in the following way:

The Radius and Interval of Convergence of a Power Series

Notice that $\sum_{n=1}^{\infty}\frac{(x-2)^n}{n}$ is a power series centered at $x = 2$, and the area of convergence of this power series is also centered at $x = 2$. This is not a coincidence. Any given power series centered at $x = c$, written $\sum_{n=0}^{\infty} a_n(x-c)^n$, clearly converges when $x = c$, because when $x = c$ we have $\sum_{n=0}^{\infty} a_n(x-c)^n = 0$. We are primarily interested in the size of the interval surrounding $x = c$ where the power series converges.

It could happen that the power series converges only for $x = c$ and it could happen that the power series converges for *all* values of x on the real axis. In the situation above, the power series $\sum_{n=1}^{\infty}\frac{(x-2)^n}{n}$ converges on the interval $|x - 2| < 1$, and we checked by hand the endpoints of this interval. In general, for a given power series $\sum_{n=0}^{\infty} a_n(x-c)^n$, there is a positive integer R such that the series converges if $|x - c| < R$ and diverges if $|x - c| > R$. This number R is called the *radius of convergence of the power series*. The interval of convergence of a power series is the exact interval (with or without endpoints) where the power series converges.

It is important to notice the distinction between the radius of convergence and the interval of convergence. In our example of $\sum_{n=1}^{\infty}\frac{(x-2)^n}{n}$, the radius of convergence is $R = 1$ and we found the interval of convergence to be $1 \le x < 3$.

Remark: Usually the ratio test can be used to determine the radius of convergence of a power series, and the ratio test is inconclusive at the endpoints of this interval. Another test should be used to calculate the convergence at the endpoints of the interval.

Let's find the radius of convergence and the interval of convergence of the power series $\sum_{n=0}^{\infty} \frac{3^n x^n}{(n+1)^2}$.

Since, for this power series, we have the terms $A_n = \frac{3^n x^n}{(n+1)^2}$, we get

$$\left|\frac{A_{n+1}}{A_n}\right| = \left|\frac{\frac{3^{n+1}x^{n+1}}{(n+1+1)^2}}{\frac{3^n x^n}{(n+1)^2}}\right| = \left|\frac{3^{n+1}x^{n+1}}{(n+2)^2} \cdot \frac{(n+1)^2}{3^n x^n}\right| = \left(\frac{n+1}{n+2}\right)^2 |3x|^1$$

Furthermore, we have

$$\lim_{n\to\infty}\left|\frac{A_{n+1}}{A_n}\right| = \lim_{n\to\infty}\left(\frac{n+1}{n+2}\right)^2 |3x|^1 = \lim_{n\to\infty} |3x|^1.$$

$$\lim_{n\to\infty}\left|\frac{A_{n+1}}{A_n}\right| = \lim_{n\to\infty} |3x| < 1 \text{ only if } |3x| < 1, \text{ that is, if } |x| < \frac{1}{3}.$$

Therefore the radius of convergence of this power series is $R = \frac{1}{3}$. In order to determine the interval of convergence, we must determine the behavior of the series at each of the endpoints of that interval, $x = \frac{1}{3}$ and $x = -\frac{1}{3}$. At those endpoints, $|3x| = 1$, so the ratio test alone is inconclusive for convergence.

If $x = \frac{1}{3}$, the series becomes $\sum_{n=0}^{\infty} \frac{3^n\left(\frac{1}{3}\right)^n}{(n+1)^2} = \sum_{n=0}^{\infty} \frac{1}{(n+1)^2}$, which is a convergent p-series.

If $x = -\frac{1}{3}$, the series becomes $\sum_{n=0}^{\infty} \frac{3^n\left(-\frac{1}{3}\right)^n}{(n+1)^2} = \sum_{n=0}^{\infty} \frac{(-1)^n}{(n+1)^2}$, which converges by the alternating series test.

Because the series converges at each of the endpoints, the interval of convergence is $-\frac{1}{3} \le x \le \frac{1}{3}$.

TAYLOR POLYNOMIAL APPROXIMATION

The real power of power series (no pun intended) lies in their ability to approximate more complicated functions. If we set:

$$f(x) = \sum_{n=0}^{\infty} a_n(x-c)^n = a_0 + a_1(x-c) + a_2(x-c)^2 + \cdots + a_n(x-c)^n + \cdots,$$

we see that a power series is like a polynomial, only it has infinitely many terms. By now we have seen many different functions in mathematics, and polynomials are some of the easiest to work with (they are continuous everywhere, their derivatives and integrals are easy to compute, etc.).

In this section, we will show how non-polynomial functions can be approximated by a power series called the *Taylor series expansion.* This means that, given any function, no matter how complicated it may be, we can always find a power series that is as close to the original function as we want, i.e., the power series *approximates* the given function.

In a way we have already seen this type of polynomial approximation. Recall that the geometric series is written as:

$$\sum_{n=0}^{\infty} r^n = 1 + r + r^2 + r^3 + \cdots + r^n + \cdots = \frac{1}{1-r}$$

In essence we have written the rational function $\frac{1}{1-r}$ as a polynomial with infinitely many terms. Let's see how we can generalize this process to find a polynomial expansion for any function.

To motivate our discussion of the Taylor and Maclaurin series, let's suppose that for a given function $f(x)$ we already have a power series expansion. If this were true then we would have

$$f(x) = \sum_{n=0}^{\infty} a_n(x-c)^n = a_0 + a_1(x-c) + a_2(x-c)^2 + \cdots + a_n(x-c)^n + \cdots$$

To determine the coefficients of this power series, we make some trivial observations. When $x = c$, we would have $f(c) = a_0$ because all the other terms in the sum would be zero. Furthermore, if we took the derivative of our function we would get

$$f'(x) = \sum_{n=0}^{\infty} na_n(x-c)^{n-1} = a_1 + 2a_2(x-c)^1 + \cdots + na_n(x-c)^{n-1} + \cdots$$

and therefore, $f'(c) = a_1$ for the exact same reasoning as before. If we took another derivative we would get

$$f''(x) = \sum_{n=0}^{\infty} n(n-1)a_n(x-c)^{n-2} = 2a_2 + 3 \cdot 2a_3(x-c)^1 \cdots + n(n-1)a_n(x-c)^{n-2} + \ldots$$

and similarly we would have $f''(c) = 2a_2$. Clearly, if we repeat this process we get $f'''(c) = 3 \cdot 2a_3$ and even more generally, we would have $f^{(n)}(c) = n!a_n$. Therefore, supposing we have such an expansion, we have a nice formula for the coefficients of the power series expansion of our function centered at the point $x = c$. Solving for a_n, we have $a_n = \frac{f^{(n)}(c)}{n!}$.

Using this motivation, we can define the Taylor and Maclaurin series of a function.

The Taylor series expansion of a function is the power series expansion of the function with coefficients discussed above and centered at the point $x = c$. For the special case when $c = 0$, we have a special name for the Taylor series expansion; it is called the Maclaurin series. The Maclaurin series of a function is the power series expansion of the function discussed above, centered at the point $x = 0$.

More formally we write,

Definition: The Taylor series of the function $f(x)$ at $x = c$ is given by

$$f(x) = \sum_{n=0}^{\infty} \frac{f^{(n)}(c)}{n!}(x-c)^n = f(c) + \frac{f'(c)}{1!}(x-c) + \frac{f''(c)}{2!}(x-c)^2 + \frac{f'''(c)}{3!}(x-c)^3 + \ldots$$

Similarly, the Maclaurin series of the function $f(x)$ is given by

$$f(x) = \sum_{n=0}^{\infty} \frac{f^{(n)}(0)}{n!}x^n = f(0) + \frac{f'(0)}{1!}x + \frac{f''(0)}{2!}x^2 + \frac{f'''(0)}{3!}x^3 + \dots$$

Note: The Maclaurin series is just the Taylor series of $f(x)$ at $c = 0$.

To stretch our legs with the newly developed Taylor and Maclaurin series, let's calculate the Taylor series at $c = 3$ of $f(x) = 10 + 5x + 3x^2$.

This function is of order 2, so there will only be at most derivatives up to order 2. We calculate

$$f'(x) = 5 + 6x$$
$$f''(x) = 6$$
$$\text{and } f^{(n)}(x) = 0 \text{ for all } n \geq 3$$

To find the Taylor series of $f(x)$ at $x = 3$, we get

$$f(x) = \sum_{n=0}^{\infty} \frac{f^{(n)}(3)}{n!}(x-3)^n = f(3) + \frac{f'(3)}{1!}(x-3) + \frac{f''(3)}{2!}(x-3)^2$$
$$= 52 + 23(x-3) + 3(x-3)^2$$

Thus, we get $f(x) = 52 + 23(x - 3) + 3(x - 3)^2$, which simplifies to $10 + 5x + 3x^2$. Furthermore, because there are only finitely many coefficients of this Taylor series, the Taylor series converges for all values of x. Therefore, the radius of convergence is infinity.

Let us now try a more interesting example.

Example:

Find the Maclaurin series of $f(x) = \sin x$.

Solution:

To calculate the Maclaurin series we must find the nth order derivatives of $\sin(x)$ and plug them into the formula provided.

$$\begin{aligned}
f(x) &= \sin x & f(0) &= \sin 0 = 0 \\
f'(x) &= \cos x & f'(0) &= \cos 0 = 1 \\
f''(x) &= -\sin x & f''(0) &= -\sin 0 = 0 \\
f'''(x) &= -\cos x \quad \text{and so,} & f'''(0) &= -\cos 0 = -1 \\
f^{(4)}(x) &= \sin x & f^{(4)}(0) &= \sin 0 = 0 \\
f^{(5)}(x) &= \cos x & f^{(5)}(0) &= \cos 0 = 1 \\
&\dots & &\dots
\end{aligned}$$

Thus, to write out the Maclaurin series we get

$$f(x) = f(0) + \frac{f'(0)}{1!}x + \frac{f''(0)}{2!}x^2 + \frac{f'''(0)}{3!}x^3 + \cdots$$
$$= x - \frac{x^3}{3!} + \frac{x^5}{5!} - \frac{x^7}{7!} + \cdots$$

To write this as a power series we get

$$\sin x = \sum_{n=0}^{\infty} (-1)^n \frac{x^{2n+1}}{(2n+1)!} = x - \frac{x^3}{3!} + \frac{x^5}{5!} - \frac{x^7}{7!} + \cdots$$

Note: the $(2n + 1)$ term ensures that we only have odd powers of x because $(2n + 1)$ is odd for every n.

MACLAURIN SERIES FOR SPECIAL FUNCTIONS

In the last section, we calculated the Maclaurin series for the sine function. In this section we will calculate the Maclaurin series of some other very important functions that we have worked with before.

Let's begin by looking at another trigonometric function, $\cos x$. To calculate the Maclaurin series we need to find a pattern in the derivatives of cosine. We calculate:

$$\begin{aligned} f(x) &= \cos x & & & f(0) &= 1 \\ f'(x) &= -\sin x & & & f'(0) &= 0 \\ f''(x) &= -\cos x & & & f''(0) &= -1 \\ f'''(x) &= \sin x & &\text{and} & f^{(3)}(0) &= 0 \\ f^{(4)}(x) &= \cos x & & & f^{(4)}(0) &= 1 \\ f^{(5)}(x) &= -\sin x & & & f^{(5)}(0) &= 0 \end{aligned}$$

Recall that the nth term of the Maclaurin series is $\frac{f^{(n)}(0)}{n!}x^n$. Therefore, the Maclaurin series for $\cos x$ will only have even order terms in x. We get

$$\cos x = \sum_{n=0}^{\infty} \frac{f^{(n)}(0)}{n!}x^n = f(0) + f'(0)x + \frac{f''(0)}{2!}x^2 + \frac{f'''(0)}{3!}x^3 + \frac{f^{(4)}(0)}{4!}x^4 + \cdots$$
$$= 1 + 0 - \frac{1}{2}x^2 + 0 + \frac{1}{4!}x^4 + \cdots$$
$$= \sum_{n=0}^{\infty} \frac{(-1)^n}{(2n)!}x^{2n}$$

What is the radius of convergence for this expression of $\cos x$? To determine this we calculate:

$$\lim_{n\to\infty} \left|\frac{A_{n+1}}{A_n}\right| = \lim_{n\to\infty} \left| \frac{\frac{(-1)^{(n+1)}}{(2(n+1))!}x^{2(n+1)}}{\frac{(-1)^n}{(2n)!}x^{2n}} \right| = \lim_{n\to\infty} \left| \frac{-x^{2n+2}}{(2n+2)!} \cdot \frac{(2n)!}{x^{2n}} \right| = \lim_{n\to\infty} \frac{-x^2}{(2n+2)(2n+1)}.$$

Thus, $\lim_{n\to\infty}\left|\frac{A_{n+1}}{A_n}\right| = 0$ for all values of x, and so the radius of convergence is the entire real line.

Note: A similar computation will show that the same result holds for the power series expansion of $\sin x$.

Now let us calculate the Maclaurin series for another very important function that is found throughout all mathematics, e^x. Not only is this function incredibly important in many diverse fields of mathematics, but the power series expansion has an incredibly striking simplicity.

As always, we calculate the nth order derivatives and look for a pattern. Recalling that the derivative of e^x is just e^x, we realize that we are done before we even start.

We have

$$\begin{aligned} f(x) &= e^x & & & f(0) &= e^0 = 1 \\ f'(x) &= e^x & &\text{and} & f'(0) &= e^0 = 1 \\ &\dots \\ f^{(n)}(x) &= e^x & & & f^{(n)}(0) &= e^0 = 1 \end{aligned}$$

And thus the Maclaurin series is as simple as it can possibly be.

$$e^x = \sum_{n=0}^{\infty}\frac{f^{(n)}(0)}{n!} = \sum_{n=0}^{\infty}\frac{x^n}{n!} = 1 + \frac{x}{1!} + \frac{x^2}{2!} + \frac{x^3}{3!} + \frac{x^4}{4!} + \dots$$

Therefore, when we let $x = 1$, we have $e = 1 + 1 + \frac{1}{2} + \frac{1}{3!} + \frac{1}{4!} + \frac{1}{5!} + \dots$

Now that we've calculated some important Maclaurin series, let's record that information:

$\frac{1}{1-x}$	$\sum_{n=0}^{\infty} x^n$	$1 + x + x^2 + x^3 + x^4 + \cdots$
e^x	$\sum_{n=0}^{\infty}\frac{x^n}{n!}$	$1 + x + \frac{x}{2!} + \frac{x}{3!} + \frac{x}{4!} + \cdots$
$\sin x$	$\sum_{n=0}^{\infty}\frac{(-1)^n}{(2n+1)!}x^{2n+1}$	$x - \frac{x^3}{3!} + \frac{x^5}{5!} + \frac{x^7}{7!} + \cdots$
$\cos x$	$\sum_{n=0}^{\infty}\frac{(-1)^n}{(2n)!}x^{2n}$	$1 - \frac{x^2}{2!} + \frac{x^4}{4!} + \frac{x^6}{6!} + \cdots$

SHORTCUTS USING TAYLOR/MACLAURIN SERIES

In the last section we calculated the Maclaurin series of some important popular functions and thus retrieved a polynomial expression for trigonometric and exponential functions. However, by now you may be asking yourself, "Apart from having a nice polynomial expression to represent some complicated function, what good are these power series expansions to us?"

Actually, they can do us a lot of good! In short, because polynomials are so easy to work with, we can substitute in the formal power series expansion for the actual function and make our lives a lot simpler. By writing a function as its Taylor series, we can greatly simplify the process of finding its derivative and antiderivative. In fact, using the power series expansion we can easily find the antiderivative of functions that previously were very difficult to find using less sophisticated techniques.

First, let's see how to use the Maclaurin series we developed in the previous section to get a new Maclaurin series. For example, to find the power series expansion for the function $f(x) = e^{-\frac{x}{2}}$, we can use what we know of e^x.

That is, because we have $e^x = \sum_{n=0}^{\infty} \frac{x^n}{n!}$, we can substitute $-\frac{x}{2}$ for the value of x to calculate the power series development of $f(x) = e^{-\frac{x}{2}}$. Using basic algebra rules, we see that

$$f(x) = e^{-\frac{x}{2}} = \sum_{n=0}^{\infty} \frac{(-\frac{x}{2})^n}{n!} = \sum_{n=0}^{\infty} \frac{\left(-1 \cdot \frac{1}{2} \cdot x\right)^n}{n!} = \sum_{n=0}^{\infty} \frac{(-1)^n \left(\frac{1}{2}\right)^n x^n}{n!} = \sum_{n=0}^{\infty} \frac{(-1)^n x^n}{2^n n!}$$

This type of substitution can greatly decrease the amount of work needed to calculate the Taylor and Maclaurin series of some functions.

For example, suppose you are asked to evaluate $\int \sin(x^2)\, dx$ as an infinite series.

Using our power series expansion of $\sin x$ this question becomes fairly routine.

To evaluate the $\sin(x^2)$ part, we use the Maclaurin series of $\sin x$ to get:

$$\begin{aligned}\sin(x^2) &= \sum_{n=0}^{\infty} \frac{(-1)^n}{(2n+1)!}(x^2)^{2n+1} \\ &= \sum_{n=0}^{\infty} \frac{(-1)^n}{(2n+1)!} x^{4n+2}\end{aligned}$$

Now since the integral of a sum is the sum of the integrals, we can calculate:

$$\begin{aligned}
\int \sin(x^2)\,dx &= \int \sum_{n=0}^{\infty} \frac{(-1)^n}{(2n+1)!} x^{4n+2}\,dx \\
&= \sum_{n=0}^{\infty} \int \frac{(-1)^n}{(2n+1)!} x^{4n+2} dx, \quad \text{because } \int \sum = \sum \int \text{ for these convergent series} \\
&= \sum_{n=0}^{\infty} \frac{(-1)^n}{(2n+1)!} \int x^{4n+2}\,dx, \quad \text{because } \frac{(-1)^n}{(2n+1)!} \text{ is just a constant} \\
&= \sum_{n=0}^{\infty} \frac{(-1)^n}{(2n+1)!} \frac{x^{4n+3}}{4n+3}, \qquad \text{because } \int x^{4n+2}\,dx = \frac{x^{4n+3}}{4n+3}
\end{aligned}$$

And we're done, thanks to the Maclaurin series.

Let's look at another example. Suppose we are asked to evaluate $\lim\limits_{x \to 0} \frac{1 - \cos x}{1 + x - e^x}$ using series development.

We get

$$\lim_{x \to 0} \frac{1 - \cos x}{1 + x - e^x} = \lim_{x \to 0} \frac{1 - \sum\limits_{n=0}^{\infty} \frac{(-1)^n}{(2n)!} x^{2n}}{1 + x - \sum\limits_{n=0}^{\infty} \frac{x^n}{n!}} = \lim_{x \to 0} \frac{1 - \left(1 + \sum\limits_{n=1}^{\infty} \frac{(-1)^n}{(2n)!} x^{2n}\right)}{1 + x - \left(1 + x + \sum\limits_{n=2}^{\infty} \frac{x^n}{n!}\right)} = \lim_{x \to 0} \frac{-\sum\limits_{n=1}^{\infty} \frac{(-1)^n}{(2n)!} x^{2n}}{-\sum\limits_{n=2}^{\infty} \frac{x^n}{n!}}$$

$$= \lim_{x \to 0} \frac{\left(\frac{x^2}{2!} - \frac{x^4}{4!} + \frac{x^6}{6!} - \ldots\right)}{\left(-\frac{x^2}{2!} - \frac{x^3}{3!} - \frac{x^4}{4!} - \frac{x^5}{5!} - \ldots\right)}$$

$$\underset{\text{Rule}}{\overset{\text{L'Hôpital's}}{=}} \lim_{x \to 0} \frac{\left(x - \frac{x^3}{3!} + \frac{x^5}{5!} - \ldots\right)}{\left(-x - \frac{x^2}{2!} - \frac{x^3}{3!} - \frac{x^4}{4!} - \ldots\right)} \underset{\text{Rule}}{\overset{\text{L'Hôpital's}}{=}} \lim_{x \to 0} \frac{\left(1 - \frac{x^2}{2!} + \frac{x^4}{4!} - \ldots\right)}{\left(-1 - x - \frac{x^2}{2!} - \frac{x^3}{3!} - \ldots\right)}$$

$$= \frac{\lim\limits_{x \to 0} \left(1 - \frac{x^2}{2!} + \frac{x^4}{4!} - \ldots\right)}{\lim\limits_{x \to 0} \left(-1 - x - \frac{x^2}{2!} - \frac{x^3}{3!} - \ldots\right)} = \frac{1}{-1} = -1$$

FUNCTIONS DEFINED BY POWER SERIES

In this section we will look at a very important function that is defined by means of an infinite power series. Surely you have seen (or can multiply by hand) the formula $(a + b)^2 = a^2 + 2ab + b^2$. Multiplying this equation by $(a + b)$ we can deduce that $(a + b)^3 = a^3 + 3a^2b + 3ab^2 + b^3$. Of course, we can continue in this manner to calculate $(a + b)^n$ for any n. Unfortunately, this method quickly becomes very computational and thus not very convenient. Fortunately, however, we have a very simple formula for $(a + b)^n$, defined as a power series.

THE BINOMIAL SERIES

Motivated by the Binomial Theorem, $(a + b)^k = \sum_{n=0}^{\infty} \binom{k}{n} a^{k-n} b^n$ where k is a positive integer, we find the binomial series:

If k is any real number and $|x| < 1$, then

$$(1+x)^k = \sum_{n=0}^{\infty} \binom{k}{n} x^n = 1 + kx + \frac{k(k-1)}{2!}x^2 + \frac{k(k-1)(k-2)}{3!}x^3 + \ldots$$

where $\binom{k}{n} = \dfrac{k(k-1)(k-2)\cdots(k-n+1)}{n!}$

Remark: This result is very different from the Binomial Theorem because the value of k can be negative and not an integer for the binomial series, whereas in the Binomial Theorem, k must be a positive integer.

As with any power series, we should question convergence. This series converges when $|x| < 1$; however, convergence at the endpoints depends entirely on the value of k. When $-1 < k \leq 0$, the binomial series converges at 1; when $k \geq 0$, it converges at both endpoints.

We can use this power series expansion for $(1 + x)^k$ in many useful ways. For example, to find the Maclaurin series and radius of convergence for $\dfrac{x}{\sqrt{4+x^2}}$. At first glance it may not be clear exactly why we should use the binomial series. With some practice these questions will make themselves more evident and their solutions will become second nature.

To use the binomial series we need to have our quotient in the proper form. Let's rearrange the quotient in a strategic way (always keeping in mind the binomial series).

$$\frac{x}{\sqrt{4+x^2}} = x(4+x^2)^{-\frac{1}{2}} = x\left(4\left(1+\frac{x^2}{4}\right)\right)^{-\frac{1}{2}} = x \cdot 4^{-\frac{1}{2}}\left(1+\frac{x^2}{4}\right)^{-\frac{1}{2}} = \frac{x}{2}\left(1+\frac{x^2}{4}\right)^{-\frac{1}{2}}$$

Now we can apply the binomial series to the second term on the right to get

$$\left(1+\frac{x^2}{4}\right)^{-\frac{1}{2}} = \sum_{n=0}^{\infty} \binom{k}{n}\left(\frac{x^2}{4}\right)^n = \sum_{n=0}^{\infty} \binom{k}{n}\left(\frac{x}{2}\right)^{2n}$$

Thus, we have

$$\frac{x}{\sqrt{4+x^2}} = \frac{x}{2} \cdot \sum_{n=0}^{\infty} \binom{k}{n} \left(\frac{x}{2}\right)^{2n} = \sum_{n=0}^{\infty} \binom{k}{n} \left(\frac{x}{2}\right)^{2n+1}$$

From the statement of the binomial series, we know that this series converges when $\left|\frac{x^2}{4}\right| < 1$. So, for this series $R = 2$.

Clearly, the binomial series is a useful tool for defining functions by power series and it should be remembered.

HOW CLOSE IS CLOSE ENOUGH? THE LAGRANGE ERROR BOUND

As we stated at the beginning of this chapter, the Taylor series expansion provides a polynomial approximation of the given function. But what exactly does this mean? Recall that for a series to converge to a finite number, as we added more and more terms together, the corresponding sum got closer and closer to the limiting value of the series. Similarly, as we add more and more terms of our Taylor series, the resulting sum function is closer and closer to the actual function we started with. We can easily see this by graphing the given function and a couple of partial sums of the Taylor series.

Let's look at the function e^x. We know that the full Maclaurin series is

$$e^x = \sum_{n=0}^{\infty} \frac{x^n}{n!} = 1 + x + \frac{x^2}{2!} + \frac{x^3}{3!} + \frac{x^4}{4!} + \frac{x^5}{5!} + \ldots$$

If we consider only the first K elements of this power series, we have the corresponding *finite* partial sums, $\sum_{n=0}^{K} \frac{x^n}{n!}$, which we call $f_K(x)$. To say that the Maclaurin series above approximates the function e^x means that $\lim_{K \to \infty} f_K(x) = e^x$. Therefore, we have

$K = 0$ $\quad f_0 = \sum_{n=0}^{0} \frac{x^n}{n!} = 1,$ $\quad$ (because $x^0 = 1, 0! = 1$)

$K = 1$ $\quad f_1 = \sum_{n=0}^{0} \frac{x^n}{n!} = 1 + x$

$K = 2$ $\quad f_2 = \sum_{n=0}^{2} \frac{x^n}{n!} = 1 + x + \frac{x^2}{2!}$

$K = 3$ $\quad f_3 = \sum_{n=0}^{3} \frac{x^n}{n!} = 1 + x + \frac{x^2}{2!} + \frac{x^3}{3!}$

...

Furthermore, if we look at the graphs of f_K, for each K, we see that the graphs begin to resemble the graph of e^x.

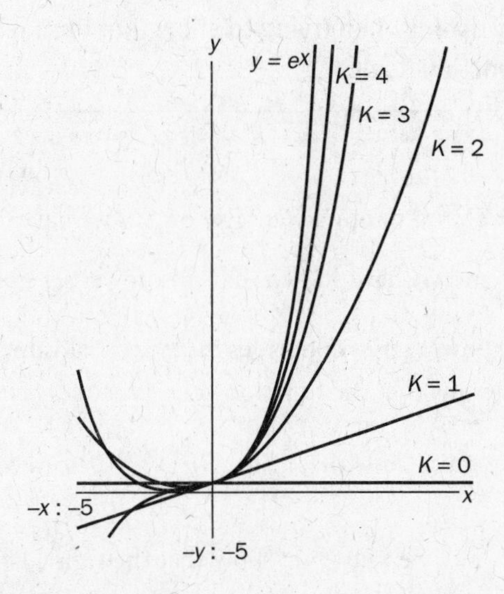

We see that $\lim_{K \to \infty} f_K(x) = e^x$. That is, the more terms we sum from the Maclaurin series, the closer we get to the actual function.

However, notice that whenever we consider a sum of finitely many terms, there is always some degree of error with the actual function we are approximating. Just as we did with infinite series, we want to be able to control the amount of error involved with our approximation. To do this, we look at the remainder terms $R_K(x) = f(x) - f_K(x)$. This difference, $R_K(x)$, is called the remainder or error of the Kth finite Taylor series. Clearly, if the remainder of the finite Taylor series of a given function goes to zero as K gets larger and larger, then the function is equal to its full Taylor series expansion on the interval where it is defined. We write:

$$\lim_{K \to \infty} R_K(x) = 0 \text{ for } |x - a| < R \quad \Rightarrow \quad f(x) = \sum_{n=0}^{\infty} \frac{f^{(n)}(a)}{n!}(x - a)^n \text{ for all } x \text{ such that } |x - a| < R$$

However, to get a bound on the actual remainder term we use a theorem much like the one we developed for infinite series.

Lagrange Error Bound for Taylor Polynomials

Suppose $|f^{(K+1)}(x)| \leq M$ when $|x - a| \leq R$. Then the remainder, $R_K(x)$, satisfies

$$R_K(x) \leq \frac{M}{(K+1)!}|x - a|^{K+1} \quad \text{for all } x \text{ such that } |x - a| \leq R.$$

In short, this theorem says that if we have an upper bound on the $(K + 1)^{\text{th}}$ derivative of $f(x)$, then we also have an upper bound on the error term of the Kth finite Taylor series of $f(x)$.

Note: This theorem is very useful when trying to show that the Taylor series of a given function is equal to the function itself.

Let's see how to use this error bound to our advantage. So far, we have calculated the Taylor series of cos *x* to be $\sum_{n=0}^{\infty} \frac{(-1)^n}{(2n)!} x^{2n}$, and we have shown that this power series has an infinite radius of convergence (meaning the power series converges for every real value of *x*). Now we will use the Lagrange Error Bound to show that the function cos *x* is actually equal to its power series $\sum_{n=0}^{\infty} \frac{(-1)^n}{(2n)!} x^{2n}$. To do this, we must show that $\lim_{K \to \infty} R_K(x) = 0$ for every value of *x*.

To use the Lagrange Error Bound, we must first show that the $(K+1)^{\text{th}}$ derivative of $f(x) = \cos x$ is bounded. Since $f^{(n+1)}(x)$ is either ± cos *x* or ± sin *x* and both sine and cosine are bounded above by 1, we have $|f^{(n+1)}(x)| \leq 1$ for all values of *x*. Therefore, the Lagrange Error Bound for Taylor polynomials tells us that $R_K(x) \leq \frac{1}{(K+1)!}|x|^{n+1}$, and thus $\lim_{K \to \infty} |R_K(x)| \leq \lim_{K \to \infty} \frac{|x|^{K+1}}{(K+1)!}$.

Now using the fact that $\lim_{k \to \infty} \frac{x^k}{k!} = 0$ (because the Taylor series of e^x converges for all values of (*x*)) we that $\lim_{K \to \infty} |R_K(x)| \leq \lim_{K \to \infty} \frac{|x|^{K+1}}{(K+1)!} = 0$.

Thus $\cos x = \sum_{n=0}^{\infty} \frac{(-1)^n}{(2n)!} x^{2n}$, because the remainder terms go to zero as *K* approaches infinity.

REVIEW QUESTIONS

1. Find the radius of convergence and interval of convergence of the power series $\sum_{n=1}^{\infty} \frac{4^n x^n}{(n+2)^2}$.

   (A) $R = \frac{1}{4}, I = \left(-\frac{1}{4}, \frac{1}{4}\right)$

   (B) $R = \frac{1}{4}, I = \left[-\frac{1}{4}, \frac{1}{4}\right]$

   (C) $R = \frac{1}{4}, I = \left(-\frac{1}{4}, \frac{1}{4}\right]$

   (D) $R = 4, I = \left[-\frac{1}{4}, \frac{1}{4}\right]$

   (E) $R = 4, I = \left(-\frac{1}{4}, \frac{1}{4}\right)$

2. Suppose $|x| < \frac{1}{2}$. Which of the following gives a power series representation for $\frac{1}{1+4x^2}$?

   (A) $\sum_{n=0}^{\infty} 4^n x^n$

   (B) $\sum_{n=0}^{\infty} 4^n x^{2n}$

   (C) $\sum_{n=0}^{\infty} (-1)^n 4^n x^{n+1}$

   (D) $\sum_{n=0}^{\infty} (-1)^n 4^n x^{2n}$

   (E) $\sum_{n=0}^{\infty} (-1)^n 4^n x^{n-1}$

3. Which of the following is the Taylor series for $f(x) = \ln x$ centered at $a = 3$?

   (A) $\ln 3 + \sum_{n=1}^{\infty} \frac{(-1)^{n+1}}{3^n n}(x-3)^n$

   (B) $\sum_{n=1}^{\infty} \frac{(x-3)^n}{3^n n!}$

   (C) $\sum_{n=1}^{\infty} \frac{(-1)^n}{n}(x-3)^n$

   (D) $\ln 3 + \sum_{n=1}^{\infty} \frac{(-1)^{n+1}}{3^n n!}(x-3)^n$

   (E) $\ln 3 + \sum_{n=1}^{\infty} \frac{3^{n+1}}{n!}(x-3)^n$

4. Using the fact that $\sin x = \sum_{n=0}^{\infty} \frac{(-1)^n}{(2n+1)!} x^{2n+1}$, what is the power series representation of $\int x^2 \sin x^3 dx$?

   (A) $\sum_{n=0}^{\infty} \frac{(-1)^n}{6(2n+1)!} x^{6(n+1)}$

   (B) $\sum_{n=0}^{\infty} \frac{x^{6n}}{(2n+1)!(n+1)}$

   (C) $\sum_{n=0}^{\infty} \frac{(-1)^n}{6(2n+1)!(n+1)} x^{6(n+1)}$

   (D) $\sum_{n=0}^{\infty} \frac{(-1)^n}{(n+1)!(n+2)} x^{6n}$

   (E) $\sum_{n=0}^{\infty} \frac{(-1)^n}{(2n+1)(n+1)!} x^{6(n+1)}$

5. Suppose $|x| < 2$. Which of the following is the correct expansion of $\frac{1}{(2+x)^3}$ as a power series?

(A) $\sum_{n=0}^{\infty} \frac{(-1)^n(n+1)x^n}{2^n}$

(B) $\sum_{n=0}^{\infty} \frac{(-1)^n(n+1)(n+2)x^n}{2^{n+4}}$

(C) $\sum_{n=0}^{\infty} \frac{(n+2)x^n}{2^n}$

(D) $\sum_{n=0}^{\infty} \frac{(-1)^n(n+1)x^{n+1}}{2^{n+3}}$

(E) $\sum_{n=0}^{\infty} \frac{(-1)^n(n+2)x^{n+2}}{2^{n+1}}$

6. What is the radius of convergence of the power series $\sum \frac{(-1)^n x^n}{\sqrt{n+3}}$?

(A) $R = \frac{1}{2}$
(B) $R = \infty$
(C) $R = 0$
(D) $R = 2\frac{1}{2}$
(E) $R = 1$

7. Using the binomial series, which power series correctly represents $\frac{x+x^2}{(1-x)^3}$?

(A) $\sum_{n=0}^{\infty} \frac{(n+1)(n+2)}{2}(x^{n+1}+x^{n+2})$

(B) $\sum_{n=0}^{\infty} \frac{(n+1)(n+2)}{2}x^{n+1}$

(C) $\sum_{n=0}^{\infty} \frac{(n+2)}{2}(x^{n+1}+x^{n+2})$

(D) $\sum_{n=0}^{\infty} (n+1)(n+2)x^{n+3}$

(E) $\sum_{n=0}^{\infty} \frac{(n+1)(n+2)}{2^{n+1}}(2x^{n+1}+x^{n+2})$

8. Which of the following is the correct Maclaurin series for $f(x) = \sin \pi x$?

(A) $\sum_{n=0}^{\infty} \frac{(-1)^{n+1}\pi^{2n}x^{2n+1}}{(2n+1)!}$

(B) $\sum_{n=0}^{\infty} \frac{(-1)^n\pi^{2n}x^{2n+1}}{(2n)!}$

(C) $\sum_{n=0}^{\infty} \frac{(-1)^n\pi^{2n}x^{2n}}{(2n)!}$

(D) $\sum_{n=0}^{\infty} \frac{(-1)^n x^{2n+1}}{(2n+1)!}$

(E) $\sum_{n=0}^{\infty} \frac{(-1)^n\pi^{2n+1}x^{2n+1}}{(2n+1)!}$

9. Let f be the function given by $f(x) = \cos x$. Which of the following is the second degree Taylor polynomial for f about $x = \frac{\pi}{3}$?

(A) $\frac{\sqrt{3}}{2} - \frac{1}{2}\left(x+\frac{\pi}{3}\right) - \frac{\sqrt{3}}{4}\left(x+\frac{\pi}{3}\right)^2$

(B) $\frac{1}{2} - \frac{\sqrt{3}}{2}\left(x+\frac{\pi}{3}\right) - \frac{1}{4}\left(x+\frac{\pi}{3}\right)^2$

(C) $\frac{\sqrt{3}}{2} - \frac{1}{2}\left(x-\frac{\pi}{3}\right) - \frac{\sqrt{3}}{4}\left(x-\frac{\pi}{3}\right)^2$

(D) $\frac{1}{2} - \frac{\sqrt{3}}{2}\left(x-\frac{\pi}{3}\right) - \frac{1}{4}\left(x-\frac{\pi}{3}\right)^2$

(E) $\frac{\sqrt{3}}{2} - \frac{1}{4}\left(x-\frac{\pi}{3}\right) - \frac{\sqrt{3}}{6}\left(x-\frac{\pi}{3}\right)^2$

10. The third-order Taylor polynomial P for a function $f(x)$ about $x = 3$ is given by $P(x) = 5 - 2(x - 3)^2 + 7(x - 3)^3$. Which of the following is true about $f(x)$?

    I. The point (3, 5) is on the graph of $f(x)$.
    II. f is concave down at $x = 3$.
    III. f has a local maximum of 5 at $x = 3$.

    (A) I and II only
    (B) I and III only
    (C) II and III only
    (D) I, II, and III
    (E) None of the above.

11. The Maclaurin series for $f(x)=x^2e^{-2x^3}$ is

    (A) $\sum_{n=0}^{\infty}\frac{x^{3n+2}}{n!}$
    (B) $\sum_{n=0}^{\infty}\frac{(-2)^n\cdot x^{3n+2}}{n!}$
    (C) $\sum_{n=0}^{\infty}\frac{(2)^n\cdot x^{3n+2}}{n!}$
    (D) $\sum_{n=0}^{\infty}\frac{(-2)^n\cdot x^{3n-2}}{n!}$
    (E) $\sum_{n=0}^{\infty}\frac{(2)^n\cdot x^{3n-2}}{n!}$

12. The third-order Taylor polynomial for $f(x)=-\frac{2}{x}$ at $x = 2$ is given by

    (A) $-1+\frac{1}{2}(x-2)-\frac{1}{2}(x-2)^2+\frac{3}{4}(x-2)^3$
    (B) $-1-\frac{1}{2}(x-2)-\frac{1}{2}(x-2)^2-\frac{3}{4}(x-2)^3$
    (C) $1-\frac{1}{2}(x-2)+\frac{1}{4}(x-2)^2-\frac{1}{8}(x-2)^3$
    (D) $-1-\frac{1}{2}(x-2)-\frac{1}{4}(x-2)^2-\frac{1}{8}(x-2)^3$
    (E) $-1+\frac{1}{2}(x-2)-\frac{1}{4}(x-2)^2+\frac{1}{8}(x-2)^3$

Question 13 requires a calculator.

13. Consider the function f, which is differentiable for all orders for all real numbers. It is known that $f(2) = 3, f'(2) = -2, f''(2) = 1$ and $f'''(2) = -1$. If a third-order Taylor polynomial at $x = 2$ is used to estimate $f(2.1)$, the result would be

    (A) 2
    (B) 2.789
    (C) 2.805
    (D) 3.191
    (E) 3.211

14. The radius of convergence of the power series $\sum_{n=0}^{\infty}\frac{n\cdot x^{2n+3}}{x^2\cdot n!}$ is

    (A) 1
    (B) 2
    (C) 3
    (D) 4
    (E) ∞

Question 15 requires a calculator.

15. A third-order Taylor polynomial centered at $x = 3$ is used to approximate a differentiable function f at $x = 3.6$. It is known that $\left|f^{(4)}(x)\right| \le 0.7$ for all $x \in [3, 3.6]$. What is the Lagrange Error Bound for the maximum error on the closed interval [3, 3.6]?

    (A) 0.00378
    (B) 0.00400
    (C) 0.00412
    (D) 0.00444
    (E) 0.00487

FREE-RESPONSE QUESTION

16. (a) Expand $f(x) = \dfrac{x}{(1-x)^2}$ as a power series.

    (b) Using part (a), find the sum of the series $\sum_{n=1}^{\infty} \frac{n}{2^n}$. Show all work.

17. (a) Find the Maclaurin series generated by $f(x) = \dfrac{\sin x}{x}$.

    (b) Determine the radius of convergence for this series.

    (c) Construct a fifth-order Taylor polynomial for $\dfrac{\sin x}{x}$ and use it to estimate $\int_0^1 \dfrac{\sin x}{x}\,dx$.

ANSWERS AND EXPLANATIONS

1. B

To calculate the radius of convergence we need to determine which values of x ensure that

$\lim_{n\to\infty}\left|\frac{a_{n+1}}{a_n}\right| < 1$, where $a_n = \frac{4^n x^n}{(n+2)^2}$. We get

$$\lim_{n\to\infty}\left|\frac{a_{n+1}}{a_n}\right| = \lim_{n\to\infty}\left|\frac{\frac{4^{n+1}x^{n+1}}{(n+1+2)^2}}{\frac{4^n x^n}{(n+2)^2}}\right| = \lim_{n\to\infty}\left|\frac{4^{n+1}x^{n+1}}{4^n x^n}\cdot\frac{(n+2)^2}{(n+1+2)^2}\right| = \lim_{n\to\infty} 4|x|\frac{(n+2)^2}{(n+3)^2} = 4|x|.$$

Thus, $\lim_{n\to\infty}\left|\frac{a_{n+1}}{a_n}\right| < 1$ precisely when $4|x| < 1$; that is, when $|x| < \frac{1}{4}$. So, $R = \frac{1}{4}$.

Now we have to check the endpoints to determine the exact interval of convergence.

If $x = \frac{1}{4}$, we have $\sum_{n=1}^{\infty}\frac{4^n\left(\frac{1}{4}\right)^n}{(n+2)^2} = \sum_{n=1}^{\infty}\frac{1}{(n+2)^2} < \sum_{n=1}^{\infty}\frac{1}{n^2} < \infty.$

If $x = -\frac{1}{4}$, we have $\sum_{n=1}^{\infty}\frac{4^n\left(-\frac{1}{4}\right)^n}{(n+2)^2} = \sum_{n=1}^{\infty}\frac{(-1)^n}{(n+2)^2} < \sum_{n=1}^{\infty}\frac{1}{(n+2)^2} < \infty.$

Therefore, both the endpoints are included. So, $I = \left[-\frac{1}{4}, \frac{1}{4}\right]$.

2. D

For this question we want to use a geometric series which states that $\sum_{n=0}^{\infty} x^n = \frac{1}{1-x}$, provided $|x| < 1$. So let's rewrite $\frac{1}{1+4x^2}$ so that it resembles the right-hand side. Note that we are given $|x| < \frac{1}{2}$, so the geometric series is convergent. We have

$$\frac{1}{1+4x^2} = \frac{1}{1-(-4x^2)} = \sum_{n=0}^{\infty}(-4x^2)^n = \sum_{n=0}^{\infty}(-1)^n 4^n (x^2)^n = \sum_{n=0}^{\infty}(-1)^n 4^n x^{2n}$$

3. A

To compute the Taylor series of $f(x)$ at the point $x = a$, we use the formula $\sum_{n=0}^{\infty}\frac{f^{(n)}(a)}{n!}(x-a)^n$. So let's get started and calculate some derivatives of $\ln x$.

$$\begin{aligned} f(x) &= \ln x, & & f(3) = \ln 3 \\ f'(x) &= \frac{1}{x}, & & f'(3) = \frac{1}{3} \\ f''(x) &= \frac{-1}{x^2}, & \text{so} \quad & f''(3) = \frac{-1}{3^2} \\ f'''(x) &= \frac{2}{x^3}, & & f'''(3) = \frac{2}{3^3} \\ &\cdots\cdots & & \cdots\cdots \\ f^{(n)}(x) &= \frac{(-1)^{n+1}(n-1)!}{x^n}, & & f^{(n)}(3) = \frac{(-1)^{n+1}(n-1)!}{3^n} \end{aligned}$$

So we get $\sum_{n=0}^{\infty} \frac{f^{(n)}(3)}{n!}(x-3)^n = f(3) + \sum_{n=1}^{\infty} \frac{\frac{(-1)^{n+1}(n-1)!}{3^n}}{n!}(x-3)^n = \ln 3 + \sum_{n=1}^{\infty} \frac{(-1)^{n+1}}{3^n n}(x-3)^n$

for the Taylor series at $x = 3$.

4. C

To compute $\int x^2 \sin x^3\, dx$ we first need to simplify $x^2 \sin x^3$. To do this we use the given information to get

$$\sin x^3 = \sum_{n=0}^{\infty} \frac{(-1)^n}{(2n+1)!}(x^3)^{2n+1} = \sum_{n=0}^{\infty} \frac{(-1)^n}{(2n+1)!} x^{6n+3}$$

Therefore,

$$x^2 \sin x^3 = x^2 \sum_{n=0}^{\infty} \frac{(-1)^n}{(2n+1)!} x^{6n+3} = \sum_{n=0}^{\infty} \frac{(-1)^n}{(2n+1)!} x^{6n+5}$$

Now, using the linearity of the integral, we get

$$\int x^2 \sin x^3\, dx = \int \sum_{n=0}^{\infty} \frac{(-1)^n}{(2n+1)!} x^{6n+5}\, dx = \sum_{n=0}^{\infty} \int \frac{(-1)^n}{(2n+1)!} x^{6n+5} dx = \sum_{n=0}^{\infty} \frac{(-1)^n}{(2n+1)!} \int x^{6n+5} dx$$

$$= \sum_{n=0}^{\infty} \frac{(-1)^n}{(2n+1)!} \frac{x^{6n+6}}{(6n+6)} = \sum_{n=0}^{\infty} \frac{(-1)^n x^{6(n+1)}}{6(2n+1)!(n+1)}$$

5. B

To answer this question we employ the binomial series:

$$(1+x)^k = \sum_{n=0}^{\infty} \binom{k}{n} x^n, \quad \text{provided } |x| < 1$$

Therefore, we have

$$\frac{1}{(2+x)^3} = (2+x)^{-3} = 2^{-3}(1+\tfrac{x}{2})^{-3} = 2^{-3} \sum_{n=0}^{\infty} \binom{-3}{n} \left(\frac{x}{2}\right)^n$$

Let's simplify the $\binom{-3}{n}$ term to get

$$\binom{-3}{n} = \frac{(-3)(-4)(-5)\ldots(-3-n+1)}{n!} = \frac{(-1)^n 3 \cdot 4 \cdot 5 \cdot \ldots \cdot (n+1)(n+2)}{n!} = \frac{(-1)^n (n+1)(n+2)}{2}.$$

Therefore, the equation above can be written as

$$\frac{1}{(2+x)^3} = 2^{-3} \sum_{n=0}^{\infty} \binom{-3}{n} \left(\frac{x}{2}\right)^n = \frac{1}{8} \sum_{n=0}^{\infty} \frac{(-1)^n (n+1)(n+2)}{2} \left(\frac{x}{2}\right)^n = \sum_{n=0}^{\infty} \frac{(-1)^n (n+1)(n+2) x^n}{2^{n+4}}.$$

6. E

$$\sum \frac{(-1)^n x^n}{\sqrt{n+3}}$$

To calculate the radius of convergence we use the ratio test yet again. We want to determine which values of x will give $\lim_{n\to\infty}\left|\frac{a_{n+1}}{a_n}\right| < 1$ where $a_n = \frac{(-1)^n x^n}{\sqrt{n+3}}$. We get

$$\lim_{n\to\infty}\left|\frac{a_{n+1}}{a_n}\right| = \lim_{n\to\infty}\left|\frac{\frac{(-1)^{n+1}x^{n+1}}{\sqrt{n+1+3}}}{\frac{(-1)^n x^n}{\sqrt{n+3}}}\right| = \lim_{n\to\infty}\left|\frac{x^{n+1}}{x^n}\cdot\frac{\sqrt{n+3}}{\sqrt{n+4}}\right| = |x|\cdot \lim_{n\to\infty}\frac{\sqrt{n+3}}{\sqrt{n+4}} = |x|$$

Therefore, this series converges for all values of x such that $|x| < 1$. So, $R = 1$.

7. A

As the question suggests, we will use the binomial series. First, let's evaluate $\frac{1}{(1-x)^3}$. We get:

$$\frac{1}{(1-x)^3} = (1-x)^{-3} = (1+(-x))^{-3} = \sum_{n=0}^{\infty}\binom{-3}{n}(-x)^n.$$

As before, let's begin by evaluating $\binom{-3}{n} = \frac{(-3)(-4)(-5)(-6)...(-3-n+1)}{n!} = \frac{(-1)^n(n+1)(n+2)}{2}$.

Therefore we have $\frac{1}{(1-x)^3} = \sum_{n=0}^{\infty}\frac{(-1)^n(n+1)(n+2)}{2}(-x)^n = \sum_{n=0}^{\infty}\frac{(n+1)(n+2)}{2}x^n$. Thus, when we multiply by $(x + x^2)$ we get

$$\frac{x+x^2}{(1-x)^3} = (x+x^2)\left(\sum_{n=0}^{\infty}\frac{(n+1)(n+2)}{2}x^n\right)$$

$$= \sum_{n=0}^{\infty}\frac{(n+1)(n+2)}{2}x^{n+1} + \sum_{n=0}^{\infty}\frac{(n+1)(n+2)}{2}x^{n+2}$$

$$= \sum_{n=0}^{\infty}\frac{(n+1)(n+2)}{2}(x^{n+1} + x^{n+2})$$

8. E

This question is probably easier than you think. It also illustrates the usefulness of power series. To make our lives simple we recall that $\sin x = \sum_{n=0}^{\infty}\frac{(-1)^n x^{2n+1}}{(2n+1)!}$. Then to find $f(x) = \sin \pi x$, we simply plug πx in for x. We get

$$\sin \pi x = \sum_{n=0}^{\infty}\frac{(-1)^n(\pi x)^{2n+1}}{(2n+1)!} = \sum_{n=0}^{\infty}\frac{(-1)^n\pi^{2n+1}x^{2n+1}}{(2n+1)!}$$

9. D

The second-order Taylor polynomial for f at $x=\dfrac{\pi}{3}$ is given by:

$$
\begin{aligned}
P_2(x) &= f\left(\frac{\pi}{3}\right)+f'\left(\frac{\pi}{3}\right)\left(x-\frac{\pi}{3}\right)+f''\left(\frac{\pi}{3}\right)\left(x-\frac{\pi}{3}\right)^2 \\
&= \cos\left(\frac{\pi}{3}\right)-\sin\left(\frac{\pi}{3}\right)\left(x-\frac{\pi}{3}\right)-\frac{\cos\left(\frac{\pi}{3}\right)}{2}\left(x-\frac{\pi}{3}\right)^2 \\
&= \frac{1}{2}-\frac{\sqrt{3}}{2}\left(x-\frac{\pi}{3}\right)-\frac{1}{4}\left(x-\frac{\pi}{3}\right)^2
\end{aligned}
$$

10. D

The third-order Taylor polynomial for f at $x = 3$ is given by:

$$f(3)+f'(3)(x-3)+\frac{f''(3)}{2!}(x-3)^2+\frac{f'''(3)}{3!}(x-3)^3$$

Thus: $P(x)=5-2(x-3)^2+7(x-3)^3=f(3)+f'(3)(x-3)+\dfrac{f''(3)}{2}(x-3)^2+\dfrac{f'''(3)}{6}(x-3)^3$

We know that $f(3) = 5$, and therefore we know that the function goes through the point (3, 5).

There is no linear term in the Taylor polynomial, which tells us that $f'(3) = 0$ and therefore (3, 5) must be a critical point.

$$\frac{f''(3)}{2}=-2\Rightarrow f''(3)=-4;$$

which tells us that f is concave down at $x = 3$ and therefore (3, 5) is a local maximum by the second derivative test.

11. B

We could apply the formula for Taylor series to this polynomial and get the answer, but this would require a great deal of work. We can simplify the process considerably by recalling that the Maclaurin series for $f(x) = e^x$ is:

$$e^x=1+x+\frac{x^2}{2!}+\frac{x^3}{3!}+\ldots=\sum_{n=0}^{\infty}\frac{x^n}{n!}$$

(you should have this memorized).

Then:

$$\begin{aligned} x^2e^{-2x^3} &= x^2\left(\sum_{n=0}^{\infty}\frac{\left(-2x^3\right)^n}{n!}\right) \\ &= x^2\left(\sum_{n=0}^{\infty}\frac{(-2)^n\cdot x^{3n}}{n!}\right) \\ &= \sum_{n=0}^{\infty}\frac{(-2)^n\cdot x^{3n+2}}{n!} \end{aligned}$$

12. E

$$f(x)=-\frac{2}{x}\Rightarrow f(2)=-1$$

$$f'(x)=\frac{2}{x^2}\Rightarrow f'(2)=\frac{1}{2}$$

$$f''(x)=-\frac{4}{x^3}\Rightarrow f''(2)=-\frac{1}{2},\ \frac{f''(2)}{2!}=-\frac{1}{4}$$

$$f'''(x)=\frac{12}{x^4}\Rightarrow f'''(2)=\frac{3}{4},\ \frac{f'''(2)}{3!}=\frac{1}{8}$$

$$\therefore P_3(x)=-1+\frac{1}{2}(x-2)-\frac{1}{4}(x-2)^2+\frac{1}{8}(x-2)^3$$

13. C

The third-order Taylor polynomial for f at $x = 2$ would be:

$$\begin{aligned} P_3(x) &= f(2)+f'(2)(x-2)+\frac{f''(2)}{2!}(x-2)^2+\frac{f'''(2)}{3!}(x-2)^3 \\ &= 3-2(x-2)+\frac{1}{2}(x-2)^2-\frac{1}{6}(x-2)^3 \end{aligned}$$

$$\begin{aligned} P_3(2.1) &= 3-2(0.1)+\frac{(.01)}{2}-\frac{(.001)}{6} \\ &= 3-0.2+0.005-0.000167 \\ &= 2.805 \end{aligned}$$

$$\therefore f(2.1)\approx 2.805$$

14. E

We can begin by simplifying the power series:

$$\sum_{n=0}^{\infty}\frac{n\cdot x^{2n+3}}{x^{2}\cdot n!}=\sum_{n=0}^{\infty}\frac{n\cdot x^{2n+3-2}}{n\cdot(n-1)!}=\sum_{n=0}^{\infty}\frac{x^{2n+1}}{(n-1)!}$$

We can now check for convergence by using the ratio test:

$$\begin{aligned}
\lim_{n\to\infty}\left|\frac{A_{n+1}}{A_n}\right| &= \lim_{n\to\infty}\left|\frac{\dfrac{x^{2(n+1)+1}}{((n+1)-1)!}}{\dfrac{x^{2n+1}}{(n-1)!}}\right| \\
&= \lim_{n\to\infty}\left|\frac{x^{2n+3}}{n!}\cdot\frac{(n-1)!}{x^{2n+1}}\right| \\
&= \lim_{n\to\infty}\left|\frac{x^{2}}{n\cdot(n-1)!}\cdot\frac{(n-1)!}{1}\right| \\
&= \lim_{n\to\infty}\frac{\left|x^{2}\right|}{n}=0 \\
\therefore \lim_{n\to\infty}\left|\frac{A_{n+1}}{A_n}\right| &= 0\ \forall x\in\mathbb{R}
\end{aligned}$$

Therefore, the series converges for all values of x, and the radius of convergence is infinity.

15. A

The Lagrange Error Bound in using a third-order polynomial centered at a (in this case 3) is given by: $\frac{M}{4!}|x-a|^{4}$ where M is the maximum possible value of the fourth derivative of f on the interval (which is given as 0.7), and x is the value at which we are trying to estimate (in this case 3.6).

Thus, the Lagrange Error Bound is: $\frac{0.7}{4!}|3.6-3|^{4}=0.00378.$

FREE-RESPONSE ANSWER

16. (a) To expand $\frac{x}{(1-x)^2}$ we use the binomial series to get

$$\frac{x}{(1-x)^2} = \frac{x}{(1+(-x))^2} = x\sum_{n=0}^{\infty}\binom{-2}{n}(-x)^n$$

$$= x\sum_{n=0}^{\infty}(-1)^n(n+1)(-x)^n = \sum_{n=0}^{\infty}(n+1)(x)^{n+1}$$

$$= \sum_{n=1}^{\infty} nx^n$$

Here we used

$$\binom{-2}{n} = \frac{(-2)(-3)(-4)...(-2-n+1)}{n!} = \frac{(-1)^n 2\cdot 3\cdot 4\cdot ...\cdot(n+1)}{n!} = (-1)^n(n+1)$$

(b) Now we can use what we have calculated to determine $\sum_{n=1}^{\infty}\frac{n}{2^n}$. If we let $x = \frac{1}{2}$, we get

$$\sum_{n=1}^{\infty}\frac{n}{2^n} = \sum_{n=1}^{\infty} n\left(\frac{1}{2}\right)^n = \frac{\frac{1}{2}}{\left(1-\frac{1}{2}\right)^2} = \frac{1}{2}\cdot 4 = 2, \text{ and we are done.}$$

17. (a) The Maclaurin series for $\sin(x)$ is known to be:

$$x - \frac{x^3}{3!} + \frac{x^5}{5!} - \frac{x^7}{7!} + ...$$

From this, we can develop:

$$\frac{\sin(x)}{x} = \frac{x - \frac{x^3}{3!} + \frac{x^5}{5!} - \frac{x^7}{7!} + ...}{x} = 1 - \frac{x^2}{3!} + \frac{x^4}{5!} - \frac{x^6}{7!} + ... = \sum_{n=0}^{\infty}\frac{(-1)^n x^{2n}}{(2n+1)!}$$

(b) We use the ratio test to determine the radius of convergence:

$$\begin{aligned}
\lim_{n\to\infty}\left|\frac{A_{n+1}}{A_n}\right| &= \lim_{n\to\infty}\left|\frac{\dfrac{(-1)^{n+1}x^{2(n+1)}}{(2n+3)!}}{\dfrac{(-1)^n x^{2n}}{(2n+1)!}}\right| \\
&= \lim_{n\to\infty}\left|\frac{(-1)^{n+1}x^{2n+2}}{(2n+3)!}\right|\cdot\left|\frac{(2n+1)!}{(-1)^n x^{2n}}\right| \\
&= \lim_{n\to\infty}\left|\frac{(-1)x^2}{(2n+3)(2n+2)}\right| \\
&= 0
\end{aligned}$$

The ratio test says that if $\lim_{n\to\infty}\left|\frac{A_{n+1}}{A_n}\right| < 1$, the series converges. In this case, we see that the limit is 0 and is independent of x, therefore, the series converges for all x and the radius of convergence is the entire real number line.

(c) The fifth-order Taylor polynomial is: $P_5(x) = 1 - \frac{x^2}{3!} + \frac{x^4}{5!}$. Using this, we have:

$$\begin{aligned}
\int_0^1 \frac{\sin x}{x}\,dx \approx \int_0^1\left(1-\frac{x^2}{6}+\frac{x^4}{120}\right)dx = \left(x-\frac{x^3}{18}+\frac{x^5}{600}\right)\Bigg|_0^1 &= \left(1-\frac{1}{18}+\frac{1}{600}\right) \\
&= \frac{1800-100+3}{1800} \\
&= \frac{1703}{1800}
\end{aligned}$$

Part Four

PRACTICE TESTS

HOW TO TAKE THE PRACTICE TESTS

The next section of this book consists of practice tests. Taking a practice AP exam gives you an idea of what it's like to answer these test questions for a longer period of time, one that approximates the real test. You'll find out which areas you're strong in, and where additional review may be required. Any mistakes you make now are ones you won't make on the actual exam, as long as you take the time to learn where you went wrong.

The practice tests in this book each include 45 multiple-choice questions and six free-response (essay) questions. You will have 105 minutes for the multiple-choice questions, a 10-minute reading period, and 90 minutes to answer the free-response questions. Before taking a practice test, find a quiet place where you can work uninterrupted for three hours. Time yourself according to the time limit at the beginning of each section. It's okay to take a short break between sections, but for the most accurate results you should approximate real test conditions as much as possible. Use the 10-minute reading period to plan your answers for the free-response questions, but don't begin writing your responses until the 10 minutes are up.

As you take the practice tests, remember to pace yourself. Train yourself to be aware of the time you are spending on each question. Try to be aware of the general types of questions you encounter, as well as being alert to certain strategies or approaches that help you to handle the various question types more effectively.

After taking a practice exam, be sure to read the detailed answer explanations that follow. These will help you identify areas that could use additional review. Even when you've answered a question correctly, you can learn additional information by looking at the answer explanation.

Finally, it's important to approach the test with the right attitude. You're going to get a great score because you've reviewed the material and learned the strategies in this book.

HOW TO COMPUTE YOUR SCORE

The practice tests are composed of multiple-choice questions and free-response questions. The multiple-choice questions are scored by an electronic scanner, while a team of trained reviewers scores the free-response questions by hand. Questions are evaluated during the month of June and exam scores are sent out to students and colleges in July. If you haven't received your score by September, contact the College Board.

SCORING THE MULTIPLE-CHOICE QUESTIONS

To compute your score on the multiple-choice portion of the two sample tests, calculate the number of questions you got right on each test, then divide by 45 to get the percentage score for the multiple choice portion of that test.

Scoring the Free-Response Questions

The readers have specific points that they want to see in each free-response question. Frame your answer in complete, coherent sentences. Make sure that you present the various components of your answer in the right order, and in an order that the readers can understand. In addition to these basic structural concerns, readers will be seeking specific pieces of information in your answer. Each piece of information that they are able to find and check off in your answer helps you toward a better score.

To figure out your approximate score for the free-response questions, look at the key points found in the sample response for each question. For each key point you include, add a point. Figure out the number of key points there are in each question, then add up the number of key points you included for each question. Divide by the total number of points available for all the free-response questions to get the percentage score for the free-response portion of that test.

Calculating Your Composite Score

Your score on the AP exam is a combination of your score on the multiple-choice portion of the exam and the free-response section. The free-response section and multiple choice section are each worth one-half of the exam score.

Add together your score on the multiple-choice portion of the exam and your approximate score on the free-response section of the exam. If your score is a decimal, then round up to a whole number. Divide by two to obtain your approximate score for each full-length exam. Remember, however, that much of this depends on how well all of those taking the AP test do. If you do better than average, your score would be higher. The numbers here are just approximations.

The approximate score range is as follows:

5 = 65 –100% (extremely well qualified)

4 = 50 – 64% (well qualified)

3 = 40 – 49% (qualified)

2 = 25 – 39% (possibly qualified)

1 = 0 – 24% (no recommendation)

If your score falls between 50 and 100, you're doing great. Keep up the good work! If your score is lower than 49, there's still hope. Keep studying and you will be able to obtain a much better score on the exam before you know it.

Good luck on the exam!

AP Calculus AB Practice Test 1 Answer Grid

1.	Ⓐ Ⓑ Ⓒ Ⓓ Ⓔ	13.	Ⓐ Ⓑ Ⓒ Ⓓ Ⓔ	25.	Ⓐ Ⓑ Ⓒ Ⓓ Ⓔ	37.	Ⓐ Ⓑ Ⓒ Ⓓ Ⓔ
2.	Ⓐ Ⓑ Ⓒ Ⓓ Ⓔ	14.	Ⓐ Ⓑ Ⓒ Ⓓ Ⓔ	26.	Ⓐ Ⓑ Ⓒ Ⓓ Ⓔ	38.	Ⓐ Ⓑ Ⓒ Ⓓ Ⓔ
3.	Ⓐ Ⓑ Ⓒ Ⓓ Ⓔ	15.	Ⓐ Ⓑ Ⓒ Ⓓ Ⓔ	27.	Ⓐ Ⓑ Ⓒ Ⓓ Ⓔ	39.	Ⓐ Ⓑ Ⓒ Ⓓ Ⓔ
4.	Ⓐ Ⓑ Ⓒ Ⓓ Ⓔ	16.	Ⓐ Ⓑ Ⓒ Ⓓ Ⓔ	28.	Ⓐ Ⓑ Ⓒ Ⓓ Ⓔ	40.	Ⓐ Ⓑ Ⓒ Ⓓ Ⓔ
5.	Ⓐ Ⓑ Ⓒ Ⓓ Ⓔ	17.	Ⓐ Ⓑ Ⓒ Ⓓ Ⓔ	29.	Ⓐ Ⓑ Ⓒ Ⓓ Ⓔ	41.	Ⓐ Ⓑ Ⓒ Ⓓ Ⓔ
6.	Ⓐ Ⓑ Ⓒ Ⓓ Ⓔ	18.	Ⓐ Ⓑ Ⓒ Ⓓ Ⓔ	30.	Ⓐ Ⓑ Ⓒ Ⓓ Ⓔ	42.	Ⓐ Ⓑ Ⓒ Ⓓ Ⓔ
7.	Ⓐ Ⓑ Ⓒ Ⓓ Ⓔ	19.	Ⓐ Ⓑ Ⓒ Ⓓ Ⓔ	31.	Ⓐ Ⓑ Ⓒ Ⓓ Ⓔ	43.	Ⓐ Ⓑ Ⓒ Ⓓ Ⓔ
8.	Ⓐ Ⓑ Ⓒ Ⓓ Ⓔ	20.	Ⓐ Ⓑ Ⓒ Ⓓ Ⓔ	32.	Ⓐ Ⓑ Ⓒ Ⓓ Ⓔ	44.	Ⓐ Ⓑ Ⓒ Ⓓ Ⓔ
9.	Ⓐ Ⓑ Ⓒ Ⓓ Ⓔ	21.	Ⓐ Ⓑ Ⓒ Ⓓ Ⓔ	33.	Ⓐ Ⓑ Ⓒ Ⓓ Ⓔ	45.	Ⓐ Ⓑ Ⓒ Ⓓ Ⓔ
10.	Ⓐ Ⓑ Ⓒ Ⓓ Ⓔ	22.	Ⓐ Ⓑ Ⓒ Ⓓ Ⓔ	34.	Ⓐ Ⓑ Ⓒ Ⓓ Ⓔ		
11.	Ⓐ Ⓑ Ⓒ Ⓓ Ⓔ	23.	Ⓐ Ⓑ Ⓒ Ⓓ Ⓔ	35.	Ⓐ Ⓑ Ⓒ Ⓓ Ⓔ		
12.	Ⓐ Ⓑ Ⓒ Ⓓ Ⓔ	24.	Ⓐ Ⓑ Ⓒ Ⓓ Ⓔ	36.	Ⓐ Ⓑ Ⓒ Ⓓ Ⓔ		

a. 17
b. 50%
d. Part A 2 questions with calculator
30 min. Part B 4 questions no calc

60 minutes
Part A
C. ~~55 min no calc~~
points are no longer deducted
for incorrect answers

a. 6
b.
c. 2, no

AP CALCULUS AB PRACTICE TEST 1

SECTION I, PART A

Time: 55 Minutes
28 Questions

NO GRAPHING CALCULATOR IS ALLOWED ON THIS PORTION OF THE EXAM

Directions: Solve the following problems, using available space for scratchwork. After examining the form of the choices, decide which one is the best of the choices given and fill in the corresponding oval on the answer sheet. No credit will be given for anything written in the test book. Do not spend too much time on any one problem.

In this test:

(1) The domain of a function f is the set of all real numbers x for which $f(x)$ is a real number, unless otherwise specified.

(2) The inverse of a trigonometric function f may be indicated using the inverse function notation f^{-1} or with the prefix "arc" (e.g., $\sin^{-1} x = \arcsin x$).

1. Given $\lim\limits_{x\to 0} \frac{1-\cos x}{x} = 0$, then $\lim\limits_{x\to 0}\left(\frac{3x^2+5\cos x-5}{2x}\right) =$

(A) 0
(B) $\frac{5}{2}$
(C) 3
(D) 5
(E) Does not exist

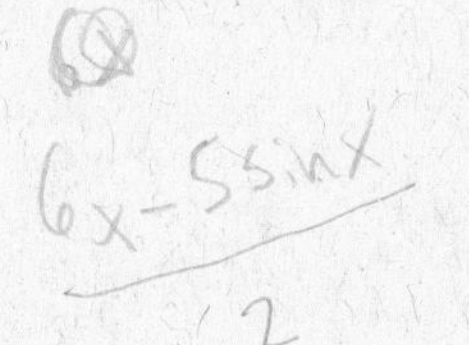

2. $f(x) = \frac{3x^2-6x-9}{x^2-x-2}$ will have vertical asymptotes at

(A) $x = 2$
(B) $x = -1$ and 2
(C) $y = 3$
(D) $x = 3$
(E) There are no vertical asymptotes.

GO ON TO THE NEXT PAGE

3. Which of the following functions grows the fastest?

(A) $a(u) = \left(\frac{1}{2}\right)^u$

(B) $b(u) = u^{100} + u^{99}$

(C) $c(u) = 4^u$

(D) $d(u) = 200e^u$

(E) $e(u) = 3^u + u^3$

4. Which of the following is true about the graph below?

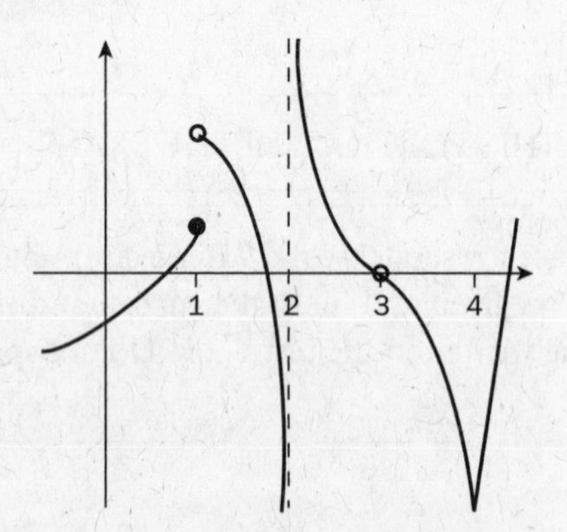

(A) An infinite discontinuity appears to occur at $x = 4$.

(B) The function does not appear to be continuous at $x = 4$.

(C) A jump discontinuity appears to occur at $x = 3$.

(D) A removable discontinuity appears to occur at $x = 2$.

(E) A jump discontinuity appears to occur at $x = 1$.

5. The cost of producing x units of a certain item is $c(x) = 2000 + 8.6x + 0.5x^2$. What is the instantaneous rate of change of c with respect to x when $x = 300$?

(A) 313.6

(B) 308.6

(C) 300.0

(D) 297.2

(E) 200.0

6. A tank holds 10,000 liters of gasoline. At the bottom of the tank, a lever can be turned to allow the gasoline to be dispensed. The tank can be emptied in exactly 40 minutes. Below is a table which gives the volume v of gasoline (in liters) which remain in the tank after t minutes of draining have taken place.

t(minutes)	0	5	10	15	20	25	30	35	40
v(liters)	4700	4100	3200	2400	2000	1400	800	500	0

During which of the following 10-minute intervals is the average rate of gasoline draining from the tank the least?

(A) $t = 0$ to $t = 10$ minutes

(B) $t = 10$ to $t = 20$ minutes

(C) $t = 15$ to $t = 25$ minutes

(D) $t = 25$ to $t = 35$ minutes

(E) $t = 30$ to $t = 40$ minutes

7. Which of the following gives the derivative of the function $f(x) = x^2$ at the point (2, 4)?

(A) $\lim_{h \to 0} \frac{(x+2)^2 - x^2}{4}$

(B) $\lim_{h \to \infty} \frac{(2+h)^2 - 2^2}{h}$

(C) $\frac{(2+h)^2 - 2^2}{h}$

(D) $\lim_{h \to 0} \frac{(2+h)^2 - 2^2}{h}$

(E) $\lim_{h \to 0} \frac{(4+h)^2 - 4^2}{h}$

GO ON TO THE NEXT PAGE

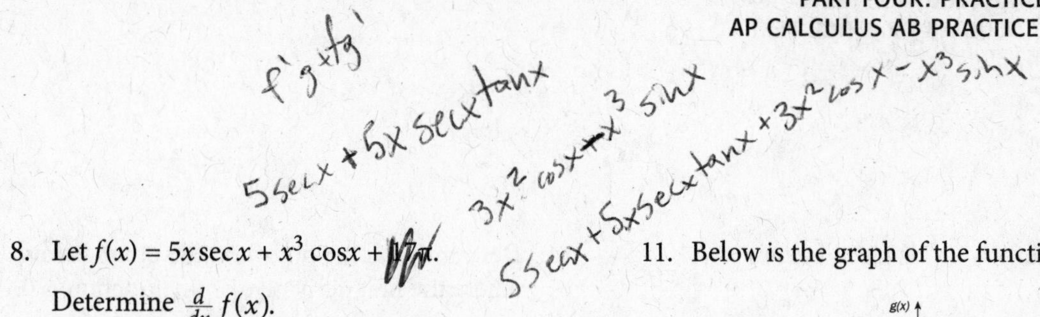

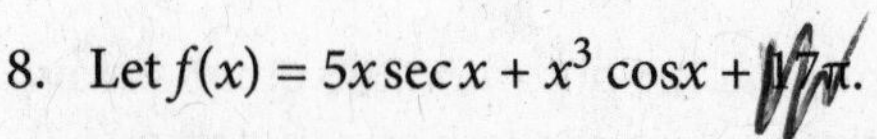

8. Let $f(x) = 5x\sec x + x^3 \cos x + 17\pi$.

   Determine $\frac{d}{dx} f(x)$.

   (A) $5 \sec x \tan x + 3x^2 \cos x + 17\pi$
   (B) $5 \sec^2 x - x^3 \sin x$
   (C) $5 \sec x \tan x - 3x^2 \sin x$
   (D) $5 \sec x + 5x\sec x \tan x + 3x^2 \cos x - x^3 \sin x$
   (E) $5 \sec x + 5x\sec x \tan x - 3x^2 \cos x + x^3 \sin x + 17\pi$

9. Find the derivative of $g(x) = 5\sin^2(6x) + 5\cos^2(6x)$ with respect to x.

   (A) $30\cos^2(6x) - 30\sin^2(6x)$
   (B) $5\cos^2(6x) - 5\sin^2(6x)$
   (C) $120\sin(6x)\cos(6x)$
   (D) 30
   (E) 0

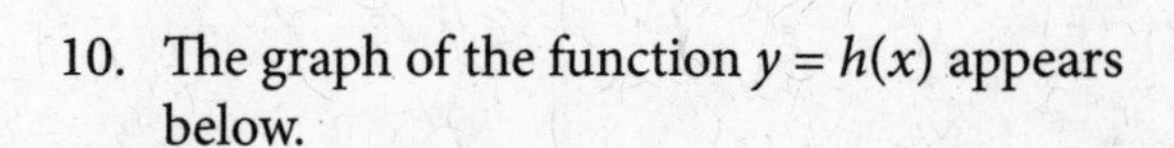

10. The graph of the function $y = h(x)$ appears below.

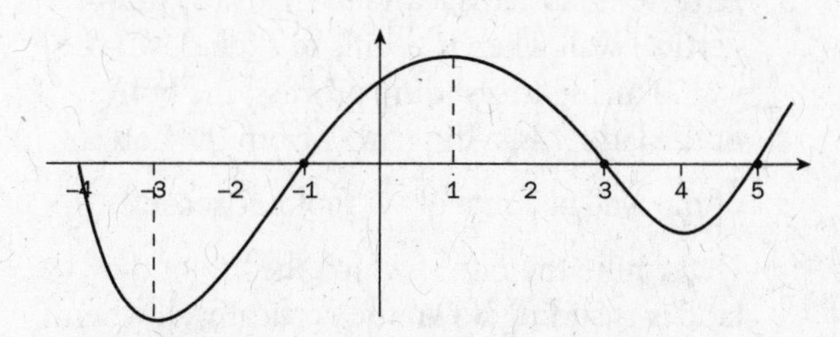

Determine the zeros of the derivative function $h'(x)$.

   (A) $x = -4, -1, 3, 5$
   (B) $x = -2, -1, 2.5$
   (C) $x = 1, 4$
   (D) $x = -3, 1, 4$
   (E) $x = -2, 2.5$

11. Below is the graph of the function $y = g(x)$.

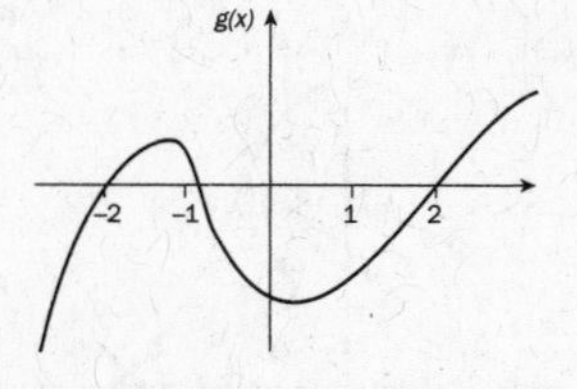

Which graph below is the graph of $y = g'(x)$?

(A)

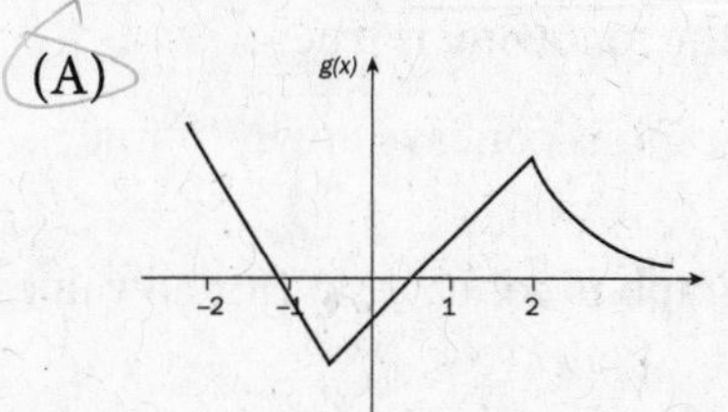

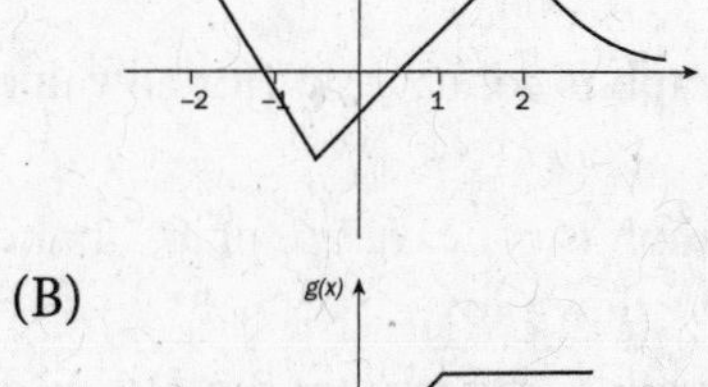

(B)

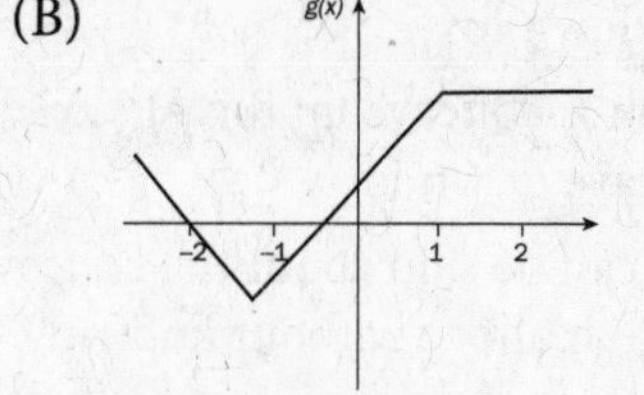

(C)

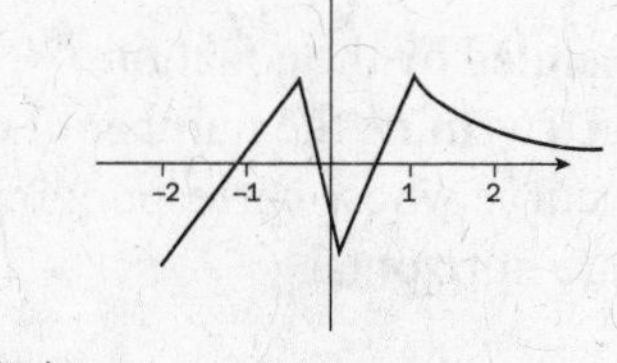

(D)

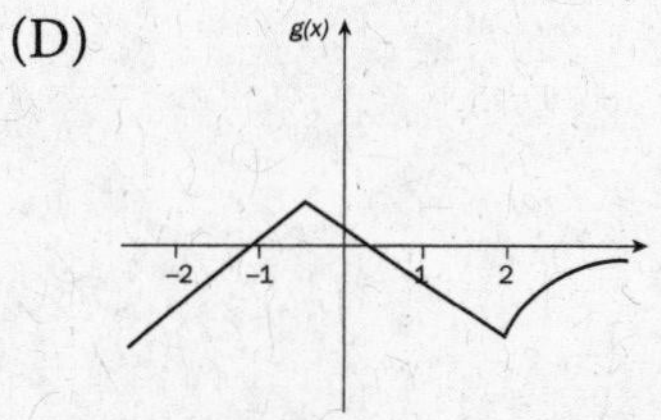

(E)

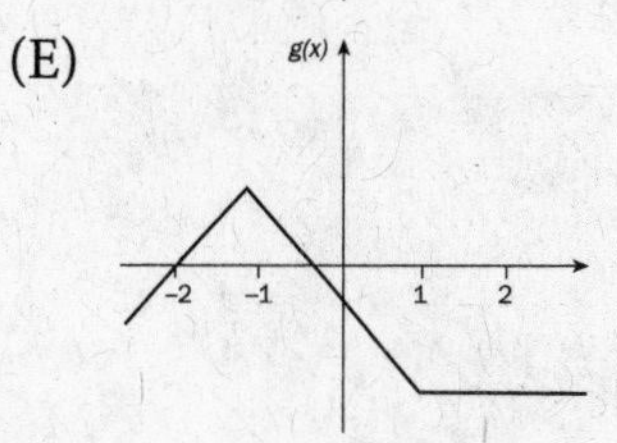

GO ON TO THE NEXT PAGE

12. The graph of $y = f(x)$ is given below.

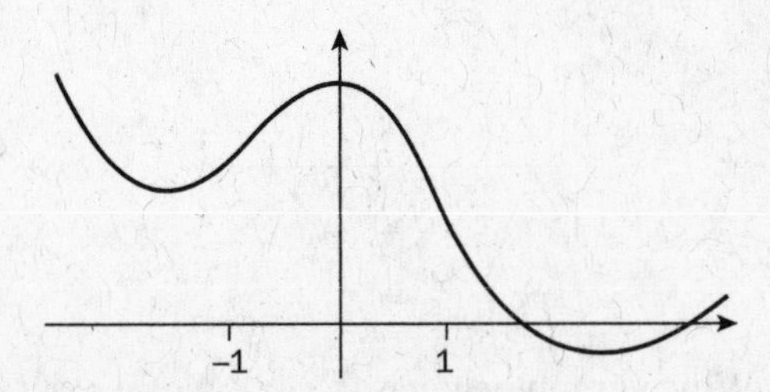

Which of the following is true?

(A) The graph is concave down (for all values of x).

(B) The graph is concave up (for all values of x).

(C) The graph is concave up for $x > 1$ and concave down for $-1 < x < 1$.

(D) The graph is concave up for $-1 < x < 1$ and concave down for $x < -1$.

(E) Nothing can be said about the concavity of the graph above without knowing the rule for the function.

13. A curve is generated by the equation $x^2 + 4y^2 = 16$. Determine the number of points on this curve whose corresponding tangent lines are horizontal.

(A) 0

(B) 1

(C) 2

(D) 3

(E) 4

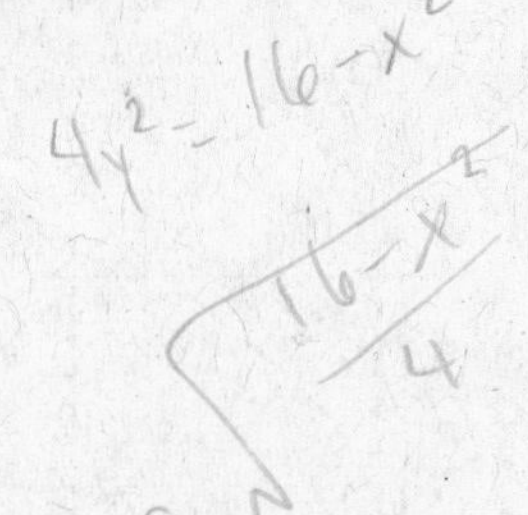

14. Determine all the points on the graph below where the first derivative of the function is 0.

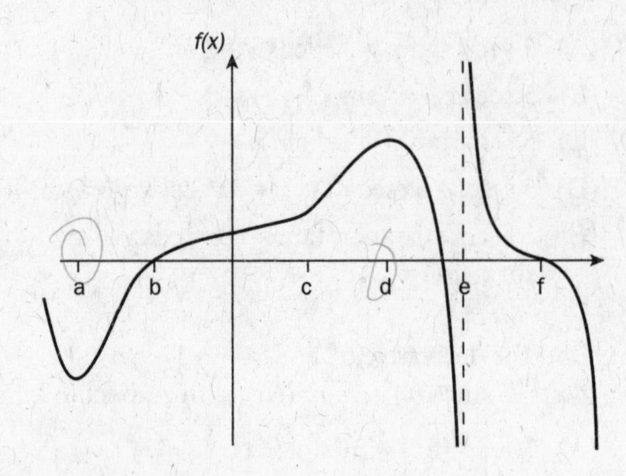

(A) a, b, e

(B) b, c, f

(C) a, d, e

(D) a, b, d, e

(E) a, d

15. A 13-foot ladder is leaning against a 20-foot vertical wall when it begins to slide down the wall. During this sliding process, the bottom of the ladder is sliding away from the bottom of the wall at a rate of $\frac{1}{2}$ foot per second. Determine the rate at which the top of the ladder is sliding down the vertical wall when the tip of the ladder is exactly 5 feet above the ground.

(A) $-\frac{6}{5}$ feet per second

(B) $\frac{5}{6}$ feet per second

(C) $-\frac{12}{13}$ feet per second

(D) -2 feet per second

(E) Not enough information is given to solve this problem.

GO ON TO THE NEXT PAGE

16. Find the solution to the differential equation $\frac{dy}{dx} = \frac{\sin x}{e^y}$, where $y\left(\frac{\pi}{4}\right) = 0$.

(A) $y = \ln\left(\cos x - \frac{\sqrt{2}}{2}\right)$

(B) $y = \ln\left(-\cos x + \frac{\sqrt{2}}{2} + 1\right)$

(C) $y = -\frac{\cos x}{e^y}$

(D) $y = \ln(-\cos x)$

(E) $y = \ln(-\cos x) + \frac{\sqrt{2}}{2}$

17. $\int_0^6 (2 - |4 - x|)dx =$

(A) 24
(B) 12
(C) 6
(D) 2
(E) −4

18. $\int (\sin(2x) + 5\tan^2 x \csc^2 x)\, dx =$

(A) $-\frac{1}{2}\cos(2x) + 5\tan x + C$

(B) $\cos(2x) - 5\cot x + C$

(C) $-\cos(2x) + 5\sec x \tan x + C$

(D) $-\frac{1}{2}\cos(2x) - 5\cot x + C$

(E) $\frac{1}{2}\sin(2x) + 5\tan x + C$

19. $\int \frac{(7+x^{\frac{2}{3}})^5}{x^{\frac{1}{3}}}\, dx =$

(A) $\frac{1}{6}(7 + x^{\frac{2}{3}})^6 + C$

(B) $\frac{1}{6}(7x^{\frac{2}{3}} + x^{\frac{5}{3}})^6 + C$

(C) $\frac{1}{4}(7 + x^{\frac{2}{3}})^6 + C$

(D) $\frac{1}{4}(7 + x^{\frac{1}{3}})^6 + C$

(E) $\frac{1}{6}(7x^{-\frac{1}{3}} + x^{\frac{2}{3}})^6 + C$

20. $\int \sin^4 x \cos x\, dx =$

(A) $\frac{1}{10}\sin^5 x \cos^2 x + C$

(B) $\frac{1}{5}\sin^5 x + C$

(C) $\frac{1}{5}\sin^5 x \cos^2 x + C$

(D) $4\sin^3 x \cos^2 x - \sin^5 x + C$

(E) $\frac{1}{5}\sin^5 x \cos x + \sin^5 x + C$

21. $\frac{d}{dt}\int_2^{t^4} e^{x^2} dx =$

(A) $e^{t^8} - e^4$

(B) $4t^3 e^{t^8} - e^4$

(C) e^{t^8}

(D) $4t^3 e^{t^8}$

(E) Cannot be determined because $\int e^{x^2}$ cannot be determined

GO ON TO THE NEXT PAGE

22. Let $g(x) = (\arccos(x^2))^5$. Then $g'(x) =$

(A) $-10\frac{(\arccos(x^2))^4}{\sqrt{1-x^2}}$

(B) $-10\frac{x(\arccos(x^2))^4}{\sqrt{1-x^4}}$

(C) $-10\frac{x(\arcsin(x^2))^4}{\sqrt{1-x^2}}$

(D) $10\frac{x(\arccos(x^2))^4}{\sqrt{1-x^2}}$

(E) $10\frac{(\arccos(x^2))^4}{\sqrt{1-x^4}}$

23. $\frac{d}{dx}(\ln(3x)5^{2x}) =$

(A) $\frac{5^{2x}}{x} + 2\ln(5)\ln(3x)5^{2x}$

(B) $\frac{5^{2x}}{3x} - 2x\ln(3x)5^{2x}$

(C) $\frac{5^{2x}}{x} - \ln(5)\ln(3x)5^{2x}$

(D) $\frac{5^{2x}}{3x} + 2\ln(3x)5^{2x}$

(E) $\frac{5^{2x}}{x} + \ln(5)\ln(3x)5^{2x}$

24. Let $u(t) = e^{3t^2} - e^{-\frac{1}{t}}$. Then $u'(t) =$

(A) $e^{t^3} + \frac{e^{-\frac{1}{t}}}{t^2}$

(B) $6te^{3t^2} + e^{-\frac{1}{t}}$

(C) $e^{3t^2} - e^{-\frac{1}{t}}$

(D) $6te^{3t^2} - \frac{e^{-\frac{1}{t}}}{t^2}$

(E) $e^{t^3} + e^{-\frac{1}{t}}$

25. $\int \frac{e^{3x}}{e^{6x}+1}\,dx =$

(A) $e^{3x^2} + x + C$

(B) $\ln(e^{6x} + 1) + C$

(C) $-\frac{1}{3}e^{-3x} - \frac{1}{6}e^{-6x} + C$

(D) $\frac{1}{6}\arctan(e^{6x}) + C$

(E) $\frac{1}{3}\arctan(e^{3x}) + C$

26. Find the volume of the solid of revolution determined by rotating the area bounded by the graphs of $f(x) = x^2$ and $g(x) = 3x$ about the x-axis.

(A) $\frac{162\pi}{5}$

(B) $\frac{27\pi}{2}$

(C) 9π

(D) 28π

(E) $\frac{240\pi}{7}$

GO ON TO THE NEXT PAGE

27. Which of the following slope fields describes the differential equation $\frac{dy}{dx} = \frac{x}{y}$?

(A)

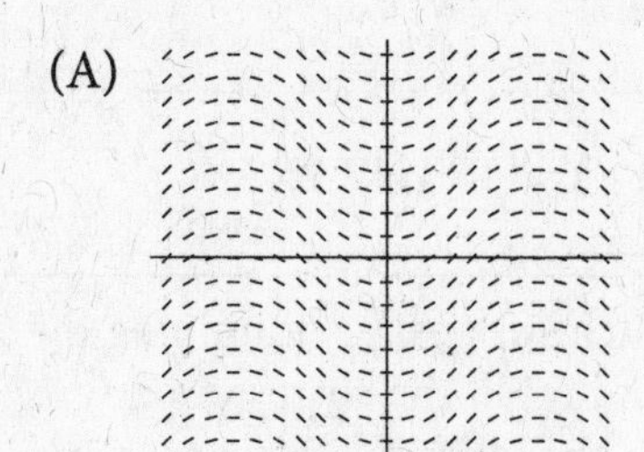

(B)

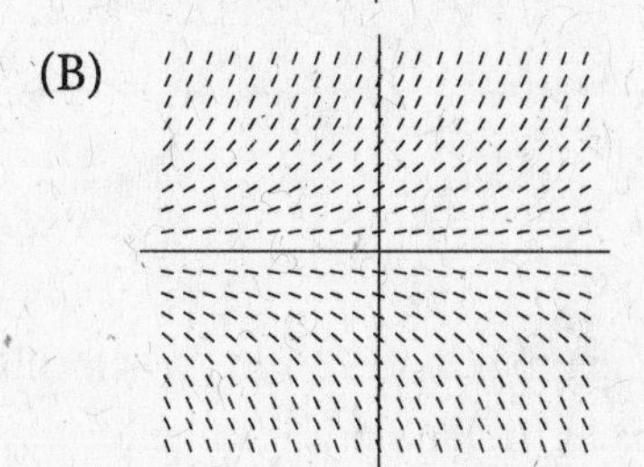

(C)

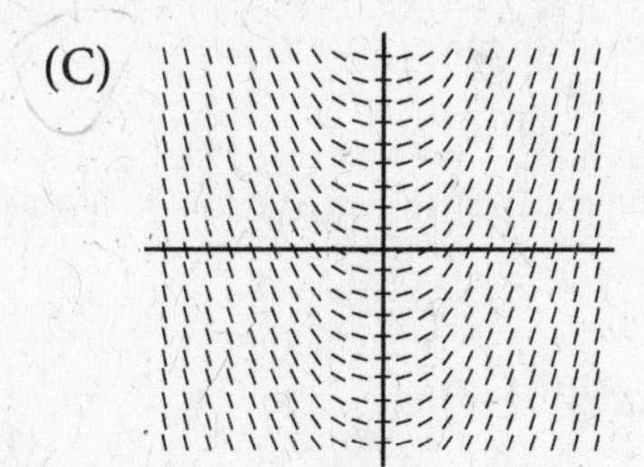

(D)

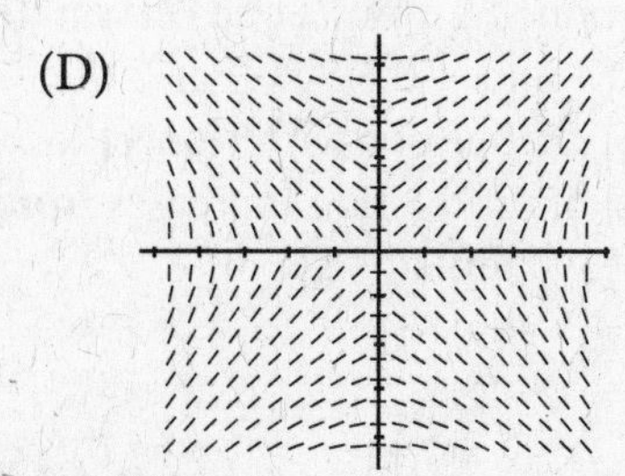

(E)

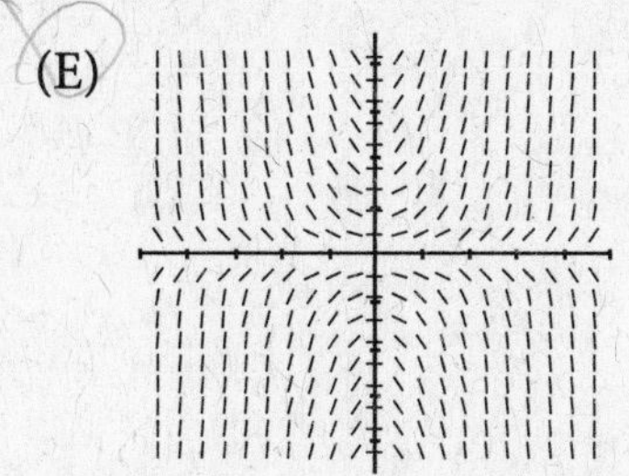

28. $\lim\limits_{x \to 5} \frac{x^2 + 2x - 35}{x^2 - 25} =$

(A) 0

(B) $\frac{6}{5}$

(C) 5

(D) 7

(E) The limit does not exist.

IF YOU FINISH BEFORE TIME IS CALLED, YOU MAY CHECK YOUR WORK ON THIS SECTION ONLY. DO NOT TURN TO ANY OTHER SECTION IN THE TEST. **STOP**

SECTION I, PART B

Time: 50 Minutes
17 Questions

A GRAPHING CALCULATOR IS ALLOWED ON THIS PORTION OF THE EXAM

Directions: Solve the following problems, using available space for scratchwork. After examining the form of the choices, decide which one is the best of the choices given and fill in the corresponding oval on the answer sheet. No credit will be given for anything written in the test book. Do not spend too much time on any one problem.

In this test:

(1) The exact numerical value of the correct answer does not always appear among the choices given. When this happens, select from among the choices the number that best approximates the numerical value.

(2) The domain of a function f is the set of all real numbers x for which $f(x)$ is a real number, unless otherwise specified.

(3) The inverse of a trigonometric function f may be indicated using the inverse function notation f^{-1} or with the prefix "arc" (e.g., $\sin^{-1} x = \arcsin x$).

29. Let $u(t) = \frac{t^2+4t-21}{t^2-9}$. Assume also that $u(t)$ is continuous for all positive real numbers. Determine $u(3)$.

(A) $\frac{5}{3}$

(B) 3

(C) $-\frac{10}{3}$

(D) 0

(E) This is not possible. The function cannot be continuous at $t = 3$.

30. The cost of producing x units of a certain item is $c(x) = 2{,}000 + 8.6x + 0.5x^2$. What is the average rate of change of c with respect to x when the level of production increases from $x = 300$ to $x = 310$ units?

(A) 313.6

(B) 310

(C) 214.2

(D) 200

(E) 10

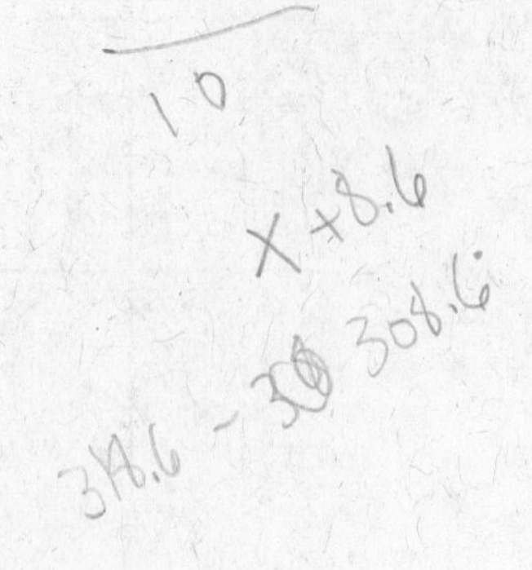

GO ON TO THE NEXT PAGE

31. Find the values of x for which the function $f(x) = x^6 + 3x^5 - \frac{15}{2}x^4 - 40x^3 - 60x^2 + 8x + 5$ has inflection points. Hint: $x^4 + 2x^3 - 3x^2 - 8x - 4 = (x^2 - 4)(x^2 + 2x + 1)$.

   (A) f has no inflection points
   (B) $x = -2, 2$
   (C) $x = -1, 0, 1$
   (D) $x = -2, -1, 2$
   (E) $x = 0$

32. Let $a(x) = \sin(\sin x)$. Find, accurate to three decimal places, $a'\left(\frac{\pi}{4}\right) =$

   (A) 0.538
   (B) −0.999
   (C) 0.009
   (D) 1.000
   (E) 0.866

33. On the interval $(0, 2\pi)$, the function $f(x) = \sin x \cos x$ has critical points at:

   (A) $x = \frac{\pi}{4}, \frac{5\pi}{4}$
   (B) $x = \frac{\pi}{4}, \frac{3\pi}{4}, \frac{5\pi}{4}, \frac{7\pi}{4}$
   (C) $x = \frac{3\pi}{4}, \frac{7\pi}{4}$
   (D) $x = \frac{\pi}{2}, \pi, \frac{3\pi}{2}$
   (E) There are no critical points for this function on the given interval.

34. Determine the slope of the normal line to the curve $x^3 + xy^2 = 10y$ at the point $(2, 1)$.

   (A) 0
   (B) 2
   (C) $-\frac{7}{3}$
   (D) $-\frac{6}{13}$
   (E) $\frac{1}{2}$

35. A spherical balloon is being inflated at a rate of 3 cubic inches per second. Determine the rate of change of the radius of the balloon when the balloon's radius is 5 inches, accurate to three decimal places. The volume of a sphere of radius r is $\frac{4}{3}\pi r^3$.

   (A) 3.000 inches per second
   (B) 1.667 inches per second
   (C) 0.010 inches per second
   (D) −2.000 inches per second
   (E) 0.120 inches per second

36. A gun is fired vertically upward from a position 100 feet above ground at an initial velocity of 400 feet per second. Determine the maximum height of the projectile. The acceleration of gravity is -32ft/sec^2.

   (A) 3,000 feet
   (B) 2,600 feet
   (C) 2,200 feet
   (D) 1,800 feet
   (E) 1,400 feet

37. The table below provides data points for the continuous function $y = h(x)$.

x	0	2	4	6	8	10
$h(x)$	9	25	30	16	25	32

Use a right Riemann sum with 5 subdivisions to approximate the area under the curve of $y = h(x)$ on the interval $[0, 10]$.

   (A) 256
   (B) 235
   (C) 210
   (D) 206
   (E) 242

GO ON TO THE NEXT PAGE

38. Using the trapezoid rule, approximate the area under the curve $f(x) = x^2 + 4$ between $x = 0$ and $x = 3$ using $n = 6$ subintervals.

(A) 18.875
(B) 20.938
(C) 21
(D) 21.125
(E) 23.375

39. The area bounded by the curves $y = x^2 + 4$ and $y = -2x + 1$ between $x = -2$ and $x = 5$ equals

(A) 86.500
(B) 86.425
(C) 86.333
(D) 86.125
(E) 86.000

40. The best approximation given below for the area of the region bounded by the graphs of $f(x) = 3x^3 - 12x^2 + 9x$ and $g(x) = -x^3 + 4x^2 - 3x$ is

(A) 12.000
(B) 12.333
(C) 12.400
(D) 12.525
(E) 12.666

41. Let R be the region in the first quadrant bounded by the graph of $y = 4 - x^2$, the x-axis, and the y-axis. Determine the value of k such that the vertical line $x = k$ cuts the region R into two smaller regions whose areas are equal.

(A) 0.520
(B) 0.695
(C) 0.875
(D) 1.000
(E) 1.135

42. The acceleration of an object at any time $t \geq 0$ is given by $a(t) = t^2 + \sqrt{t+9} + e^{-t}$. The velocity of the object at time $t = 0$ is 4 and the position of the object at time $t = 0$ is 5. Then the position of the object at time $t = 7$ is approximately equal to

(A) −14.714
(B) 15.507
(C) 103.668
(D) 246.752
(E) 321.351

43. The average value of the function $f(x) = \frac{2x}{x^2 - 4}$ on the interval [5, 8] is approximately

(A) 0.350
(B) 1.050
(C) 0.743
(D) 0.248
(E) 0.201

GO ON TO THE NEXT PAGE

44. Which of the following is a solution of $\frac{dy}{dx} = \frac{5}{y(x+4)}$ given that $y(e-4) = 0$?

(A) $y = (t+4)^2 - e^2$

(B) $y^2 = 5\ln|x+4| - 5$

(C) $y = 2\sqrt{t+4} - 2\sqrt{e}$

(D) $y^2 = 10\ln|x+4| - 10$

(E) No solution exists.

45. Which of the following is true of the function $f(x) = \sqrt{x^2+1}$?

(A) $\lim_{x\to\infty}(f(x)-x) = 0$ and $\lim_{x\to-\infty}(f(x)-x) = 0$

(B) $\lim_{x\to\infty}(f(x)+x) = 0$ and $\lim_{x\to-\infty}(f(x)-x) = 0$

(C) $\lim_{x\to\infty}(f(x)-x) = 0$ and $\lim_{x\to-\infty}(f(x)+x) = 0$

(D) $\lim_{x\to\infty}(f(x)+x) = 0$ and $\lim_{x\to-\infty}(f(x)+x) = 0$

(E) None of the above

IF YOU FINISH BEFORE TIME IS CALLED, YOU MAY CHECK YOUR WORK ON THIS SECTION ONLY. DO NOT TURN TO ANY OTHER SECTION IN THE TEST.

STOP

SECTION II

Time: 1 hour and 30 minutes
6 Problems
50 percent of total grade

Directions: Solve the following problems, using available space for scratchwork. Show how you arrived at your answer.

You may wish to look over the problems before starting to work on them; on the actual test, it is not expected that everyone will be able to complete all parts of all problems. All problems are given equal weight, but the individual parts of a particular problem are not necessarily given equal weight. You should not spend too much time on any one problem.

- You should write out all your work for each part. On the actual test, you will do this in the space provided in the test booklet. Be sure to write clearly and legibly. If you make a mistake, you can save time by crossing it out rather than trying to erase it. Erased or crossed-out work will not be graded.
- Show all your work. Clearly label any functions, graphs, tables, or other objects that you use. On the actual exam, you will be graded on the correctness and completeness of your methods as well as your answers. Answers without any supporting work may not receive credit.
- Justifications (i.e., the request that you "justify your answer") require that you give mathematical (non-calculator) reasons.
- Work must be expressed in standard mathematical notation, not calculator syntax.
- Unless otherwise specified, answers (numeric or algebraic) need not be simplified.
- If you use decimal approximations in calculations, the readers of the actual exam will grade you on accuracy. Unless otherwise specified, your final answers should be accurate to three places after the decimal point.
- Unless otherwise specified, the domain of function f is the set of all real numbers x for which $f(x)$ is a real number.

GO ON TO THE NEXT PAGE

PART A

Time: 30 Minutes
2 Problems

A GRAPHING CALCULATOR IS ALLOWED ON THIS PORTION OF THE EXAM

1. The rate of change of charge passing into a battery is modeled by the function $C(t) = 10 + 6\sin\left(\frac{t^2}{3}\right)$, for $t \geq 0$. $C(t)$ has units in coulombs per hour and t has units in hours. At $t = 0$ the battery is empty of any charge.

   (a) Is the amount of charge in the battery increasing or decreasing at $t = 4$? Give a reason for your answer.

   (b) Is the rate of change of charge in the battery increasing or decreasing at $t = 4$? Give a reason for your answer.

   (c) What is the average rate of change of the charge between $t = 2$ and $t = 10$?

   (d) The battery can hold a charge of 160 coulombs. How long, in hours, does it take to charge the battery?

2. Let f and g be the functions given by the equations $f(x) = \sqrt{x}$ and $g(x) = 1 - \cos\left(\frac{\pi x}{2}\right)$. Let A be the region in the first quadrant enclosed by the graphs of f and g, and let B be the region in the first quadrant enclosed by the graph of f, the y-axis, and the line $y = 1$, as shown in the figure below.

   (a) Find the area of A.

   (b) Find the volume of the solid generated when A is rotated about the x-axis.

   (c) Find the volume of the solid generated when B is rotated about the y-axis.

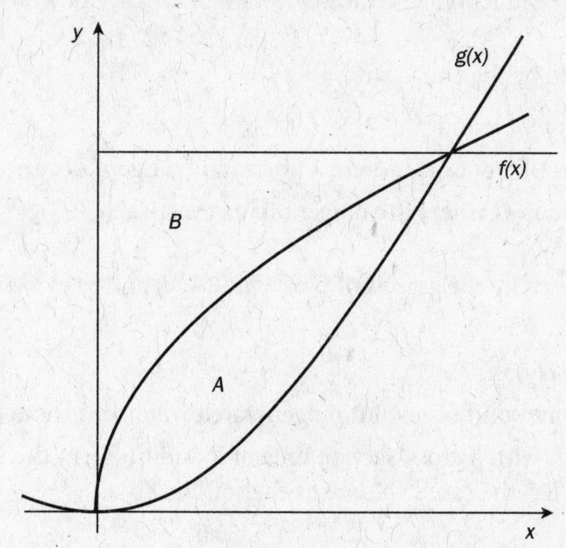

GO ON TO THE NEXT PAGE

PART B

Time: 60 Minutes
4 Problems

NO GRAPHING CALCULATOR IS ALLOWED ON THIS PORTION OF THE EXAM

Note: If you have extra time, you can go back and work on Part A of Section II, but you cannot use a calculator to complete your work at this time.

3. The graph of the function $f'(x)$ appears below, where $f'(x)$ is the derivative of some function $f(x)$. The domain of $f'(x)$ here is the set of real numbers x satisfying the inequalities $-6 < x < 6$.

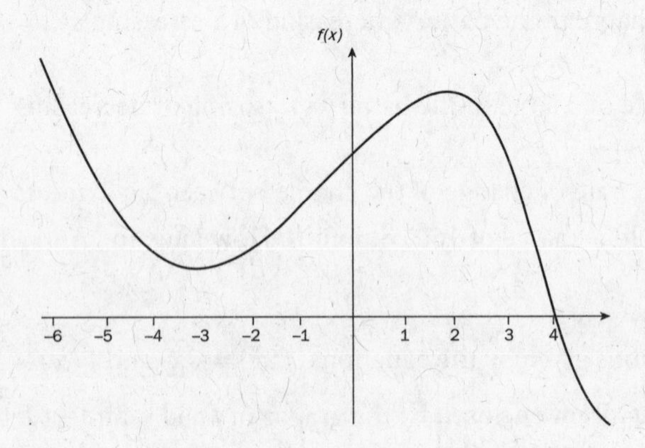

   (a) Determine all points at which f has a relative extremum (if any exist) and state whether each is a maximum or minimum. Justify your answers.
   (b) Determine all points of inflection for f. Justify your answers.
   (c) Determine whether the function f is concave upward or concave downward on the interval $(-6, -3)$.

4. Let C be the curve defined by the equation $xy^2 + y^3 = 5$.
   (a) Find $\frac{dy}{dx}$ as a function of x and y.
   (b) Determine all points (if any exist) where a horizontal tangent occurs.
   (c) Determine the point on C where the tangent line is vertical.

5. Let R be the region enclosed by the graphs of $f(x) = ax(2 - x)$ and $g(x) = ax$ for some positive real number a.
   (a) Find the area of the region R.
   (b) Find the volume of the solid of revolution generated when R is rotated about the x-axis.
   (c) Assume a solid exists with a cross-section area of R and uniform thickness π. Find the value of a for which this solid has the same volume as the solid in (b).

GO ON TO THE NEXT PAGE

6. Consider the differential equation $\frac{dy}{dx} = x^3(y-1)$. Let $f(x)$ be a solution to the differential equation satisfying $f(2) = 2$.

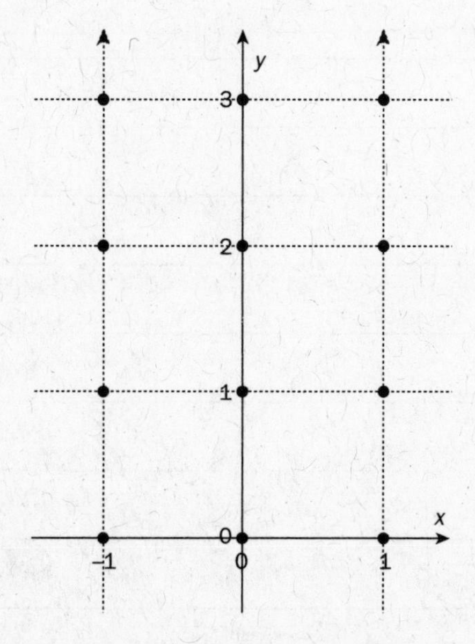

(a) On the axes provided, sketch a slope field for the given differential equation at the 12 points indicated.

(b) Write an equation for the tangent line to the graph of $f(x)$ at $x = 2$.

(c) Find the solution $y = f(x)$ to the given differential equation with the initial condition $f(2) = 2$.

GO ON TO THE NEXT PAGE

IF YOU FINISH BEFORE TIME IS CALLED, YOU MAY CHECK YOUR WORK ON THIS SECTION ONLY. DO NOT TURN TO ANY OTHER SECTION IN THE TEST.

AP Calculus AB Practice Test 1 **Answer Key**

1. A
2. A
3. C
4. E
5. B
6. E
7. D
8. D
9. E
10. D
11. A
12. C
13. C
14. E
15. A
16. B
17. D
18. A
19. C
20. B
21. D
22. B
23. A
24. D
25. E
26. A
27. D
28. B
29. A
30. A
31. B
32. A
33. B
34. D
35. C
36. B
37. A
38. D
39. C
40. B
41. B
42. E
43. A
44. D
45. C

ANSWERS AND EXPLANATIONS

1. A

$$\lim_{x\to0}\left(\frac{3x^2+5\cos x-5}{2x}\right)=\lim_{x\to0}\frac{3x^2}{2x}+\lim_{x\to0}\frac{5\cos x-5}{2x}$$
$$=\lim_{x\to0}\frac{3x}{2}+\lim_{x\to0}\frac{5(\cos x-1)}{2x}$$
$$=0+\lim_{x\to0}\frac{5}{2}\left(\frac{\cos x-1}{x}\right)$$
$$=\lim_{x\to0}\frac{5}{2}(0)$$
$$=0$$

2. A

First, note that answer (C) is the *horizontal* asymptote for $f(x)$. Avoid confusing horizontal and vertical asymptotes in these problems.

Second, after factoring the denominator, we know that -1 and 2 are the roots of the expression in the denominator. However, note that -1 is also a root of the numerator, while 2 is not a root of the numerator. Therefore, $x = -1$ is *not* a vertical asymptote for $f(x)$. The only vertical asymptote for $f(x)$ is $x = 2$.

3. C

We know that exponential functions (with bases greater than 1) grow faster than any power (polynomial) functions. Therefore, answer (B) cannot be the correct answer as $b(u)$ is a polynomial function. Next, we see that the other functions are exponential functions with bases of $\frac{1}{2}$, 4, e, and 3 respectively. We simply need to ask which of these bases is largest (no other facts about the functions are important, including the size of the coefficient 200 in $d(u)$). Recalling that $e \approx 2.7$, we know that the largest of the bases is 4, so the answer is (C).

4. E

The discontinuity at $x = 2$ is an infinite discontinuity. The discontinuity at $x = 3$ is a removable discontinuity. The function is actually continuous at $x = 4$.

5. B

There are several ways to approach this problem. The preferred approach is to calculate the derivative of the function and finding the slope (rate of change) at the point specified.

$$\frac{d}{dx}c(x)=\frac{d}{dx}(2000+8.6x+0.5x^2)=8.6+x$$

Substitute in 300 for x, and you will obtain the same result: $8.6 + 300 = 308.6$.

Another approach is to approximate the average rate of change of c from $x = 300$ to $x = 300 + h$ where h is some small quantity (like $h = 1$, $h = 0.1$, $h = 0.001$, and so on). However, this method is not 100 percent accurate. Another approach is to calculate the limit via algebraic manipulation.

$$\begin{aligned}\lim_{x\to300}\frac{c(x)-c(300)}{x-300} &= \lim_{x\to300}\frac{2000+8.6x+0.5x^2-49{,}580}{x-300}\\ &= \lim_{x\to300}\frac{0.5x^2+8.6x-47{,}580}{x-300}\\ &= \lim_{x\to300}\frac{(x-300)(0.5x+158.6)}{x-300}\\ &= \lim_{x\to300} 0.5x+158.6\\ &= 308.6\end{aligned}$$

6. E

Note that the gasoline drains at an average rate of $\frac{v(40)-v(30)}{10-0} = \frac{0-800}{10} = -80$ liters per minute during the last 10 minutes. Moreover, this is the smallest of the average rates in the choices, so the answer must be (E).

7. D

This is simply the definition of the derivative at a point.

8. D

$$\begin{aligned}f'(x) &= 5x(\sec x\tan x) + \sec x(5) + x^3(-\sin x) + \cos x(3x^2)\\ &= 5x\sec x\tan x + 5\sec x - x^3\sin x + 3x^2\cos x\\ &= 5\sec x + 5\sec x\tan x + 3x^2\cos x - x^3\sin x\end{aligned}$$

9. E

Although one can attack this problem by actually calculating the derivatives as the first step in the solution, it is much easier to first remember a trigonometric identity that will simplify the problem greatly. Namely, we remember that

$\sin^2\theta + \cos^2\theta = 1$ for any angle θ. Thus, we know that

$g(x) = 5\sin^2(6x) + 5\cos^2(6x) = 5(\sin^2(6x) + \cos^2(6x)) = 5(1) = 5$

for all x. That means that $g(x)$ is a constant function, and therefore its derivative is zero.

10. D

The zeros of the derivative function $h'(x)$ are exactly those values of x at which $h(x)$ reaches its (relative) minimum or maximum values. In this case, this occurs at $x = -3$ (a relative minimum), $x = 1$ (a relative maximum), and $x = 4$ (a relative minimum).

11. A

The solution should promote good test-taking skills. Start by noticing that the slope is positive from $-\infty$ to -1. Thus, $f'(x)$ must be above the x-axis. This will rule out choices (C), (D), and (E). Next, notice that the graph has horizontal tangent lines around $x = -1$ and $x = 0.5$. $f'(x)$ must equal zero at these points. Therefore, choice (A) is the answer. Note that the function $g(x)$ appears to be increasing for x from $-\infty$ to -1.1 or so. This means that the graph of $g'(x)$ needs to be above the x-axis on the interval $(-\infty,-1.1)$. Similarly, the function $g(x)$ appears to be increasing on the interval $(0.5,\infty)$ (again approximating this value 0.5). Thus, the graph of $g'(x)$ needs to be above the x-axis on the interval $(0.5,\infty)$ Finally, the graph of $g(x)$ appears to be decreasing on $(-1.1,0.5)$ so the graph of $g'(x)$ must be below the x-axis on this interval. The only choice which satisfies these criteria is (A).

12. C

Although it may be difficult to ascertain, the graph appears to be a "cup" for both $x < -1$ and $x > 1$, and is therefore concave up over these two intervals. Its shape is that of an upside-down cup for $-1 < x < 1$ where it is concave down. If you are having trouble, look for relative minimums and maximums. Remember that relative minimums occur in concave up portions of the graph, and relative maximums occur on concave down portions of the graph.

13. C

There are two ways to obtain this answer. The first is to realize that the graph of this equation is an ellipse with center (0,0) and vertices at the points (4,0), (−4,0), (0,2), and (0,−2). Once drawn, one can easily see that there are exactly two points whose tangent lines are horizontal (the points (0,2) and (0,−2)).

The second approach is to find critical points via the derivative. From the equation, we see that $2x + 8y\frac{dy}{dx} = 0$, so $\frac{dy}{dx} = -\frac{x}{4y}$. Setting this derivative equal to 0 shows that all points which satisfy $x = 0$ will have horizontal tangent lines. Substituting $x = 0$ into the original equation, we have $4y^2 = 16$, or $y = \pm 2$. Thus, we have determined that horizontal tangent lines occur at (0,2) and (0,−2).

14. E

Note that the point e is not a point where the first derivative is zero. Indeed, it appears that the first derivative does not exist at the point. Points b, c, and f may be inflection points, but then the second derivative would be zero, not necessarily the first derivative.

15. A

Let the horizontal distance from the bottom of the wall to the bottom of the ladder be x and let the vertical distance from the top of the ladder to the ground be y. Then, by the Pythagorean Theorem, we know that

$x^2 + y^2 = 13^2$. We also know that $\frac{dx}{dt} = \frac{1}{2}$ and we want to find $\frac{dy}{dt}$. We implicitly differentiate our equation to obtain

$$2x\frac{dx}{dt} + 2y\frac{dy}{dt} = 0.$$

At the time when $y = 5$, we know that $x = 12$ (again thanks to the Pythagorean Theorem and the fact that the ladder is always 13 feet long). Substituting all this information yields

$$2(12)\left(\frac{1}{2}\right) + 2(5)\frac{dy}{dt} = 0,$$

which gives answer (A) upon solving for $\frac{dy}{dt}$.

16. B

By cross-multiplying, we see that $e^y dy = \sin x\, dx$. Integrating both sides,

$$\int e^y\, dy = e^y = \int \sin x\, dx = -\cos x + C, \text{ or } y = \ln(-\cos x + C).$$

Plugging in our initial condition,

$$y\left(\frac{\pi}{4}\right) = 0 = \ln\left(-\cos\frac{\pi}{4}\right) + C = \ln\left(-\frac{\sqrt{2}}{2} + C\right).$$

Because $\ln(1) = 0$, we find $C = \frac{\sqrt{2}}{2} + 1$, so

$$y = \ln\left(-\cos x + \frac{\sqrt{2}}{2} + 1\right)$$

17. D

The key to this problem is to draw the graph of the integrand function, $2 - |4 - x|$. The graph is as follows:

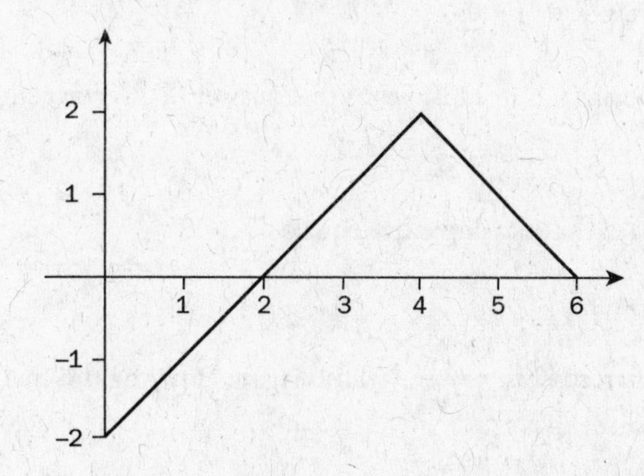

Now we see that the integral in question is simply the signed sum of the areas of two triangles. Note that the left triangle, the one that appears below the x-axis, has the area of 2. But its contribution to the integral is -2 because this triangular area appears below the x-axis. The area of the other triangle is 4, so the integral in question equals $-2 + 4 = 2$.

18. A

First, we note that an antiderivative of $\sin(2x)$ is $-\frac{1}{2}\cos(2x)$. So at this point, the only viable answers are (A) and (D). Next, we have

$$5\tan^2 x \csc^2 x = 5\frac{\sin^2 x}{\cos^2 x}\frac{1}{\sin^2 x} = 5\frac{1}{\cos^2 x} = 5\sec^2 x.$$

An antiderivative of $\sec^2 x$ is $\tan x$, so the correct answer must be (A).

19. C

Note that a u-substitution is in order here. Namely, let $u = 7 + x^{2/3}$. Then $du = \frac{2}{3}x^{-1/3}dx$. Now we rewrite the original integral as

$$\frac{3}{2}\int (7 + x^{2/3})^5 \times \frac{2}{3}x^{-1/3}\,dx.$$

Rewritten in terms of u, we have

$$\frac{3}{2}\int u^5\,du = \frac{3}{2} \times \frac{u^6}{6} + C.$$

Converted back in terms of x, our final answer is

$$\frac{1}{4}(7 + x^{2/3})^6 + C.$$

20. B

Let $u = \sin x$ so that $du = \cos x\,dx$. Then we have

$$\int \sin^4 x \cos x\,dx = \int u^4\,du = \frac{u^5}{5} + C = \frac{1}{5}\sin^5 x + C.$$

21. D

Thanks to the Fundamental Theorem of Calculus, we know that the answer is

$$e^{(t^4)^2}\frac{d}{dt}(t^4) - e^{2^2}\frac{d}{dt}(2) = e^{t^8}(4t^3) - 0$$

because the derivative of 2, a constant, is 0. This means that answer (D) is correct.

22. B

We first apply the chain rule to see that the derivative equals

$$5\left(\arccos\left(x^2\right)\right)^4 \frac{d}{dx}\arccos\left(x^2\right).$$

Next, note that the derivative of $\arccos x$ is $-\frac{1}{\sqrt{1-x^2}}$. Thus, again applying the chain rule to handle the argument of x^2 in the original problem, we see that

$$\frac{d}{dx}\arccos(x^2) = -2x\frac{1}{\sqrt{1-x^4}}.$$

Combining these quantities yields answer (B).

23. A

The product rule is necessary in the solution of this problem, along with the chain rule in two places. The solution is as follows:

$$\begin{aligned}\frac{d}{dx}(\ln(3x)5^{2x}) &= 5^{2x}\frac{d}{dx}(\ln(3x)) + \ln(3x)\frac{d}{dx}(5^{2x})\\ &= 5^{2x}\frac{1}{3x}\frac{d}{dx}(3x) + \ln(3x)5^{2x}\ln(5)\frac{d}{dx}(2x)\\ &= 5^{2x}\frac{3}{3x} + \ln(3x)5^{2x}2\ln(5)\\ &= \frac{5^{2x}}{x} + 2\ln(5)\ln(3x)5^{2x}\end{aligned}$$

24. D

The chain rule must be used in this problem. We must also remember that $\frac{d}{dt}e^{f(t)} = e^{f(t)}\frac{d}{dt}(f(t))$. Because the derivative of $3t^2$ with respect to t is $6t$, and because the derivative of $-\frac{1}{t}$ with respect to t is $\frac{1}{t^2}$, we see that

$$u(t) = e^{3t^2} \times 6t - e^{-1/t} \times \frac{1}{t^2} = 6te^{3t^2} - \frac{e^{-1/t}}{t^2}.$$

25. E

We first rewrite the integral in question as $\int \frac{e^{3x}}{(e^{3x})^2+1}dx$. This new version of the problem gives us a hint to attempt the u-substitution $u = e^{3x}$, $du = 3e^{3x}$. This implies that the integral is equal to $\frac{1}{3}\int \frac{du}{u^2+1}$ which equals $\frac{1}{3}\arctan(u) + C$. Rewriting in terms of x, we have the final answer: $\frac{1}{3}\arctan(e^{3x}) + C$.

26. A

First, we see that the graph of $f(x)$ is always below the graph of $g(x)$ from $x = 0$ to $x = 3$ (where the two graphs intersect). We then see that the volume in question can be determined using the washer method. The typical washer will have area $\pi((3x)^2 - (x^2)^2)$. Therefore, the volume in question equals

$$\int_0^3 \pi((3x)^2 - (x^2)^2)dx.$$

Note that

$$\int \pi((3x)^2 - (x^2)^2)\,dx = \pi\int (9x^2 - x^4)\,dx = \pi\left[3x^3 - \frac{x^5}{5}\right]$$

Now we evaluate from $x = 0$ to $x = 3$ using the Fundamental Theorem of Calculus and obtain

$$\pi\left(81 - \frac{243}{5}\right) = \frac{162\pi}{5}.$$

27. D

There are at least two different approaches to this problem. One is to solve the differential equation in question. Because it is a first order, separable differential equation, this is straightforward. First, rewrite the original differential equation as $y\,dy = x\,dx$. Then integrate both sides to obtain $\frac{y^2}{2} = \frac{x^2}{2} + C$ or $\frac{y^2}{2} - \frac{x^2}{2} = C$ for some constant C. Next, we must ask what the graph of such an equation looks like. In this case, the answer would be a hyperbola. It should be clear that answer (D) most closely resembles a family of hyperbolas.

You may also choose the following alternate approach. For the differential equation $\frac{dy}{dx} = \frac{x}{y}$, it is the case that the slope at each point (a, a) where a is an integer must be one. Hence, in the slope field, the line segments corresponding to the points (1,1), (2,2), (3,3), and so on must all have a slope of one. The only answer choice which satisfies this criterion is (D).

28. B

Note that simple substitution cannot be applied here because the denominator would then equal 0. However, because the numerator would also equal 0 on simple substitution, we must factor both the numerator and the denominator of the expression to see if we can cancel anything.

$$\frac{x^2 + 2x - 35}{x^2 - 25} = \frac{(x-5)(x+7)}{(x-5)(x+5)} = \frac{x+7}{x+5}$$

With the factors of $x - 5$ now removed from both the numerator and the denominator, we can now apply simple substitution to obtain the limit:

$$\begin{aligned}\lim_{x\to 5} \frac{x^2 + 2x - 35}{x^2 - 25} &= \lim_{x\to 5} \frac{x+7}{x+5} \\ &= \frac{5+7}{5+5} \\ &= \frac{12}{10} \\ &= \frac{6}{5}\end{aligned}$$

29. A

Note that $u(t) = \frac{(t+7)(t-3)}{(t+3)(t-3)} = \frac{t+7}{t+3}$ for all values of $t \neq \pm 3$. Because $u(t)$ is continuous for all positive real numbers, we know that $\lim\limits_{t\to 3} u(t)$ must exist and equal the value of $u(3)$. Thus, $u(3) = \lim\limits_{t\to 3} \frac{t+7}{t+3} = \frac{10}{6} = \frac{5}{3}$.

30. A

In order to calculate the average rate of change, we must find $\frac{c(310)-c(300)}{310-300}$. This is given by

$$\begin{aligned}\frac{c(310)-c(300)}{310-300} &= \frac{(2{,}000+8.6(310)+0.5(310)^2)-(2{,}000+8.6(300)+0.5(300)^2)}{10}\\ &= \frac{3{,}136}{10}\\ &= 313.6\end{aligned}$$

31. B

In order to find possible inflection points for $f(x)$, we must calculate $f''(x)$. In this case, $f''(x) = 30x^4 + 60x^3 - 90x^2 - 240x - 120$, which is factorable as

$$f\ (x) = 30(x^4 + 2x^3 - 3x^2 - 8x - 4) =$$
$$30(x^2 - 4)(x^2 + 2x + 1) = 30(x^2 - 4)(x + 1)^2$$

At this point, there are three values of x which make $f''(x) = 0$. These are $x = -2, -1, 2$. But note that $x = -1$ is not an inflection point because the sign of $f''(x)$ does not change for values of x on either side of (and close to) $x = -1$.

32. A

This problem is an application of the chain rule. Letting $u(x) = \sin x$, so that $a = \sin(u)$, we see that $\frac{da}{dx} = \frac{da}{du} \times \frac{du}{dx}$. Then we have $\frac{da}{du} = \cos u = \cos(\sin x)$ and $\frac{du}{dx} = \cos x$. Thus, the derivative is the product of these two, $\cos(\sin x)\cos x$. Now we simply evaluate this function at $\frac{\pi}{4}$, yielding 0.5376.

33. B

We must solve $f'(x) = 0$ to find the critical points. There are two approaches to doing so. One is to simply determine $f'(x)$ as it is given to us (using the product rule). Then we have $f'(x) = \cos^2 x - \sin^2 x$. We can then factor this to obtain $f'(x) = (\cos x - \sin x)(\cos x + \sin x)$. Then the critical points occur when either $\cos x = \sin x$ or when $\cos x = -\sin x$. Solving these two equations yields (B).

The second approach is to first use a trigonometric identity, which tells us that $f(x) = \frac{1}{2}\sin 2x$. Then $f'(x) = \cos 2x$, so we only have to solve $\cos 2x = 0$. This again yields (B).

Of course, with a calculator, one can plot the graph and visually determine the critical points from the locations of the relative extrema.

34. D

Using implicit differentiation, we have

$$3x^2 + 2xy\frac{dy}{dx} + y^2 = 10\frac{dy}{dx}.$$

Solving for $\frac{dy}{dx}$ yields

$$\frac{dy}{dx} = \frac{3x^2 + y^2}{10 - 2xy}.$$

Thus, the slope of the tangent line at the point (2, 1) is given by $\frac{dy}{dx} = \frac{13}{6}$. The slope of the normal line at that same point is the negative reciprocal of the slope of the tangent line.

35. C

First, note that the balloon is increasing in size, so the rate of change of the radius must be positive. Thus, answer (D) is not viable. Next, note that the rate at which the balloon is being inflated is exactly the rate of change of the volume of the balloon. Hence, we need an equation that relates the volume of a sphere with its radius. Such an equation is

$$V = \frac{4}{3}\pi r^3.$$

Implicit differentiation yields

$$\frac{dV}{dt} = 4\pi r^2\frac{dr}{dt}.$$

The problem is asking for $\frac{dr}{dt}$ when $r = 5$. Moreover, we know that $\frac{dV}{dt} = 3$. Substituting this information yields $3 = 4\pi(5)^2\frac{dr}{dt}$ or $\frac{dr}{dt} = \frac{3}{100\pi}$. Using a calculator, we find this is approximately 0.0095 inches per second.

36. B

We know that the acceleration due to gravity is –32 feet per second. Thus, the acceleration of the projectile is $a(t) = -32$. The velocity of the object is $v(t) = -32t + 400$, because the initial velocity is 400 feet per second. The projectile will reach its maximum height when $v(t) = 0$. In this case, $t = 12.5$ seconds after the initial firing. Next, we know that $h(t) = -16t^2 + 400t + 100$ where $h(t)$ is the height of the projectile above the ground exactly t seconds after firing. Thus, our answer is $h(12.5)$, which is 2,600 feet.

37. A

In order to compute this approximation correctly, we must first note that the width of any rectangle in this problem is $\frac{10-0}{5} = 2$. Moreover, the sum in question is $h(2) + h(4) + h(6) + h(8) + h(10) = 25 + 30 + 16 + 25 + 32 = 128$. Therefore, the approximate area is $2 \times 128 = 256$.

38. D

First, note that 21 is the exact value of the area in question, but this is not what is requested, so this is not the correct answer. Note that the other answer choices are the results of approximating in different ways (such as using left-hand endpoints, right-hand endpoints, Simpson's rule, and so on). So great care must be taken in choosing the correct answer here.

The trapezoid rule in this case yields

$$\frac{3-0}{2\times 6}\left(f(0)+2f(0.5)+2f(1)+2f(1.5)+2f(2)+2f(2.5)+f(3)\right),$$

which simplifies to 21.125.

39. C

Our first task is to determine the graphical relationship between the two curves, $y = x^2 + 4$ and $y = -2x + 1$. Using a graphing calculator, we can quickly see that $y = x^2 + 4$ is always above $y = -2x + 1$ on the interval $[-2, 5]$. Therefore, the area in question is found by computing the value of

$$\int_{-2}^{5}(x^2+4-(-2x+1))dx = \int_{-2}^{5}(x^2+2x+3)dx.$$

This can be done by hand via antidifferentiation (or by calculator if desired). The final answer is $\frac{259}{3}$, which is approximately 86.3333.

40. B

Plotting $f(x)$ and $g(x)$ on a calculator, we see that the area bounded by the two graphs is restricted over the two intervals $[0,1]$ and $[1,3]$.

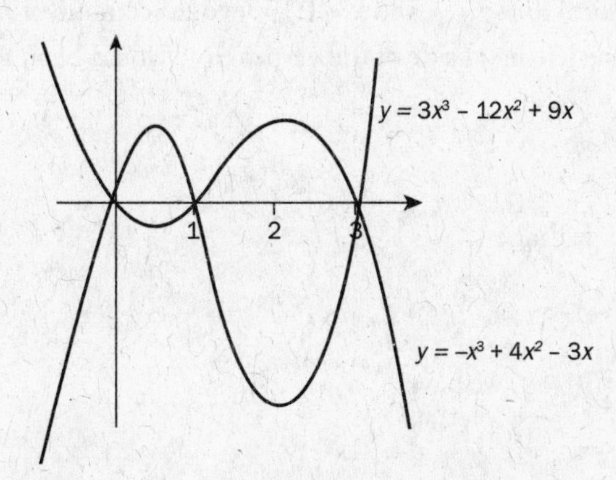

Moreover, we can determine which of the two graphs is above the other in these two intervals. On the interval $[0,1]$, the graph of $f(x)$ is above the graph of $g(x)$ while on the interval $[1,3]$ the roles are reversed. Thus, the area in question is the following sum of two integrals:

$$\int_{0}^{1}((3x^3-12x^2+9x)-(-x^3+4x^2-3x))dx + \int_{1}^{3}((-x^3+4x^2-3x)-(3x^3-12x^2+9x))dx.$$

This simplifies as

$$\int_0^1 (4x^3 - 16x^2 + 12x)dx + \int_1^3 (-4x^3 + 16x^2 - 12x)dx.$$

The first integral equals $\frac{5}{3}$ while the second equals $\frac{32}{3}$, so their sum is $\frac{37}{3}$. The best approximation above is then given by 12.333.

41. B

Assume that we have such a line $x = k$ that cuts the region R into two regions of equal area. Then we know that

$$\int_0^k (4 - x^2)dx = \int_k^2 (4 - x^2)dx.$$

(This follows from the definition of the integral and the fact that the two areas are equal.) After performing the integration and simplifying the results, we have the following equality:

$$4k - \frac{k^3}{3} = \frac{16}{3} - 4k + \frac{k^3}{3}.$$

We now need to solve this equality for k, the desired quantity in this problem. Further simplification of the equality above gives $k^3 - 12k + 8 = 0$.

Using a calculator, the root in question can be found in a number of ways, and is approximately equal to 0.695.

Note that answers (D) and (E) really do not make any sense. It should be clear from a quick sketch of $y = 4 - x^2$ in the first quadrant that the vertical lines $x = 1$ and $x = 1.135$ would certainly not produce two smaller areas of equal area. Indeed, the area on the left-hand side of the vertical line would be much larger than the area on the right-hand side.

42. E

First, because $a'(t) = v(t)$, we know that

$$v(t) = \frac{t^3}{3} + \frac{2}{3}(t + 9)^{\frac{3}{2}} - e^{-t} + C$$

via antidifferentiation. Because $v(0) = 4$, we have

$$4 = \frac{2}{3}(9)^{\frac{3}{2}} - 1 + C.$$

Simplifying this equality and solving for C gives $C = -13$. Thus, we know that $v(t) = \frac{t^3}{3} + \frac{2}{3}(t + 9)^{\frac{3}{2}} - e^{-t} - 13$. Next, the position function $x(t)$ is related to the velocity by $x'(t) = v(t)$.

Thus, $x(t) = \frac{t^4}{12} + \frac{4}{15}(t + 9)^{\frac{5}{2}} + e^{-t} - 13t + C$. Because $x(0) = 5$, we know that $5 = \frac{4}{15}(9)^{\frac{5}{2}} + 1 + C$.

Solving for C yields $C = -\frac{304}{5}$. Therefore, we know that the position of the object at any time t is given by

$$x(t) = \frac{t^4}{12} + \frac{4}{15}(t + 9)^{5/2} + e^{-t} - 13t - \frac{304}{5}.$$

Now we simply evaluate this quantity at $t = 7$ and have our final answer:

$$x(7) = \frac{7^4}{12} + \frac{4}{15}(7+9)^{5/2} + e^{-7} - 13(7) - \frac{304}{5} = 321.351.$$

43. A

The average value of the function $f(x)$ on the interval $[a, b]$ is given by

$$\frac{1}{b-a}\int_a^b f(x)dx.$$

So in this problem, we must compute

$$\frac{1}{8-5}\int_5^8 \frac{2x}{x^2-4}dx.$$

Note that $\int \frac{2x}{x^2-4}dx = \ln(x^2-4) + C$ via u-substitution. Therefore, the average value is equal to $\frac{1}{3}(\ln(8^2-4) - \ln(5^2-4))$, which is approximately 0.3499.

44. D

We note that the differential equation supplied in the problem is separable and may be rewritten as

$$y\,dy = \frac{5}{x+4}dx.$$

Integrating both sides of this differential equation yields

$$\frac{y^2}{2} = 5\ln|x+4| + C$$

or

$$y^2 = 10\ln|x+4| + C$$

for some constant C. Because $y(e-4) = 0$, we know that $0 = 10\ln(e-4+4) + C = 10\ln(e) + C = 10 + C$. Thus, $C = -10$. Hence, we know that $y^2 = 10\ln|x+4| - 10$ and our answer is (D).

45. C

For positive or negative values of x of very large magnitude, $\sqrt{x^2+1}$ is very close to $\sqrt{x^2}$, which is equal to $|x|$. Remember, $|x| = x$ is x is a positive number, and $|x| = -x$ if x is a negative number.

As $x \to \infty$, x will be taking on the values of positive numbers.

Therefore, $\lim_{x\to\infty} \sqrt{x^2+1} - x = \lim_{x\to\infty} \sqrt{x^2} - x = \lim_{x\to\infty} |x| - x = \lim_{x\to\infty} (x - x) = 0.$

As $x \to -\infty$, x will be taking on the values of negative numbers.

Therefore, $\lim_{x\to-\infty} \sqrt{x^2+1} + x = \lim_{x\to-\infty} \sqrt{x^2} + x = \lim_{x\to-\infty} |x| + x = \lim_{x\to-\infty} (-x + x) = 0$

FREE-RESPONSE ANSWERS

1. (a) At $t = 4$, $C(t) = 5.12 > 0$, so the amount of charge is increasing.

   (b) $C\ (t) = 0 + 6\cos\left(\frac{t^2}{3}\right)\left(\frac{2t}{3}\right) = 4t\cos\left(\frac{t^2}{3}\right)$.

   At $t = 4, C'(t) = 9.309 > 0$, so the rate of change of the charge in the battery is increasing at $t = 4$.

   (c) The average rate of change of charge in the battery is given by

   $$\frac{1}{10-2}\int_2^{10} C(t)\,dt = \frac{1}{10-2}\int_2^{10} 10 + 6\sin\left(\frac{t^2}{3}\right)dt = 10.2640 \quad \text{coulombs per hour.}$$

   (d) We need to solve

   $$160 = \int_0^s C(t)\,dt = \int_0^s 10 + 6\sin\left(\frac{t^2}{3}\right)dt.$$

   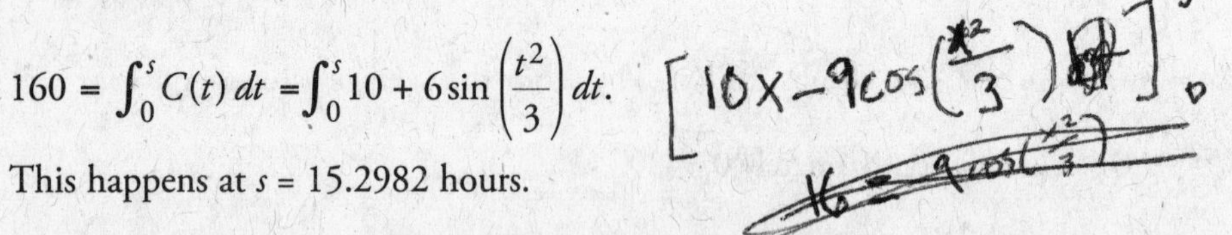

   This happens at $s = 15.2982$ hours.

   For calculator-based problems, it is enough to write down the integral you put into the calculator followed by the answer it spits out. If you make any algebraic simplifications, though, or separate your integral into more than one piece, make sure to illustrate that in your answer. You will lose points if you lose significant digits by rounding two integrals and then summing them. Similarly, do not write more than is needed. If you had your calculator return $C'(t) = 9.309$ for part (b), it is not necessary to do the explicit derivation yourself. Finally, in part (d), make the calculator do all the work. Be familiar with your calculator's solve features and know how to use them for problems like this. Never forget your units!

2. (a) The area of A is given by the integral $\int_0^1 f(x) - g(x)\,dx = \int_0^1 \sqrt{x} - 1 - \left(\cos\left(\frac{\pi x}{2}\right)\right)dx = 0.3033.$

   (b) The volume of the solid generated when A is rotated about the x-axis is given by the integral

   $$\pi\int_0^1 (f(x))^2 - (g(x))^2\,dx = \pi\int_0^1 x - \left(1 - \cos\left(\frac{\pi x}{2}\right)\right)^2 dx = 0.8584.$$

   (c) The volume of the solid generated when B is rotated about the y-axis is given by the integral

   $\int_0^1 \pi(F(y))^2 dy = \pi\int_0^1 y^4 dy = 0.6283.$ Here $F(y) = y^2$ gives the length of x as a function of y.

   Note: For part (b) we integrate using the *washer method*. For part (c) it suffices to use the *cylinder method*. Don't forget to use the inverse function, however. Remember, when we rotate around the y-axis, we integrate with respect to the variable y, so we need to integrate a function of y.

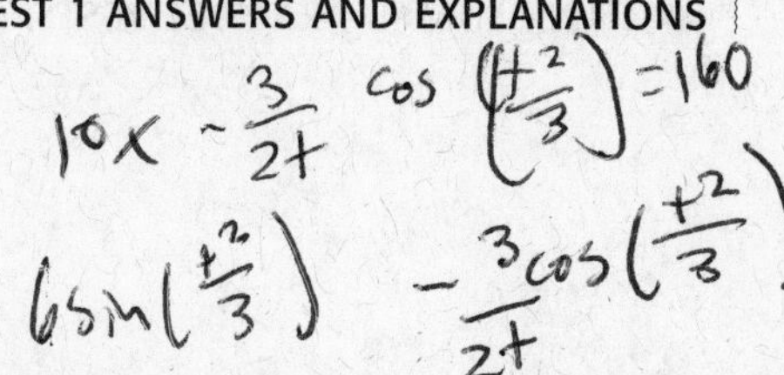

3. (a) f' changes sign only at $x = 4$, so f has a relative extremum only there. Because f' changes from positive to negative there, it is a local max.

   (b) The inflection points of f are the relative extrema of f'. These occur at $x = -3, 2$.

   (c) Because f' is decreasing on $[-6, -3]$, f is concave down on that interval.

4. (a) Implicit differentiation gives:

$$y^2 + x\left(2y\frac{dy}{dx}\right) + 3y^2\frac{dy}{dx} = 0.$$

   Solving for $\frac{dy}{dx}$:

$$x\left(2y\frac{dy}{dx}\right) + 3y^2\frac{dy}{dx} = -y^2$$

$$\frac{dy}{dx}(2xy + 3y^2) = -y^2$$

$$\frac{dy}{dx} = \frac{-y^2}{(2xy + 3y^2)}$$

   (b) Horizontal tangents occur when $\frac{dy}{dx} = 0$. By (a), this could only occur if $y = 0$. Plugging $y = 0$ into the equation for C gives $0 = 5$, so there are no points on the curve with $y = 0$ and no horizontal tangents.

   (c) Vertical tangencies can only occur if $\frac{dy}{dx}$ is undefined. The only possibilities are if

$$(2xy + 3y^2) = 0$$

$$y(2x + 3y) = 0$$

   Because, by (b), $y \neq 0$, we have

   $(2x + 3y) = 0$ or $x = -\frac{3}{2}y$. Because we are given that there *is* a vertical tangency, this must be it. Plugging this back into the equation for C gives:

$$-\frac{3}{2}y \cdot y^2 + y^3 = 5$$

$$-\frac{3}{2}y^3 + y^3 = 5$$

$$-\frac{3}{2}y^3 = 5$$

$$y = \sqrt[3]{-10}.$$

So $x = -\frac{3}{2}y = -\frac{3}{2}\sqrt[3]{-10} = \frac{3}{2}\sqrt[3]{10}$ and

$(x, y) = \left(\frac{3}{2}\sqrt[3]{10}, \sqrt[3]{-10}\right)$

Note: Without knowing that a vertical tangency exists in part (c), the question is more delicate. We have shown that if there is a vertical tangency, it must be at $(x, y) = \left(\frac{3}{2}\sqrt[3]{10}, \sqrt[3]{-10}\right)$, but there's more work to be done to show that this is indeed a vertical tangency.

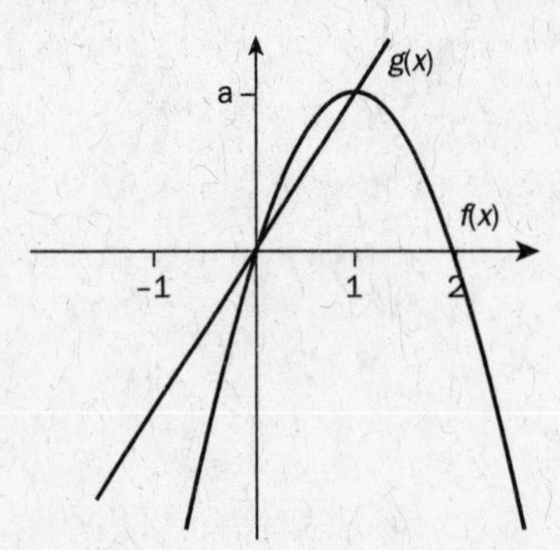

5. (a)

$$f(x) = g(x)$$
$$ax(2 - x) = ax$$
$$x = 0 \quad \text{or} \quad 2 - x = 1$$
$$x = 1$$

Therefore, the area of the region is given by

$$\int_0^1 f(x) - g(x)\,dx = \int_0^1 (2ax - ax^2) - ax\,dx = \int_0^1 ax - ax^2\,dx = \frac{ax^2}{2} - \frac{ax^3}{3}\Bigg|_0^1 = \frac{a}{2} - \frac{a}{3} = \frac{a}{6}$$

(b) The volume of revolution (about the x-axis) is given by

$$\int_0^1 \pi f^2(x) - \pi g^2(x)\,dx = \int_0^1 \pi(2ax - ax^2)^2 - \pi(ax)^2\,dx = \pi\int_0^1 4a^2x^2 - 4a^2x^3 + a^2x^4 - a^2x^2dx$$

$$= \pi\ a^2x^3 - a^2x^4 + \frac{a^2x^5}{5}\Bigg|_0^1 = \pi\ a^2 - a^2 + \frac{a^2}{5} - 0 = \frac{\pi a^2}{5}$$

(c) The volume created by the cross-section R and thickness a is $\left(\frac{a}{6}\right)\pi = \frac{\pi a}{6}$. We need to find the positive value of a for which $\frac{\pi a}{6} = \frac{\pi a^2}{5}$.

Therefore, for $a = \frac{5}{6}$ the volumes are equal.

6. (a)

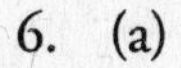

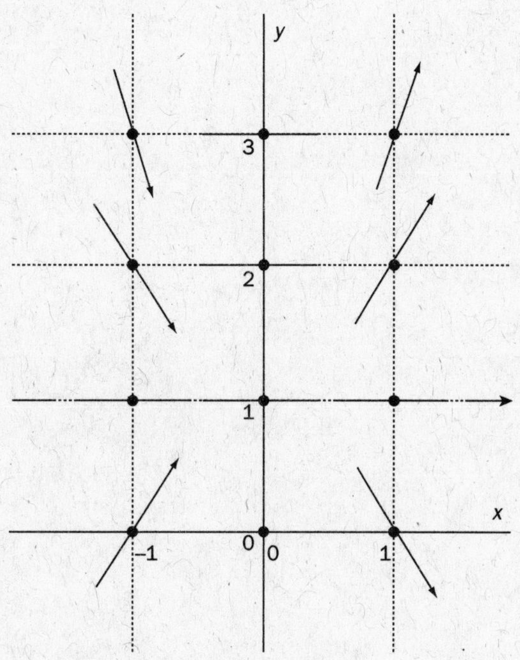

x	y	$\frac{dy}{dx} = x^3(y - 1)$
–1	0	1
–1	1	0
–1	2	–1
–1	3	–2
0	0	0
0	1	0
0	2	0
0	3	0
1	0	–1
1	1	0
1	2	1
1	3	2

(b) Because $f(x)$ is a solution to the differential equation and $f(2) = 2$, the slope of the tangent line to the graph of f is $\frac{dy}{dx} = x^3(y-1) = 2^3(2-1) = 8$. The equation for the tangent line is $(y - 2) = 8(x - 2)$.

(c) Solving:

$$\frac{dy}{dx} = x^3(y-1)$$

$$\frac{dy}{y-1} = x^3 dx$$

$$\int \frac{dy}{y-1} = \int x^3 dx$$

$$\ln(y-1) = \frac{1}{4}x^4 + C$$

$$y - 1 = e^{\frac{1}{4}x^4 + C}$$

$$y = e^{\frac{1}{4}x^4 + C} + 1$$

To find C:

$$y = e^{\frac{1}{4}x^4 + C} + 1$$

$$2 = e^{\frac{1}{4}2^4 + C} + 1$$

$$1 = e^{4+C}$$

$$C = -4$$

So our particular solution is given by the equation $f(x) = e^{\frac{1}{4}x^4 - 4} + 1$.

Note: For part (b), because we know the point we are looking at, the slope is just given by the slope field, namely, $\frac{dy}{dx} = x^3(y-1) = 2^3(2-1) = 8$. Use any form for the equation of the line, and don't bother to simplify because there's a small chance you could make an error and lose points. You won't lose points for leaving the equation for the tangent line unsimplified. For part (c), always remember to separate the variables before integrating. Even if you have the correct answer, you will lose points if you fail to separate the variables first.

AP CALCULUS AB PRACTICE TEST 1 CORRELATION CHART

Multiple Choice Question	Review Chapter: Subsection of Chapter
1	Chapter 7: Evaluating Limits Algebraically and Some Important Limits
2	Chapter 8: Vertical Asymptotes–Understanding Infinite Limits Graphically
3	Chapter 8: Rates of Growth
4	Chapter 9: Continuity and Limits
5	Chapter 13: The Slope of a Curve at a Point
6	Chapter 11: A Physical Interpretation of the Derivative: Speed
7	Chapter 12: Computation of Derivatives
8	Chapter 12: Rules for Computing Derivatives Chapter 12: Derivatives of Trigonometric Functions
9	Chapter 12: Rules for Computing Derivatives Chapter 12: Derivatives of Trigonometric Functions
10	Chapter 14: Corresponding Characteristics of the Graphs of f and f'
11	Chapter 14: Corresponding Characteristics of the Graphs of f and f'
12	Chapter 14: The Derivative As a Function
13	Chapter 12: Implicit Differentiation
14	Chapter 14: Critical Points and Local Extrema
15	Chapter 16: Related Rates
16	Chapter 16: Differential Equations
17	Chapter 17: Finding the Area Under a Curve
18	Chapter 21: Computing Antiderivatives Directly
19	Chapter 21: When F Is Complicated: The Substitution Game
20	Chapter 21: When F Is Complicated: The Substitution Game
21	Chapter 20: What's So Fundamental? The Connection Between Integrals and Derivatives
22	Chapter 12: The Derivative of Inverse Functions
23	Chapter 12: Rules for Computing Derivatives
24	Chapter 12: Rules for Computing Derivatives Chapter 12: Exponential and Logarithmic Functions

Multiple Choice Question	Review Chapter: Subsection of Chapter
25	Chapter 21: When *F* Is Complicated: The Substitution Game
26	Chapter 18: Solids of Revolution
27	Chapter 16: Differential Equations
28	Chapter 7: Evaluating Limits Algebraically
29	Chapter 9: Continuity and Limits
30	Chapter 11: A Physical Interpretation of the Derivative: Speed
31	Chapter 15: Inflection Points
32	Chapter 12: Rules for Computing Derivatives Chapter 12: Derivatives of Trigonometric Functions
33	Chapter 14: Critical Points and Local Extrema
34	Chapter 12: Implicit Differentiation Chapter 13: The Slope of a Curve at a Point
35	Chapter 16: Related Rates
36	Chapter 16: Interpreting the Derivative As a Rate of Change
37	Chapter 17: Riemann Sums
38	Chapter 6: Function Operations in Action on the AP Exam
39	Chapter 18: Area Bounded by Curves
40	Chapter 18: Area Bounded by Curves
41	Chapter 20: The Fundamental Theorem of Calculus (FTC)
42	Chapter 22: Accumulation Functions and Antiderivatives
43	Chapter 18: Average Value of a Function
44	Chapter 22: Separable Differential Equations
45	Chapter 8: Horizontal Asymptotes

Free Response Question	Review Chapter: subsection of chapter
1(a)	Chapter 6: Evaluating a Function
1(b)	Chapter 16: Interpreting the Derivative as a Rate of Change
1(c)	Chapter 18: Average Value of a Function
1(d)	Chapter 20: The Fundamental Theorem of Calculus (FTC)
2(a)	Chapter 18: Area Bounded by Curves
2(b)	Chapter 18: Solids of Revolution
2(c)	Chapter 18: Solids of Revolution
3(a)	Chapter 14: Critical Points and Local Extrema
3(b)	Chapter 15: Inflection Points
3(c)	Chapter 15: Concavity and the Sign of f''
4(a)	Chapter 12: Implicit Differentiation
4(b)	Chapter 14: Critical Points and Local Extrema
4(c)	Chapter 14: Critical Points and Local Extrema
5(a)	Chapter 18: Area Bounded by Curves
5(b)	Chapter 18: Solids of Revolution
5(c)	Chapter 18: Volumes of Solids
6(a)	Chapter 16: Differential Equations
6(b)	Chapter 13: The Slope of a Curve at a Point
6(c)	Chapter 22: Separable Differential Equations

AP Caclulus AB Practice Test 2 Answer Grid

1.	Ⓐ Ⓑ Ⓒ Ⓓ Ⓔ	13.	Ⓐ Ⓑ Ⓒ Ⓓ Ⓔ	25.	Ⓐ Ⓑ Ⓒ Ⓓ Ⓔ	37.	Ⓐ Ⓑ Ⓒ Ⓓ Ⓔ
2.	Ⓐ Ⓑ Ⓒ Ⓓ Ⓔ	14.	Ⓐ Ⓑ Ⓒ Ⓓ Ⓔ	26.	Ⓐ Ⓑ Ⓒ Ⓓ Ⓔ	38.	Ⓐ Ⓑ Ⓒ Ⓓ Ⓔ
3.	Ⓐ Ⓑ Ⓒ Ⓓ Ⓔ	15.	Ⓐ Ⓑ Ⓒ Ⓓ Ⓔ	27.	Ⓐ Ⓑ Ⓒ Ⓓ Ⓔ	39.	Ⓐ Ⓑ Ⓒ Ⓓ Ⓔ
4.	Ⓐ Ⓑ Ⓒ Ⓓ Ⓔ	16.	Ⓐ Ⓑ Ⓒ Ⓓ Ⓔ	28.	Ⓐ Ⓑ Ⓒ Ⓓ Ⓔ	40.	Ⓐ Ⓑ Ⓒ Ⓓ Ⓔ
5.	Ⓐ Ⓑ Ⓒ Ⓓ Ⓔ	17.	Ⓐ Ⓑ Ⓒ Ⓓ Ⓔ	29.	Ⓐ Ⓑ Ⓒ Ⓓ Ⓔ	41.	Ⓐ Ⓑ Ⓒ Ⓓ Ⓔ
6.	Ⓐ Ⓑ Ⓒ Ⓓ Ⓔ	18.	Ⓐ Ⓑ Ⓒ Ⓓ Ⓔ	30.	Ⓐ Ⓑ Ⓒ Ⓓ Ⓔ	42.	Ⓐ Ⓑ Ⓒ Ⓓ Ⓔ
7.	Ⓐ Ⓑ Ⓒ Ⓓ Ⓔ	19.	Ⓐ Ⓑ Ⓒ Ⓓ Ⓔ	31.	Ⓐ Ⓑ Ⓒ Ⓓ Ⓔ	43.	Ⓐ Ⓑ Ⓒ Ⓓ Ⓔ
8.	Ⓐ Ⓑ Ⓒ Ⓓ Ⓔ	20.	Ⓐ Ⓑ Ⓒ Ⓓ Ⓔ	32.	Ⓐ Ⓑ Ⓒ Ⓓ Ⓔ	44.	Ⓐ Ⓑ Ⓒ Ⓓ Ⓔ
9.	Ⓐ Ⓑ Ⓒ Ⓓ Ⓔ	21.	Ⓐ Ⓑ Ⓒ Ⓓ Ⓔ	33.	Ⓐ Ⓑ Ⓒ Ⓓ Ⓔ	45.	Ⓐ Ⓑ Ⓒ Ⓓ Ⓔ
10.	Ⓐ Ⓑ Ⓒ Ⓓ Ⓔ	22.	Ⓐ Ⓑ Ⓒ Ⓓ Ⓔ	34.	Ⓐ Ⓑ Ⓒ Ⓓ Ⓔ		
11.	Ⓐ Ⓑ Ⓒ Ⓓ Ⓔ	23.	Ⓐ Ⓑ Ⓒ Ⓓ Ⓔ	35.	Ⓐ Ⓑ Ⓒ Ⓓ Ⓔ		
12.	Ⓐ Ⓑ Ⓒ Ⓓ Ⓔ	24.	Ⓐ Ⓑ Ⓒ Ⓓ Ⓔ	36.	Ⓐ Ⓑ Ⓒ Ⓓ Ⓔ		

AP CALCULUS AB PRACTICE TEST 2

SECTION I, PART A

Time: 55 Minutes
28 Questions

NO GRAPHING CALCULATOR IS ALLOWED ON THIS PORTION OF THE EXAM

Directions: Solve the following problems, using available space for scratchwork. After examining the form of the choices, decide which one is the best of the choices given and fill in the corresponding oval on the answer sheet. No credit will be given for anything written in the test book. Do not spend too much time on any one problem.

In this test:

(1) The domain of a function f is the set of all real numbers x for which $f(x)$ is a real number, unless otherwise specified.

(2) The inverse of a trigonometric function f may be indicated using the inverse function notation f^{-1} or with the prefix "arc" (e.g., $\sin^{-1}x = \arcsin x$).

1. Suppose that the function f satisfies $f'(x) = 3x^2 - \sin \pi x$. Then the slope of the line tangent to the graph of f at the point $x = 2$ is

(A) 12
(B) $8 - \frac{1}{\pi}$
(C) 7
(D) $12 - \pi$
(E) 24

2. Consider a continuous function f with the properties that f is concave up on the interval $[-1, 3]$ and concave down on the interval $[3, 5]$. Which of the following statements is true?

(A) $f''(2) > 0$ and $f''(4) < 0$.
(B) $f''(2) < 0$ and $f''(4) > 0$.
(C) $f''(3) > 0$ and $x = 3$ is a point of inflection of f.
(D) Both (A) and (C)
(E) Both (B) and (C)

GO ON TO THE NEXT PAGE

3. An object moving in a straight line has a velocity given by the equation $v(t) = 4t + e^{t-2}$. At time $t = 2$ the object's position, $y(t)$, is given by $y(2) = 3$. The function, $y(t)$, describing the object's position for any time $t > 0$ is

(A) $y(t) = 9t - 15$

(B) $y(t) = 2t^2 + e^{t-2} + 9$

(C) $y(t) = 9t + 9$

(D) $y(t) = 2t^2 + e^{t-2} - 6$

(E) $y(t) = 9t - 6$

4. The graph of a function, f, is given below. Based on the graph, which of the following statements is true?

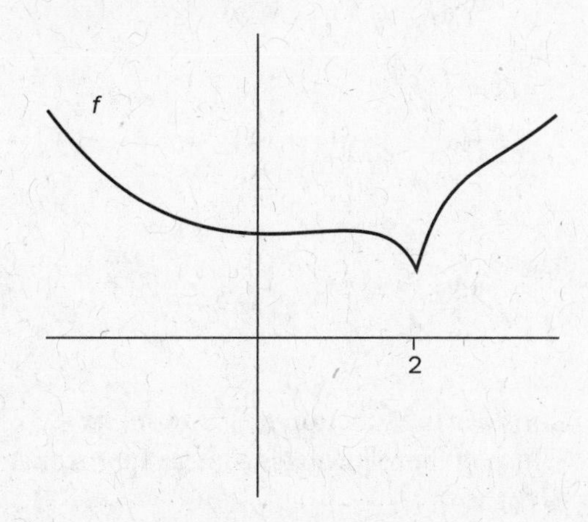

(A) f is continuous and differentiable everywhere, and $f'(2) = 0$.

(B) f is continuous everywhere and differentiable everywhere except at the point $x = 2$.

(C) f is discontinuous at the point $x = 2$, and f is differentiable everywhere except at the point $x = 2$.

(D) f is discontinuous at the point $x = 2$, f is differentiable everywhere, and $f'(2) = 0$.

(E) None of the above

5. Which of the following is a necessary and sufficient condition for a function, f, to be continuous at the point $x = 4$?

(A) $\lim_{x \to 4} f(x)$ exists.

(B) $\lim_{x \to 4} f(x)$ does not exist.

(C) f is differentiable at $x = 4$.

(D) f is defined on an open interval that contains $x = 4$.

(E) $\lim_{x \to 4} f(x) = f(4)$.

6. The derivative of $f(x) = \frac{x^3 \sin x}{x^2 + 3}$ is given by which of the following?

(A) $f(x) = \frac{3x \cos x}{2}$

(B) $f(x) = \frac{3x^2 \sin x + x^3 \cos x}{x^2 + 3} - \frac{2x^4 \sin x}{(x^2 + 3)^2}$

(C) $f(x) = \frac{3x \sin x + x^2 \cos x}{2}$

(D) $f(x) = \frac{(3x^2 \cos x)(x^2 + 3) - 2x^4 \sin x}{(x^2 + 3)^2}$

(E) $f(x) = \frac{3x^2 \sin x + x^3 \cos x}{x^2 + 3} - \frac{(2x + 3)x^3 \sin x}{(x^2 + 3)^2}$

7. If $f(x) = \sin x + 4x - e^{-}x$, then an antiderivative of f is

(A) $F(x) = \cos x + 4 + e^{-x}$

(B) $F(x) = \cos x + 2x^2 + e^{-x}$

(C) $F(x) = -\cos x + 2x^2 + e^{-x}$

(D) $F(x) = -\cos x + 4 - e^{-x}$

(E) $F(x) = -\cos x + 2x^2 - e^{-x}$

GO ON TO THE NEXT PAGE

8. "f is continuous at $x = a$ and $\lim\limits_{x \to a} f(x) = L$" is best represented by which of the following illustrations?

(A)

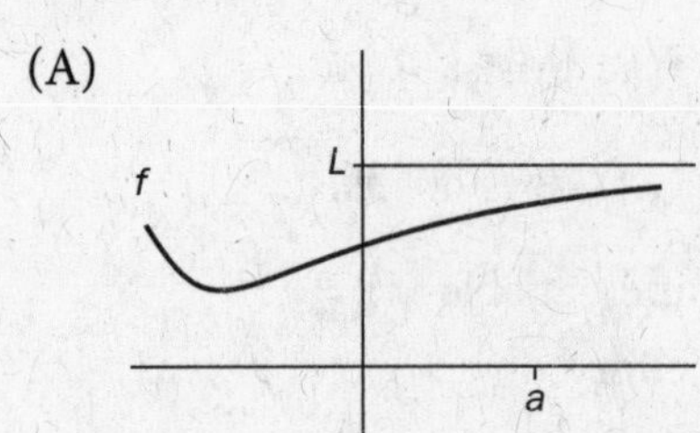

(B)

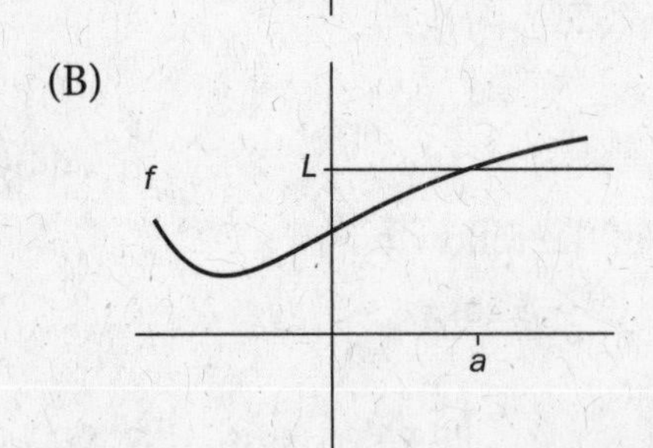

(C)

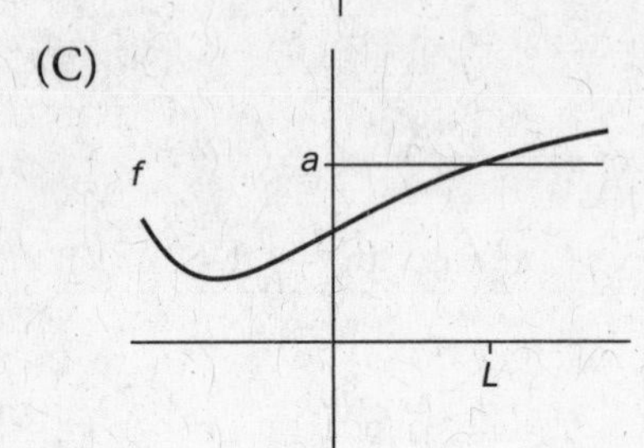

(D)

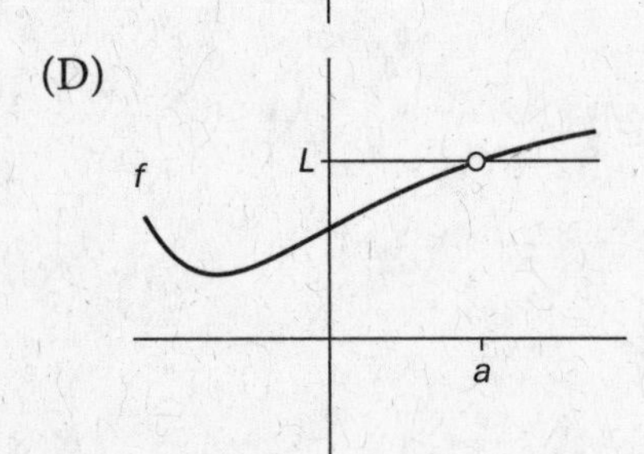

(E)

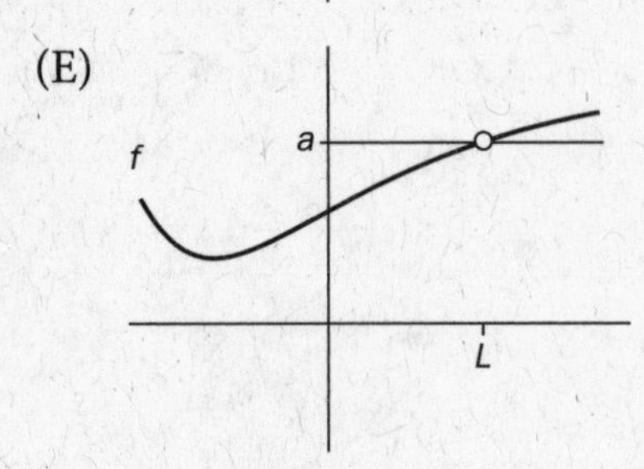

9. Evaluate $\int_{-1}^{2} (3x^2 - 4x + 2)\,dx$.

(A) -2
(B) 14
(C) 9
(D) 18
(E) 21

10. If $x^2 - xy + y^3 = 8$, then $\frac{dy}{dx} =$

(A) $\frac{2x}{y + 3y^2}$

(B) $\frac{-x^2 - 2xy^2 - 8}{(y^2 - x)^2}$

(C) 2

(D) $\frac{8 - x^2 + y - 4y^2}{y^2 - x}$

(E) $\frac{y - 2x}{3y^2 - x}$

11. Consider the function $f(x) = x^2 \sin \pi x - 3x$. Which of the following is a linear approximation to f at $x = 1$?

(A) $f(x) = (2x \sin \pi x + x^2\pi \cos \pi x - 3)x - 1$
(B) $f(x) = -x - 1$
(C) $f(x) = -\pi x + \pi$
(D) $f(x) = (-\pi - 3)x + \pi$
(E) $f(x) = (2x \sin \pi x + x^2\pi\cos \pi x - 3)x + \pi$

GO ON TO THE NEXT PAGE

12. The asymptotic behavior of the function $f(x) = \frac{1}{\sqrt{x}} + 3, x > 0$ is best described by which of the following?

(A) f has a horizontal asymptote at $y = 3$ and a vertical asymptote at $x = 0$.
(B) f has a horizontal asymptote at $y = 0$ and a vertical asymptote at $x = 3$.
(C) f has a horizontal asymptote at $y = 0$ and a vertical asymptote at $x = 0$.
(D) f has a horizontal asymptote at $x = 3$ and a vertical asymptote at $y = 0$.
(E) f has a horizontal asymptote at $x = 0$ and a vertical asymptote at $y = 3$.

13. Let $f(x) = \pi\sec^2 \pi x - 1$. Which of the following statements is true?

(A) An antiderivative of f is $F(x) = \tan\pi x$ and $\int_1^3 f(x)\,dx = -2$.
(B) An antiderivative of f is $F(x) = \tan\pi x$ and $\int_1^3 f(x)\,dx$ is undefined.
(C) An antiderivative of f is $F(x) = \tan\pi x + 3$ and $\int_1^3 f(x)\,dx = -2$.
(D) An antiderivative of f is $F(x) = \tan\pi x - x$ and $\int_1^3 f(x)\,dx = -2$.
(E) An antiderivative of f is $F(x) = \tan\pi x - x + 3$ and $\int_1^3 f(x)\,dx$ is undefined.

14. Let $f(x) = x\sin\left(\frac{\pi}{2}\sqrt{x}\right)$. Which of the following statements is a conclusion of the Mean Value Theorem?

(A) There exists c in (1,4) such that $f(c) = -\frac{1}{3}$.
(B) There exists c in (0,1) such that $f(c) = -\frac{1}{3}$.
(C) There exists c in (0,1) such that $f'(c) = 3$.
(D) There exists c in (1,4) such that $f'(c) = 3$.
(E) There exists c in (0,1) such that $f(c) = -\frac{1}{2}$.

15. Compute the derivative of $f(x) = \ln x - \sin x + \arctan x + 2^x, x > 0$.

(A) $f(x) = \frac{1}{x} - \cos x + \frac{1}{1+x^2} + x2^x$
(B) $f(x) = \frac{1}{x} - \cos x + \frac{1}{1-x^2} + x2^x$
(C) $f(x) = \frac{1}{x} + \cos x + \frac{1}{1-x^2} + (\ln 2)2^x$
(D) $f(x) = \frac{1}{x} - \cos x + \frac{1}{1+x^2} + (\ln 2)2^x$
(E) $f(x) = \frac{1}{x} + \cos x + \frac{1}{1+x^2} + (\ln 2)2^x$

16. An apple farmer can produce 600 apples from each of his apple trees if no more than 20 trees are planted. If he plants more than 20 trees, the yield per tree will decrease. In fact, he figures that for each extra tree he plants, his yield per tree will decrease by 15 apples. How many trees should he plant to obtain the maximum number of apples?

(A) 450
(B) 40
(C) 200
(D) 10
(E) 30

GO ON TO THE NEXT PAGE

17. Consider the function $f(x)=\begin{cases} x^2 \text{ if } 0 \le x < 1 \\ 0 \text{ if } 1 \le x \le 2 \end{cases}$.

Which of the following is true?

(A) f attains an absolute maximum value of 1.

(B) f attains an absolute minimum value of 0.

(C) f attains an absolute maximum value of 1 somewhere on the interval [0, 2].

(D) f does not attain an absolute minimum value.

(E) Both (A) and (C)

18. An equivalent representation of the definite integral $\int_1^3 2x\cos(x^2)dx$ is

(A) $\int_1^3 \cos u\,du$

(B) $\int_1^9 \cos u\,du$

(C) $\int_1^{\sqrt{3}} \cos u\,du$

(D) $\int_1^9 2\sqrt{u}\cos u\,du$

(E) $\int_1^{\sqrt{3}} 2\sqrt{u}\cos u\,du$

19. If the function $f(t) = \cos t - t^2 + 4t$ represents the position of a particle in meters after t seconds, $t \ge 0$, then the instantaneous rate of change of the particle at time $t = 2$ seconds is

(A) $\cos 2$ m/s

(B) $\frac{1}{2}\cos 2 + 2$ m/s

(C) $-\sin 2$ m/s

(D) 4 m/s

(E) $-\frac{1}{2}\sin 4 - 2$ m/s

20. Compute the limit $\lim_{x\to\infty} \frac{x^2 - 3x + 7}{\sqrt{4x^4 - 3x^3 + 2x^2}}$.

(A) The limit does not exist.

(B) 1

(C) $\frac{1}{2}$

(D) $\frac{1}{4}$

(E) 0

21. A spherical balloon is being filled with water so that its volume increases at a rate of 100 cm^3/s. How fast is the radius of the balloon increasing when the diameter is 50 cm? The volume of a sphere of radius r is $\frac{4}{3}\pi r^3$.

(A) $\frac{1}{100\pi}$ cm/s

(B) $\frac{1}{25\pi}$ cm/s

(C) $\frac{1}{50\pi}$ cm/s

(D) $\frac{1}{75\pi}$ cm/s

(E) There is not enough information to determine the answer.

22. The instantaneous rate of change of the function, $f(x) = |x|$, at $x = 0$ is

(A) nonexistent

(B) $\lim_{\Delta x\to 0} \frac{f(\Delta x) - f(0)}{\Delta x}$

(C) $\lim_{\Delta x\to 0} \frac{|\Delta x|}{\Delta x}$

(D) Both (B) and (C)

(E) None of the above

GO ON TO THE NEXT PAGE

23. The derivative $f'(x)$ of the function $f(x) = x^2$ may be expressed as a limit by

(A) $\lim_{\Delta x \to 0} \frac{(x - \Delta x)^2 - x^2}{\Delta x}$

(B) $\lim_{\Delta x \to 0} \frac{(x + \Delta x)^2 - x^2}{\Delta x}$

(C) $\lim_{\Delta x \to 0} \frac{x^2 - (x - \Delta x)^2}{\Delta x}$

(D) Both (B) and (C)

(E) Both (A) and (B)

24. If the rate of change of a quantity over the interval $[-1, 5]$ is given by $f'(x) = xe^{x^2}$, then the total change of the quantity over the interval $[-1, 5]$ is

(A) $e^{25} - e$

(B) $5e^{25} + e$

(C) $\frac{5}{2}e^{25} + \frac{1}{2}e$

(D) $51e^{25} - 3e$

(E) $\frac{1}{2}e^{25} - \frac{1}{2}e$

25. At time $t = 0$ hours, a car that is traveling in a straight line at 30 mi/hr begins accelerating at the rate of 2 mi/hr^2, and is 31 miles from its destination. How much time must elapse before the car reaches its destination?

(A) More than 1 hour, 2 minutes

(B) Less than 58 minutes

(C) 1 hour, 2 minutes

(D) 58 minutes

(E) 1 hour

26. Consider the function,

$$f(t) = \begin{cases} t^2 + 1 \text{ if } -1 \le t < 1 \\ -t + 1 \text{ if } 1 \le t < 2 \\ -1 \text{ if } t > 2 \end{cases}$$

The points in the domain of f at which f is continuous are

(A) $(-1,1), (1,\infty)$

(B) $[-1,1), [1,\infty)$

(C) $(-1,1), (1,2), (2,\infty)$

(D) $[-1,1), (1,2), (2,\infty)$

(E) $(-1,2), (2,\infty)$

27. Consider the graph of the function $f(x) = \sqrt[3]{x}$. Which of the following is true?

(A) f has a horizontal tangent at $x = 0$.

(B) f has a vertical tangent at $x = 0$.

(C) The slope of the tangent to the curve is increasing on the interval $(-1, 1)$.

(D) Both (A) and (C)

(E) Both (B) and (C)

28. Consider the curve described by the equation $x^2 - 4xy - y^2 = -4$. Calculate $\frac{dy}{dx}$ at the point $(x, y) = (3, 1)$.

(A) $\frac{1}{7}$

(B) $\frac{7}{2}$

(C) $\frac{3}{2}$

(D) $\frac{2}{3}$

(E) $\frac{2}{7}$

IF YOU FINISH BEFORE TIME IS CALLED, YOU MAY CHECK YOUR WORK ON THIS SECTION ONLY. DO NOT TURN TO ANY OTHER SECTION IN THE TEST. STOP

SECTION I, PART B

Time: 50 Minutes
17 Questions

A GRAPHING CALCULATOR IS ALLOWED ON THIS PORTION OF THE EXAM

Directions: Solve the following problems, using available space for scratchwork. After examining the form of the choices, decide which one is the best of the choices given and fill in the corresponding oval on the answer sheet. No credit will be given for anything written in the test book. Do not spend too much time on any one problem.

In this test:

(1) The exact numerical value of the correct answer does not always appear among the choices given. When this happens, select from among the choices the number that best approximates the numerical value.

(2) The domain of a function f is the set of all real numbers x for which $f(x)$ is a real number, unless otherwise specified.

(3) The inverse of a trigonometric function f may be indicated using the inverse function notation f^{-1} or with the prefix "arc" (e.g., $\sin^{-1} x = \arcsin x$).

29. If the derivative of f is $f'(x) = \ln(x^2 + 1) - 2$, at which of the following values of x does f have an absolute minimum value?

(A) 2.53
(B) 0
(C) 2.72
(D) 4.44
(E) 1.72

30. The graph of f is shown in the figure below. If $\int_0^3 f(x)dx = 3.5$ and $F'(x) = f(x)$, then $F(4) - F(0) =$

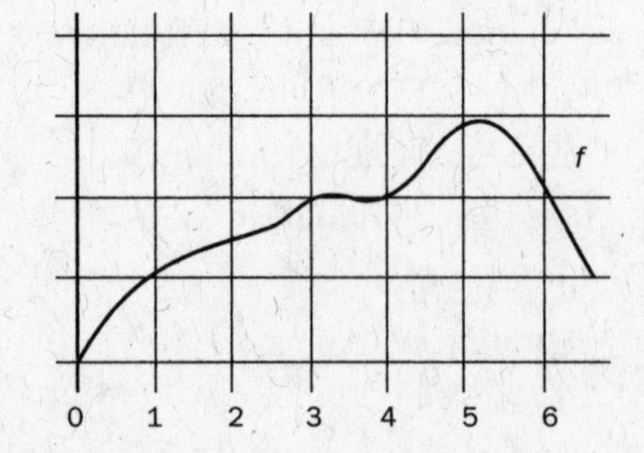

(A) 6.5
(B) 1.5
(C) 2.5
(D) 5.5
(E) 4.5

GO ON TO THE NEXT PAGE

31. Let $f''(x) = \sin x^2$. Which of the following three statements are true?

   I. f is concave up on (0, 1.77) and (2.5, 3.06).

   II. f is concave down on (1.78, 2.50).

   III. f' is increasing on (0, 1.77).

   (A) I and II only
   (B) I and III only
   (C) I, II, and III
   (D) II and III only
   (E) III only

32. Suppose that a differential equation, $y' = f(x, y)$, has the slope field shown below. Which of the following could be one possible solution of the differential equation?

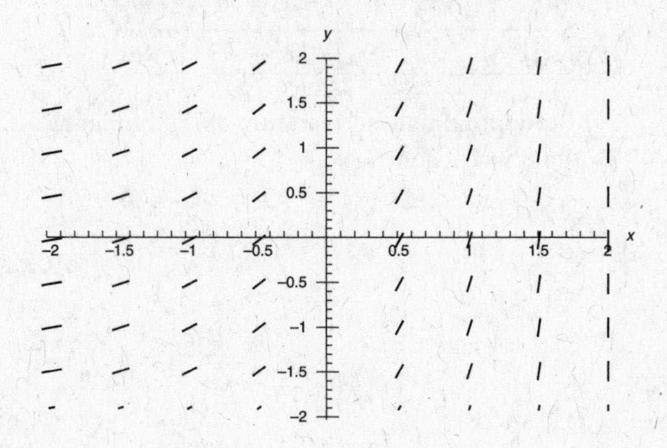

   (A) xe^x
   (B) $2e^{\frac{x^2}{2}}$
   (C) e^x
   (D) $4e^{\frac{x^2}{2}}$
   (E) $3e^{\frac{x^3}{2}}$

33. Consider the function $P(t) = 500e^t - 500\ln(t + 1)$ for $t \geq 0$. Which of the following statements is true?

   (A) $P(t)$ increases, then decreases, then increases indefinitely as $t \to \infty$.
   (B) $P(t)$ decreases indefinitely as $t \to \infty$.
   (C) $P(t)$ increases indefinitely as $t \to \infty$.
   (D) $P(t)$ decreases, then increases indefinitely as $t \to \infty$.
   (E) $P(t)$ increases, then decreases indefinitely as $t \to \infty$.

34. Suppose that f is continuous and has the properties that $f'(x) > 0$ on $(-1, 0)$, $f'(x) \geq 0$ on $(0, \infty)$, $f'(x) = 0$ at $x = -1$ and at $x = 0$, and $f'(x) < 0$ for $x < -1$. Which of the following is true of f?

   (A) f has a local maximum at $x = -1$ and a local minimum at $x = 0$.
   (B) f has a local minimum at $x = -1$ and a local maximum at $x = 0$.
   (C) f has a local minimum at $x = -1$ and no local extreme value at $x = 0$.
   (D) f has a local maximum at $x = -1$ and no local extreme value at $x = 0$.
   (E) f has no local extreme value at $x = -1$ and a local minimum at $x = 0$.

GO ON TO THE NEXT PAGE

35. The speeds of a bicyclist at various times t are given in the table below.

Minutes	0	1	2	3	4	5	6
Miles/hr	0	20	40	45	35	20	5

Assume that the bicyclist's acceleration is positive on (0, 3) and negative on (3, 6). If at $t = 3$ minutes, the bicycle has traveled 1.25 miles, then at $t = 4$ minutes, which of the following could represent the total distance traveled by the bicyclist?

(A) 1.5 miles
(B) 1.9 miles
(C) 1.25 miles
(D) 2 miles
(E) 1.8 miles

36. Let $g(x) = \int_{\frac{1}{2}}^{x} \sqrt{t^3+1}\,dt$. Then $g'(x) =$

(A) $\int_{\frac{1}{2}}^{1} \sqrt{t^3+1}\,dt>$

(B) $\int_{\frac{1}{2}}^{x} \frac{3t^2}{2\sqrt{t^3+1}}\,dt$

(C) $\sqrt{x^3+1} - \frac{3\sqrt{2}}{4}$

(D) $\sqrt{x^3+1}$

(E) 0

37. Consider the functions, $f(x) = \sin\frac{1}{x}$, $x \neq 0$, and $g(x) = x\sin\frac{1}{x}$, $x \neq 0$. Which of the following describes the behavior of f and g as $x \to 0$?

(A) $\lim_{x\to 0} f(x) = 0$ and $\lim_{x\to 0} g(x) = 0$.

(B) $\lim_{x\to 0} f(x)$ and $\lim_{x\to 0} g(x)$ do not exist.

(C) $\lim_{x\to 0} f(x) = 0$ and $\lim_{x\to 0} g(x)$ does not exist.

(D) $\lim_{x\to 0} f(x)$ does not exist and $\lim_{x\to 0} g(x) = 0$.

(E) $\lim_{x\to 0} f(x) = \infty$ and $\lim_{x\to 0} g(x) = 0$.

38. The speeds of a runner in feet per second as a function of time is given in the table below.

t	0	1	2	3	4
s(t)	5	10	12	11	9

The approximate acceleration of the runner at time $t = 2$ seconds is

(A) 1 ft/s^2
(B) -1 ft/s^2
(C) 2 ft/s^2
(D) $-\frac{1}{2}$ ft/s^2
(E) $\frac{1}{2}$ ft/s^2

39. If a differentiable function, f, is increasing on (0, 3), then for $x \in (0,3)$,

(A) $f'(x) > 0$
(B) $f''(x) > 0$
(C) $f''(x) < 0$
(D) Both (A) and (B)
(E) Both (A) and (C)

GO ON TO THE NEXT PAGE

40. Let $y = \frac{1}{3}y$ be the differential equation that approximately models the growth of a population of rabbits. At time $t = 0$ years, there are 50 rabbits. Approximately how many rabbits are there after nine years?

(A) A whole number of rabbits close to 150
(B) A whole number of rabbits close to $e^3 + 49$
(C) 50 rabbits
(D) A whole number of rabbits close to $50e^3$
(E) A whole number of rabbits close to $150e^{\frac{1}{3}}$

41. The graph of a function, f, is shown below. What can be deduced about the function from its graph?

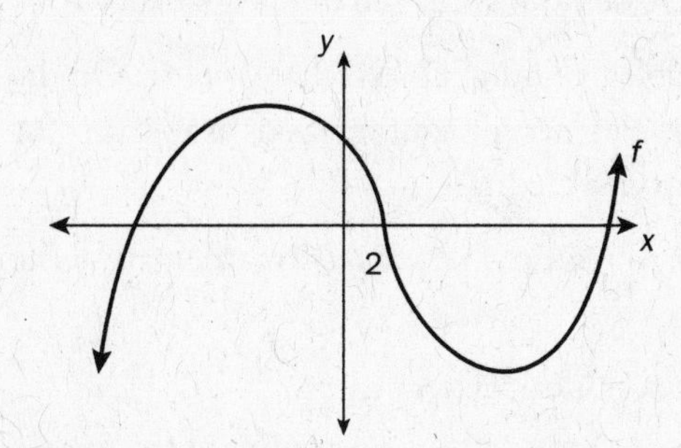

(A) $f''(x) < 0$ for $x \in (-\infty, 2)$
(B) $f''(2) = 0$
(C) $f''(x) > 0$ for $x \in (2, \infty)$
(D) $x = 2$ is a point of inflection for f.
(E) All of the above

42. Suppose that $f'(x) > 0$ on $(-\infty, -1)$, $f'(x) < 0$ on $(-1, 1)$, and $f'(x) > 0$ on $(1, \infty)$. Which of the following functions could be f?

(A) $f(x) = x^2 - 2x$
(B) $f(x) = x^3 - 3x$
(C) $f(x) = x^2 + 2x$
(D) $f(x) = x^4 - 4x$
(E) $f(x) = x^3 + 3x$

43. Let $f(x) = \int_1^x (4t^2 - \sin t)\,dt$. Then,

(A) $f(-1)$ is undefined.
(B) $f(-1) = -\frac{2}{3} - 2\cos 1$
(C) $f(-1) = -\frac{8}{3}$
(D) $f(-1) = -\frac{8}{3} - 2\cos 1$
(E) $f(-1) = 0$

44. Suppose that a function, f, that is continuous on its domain, has the additional properties that

I. $f''(x) > 0$ for $x > 3$
II. $f''(x) < 0$ for $x < 3$
III. $f'(x) < 0$ for $x > 3$
IV. $f'(x) < 0$ for $x < 3$
V. $f(3.1) > f(2.9)$

Which of the following could be a property of f?

(A) f has a vertical asymptote at $x = 3$.
(B) f is decreasing at $x = 3$.
(C) f is continuous at $x = 3$.
(D) $x = 3$ is a point of inflection for f.
(E) All of the above

45. Find $\lim_{h \to 0} h^{-1}[(x+h)^9 \sin(x+h)^2 - x^9 \sin x^2]$

(A) $x^8(2x\sin x\cos x + 9\sin x^2)$
(B) 0
(C) $x^8(2x^2\cos x^2 + 9\sin x^2)$
(D) ∞
(E) The limit does not exist.

IF YOU FINISH BEFORE TIME IS CALLED, YOU MAY CHECK YOUR WORK ON THIS SECTION ONLY. DO NOT TURN TO ANY OTHER SECTION IN THE TEST. **STOP**

SECTION II

Time: 1 hour and 30 minutes
6 Problems
50 percent of total grade

Directions: Solve the following problems, using available space for scratchwork. Show how you arrived at your answer.

You may wish to look over the problems before starting to work on them; on the actual test, it is not expected that everyone will be able to complete all parts of all problems. All problems are given equal weight, but the individual parts of a particular problem are not necessarily given equal weight. You should not spend too much time on any one problem.

- You should write out all your work for each part. On the actual test, you will do this in the space provided in the test booklet. Be sure to write clearly and legibly. If you make a mistake, you can save time by crossing it out rather than trying to erase it. Erased or crossed-out work will not be graded.
- Show all your work. Clearly label any functions, graphs, tables, or other objects that you use. On the actual exam, you will be graded on the correctness and completeness of your methods as well as your answers. Answers without any supporting work may not receive credit.
- Justifications (i.e., the request that you "justify your answer") require that you give mathematical (non-calculator) reasons.
- Work must be expressed in standard mathematical notation, not calculator syntax.
- Unless otherwise specified, answers (numeric or algebraic) need not be simplified.
- If you use decimal approximations in calculations, the readers of the actual exam will grade you on accuracy. Unless otherwise specified, your final answers should be accurate to three places after the decimal point.
- Unless otherwise specified, the domain of function f is the set of all real numbers x for which $f(x)$ is a real number.

GO ON TO THE NEXT PAGE

PART A

Time 30 Minutes
2 Problems

A GRAPHING CALCULATOR IS ALLOWED ON THIS PORTION OF THE EXAM

1. A particle moves along the x-axis so that its velocity v at time t, for $0 \le t \le 6$, is given by $v(t) = 4\sin\left(\frac{t^2}{2} - 2t + 2\right)$. At time $t = 0$, the particle is at $x = 1$.

   (a) Find the velocity and acceleration of the particle at time $t = 4$.
   (b) Is the speed of the particle increasing or decreasing at time $t = 4$? Give a reason for your answer.
   (c) Find the time when the particle first changes direction.
   (d) Describe the movement of the particle near time $t = 2$.

2. A ski resort uses a snow machine to control the snow level on a ski slope. Over a 24-hour period the volume of snow added to the slope per hour is modeled by the equation $S(t) = 24 - t\sin^2\left(\frac{t}{14}\right)$. The rate that the snow melts is modeled by the equation $M(t) = 10 + 8\cos\left(\frac{t}{3}\right)$. Both $S(t)$ and $M(t)$ have units of cubic yards per hour and t is measured in hours for $0 \le t \le 24$. At time $t = 0$, the slope holds 50 cubic yards of snow.

   (a) Compute the total volume of snow added to the mountain over the first 6-hour period.
   (b) Determine the first time that the instantaneous rate of change of the volume of snow is 0.
   (c) Is the volume of snow increasing or decreasing at time $t = 4$?
   (d) Suppose the snow machine is turned off at time $t = 6$. At what time will all the snow be melted?

GO ON TO THE NEXT PAGE

PART B

Time 60 Minutes
4 Problems

NO GRAPHING CALCULATOR IS ALLOWED ON THIS PORTION OF THE EXAM

Note: If you have extra time, you can go back and work on Part A of Section II, but you cannot use a calculator to complete your work at this time.

3. Let f be a function defined on the closed interval $[-5, 5]$. The graph of f consists of three line segments and is shown below. Let g be the function given by $g(x) = \int_{-3}^{x} f(t)\,dt$.

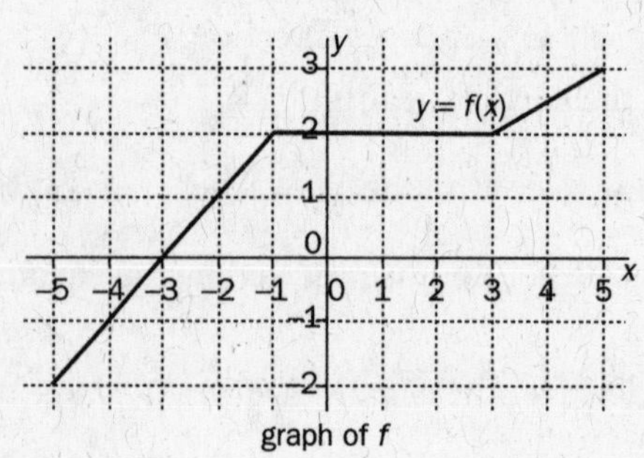

graph of f

(a) Find the values of $g(-5)$, $g(-3)$, $g(4)$.

(b) Write an equation for the tangent line to the graph of g at $x = 4$.

(c) Write a piecewise defined function for $g''(x)$.

GO ON TO THE NEXT PAGE

4. Consider the differential equation $\frac{dy}{dx} = \frac{x^2 - 1}{y}$.

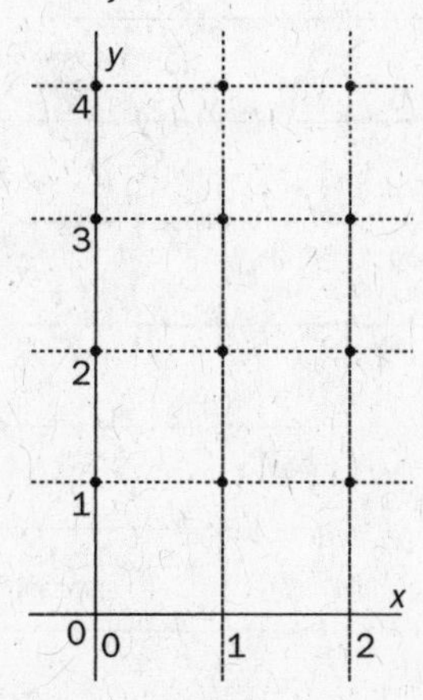

(a) On the axes provided, sketch a slope field for the given differential equation at the 12 points indicated.
(b) Let $y = f(x)$ be the particular solution to the given differential equation with the initial condition $f(3) = 4$. Write an equation for the tangent line to the graph of f at $x = 3$.
(c) Find the particular solution to the given differential equation with the initial condition $f(3) = 4$.

5. A circular oil slick of uniform thickness contains 100 cm^3 of oil.

(a) The volume of the oil remains constant. Use the equation for the volume of a cylinder to relate the thickness to the radius.
(b) As the oil spreads, the thickness is decreasing at the rate of 0.01 cm/min. At what rate is the radius of the slick increasing when the diameter is 20 cm?
(c) Suppose that after 1 minute, the diameter of the oil slick is 20 cm. How long will it take for the diameter of the slick to reach 40 cm?

Time t (days)	0	1	4	5	8	10
Mass $M(t)$(grams)	0	4	15	18	26	31

6. A crystal being grown in a lab has its mass recorded daily over a period of 10 days. The table above gives a sample of the mass of the crystal at selected days during its growth, where t represents the amount of time the crystal has been growing in days, and $M(t)$ is a twice-differentiable function representing the mass of the crystal in grams at time t.

(a) Estimate $M'(9)$. Show the work that leads to your answer. Indicate units of measure.
(b) Write an integral expression representing the average mass of the crystal over the 10-day period.
(c) Estimate the value of the above integral using a right Riemann sum with two subintervals of equal length.
(d) Is the data in the table consistent with the statement that $M''(t) < 0$ for every t in the interval $0 < t < 10$? Explain your answer.

GO ON TO THE NEXT PAGE

IF YOU FINISH BEFORE TIME IS CALLED, YOU MAY CHECK YOUR WORK ON THIS SECTION ONLY. DO NOT TURN TO ANY OTHER SECTION IN THE TEST. STOP

AP Calculus AB Practice Test 2 **Answer Key**

1. D
2. A
3. D
4. B
5. E
6. B
7. C
8. B
9. C
10. E
11. D
12. A
13. E
14. A
15. D
16. E
17. B
18. B
19. C
20. C
21. B
22. A
23. D
24. E
25. E
26. D
27. B
28. A
29. A
30. D
31. C
32. E
33. C
34. C
35. B
36. D
37. D
38. E
39. A
40. D
41. E
42. B
43. C
44. A
45. C

ANSWERS AND EXPLANATIONS

1. D

For this problem, you need to recall that the slope of a tangent line to the graph of a function at a point is the derivative of the function at that point. With that knowledge, it is easy to see that the answer may be obtained by substituting 2 for x into the expression $f'(x)$.

Here, $f(x) = 3x^2 - \sin \pi x$.

Then $f'(x) = 3(2x) - (\cos \pi x)(\pi) = 6x - \pi\cos \pi x$.

Thus, $f'(x) = 6x - \pi\cos \pi x$.

Let's substitute 2 for x into $f'(x) = 6x - \pi\cos \pi x$.

Then $f'(2) = 6(2) - \pi\cos(\pi(2)) = 12 - \pi\cos 2\pi = 12 - \pi(1) = 12 - \pi$.

Thus, $f'(2) = 12 - \pi$.

Be sure to read the question carefully and differentiate carefully.

2. A

This question requires you to know that a positive second derivative on an interval corresponds to the function being concave up on that interval, while a negative second derivative on an interval corresponds to the function being concave down on that interval. With this knowledge, choices (B) and (E) may both be eliminated. To finish answering the question, you still need to understand what is happening at the point $x = 3$. From the question, it is apparent that the concavity of the function changes at that point. You may remember that such a point is called a point of inflection. If you know that $f''(3) = 0$ is equivalent to the point $x = 3$ being a point of inflection for f, you can eliminate choices (C) and (D). If you don't know this, you may be tempted to choose (C) or (D). But if you can think of a function you know that has the properties described in the question, for example, $f(x) = -(x - 3)^3$, and if you check the second derivative at $x = 3$, you'll get 0. Thus by thinking of a simple example, you can eliminate (C) and (D), and be left with the correct answer (A) as the only remaining choice. Having in mind some examples of simple functions and expressions can be handy in eliminating incorrect answers.

3. D

To obtain the position function from the velocity, you must integrate $v(t)$, and only then can you legitimately use the object's position at $t = 2$ to obtain the constant term.

$$y(t) = \int v(t)\,dt = \int (4t + e^{t-2})\,dt = 2t^2 + e^{t-2} + C$$

Now plug in the initial condition to find C:

$y(2) = 3 \Leftrightarrow 2(4) + e^0 + C = 9 + C = 3$.

Therefore, $C = -6$

$y(t) = 2t^2 + e^{t-2} - 6$.

4. B

This question is testing your knowledge of what continuity and differentiability mean graphically. Continuity is the more intuitive concept: You don't need to take your pencil off the page to trace the graph where it is continuous. Armed with this knowledge, you can immediately eliminate choices (C) and (D). Remember that the derivative of a function at a point corresponds to the slope of the tangent line to the graph at that point. There is no tangent line to the function at the point $x = 2$, but there is a tangent line at every other point. From this, it becomes clear that choice (B) can be the only true statement. Don't be fooled into thinking that $f'(2) = 0$ just because the graph has a minimum there. It would have to be differentiable there too, and it is not. You should use easily recalled knowledge to eliminate some of the answers at the outset, but don't be fooled by the answer choices that remain.

5. E.

This question is testing your understanding of what it means for a function to be continuous at a point, in terms of limits. The words "necessary and sufficient" mean that the correct answer has to both imply and be implied by the continuity of f at $x = 4$. In other words the correct answer is equivalent to the continuity of f at $x = 4$. Since differentiability at a point is a stronger condition than continuity at a point, choice (C) can be eliminated. Choice (D) is certainly implied by the continuity of f at $x = 4$, but it is not enough to imply continuity, so it may also be eliminated. Of the remaining choices, (A), (B), and (E), the one that seems not to fit is (B), because a limit not existing at a point implies the opposite of continuity. Of the remaining two choices, you might recognize choice (E) as the definition of continuity of a function at the point $x = 4$. Systematically eliminating incorrect answer choices can be a sound strategy for honing in on the correct answer.

6. B

Here, you are supposed to compute the derivative of the given function, which is a quotient. If the first

thought that jumps into your mind is "quotient rule," you're on the right track! Recall the quotient rule:

$\left(\frac{a}{b}\right)' = \frac{a'b - b'a}{b^2}$. Therefore, for $f(x) = \frac{x^3 \sin x}{x^2 + 3}$:

$$f'(x) = \frac{(3x^2 \sin x + x^3 \cos x)(x^2 + 3) - (2x)(x^3 \sin x)}{(x^2 + 3)^2}$$

$$= \frac{3x^2 \sin x + x^3 \cos x}{x^2 + 3} - \frac{2x^4 \sin x}{(x^2 + 3)^2}$$

7. C

When you see the word *antiderivative,* you should be thinking you may need to integrate. In this case, you simply need to integrate the given function. The function has three parts that may be integrated separately. Take care to integrate each part accurately. A slip-up on any one of them could cause you to choose the wrong answer. The antiderivative of $\sin x$ is $-\cos x$, the antiderivative of $4x$ is $2x^2$, and the antiderivative of $-e^{-x}$ is e^{-x}. Adding these together yields (C). Break the problem up into smaller pieces, and double-check your answer for each piece before combining them together to find the solution.

8. B

You are given an expression that tells you that a function is continuous at a given point, and that the limit of the function at that point is L. Since the expression has two pieces, you can break it down and analyze each part separately.

The part about continuity is intuitive: If the function is continuous at $x = a$, then it will look continuous there. So, you can eliminate (D) and (E). You might notice as you are perusing the answer choices, that some of them have $x = a$ appearing along the x-axis, while others have $y = a$ along the y-axis. Those latter choices can't be right! So you can eliminate (C) as well. Of the remaining two choices, (A) shows a function whose limit is L as x tends to infinity, while (B) correctly illustrates the limit as x tends to a. Examine the answer choices before trying to solve the problem, and eliminate those which are obviously incorrect.

9. C

$$\int_{-1}^{2} (3x^2 - 4x + 2)dx = x^3 - 2x^2 + 2x \Big|_{-1}^{2}$$

$$= 8 - 8 + 4 - (-1 - 2 - 2) = 9$$

You must evaluate the definite integral. If you differentiate instead of integrate, and then plug in the limits of integration, you'll wind up with answer (D). Other mistakes, like not keeping track of your signs properly, will lead to other incorrect answers. Answer choice (E) arises out of not computing the antiderivative correctly. Carefully taking the antiderivative of the integrand and then plugging in the limits of integration will yield the correct answer, (C).

10. E

Differentiating both sides of the equation gives

$$2x - y - x\frac{dy}{dx} + 3y^2\frac{dy}{dx} = 0$$

$$(3y^2 - x)\frac{dy}{dx} = (y - 2x)$$

$$\frac{dy}{dx} = \frac{(y-2x)}{(3y^2 - x)}$$

This is an exercise in implicit differentiation. You must remember when differentiating both sides of this equation, to treat y as a function of x, whereby both the product rule and the chain rule will apply. If you forget to apply these rules, in essence treating y as a constant, you will arrive at one of the incorrect answers. Remember when implicitly differentiating that y is implicitly a function of x.

11. D

$$f(x) = x^2 \sin \pi x - 3x$$

$$f'(x) = 2x \sin \pi x + \pi x^2 \cos \pi x - 3$$

$$f'(1) = \pi(-1) - 3 = -\pi - 3$$

This problem asks you to find a linear approximation to a given function f at the point $x = 1$. A keyword here is "linear." If you know what a linear function looks like, you should be able to eliminate (A) and (E) at the outset, since they are not linear. After that, you will need to know that the tangent line to the function at the given point provides a linear approximation to the function at that point. Since the answer choices (B), (C), and (D) all have different slopes, it suffices to figure out the slope of the tangent line to f at $x = 1$. In other words, finding $f'(1) = -\pi - 3$ will narrow things down to the correct answer, (D). Don't do more work than is needed to obtain the correct solution.

12. A

This question is testing your knowledge of the asymptotic behavior of functions. The particular function given is not difficult to visualize, so starting with a graph will go a long way with this problem. Observe that the domain of the function is $x > 0$. As x approaches 0, the function value increases to infinity. This corresponds to a vertical asymptote at $x = 0$. With this one fact, you can eliminate (B), (D), and (E). (Choices (D) and (E) don't make sense anyway; you can't have a vertical asymptote at a horizontal line!) The next step is to look at the two remaining answers. The function will either have a horizontal asymptote at $y = 3$ or $y = 0$. The horizontal asymptote, by definition, must be the limit of the function as x tends to infinity. A graph of the function tells the whole story:

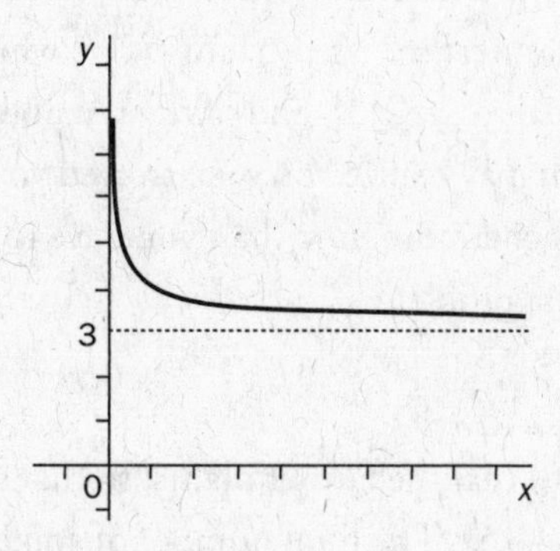

The asymptote occurs at $y = 3$, so the answer must be (A). A picture is worth a thousand words.

13. E

Every once in a while, a question will be much trickier than it appears to be at first. This is one such question. You are given a function and you are asked to find an antiderivative and to evaluate the integral of the function with set limits of integration. Even if you wanted to start by integrating, you'd have to find the antiderivative first, so finding the antiderivative is a good place to start. You could even work backward, differentiating the antiderivative candidates in the answer choices, and seeing which ones give the original function back. Only answer choices (D) and (E) do that. Moving on to the definite integral, there are two possible choices: either the integral equals -2 or it is undefined. If you simply plug the limits of integration into the antiderivative, you do get -2, but is that the correct answer? In order for this integral to be evaluated this way (using the Fundamental Theorem of Calculus), the integrand must be a continuous function over the subject interval. The given function is not continuous on the interval $[1, 3]$, so (D) is not the correct choice, leaving only (E). Keep your eye out for tricky questions that might require you to remember the assumptions underlying a procedure.

14. A

This question is testing your ability to recall the Mean Value Theorem. If you can't recall it, you may have difficulties with this one. The Mean Value Theorem says, in essence, that in a given interval, there is some point such that the derivative of the function at that point equals the slope of the secant whose endpoints correspond to the endpoints of the interval. So to check which answer is the correct one, just start going down the list of answer choices. The slope of the secant in answer choices (A) and (D) is $\frac{f(4) - f(1)}{4 - 1} = -\frac{1}{3}$, so (D) can be eliminated and (A) looks like the right choice. For (B), (C), and (E), the slope of the secant is 1, and none of these answer choices give 1 as a possible value of $f'(c)$, so you can eliminate all of them. The only viable answer is (A). Don't hesitate to evaluate expressions using each of the answer choices in turn.

15. D

This is a problem that may be broken up into four separate pieces. For each piece, you must compute the derivative, then add them together when you are finished. The answer choices are designed to capture mistakes in any one of the given pieces, so be very careful. The last two pieces are probably more difficult, so differentiating the first two pieces first can help eliminate answer choices. If you differentiate these correctly, you'll be able to eliminate (C) and (E). If you don't know how to differentiate one or both of the remaining terms, here's a trick you can use for the last term: 2^x is equal to $e^{x \ln 2}$, which is easy to differentiate using the chain rule. The result, $(\ln 2)e^{x \ln 2}$, is equal to $(\ln 2)2^x$, and of the remaining answer choices, only one of them offers this possibility, so you're done. You may realize you've narrowed your answers down to one choice before you expected to. If this happens, don't spend more time than you have to by continuing to solve the problem. Mark the answer and move on to the next problem.

16. E

Let x = number of extra trees (beyond 20) that the farmer plants, so that the number of trees planted is $x + 20$. Then, the yield of apples per tree is $600 - 15x$. The total number of apples that can be picked, T, is the product of these two values:

$$\begin{aligned} T &= (x + 20)(600 - 15x) \\ &= 600x - 15x^2 + 12000 - 300x \\ &= 300x - 15x^2 + 12000 \end{aligned}$$

We want to find the maximum value of T, which requires that we compute T' and set it equal to zero:

$T' = 300 - 30x = 0 \Rightarrow x = 10$ extra trees planted. So the number of trees the farmer should plant is $x + 20 = 30$ trees. Note that the second derivative is -30, which is negative. Thus this amount is a maximum.

17. B

This is another problem in which visualizing the graph of the given function proves extremely helpful. The function is defined piecewise, and each piece is simple. The answer choices are concerned with absolute extrema. The highest value on the graph is almost 1 (since the function jumps to 0 at $x = 1$), and the lowest value is 0. At first glance, it may seem that all of the choices (A), (B), and (C) are correct, but the keyword here is "attains." Choice (A) is incorrect because the maximum value of 1 is never attained by a particular x, and similar reasoning eliminates (C). But f does attain its absolute minimum value of 0, so (B) must be the correct choice. Visualize the graphs of functions whenever possible.

18. B

This is an exercise in the correct application of u-substitution in a definite integral. The obvious choice is to let $u = x^2$, in which case $du = 2x\,dx$. Once you have the correct integrand, you can eliminate choices (D) and (E).

Then you have to make sure you derive the limits of integration correctly; forgetting to do this might lead you to the incorrect answer. The original limits are in terms of x. Now you are differentiating with respect to u, so the new limits should reflect this. Simply plug the original limits for x into the expression $u = x^2$ to get the new limits, 1 and 9, and the correct answer follows. When performing a u-substitution, don't forget to include your limits of integration.

19. C

This question is testing your understanding of the representation of the derivative of a function as an instantaneous rate of change. When you see the words, "instantaneous rate of change," hopefully you will think, "derivative." From there, it's easy; just differentiate and plug in the point $t = 2$ to arrive at the correct answer.

20. C

To compute the limit of the given expression as x approaches infinity, you need to divide both the numerator and the denominator by the highest order term, which is x^2. In the denominator, since a square root appears, divide by $\sqrt{x^4}$ so that everything will stay under the square root. Then when you take the limit, all the terms except the highest order terms will go to 0. You are left with 1 in the numerator and $\sqrt{4} = 2$ in the denominator. The limit is $\frac{1}{2}$. When computing the limit of a rational function, divide everything through by the highest ordered term.

21. B

This is a related-rates problem. You need to know the formula for the volume of a sphere, which is $V = \frac{4}{3}\pi r^3$. It is given that the volume increases at a rate of 100 cm^3/s. Here, you also need to know that the volume's rate of increase can be described by $\frac{dV}{dt}$, to arrive at the equation $\frac{dV}{dt} = 100 \text{ cm}^3/\text{s}$. Now you need to relate the rate of increase of the volume to the rate of increase of the radius (which you are trying to find) by using the formula for the volume, and thinking of both v and r as functions of t. Differentiating both sides of the volume equation with respect to t yields $\frac{dV}{dt} = 4\pi r^2 \frac{dr}{dt}$.

Since you are given that $\frac{dV}{dt} = 100 \text{ cm}^3/\text{s}$, you have $100 = 4\pi r^2 \frac{dr}{dt}$, which is the equation you need to solve the problem. When the diameter is 50 cm, the radius is 25 cm, so $\frac{dr}{dt} = \frac{100}{4\pi(25)^2} = \frac{1}{25\pi}$ cm/s.

22. A

This question asks about the instantaneous rate of change of the absolute value function at $x = 0$. If you know that the instantaneous rate of change of a function at a point is equal to the derivative of the function at that point, then this problem is straightforward. The absolute value function is not differentiable at $x = 0$, so the instantaneous rate of change is nonexistent. Your task is to recognize this. The remaining answer choices offer versions of the definition of the derivative, involving a limit that, in the case of this function, does not exist. So those answer choices are nonsensical.

23. D

Here, you are being tested on your understanding of the definition of the derivative in terms of limits. The first three answer choices look very similar to each other, and the last two offer combinations of those three. So you must examine the answer choices very carefully. Look for similarities and differences. Look for ways in which one answer choice contradicts another. In this case, answers (A) and (C) cannot both be correct, because one is the negative of the other. But that doesn't eliminate any of the answers. Here's a clever idea: Compute the limits in (A), (B), and (C), and see which ones give the right answer, since you know what the derivative of $f(x) = x^2$ should be. Expanding the numerator allows you to cancel some terms in each of the limits, so that in (A) you obtain a limit of $-2x$, in (B) you obtain a limit of $2x$, and in (C) you obtain a limit of $2x$. Both (B) and (C) are correct, so the right answer must be (D). It can be helpful to start with what you know, and work backward.

24. E

The key to this question is to recognize that obtaining the total change of the quantity over the interval amounts to integrating the instantaneous rate of change, or $f'(x)$, over that interval. To do this, you find the antiderivative of $f'(x)$ first. Using u-substitution, let $u = x^2$, then $du = 2x\,dx$. So, $\int_{-1}^{5} xe^{x^2}\,dx = \frac{1}{2}e^{x^2}\Big|_{-1}^{5} = \frac{e^{25}}{2} - \frac{e}{2}$

Various mistakes could lead you to the wrong answer, like plugging the interval into $f'(x)$ instead of $f(x)$, so be careful. A rate of change corresponds to the derivative of a function, while total change corresponds to the function itself.

25. E

This is a problem dealing with position, velocity, and acceleration. The best strategy for this problem is to begin writing down everything you know. If f, v, and a represent position, velocity, and acceleration, respectively, then you know that $v(0) = 30$ mi/hr, and you can assume that $f(0) = 0$. You also know that $a(t) = 2$ mi/hr^2 for all $t > 0$. You need to find t in hours such that $f(t) = 31$ miles. To get the position function, you must integrate $a(t)$ twice, each time using the initial conditions to find your constant term. The first integration yields $v(t) = 2t + 30$ mi/hr. Integrating again yields $f(t) = t^2 + 30t$ miles. Now you need to find t that solves the equation $t^2 + 30t - 31 = 0$. This quadratic equation factors nicely as $(t + 31)(t - 1) = 0$. The positive solution, $t = 1$ hour, gives the correct answer. When units are involved, keep careful track of them all along the way.

26. D

This question is testing your ability to distinguish between points where a function is continuous and where it is not. The first observation to be made is that the function is continuous on the open intervals corresponding to each of the three pieces of the function. Next, observe that the point $t = 2$ is not even in the domain of the function, so answer choices (A) and (B) may be eliminated. The remaining points in question are $t = 1$ and $t = -1$. At $t = 1$, $f(t) = 0$, while the limit of $f(t)$ from the left as t approaches 1 is 2. So $t = 1$ is not a point of continuity of the function, and (E) may be eliminated. The last point, $t = -1$, presents a technical question:

Is continuity well defined at the boundary of the domain of a function? The answer is yes, so the point $t = -1$ should be included, which yields (D) as the correct answer.

27. B

This question is testing your ability to visualize a simple function and analyze its tangents. The first step is to visualize the function. If you make the observation that $f(x) = \sqrt[3]{x}$ is the inverse of the function $g(x) = x^3$, then the graph is just the graph of $g(x)$ flipped over the line $y = x$. Now that you are visualizing the function, it is easy to glean information about its tangents. At $x = 0$, the function has a vertical tangent, so (B) is correct. Choice (C) is incorrect since the slope becomes increasingly steep as it approaches 0, and then decreases as it moves away from 0.

28. A

This is an implicit differentiation problem. Remember, treat y as a function of x and you'll be fine. Differentiating both sides of the equation requires the product rule, and yields $2x - 4xy' - 4y - 2yy' = 0$. Isolating y', you obtain $y' = \frac{-2x + 4y}{-4x - 2y}$. Now just plug in the given point to obtain the answer. This problem could be much more difficult if you approach it by trying to isolate y first. Implicit differentiation is a useful tool for finding the derivative when the function is defined implicitly.

29. A

You are given the derivative of a function and asked to find the approximate value of x where the function has an absolute minimum value. You could use the derivative test, setting the given derivative equal to 0 to obtain the only critical points, $x = \pm\sqrt{e^2 - 1}$. To test whether each of these points is a max or a min, you could use the second derivative test, where $f''(x) = \frac{2x}{x^2 + 1}$. You can see that $f''(x)$ is positive at $x = \sqrt{e^2 - 1}$, so this point is a local minimum. However, the problem asked you to find an absolute minimum. You'll need to check any endpoints. But the derivative is defined everywhere, so there are no endpoints to check. So the local minimum you found must also be an absolute minimum. You can translate this to a number using your calculator to get $x \approx 2.53$.

30. D

For this problem, you will need to recognize that the function F is the antiderivative of f, so that the expression $F(4) - F(0)$ is, by the Fundamental Theorem of Calculus, equal to $\int_0^4 f(x)dx$. You are given the graph of f, as well as the value of $\int_0^3 f(x)dx$. Recognizing that this integral represents the area under the graph between $x = 0$ and $x = 3$, and that the remaining part of the integral is the area under the graph between $x = 3$ and $x = 4$, you can see from the graph that the remaining piece has area 2. Therefore, the total area is 5.5.

31. C

For this problem, you are given a second derivative of a function, and asked to consider which of the three statements about the function are true. The first two statements deal with where the function is concave up and where it is concave down. A differentiable function is concave up on an interval if its second derivative is greater than 0 on that interval. It is concave down on an interval if its second derivative is less than 0 on that interval. So for the first two statements, you need to check the sign of $f''(x)$ on the given intervals. You could use your calculator to do so. For statement I, when x is in the first interval, x^2 lies within $(0, \pi)$, so $f''(x) > 0$ on the first interval. When x is in the second interval, x^2 lies within $(2\pi, 3\pi)$, so $f''(x) > 0$ on the second interval as well. Thus, statement I is true. For statement II, when x is in the given interval, x^2 lies within $(\pi, 2\pi)$, so $f''(x) < 0$ on that interval. This implies that f is concave down on the interval, so that statement II is also true. Statement III considers the same interval as the first interval in statement I. But now under consideration is whether or not $f'(x)$ is increasing on that interval. If a function's derivative is greater than 0 on an interval, then the function is increasing on that interval. You already know that $f''(x) > 0$ on that interval, so $f'(x)$ must be increasing there, and statement III must be true. Thus all three statements are true.

32. E

Here's a question about differential equations and their slope fields. A slope field is shown to you, and your job is to figure out what function could be a solution of the corresponding differential equation. If you graph each of the functions in the answer choices, you will see that only choice (E) fits the given slope field. The other choices either cross the x-axis or don't exhibit the requisite behavior of the function in some other way.

33. C

For this question, you are given a function and asked to consider how the function increases and decreases as t tends to infinity, $t \geq 0$. Here's another problem for which graphing the function could help you to figure out the solution. You could also check yourself by computing the derivative, $P'(t) = 500e^t - \frac{500}{t+1}$, and checking to see where it is positive (increasing) and where it is negative (decreasing). Since $e^t \geq \frac{1}{t+1}$ for all $t \geq 0$, $P(t)$ must increase indefinitely as t tends to infinity.

34. C

For this problem, the first step is to consider what the function f might look like. It is increasing on $(-1, 0)$, non-decreasing on $(0, \infty)$, has critical points at $x = -1$ and $x = 0$, and is decreasing on $(-\infty, -1)$. If you draw such a function, you will see that it must have a local minimum at $x = -1$, and no local extreme value at $x = 0$. This is enough to pinpoint the correct answer.

35. B

In this problem, the assumption that the bicyclist's acceleration is negative on $(3, 6)$ means that the speed is decreasing on that interval. Your goal is to figure out what distance the bicycle could have traveled after four minutes, given that it travels 1.25 miles after three minutes. Since the velocity is decreasing after three minutes, the total distance traveled can't be more than $1.25 + \frac{45}{60} = 2$ miles. Also, the total distance traveled can't be less than $1.25 + \frac{35}{60} \approx 1.83$ miles. If you look for the answer choice that lies between 1.83 miles and 2 miles, you'll find that there's only one such answer.

36. D

This question is testing your understanding of the Fundamental Theorem of Calculus. You are given a function defined in terms of an integral whose upper limit of integration is the argument of the function. You are asked to compute the derivative of the function. The Fundamental Theorem of Calculus says that the derivative will be the integrand evaluated at x. Thus, $g'(x) = \sqrt{x^3 + 1}$.

37. D

You are asked to analyze the behavior of two functions, so it is wise to graph each one on your graphing calculator. If you use the zoom feature on your calculator, you can analyze what is happening as x tends to 0 for each function. It appears that $f(x)$ keeps oscillating between -1 and 1 more and more rapidly, whereas $g(x)$, while also oscillating, decreases in magnitude to 0. Thus, $f(x)$ doesn't have a limit as x goes to 0, whereas $g(x)$ tends to 0 as x goes to 0.

38. E

The best approximation of the runner's acceleration at $t = 2$ seconds is the slope of the secant between $t = 1$ and $t = 3$. (Secants between $t = 1$ and $t = 2$ or between $t = 2$ and $t = 3$ would be better at approximating the acceleration at $t = 1.5$ and $t = 2.5$ seconds respectively.) The slope of the secant between $t = 1$ and $t = 3$ seconds is $\frac{f(3) - f(1)}{3 - 1} = \frac{11 - 10}{2} = \frac{1}{2}$ ft/s^2.

39. A

For this problem, you need to know that if a function is increasing on an interval, then its derivative is greater than 0 on that interval. The second derivative could be either positive or negative, so the remaining answer choices may be excluded.

40. D

This differential equation says that the rate of growth of the rabbits is proportional to the number of rabbits. Using the initial condition and separation of variables, the solution to this differential equation is:

$$y' = \frac{dy}{dt} = \frac{1}{3}y$$

$$\frac{dy}{y} = \frac{1}{3}dt$$

$$\int \frac{dy}{y} = \int \frac{1}{3}\, dt$$

$$\ln y = \frac{1}{3}t + C$$

$$y = e^{\frac{1}{3}t+C} = e^{\frac{1}{3}t} \cdot e^{C} = Ae^{\frac{1}{3}t}$$

$$y(0) = Ae^{\frac{1}{3}(0)} = A = 50$$

Thus, $y = 50e^{\frac{1}{3}t}$.

Now, plug $t = 9$ years into this equation to arrive at approximately $50e^3$ or 1,004 rabbits after nine years.

41. E

From the graph of the function, you can deduce that it is concave down to the left of $x = 2$ and concave up to the right. Thus, $x = 2$ is a point of inflection, and $f''(2) = 0$. Where the function is concave down, $f''(x) < 0$, and where it is concave up, $f''(x) > 0$. So, all of the answer choices are true, and the answer is "all of the above."

42. B

For this problem, it will help to try to visualize the graph of *f*. The information given implies that *f* is increasing on $(-\infty, -1)$, decreasing on $(-1, 1)$, and increasing again on $(1, \infty)$. You are given various answer choices that represent choices for *f*. Use your graphing calculator to graph these functions and check to see which one satisfies the given conditions. Answer choices (A), (C), and (D) don't fit because their orders are even with a positive leading coefficient, so they must be decreasing in some interval $(-\infty, a)$, not increasing. Answer choice (E) doesn't work because it is always increasing. But answer choice (B) fits the information perfectly.

43. C

To compute $f(-1)$, you replace the x in the integral's limit of integration with -1, and compute the integral if it exists. Since the integrand is continuous, and has antiderivative $F(t) = \frac{4t^3}{3} + \cos t$, the definite integral is equal to $F(-1) - F(1) = -\frac{4}{3} + \cos(-1) - \frac{4}{3} + \cos(1) = -\frac{8}{3}$. The last equality was obtained with the help of the identity $\cos(-a) = \cos(a)$.

44. A

For this problem, you are given five properties for a function, *f*, that is continuous on its domain. Note that you are not told the domain of the function. In particular, if you look at the properties I–V, it becomes evident that $x = 3$ might not be in the domain of *f*. Property I implies that *f* is concave up for $x > 3$. Property II implies that *f* is concave down for $x < 3$. Property III implies that *f* is decreasing for $x > 3$. Property IV implies that *f* is decreasing for $x < 3$. Finally, property V, when combined with the other properties, implies that *f* is greater immediately to the right of $x = 3$ than it is immediately to the left of $x = 3$. If you can come up with an example of a function that satisfies all of these properties, and that is not defined at $x = 3$, then you can eliminate answer choices (B), (C), (D), and (E). The function $f(x) = \frac{1}{x-3}$ is one such function. In fact, this function has the property that it has a vertical asymptote at $x = 3$. Since the question asks which of the answer choices could be a property of *f*, (A) is the only possible correct answer.

45. C

By putting h^{-1} into the denominator, we see that this just becomes the limit equation for the derivative of $x^9 \sin x^2$. Therefore, we just need to use the product rule to solve:

$$\begin{aligned} f(x) &= x^9 \sin x^2 \\ f'(x) &= x^9 \cos x^2 (2x) + \sin x^2 (9x^8) \\ &= 2x^{10} \cos x^2 + 9x^8 \sin x^2 \\ &= x^8 (2x^2 \cos x^2 + 9 \sin x^2) \end{aligned}$$

FREE-RESPONSE ANSWERS

1. (a) Velocity is given by $v(t) = 4\sin\left(\frac{t^2}{2} - 2t + 2\right)$. Acceleration is given by

$v'(t) = 4\cos\left(\frac{t^2}{2} - 2t + 2\right)\left(\frac{2t}{2} - 2\right)$. Thus the velocity at $t = 4$ is $v(4) = 3.637$ unit/sec

and the acceleration is $v'(4) = -3.329$ units/sec^2.

(b) At $t = 4$, $v(t)$, is positive and $v'(t)$, is negative so the speed of the particle is decreasing. Speed is the absolute value of velocity, so since velocity is positive, speed is equal to velocity and the rate of change of speed is the rate of change of velocity.

(c) The particle changes direction when $v(t)$ changes sign. In part (d), there's a sketch of the graph of $v(t)$; $v(t)$ first changes sign at $t = 2$.

(d)

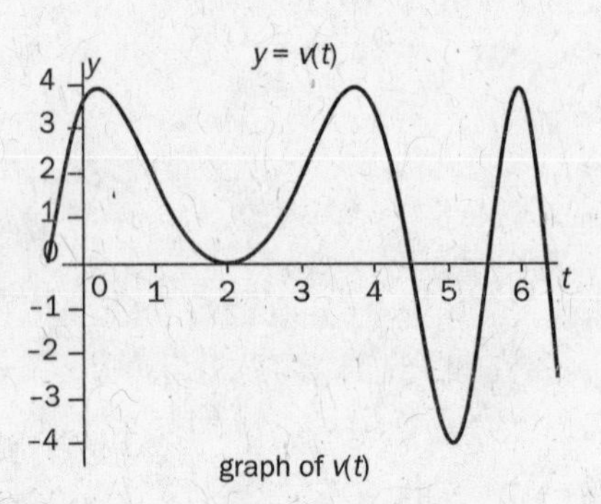

Near $t = 2$, the graph of the velocity of the particle decreases to 0 and then begins increasing again. Before $t = 2$, the particle is in the positive part of the x-axis, moving in the positive direction, but slowing down. As t approaches 2, the particle stops. At $t = 2$, the particle is stopped. After $t = 2$, the particle begins moving in the positive direction again. Informally, the particle is moving, slowing down, stops for an instant, and then speeds back up again.

When answering questions that ask for a description of the system an equation is modeling, always write clearly and in complete sentences. (In fact, you should always write your answers as though you were explaining the answer to someone, and always in complete sentences.) You might get most of the credit for an answer such as "it stopped and then started again," but then again, you might not. For part (B), and similar problems, questions often come up where the velocity is negative *and* the acceleration is negative. In these cases, the speed is increasing, since negative velocity and acceleration mean that velocity is becoming more negative, and so the particle is moving faster and faster in the negative direction. If velocity were negative but acceleration positive, then the speed would be decreasing. For part (c), make sure you look at the graph of the function before you ask the calculator to find the zero: If not, you might end up with $t = 2$ as the first time the particle changes direction.

2. The total volume of snow added to the mountain over the first 6-hour period is given by the integral $\int_0^6 S(t)dt = \int_0^6 24 - t\sin^2\left(\frac{t}{14}\right)dt = 142.4132 \text{ yds}^3$.

(e) The first time that the instantaneous rate of change of the volume of snow is 0 occurs the first time $M(t) = S(t)$, at $t = 15.0404$ hours.

(f) At time $t = 4$, the rate of change of the volume of snow is given by $S(4) - M(4) = 11.8004 \text{ yds}^3/\text{hr}$. This is positive so the volume of snow is increasing.

(g) At time $t = 6$, the amount of snow added to the mountain was 142.4132 yds^3, bringing the total amount of snow needing to be melted to 192.4132 yds^3. To determine the time when all the snow is melted we solve the equality

$\int_0^S M(t)dt = \int_0^S 10 + 8\cos\left(\frac{t}{14}\right)dt = 192.4132$. This occurs at $S = 11.204$ hours.

For part (a), read the question carefully and make sure you understand what it's asking. (For example, it's not asking for the actual volume of snow on the mountain at $t = 6$.) Remember, since $M(t)$ measures the rate the snow melts, it contributes negatively to the rate of change of the volume of the snow. Again, don't forget your units!

3. (a)

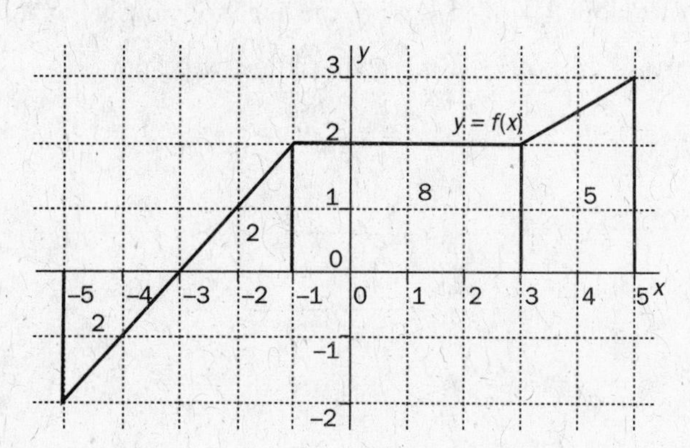

$g(-5) = 2$

$g(-3) = 0$

$g(4) = 12.25$

(b) By the Fundamental Theorem of Calculus, $g'(x) = f(x)$ so the tangent line is given by the equation $y - g(4) = g'(4)(x - 4)$, giving $y - 12.25 = 2\frac{1}{2}(x - 4)$.

(c) By the Fundamental Theorem of Calculus, $g'(x) = f(x)$, and $g''(x) = f'(x)$.

$$f'(x) = \begin{cases} 1 \text{ if } -5 \leq x < -1 \\ 0 \text{ if } -1 < x < 3 \\ \frac{1}{2} \text{ if } 3 < x \leq 5 \end{cases} \quad \text{so} \quad g''(x) = \begin{cases} 1 \text{ if } -5 \leq x < -1 \\ 0 \text{ if } -1 < x < 3 \\ \frac{1}{2} \text{ if } 3 < x \leq 5 \end{cases}$$

Be very careful about computing $g(-5)$ in particular. Using the definition $g(x) = \int_{-3}^{x} f(t)\,dt$, we compute $g(-5) = \int_{-3}^{-5} f(t)\,dt = -\int_{-5}^{-3} f(t)\,dt = -(-2) = 2$.

An intuitive way to see this is by thinking of f as the derivative of g. Since f is negative, g is *decreasing* on the interval $[-5,-3]$. Since $g(-3) = 0$, the value of g at $x = -5$ must be positive. When computing

$$g''(x) = f'(x) = \begin{cases} 1 \text{ if } -5 \leq x < -1 \\ 0 \text{ if } -1 < x < 3 \\ \frac{1}{2} \text{ if } 3 < x \leq 5 \end{cases}$$

we just compute the slopes of each line segment in the graph of f. On $-5 \leq x < -1$, the line has a vertical shift of $2 - (-2) = 4$, giving a line with slope 1. On $-1 < x < 3$, the line is obviously flat and has a slope of 0. On $3 < x \leq 5$, the line has a vertical shift of $3 - 2 = 1$, giving a line with slope $\frac{1}{2}$.

4. (a)

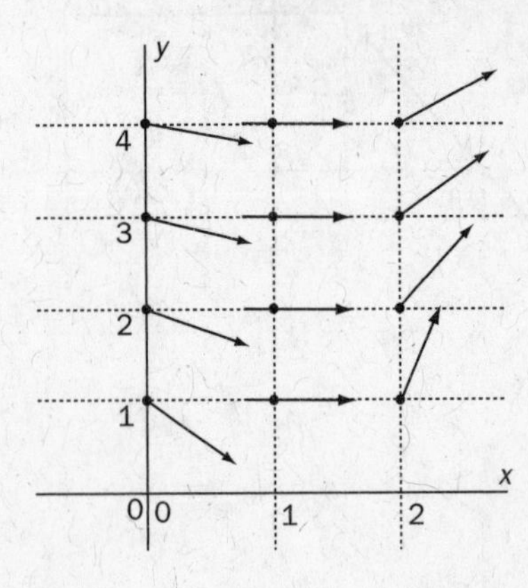

x	y	$\frac{dy}{dx} = \frac{x^2 - 1}{y}$
0	1	-1
0	2	$-\frac{1}{2}$
0	3	$-\frac{1}{3}$
0	4	$-\frac{1}{4}$
1	1	0
1	2	0
1	3	0
1	4	0
2	1	3
2	2	$\frac{3}{2}$
2	3	1
2	4	$\frac{3}{4}$

(b) Since $f(x)$ is a solution to the differential equation and $f(3) = 4$, the slope of the tangent line to the graph of f is $\frac{dy}{dx} = \frac{x^2 - 1}{y} = \frac{9 - 1}{4} = 2$. The equation for the tangent line is $(y - 4) = 2(x - 3)$.

(c) Solving:

$$\frac{dy}{dx} = \frac{x^2 - 1}{y}$$

$$y\,dy = (x^2 - 1)\,dx$$

$$\int y\,dy = \int x^2 - 1\,dx$$

$$\frac{1}{2}y^2 = \frac{1}{3}x^3 - x + C$$

$$y = \pm\sqrt{\frac{2}{3}x^3 - 2x + 2C}$$

To compute C, use $\frac{1}{2}y^2 = \frac{1}{3}x^3 - x + C$. Thus $\frac{1}{2}(4^2) = \frac{1}{3}(3^3) - 3 + C$, and $C = 2$. Finally, our particular solution is given by the equation $f(x) = \sqrt{\frac{2}{3}x^3 - 2x + 4}$. (Since $f(3) = 4$ we take the positive root.)

For part (b), because we know the point we are looking at, the slope is just given by the slope field, namely $\frac{dy}{dx} = \frac{x^2 - 1}{y} = \frac{3^2 - 1}{4} = 2$. Use any form for the equation of the line, and don't bother to simplify since there's a small chance you could make an error and lose points. You won't lose points for leaving the equation for the tangent line unsimplified. For part (c), always remember to separate the variables before integrating. Even if you have the correct answer, you will lose points if you fail to separate the variables first. Finally, in the computation of C, we use the simpler of the two equivalent formulations of the solution to ease calculations, but we need the more complicated one in order to give a final equation. Note also that if our initial condition was $f(3) = -4$, we would have ended up with the same solution only using the *negative* root.

5. (a) For a cylinder, $Vol = \pi r^2 h$. Thus

$$100 = \pi r^2 h$$

$$h = \frac{100}{\pi r^2}$$

(b) We are given that $\frac{dh}{dt} = -0.01$ cm/min. By (a), we have $\frac{dh}{dt} = -\frac{200}{\pi} r^{-3} \frac{dr}{dt}$, and we want $\frac{dr}{dt}$ at $r = 10$ cm. Solving:

$$\frac{dr}{dt} = \frac{dh}{dt}\left(-\frac{\pi r^3}{200}\right) = -0.01\left(-\frac{\pi r^3}{200}\right) = 0.01\left(-\frac{\pi r^3}{200}\right).$$

At $r = 10$, we have

$$\frac{dr}{dt} = 0.01\left(\frac{\pi(10^3)}{200}\right) = 0.05\pi.$$

Thus, at $r = 10$ cm, the radius is increasing at 0.05π cm/min.

(c) At $r = 10$, $h(1) = \frac{100}{\pi r^2} = \frac{100}{\pi 100} = \frac{1}{\pi}$. At $r = 20$, $h(t) = \frac{100}{\pi r^2} = \frac{100}{\pi 400} = \frac{1}{4\pi}$.

Applying the Fundamental Theorem of Calculus,

$$h(t) = \int_0^t h(s)\,ds = \int_0^1 h(s)\,ds + \int_1^t h(s)\,ds.$$

Since we know that $h'(t) = -0.01$ and $h(1) = \frac{1}{\pi}$,

$$h(t) = \frac{1}{\pi} - 0.01\int_1^t ds = \frac{1}{\pi} - 0.01(t-1)$$

Finally, when $r = 20$, $h(t) = \frac{1}{4\pi}$. So we need to find t so that $h(t) = \frac{1}{4\pi}$.

$$\frac{1}{4\pi} = h(t) = \frac{1}{\pi} - 0.01\int_1^t ds = \frac{1}{\pi} - 0.01(t-1)$$

$$\frac{1}{4\pi} - \frac{1}{\pi} = -0.01(t-1)$$

$$t = \frac{1}{0.01}\left(\frac{1}{\pi} - \frac{1}{4\pi}\right) + 1 = 100\frac{3}{4\pi} + 1 = \frac{75+\pi}{\pi}$$

For part (c), you can also solve the problem by finding and integrating $\frac{dr}{dt}$, which is somewhat messier.

6. (a) We estimate $M'(9)$ by the secant line between $t = 8$ and $t = 10$. $M(9) \approx \frac{M(10) - M(8)}{10 - 8} = \frac{5}{2}$ grams/day.

(b) The average mass is given by the integral $\frac{1}{10}\int_0^{10} M(t)dt$.

(c) The right Riemann sum with intervals [0,5] and [5,10] is given by the expression

$(5 - 0)M(5) + (10 - 5)M(10) = 5 \cdot 18 + 5 \cdot 31 = 90 + 155 = 245$

(d)

Time t (days)	0	1	4	5	8	10
Mass $M(t)$ (grams)	0	4	15	18	26	31
Mass change ΔM	4	11	3	3	5	?
Average Mass Change $\Delta M/day$	4	$\frac{11}{3} = 3\frac{2}{3}$	3	$\frac{8}{3} = 2\frac{2}{3}$	$\frac{5}{2} = 2\frac{1}{2}$	?

The estimates of $M'(t)$ above are decreasing, so our estimates of $M''(t)$ would be negative. The data in the table is consistent with the statement that $M''(t) < 0$ for every t in the interval $0 < t < 10$.

Don't forget that the average value of a function can be computed by taking the integral and then dividing by the size of the interval of integration. Don't forget to divide (in this case by 10). Part (d) is a really neat question. Use the data in the table to estimate $\frac{dM}{dt}$ and then see if these estimates can tell you something. (For example, if our estimates of $\frac{dM}{dt}$ increased somewhere and decreased somewhere, we could use the Intermediate Value Theorem and the Mean Value Theorem (or our intuition) to show that $M''(t)$ would have to be 0 somewhere.)

AP CALCULUS AB PRACTICE TEST 2 CORRELATION CHART

Multiple Choice Question	Review Chapter Subsection of Chapter
1	Chapter 13: The Slope of a Curve at a Point
2	Chapter 15: Concavity and the Sign of f''
3	Chapter 22: Accumulation Functions and Antiderivatives
4	Chapter 11: The Relationship Between Differentiability and Continuity
5	Chapter 9: Continuity and Limits
6	Chapter 12: Rules for Computing Derivatives
7	Chapter 21: Computing Antiderivatives Directly
8	Chapter 9: Continuity and Limits
9	Chapter 20: What's So Fundamental? The Connection Between Integrals and Derivatives
10	Chapter 12: Implicit Differentiation
11	Chapter 13: Local Linear Approximation
12	Chapter 8: Vertical Asymptotes – Understanding Infinite Limits Graphically Chapter 8: Horizontal Asymptotes
13	Chapter 21: Computing Antiderivatives Directly Chapter 20: What's So Fundamental? The Connection Between Integrals and Derivatives
14	Chapter 14: The Mean Value Theorem
15	Chapter 12: Rules for Computing Derivatives Chapter 12: Derivatives of Trigonometric Functions Chapter 12: The Derivative of Inverse Functions Chapter 12: Exponential and Logarithmic Functions
16	Chapter 16: Optimization
17	Chapter 14: Absolute Extrema
18	Chapter 21: When *F* Is Complicated: The Substitution Game
19	Chapter 13: The Slope of a Curve at a Point
20	Chapter 8: Limits Approaching Infinity
21	Chapter 16: Related Rates
22	Chapter 11: The Relationship Between Differentiability and Continuity
23	Chapter 12: Computation of Derivatives
24	Chapter 22: Accumulation Functions and Antiderivatives
25	Chapter 22: Accumulation Functions and Antiderivatives

Multiple Choice Question	Review Chapter Subsection of Chapter
26	Chapter 9: Continuous Functions
27	Chapter 16: Graphing Using Derivatives
28	Chapter 12: Implicit Differentiation
29	Chapter 14: Absolute Extrema
30	Chapter 20: What's So Fundamental? The Connection Between Integrals and Derivatives Chapter 20: Interpreting the Fundamental Theorems Graphically
31	Chapter 16: Graphing Using Derivatives
32	Chapter 16: Differential Equations
33	Chapter 14: Increasing and Decreasing Behavior of f and the Sign of f'
34	Chapter 14: Critical Points and Local Extrema
35	Chapter 22: Accumulation Functions and Antiderivatives
36	Chapter 20: What's So Fundamental? The Connection Between Integrals and Derivatives
37	Chapter 23: Approximating the Area Under the Curve Using Riemann Sums
38	Chapter 23: Trapezoidal Approximation
39	Chapter 14: Increasing and Decreasing Behavior of f and the Sign of f'
40	Chapter 22: Separable Differential Equations
41	Chapter 16: Graphing Using Derivatives
42	Chapter 14: Corresponding Characteristics of the Graphs of f and f'
43	Chapter 21: Computing Antiderivatives Directly
44	Chapter 16: Graphing Using Derivatives
45	Chapter 12: Computation of Derivatives

Free Response Question	Review Chapter subsection of chapter
1(a)	Chapter 6: Evaluating a Derivative at a Point
1(b)	Chapter 16: Interpreting the Derivative As a Rate of Change
1(c)	Chapter 16: Interpreting the Derivative As a Rate of Change
1(d)	Chapter 16: Interpreting the Derivative As a Rate of Change Chapter 11: A Physical Interpretation of the Derivative: Speed
2(a)	Chapter 18: Accumulated Change Chapter 22: Accumulation Functions and Antiderivatives
2(b)	Chapter 14: Critical Points and Local Extrema
2(c)	Chapter 16: Interpreting the Derivative As a Rate of Change
2(d)	Chapter 18: Accumulated Change Chapter 22: Accumulation Functions and Antiderivatives
3(a)	Chapter 23: Approximating the Area Under Functions Given Graphically or Numerically
3(b)	Chapter 13: The Slope of a Curve at a Point
3(c)	Chapter 20: The Fundamental Theorem of Calculus (FTC)
4(a)	Chapter 16: Differential Equations
4(b)	Chapter 13: The Slope of a Curve at a Point
4(c)	Chapter 22: Separable Differential Equations
5(a)	Chapter 16: Related Rates
5(b)	Chapter 16: Related Rates
5(c)	Chapter 16: Related Rates Chapter 20: The Fundamental Theorem of Calculus (FTC)
6(a)	Chapter 13: Approximating Derivatives Using Tables and Graphs
6(b)	Chapter 18: Average Value of a Function
6(c)	Chapter 23: Approximating the Area Under the Curve Using Riemann Sums
6(d)	Chapter 13: Approximating Derivatives Using Tables and Graphs Chapter 15: Second Derivatives

Calculus AB Practice Test 3
Answer Grid

1.	Ⓐ Ⓑ Ⓒ Ⓓ Ⓔ	13.	Ⓐ Ⓑ Ⓒ Ⓓ Ⓔ	25.	Ⓐ Ⓑ Ⓒ Ⓓ Ⓔ	37.	Ⓐ Ⓑ Ⓒ Ⓓ Ⓔ
2.	Ⓐ Ⓑ Ⓒ Ⓓ Ⓔ	14.	Ⓐ Ⓑ Ⓒ Ⓓ Ⓔ	26.	Ⓐ Ⓑ Ⓒ Ⓓ Ⓔ	38.	Ⓐ Ⓑ Ⓒ Ⓓ Ⓔ
3.	Ⓐ Ⓑ Ⓒ Ⓓ Ⓔ	15.	Ⓐ Ⓑ Ⓒ Ⓓ Ⓔ	27.	Ⓐ Ⓑ Ⓒ Ⓓ Ⓔ	39.	Ⓐ Ⓑ Ⓒ Ⓓ Ⓔ
4.	Ⓐ Ⓑ Ⓒ Ⓓ Ⓔ	16.	Ⓐ Ⓑ Ⓒ Ⓓ Ⓔ	28.	Ⓐ Ⓑ Ⓒ Ⓓ Ⓔ	40.	Ⓐ Ⓑ Ⓒ Ⓓ Ⓔ
5.	Ⓐ Ⓑ Ⓒ Ⓓ Ⓔ	17.	Ⓐ Ⓑ Ⓒ Ⓓ Ⓔ	29.	Ⓐ Ⓑ Ⓒ Ⓓ Ⓔ	41.	Ⓐ Ⓑ Ⓒ Ⓓ Ⓔ
6.	Ⓐ Ⓑ Ⓒ Ⓓ Ⓔ	18.	Ⓐ Ⓑ Ⓒ Ⓓ Ⓔ	30.	Ⓐ Ⓑ Ⓒ Ⓓ Ⓔ	42.	Ⓐ Ⓑ Ⓒ Ⓓ Ⓔ
7.	Ⓐ Ⓑ Ⓒ Ⓓ Ⓔ	19.	Ⓐ Ⓑ Ⓒ Ⓓ Ⓔ	31.	Ⓐ Ⓑ Ⓒ Ⓓ Ⓔ	43.	Ⓐ Ⓑ Ⓒ Ⓓ Ⓔ
8.	Ⓐ Ⓑ Ⓒ Ⓓ Ⓔ	20.	Ⓐ Ⓑ Ⓒ Ⓓ Ⓔ	32.	Ⓐ Ⓑ Ⓒ Ⓓ Ⓔ	44.	Ⓐ Ⓑ Ⓒ Ⓓ Ⓔ
9.	Ⓐ Ⓑ Ⓒ Ⓓ Ⓔ	21.	Ⓐ Ⓑ Ⓒ Ⓓ Ⓔ	33.	Ⓐ Ⓑ Ⓒ Ⓓ Ⓔ	45.	Ⓐ Ⓑ Ⓒ Ⓓ Ⓔ
10.	Ⓐ Ⓑ Ⓒ Ⓓ Ⓔ	22.	Ⓐ Ⓑ Ⓒ Ⓓ Ⓔ	34.	Ⓐ Ⓑ Ⓒ Ⓓ Ⓔ		
11.	Ⓐ Ⓑ Ⓒ Ⓓ Ⓔ	23.	Ⓐ Ⓑ Ⓒ Ⓓ Ⓔ	35.	Ⓐ Ⓑ Ⓒ Ⓓ Ⓔ		
12.	Ⓐ Ⓑ Ⓒ Ⓓ Ⓔ	24.	Ⓐ Ⓑ Ⓒ Ⓓ Ⓔ	36.	Ⓐ Ⓑ Ⓒ Ⓓ Ⓔ		

AP CALCULUS AB PRACTICE TEST 3

SECTION I, PART A

Time 55 Minutes
28 Questions

NO GRAPHING CALCULATOR IS ALLOWED ON THIS PORTION OF THE EXAM

Directions: Solve the following problems, using available space for scratchwork. After examining the form of the choices, decide which one is the best of the choices given and fill in the corresponding oval on the answer sheet. No credit will be given for anything written in the test book. Do not spend too much time on any one problem.

In this test:

(1) The domain of a function f is the set of all real numbers x for which $f(x)$ is a real number, unless otherwise specified.

(2) The inverse of a trigonometric function f may be indicated using the inverse function notation f^{-1} or with the prefix "arc" (e.g., $\sin^{-1} x = \arcsin x$).

1. Suppose $f(3) = 5$ and $f'(x) = \dfrac{x^2 - 3}{x}$. Using a tangent line approximation, the value of $f(3.1)$ is best approximated by

(A) 4.8
(B) 4.9
(C) 5
(D) 5.1
(E) 5.2

GO ON TO THE NEXT PAGE

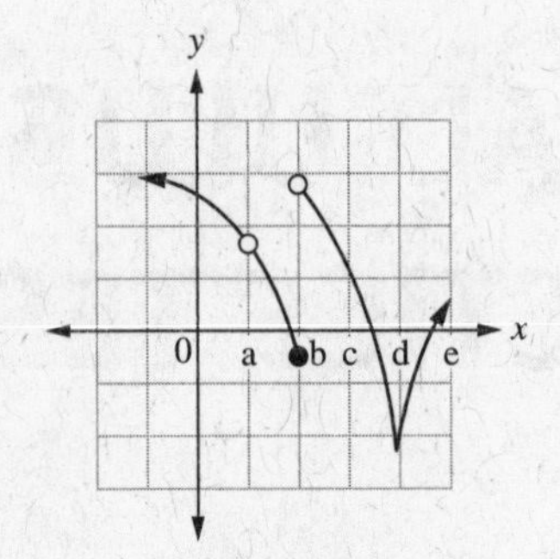

2. The graph of the function f is given above. Which of the following statements is false?

(A) $\lim_{x \to e} f(x)$ exists

(B) $\lim_{x \to b} f(x)$ exists

(C) $\lim_{x \to c} f(x)$ exists

(D) $\lim_{x \to d} f(x)$ exists

(E) $\lim_{x \to e} f(x)$ exists

3. What is the absolute minimum value of the function f defined as $f(x) = x \ln x$?

(A) $-\frac{1}{e}$

(B) 0

(C) $\frac{1}{e}$

(D) 1

(E) e

4. Let f and g be functions such that $f(2) = 4, f'(2) = 6, g(2) = 2$, and $g'(2) = -3$. If $h(x) = f(g(x))$, then $h'2) =$

(A) -18

(B) -3

(C) 0

(D) 6

(E) 12

5. For $y = x \cos x$, the derivative is $y' =$

(A) $x(\cos x - \sin x)$

(B) $\cos x - x \sin x$

(C) $\cos x + x \sin x$

(D) $\cos x - \sin x$

(E) $x(\sin x - \cos x)$

6. Let f be a function whose derivative is given by $f'(x) = \dfrac{6x}{(x^2+3)^2}$. On which of the following intervals of x is the graph of f concave upward?

(A) $x < 0$

(B) $x > 0$

(C) $-1 < x < 1$

(D) $x < -1$ and $x > 1$

(E) f is never concave upward.

7. If f is continuous at $x = 4$, which of the following must be true?

I. $\lim_{x \to 4} f(x)$ exists.

II. $f(4)$ exists.

III. $f'(4)$ exists.

(A) I only

(B) II only

(C) III only

(D) I and II only

(E) I, II, and III

GO ON TO THE NEXT PAGE

x	0	4	6	9
$h(x)$	−1	4	1	5

8. The continuous function h has values as given in the table above. The approximate value of $\int_0^9 h(x)\,dx$ using trapezoids with the three sub-intervals indicated by the data in the table is

(A) 7
(B) 17
(C) 20
(D) 28
(E) 33

9. $\lim\limits_{h\to 0}\dfrac{\sqrt[3]{1+h}-\sqrt[3]{1}}{h}$ gives

(A) 0
(B) $\frac{1}{3}$
(C) $\frac{2}{3}$
(D) 1
(E) Nonexistent

10. At any time t, the velocity of a particle moving along the y-axis is given by $4t - t^2$. The total distance traveled by the particle from time $t = 0$ to $t = 3$ is

(A) 3
(B) 9
(C) 15
(D) 18
(E) 21

11. Using the substitution $u = x^2$, which of the following integrals is equivalent to $\int_0^2 xe^{x^2}\,dx$?

(A) $\frac{1}{2}\int_0^2 e^u\,du$
(B) $\int_0^2 e^u\,du$
(C) $\frac{1}{2}\int_0^2 ue^u\,du$
(D) $\int_0^4 e^u\,du$
(E) $\frac{1}{2}\int_0^4 e^u\,du$

12. $\int \sec^2 x\,dx =$

(A) $\tan x + C$
(B) $\sec x + C$
(C) $\dfrac{\sec^3 x}{3} + C$
(D) $\sec x \tan x + C$
(E) $\dfrac{\sec^3 x}{3} + C$

13. A particle travels along the x-axis so that its velocity is given by $v(t) = 2\cos t$ for $t \geq 0$. The position of the particle at $t = \frac{\pi}{2}$ is 5. The position of the particle at any time t is given by

(A) $x(t) = 2\sin t + 5$
(B) $x(t) = -2\sin t + 7$
(C) $x(t) = 2\sin t + 3$
(D) $x(t) = -2\cos t + 5$
(E) $x(t) = -2\cos t + 7$

GO ON TO THE NEXT PAGE

14. A curve given by the equation $x^3 + xy = 8$ has slope given by $\frac{dy}{dx} = \frac{-3x^2 - y}{x}$. The value of $\frac{d^2y}{dx^2}$ at the point where $x = 2$ is

    (A) -6
    (B) -3
    (C) 0
    (D) 4
    (E) undefined

15. Consider the function given by $f(x) = 27x - x^3$. The function f is decreasing on the interval(s)

    (A) $[-3, 3]$ only
    (B) $[0, 3]$ only
    (C) $[0, \infty)$ only
    (D) $\left[-3\sqrt{3}, 3\sqrt{3}\right]$ only
    (E) $(-\infty, -3]$ and $[3, \infty)$

16. The function f has the property that $f(x)$, $f'(x)$, and $f''(x)$ are positive for all $x > 0$. Which of the following could be the graph of f?

    (A)

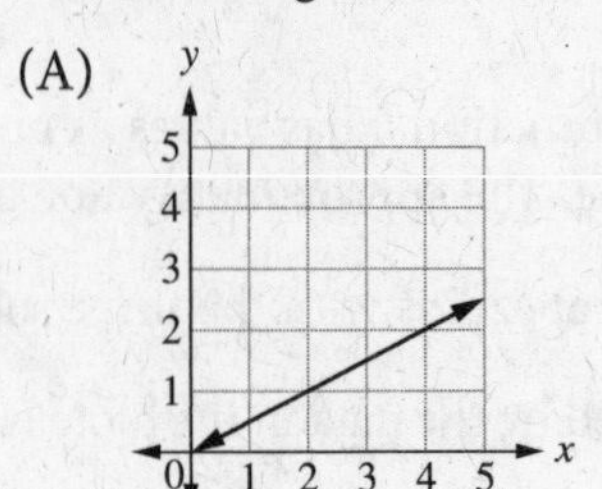

    (B)

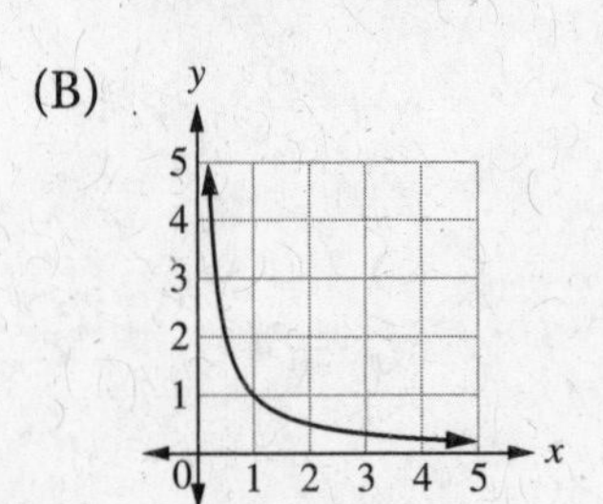

    (C)

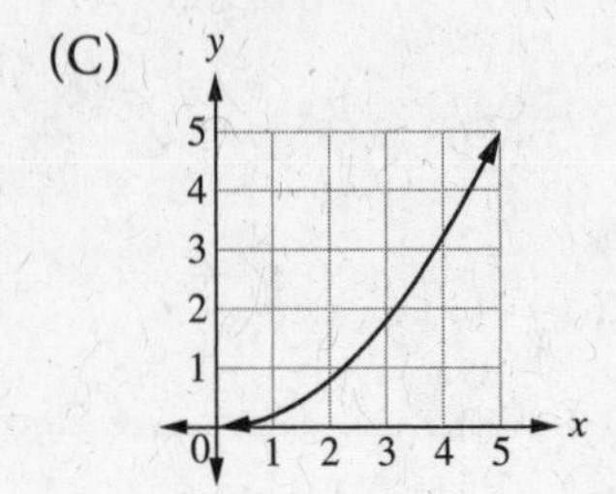

    (D)

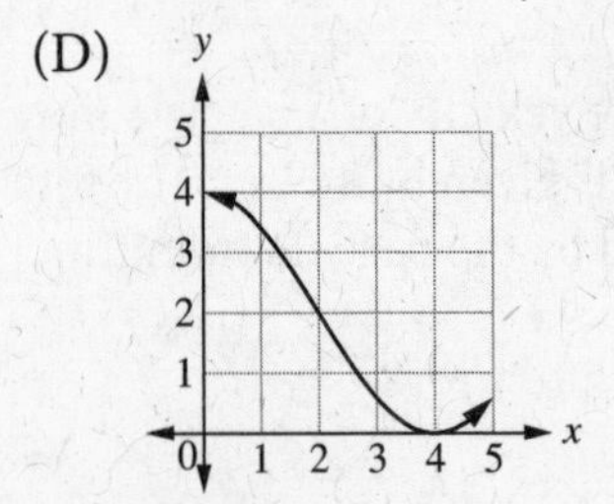

    (E)

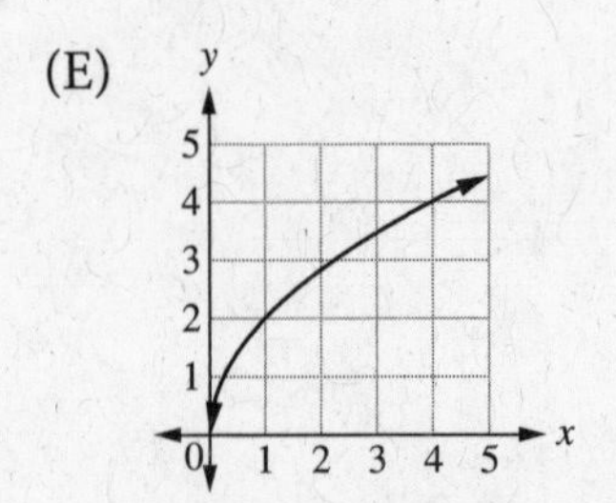

GO ON TO THE NEXT PAGE

17. 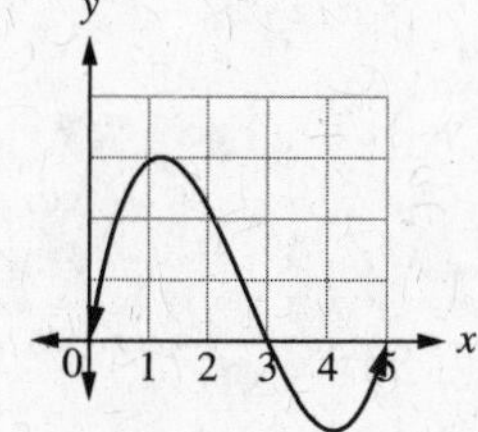

The graph of g is shown above. The area of the region between g and the x-axis on the interval [0, 3] is 9. The area of the region between g and the x-axis on the interval [3, 5] is 2. The value of $\int_0^5 g(x)dx$ is

(A) 5
(B) 7
(C) 9
(D) 11
(E) 18

18. Consider the function f given by $f(x)=\dfrac{x}{x+3}$. The line tangent to f is parallel to the line $x - 3y = 3$ at which of the following locations?

(A) $x = 0$ only
(B) $x = 3$ only
(C) $x = -2$ only
(D) $x = 0$ and $x = -6$
(E) $x = -4$ and $x = -2$

19. $\dfrac{d}{dx}\int_1^{x^3}\ln(5+t^2)dt=$

(A) $\text{In}(5 + x^6)$
(B) $\text{In}(5 + x^2)$
(C) $3x^2\,\text{In}(5 + x^6)$
(D) $\dfrac{1}{5+x^6}$
(E) $\dfrac{3x^2}{5+x^6}$

20. Which of the following functions has the line $y = 3$ as a horizontal asymptote?

I. $y=\dfrac{\sqrt{9x^2+1}}{x-2}$

II. $y=\dfrac{3x}{1-x}$

III. $y=\dfrac{9x^2-2}{2+3x^2}$

(A) I only
(B) III only
(C) I and II only
(D) I and III only
(E) I, II, and III

21. Given that $f(x) = 5x^2 - 4 + \ln x$, the value of $f'(1)$ is

(A) 7
(B) 8
(C) 9
(D) 10
(E) 11

22. The position of a particle moving along the x-axis is given by $x(t) = e^{2t} - e^t$ for all $t \geq 0$. When the particle is at rest, the acceleration of the particle is

(A) $\dfrac{1}{2}$
(B) $\dfrac{1}{4}$
(C) $\ln\dfrac{1}{2}$
(D) 2
(E) 4

GO ON TO THE NEXT PAGE

23. Let g be the function defined as

$$g(x)=\begin{cases}\dfrac{x^2-9}{x-3} & \text{if } x \neq 3\\ k & \text{if } x=3\end{cases}$$

If g is continuous for all x, then the value of k is

(A) 0
(B) 2
(C) 3
(D) 4
(E) 6

24. The position of a particle at any time t as it moves on the y-axis is given by $y(t) = t^2 - jt + k$, where j and k are nonzero constants. At what time(s) t is the particle at rest?

(A) $t = 0$ only
(B) $t = 2j$ only
(C) $t = j - k$ olny
(D) $t=\dfrac{j}{2}$ only
(E) $t = 0$ and $t = j - k$ only

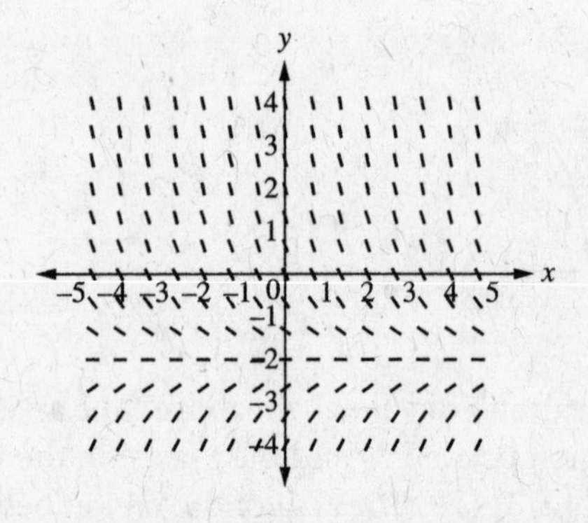

25. Shown above is a slope field for which of the following differential equations?

(A) $\dfrac{dy}{dx}=x+2$

(B) $\dfrac{dy}{dx}=y+2$

(C) $\dfrac{dy}{dx}=-y-2$

(D) $\dfrac{dy}{dx}=\dfrac{y+2}{x}$

(E) $\dfrac{dy}{dx}=\dfrac{y+2}{x^2}$

26. Let a function f be defined as $f(x) = x^3 - 2x - 4$ for $x \geq 1$. Let $g(x)$ be the inverse function of $f(x)$ and note that $f(2) = 0$. The value of $g'(0) =$

(A) $\dfrac{1}{10}$
(B) 1
(C) 4
(D) −2
(E) $-\dfrac{1}{2}$

GO ON TO THE NEXT PAGE

27. The area of the region in the first quadrant bounded by the graph of $f(x)=\dfrac{\ln x}{x}$ and the lines $x = 1$ and $x = e$ is

    (A) $\dfrac{1}{3}$

    (B) $\dfrac{1}{2}$

    (C) 1

    (D) e

    (E) 3

28. Let $y = (x^5 + \sin x)^4$. The derivative is $\dfrac{dy}{dx} =$

    (A) $4(5x^4 + \cos x)^3$

    (B) $4(x^5 + \sin x)^3 (5x^4 - \cos x)$

    (C) $4(x^5 + \sin x)^3$

    (D) $4(x^5 + \sin x)^3 (5x^4 + \text{cox } x)$

    (E) $4(5x^4 + \cos x)^3$

IF YOU FINISH BEFORE TIME IS CALLED, YOU MAY CHECK YOUR WORK ON THIS SECTION ONLY. DO NOT TURN TO ANY OTHER SECTION IN THE TEST. **STOP**

SECTION I, PART B

Time: 50 Minutes
17 Questions

A GRAPHING CALCULATOR IS ALLOWED ON THIS PORTION OF THE EXAM

Directions: Solve the following problems, using available space for scratchwork. After examining the form of the choices, decide which one is the best of the choices given and fill in the corresponding oval on the answer sheet. No credit will be given for anything written in the test book. Do not spend too much time on any one problem.

In this test:

(1) The exact numerical value of the correct answer does not always appear among the choices given. When this happens, select from among the choices the number that best approximates the numerical value.

(2) The domain of a function f is the set of all real numbers x for which $f(x)$ is a real number, unless otherwise specified.

(3) The inverse of a trigonometric function f may be indicated using the inverse function notation f^{-1} or with the prefix "arc" (e.g., $\sin^{-1} x = \arcsin x$).

29. Suppose R is the region in the first quadrant enclosed by the graphs of $y = x^3$ and $y = \sqrt{x}$. What is the volume of the solid of revolution that is generated when R is revolved about the y-axis?

(A) 0.357
(B) 0.400
(C) 1.122
(D) 1.257
(E) 1.309

30. Suppose a function f is continuous for all x. If f has a relative minimum at $(-2, -3)$ and a relative maximum at $(4, 5)$, which of the following statements must be true?

(A) The graph of f has an absolute maximum at $(4, 5)$.
(B) The graph of f is increasing for all x in $-2 \le x \le 4$.
(C) There exists some c in the interval $-2 \le x \le 4$ such that $f(c) = 0$.
(D) There exists some c in the interval $-2 \le x \le 4$ such that $f'(c) = 0$.
(E) There exists some c in the interval $-2 \le x \le 4$ such that at $x = c$ the sign of $f''(X)$ changes from positive to negative.

GO ON TO THE NEXT PAGE

31. Let f be a continuous and differentiable function for all real numbers. The only critical points for f are located at $x = -3$ and $x = 5$. If $f''(x) = 2x - 2$, which of the following must necessarily be true?

I. f has a relative maximum at $x = -3$.
II. f has a relative maximum at $x = 5$.
III. f has a point of inflection at $x = 1$.

(A) I only
(B) II only
(C) III only
(D) I and III only
(E) II and III only

x	0	3	6	9	12	15	18
$g(x)$	−4	−2	3	4	9	5	1

32. Suppose $g(x)$ is a continuous function. A table of selected values of $g(x)$ is shown above. The approximate value of $\int_0^{18} g(x)dx$ using a midpoint Riemann sum with three subintervals of equal length is

(A) 48
(B) 42
(C) 39
(D) 24
(E) 21

33. The growth of a population P is modeled by the differential equation $\frac{dP}{dt} = 0.713P$. If the population is 2 at $t = 0$, then the population at $t = 5$ is

(A) 3.565
(B) 5.565
(C) 20.810
(D) 50.810
(E) 70.679

34. Suppose $f'(x) = \frac{-x^3 + 4x - 2}{\sqrt{x^2 + 1}}$. Which of the following statements must be true?

I. f has a relative minimum at $x = -0.768$.
II. f has a relative minimum at $x = 0.539$.
III. f has a relative minimum at $x = 1.675$.

(A) I only
(B) II only
(C) III only
(D) I and II only
(E) I and III only

GO ON TO THE NEXT PAGE

35. Let $f(x) = x^3 + 2x - 5$. What is the x-coordinate of a point where the instantaneous rate of change of f is the same as the average rate of change of f on the interval $-1 < x < 1$?

(A) $\frac{\sqrt{3}}{3}$

(B) $\frac{1}{2}$

(C) 0

(D) $\frac{1}{3}$

(E) $\sqrt{3}$

36. The rate of fuel consumption $r(t)$, in gallons per hour, for a truck is given by $r(t) = \sin t - t \cos t + 3$, where t is time measured in hours for $0 \le t \le 5$. How many gallons of fuel does the truck use from $t = 0$ hours to $t = 5$ hours?

(A) 2.377

(B) 15.713

(C) 19.333

(D) 21.227

(E) 24.333

37. The radius of a circle is increasing at a rate of $\frac{5}{2}$ centimeters per minute. At the instant when the area of the circle is 16π square centimeters, what is the rate of increase in the area of the circle, in square centimeters per minute?

(A) 20π

(B) 16π

(C) 8π

(D) $20\pi^2$

(E) 4

38. What is the slope of the curve $y = 3^{\sin x} - 2$ at its first positive x-intercept?

(A) 0.683

(B) 1.643

(C) 1.705

(D) 1.805

(E) 2

39. A particle moves along the y-axis in such a way that its position at time t is given by $s(t)$ and its velocity at time t is given by $v(t)$. Which of the following calculates the average velocity of the particle from $t = 0$ to $t = 24$?

I. $\frac{1}{24}\int_0^{24} v(t)\,dt$

II. $\frac{v(0)+v(24)}{2}$

III. $\frac{s(24)-s(0)}{24}$

(A) I only

(B) II only

(C) III only

(D) I and II only

(E) I and III only

GO ON TO THE NEXT PAGE

40. A company sells graphing calculators at a rate that is modeled by $S(t) = 97 + 63\ \text{In}(t^2 + 1)$, where $S(t)$ is measured in thousands of calculators sold per year and t is in years from the beginning of 2010. To the nearest calculator, how many calculators were sold during the four-year period from January 1, 2010, to January 1, 2014?

(A) 97,000
(B) 275,492
(C) 505,568
(D) 765,023
(E) 953,618

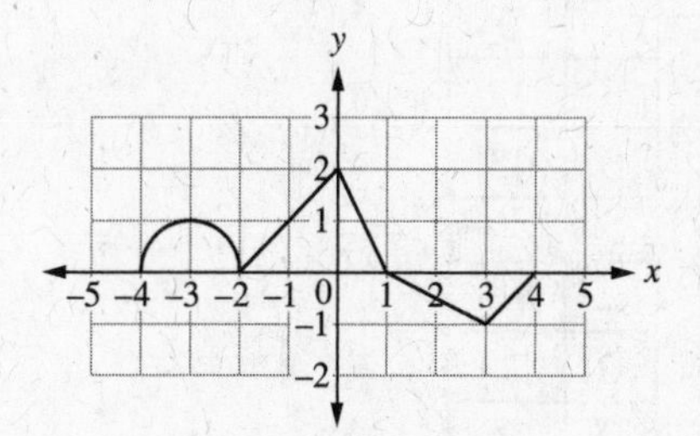

41. Let f be a function defined on $-4 < x < 4$ and consisting of a semi-circle and four line segments, as shown above. Suppose g is the function defined by $g(x) = \int_0^x f(t)dt$. The function g is decreasing for

(A) $1 \leq x < 4$ only
(B) $1 \leq x \leq 3$ only
(C) $0 \leq x \leq 3$ only
(D) $-4 < x \leq 0$ only
(E) $-3 \leq x \leq -2$ and $0 \leq x \leq 3$ only

42. A stream flows into a lake at a rate of $R(t) = \sin\frac{\pi t}{12} + 40t$ gallons per hour. A pump is used to remove water from the lake at a rate of 500 gallons per hour. At the beginning of a certain day, the lake contains 30,000 gallons of water. How much water is in the lake after 24 hours?

(A) 11,020
(B) 11,520
(C) 29,520
(D) 41,020
(E) 41,520

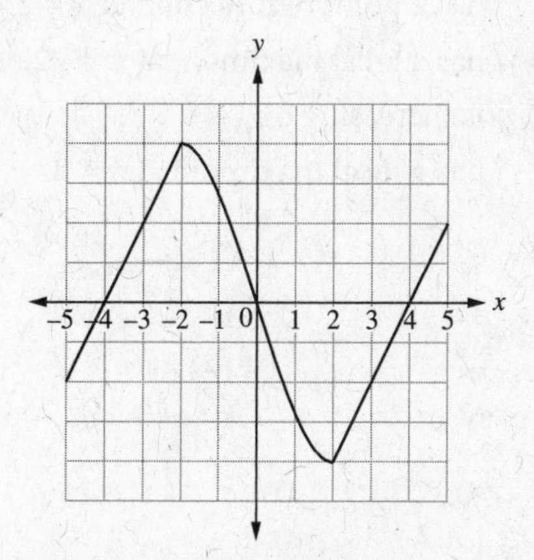

43. If the graph shown above represents $f'(x)$, the first derivative of a function $y = f(x)$, then the function f is concave down for

(A) $-2 < x < 0$
(B) $-2 < x < 2$
(C) $x < -4$ and $0 < x < 4$
(D) $-4 < x < 0$ and $x > 4$
(E) No intervals of x

GO ON TO THE NEXT PAGE

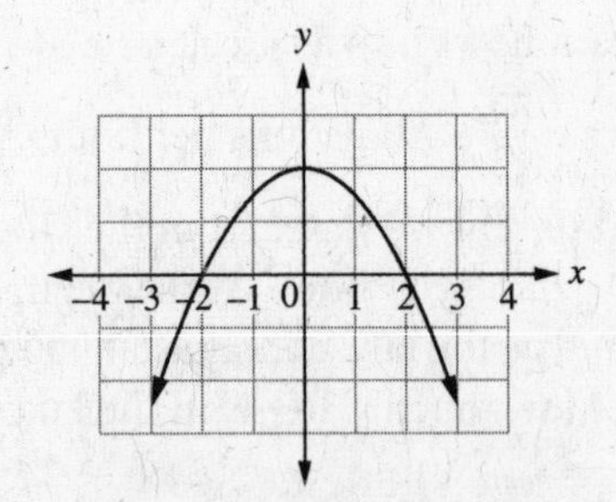

NOTE: Graph is of $f'(x)$

44. The graph of the derivative of $f(x)$ is shown above. Which of the following statements is true?

(A) f has a local maximum at $x = 0$.
(B) f has a point of inflection at $x = 2$.
(C) f has a local maximum at $x = -2$.
(D) f is increasing on $(-\infty, 0)$.
(E) f has a local maximum at $x = 2$.

45. Suppose f is a function such that $f(x) > 0$ and $f'(x) > 0$ but $f''(x) < 0$. Which of the following could represent a table of select values for f?

I.

x	f(x)
0	2
2	5
4	9
6	14
8	20

II.

x	f(x)
0	2
2	5
4	9
6	12
8	13

III.

x	f(x)
0	1
2	8
4	13
6	16
8	18

(A) I only
(B) II only
(C) III only
(D) II and III only
(E) None of the above

IF YOU FINISH BEFORE TIME IS CALLED, YOU MAY CHECK YOUR WORK ON THIS SECTION ONLY. DO NOT TURN TO ANY OTHER SECTION IN THE TEST. **STOP**

SECTION II

Time: 1 hour and 30 minutes
6 Problems
50 percent of total grade

Directions: Solve the following problems, using available space for scratchwork. Show how you arrived at your answer.

You may wish to look over the problems before starting to work on them; on the actual test, it is not expected that everyone will be able to complete all parts of all problems. All problems are given equal weight, but the individual parts of a particular problem are not necessarily given equal weight. You should not spend too much time on any one problem.

- You should write out all your work for each part. On the actual test, you will do this in the space provided in the test booklet. Be sure to write clearly and legibly. If you make a mistake, you can save time by crossing it out rather than trying to erase it. Erased or crossed-out work will not be graded.
- Show all your work. Clearly label any functions, graphs, tables, or other objects that you use. On the actual exam, you will be graded on the correctness and completeness of your methods as well as your answers. Answers without any supporting work may not receive credit.
- Justifications (i.e., the request that you "justify your answer") require that you give mathematical (non-calculator) reasons.
- Work must be expressed in standard mathematical notation, not calculator syntax.
- Unless otherwise specified, answers (numeric or algebraic) need not be simplified.
- If you use decimal approximations in calculations, the readers of the actual exam will grade you on accuracy. Unless otherwise specified, your final answers should be accurate to three places after the decimal point.
- Unless otherwise specified, the domain of function f is the set of all real numbers x for which $f(x)$ is a real number.

GO ON TO THE NEXT PAGE

PART A

Time: 30 Minutes

2 Problems

A GRAPHING CALCULATOR IS ALLOWED ON THIS PORTION OF THE EXAM

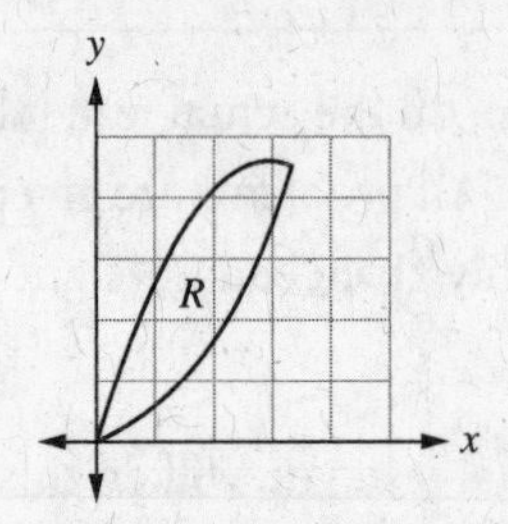

1. Let R be the region in the first quadrant bounded by the graphs of $f(x) = 5x - x^2$ and $g(x) = 2^x - 1$ as shown in the figure above.
   (a) Find the area of region R.
   (b) Let k be the number such that the vertical line $x = k$ divides the area of region R exactly in half, forming two separate regions of equal area. Write, but do not solve, an equation involving an integral that can be used to find the value of k.
   (c) Find the volume of the solid generated when R is rotated about the x-axis.
   (d) Suppose region R is the base of a solid whose cross sections cut by a plane perpendicular to the x-axis are squares. Write, but do not evaluate, an integral expression that gives the volume of the solid.

2. A particle moves along the y-axis so that its velocity v at time t is given by $v(t) = t \sin t$, where $t \geq; 0$. At time $t = 0$, the particle is at position $y = 2$.
   (a) Find the velocity and the acceleration of the particle at time $t = 4$.
   (b) Is the speed of the particle increasing or decreasing at time $t = 4$? Explain the reason for your answer.
   (c) Find the total distance traveled by the particle from time $t = 0$ to time $t = 5$.
   (d) Find the average velocity of the particle for the time interval from time $t = 0$ to time $t = 5$.
   (e) Find the position of the particle at time $t = 5$ Show the work that leads to your answer.

GO ON TO THE NEXT PAGE

PART B

Time: 60 Minutes
4 Problems

NO GRAPHING CALCULATOR IS ALLOWED ON THIS PORTION OF THE EXAM

Note: If you have extra time, you can go back and work on Part A of Section II, but you cannot use a calculator to complete your work at this time.

3. A 13-foot ladder that is leaning against the wall of a building is pulled away from the wall. At the instant when the base of the ladder is 5 feet from the wall, the base is moving at a rate of 3 feet per second.
   (a) At what rate is the top of the ladder sliding down the wall at the instant when the base of the ladder is 5 feet from the wall? Indicate units of measure.
   (b) Let θ be the angle between the ladder and the ground. At what rate is this angle changing at the instant when the base of the ladder is 5 feet from the wall? Indicate units of measure.
   (c) The wall of the building, the ladder, and the ground form a right triangle. At what rate is the area of the triangle changing at the instant when the base of the ladder is 5 feet from the wall? Indicate units of measure.

x	0	2	3	7	9
$f(x)$	−10	−2	0	6	8

4. Let f be a twice differentiable and strictly increasing function for all real numbers. Values of f at selected values of x are shown in the table above.
   (a) Approximate $f'(5)$. Show the work that supports your answer.
   (b) Suppose $f'(9) = \frac{4}{5}$. Write an equation of the line tangent to the graph of f at the point where $x = 9$. Use the tangent line to approximate $f(9.1)$.
   (c) Determine the value of $\int_0^9 2f'(x)\,dx$. Show the work that supports your answer.
   (d) Approximate $\int_0^9 f(x)\,dx$ with a Riemann sum, using the right endpoints of the four subintervals indicated by the data in the table. Is this numerical approximation greater than the value of $\int_0^9 f(x)\,dx$? Explain your reasoning.
   (e) Is there some x in $0 < x < 9$ such that $f'(x) = 2$? Justify your answer.

GO ON TO THE NEXT PAGE

5. Consider the curve given by $x^3 + xy = 16$.

(a) Show that $\frac{dy}{dx} = -\frac{3x^2 + y}{x}$.

(b) Find the coordinates of any point on the curve where the tangent line is vertical.

(c) Find the value of $\frac{d^2y}{dx^2}$ at the point $(-2, -12)$.

(d) Does the curve have a relative minimum, a relative maximum, or neither at $x = -2$? Justify your answer using $\frac{dy}{dx}$ and $\frac{d^2y}{dx^2}$.

6. The rate of change of the population $P(t)$ of a herd of deer is given by $\frac{dP}{dt} = \frac{1}{4}(1200 - P)$, where t is measured in years. When $t = 0$, the population P is 200.

(a) Write an equation of the line tangent to the graph of P at $t = 0$. Use the tangent line to P in order to approximate the population of the herd after 2 years.

(b) Show that $\frac{d^2P}{dt^2} = -\frac{1}{16}(1200 - P)$.

(c) Does the approximation for the population of the herd as found in part (a) overestimate or underestimate the actual population of the herd at time $t = 2$? Support your answer using $\frac{d^2P}{dt^2}$.

(d) Using separation of variables, find the particular solution $y = P(t)$ to the differential equation $\frac{dP}{dt} = \frac{1}{4}(1200 - P)$ with the initial population $P(0) = 200$.

IF YOU FINISH BEFORE TIME IS CALLED, YOU MAY CHECK YOUR WORK ON THIS SECTION ONLY. DO NOT TURN TO ANY OTHER SECTION IN THE TEST. **STOP**

AP Calculus AB Practice Test 3 **Answer Key**

1. E
2. B
3. A
4. A
5. B
6. C
7. D
8. C
9. B
10. B
11. E
12. A
13. C
14. C
15. E
16. C
17. B
18. D
19. C
20. D
21. E
22. A
23. E
24. D
25. C
26. A
27. B
28. D
29. D
30. C
31. D
32. B
33. E
34. B
35. A
36. D
37. A
38. C
39. E
40. D
41. A
42. C
43. B
44. E
45. C

ANSWERS AND EXPLANATIONS

1. E

Since $f(3)=5$, we know the coordinates of the point of tangency. To write the equation of the tangent line, we need the value of the slope at $x=3$. From $f'(x)=\frac{x^2-3}{x}$, we compute that $f'(3)=\frac{3^2-3}{3}=2$. Thus, the equation of the tangent line is $y-5=2(x-3)$ or $y=2(x-3)+5$. When $x=3.1$, $y=2(3.1-3)+5=2(0.1)+5=5.2$. This means $f(3.1)\approx 5.2$.

2. B

$\lim_{x\to b} f(x)$ does not exist since $\lim_{x\to b^-} f(x) \neq \lim_{x\to b^+} f(x)$. Thus, statement (B) is false.

3. A

To find the extrema of f, we will use $f'(x)=0$ to find the critical points. For $f(x)=x \ln x$, $f'(x)=x\left(\frac{1}{x}\right)+\ln x=1+\ln x=0$. Solving for x gives $\ln x=-1$ or $x=e^{-1}=\frac{1}{e}$. Checking the sign of $f'(x)$ shows that $f'(x)<0$ for $x<\frac{1}{e}$ and $f'(x)>0$ for $x>\frac{1}{e}$, so there is an absolute minimum at $x=\frac{1}{e}$. The absolute minimum is $f\left(\frac{1}{e}\right)=\frac{1}{e}\ln\left(\frac{1}{e}\right)=\frac{1}{e}\left(\ln e^{-1}\right)=-\frac{1}{e}$.

4. A

Differentiating $h(x)=f(g(x))$ using the chain rule gives $h'(x)=f'(g(x))\cdot g'(x)$. Evaluating at $x=2$, we get

$$\begin{aligned} h'(2) &= f'(g(2))\cdot g'(2) \\ &= f'(2)\cdot g'(2)=6\cdot -3=-18 \end{aligned}$$

5. B

Using the product rule on $y=x\cos x$, the derivative is

$$\begin{aligned} y' &= x(-\sin x)+\cos x \\ &= -x\sin x+\cos x \\ &= \cos x - x\sin x \end{aligned}$$

6. C

For $f'(x) = \frac{6x}{(x^2+3)^2}$, we use the quotient rule to get.

$$f''(x) = \frac{(x^2+3)^2(6) - 6x(2)(x^2+3)(2x)}{(x^2+3)^4}$$
$$= \frac{(x^2+3)[6x^2+18-12x(2x)]}{(x^2+3)^4}$$
$$= \frac{6x^2+18-24x^2}{(x^2+3)^3}$$
$$= \frac{-18x^2+18}{(x^2+3)^3} = \frac{-18(x^2-1)}{(x^2+3)^3}$$

To find the possible points of inflection, we consider $f''(x) = 0$. This means $x^2 - 1 = 0$, which happens when $x^2 = 1$ or $x^2 = \pm 1$.

We will check the sign of f'' on each side of $x = \pm 1$ to determine when $f'' > 0$.

$x < -1$	$-1 < x < 1$	$x > 1$
−	+	−

From the sign chart above, f is concave upward on $-1 < x < 1$ since $f'' > 0$ on $-1 < x < 1$.

7. D

By definition of continuity, if f is continuous at $x = 4$, then $\lim_{x \to 4} f(x) = f(4)$. Thus, statement I and statement II are both true because $\lim_{x \to 4} f(x)$ exists and $f(4)$ is defined due to the definition of continuity.

A function can be continuous without also being differentiable. Thus, even though we know that f is continuous, we do not have enough information to know if f is also differentiable. While statement III could be true, we do NOT know that it must be true. That makes the answer statements I and II only.

8. C

Since the data on the table is spaced unevenly, each trapezoid must be dealt with individually.

$$\int_0^9 h(x)\,dx \approx \frac{1}{2}(-1+4)(4) + \frac{1}{2}(4+1)(2) + \frac{1}{2}(1+5)(3)$$
$$= \frac{1}{2}(3)(4) + \frac{1}{2}(5)(2) + \frac{1}{2}(6)(3)$$
$$= (3)(2) + 5 + (3)(3) = 20$$

9. B

The definition of the derivative of f at $x = a$ is $f'(a) = \lim_{h\to 0}\frac{f(a+h)-f(a)}{h}$. This means that $\lim_{h\to 0}\frac{\sqrt[3]{1+h}-\sqrt[3]{1}}{h}$ gives $f'(1)$ for $f(x)=\sqrt[3]{x}$. If $f(x)=\sqrt[3]{x}$, then $f'(x)=\frac{1}{3}x^{-\frac{2}{3}}$ and $f'(1)=\frac{1}{3}(1)^{-\frac{2}{3}}=\frac{1}{3}$

Thus, $\lim_{h\to 0}\frac{\sqrt[3]{1+h}-\sqrt[3]{1}}{h}=\frac{1}{3}$

10. B

Since $v(t) = 4t - t^2 > 0$ for $0 < t < 3$, the total distance traveled by the particle from time $t = 0$ to $t = 3$ is given by

$$\int_0^3 \left(4t-t^2\right)dt = \left[\frac{4t^2}{2}-\frac{t^3}{3}\right]_0^3 = \left[2t^2-\frac{t^3}{3}\right]_0^3$$

$$= \left[2(3)^2-\frac{3^3}{3}-0\right] = 18-9=9.$$

11. E

The substitution $u = x^2$ gives $du = 2x\,dx$ or $\frac{1}{2}du = x\,dx$. When performing a u-substitution on the integral, it is important to remember also to convert the limits of integration from the given interval of x values to the corresponding value of u. If $x = 0$, then $u = 0$. If $x = 2$, then $u = 4$.

This allows the integral to be rewritten as follows $\int_0^2 xe^{x^2}\,dx = \frac{1}{2}\int_0^4 e^u\,du.$

12. A

$$\int \sec^2 x\,dx = \tan x + C$$

13. C

The position is $x(t) = \int v(t)\,dt = \int 2\cos t\,dt = 2\sin t + C$. The value of C can be found by using the initial condition $x\left(\frac{\pi}{2}\right)=5$. Substituting into the position equation, $x\left(\frac{\pi}{2}\right)=2\sin\left(\frac{\pi}{2}\right)+C=5$.

Thus, $2(1) + C = 5$ or $C = 3$. Now, the position is given by $x(t) = 2\sin t + 3$.

14. C

Substituting $x = 2$ into the original equation gives $2^3 + 2y = 8$ or $2y = 0$ so $y = 0$. To find $\frac{d^2y}{dx^2}$ from $\frac{dy}{dx}=\frac{-3x^2-y}{x}$, use the Quotient Rule. Using the quotient rule,

$$\frac{d^2y}{dx^2}=\frac{x\left(-6x-\frac{dy}{dx}\right)-\left(-3x^2-y\right)}{x^2}$$
$$=\frac{x\left(-6x-\frac{-3x^2-y}{x}\right)-\left(-3x^2-y\right)}{x^2}$$
$$=\frac{-6x^2+3x^2+y+3x^2+y}{x^2}$$

At the point (2, 0),

$$\frac{d^2y}{dx^2}=\frac{-6(2)^2+3(2)^2+0+3(2)^2+0}{(2)^2}=$$
$$\frac{-24+12+12}{4}=0$$

15. E

The function $f'(x)=27x-x^3$ gives $f'(x)=27-3x^2$. Using this, we can find the critical points of f.

$$f'(x)=27-3x^2=3(9-x^2)=C$$
$$x=\pm 3$$

For f to be decreasing, we need $f'<0$. This occurs on the intervals $[-\infty,-3]$ and $[3,\infty]$.

16. C

First, $f(x)>0$ implies that f is above the x-axis. All of the choices meet this condition.

Second, $f'(x)>0$ implies that f is increasing for all x. This eliminates choices (B) and (D) which are both decreasing.

Third, $f''(x)>0$ implies that f is concave up for all x. This eliminates choice (A) which is linear and does not have concavity and choice (E) which is concave down. Thus, the correct answer is choice (C) which is both increasing and concave up.

17. B

Integrals give a "net" area or signed area, so the area for the region on [0, 3] will be treated as positive area since it is above the x-axis and the region on [3, 5] will be treated as negative area since it is below the x-axis.

Thus, the value of $\int_0^5 g(x)\,dx=9-2=7$.

18. D

To find the slope of the line $x-3y=3$, we rewrite the line as $3y=x-3$ or $y=\frac{1}{3}x-1$.

Thus, the desired slope is $m=\frac{1}{3}$. To find the slope of *f*, we differentiate $f(x)=\frac{x}{x+3}$ use the quotient rule to get $f'(x)=\frac{(x+3)(1)-x(1)}{(x+3)^2}=\frac{3}{(x+3)^2}$. Setting $f'(x)=\frac{1}{3}$ gives $\frac{3}{(x+3)^2}=\frac{1}{3}$.

Solving for x gives $9=(x+3)^2$ which in turn gives $x+3=\pm 3$. This gives $x=-3\pm 3$ so $x=-6$ or $x=0$.

19. C

Using the Second Fundamental Theorem of Calculus,

$$\begin{aligned}\frac{d}{dx}\int_1^{x^3}\ln(5+t^2)\,dt &= \ln(5+(x^3)^2)\cdot(3x^2)\\ &= 3x^2\ln(5+x^6).\end{aligned}$$

20. D

Horizontal asymptotes occur if the function approaches a particular y value as $x\to\infty$ or $x\to-\infty$.

For choice I, $\lim\limits_{x\to\infty}\frac{\sqrt{9x^2+1}}{x-2}=\frac{3}{1}=3$, so $y=3$ is a horizontal asymptote.

For choice II, $\lim\limits_{x\to\infty}\frac{3x}{1-x}=\frac{3}{-1}=-3$ and $\lim\limits_{x\to-\infty}\frac{3x}{1-x}=\frac{-3}{1}=-3$, so $y=-3$ is a horizontal asymptote but not $y=3$.

For choice III, $\lim\limits_{x\to\infty}\frac{9x^2-2}{2+3x^2}=\frac{9}{3}=3$ so $y=3$ is a horizontal asymptote.

Thus, I and III have $y=3$ as horizontal asymptotes.

21. E

Given that $f(x)=5x^2-4+\ln x$, we find that $f'(x)=10x+\frac{1}{x}$. This gives that $f'(1)=10+1=11$.

22. A

Position is $x(t)=e^{2t}-e^t$. Thus, velocity is $v(t)=2e^{2t}-e^t$ and acceleration is $a(t)=4e^{2t}-e^t$.

The particle is at rest when $v(t)=2e^{2t}-e^t=0$. Factoring gives $e^t(2e^t-1)=0$ The first factor will never equal 0. For the second factor, $2e^t=1$. Solving for t gives $e^t=\frac{1}{2}$ or $t=\ln\frac{1}{2}$. The acceleration at this time is

$$\begin{aligned}a\left(\ln\frac{1}{2}\right) &= 4e^{2\ln\frac{1}{2}}-e^{\ln\frac{1}{2}} = 4e^{\ln\frac{1}{4}}-e^{\ln\frac{1}{2}}\\ &= 4\left(\frac{1}{4}\right)-\frac{1}{2}=1-\frac{1}{2}=\frac{1}{2}.\end{aligned}$$

23. E

$$\lim_{x\to 3} g(x) = \lim_{x\to 3}\frac{x^2-9}{x-3} = \lim_{x\to 3}(x+3) = 6$$

For g to be continuous, we need $g(3) = k = 6$.

24. D

If the position is $y(t) = t^2 - jt + k$, then velocity is $v(t) = 2t - j$. The particle is at rest when $v(t) = 0$.

Solving $v(t) = 2t - j = 0$ give $t = \frac{j}{2}$.

25. C

The sample tangent line pieces all have the same slope for a given value of *y*, so the slope only depends on *y*. Thus, the differential equation should only be in terms of the variable *y*. This eliminates all choices except (B) and (C).

The slope field has horizontal tangent lines at $y = -2$, so the differential equation must be one that has a value of 0 when $y = -2$. Unfortunately, that is true for both (B) and (C) so that doesn't help narrow down the choices.

Another characteristic to check is the sign of the slope. For choice (B), $\frac{dy}{dx} = y + 2 > 0$ when $y > -2$. This is not consistent with the slope field, so (B) must be eliminated. Thus, the answer is choice (C).

26. A

Since *f* and *g* are inverses of each other, we know that $g'(x) = \frac{1}{f'(g(x))} = \frac{1}{f'(y)}$.

The function $f(x) = x^3 - 2x - 4$ has derivative $f'(x) = 3x^2 - 2$.

Since (2, 0) is a point on *f*, we know that the corresponding point on *g* is (0, 2).

Thus,

$$g'(0) = \frac{1}{f(2)} = \frac{1}{3(2)^2 - 2} = \frac{1}{10}.$$

27. B

The area is $A = \int_1^e \frac{\ln x}{x}\,dx$. Using the substitution $u = \ln x$ and $du = \frac{1}{x}dx$, the integral becomes $A = \int_0^1 u\,du = \left[\frac{u^2}{2}\right]_0^1 = \frac{1}{2} - 0 = \frac{1}{2}$.

28. D

Given, $y = (x^5 + \sin x)^4$, the chain rule gives the derivative as $\frac{dy}{dx} = 4(x^5 + \sin x)^3(5x^4 + \cos x)$.

29. D

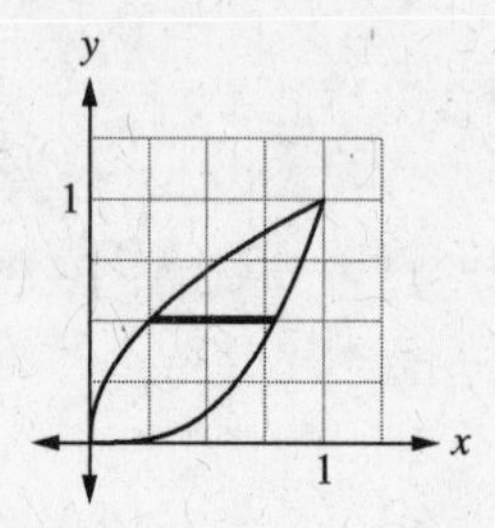

When dealing with solids of revolution, we always slice the region with rectangles that are perpendicular to the axis of revolution. In this case, that means using rectangles that are perpendicular to the *y*-axis, as shown above. The graphs intersect at the point (1,1), so the rectangles will range from [0, 1] along the *y*-axis.

This type of rectangle will generate washers with a thickness of Δy. For the washer method, the outer radius R and the inner radius r are both calculated as $x_{big} - x_{little}$, so we need to solve for the x value in each of the equations of the graphs.

From $y = x^3$, we get $x = \sqrt[3]{y}$. From $y = \sqrt{x}$, we get $x = y^2$.

Thus, $R = x_{big} - x_{little} = \sqrt[3]{y} - 0 = \sqrt[3]{y}$ and $r = x_{big} - x_{little} = y^2 - 0 = y^2$.

Using the washer method, the volume is

$$V = \pi \int_0^1 \left(R^2 - r^2\right) dy$$

$$= \pi \int_0^1 \left(\left(\sqrt[3]{y}\right)^2 - \left(y^2\right)^2\right) dy = 0.400\pi = 1.257.$$

30. C

Choice (A): The function *f* does not have to have an absolute extrema. That is only guaranteed if the function is continuous on a closed interval, so the statement does not have to be true.

Choice (B): The graph of *f* could be increasing for all x in $-2 \le x \le 4$. but it does not have to be. There could be other extrema and thus other turning points between $x = -2$ and $x = 4$.

Choice (C): Since 0 is an intermediate value between $f(-2)$ and $f(4)$, this statement must be true since the Intermediate Value Theorem guarantees that there exists a c in the interval $-2 \le x \le 4$ such that $f(c) = 0$.

Choice (D): The function f is continuous for all x, but that does not guarantee that the function is differentiable. Since f does not have to be differentiable, f does not have to be smooth. Thus, the graph of f could look like the graph below and there is not a point at which $f'(c) = 0$. If that is the case, the statement does not have to be true.

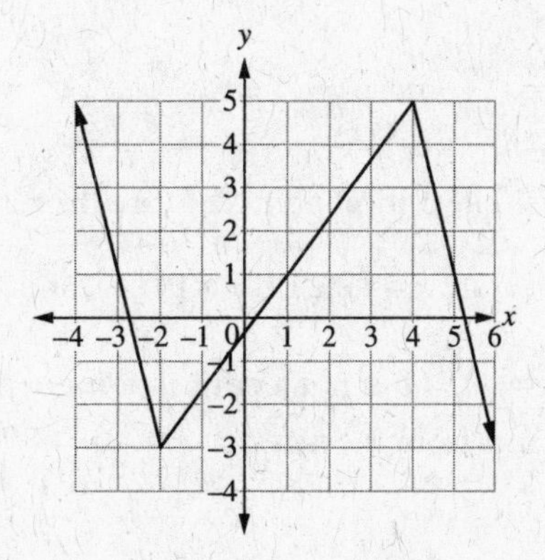

Choice (E): As shown in the graph above, it is possible that f could be piecewise linear and thus not have a change in concavity. If that is the case, then there would not be a sign change in f'' for the interval $-2 \leq x \leq 4$. Thus, although this statement could be true, it does not have to be true.

31. D

Extrema occur at critical points. Since f'' is provided, we will use the Second Derivative Test.

Using the Second Derivative Test on $x = -3$, $f''(-3) = 2(-3) - 2 = -8 < 0$ so $x = -3$ gives a relative maximum. Thus, statement I is true.

Using the Second Derivative Test on $x = 5$, $f''(5) = 2(5) - 2 = 8 > 0$ so $x = 5$ gives a relative minimum. This means statement II is false.

There is a possible inflection point at $x = 1$ since $f''(x) = 0$ when $x = 1$. The actual change in concavity can be verified by noting that $f'' < 0$ for $x < 1$ and $f'' > 0$ for $x > 1$. This means that statement III is true.

32. B

Three uniform subintervals on the interval [0, 18] make $\Delta x = \dfrac{18-0}{3} = 6$.

This makes the subintervals [0, 6], [6, 12], and [12, 18] with midpoints at $x = 3$, $x = 9$, and $x = 15$.

Now

$$\int_0^{18} g(x)\,dx \approx g(3)\Delta x + g(9)\Delta x + g(15)\Delta x = \Delta x(-2 + 4 + 5) = 6(7) = 42.$$

33. E

This differential equation is of the form $\frac{dy}{dt} = ky$, where the rate of change is proportional to the amount present. This means that the population's growth follows the Law of Exponential Change, so we know $P = P_0 e^{kt}$. Using $P_0 = 2$ and, $k = 0.713$, this model becomes $P = 2e^{0.713t}$. At $t = 5$, the population is $P = 2e^{0.713(5)} = 70.679$.

34. B

The extrema of f occur at critical points. The critical points of f are the zeroes of $f'(x)$. From the graph of $f'(x) = \frac{-x^3 + 4x - 2}{\sqrt{x^2 + 1}}$, the zeroes of $f'(x)$ are $x = -2.214$, $x = 0.539$, and $x = 1.675$.

Thus, statement I is eliminated since $x = -0.768$ is not a critical point of f. (If you were tempted by this choice, the problem and possible confusion comes from the fact that $x = -0.768$ is the location of a relative minimum on the graph of $f'(x)$.)

Statement II is true. It can be seen that f has a relative minimum at $x = 0.539$ since $f'(x)$ goes from negative to positive (since the graph of $f'(x)$ goes from below the x-axis to above the x-axis).

Statement III is false. Rather than a minimum, f has a relative maximum at $x = 1.675$ since $f'(x)$ goes from positive to negative (since the graph of $f'(x)$ goes from above the x-axis to below the x-axis).

Thus, the only statement that is true is statement II.

35. A

The instantaneous rate of change of $f(x)$ is $f'(x) = 3x^2 + 2$. The average rate of change of $f(x)$ on $-1 < x < 1$ is $\frac{f(1) - f(-1)}{1 - (-1)} = \frac{-2 - (-8)}{2} = \frac{6}{2} = 3$.

We are looking for the point where $f'(x) = 3$. Solving $3x^2 + 2 = 3$ gives $x^2 = \frac{1}{3}$ or $x = \pm\frac{1}{\sqrt{3}} = \pm\frac{\sqrt{3}}{3}$. Since the answer choices do not include $x = -\frac{\sqrt{3}}{3}$, we choose $x = \frac{\sqrt{3}}{3}$.

36. D

The integral of the rate of fuel consumption over an interval will give the amount of fuel consumed over that interval.

$$\int_0^5 r(t)\,dt = \int_0^5 (\sin t - t\cos t + 3)\,dt = 21.227$$

The units in this situation support this fact as the integral is summing (gallons/hour)*hours, thus giving a quantity that is measured in gallons.

37. A

From the given information, we know $\frac{dr}{dt}=\frac{5}{2}$. When $A=\pi r^2=16\pi$, the radius is $r=4$.

We want to find $\frac{dA}{dt}$ at this instant when $r=4$. Starting with the area equation $A=\pi r^2$, we differentiate to get $\frac{dA}{dt}=2\pi r\frac{dr}{dt}$.

Substituting $\frac{dr}{dt}=\frac{5}{2}$ and $r=4$, we get $\frac{dA}{dt}=2\pi(4)\frac{5}{2}=20\pi$.

38. C

The first positive x-intercept of $y=3^{\sin x}-2$ is $x=0.6827$. To find the slope at this value of x, we will need to find the value of the derivative at $x=0.6827$. This can be done using the numerical derivative feature of the calculator. This gives $y'|_{x=0.6827}=1.705$.

39. E

Statement I is the definition of average value of a function, so it calculates the average velocity on the interval [0, 24].

Statement II is finding the average value of $v(t)$ at time $t=0$ and $t=24$, but only at these two values of time. It does not find the average value over the entire time interval.

Statement III is finding the rate of change of position with respect to time on the interval [0, 24].

Thus, only statements I and III calculate the average velocity on [0, 24].

40. D

The integral of the rate of sales over an interval will give the amount of sales over that interval.

$$\int_0^4 S(t)\,dt = \int_0^4 [97+63\ln(t^2+1)]\,dt = 765.023$$

The units for this answer are in thousands of calculators since the integral is summing up (thousands of calculators/year)*years. This makes the answer (765.023)(1000)= 765,023.

41. A

We want g decreasing, so we want $g'(x)<0$. Using the Second Fundamental Theorem of Calculus on $g(x)=\int_0^x f(t)\,dt$ tells us that $g'(x)=f(x)$. The given graph is the graph of f, so it is the graph of $g'(x)$. The graph of $g'(x)=f(x)$ is below the x-axis and thus negative on the interval $1<x<4$, so g is decreasing on $1\le x<4$.

42. C

The amount of water that flows into the lake is calculated as $\int_0^{24} R(t)\,dt = \int_0^{24}\left(\sin\frac{\pi t}{12}+40t\right)dt = 11520$ gallons.

The amount of water pumped out of the lake over the 24-hour period is (500)(24)=12000. Thus, the amount of the water in the lake after 24 hours is computed as $30{,}000 + 11{,}520 - 12{,}000 = 29{,}520$.

43. B

By definition, f is concave down when f' is decreasing. From the graph, this is the case for $-2 < x < 2$.

Using a different approach, typically f'' is used to analyze concavity. In this case, since the given graph is of f', the value of f'' is the slope of f'. For concave down, we need $f'' < 0$ which mean we need f' to be decreasing. In this case, f' is decreasing on $-2 < x < 2$ so f''(the slope of f') is negative.

44. E

From the graph of $f'(x)$, the critical points of f are $x = -2$ and $x = 2$. To test the critical points to determine the type of extrema, we analyze the sign of $f'(x)$. The sign of f' can be determined based on whether the given graph is above or below the x-axis. There is a relative minimum at $x = -2$ since f' changes from negative to positive. There is a relative maximum at $x = 2$ since f' changes from positive to negative.

45. C

Since $f(x) > 0$ and $f'(x) > 0$, f has positive output values and is increasing. All three choices meet that requirement.

Since $f''(x) < 0$, then f is concave down and f' is decreasing. Table I is not consistent with this condition since the successive difference in values of f goes from 3 to 4 to 5 to 6, indicating that the slopes of f are increasing rather than decreasing. Table II is not consistent with this condition since the successive difference in values of f goes from 3 to 4 to 3 to 1, indicating that the slopes of f are both increasing and decreasing rather than just decreasing.

Table III is consistent with this condition since the successive difference in values of f goes from 7 to 5 to 3 to 2, indicating that the slopes of f are decreasing.

Free Response Answers

1. (a) The points of intersection of $f(x)$ and $g(x)$ are (0,0) and (2.8354, 6.137). Let $a = 2.8354$. Special Note on Decimals: Answers using decimal approximations from the calculator should always be given to at least three places after the decimal, rounded or truncated. However, rounding or truncating should only take place on final answers. Since this value is an intermediate answer that will be used in subsequent computations, store the point of intersection of the two functions in the memory of the calculator in its exact form as $a = 2.8354\ldots$ to avoid approximation errors. At the very least, carry an extra digit or two beyond the necessary three places after the decimal to ensure accuracy in the final answer.

   The upper boundary of R is $f(x)$ and the lower boundary is $g(x)$, so the typical rectangle has a height given as $h = f(x) - g(x)$.

   Thus, the area of the region is $A = \int_0^a (f(x) - g(x))\,dx = \int_0^a (5x - x^2 - (2^x - 1))\,dx = 6.481.$

   (b) To find the area of the half of the region before the vertical line $x = k$, we change the upper limit of integration on the integral to k to give $A_{half} = \int_0^k (f(x) - g(x))\,dx.$

   Half of the area of region R is $\dfrac{6.481}{2} = 3.240$ or 3.241.

   Equating these two, the integral equation that could be used to find k is
   $\int_0^k (f(x) - g(x))\,dx = \dfrac{6.481}{2}$ or $\int_0^k (f(x) - g(x))\,dx = 3.240$ or 3.241.

   (c) For solids of revolution, the slices are to be taken perpendicular to the axis of revolution. Thus, the typical rectangle will be perpendicular to the x-axis. This type of rectangle will generate washers with a thickness of Δx, so the outer radius R and the inner radius r are both calculated as $y_{big} - y_{little}$.

   Thus, $R = y_{big} - y_{little} = 5x - x^2 - 0 = 5x - x^2$ and $r = y_{big} - y_{little} = 2^x - 1 - 0 = 2^x - 1$.

   Using the washer method, the volume is

$$V = \pi \int_0^a (R^2 - r^2)\,dx$$

$$= \pi \int_0^a ((f(x))^2 - (g(x))^2)\,dx$$

$$= \pi \int_0^a ((5x - x^2)^2 - (2^x - 1)^2)\,dx = 0.400\pi = 1.257$$

$$= 43.875\pi \text{ or } 43.876\pi.$$

   (137.840 or 137.841)

(d) To find the volume of a solid with known cross sections, we use $V = \int_0^a A(x)\,dx$. For cross sections that are squares, the area is $A(x) = s^2$. The side of the square is embedded into the region R, so the length of the side is $s = y_{big} - y_{title} = 5x - x^2 - (2^x - 1)$. Thus, $A(x) = (5x - x^2 - (2^x - 1))^2 = (5x - x^2 - 2^x + 1)^2$.

The expression that will find the volume of the solid is $V = \int_0^a A(x)\,dx = \int_0^a (5x - x^2 - 2^x + 1)^2\,dx$.

2. (a) Evaluating the given $v(t)$ function at $t = 4$, $v(4) = (4)\sin(4) = -3.027$.

Finding $a(t)$ by differentiating $v(t)$ would require using the product rule. However, we only need to find the value of the derivative at a specific point. This can be accomplished using the numeric derivative feature of a graphing calculator.

Using the calculator, $a(4) = v'(4) = -3.371$.

(b) As a general rule, when v and a have the same sign, speed is increasing. When v and a have different signs, speed is decreasing. Since $v(4) < 0$ and $a(4) < 0$, the speed is increasing at time $t = 4$.

(c) The total distance traveled on $[a, b]$ is found by using $\int_a^b |v(t)|\,dt$.

In this case, the total distance traveled is $\int_0^5 |t \sin t|\,dt = 8.660$.

(d) By definition, the average value of f on $[a, b]$ is given by $\frac{1}{b-a}\int_a^b f(x)\,dx$.

In this case, $\frac{1}{5-0}\int_0^5 t \sin t\,dt = \frac{1}{5}(-2.377) = -0.475$.

(e) The integral $\int_0^5 v(t)\,dt$ finds the displacement of the particle for the time interval [0, 5]. Adding this to the initial position should find the position at time $t = 5$.

Thus, $y(5) = y(0) + \int_0^5 v(t)\,dt = 2 + (-2.377) = -0.377$.

3. (a) The physical situation can be represented as a right triangle, as shown below.

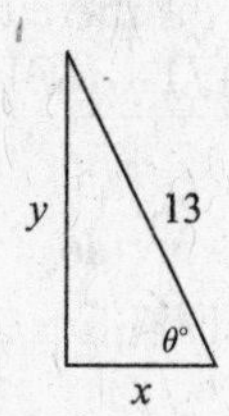

It is given that $\frac{dx}{dt} = 3$ when $x = 5$ when $x = 5$. We want to find $\frac{dy}{dt}$.

A basic equation to relate the relevant variables is the Pythagorean Theorem, which gives us $x^2 + y^2 = 13^2$. We note that when $x = 5$, the value of y is $y = 2$ using the Pythagorean relationship.

To get a $\frac{dy}{dt}$ to result, we must differentiate the Pythagorean Equation with respect to t.

This gives $2x\frac{dx}{dt} + 2y\frac{dy}{dt} = 0$.

Substituting the given values, we get $2(5)(3) + 2(12)\frac{dy}{dt} = 0$.

Solving for $\frac{dy}{dt}$, we find that $12\frac{dy}{dt} = -15$ or $\frac{dy}{dt} = -\frac{15}{12} = -\frac{5}{4}$ feet/second.

(b) In order to find $\frac{d\theta}{dt}$, we need an equation that involves θ. In the given right triangle, we choose cosine since this will involve x and the constant hypotenuse.

Differentiating the equation $\cos\theta = \frac{x}{13}$ gives $-\sin\theta \cdot \frac{d\theta}{dt} = \frac{1}{13} \cdot \frac{dx}{dt}$.

From the original triangle situation, we know that $\sin\theta = \frac{12}{13}$ at the instant when $x = 5$.

Now $-\sin\theta \cdot \frac{d\theta}{dt} = \frac{1}{13} \cdot \frac{dx}{dt}$ becomes $-\frac{12}{13} \cdot \frac{d\theta}{dt} = \frac{3}{13}$.

This gives $\frac{d\theta}{dt} = \frac{\frac{3}{13}}{-\frac{12}{13}} = -\frac{1}{4}$ radians per second.

(c) The area of the right triangle is $A = \frac{1}{2}bh = \frac{1}{2}xy$. We want to find $\frac{dA}{dt}$, so we differentiate the area equation with respect to t. Using the product rule, we get $\frac{dA}{dt} = \frac{1}{2}\left(x\frac{dy}{dt} + y\frac{dx}{dt}\right)$.

Substituting the given values, $\frac{dA}{dt} = \frac{1}{2}\left(5 \cdot -\frac{5}{4} + 12 \cdot 3\right)$ square feet per second. On the AP Exam, there is no requirement to simplify an answer, so the answer can be left in this form. In any case, we do need to make sure that the answer does include the units.

4. (a) The instantaneous rate of change $f'(5)$ can be approximated by using average slope on an interval as close to $x = 5$ as possible. The data points on the table that are closest to $x = 5$ are at $x = 3$ and $x = 7$. Forming a difference quotient, $f'(5) \approx \frac{f(7) - f(3)}{7-3} = \frac{6-0}{4} = \frac{3}{2}$.

NOTE: Since the difference quotient only provides an approximation, it is important to use the symbol $\approx$ instead of the standard $=$ symbol. This is also true in parts (b) and (d) as well.

(b) From the table, the point of tangency is $f(9) = 8$. The slope is $f'(9) = \frac{4}{5}$. Thus, the equation of the line is $y - 8 = \frac{4}{5}(x-9)$ or $y = \frac{4}{5}(x-9)+8$.

At $x = 9.1$, the tangent line has the value $y = \frac{4}{5}(9.1-9)+8 = \frac{4}{5}\left(\frac{1}{10}\right)+8 = 8\frac{4}{50} = 8\frac{2}{25}$ or 8.08. Thus, $f(9.1) \approx 8.08$.

(c) According to the Fundamental Theorem of Calculus, $\int_0^9 2f'(x)\,dx = 2[f(9) - f(0)]$. These values can be found from the table.

Thus, this gives $\int_0^9 2f'(x)\,dx = 2[f(9) - f(0)] = 2[8-(-10)] = 2[18] = 36$.

(d) The right endpoints of the subintervals are $x = 2$, $x = 3$, $x = 7$, and $x = 9$ and the widths of the subintervals are 2, 1, 4, and 2, respectively. Using these to form a right Riemann sum gives

$$\int_0^9 f(x)\,dx \approx f(2)(2) + f(3)(1) + f(7)(4) + f(9)(2)$$

$$= (-2)(2)+(0)(1)+(6)(4)+(8)(2) = -4+0+24+16 = 36.$$

Since f is strictly increasing, rectangles formed using right endpoints will extend above the curve. Thus, this numerical approximation is greater than the actual value of $\int_0^9 f(x)\,dx$.

(e) Since f is differentiable on $0 < x < 9$, we can apply the Mean Value Theorem on this interval. According to the Mean Value Theorem, that means there exists a c in $0 < x < 9$ such that

$$f'(c) = \frac{f(9) - f(0)}{9-0} = \frac{8-(-10)}{9} = \frac{18}{9} = 2.$$

Thus, the answer is yes, there is a value of x in $0 < x < 9$ in such that $f'(x) = 2$.

5. (a) Differentiating each side of the equation $x^3 + xy = 16$ with respect to x and using the product rule gives

$$3x^2 + \left(x\frac{dy}{dx} + y\right) = 0.$$

$$x\frac{dy}{dx} = -3x^2 - y$$

$$\frac{dy}{dx} = \frac{-3x^2 - y}{x} = -\frac{3x^2 + y}{x}.$$

(b) A vertical tangent line requires that $\frac{dy}{dx} = -\frac{3x^2 + y}{x}$ is undefined.

This occurs when $x = 0$.

To find the other coordinate, substitute this value into the original equation $x^3 + xy = 16$ to get $0^3 + 0 \cdot y = 16$ or $0 = 16$. Since this is not possible, there is not such a point where $x = 0$. We thus conclude that there are no vertical tangent lines to the curve.

(c) Using the quotient rule and the chain rule to differentiate $\frac{dy}{dx} = -\frac{3x^2 + y}{x}$ gives

$$\frac{d^2y}{dx^2} = -\frac{x\left(6x + \frac{dy}{dx}\right) - (3x^2 + y)(1)}{x^2}.$$

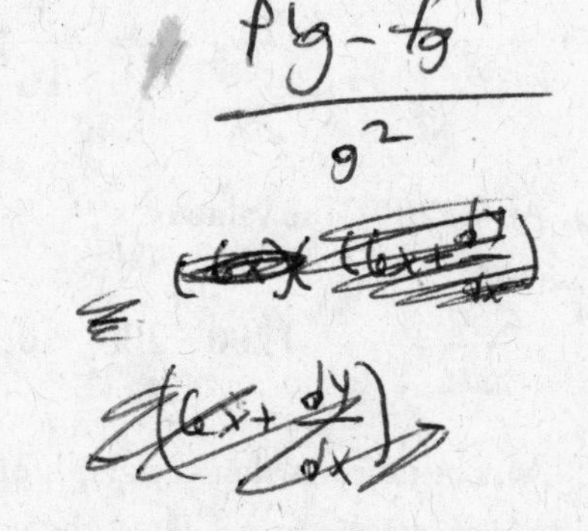

At (–2, –12), the value of $\frac{dy}{dx}$ is

$$\frac{dy}{dx} = -\frac{3(-2)^2 + (-12)}{-2} = -\frac{3(4) - 12}{-2} = 0.$$

At (–2, –12), the value of $\frac{d^2y}{dx^2}$ is

$$\frac{d^2y}{dx^2} = -\frac{(-2)(6(-2) + 0) - (3(-2)^2 + (-12))}{(-2)^2}$$

$$= -\frac{(-2)(-12) - (3(4) - 12)}{4} = -\frac{24}{4} = -6.$$

(d) As found in part (c), at (–2, –12), the value of $\frac{dy}{dx}$ is $\frac{dy}{dx} = 0$ so $x = -2$ is a critical point on the curve.

From part (c) we know $\frac{d^2y}{dx^2} = -6 < 0$. Using the Second Derivative Test to analyze this critical point, we conclude that there is a relative maximum at $x = -2$ since $\frac{dy}{dx} = 0$ and $\frac{d^2y}{dx^2} < 0$.

6. (a) The point of tangency is (0, 200). At the point (0, 200), the slope is

$$\frac{dP}{dt}=\frac{1}{4}(1200-200)=\frac{1000}{4}=250.$$

Using point-slope form, the equation of the tangent line is $y - 200 = 250(t - 0)$ or $y = 250t + 200$. When $t = 2$, the tangent line approximation gives $y = 250(2) + 200 = 700$.

(b) Starting with

$$\frac{dP}{dt}=\frac{1}{4}(1200-P),$$

we get

$$\frac{d^2P}{dt^2}=\frac{1}{4}\;0-\frac{dP}{dt}\quad=-\frac{1}{4}\frac{dP}{dt}$$

$$=-\frac{1}{4}\left(\frac{1}{4}(1200-P)\right)$$

$$=-\frac{1}{16}(1200-P).$$

(c) At (0, 200), the value of $\frac{d^2P}{dt^2}$ is

$$\frac{d^2P}{dt^2}=-\frac{1}{16}(1200-200)<0$$

which indicates that the graph of P is concave down and so the tangent line to the curve at $P = 200$ lies above the curve. Thus, the value of the approximation overestimates the actual value of P.

(d) Using separation of variables, find the particular solution $y = P(t)$ to the differential equation

$$\frac{dP}{dt}=\frac{1}{4}(1200-P)$$

with the initial population $P(0) = 200$.

The first step in solving a differential equation must be to separate the variables. If this is not done or at least attempted, you typically will not be eligible for any partial credit on this entire part of the problem.

For the differential equation $\frac{dP}{dt} = \frac{1}{4}(1200 - P)$,

separating variables gives $\frac{1}{1200 - P}dP = \frac{1}{4}dt$.

Integrating both sides is the next step. The antiderivative of the left side can be determined by inspection or can be handled more thoroughly with a *u*-substitution where $u = 1200 - P$.

$$-\ln|1200 - P| = \frac{1}{4}t + C$$

Next we isolate *P*:

$$\ln|1200 - P| = -\frac{1}{4}t - C$$

$$|1200 - P| = e^{-\frac{1}{4}t - C}$$

$$|1200 - P| = e^{-\frac{1}{4}t}e^{-C}$$

$$|1200 - P| = Ce^{-\frac{1}{4}t}$$

$$1200 - P = Ce^{-\frac{1}{4}t}$$

The absolute values can be dropped since *C* can be positive or negative as necessary to handle the issue of the sign. $-P = Ce^{-\frac{1}{4}t} - 1200 \quad P = -Ce^{-\frac{1}{4}t} + 1200$

To find the value of the constant of integration, substitute the initial condition (0, 200)

$$200 = -Ce^{-\frac{1}{4}(0)} + 1200$$

$$C = 1000$$

This gives the final solution as $P = -1000e^{-\frac{1}{4}t} + 1200$.

AP CALCULUS AB PRACTICE TEST 3 CORRELATION CHART

Multiple Choice Question	Review Chapter: Subsection of Chapter
1	Chapter 13: Local Linear Approximation
2	Chapter 7: Basic Definitions and Understanding Limits Graphically
3	Chapter 14: Absolute Extrema Chapter 12: Rules for Computing Derivatives Chapter 12: Exponential and Logarithmic Functions
4	Chapter 12: Chain Rule
5	Chapter 12: Rules for Computing Derivatives Chapter 12: Derivatives of Trigonometric Functions
6	Chapter 15: Concavity and the Sign of f''
7	Chapter 11: The Relationship Between Differentiability and Continuity
8	Chapter 23: Trapezoidal Approximation
9	Chapter 11: A Geometric Interpretation: The Derivative as a Slope
10	Chapter 18: Distance Traveled by a Particle Along a Line
11	Chapter 21: When F Is Complicated: The Substitution Game
12	Chapter 21: Computing Antiderivatives Directly
13	Chapter 22: Accumulation Functions and Antiderivatives
14	Chapter 12: Implicit Differentiation
15	Chapter 14: Increasing and Decreasing Behavior of f and the Sign of f'
16	Chapter 14: Increasing and Decreasing Behavior of f and the Sign of f' Chapter 15: Concavity and the Sign of f''
17	Chapter 17: Properties of the Definite Integral
18	Chapter 13: The Slope of a Curve at a Point Chapter 12: Rules for Computing Derivatives
19	Chapter 20: The Fundamental Theorem of Calculus (FTC)
20	Chapter 8: Horizontal Asymptotes
21	Chapter 12: Derivative of x^n Chapter 12: Exponential and Logarithmic Functions
22	Chapter 12: Exponential and Logarithmic Functions Chapter 16: Interpreting the Derivative as a Rate of Change

Multiple Choice Question	Review Chapter: Subsection of Chapter
23	Chapter 9: Continuity and Limits
24	Chapter 16: Interpreting the Derivative as a Rate of Change
25	Chapter 16: Differential Equations
26	Chapter 12: The Derivative of Inverse Functions
27	Chapter 17: Finding the Area Under a Curve
28	Chapter 12: Chain Rule
29	Chapter 18: Solids of Revolution
30	Chapter 14: Absolute Extrema Chapter 14: Increasing and Decreasing Behavior of f and the Sign of f' Chapter 9: Properties of Continuous Functions Chapter 14: The Mean Value Theorem Chapter 15: Inflection Points
31	Chapter 14: Critical Points and Local Extrema Chapter 15: Inflection Points
32	Chapter 23: Approximating the Area Under Functions Given Graphically or Numerically
33	Chapter 22: Separable Differential Equations
34	Chapter 14: Critical Points and Local Extrema
35	Chapter 14: The Mean Value Theorem
36	Chapter 17: Rates of Change and the Definite Integral
37	Chapter 16: Related Rates
38	Chapter 13: The Slope of a Curve at a Point
39	Chapter 11: A Physical Interpretation of the Derivative: Speed Chapter 18: Average Value of a Function
40	Chapter 17: Rates of Change and the Definite Integral
41	Chapter 18: Accumulated Change Chapter 20: What's So Fundamental? The Connection Between Integrals and Derivatives
42	Chapter 17: Rates of Change and the Definite Integral
43	Chapter 15: Concavity and the Sign of f''
44	Chapter 14: Corresponding Characteristics of the Graphs of f and f'
45	Chapter 14: Increasing and Decreasing Behavior of f and the Sign of f' Chapter 15: Concavity and the Sign of f''

Free Response Question	Review Chapter: Subsection of Chapter
1(a)	Chapter 18: Area Between Two Curves
1(b)	Chapter 18: Area Between Two Curves
1(c)	Chapter 18: Solids of Revolution
1(d)	Chapter 18: Volumes of Solids
2(a)	Chapter 16: Interpreting the Derivative As a Rate of Change
2(b)	Chapter 16: Interpreting the Derivative As a Rate of Change
2(c)	Chapter 18: Distance Traveled by a Particle Along a Line
2(d)	Chapter 18: Average Value of a Function
2(e)	Chapter 18: Distance Traveled by a Particle Along a Line
3(a)	Chapter 16: Related Rates
3(b)	Chapter 16: Related Rates
3(c)	Chapter 16: Related Rates
4(a)	Chapter 13: Approximating Derivatives Using Tables and Graphs
4(b)	Chapter 13: Local Linear Approximation
4(c)	Chapter 20: The Fundamental Theorem of Calculus
4(d)	Chapter 23: Approximating the Area Under the Curve Using Riemann Sums Chapter 23: Approximating the Area Under Functions Given Graphically or Numerically
4(e)	Chapter 14: The Mean Value Theorem
5(a)	Chapter 12: Implicit Differentiation
5(b)	Chapter 12: Implicit Differentiation
5(c)	Chapter 12: Implicit Differentiation
5(d)	Chapter 14: Critical Points and Local Extrema
6(a)	Chapter 13: Local Linear Approximation
6(b)	Chapter 12: Implicit Differentiation
6(c)	Chapter 15: Concavity and the Sign of f''
6(d)	Chapter 22: Separable Differential Equations

AP Calculus BC Practice Test 1 Answer Grid

1. Ⓐ Ⓑ Ⓒ Ⓓ Ⓔ
2. Ⓐ Ⓑ Ⓒ Ⓓ Ⓔ
3. Ⓐ Ⓑ Ⓒ Ⓓ Ⓔ
4. Ⓐ Ⓑ Ⓒ Ⓓ Ⓔ
5. Ⓐ Ⓑ Ⓒ Ⓓ Ⓔ
6. Ⓐ Ⓑ Ⓒ Ⓓ Ⓔ
7. Ⓐ Ⓑ Ⓒ Ⓓ Ⓔ
8. Ⓐ Ⓑ Ⓒ Ⓓ Ⓔ
9. Ⓐ Ⓑ Ⓒ Ⓓ Ⓔ
10. Ⓐ Ⓑ Ⓒ Ⓓ Ⓔ
11. Ⓐ Ⓑ Ⓒ Ⓓ Ⓔ
12. Ⓐ Ⓑ Ⓒ Ⓓ Ⓔ
13. Ⓐ Ⓑ Ⓒ Ⓓ Ⓔ
14. Ⓐ Ⓑ Ⓒ Ⓓ Ⓔ
15. Ⓐ Ⓑ Ⓒ Ⓓ Ⓔ
16. Ⓐ Ⓑ Ⓒ Ⓓ Ⓔ
17. Ⓐ Ⓑ Ⓒ Ⓓ Ⓔ
18. Ⓐ Ⓑ Ⓒ Ⓓ Ⓔ
19. Ⓐ Ⓑ Ⓒ Ⓓ Ⓔ
20. Ⓐ Ⓑ Ⓒ Ⓓ Ⓔ
21. Ⓐ Ⓑ Ⓒ Ⓓ Ⓔ
22. Ⓐ Ⓑ Ⓒ Ⓓ Ⓔ
23. Ⓐ Ⓑ Ⓒ Ⓓ Ⓔ
24. Ⓐ Ⓑ Ⓒ Ⓓ Ⓔ
25. Ⓐ Ⓑ Ⓒ Ⓓ Ⓔ
26. Ⓐ Ⓑ Ⓒ Ⓓ Ⓔ
27. Ⓐ Ⓑ Ⓒ Ⓓ Ⓔ
28. Ⓐ Ⓑ Ⓒ Ⓓ Ⓔ
29. Ⓐ Ⓑ Ⓒ Ⓓ Ⓔ
30. Ⓐ Ⓑ Ⓒ Ⓓ Ⓔ
31. Ⓐ Ⓑ Ⓒ Ⓓ Ⓔ
32. Ⓐ Ⓑ Ⓒ Ⓓ Ⓔ
33. Ⓐ Ⓑ Ⓒ Ⓓ Ⓔ
34. Ⓐ Ⓑ Ⓒ Ⓓ Ⓔ
35. Ⓐ Ⓑ Ⓒ Ⓓ Ⓔ
36. Ⓐ Ⓑ Ⓒ Ⓓ Ⓔ
37. Ⓐ Ⓑ Ⓒ Ⓓ Ⓔ
38. Ⓐ Ⓑ Ⓒ Ⓓ Ⓔ
39. Ⓐ Ⓑ Ⓒ Ⓓ Ⓔ
40. Ⓐ Ⓑ Ⓒ Ⓓ Ⓔ
41. Ⓐ Ⓑ Ⓒ Ⓓ Ⓔ
42. Ⓐ Ⓑ Ⓒ Ⓓ Ⓔ
43. Ⓐ Ⓑ Ⓒ Ⓓ Ⓔ
44. Ⓐ Ⓑ Ⓒ Ⓓ Ⓔ
45. Ⓐ Ⓑ Ⓒ Ⓓ Ⓔ

AP CALCULUS BC PRACTICE TEST 1

SECTION I, PART A

Time: 55 Minutes
28 Questions

NO GRAPHING CALCULATOR IS ALLOWED ON THIS PORTION OF THE EXAM

Directions: Solve the following problems, using available space for scratchwork. After examining the form of the choices, decide which one is the best of the choices given and fill in the corresponding oval on the answer sheet. No credit will be given for anything written in the test book. Do not spend too much time on any one problem.

In this test:

(1) The domain of a function f is the set of all real numbers x for which $f(x)$ is a real number, unless otherwise specified.

(2) The inverse of a trigonometric function f may be indicated using the inverse function notation f or with the prefix "arc" (e.g., $\sin^{-1} x = \arcsin x$).

1. $f(x) = \sin(x^2 + 1)$. Compute $f'(x)$.

(A) $(2x + 1)\cos(x^2 + 1)$
(B) $2x\cos(x^2 + 1)$
(C) $2x\sin(x^2 + 1)$
(D) $\cos(x^2 + 1)$
(E) $\sin(x^2 + 1)\cos(x^2 + 1)$

2. $f(x) = \int_0^x g(t)\,dt$ and $h(x) = x^2$. Which expression gives $\frac{d}{dx}(f(h(x)))$?

(A) $2x \cdot f(x^2)$
(B) $f(x^2)$
(C) $g(x^2)$
(D) $2x \cdot g(x)$
(E) $2x \cdot g(x^2)$

GO ON TO THE NEXT PAGE

3. For what values of p does the following series converge?

$$\sum_{n=1}^{\infty} \frac{1}{n^{2p}}$$

(A) $p > 0$

(B) $p \geq 1$

(C) $p > 2$

(D) $p > \frac{1}{2}$

(E) The series converges for all p.

4. Below is a table of values for f. Estimate $\int f(x)\,dx$ by using a right Riemann sum over the intervals [1, 6], [6, 8], and [8, 12].

x	1	6	8	12
$f(x)$	21	15	29	37

(A) 201

(B) 236

(C) 276

(D) 281

(E) 302

5. The Taylor series for $\sin x$ centered at $x = 0$ begins $x - \frac{x^3}{3!} + \frac{x^5}{5!} - \frac{x^7}{7!} + \cdots$. What is the coefficient of x^3 in the Taylor series for $x \sin x^2$ centered at $x = 0$?

(A) 1

(B) $\frac{2}{3}$

(C) $\frac{1}{2}$

(D) $\frac{1}{3}$

(E) $\frac{1}{6}$

6. Solve the separable differential equation $\frac{dy}{dx} = 2xy\cos(x^2)$.

(A) $y = \ln(\sin(x^2)) + C$

(B) $y^2 = 2\sin(x^2) + C$

(C) $y = Ce^{\cos(x^2)}$

(D) $y = e^{\sin(x^2)} + C$

(E) $y = Ce^{\sin(x^2)}$

7. A curve is given by the equation $g(x) = x^3 - 2x + 1$. At what point does the tangent to the curve at (2,5) intersect the line $y = x - 5$?

(A) $(1,-4)$

(B) $\left(\frac{10}{9}, \frac{-35}{9}\right)$

(C) $\left(\frac{5}{3}, \frac{-10}{3}\right)$

(D) $\left(\frac{3}{2}, \frac{-7}{2}\right)$

(E) The two lines do not intersect.

8. Compute $\int (2x + 3)^4\,dx$.

(A) $(x^2 + 3)^4 + C$

(B) $\frac{1}{5}(x^2 + 3)^5 + C$

(C) $\frac{3}{5}(x + 3)^5 + C$

(D) $\frac{1}{10}(2x + 3)^5 + C$

(E) $\frac{1}{5}(2x + 3)^5 + C$

GO ON TO THE NEXT PAGE

9. A spherical balloon is inflating at a rate that is directly proportional to the square of its radius. Which differential equation describes this?

(A) $\frac{dV}{dt} = cr^2$

(B) $\frac{dV}{dt} = cr$

(C) $\left(\frac{dV}{dt}\right)^2 = r + C$

(D) $\frac{dV}{dt} = r^2 + C$

(E) $\frac{dV}{dt} = \frac{c}{r^2}$

10. Which of the following integrals gives the area of the shaded region?

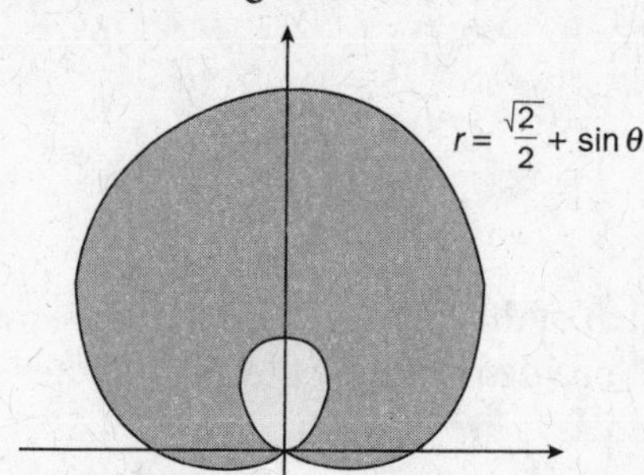

(A) $\int_{-\frac{\pi}{4}}^{\frac{5\pi}{4}} \frac{1}{2}\left(\frac{\sqrt{2}}{2} + \sin\theta\right)^2 d\theta$

(B) $\int_{0}^{2\pi} \frac{1}{2}\left(\frac{\sqrt{2}}{2} + \sin\theta\right)^2 d\theta$

(C) $\int_{-\frac{\pi}{4}}^{\frac{5\pi}{4}} \left(\frac{\sqrt{2}}{2} + \sin\theta\right) d\theta$

(D) $\int_{-\frac{\pi}{4}}^{\frac{5\pi}{4}} \frac{1}{2}\left(\frac{\sqrt{2}}{2} + \sin\theta\right)^2 d\theta - \int_{\frac{5\pi}{4}}^{\frac{7\pi}{4}} \frac{1}{2}\left(\frac{\sqrt{2}}{2} + \sin\theta\right)^2 d\theta$

(E) $\int_{\frac{5\pi}{4}}^{\frac{7\pi}{4}} \frac{1}{2}\left(\frac{\sqrt{2}}{2} + \sin\theta\right)^2 d\theta - \int_{-\frac{\pi}{4}}^{\frac{5\pi}{4}} \frac{1}{2}\left(\frac{\sqrt{2}}{2} + \sin\theta\right)^2 d\theta$

11. $g(x) = \int_0^{x^3} \sin(t^2)dt$. What is $g'(x)$?

(A) $2x\cos(x^2)$

(B) $2x^3\cos(x^6)$

(C) $\sin(x^6)$

(D) $3x^2\sin(x^6)$

(E) $2x^3\sin(x^6)$

12. $\lim_{x\to 0} \frac{\ln(x+1) - x\cos x}{x^2 + x}$

(A) ∞

(B) 2

(C) $\frac{2}{5}$

(D) 0

(E) $-\frac{1}{5}$

13. Ford is water-skiing around Cartesian Lake, towed by a motorboat that is following a path parameterized by the equations

$$y(t) = t^3 + 3t^2 + 4$$
$$x(t) = 2t^2 + 2$$

At time $t = 2$ Ford lets go and continues in a straight line. What is the slope of that line?

(A) 3

(B) 2

(C) 1

(D) $\frac{1}{2}$

(E) $\frac{1}{3}$

GO ON TO THE NEXT PAGE

14. Compute $\int \frac{x+3}{(x+1)x}\,dx$.

(A) $-2\ln|x+1| + 3\ln|x| + C$

(B) $-\frac{1}{2(x+1)^2} + C$

(C) $-2\ln|x+1| + 3\ln|x| + C$

(D) $\ln|x(x+1)| + C$

(E) $3\ln|x+1| + \ln|x| + C$

15.

x	3	6	7	12	17
$f(x)$	3	−1	−2	5	14

The table above gives some values of a continuous function f. Which of the following must be true?

(A) f is differentiable.

(B) For some $3 \le x \le 12$, $f'(x) = 0$.

(C) $f(x) \le 14$

(D) f has exactly one zero.

(E) f has at least two zeros.

16. Which of the following sequences converge?

I. $\sum_{n=0}^{\infty} \frac{n+1}{e^n}$

II. $\sum_{n=1}^{\infty} \frac{e^n}{4^n}$

III. $\sum_{n=0}^{\infty} \sin \frac{\pi}{2} - \frac{1}{n}$

(A) I only

(B) II only

(C) I and II

(D) II and III

(E) I, II, and III

17. Compute $\int_1^{\infty} \frac{3}{x^2}\,dx$.

(A) −1

(B) 1

(C) 3

(D) $\frac{3}{2}$

(E) undefined

18. Compute $\int xe^{-3x}\,dx$.

(A) $-\frac{1}{3}e^{-3x^2} + C$

(B) $-\frac{1}{3}xe^{-3x} + C$

(C) $-\frac{1}{6}x^2e^{-3x} + C$

(D) $-\frac{1}{3}xe^{-3x} - \frac{1}{9}e^{-3x} + C$

(E) $\frac{1}{9}xe^{-3x} - \frac{1}{3}e^{-3x} + C$

19. Which of the following differential equations could be described by the given vector (slope) field?

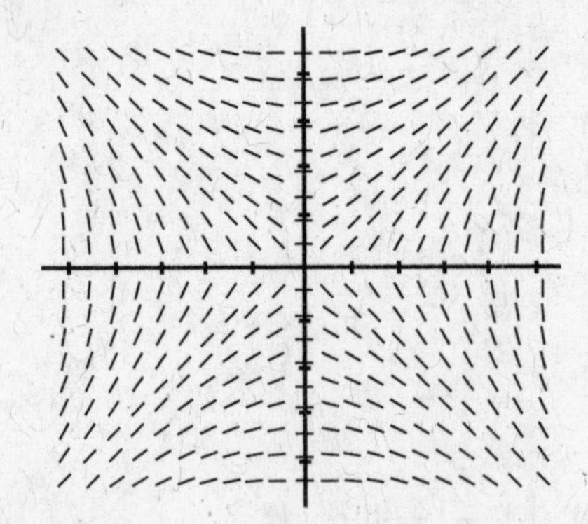

(A) $y' = x$

(B) $y' = xy$

(C) $y' = x^2y$

(D) $y' = \frac{x}{y}$

(E) $y' = \frac{y^2}{x}$

GO ON TO THE NEXT PAGE

20. A population of voles is described by the differential equation $\frac{dP}{dt} = 0.6P\left(1 - \frac{P}{400}\right)$ with initial condition $P(0) = a$. For which values of a will $\lim_{x\to\infty} P(x) = 400$?

(A) $a = 0$
(B) $a = 400$
(C) $a \geq 400$
(D) $0 \leq a \leq 400$
(E) $a > 0$

21.

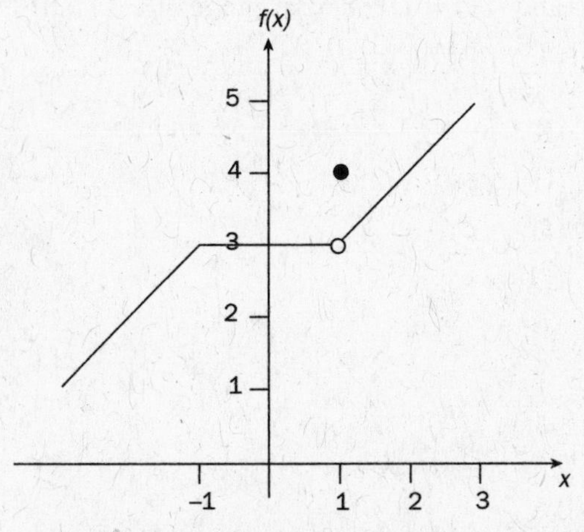

The graph of f is given above.

What is $f\left(\lim_{h\to 2} g(h)\right)$ if $g(h) = \frac{h^2 - 3h + 2}{h - 2}$?

(A) 1
(B) 2
(C) 3
(D) 4
(E) 5

22. What is the coefficient of x^2 in the Taylor expansion of $\ln(3x^2 + 1)$ centered $x = 0$?

(A) 6
(B) 3
(C) $\frac{3}{2}$
(D) 0
(E) $-\frac{1}{2}$

23. The slope field below describes a differential equation. If $y(x)$ is a solution to that differential equation satisfying $y(3) = 0$, what can be said about $\lim_{x\to\infty} y(x)$?

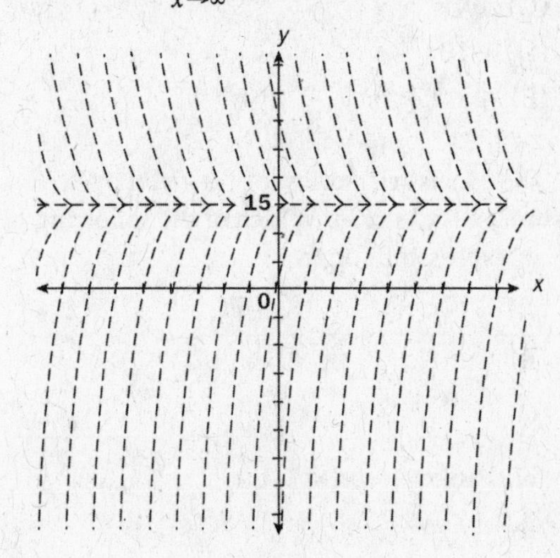

(A) The limit exists and is equal to 0.
(B) The limit exists and is equal to 3.
(C) The limit exists and is equal to 15.
(D) The function grows without bound (i.e., approaches infinity) and the limit does not exist.
(E) The function is bounded but the limit does not exist.

24. Consider the differential equation given by $y' = xy$, with initial condition $y(0) = 1$. Using Euler's method starting at $x = 0$ with step size 1, what is the approximate value of $y(2)$?

(A) 0
(B) 1
(C) 2
(D) 4
(E) 6

GO ON TO THE NEXT PAGE

25. $f(x) = \sin(2\pi e^{3x} - 4\pi x)$. What is $f'(0)$?

(A) 2π

(B) -1

(C) 0

(D) -10π

(E) $2\pi e^3 - 4\pi$

26. The Maclaurin series for f is given by $x + x^3 + x^5 + x^7 \cdots$. Which of the following is an expression for f?

(A) $e^{x^2} - 1$

(B) $\sin(3x)$

(C) $\dfrac{x}{1 - x^2}$

(D) $\sin(x) - \cos(x^2)$

(E) $\dfrac{1}{1 + x^2} - 1$

27. Evaluate $\sum_{n=1}^{\infty} \dfrac{(-3)^{n+1}}{4^n}$.

(A) $\dfrac{4}{7}$

(B) $\dfrac{16}{7}$

(C) 1

(D) $\dfrac{9}{7}$

(E) $\dfrac{1}{7}$

28. A function f satisfies $f'(x) = \dfrac{\sin x + 1}{x}$ for $x > 0$ and $f(1) = 0$. For what range of x is $f > 0$?

(A) $\{x > 0 \mid x \neq k\pi\}$ k an integer

(B) $\left\{x > 0 \mid x \neq k\pi + \dfrac{\pi}{2}\right\}$ k an integer

(C) $x > 0$

(D) $x > 1$

(E) $x < 1$

IF YOU FINISH BEFORE TIME IS CALLED, YOU MAY CHECK YOUR WORK ON THIS SECTION ONLY. DO NOT TURN TO ANY OTHER SECTION IN THE TEST. **STOP**

SECTION I, PART B

Time: 50 Minutes
17 Questions

A GRAPHING CALCULATOR IS ALLOWED ON THIS PORTION OF THE EXAM

Directions: Solve the following problems, using available space for scratchwork. After examining the form of the choices, decide which one is the best of the choices given and fill in the corresponding oval on the answer sheet. No credit will be given for anything written in the test book. Do not spend too much time on any one problem.

Note: For the actual test, no scrap paper is provided.

In this test:

(1) The exact numerical value of the correct answer does not always appear among the choices given. When this happens, select from among the choices the number that best approximates the numerical value.

(2) The domain of a function *f* is the set of all real numbers *x* for which *f*(*x*) is a real number, unless otherwise specified.

(3) The inverse of a trigonometric function *f* may be indicated using the inverse function notation f^{1} or with the prefix "arc" (e.g., $\sin^{-1} x = \arcsin x$).

x	$f(x)$	$f'(x)$	$g(x)$	$g'(x)$
1	1	3	1	5
3	5	1	5	3
5	3	5	3	1

29. If $h(x) = g(f(x))$, what is $h'(1)$?

(A) 1
(B) 3
(C) 9
(D) 15
(E) 25

GO ON TO THE NEXT PAGE

30. What is $\lim_{x\to 4} \ln(f(x))$?

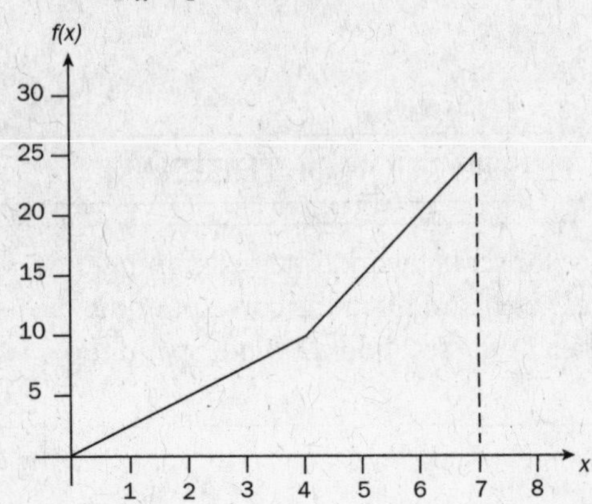

(A) 1.386

(B) 2.303

(C) 2.125

(D) 2.708

(E) 1.681

31. A particle travels along the x-axis with velocity given by $v(t) = (1 - t^2)\sin t$. If the particle is at $x = 2$ when $t = 0$, where is the particle at $t = 5$?

(A) 22.116

(B) 18.230

(C) 16.115

(D) 19.411

(E) 20.830

32. For the first six seconds of driving, a car accelerates at a rate of $a(t) = 10\sin\left(1+\frac{t^2}{10}\right)$ m/sec^2. Which expression gives the velocity of the car when it first begins to decelerate?

(A) $\int_0^{0.755} 10\sin\left(1+\frac{t^2}{10}\right)dt$

(B) $\int_0^{3.830} 100\sin^2\left(1+\frac{t^2}{10}\right)dt$

(C) $\int_0^{1.715} 2t\cos\left(1+\frac{t^2}{10}\right)dt$

(D) $\int_0^{4.627} 10\sin\left(1+\frac{t^2}{10}\right)dt$

(E) $\int_0^{3.830} 10\sin\left(1+\frac{t^2}{10}\right)dt$

GO ON TO THE NEXT PAGE

33. With respect to the intervals [1, 3] and [3, 5], the integral $\int_{1}^{5} f(x)\,dx$ is over-approximated by a left Riemann sum, and under-approximated by a right Riemann sum. Which of the following could be the graph of *f*?

(A)

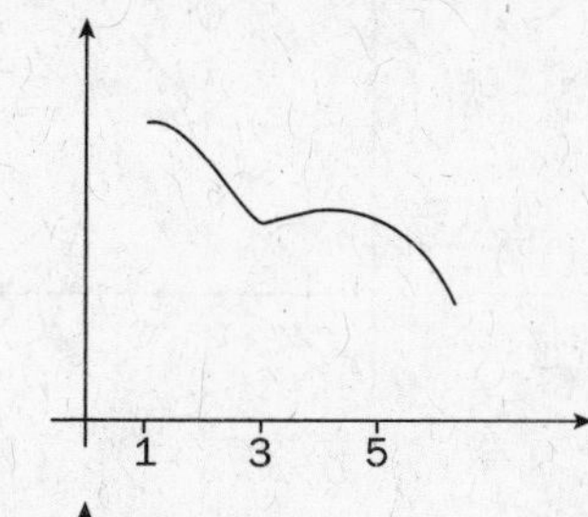

(B)

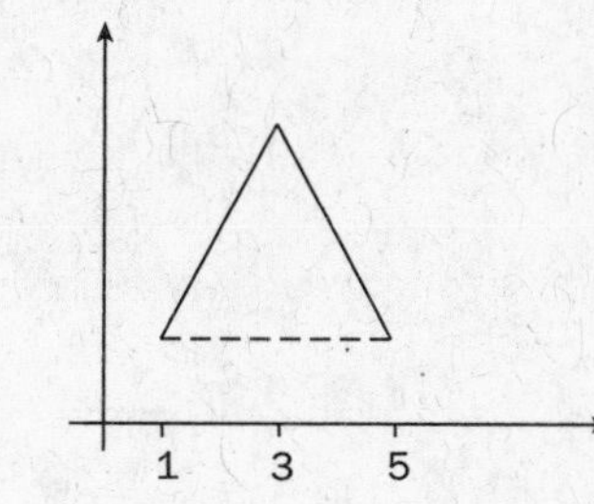

(C)

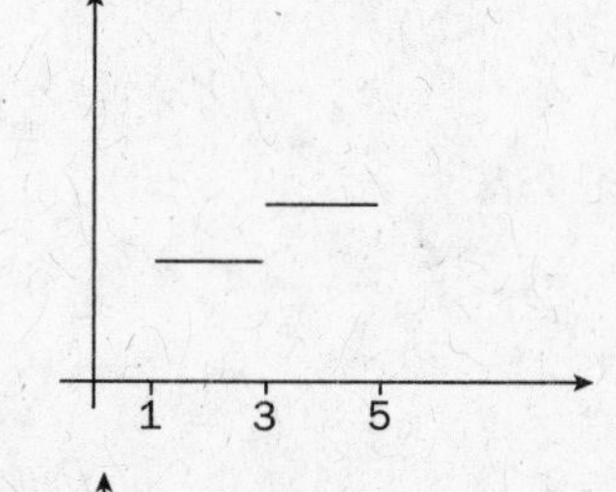

(D)

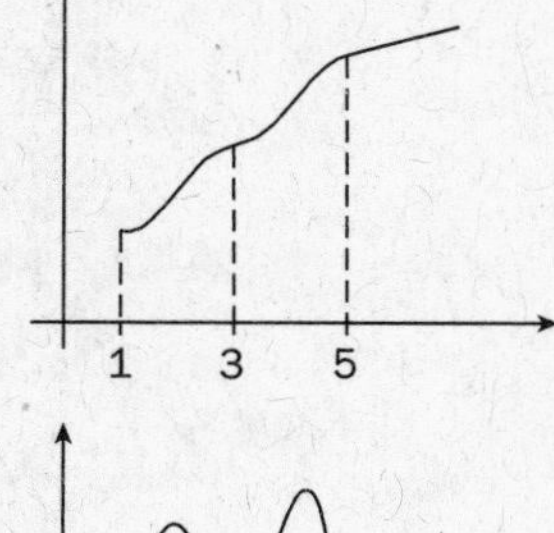

(E)

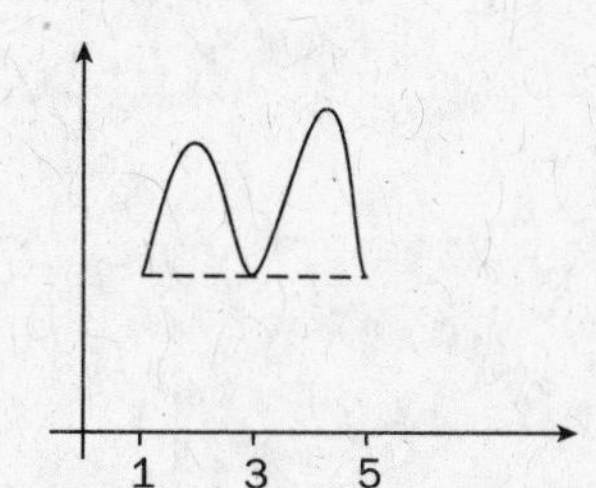

34. Let *R* be the region bounded by $\cos \pi x$ and the *x*-axis, over the interval $\left[-\frac{1}{2}, \frac{1}{2}\right]$. Compute the volume of a solid whose base is *R*, and whose cross section parallel to the *y*-axis is a rectangle with height equal to1.5 times the length of the base.

(A) 2

(B) 1.571

(C) 1

(D) 0.75

(E) 0.637

35. The movement of a particle in the *xy*-plane is given by $(x(t), y(t))$ where $x(t) = e^t + 1$ and $y(t) = t^2 + 2$. What is the speed of the particle at time $t = 2$?

(A) 3.375

(B) 8.402

(C) 14.389

(D) 33.314

(E) 70.598

36. For $t > 0$, the volume of a cloud is changing at a rate of $r(t) = \frac{t}{4} + \frac{\sin(3t)}{t}$. What is the rate of growth of the cloud when the cloud is shrinking the fastest?

(A) 1.456

(B) 0.696

(C) −1.456

(D) −0.282

(E) −3.647

GO ON TO THE NEXT PAGE

37. The graph of f' is given below. If $f(0) = 0$, which of the following must be true?

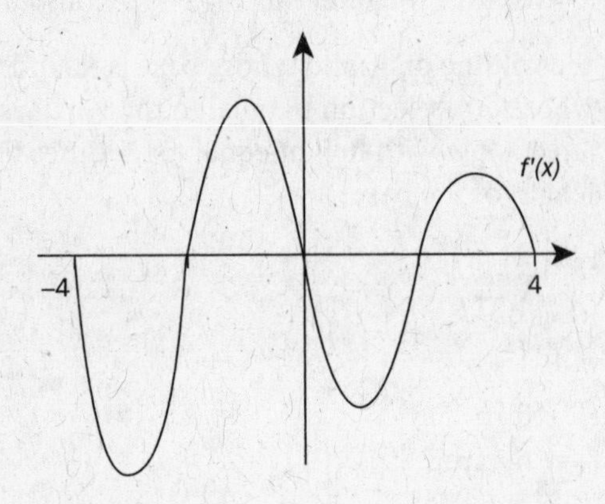

(A) $f(-4) = -f(4)$
(B) $f(x)$ has a local minimum at $x = 0$
(C) $f(x)$ is increasing on $[-4, 4]$
(D) $f(x) \geq 0$ on $[-4, 0]$
(E) $f(x) \leq 0$ on $[-4, 4]$

38. A tree grows at a rate of $g(x) = (x^2 + x\cos(3x))0.9^{\frac{x^2+2}{6}}$ feet per year, with time measured in years. How tall is the tree after eight years? Round your answer to the nearest foot.

(A) 98 ft
(B) 87 ft
(C) 72 ft
(D) 47 ft
(E) 2 ft

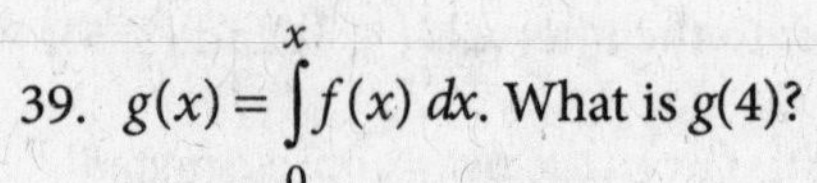
39. $g(x) = \int_0^x f(x)\,dx$. What is $g(4)$?

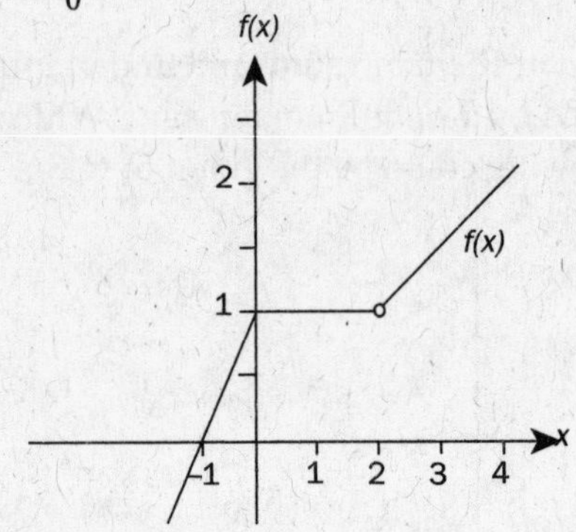

(A) 2
(B) 3
(C) 5
(D) 6
(E) 8

GO ON TO THE NEXT PAGE

40. The degree four Taylor olynomial of f centered at $x = 3$ is given by $P(x) = 2 + (x-3) + 3(x-3)^2 + \frac{4}{3}(x-3)^3 - \frac{1}{4}(x-3)^4$.
What is $f''(3)$?

(A) $\frac{1}{2}$
(B) $\frac{4}{3}$
(C) 3
(D) 6
(E) 8

41. A rectangle is drawn in the xy-plane with base on the x-axis and with corner points on the graph of $y = 2x - x^2$. What is the maximum area possible for the rectangle?

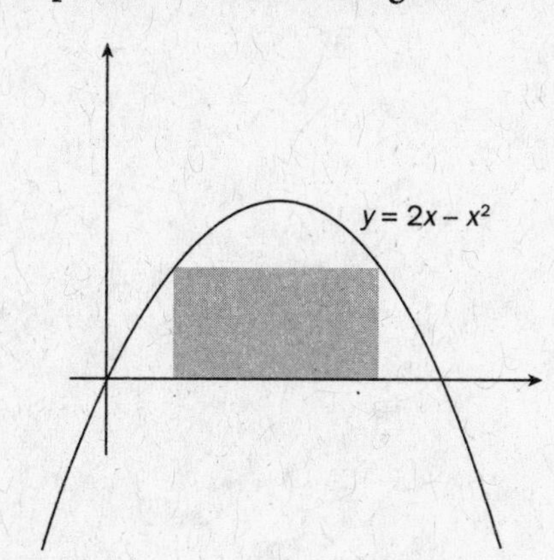

(A) 1.155
(B) 1.434
(C) 2.101
(D) 0.770
(E) 3.309

42.

x	0	4	7	12
$f(x)$	3	7	4	14

f is continuous and differentiable on [0, 12]. Which of the following must be true?

(A) $f > 0$ on [0,12]
(B) $f' > 0$ on [0,12]
(C) $f''(4) < 0$
(D) f has a local max at 4
(E) $f'(c) = 0$ for some c in [0,12]

43. On the interval [0,4], the function is given by the equation $f(x) = e^x + x(1 + \sin(2\pi x))$. At what point in [0,4] does the average rate of change equal the instantaneous rate of change?

(A) 4.432
(B) 3.398
(C) 2.292
(D) 1.608
(E) 1.069

GO ON TO THE NEXT PAGE

44. Below is a graph of the function f'. Which of the following describes the set of extrema of f 000and the inflection points of f?

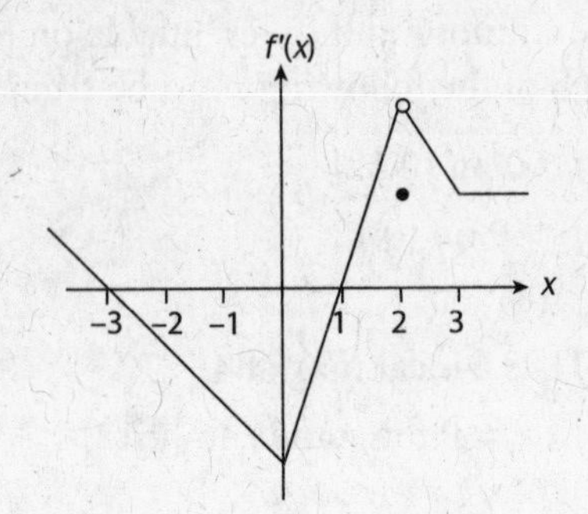

(A) f has extrema at $x = -3, 1$ and an inflection point at $x = 2$.
(B) f has extrema at $x = -3, 1$ and an inflection point at $x = 0$.
(C) f has extrema at $x = -3, 1$ and inflection points at $x = 0, 2$.
(D) f has extrema at $x = 0, 3$ and inflection points at $x = 0, 2$.
(E) f has extrema at $x = 0, 3$ and inflection points at $x = 0, 2, 3$.

45. Using a right Riemann sum over the given intervals, estimate $\int_{5}^{35} f(t)dt$.

x	5	13	22	27	35
$f(x)$	44	12	13	17	22

(A) 730
(B) 661
(C) 564
(D) 474
(E) 325

IF YOU FINISH BEFORE TIME IS CALLED, YOU MAY CHECK YOUR WORK ON THIS SECTION ONLY. DO NOT TURN TO ANY OTHER SECTION IN THE TEST. **STOP**

SECTION II

Time: 1 hour and 30 minutes
6 Problems
50 percent of total grade

Directions: Solve the following problems, using available space for scratchwork. Show how you arrived at your answer.

You may wish to look over the problems before starting to work on them; on the actual test, it is not expected that everyone will be able to complete all parts of all problems. All problems are given equal weight, but the individual parts of a particular problem are not necessarily given equal weight. You should not spend too much time on any one problem.

- You should write out all your work for each part. On the actual test, you will do this in the space provided in the test booklet. Be sure to write clearly and legibly. If you make a mistake, you can save time by crossing it out rather than trying to erase it. Erased or crossed-out work will not be graded.
- Show all your work. Clearly label any functions, graphs, tables, or other objects that you use. On the actual exam, you will be graded on the correctness and completeness of your methods as well as your answers. Answers without any supporting work may not receive credit.
- Justifications (i.e., the request that you "justify your answer") require that you give mathematical (non-calculator) reasons.
- Work must be expressed in standard mathematical notation, not calculator syntax.
- Unless otherwise specified, answers (numeric or algebraic) need not be simplified.
- If you use decimal approximations in calculations, the readers of the actual exam will grade you on accuracy. Unless otherwise specified, your final answers should be accurate to three places after the decimal point.
- Unless otherwise specified, the domain of function f is the set of all real numbers x for which $f(x)$ is a real number.

GO ON TO THE NEXT PAGE

PART A

Time: 30 Minutes
2 Problems

A GRAPHING CALCULATOR IS ALLOWED ON THIS PORTION OF THE EXAM

1. A city was founded in 1900 with an initial population of 4,000. The rate of change in population is modeled by the equation $P'(t) = \ln(1+5t^2)$, where P is population in thousands and t is measured in years since 1900.

   (a) Using the model, find and solve an integral formula to estimate the population of the city in the year 1950.

   (b) Using the model, find and solve an integral formula to determine during which year the population reaches 1 million.

   (c) Use Euler's method, starting at $t = 0$ and with a step size of 25 years, to estimate the population of the city in the year 2000.

2. The region R is bounded by the curves $y_1 = \sin x$ and $y_2 = 1 - \cos x$.

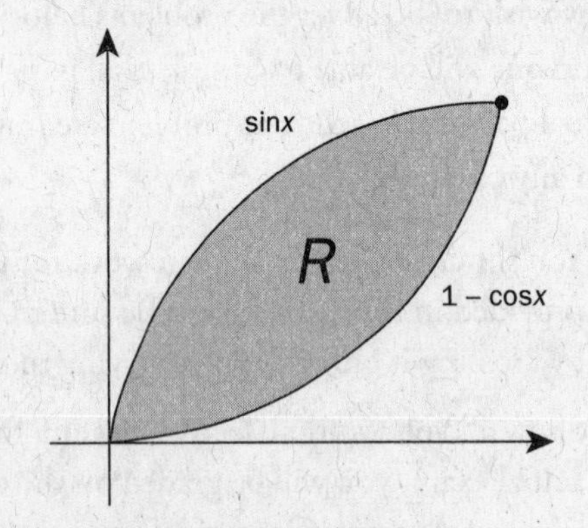

   (a) Find the area of R.

   (b) Find the volume of the solid obtained by rotating the region R about the x-axis.

   (c) Compute the volume of the solid with base R and height (as a function of x) equal to the square of the length of the base $|y1 - y2|$.

GO ON TO THE NEXT PAGE

PART B

60 Minutes
4 Problems

NO GRAPHING CALCULATOR IS ALLOWED ON THIS PORTION OF THE EXAM

Note: If you have extra time, you can go back and work on Part A of Section II, but you cannot use a calculator to complete your work at this time.

3. The position of a particle moving in the xy-plane is determined by the curve $(x(t), y(t))$, $x'(t) = 3e^{3t}$, and $y'(t) = 2t - 4$. At $t = 0$, the particle is at the point (0,0).

   (a) Compute the speed and the acceleration vector of the particle at $t = 0$.

   (b) Compute the x- and y-coordinates of the particle at $t = 4$.

   (c) Describe the behavior of the x- and y-coordinates as $t \to -\infty$ and as $t \to +\infty$.

4. A curve satisfies the equation $y^2 + 2x = x^2 + 2xy + 1$.

   (a) Show that $\frac{dy}{dx} = \frac{x + y - 1}{y - x}$.

   (b) Determine the points of intersection of the curve with the line $y = 2$, and find the slope of the tangent line to the curve at the point with positive x-coordinates.

   (c) Determine all points on the curve with a horizontal tangent line.

GO ON TO THE NEXT PAGE

5. The shape of an anthill over time is modeled by a cone with fixed base and increasing height. The population that the anthill can hold is directly proportional to its volume and is given by $P(t) = 5r^2h$, where the population is measured in hundreds. As the anthill grows, the population changes at a rate of $\frac{t+1}{h}$ hundreds of ants per day.

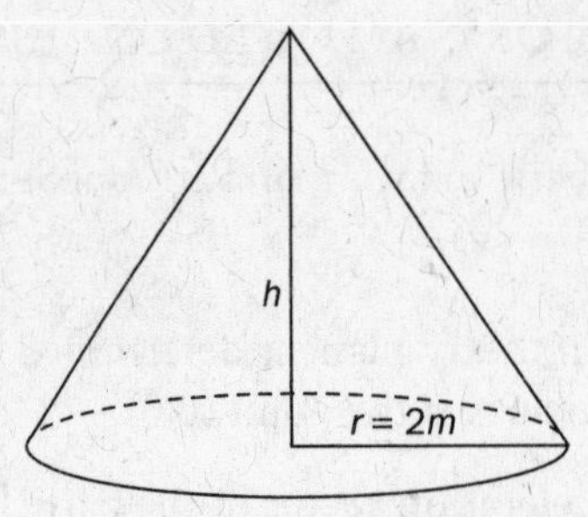

(a) Show that $\frac{dh}{dt} = \frac{t+1}{20h}$.

(b) Solve the differential equation for the initial condition $h(0) = 0$.

(c) When will the population of ants reach 4,000?

6. The derivatives of $f(x)$ at $x = 0$ are given by $f^{(n)}(0) = \frac{(n+1)!}{3^n}$.

(a) Show that the Taylor series for $f(x)$ at $x = 0$ is $\sum_{n=0}^{\infty} \frac{(n+1)x^n}{3^n}$.

(b) What is the radius of convergence of the series?

(c) Integrate the series term by term to get a power series for $\int f(x)\,dx$.

(d) Use the series from part (c) to show that $\frac{3}{1-\frac{x}{3}}$ is the antiderivative for $f(x)$ with $C = 3$.

(Note: The power series for $\frac{1}{1-x} = \sum_{n=0}^{\infty} x^n$.)

IF YOU FINISH BEFORE TIME IS CALLED, YOU MAY CHECK YOUR WORK ON THIS SECTION ONLY. DO NOT TURN TO ANY OTHER SECTION IN THE TEST.

STOP

AP Calculus BC Practice Test 1 **Answer Key**

1. B
2. E
3. D
4. D
5. A
6. E
7. B
8. D
9. A
10. D
11. D
12. D
13. A
14. C
15. E
16. C
17. C
18. D
19. D
20. E
21. D
22. B
23. C
24. C
25. A
26. C
27. D
28. D
29. D
30. B
31. E
32. D
33. A
34. D
35. B
36. D
37. E
38. B
39. C
40. D
41. D
42. E
43. B
44. B
45. D

ANSWERS AND EXPLANATIONS

1. B

We're asked to compute the derivative of $f(x) = \sin(x^2 + 1)$. To do this we need to use the chain rule. We view f as a composition of two simpler functions—$g_1(x) = \sin x$ and $g_2(x) = x^2 + 1$—that have easy derivatives. Write $f(x) = g_1(g_2(x))$, which is by the chain rule $f'(x) = g'_1(g_2(x)) \cdot g'_2(x)$, giving us $f'(x) = \cos(x^2 + 1)(2x)$. Don't get in a hurry and answer with $f'(x) = g'_1(g_2(x)) = \cos(x^2 + 1)$. Another easy mistake to make is writing $f'(x) = g'_1(g'_2(x)) = \cos(2x)$.(but luckily, this isn't a possible answer here).

2. E

This problem combines the chain rule with the Fundamental Theorem of Calculus. First, use the chain rule symbolically to get $\frac{d}{dx}(f(h(x))) = f'(h(x))h'(x)$. Now we need to use the FTC: $f(x) = \int_0^x g(t)\,dt$ so $f'(x) = g(x)$.(Remember, the derivative of an antiderivative is just the original function again.) Lastly we use $h(x) = x^2$ to compute. We put all this together to see that $\frac{d}{dx}(f(h(x)) = f'(h(x))h'(x) = g(x^2)2x$. You may also confront a similar problem in the form of "compute $f'(x)$ if $f(x) = \int_0^{x^2} g(t)\,dt$." To do such a computation, just follow the steps above.

3. D

In general, the p-series $\sum_{n=1}^{\infty} \frac{1}{n^p}$ converges for $p > 1$. (Remember, $\sum_{n=1}^{\infty} \frac{1}{n}$ does *not* converge, but $\sum_{n=1}^{\infty} \frac{1}{n^2}$ does.) We can use this to determine when $\sum_{n=1}^{\infty} \frac{1}{n^{2p}}$ converges. In other words, we know that $\sum_{n=1}^{\infty} \frac{1}{n^{2p}}$ converges for $2p > 1$ or $p > \frac{1}{2}$.

4. D

To compute a right Riemann sum, we approximate the area under the curve with a set of rectangles whose height is determined by the value of the function on the right endpoint. For the first interval, [1,6], the height of the rectangle is $f(6) = 15$ and the base has length $6 - 1 = 5$. Similarly, for the second interval, [6,8], the rectangle has height $f(8) = 29$ and base $8 - 6 = 2$, and the last interval, [8,12], has a rectangle with height $f(12) = 37$ and base $12 - 8 = 4$. To get the right Riemann sum we add the areas of the rectangles:

$$f(6)(6 - 1) + f(8)(8 - 6) + f(12)(12 - 8) = 15 \cdot 5 + 29 \cdot 2 + 37 \cdot 4 = 281$$

5. A

Though our function, $x \sin(x^2)$, is itself fairly complicated, we don't need to compute all the derivatives in order to determine the power series. We know that a power series for $\sin x$ is given by $\left(x - \frac{x^3}{3!} + \frac{x^5}{5!} - \frac{x^7}{7!} + \cdots\right)$. This will give us a power series for $\sin(x^2)$ by composition:

$$\left((x^2) - \frac{(x^2)^3}{3!} + \frac{(x^2)^5}{5!} - \frac{(x^2)^7}{7!} + \cdots\right) = \left(x^2 - \frac{x^6}{3!} + \frac{x^{10}}{5!} - \frac{x^{14}}{7!} + \cdots\right).$$

Finally, to get a power series for $x \sin(x^2)$, we just multiply the power series for x (which is just the series $x + 0 + 0 + \cdots + 0 + \cdots$) and $\sin(x^2)$, namely $x\left(x^2 - \frac{x^6}{3!} + \frac{x^{10}}{5!} - \frac{x^{14}}{7!} + \cdots\right) = x^3 - \frac{x^7}{3!} + \frac{x^{11}}{5!} - \frac{x^{15}}{7!} + \cdots$.

The coefficient of x^3 is 1.

6. E

We're given that it's a separable differential equation, so make sure to separate first and then integrate. We start with $\frac{dy}{dx} = 2xy\cos(x^2)$. By separating we get $\frac{dy}{y} = 2x\cos(x^2)dx$. Integrating the left-hand side gives $\ln(y) + C$ and integrating the right-hand side gives $\sin(x^2) + C$. We need to combine these in order to write the answer as a function of x. Combine the constants and write $\ln(y) = \sin(x^2) + C_1$. Exponentiate each side to get $y = e^{\sin(x^2) + C_1} = e^{\sin(x^2)} \cdot e^{C_1} = C_2 e^{\sin(x^2)}$. Here $C_2 = eC_1$. Remember, it doesn't matter what we call the constant, and $e^{\text{some constant}}$ is just another constant.

7. B

First, notice we're looking for the intersection of the *tangent line,* not the intersection with the graph itself (i.e., don't plug $y = x - 5$ into the equation for the curve). Our solution has two parts. First, use the derivative to write an equation for the tangent line to the curve at the point (2,5), and second, compute the intersection of this line and the line $y = x - 5$. Our curve is given by the equation and the derivative is $g'(x) = 3x^2 - 2$. Thus the slope of the tangent line is $g'(2) = 10$ and the equation of the tangent line (in point-slope format) is $y - 5 = 10(x - 2)$. To see where this intersects $y = x - 5$, we solve the linear system (in this case just substitute $y = x - 5$ into $y - 5 = 10(x - 2)$ and solve).

$$(x - 5) - 5 = 10(x - 2)$$
$$x - 10 = 10x - 20$$
$$10 = 9x$$

Finally solving, we see $x = \frac{10}{9}$ and $y = x - 5 = \frac{10}{9} - \frac{45}{9} = -\frac{35}{9}$.

8. D

We are asked to compute $\int (2x + 3)^4 dx$. First, don't make the silly mistake of forgetting to *u*-substitute and quickly write down $\frac{1}{5}(2x + 3)^5 + C$. To solve, we must integrate using *u*-substitution. The obvious substitution is $u = 2x + 3$ and $du = 2dx$. Substituting we get

$$\int (2x + 3)^4 \, dx = \int (u)^4 \frac{du}{2}$$

$$\frac{1}{2}\frac{u^5}{5} + C = \frac{u^5}{10} + C = \frac{1}{10}(2x + 3)^5 + C$$

9. A

We are told to write an equation to model the statement "a spherical balloon is inflating at a rate that is directly proportional to the square of its radius." First, let's translate the pieces of the statement. Let's choose that V represents the volume of the balloon, t represents time, and r the radius. The rate of inflation is just the rate of change of volume: $\frac{dV}{dt}$. We are given that this is directly proportional to the square of the radius: r^2. This translates directly into $\frac{dV}{dt} = cr^2$. That is to say, "directly proportional to" means that it is a constant multiple, c, of the other. Let's examine the rest of the possible answers.

(B) $\frac{dV}{dt} = cr$. Here the rate of inflation is directly proportional to just the radius, not its square.

(C) $\left(\frac{dV}{dt}\right)^2 = r + C$. Here the square of the rate of inflation is exactly the radius plus a constant.

(D) $\frac{dV}{dt} = r^2 + C$. Here the rate of inflation is exactly the square of the radius plus a constant. We're pretty close here, but don't be fooled. This isn't the description we're looking for.

(E) $\frac{dV}{dt} = \frac{c}{r^2}$. Here we are close again, but now the rate of inflation is *inversely proportional* to the square of the radius.

Only (A) says that the rate of change of volume, V, is a constant times the radius squared.

10. D

To start with, how would we compute the total area enclosed by the outer curve?

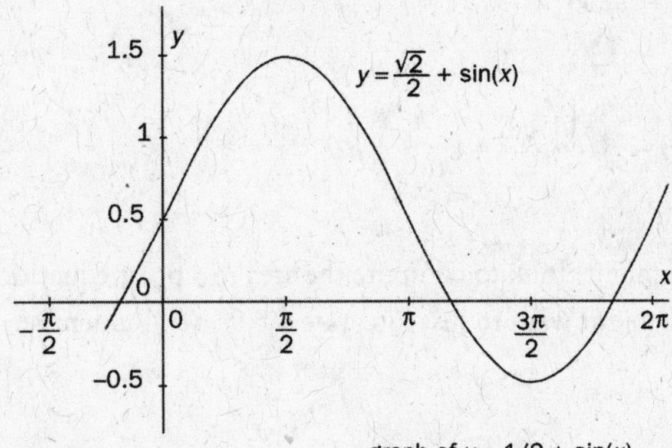

graph of $y = 1/2 + \sin(x)$

Here's a graph of $y = \frac{\sqrt{2}}{2} + \sin x$. Notice that it's positive except on the interval $\left(\frac{5\pi}{4}, \frac{7\pi}{4}\right)$.

Plotting a few points if necessary, we see that the outer curve is given by $r = \frac{\sqrt{2}}{2} + \sin\theta$ on the interval $\left(-\frac{\pi}{4}, \frac{5\pi}{4}\right)$, i.e., exactly where the radius is positive. (Note, we can see by inspecting the graph that because the positive values of $y = \frac{\sqrt{2}}{2} + \sin x$ are much larger than the negative values, the points on the graph with positive radius correspond to the *outer curve*.) To compute the area of this region we need to compute

$$\int_{-\frac{\pi}{4}}^{\frac{5\pi}{4}} \frac{1}{2}[f(\theta)]^2\, d\theta = \int_{-\frac{\pi}{4}}^{\frac{5\pi}{4}} \frac{1}{2}\left(\frac{\sqrt{2}}{2} + \sin\theta\right)^2 d\theta$$

Just as above, the area of the inner shell is given by the integral

$$\int_{\frac{5\pi}{4}}^{\frac{7\pi}{4}} \frac{1}{2}[f(\theta)]^2\, d\theta = \int_{\frac{5\pi}{4}}^{\frac{7\pi}{4}} \frac{1}{2}\left(\frac{\sqrt{2}}{2} + \sin\theta\right)^2 d\theta$$

Finally, to compute the area we just take the difference. We need the polar integral over $\left(-\frac{\pi}{4}, \frac{5\pi}{4}\right)$ minus the polar integral over $\left(\frac{5\pi}{4}, \frac{7\pi}{4}\right)$.

$$\text{Area} = \int_{\frac{-\pi}{4}}^{\frac{5\pi}{4}} \frac{1}{2}\left(\frac{\sqrt{2}}{2} + \sin\theta\right)^2 d\theta - \int_{\frac{5\pi}{4}}^{\frac{7\pi}{4}} \frac{1}{2}\left(\frac{\sqrt{2}}{2} + \sin\theta\right)^2 d\theta$$

Note: If you are in a hurry you might think to compute the area in a polar equation just as you would for a standard function. That is, you might want to just integrate $\int r(\theta)\, d\theta$. Remember, the area formula in polar coordinates is:

$$A = \int_{\theta_1}^{\theta_2} \frac{1}{2}[f(\theta)]^2\, d\theta$$

11. D

We are given $g(x) = \int_0^{x^3} \sin(t^2)dt$ and asked to compute $g'(x)$. Just like in question 2, we need to split this into two functions to which we can apply the chain rule. Using the Fundamental Theorem of Calculus we start by letting $F(x) = \int_0^x \sin(t^2)\, dt$. Notice then that $g(x) = F(x^3)$.

Now we proceed just as before. First, using the chain rule we have $g'(x) = F'(x^3)3x^2$. Using FTC again we show that $f'(x) = \sin(x^2)$. Finally, we plug this into the formula for g' to get $g'(x) = F'(x^3)3x^2 = 3x^2 \sin(x^6)$.

12. D

As always, the first thing you do when asked to compute a limit is to evaluate the function to see if it's obviously undefined or infinity. Here we have $\frac{\ln(0+1) - 0\cos(0)}{0^2 + 0} = \frac{0-0}{0}$, and because both the top and bottom evaluate to 0 we can use L'Hôpital's Rule. Remember, L'Hôpital's Rule says that if the limit of both the numerator and denominator is 0, then you have $\lim_{x \to a} \frac{f(x)}{g(x)} = \lim_{x \to a} \frac{f'(x)}{g'(x)}$.

Applying L'Hôpital's Rule we have

$\lim_{x \to 0} \frac{\ln(x+1) - x\cos x}{x^2 + x} = \lim_{x \to 0} \frac{\frac{1}{x+1} - (\cos x - x\sin x)}{2x+1}$. Evaluating the right-hand side gives

$\frac{\frac{1}{0+1} - (\cos(0) - 0\sin(0))}{2 \cdot 0 + 1} = \frac{1 - (1-0)}{1} = 0$. The trick here, as is the trick with all L'Hôpital questions, is to work carefully and evaluate before simplifying the expressions. This will help to eliminate any algebraic mistakes. Just make sure to keep track of signs.

13. A

At the heart of this word problem is a question about tangent lines to parametric curves. We are given a parametric curve in the *xy*-plane and asked to compute the slope of the tangent line at $t = 2$. In general, the trajectory of a point traveling along a curve given by a parametric equation is the vector $(x'(t), y'(t))$, which points along a line with slope $\frac{y'(t)}{x'(t)}$.

The equations for the curve are $\begin{array}{l} y(t) = t^3 + 3t^2 + 4 \\ x(t) = 2t^2 + 2 \end{array}$. The general formula for the slope of the tangent line to a parametric curve is $\frac{y'(t)}{x'(t)}$ (this is just the infinitesimal version of "rise over run"). First we compute derivatives $\begin{array}{l} y'(t) = 3t^2 + 6t \\ x'(t) = 4t \end{array}$. Now, we want the slope at $t = 2$: $\frac{y'(2)}{x'(2)} = \frac{3 \cdot 2^2 + 6 \cdot 2}{4 \cdot 2} = \frac{12 + 12}{8} = 3.$

14. C

In order to integrate this rational function we need to first compute the partial fraction decomposition. We start with our function $\frac{x + 3}{(x + 1)x}$ and try to write it as a sum: $\frac{A}{x + 1} + \frac{B}{x}$. To solve this we add the two fractions by first forming their common denominator and then comparing that with our original equation.

$$\frac{x + 3}{(x + 1)x} = \frac{A}{x + 1} + \frac{B}{x} = \frac{Ax}{(x + 1)x} + \frac{B(x + 1)}{x(x + 1)}$$

$$\frac{x + 3}{(x + 1)x} = \frac{x(A + B) + 1(B)}{(x + 1)x}$$

By comparing the coefficients of the numerator of both sides we have $\begin{array}{l} A + B = 1 \\ B = 3 \end{array}$, and so $A = -2$. Finally we see that $\frac{x + 3}{(x + 1)x} = \frac{-2}{x + 1} + \frac{3}{x}$.

Now we integrate:

$$\int \frac{x + 3}{(x + 1)x}\,dx = \int \frac{-2}{x + 1} + \frac{3}{x}\,dx = \int \frac{-2}{x + 1}\,dx + \int \frac{3}{x}\,dx$$

$$= -2\int \frac{1}{x + 1}\,dx + 3\int \frac{1}{x}\,dx = -2\ln|x + 1| + 3\ln|x| + C$$

15. E

We're given a table of values for a function f and told that the function is *continuous*. First, we can immediately rule out answers (A) and (B), which need not be true because f need not be differentiable. You may be tempted to consider (B) because it looks like an answer to a Mean Value Theorem question, but we are only told that the function is continuous, so MVT doesn't apply. For answer (C), we aren't assuming that the function is *increasing* everywhere, and nothing about continuity implies that f can't be bigger than 14. Looking at the last two options should make you think this is an Intermediate Value Theorem question (and it is), so we need to closely evaluate the minimum number of zeros guaranteed by IVT. Choice (D) isn't exactly true. IVT says that we can, at the very least, find two distinct zeros, but there could certainly be more. Choice (E) is indeed true. f crosses the x-axis between 3 and 6, and again between 7 and 12, because our function changes sign between each of the two pairs of points.

16. C

First notice that $\sum_{n=0}^{\infty} \sin\left(\frac{\pi}{2} - \frac{1}{n}\right)$ diverges. To see this, observe that $\lim_{t\to\infty} \sin\left(\frac{\pi}{2} - \frac{1}{t}\right) = \sin\left(\frac{\pi}{2} - 0\right) = 1$ and so the series cannot converge (always remember that the first condition to check for series convergence is whether the limit of the summands is 0. $\lim_{n\to\infty} a_n = 0$ in order for $\sum_{n=0}^{\infty} a_n$ to converge). Notice that this immediately rules out answer choices (D) and (E). For I, we need to remember the order of growth of functions. Because we're comparing polynomials and exponential functions, this is relatively simple; exponentials dominate *any polynomial*, so $\sum_{n=1}^{\infty} \frac{n+1}{e^n}$ converges. For II, notice that $\sum_{n=1}^{\infty} \frac{e^n}{4^n} = \sum_{n=1}^{\infty} \left(\frac{e}{4}\right)^n$, so this is a geometric series with base $\frac{e}{4}$. $e \approx 2.71$ Thus $\frac{e}{4} < 1$ and the series converges. (Remember, a geometric series converges if and only if the base has absolute value less than 1.)

17. C

This question asks us to compute an improper integral, which means that we will eventually need to compute more limits. To start, though, just find an antiderivative:

$$\int_1^{\infty} \frac{3}{x^2}\,dx = 3\int_1^{\infty} x^{-2}\,dx = -3x^{-1}\Big|_1^{\infty} = \left(\lim_{x\to\infty} - 3x^{-1}\right) - (-3\cdot 1) = -3\left(\lim_{x\to\infty} \frac{1}{x}\right) + 3 = 0 + 3 = 3$$

In order to evaluate, we take the limit of the antiderivative as x goes to infinity (in place of evaluating the function at infinity), as above.

18. D

Here we are given a relatively straightforward indefinite integral to compute. First look for a possible u-substitution. However, in this case no such substitution exists. That leads us next to integration by parts. (This is usually the next step for functions involving e^x, a popular element in integration-by-parts questions.)

First, recall the mnemonic: $\int u\,dv = uv - \int v\,du$. The tactic involves simplifying one part of the function, u, (by differentiating) at the expense of complicating the other, dv, (by integrating). Because e^x integrated relatively easily, let's try $\begin{array}{l} u = x \\ dv = e^{-3x}dx \end{array}$ which then makes $\begin{array}{l} du = dx \\ v = \int e^{-3x}\,dx = \frac{-1}{3}e^{-3x} \end{array}$.

Plugging this into the formula above gives

$$\int xe^{-3x}\,dx = x\frac{-1}{3}e^{-3x} - \int \frac{-1}{3}e^{-3x}\,dx = \frac{-xe^{-3x}}{3} + \frac{1}{3}\int e^{-3x}\,dx$$

$$= \frac{-xe^{-3x}}{3} + \frac{1}{3}\cdot\frac{-1}{3}e^{-3x} + C = -\frac{1}{3}xe^{-3x} - \frac{1}{9}e^{-3x} + C.$$

19. D

We're given a slope field and asked which of the possible differential equations listed it could correspond to. Let's look at each choice individually.

(A) $y' = x$. First, notice that the slope field is *not* independent of y, and so this is not a possible choice. If you look along the vertical line $x = 4$, for example, the slope increases as the y-coordinate increases.

(B) $y' = xy$. This differential equation is defined everywhere. In particular, there are no *vertical* slopes in a slope field corresponding to $y' = xy$. If you look at the slope field given, it has vertical slopes near or along the x-axis. (This should give you a better hint at what the solution should be.)

(C) $y' = x^2y$. The same issue about vertical slopes applies here as well. Also, notice that this differential equation is symmetric across the y-axis. By this we mean that the slope of the curve through the point (x_0, y_0) is the same as the slope of the curve through $(-x_0, y_0)$. If you look at the slope field we are given, it appears that the slope at $(-x_0, y_0)$ is actually the *negative* of the one at (x_0, y_0).

(D) $y' = \frac{x}{y}$. If we look at the problems we've submitted for answers (A) through (C), none of these apply to answer (D). Comparing with the slope field given, we notice that this differential equation is undefined where $y = 0$, i.e., along the x-axis. It is 0 where $x = 0$, i.e., along the y-axis. These two conditions alone rule out all other answer choices. To be complete, though, let's also examine answer choice (E).

(E) $y' = \frac{y^2}{x}$. If we weren't paying close enough attention, this might seem like a valid option, and in fact, we should look closely before making a decision. The thing to notice is that the vertical slopes for this differential equation, correspond to points with $x = 0$ and $y \neq 0$, i.e., along the y-axis (except possibly the origin). Easier to notice, though, is that the horizontal slopes, the points with $y' = 0$, all fall on the x-axis. This is enough to rule out this answer choice.

In general, when solving this type of problem, isolate the lines that have either vertical slopes (y' is undefined) or horizontal slopes ($y' = 0$) and compare these with the differential equations given. This is usually enough. If this isn't enough to isolate a solution, move on to the symmetries of the slope field.

20. E

First off, we need to recognize that we are modeling a population and $P(0) = a$ is the number of creatures we are starting with. Common sense should tell you that if we start with a population of 0, it will always stay 0. Looking at the model $\frac{dP}{dt} = 0.6P\left(1 - \frac{P}{400}\right)$, if $P = 0$, then $\frac{dP}{dt} = 0$, and so, as we'd expect, there is no population change and the population stays at 0. This should immediately rule out answer choices (A) and (D). Looking closer, $\frac{dP}{dt} = 0.6P\left(1 - \frac{P}{400}\right)$ is well behaved everywhere and if $P > 400$ then the population is decreasing, because $\frac{dP}{dt} < 0$.

If $P < 400$, then the population is increasing because $\frac{dP}{dt} > 0$. Finally, if $P = 400$, then $\frac{dP}{dt} = 0$, and again we have another constant solution. These clues all point to answer choice (E).

In fact, the differential equation given is a *logistic differential equation* and limits to 400 for *every* nonzero initial condition. Below is a graph of the slope field given by $\frac{dP}{dt} = 0.6P\left(1 - \frac{P}{400}\right)$.

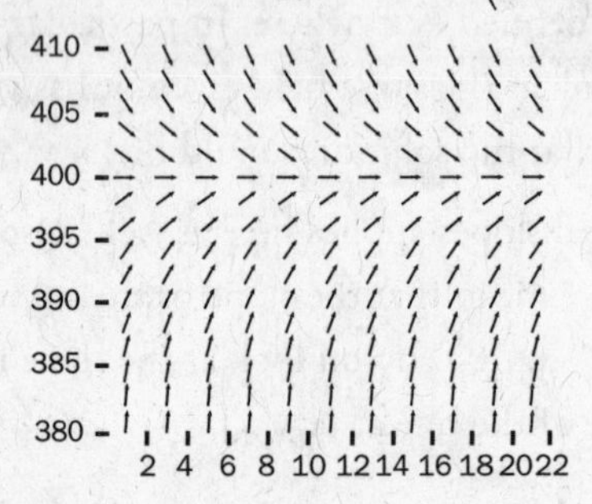

21. D

This question is asking about the composition of two functions. First, we know that $g(h) = \frac{h^2 - 3h + 2}{h - 2}$ is discontinuous at $h = 2$, but the limit of $g(h)$ can still exist as h approaches 2. To find $\lim_{h\to 2} g(h)$, let's factor the numerator of $g(h)$: $g(h) = \frac{(h-1)(h-2)}{h-2}$. We can simplify the limit expression by canceling out the common factor in the denominator and numerator, leaving us with $\lim_{h\to 2}(h - 1) = 2 - 1 = 1$. We can now substitute this into $f(x)$ to get $f\left(\lim_{h\to 2} g(h)\right) = f(1)$. If we look at the graph of the function $f(x)$, we see that there is a point discontinuity at $x = 1$ that passes through 3, but $f(x)$ does have an exact solution at this point: $f(1) = 4$. Therefore the correct answer is choice (D), 4. Be careful not to confuse the two functions. You may have been tempted to pick choice (C), which is $\lim_{x\to 1} f(x)$, but the question is asking for the value of $f(x)$ at $x = 1$, not its limit.

22. B

We are asked to compute the coefficient of x^2 in the Taylor expansion of f. First, remember that the Taylor series is given by the formula $f(a) + f'(a)x + \frac{f''(a)}{2}x^2 + \frac{f'''(a)}{3!} + \cdots$. The coefficient of x^2 is then given by the formula $\frac{f''(a)}{2}$. (Don't forget the coefficient of $\frac{1}{2}$!) We are told that $f(x) = \ln(3x^2 + 1)$. Now we just need to compute $f'(x)$ and $f''(x)$.

We compute $f'(x) = \frac{1}{3x^2+1}(6x) = \frac{6x}{3x^2+1}$ using the chain rule with functions $h(x) = \ln(x)$ and $g(x) = 3x^2 + 1$. Now, $f(x) = h(g(x))$, so $f'(x) = h'(g(x)) \cdot g'(x)$, which by plugging in $h'(x) = \frac{1}{x}$ and $g'(x) = 6x$ gives the formula. To compute $f''(x)$, we use the quotient rule to differentiate $f'(x)$. Remember, the mnemonic for the quotient rule is "low-dee-high minus high-dee-low over low-low." That is, writing $f'(x) = \frac{p(x)}{q(x)}$, we can compute f'' by $f'(x) = \frac{q(x)p'(x) - p(x)q'(x)}{(q(x))^2}$. For the function we are given, this evaluates as $f''(x) = \frac{(3x^2+1)6 - 6x(6x)}{(3x^2+1)^2}$. Now, in order to save some time and ensure we don't make an algebraic mistake simplifying the expression for f'', compute $f''(0)$ now, using the formula we've just found. $f''(0) = \frac{(3 \cdot 0^2 + 1)6 - 6 \cdot 0(6 \cdot 0)}{(3 \cdot 0^2 + 1)^2} = \frac{6}{1} = 6$. Finally, just as above, the coefficient of x^2 in the Taylor expansion is given by $\frac{f''(0)}{2} = \frac{6}{2} = 3$.

23. C

By visually estimating a flow line, you can see any solution has the line $y = 15$ as a horizontal asymptote. Thus *any* integral curve of the slope field approaches $y = 15$ as $x \to \infty$. In particular, $\lim_{x\to\infty} y(x) = 15$.

24. C

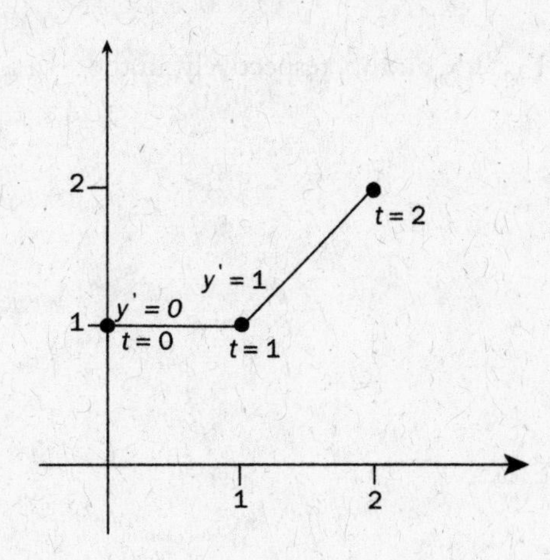

We are given the differential equation $y' = xy$ with the initial condition $y(0) = 1$, and asked to use Euler's method starting at $x = 0$ with step size 1, to approximate the value of $y(2)$. Euler's method works by approximating a function with its derivative. We start at the point (0,1) (remember our initial condition of $y(0) = 1$) and move in the direction given by $y' = xy$ for one step. At the point (0,1), $y' = 0 \cdot 1 = 0$. The slope is 0 so we move one step to the right along the line through (0,1) with slope 0. We're now at the point (1,1). Here $y' = 1 \cdot 1 = 1$, so we move one step to the right alone the line through (1,1) with slope 1 and end up at the point (2,2). This is the final x-value and so our approximation of $y(2)$ is 2.

25. A

This is just a direct computation. We are asked to compute the derivative of a function $f(x) = \sin(2\pi e^{3x} - 4\pi x)$. This involves a seemingly complicated chain rule. Let's break it into pieces to see it easier. Write $f(x) = h(g(x))$, where $h(x) = \sin x$ and $g(x) = 2\pi e^{3x} - 4\pi x$. Then by the chain rule $f'(x) = h'(g(x)) \cdot g'(x)$, so let's compute the derivatives of each of the pieces and plug them in to get a final formula. $h(x) = \sin x$ so $h'(x) = \cos x$. Similarly, $g(x) = 2\pi e^{3x} - 4\pi x$ so $g'(x) = 2\pi e^{3x} \cdot 3 - 4\pi = 6\pi e^{3x} - 4\pi$. Plugging this into the chain rule formula gives $f'(x) = h'(g(x)) \times g'(x) = \cos(2\pi e^{3x} - 4\pi x)(6\pi e^{3x} - 4\pi)$. Evaluating at $x = 0$ gives $f'(0) = \cos(2\pi e^{3 \cdot 0} - 4\pi 0)(6\pi e^{3 \cdot 0} - 4\pi) = \cos(2\pi)(2\pi) = 2\pi$.

26. C

We're given a series, told that it's the Maclaurin series for a function, and asked to find that function. The most effective way to answer this question is through the process of elimination, using guess and check techniques. To start, we're told that this is the Maclaurin series, i.e., the Taylor series about 0. Thus we know the values for the function and all its derivatives at $x = 0$. The series is $x + x^3 + x^5 + x^7 + x^9 + \ldots$ so we know that

$f(x) = 0$

$f'(x) = 1$

$f''=(x)$

because these are the coefficients of 1 (x^0), x, and x^2, respectively, in the series. Let's start by looking at the values of the functions at $x = 0$.

(A) $e^{0^2} - 1 = 1 - 1 = 0$

(B) $\sin(3 \cdot 0) = 0$

(C) $\dfrac{0}{1 - 0^2} = 0$

(D) $\sin 0 - \cos(0^2) = -1$

(E) $\dfrac{1}{1 + 0^2} - 1 = 0$

That rules out answer choice (D). Now let's try the first derivative:

$f(x)$	$f'(x)$	$f'(0)$
$e^{x^2}-1$	$e^{x^2}\cdot 2x$	$e^{0^2}\cdot 2\cdot 0=0$
$\sin(3x)$	$3\cos(3x)$	$3\cos(3\cdot 0)=3$
$\frac{x}{1-x^2}$	$\frac{(1-x^2)1+x(2x)}{(1-x^2)}$	$\frac{(1-0^2)1-0(2\cdot 0)}{(1-0^2)^2}=\frac{1}{1}=1$
~~$\sin(x)-\cos(x^2)$~~		
$\frac{1}{1+x^2}-1=(1+x^2)^{-1}-1$	$-1(1+x^2)^{-2}\cdot 2x$	$-1(1+0^2)^{-2}\cdot 2\cdot 0=0$

This rules out all choices other than $\frac{x}{1-x^2}$. Computing the derivatives of $\frac{x}{1-x^2}$ and $\frac{1}{1+x^2}-1$ can be time consuming sometimes, so let's look at these a little closer to see why we get the answer we do. Now, we should know the formula $\frac{1}{1-x}=\sum_{n=0}^{\infty}x^n$. This is just the formula for summing a geometric series. This will give us a formula for $\frac{1}{1+x^2}$: $\frac{1}{1-(-x^2)}=\sum_{n=0}^{\infty}(-x^2)^n=\sum_{n=0}^{\infty}(-1)^n x^{2n}$. Even subtracting off 1, we don't get anything that even remotely resembles $x+x^3+x^5+x^7+\cdots$. However, using this technique for $\frac{1}{1-x^2}$ gives $\frac{1}{1-x^2}=\sum_{n=0}^{\infty}(x^2)^n=\sum_{n=0}^{\infty}x^{2n}$. This gives us a power series for $\frac{x}{1-x^2}=x\frac{1}{1-x^2}=x\left(\sum_{n=0}^{\infty}x^{2n}\right)=\sum_{n=0}^{\infty}x^{2n+1}$. This is looking better. In fact, we've got it. In a different notation we have the Maclaurin series for $\frac{1}{1-x}=1+x+x^2+x^3+x^4+\cdots$.

So a power series for $\frac{1}{1-x^2}=1+x^2+x^4+x^6+x^8+\cdots$

and $\frac{x}{1-x^2}=x\left(\frac{1}{1-x^2}\right)=x(1+x^2+x^4+x^6+x^8\cdots)=x^3+x^5+x^7+x^9\cdots$. If you are worried that all we have is *some* power series for our function, and that we might get a different one by constructing the

Taylor series using all of the derivatives, just remember that the Taylor series for a power series is the power series itself (just take the derivatives); so as soon as we have any power series, we have the Taylor (and Maclaurin) series.

27. D

For this question we do something akin to what we did for problem 26; we modify our given series until we have a geometric series that we know how to solve. For example, $\sum_{n=1}^{\infty} \frac{(-3)^{n+1}}{4^n} = -3\sum_{n=1}^{\infty} \frac{(-3)^n}{4^n}$. We know how to compute this one: $\sum_{n=1}^{\infty} \frac{(-3)^n}{4^n} = \sum_{n=1}^{\infty} \left(-\frac{3}{4}\right)^n = \frac{-\frac{3}{4}}{1-\left(-\frac{3}{4}\right)} = \frac{-\frac{3}{4}}{\frac{7}{4}} = -\frac{3}{4}\cdot\frac{4}{7} = -\frac{3}{7}$. Plugging this into the formula above gives $\sum_{n=1}^{\infty} \frac{(-3)^{n+1}}{4^n} = -3\sum_{n=1}^{\infty} \frac{(-3)^n}{4^n} = -3\cdot\frac{-3}{7} = \frac{9}{7}$.

28. D

Notice that because $\sin x$ is always greater than (or equal to) -1, the derivative is always non-negative for $x > 0$: $f'(x) = \frac{\sin x + 1}{x} \geq 0$, which implies that $f(x)$ is always increasing for $x > 0$. Because we're given that $f(1) = 0$, we can conclude that $f(x) \geq 0$ for $x > 1$. In order to show that $f(x) > 0$, we just need to notice that at $x = 1$, the derivative is larger than 0: $f'(1) = \frac{\sin(1) + 1}{1} > 0$, and so the function is immediately increasing. Thus immediately after $x = 1$, the function becomes greater than 0. Because the function is non-decreasing, it remains greater than 0 forever. Also, because the function is increasing, we know that $f(x) \leq 0$ for $x < 1$.

29. D

We start with the chain rule. $h(x) = g(f(x))$ so $h'(x) = g'(f(x))f'(x)$. In order to compute $h'(1)$, we just find the appropriate values from the table. First we use that $f(1) = 1$ and $f'(1) = 3$, so $h'(1) = g'(f(1))f'(1) = g'(1)\cdot 3$. Now we need $g'(1) = 5$, and $h'(1) = g'(f(1))f'(1) = g'(1)\cdot 3 = 5\cdot 3 = 15$.

30. B

Just as in problem 21, we can see from the graph of f that it's continuous at $x = 4$ and that $f(4) = 10$. Because f is continuous at $x = 4$ and $\ln x$ is continuous at $x = 10$, the composition is continuous and the limit is just the evaluation $\lim_{x\to 4} \ln(f(x)) = \ln(f(4)) = \ln(10) = 2.303$.

31. E

To compute the position we integrate velocity.

$$p(5) = \int_0^5 v(t)dt + p(0) = \int_0^5 (1-t^2)\sin t dt + 2 = 18.830 + 2 = 20.830$$

Because we have access to calculators, make the calculator do all the work. Don't forget to add in the initial position, though! Just integrating velocity gives you *displacement* and tells you how far (and in which direction) the particle has moved. In order to determine the final position, we need to know the initial position (and then the displacement).

32. D

Implicit in the statement "for the first six seconds of driving" is the idea that the car is starting from a standing position, that is, the initial velocity is 0. In order to determine the velocity when the car *first begins to decelerate* we need to know at what time that happens. Graphing acceleration shows that the first time acceleration becomes negative occurs at $t = 4.627$ seconds. Velocity is then given by integrating acceleration from $t = 0$ to $t = 4.627$, answer choice (D).

33. A

Notice that for answers (B) and (C), either the left or the right Riemann sum gives the actual area. Option (D) is increasing instead of decreasing and (E), the left Riemann sum, under-approximates it. Only (A) satisfies the conditions.

34. D

We compute the volume of the solid by integrating the areas of the rectangular slices that lie over slices of the region R. We slice parallel to the y-axis. The rectangle that lies over the line $x = x_0$, has a base of length $\cos \pi x_0$ and so has a height of $\frac{3}{2}\cos \pi x_0$. This tells us the area of the rectangle is $b \cdot h = \cos \pi x_0 \cdot \frac{3}{2}\cos \pi x_0 = \frac{3}{2}\cos^2 \pi x_0$. Because the slices are parameterized by their x-coordinate, to compute the volume we integrate the area of each slice as a function of x. We are given the interval of integration in the statement of the question. All that's left is to plug it into the calculator and integrate:

$$\int_{-\frac{1}{2}}^{\frac{1}{2}} \frac{3}{2}\cos^2 \pi x \, dx = 0.75$$

35. B

Remember, for a particle traveling along a parametric curve, the speed of the particle is given by the equation $\sqrt{x'(t)^2 + y'(t)^2}$ (this is just the length of the tangent vector). For the parametric equation we were given, $x'(t) = e^t$ and $y'(t) = 2t$. To compute the speed at $t = 2$, we just plug in to get $\sqrt{x'(2)^2 + y'(2)^2} = \sqrt{(e^2)^2 + (2 \cdot 2)^2} = 8.402$.

36. D

To find the time that the cloud is shrinking the fastest, we need to determine where the rate of change of volume is *most negative,* i.e., at the minimum of $r(t)$. Use the calculator directly to find this is at $t = 1.456$, which gives a rate $r(t)$ of -0.282.

37. E

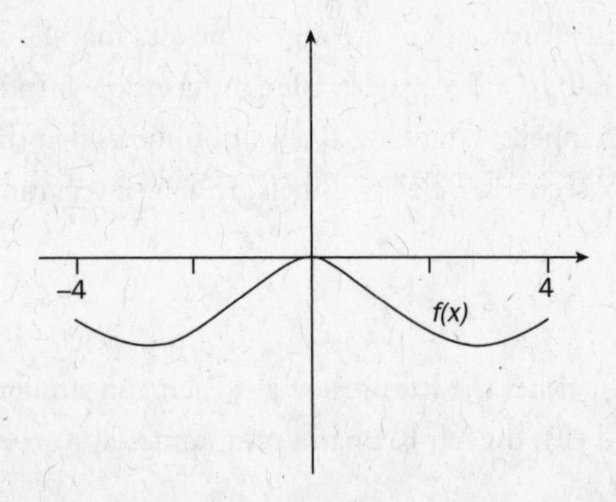

The picture above is an estimate of the graph of $f(x)$. Only answer (E) is satisfied. Let's look at each answer choice separately to see where they fail.

(A) $f(-4) = -f(4)$. This looks like it might be true, but be careful. $f'(x)$ looks like it's an odd function so the area under the graph of $f'(x)$ between –4 and 0 is the negative of the area from 0 to 4. Using the Fundamental Theorem we have $f(0) - f(-4) = \int_{-4}^{0} f'(x)\,dx = -\int_{0}^{4} f'(x)\,dx = -(f(4) - f(0))$. Simplifying this expression:

$$\begin{aligned} f(0) - f(-4) &= -(f(4) - f(0)) \\ f(0) - f(-4) &= -f(4) + f(0) \\ -f(-4) &= -f(4) \\ f(-4) &= f(4) \end{aligned}$$

(B) $f(x)$ has a local minimum at $x = 0$. Because the graph of $f'(x)$ changes from *positive* to *negative* at $x = 0$, $f(x)$ has a local *maximum* at $x = 0$, not a local minimum.

(C) $f(x)$ is increasing on $[-4,4]$. Because $f'(x)$ is negative on the intervals $[-4,-2]$ and $[0,2]$, it is decreasing here, not increasing.

(D) $f(x) \geq 0$ on $[-4,0]$. We are given $f(0) = 0$. Because $f'(x)$ is positive on the interval $[-2,0]$, f is increasing on that interval. Because it increases to 0 at $x = 0$, the function must be negative before that.

(E) $f(x) \leq 0$ on $[-4,4]$. In fact, if we continue with the line of reasoning above, because the area below the graph of $f'(x)$ on $[-4,-2]$ is visibly larger than the region on $[-2,0]$, the function must be less than 0 on the whole interval $[-4,0]$. Turning this logic around, we see that the function decreases again on the interval $[0,2]$; even though it increases over the interval $[2,4]$, it doesn't increase enough to make the function positive. This is our answer.

38. B

Again, total change is the integral of the rate of change. Because we measure height beginning with the sprout (i.e., height is 0 feet), the height of the tree is just given by $\int_0^8 (x^2 + x\cos(3x))0.9^{\frac{x^2+2}{6}}\,dx = 86.882$. Plug it into the calculator and compute. Round to 87.

39. C

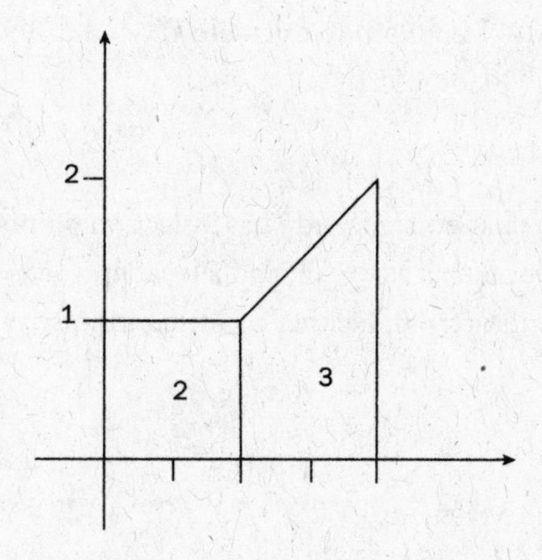

40. D

The coefficient of $(x-3)^2$ in the degree four Taylor polynomial (in fact, in any degree Taylor polynomial) $P(x)$ is equal to $\frac{f''(3)}{2!}$. In this case the coefficient is 3, so $\frac{f''(3)}{2} = 3$, or $f''(3) = 2 \cdot 3 = 6$.

41. D

First we need to determine an equation for the area of the rectangle as a function of either x or y (but not both). For a given point (x, y) on the graph of $y = 2x - x^2$, the base of the rectangle has length $(2 - x) - x = 2 - 2x$ for $x \leq 1$, and height $f(x) = 2x - x^2$. So the area of the rectangle (as a function of x) is $A(x) = (2 - 2x)(2x - x^2) = 2x^3 - 6x^2 + 4x$. To compute the maximum, either plug directly into the calculator (definitely the faster option) or compute a zero of the derivative: $A'(x) = 6x^2 - 12x + 4$. One of the roots is $1 - \frac{\sqrt{3}}{3}$, and plugging this into A gives $A\left(1 - \frac{\sqrt{3}}{3}\right) = A(0.423) = 0.770$.

The other root corresponds to values of $x \geq 1$. In this case our formula for the base of the rectangle produces a negative value. We need to adjust the length of the rectangle (for $x > 1$) to $2x - 2$. The corresponding solution for $x = 1 + \frac{\sqrt{3}}{3}$ results in $A\left(1 + \frac{\sqrt{3}}{3}\right) = A(1.577) = 0.770$.

42. E

The Intermediate Value Theorem says that somewhere in the interval [0,4] there's a c with $f(c) = 4$. The Mean Value Theorem then says that somewhere on $[c,7]$, there's a d with $f'(d) = 0$. Informally, on [0,4], f increased from 3 to 7, then on [4,7], f decreased from 7 to 4. So there was a local max somewhere on the interval [0,7].

43. B

On the interval [0, 4], the average rate of change is given by the formula $\frac{f(4) - f(0)}{4 - 0} = \frac{e^4 + 4 - 1}{4} =$ 14.39953. The instantaneous rate of change is always just the derivative and is given by the equation $f'(x) = e^x + (1 + \sin(2\pi x)) + x \cdot \cos(2\pi x) \cdot 2\pi = e^x + (1 + \sin(2\pi x)) + 2\pi x \cos(2\pi x)$. Use your calculator's solver feature (or find the zero of $f'(x) - 14.39953$) to calculate $f'(x) = 14.39953$ at $x = 3.398$. Other possible answers are 1.864, 2.212, 2.740, or 3.551.

44. B

Remember, f has extrema wherever f' changes sign, and f has inflection points at the extrema of f'. Looking at the graph of f', we see that f has a local max at $x = -3$, and a local min at $x = 1$. f has one inflection point—the local min of f' which occurs at $x = 0$. Neither the discontinuity at $x = 2$ nor the point $x = 3$ is an inflection point.

45. D

The right Riemann sum is given by $(13 - 5)f(13) + (22 - 13)f(22) + (27 - 22)f(27) + (35 - 27)f(35) =$ $(13 - 5)12 + (22 - 13)13 + (27 - 22)17 + (35 - 27)22 = 474$.

FREE-RESPONSE ANSWERS

1. (a) By the Fundamental Theorem of Calculus $P(t) = \int_0^t P'(s)\, ds + P(0) = \int_0^t \ln(1 + 5s^2)\, ds + P(0)$

Calculating at $t = 50$ gives $P(50) = 373.075 + P(0)$. So the population in 1950 is approximately 377,075.

(b) By the formula given we need to solve

$1000 = P(t) = \int_0^t P'(s)\, ds + P(0) = \int_0^t \ln(1 + 5s^2)\, ds + P(0)$ Calculating, this occurs $t \approx 110.314$.

Further,

$$\int_0^{110} \ln(1 + 5t^2)dt + P(0) \approx 997,$$

$$\int_0^{111} \ln(1 + 5t^2)dt + P(0) \approx 1008$$

So the population reaches 1 million between $t = 110$ and $t = 111$, that is, in the year 2010.

(c) The starting point is the initial population $t = 0$, $P_0 = 4$.

$t = 0$	$P_1 = 4 + 25 \cdot P'(0)$
$t = 25$	$P_2 = P_1 + 25 \cdot P'(25)$
$t = 50$	$P_3 = P_2 + 25 \cdot P'(50)$
$t = 75$	$P_4 = P_3 + 25 \cdot P'(75)$

$P = 4 + 25 \cdot (P'(0) + P'(25) + P'(50) + P'(75)) = 697.138$

So using Euler's method, the population in 2000 is 697,138.

Once you know what the integral needs to be for parts (a) and (b), make the calculator do all the work. The less you do by hand, the less likely you are to make a silly mistake. For part (b), use your calculator's solve feature or plot the integral. Remember, for Euler's method, we estimate the graph of the integral by a sequence of line segments where the slope is determined by the derivative of the function at the starting point of the line segment.

2. (a) To compute area we need to solve $\sin x = (1 - \cos x)$. The two solutions that fit the graph are $x = 0$ and $x = \frac{\pi}{2}$. Then the area of R is given by the integral Area $\int_0^{\frac{\pi}{2}} \sin x - (1 - \cos x)\, dx = 0.429$.

   (b) We have the interval of integration from part (a). The volume of the region generated when R is revolved about the x-axis is given by the integral $\pi \int_0^{\frac{\pi}{2}} (\sin x)^2 - (1 - \cos x)^2\, dx = 1.348$.

   (c) To compute the volume we integrate the area of slices parallel to the y-axis. Each rectangular slice has height $h = b^2$, and base $b(x) = \sin x - (1 - \cos x)$. Thus the area of each slice is $bh = b^3 = (\sin x - (1 - \cos x))^3$. The volume of the solid is given by the integral

$$\int_0^{\frac{\pi}{2}} bh\, dx = \int_0^{\frac{\pi}{2}} b^3\, dx = \int_0^{\frac{\pi}{2}} (\sin x - (1 - \cos x))^3\, dx = 0.050.$$

The first thing we need to do is compute the intersection points; $y = \sin(x)$ and $y = 1 - \cos(x)$ obviously intersect at $x = 0$. Use your calculator to find the second intersection point, $\left(\frac{\pi}{2}, 1\right)$. You can use the decimal approximation, if you like, but use the calculator to save the coordinates of the point, or record eight or nine significant figures, for accuracy. You can see the area integral easily by the following:

$$\int_0^{\frac{\pi}{2}} \sin x - (1 - \cos x)\, dx = (-\cos x - x + \sin x)\Big|_0^{\frac{\pi}{2}} = \left(0 - \frac{\pi}{2} + 1\right) - (-1 - 0 + 0)$$

$$= 2 - \frac{\pi}{2} = 0.429.$$

To compute the volume of the solid of rotation, we use the washer method. Again, make the calculator do all the work. For part (c), we determine the answer by integrating parallel cross sections. Draw a picture or do whatever you need to do to see what the picture needs to be, but the general idea is always the same. They give you the information you need to determine the area of the cross section as a function of x. Integrate this function over the appropriate region. The only trick is in finding the formula.

3. (a) The speed of the particle at time t is the length of the tangent vector of velocity. At $t = 0$ this is:

$$\sqrt{x(0)^2 + y(0)^2} = \sqrt{(3e^{3(0)})^2 + (2(0) - 4)^2} = \sqrt{(3(1))^2 + (-4)^2} = \sqrt{25} = 5$$

At a general time t, the acceleration vector is given by the pair $(x''(t), y''(t))$. We compute that and $x''(t) = 9e^3$ and $y''(t) = 2$. At $t = 0$ we see that the acceleration vector is given by $(9e^{3t}, 2) = (9e^0, 2) = (9, 2)$.

(b) The x- and y-coordinates are computed by integrating the velocity vector equations and taking into account the initial conditions of the position. Here, the initial condition is at $t = 0$ and position (0,0).

$$x(t) = \int_0^t x'(t)\,dt = \int_0^t 3e^{3t}\,dt = e^{3t} - x(0)$$

$$y(t) = \int_0^t y'(t) = \int_0^t 2t - 4 = t^2 - 4t - y(0)$$

At $t = 4$, we compute the x- and y-coordinates as:

$x(4) = e^{3(4)} - 0 = e^{12}$

$y(4) = 4^2 - 4(4) - 0 = 0$

(c) To describe the behavior of the x- and y-coordinates as $t \to -\infty$ and as $t \to +\infty$, we should think about what happens when we input very small or very large values for t into the equations for $x(t)$ and $y(t)$. For very small, or increasingly negative, values of t, we see that the x-coordinate approaches 0 while the y-coordinate approaches infinity. Thus, there is a vertical asymptote at $x = 0$ as $t \to -\infty$. For very large values of t, both the $x(t)$ and $y(t)$ equations go to infinity. Therefore, the x- and y-coordinates are unbounded as $t \to \infty$.

4. (a) We solve by implicit differentiation.

$$\frac{d}{dx}(y^2 + 2x = x^2 + 2xy + 1)$$

$$\frac{d}{dx}(y^2 + 2x) = \frac{d}{dx}(x^2 + 2xy + 1)$$

$$2y\frac{dy}{dx} + 2 = 2x + 2y + 2x\frac{dy}{dx}$$

$$\frac{dy}{dx}(2y - 2x) = 2x + 2y - 2$$

$$\frac{dy}{dx} = \frac{2}{2} \cdot \frac{x + y - 1}{y - x}$$

(b) We solve by plugging $y = 2$ into the equation for the curve.

$$2^2 + 2x = x^2 + 2x(2) + 1$$

$$0 = x^2 + 4x - 2x + 1 - 4$$

$$0 = x^2 + 2x - 3$$

$$0 = (x + 3)(x - 1)$$

So the points of intersection are (1,2) and (−3,2). The slope of the tangent line at (1,2) is

$$\frac{dy}{dx} = \frac{1 + 2 - 1}{2 - 1} = 2$$

(c) A horizontal tangent line occurs where:

$$0 = \frac{dy}{dx} = \frac{x + y - 1}{y - x}$$

$$0 = x + y - 1$$

$$y = 1 - x$$

Plugging this into the equation for the curve gives:

$$(1 - x)^2 + 2x = x^2 + 2x(1 - x) + 1$$

$$1 - 2x + x^2 + 2x = x^2 + 2x(1 - x) + 1$$

$$x^2 + 1 = x^2 + 2x(x - 1) + 1$$

$$0 = 2x(x - 1)$$

So $x = 0,1$. Because $y = 1 - x$, we have that at $x = 0$, $y = 1$, and at $x = 1$, $y = 0$.

So the points with horizontal tangencies are (0,1) and (1,0).

Check that these all satisfy $\frac{dy}{dx} = 0$.

For (0,1), $\frac{dy}{dx} = \frac{0 + 1 - 1}{1 - 0} = \frac{0}{1} = 0.$

For (1,0), $\frac{dy}{dx} = \frac{1+0-1}{0-1} = \frac{0}{-1} = 0.$

Thus, the points on the curve with horizontal tangencies are indeed (0,1) and (1,0).

For part (a), always remember that because y is a function of x, taking the derivative of y^2 involves a chain rule: $\frac{d}{dx}(y^2) = 2(y)\frac{d}{dx}(y) = 2y\frac{dy}{dx}$. Similarly, taking the derivative of $2xy$ involves the product rule: $\frac{d}{dx}(2xy) = 2\frac{d}{dx}(x)y + 2x\frac{d}{dx}(y) = 2 \cdot 1 \cdot y + 2x\frac{dy}{dx}$.

For part (c), always remember to *double-check* all points you get through an algebraic solution. It's possible that during an algebraic manipulation, you chose the sign of a square root or assumed something was nonzero and divided by it. There are many ways to lose small pieces of vital information. Always go back and check that your answer makes sense.

5. (a) We are told that the area of the base remains constant, so only p and h vary with time. We are told that $\frac{dP}{dt} = \frac{t+1}{h}$. Therefore,

$$\frac{dP}{dt} = 5r^2 \frac{dh}{dt} = 20\frac{dh}{dt},$$ because r is a constant 2 meters.

Plugging in $\frac{dP}{dt} = \frac{t+1}{h}$ gives

$$20\frac{dh}{dt} = \frac{t+1}{h}$$

$$\frac{dh}{dt} = \frac{t+1}{20h}$$

(b) We solve $\frac{dh}{dt} = \frac{t+1}{20h}$ with initial condition that at $t = 0$, $h = 0$.

Separating $\frac{dh}{dt} = \frac{t+1}{20h}$ gives $20h\,dh = (t+1)dt$.

Integrating:

$$\int 20h\,dh = \int t + 1\,dt$$

$$10h^2 = \frac{t^2}{2} + t + C$$

$$h(t) = \sqrt{\frac{1}{10}\ \frac{t^2}{2} + t + C}$$

To solve for the constant we use $10h^2 = \frac{t^2}{2} + t + C$, giving $10 \cdot 0^2 = \frac{0^2}{2} + 0 + C$ and $C = 0$.

Thus the solution is given by the equation $h(t) = \sqrt{\frac{1}{10}\left(\frac{t^2}{2} + t\right)}$.

(c) Because P is given in units of hundreds of ants, we need to solve for $P = 40$.

$40 = P(t) = 5r^2h(t) = 20h$, so $h = 2$. Using the equations from part (b),

$$10 \cdot 2^2 = \frac{t^2}{2} + t$$

$$80 = t^2 + 2t$$

$$0 = t^2 + 2t - 80 = (t-8)(t+10)$$

So, because $t > 0$, $t = 8$ days.

Part (a) is a straightforward related-rates problem. The only thing to note is that because the radius is constant, you do not need to take its derivative. (Even if you do, because $\frac{dr}{dt} = 0$, you get the same answer.) For part (b), when we finally solve for h as a function of t, we make a choice of the positive square root. This is necessary in this case, because we expect the height of the anthill always to be positive. If we were doing these same computations for a function that is negative, we would need to choose the negative root. This is why the shortcut used in part (c) is legitimate, because we would need to undo these steps anyway to solve $h = 2 = \sqrt{\frac{1}{10}\left(\frac{t^2}{2} + t\right)}$.

6. (a) The formula for the Taylor series for $f(x)$ about $x = 0$ is

$$\sum_{n=0}^{\infty} \frac{f^{(n)}(0)}{n!} x^n = \sum_{n=0}^{\infty} \frac{(n+1)!}{n!3^n} x^n = \sum_{n=0}^{\infty} \frac{(n+1)}{3^n} x^n$$

We just plug the formula $f^{(n)}(0) = \frac{(n+1)!}{3^n}$ into the formula for the general term of the Taylor series: $\frac{f^{(n)}(0)}{n!} x^n$. Don't forget the $n!$ in the denominator! Because there's already a factorial given in the formula for $f^{(n)}(0)$ you might overlook it.

(b) We can determine the radius of convergence for this series using the ratio test.

$$\lim_{x\to\infty} \left|\frac{a_{n+1}}{a_n}\right| = \left|\lim_{x\to\infty} \frac{\frac{(n+2)}{3^{n+1}} x^{n+1}}{\frac{(n+1)}{3^n} x^n}\right| = \lim_{x\to\infty} \left|\frac{3^n (n+2)}{3^{n+1}(n+1)} x\right| = \lim_{x\to\infty} \left|\frac{n+2}{n+1} \frac{x}{3}\right| = \lim_{x\to\infty} \frac{(n+2)}{(n+1)}\left|\frac{x}{3}\right| = \left|\frac{x}{3}\right|$$

The series will diverge when the above limit is greater than 1 and will converge when the limit is less than 1. Therefore the series will converge for all x such that $\left|\frac{x}{3}\right| < 1 \Rightarrow |x| < 3$. The radius of convergence for this series is 3. Because we are not asked for the interval of convergence, we do not have to test the endpoints.

(c) Integrating the general term of the Taylor series of f gives

$$\int \frac{(n+1)}{3^n} x^n \, dx = \frac{n+1}{3^n} \frac{x^{n+1}}{n+1} + C = \frac{x^{n+1}}{3^n} + C.$$

A term-by-term integration of the Taylor series of f then gives the series

$$\sum_{n=0}^{\infty} \int \frac{(n+1)x^n}{3^n} \, dx = \sum_{n=0}^{\infty} \frac{x^{n+1}}{3^n} + C$$

(d) We are given $\frac{1}{1-x} = \sum_{n=0}^{\infty} x^n$. Then the power series for $\frac{3}{1-\frac{x}{3}}$ is

$$3\frac{1}{1-\frac{x}{3}} = 3\sum_{n=0}^{\infty}\left(\frac{x}{3}\right)^n = \sum_{n=0}^{\infty}\frac{x^n}{3^{n-1}} = \sum_{n=0}^{\infty}\frac{x^{n+1}}{3^n} + 3$$

In part (c), Taylor's theorem tells you that on the interval of convergence, you can interchange a function and its Taylor series in almost any situation. In particular, suppose that $g'(x) = f(x)$, and so $g^{(n+1)}(x) = f^{(n)}(x)$. Then $\int \frac{f^{(n)}(0)}{n!}x^n = \frac{f^{(n)}(0)}{n!}\cdot\frac{x^{n+1}}{n+1} = \frac{g^{(n+1)}}{(n+1)!}x^{n+1}$. So integrating the Taylor series for f gives you the Taylor series for g, up to the constant. In fact, this tells you some information for part (d) as well. This means that the Taylor series for any power series is just the power series itself. You can also answer part (d) by going the other direction. Start with the antiderivative and show that it is the same as the power series for $\frac{3}{1-\frac{x}{3}}$, as follows:

$$\sum_{n=0}^{\infty}\frac{x^{n+1}}{3^n} + 3 = 3\sum_{n=0}^{\infty}\frac{x^{n+1}}{3^{n+1}} + 3 = 3\left(\sum_{n=1}^{\infty}\frac{x^n}{3^n} + 1\right) = 3\left(\sum_{n=0}^{\infty}\frac{x^n}{3^n}\right) = 3\left(\frac{1}{1-\left(\frac{x}{3}\right)}\right) = \frac{3}{1-\frac{x}{3}}$$

AP CALCULUS BC PRACTICE TEST 1 CORRELATION CHART

Multiple Choice Question	Review Chapter: Subsection of Chapter
1	Chapter 12: Chain Rule Chapter 12: Derivatives of Trigonometric Functions
2	Chapter 12: Chain Rule Chapter 20: What's So Fundamental? The Connection Between Integrals and Derivatives
3	Chapter 25: Comparing Series to Test for Convergence or Divergence
4	Chapter 23: Approximating the Area Under the Curve Using Riemann Sums
5	Chapter 26: Taylor Polynomial Approximation
6	Chapter 22: Separable Differential Equations
7	Chapter 13: The Slope of a Curve at a Point
8	Chapter 21: When *F* Is Complicated: The Substitution Game
9	Chapter 16: Related Rates Chapter 16: Differential Equations
10	Chapter 10: Polar Functions Chapter 18: Area Bounded by Curves
11	Chapter 20: What's So Fundamental? The Connection Between Integrals and Derivatives
12	Chapter 16: L'Hôpital's Rule
13	Chapter 10: Parametric Functions Chapter 12: Derivatives of Parametric, Polar, and Vector Functions
14	Chapter 21: Antiderivatives by Simple Partial Fractions
15	Chapter 9: Properties of Continuous Functions
16	Chapter 8: Rates of Growth Chapter 24: To Converge or Not to Converge Chapter 25: The Properties of Series Chapter 25: The Geometric Series with Applications
17	Chapter 21: Antiderivatives by Improper Integrals
18	Chapter 21: Integration by Parts
19	Chapter 16: Differential Equations
20	Chapter 22: Solving Logistic Differential Equations and Using Them in Modeling
21	Chapter 7: Evaluating Limits Algebraically Chapter 9: Continuity and Limits

Multiple Choice Question	Review Chapter: Subsection of Chapter
22	Chapter 26: Taylor Polynomial Approximation
23	Chapter 16: Differential Equations Chapter 8: Horizontal Asymptotes
24	Chapter 16: Differential Equations Via Euler's Method
25	Chapter 12: Chain Rule Chapter 12: Rules for Computing Derivatives Chapter 12: Derivatives of Trigonometric Functions
26	Chapter 26: Taylor Polynomial Approximation Chapter 26: Maclaurin Series for Special Functions
27	Chapter 25: The Geometric Series with Applications
28	Chapter 14: Increasing and Decreasing Behavior of f and the Sign of f'
29	Chapter 12: Chain Rule
30	Chapter 7: Evaluating Limits Computationally
31	Chapter 18: Distance Traveled by a Particle Along a Line
32	Chapter 22: Accumulation Functions and Antiderivatives
33	Chapter 23: Approximating the Area Under the Curve Using Riemann Sums
34	Chapter 18: Volumes of Solids
35	Chapter 10: Parametric Functions Chapter 16: Interpreting the Derivative As a Rate of Change
36	Chapter 16: Optimization
37	Chapter 16: Graphing Using Derivatives
38	Chapter 22: Accumulation Functions and Antiderivatives
39	Chapter 23: Approximating the Area Under Functions Given Graphically or Numerically
40	Chapter 26: Taylor Polynomial Approximation
41	Chapter 16: Optimization
42	Chapter 9: Properties of Continuous Functions
43	Chapter 14: The Mean Value Theorem
44	Chapter 14: Critical Points and Local Extrema Chapter 14: Corresponding Characteristics of the Graphs of f and f' Chapter 15: Inflection Points
45	Chapter 23: Approximating the Area Under the Curve Using Riemann Sums

Free Response Question	Review Chapter: Subsection of Chapter
1(a)	Chapter 20: What's So Fundamental? The Connection Between Integrals and Derivatives Chapter 22: Accumulation Functions and Antiderivatives
1(b)	Chapter 22: Accumulation Functions and Antiderivatives
1(c)	Chapter 16: Differential Equations Via Euler's Method
2(a)	Chapter 18: Area Bounded by Curves
2(b)	Chapter 18: Solids of Revolution
2(c)	Chapter 18: Volumes of Solids
3(a)	Chapter 16: Interpreting the Derivative as a Rate of Change
3(b)	Chapter 21: Computing Antiderivatives Directly
3(c)	Chapter 10: Parametric Functions Chapter 8: Vertical Asymptotes—Understanding Infinite Limits Graphically Chapter 8: Horizontal Asymptotes
4(a)	Chapter 12: Implicit Differentiation
4(b)	Chapter 13: The Slope of a Curve at a Point
4(c)	Chapter 13: The Slope of a Curve at a Point Chapter 14: Critical Points and Local Extrema
5(a)	Chapter 16: Related Rates Chapter 22: Differential Equations
5(b)	Chapter 22: Separable Differential Equations
5(c)	Chapter 22: Differential Equations
6(a)	Chapter 26: Taylor Polynomial Approximation
6(b)	Chapter 25: The Ratio Test for Convergence and Divergence
6(c)	Chapter 26: Shortcuts Using Taylor/Maclaurin Series
6(d)	Chapter 26: Shortcuts Using Taylor/Maclaurin Series

AP Calculus BC Practice Test 2 Answer Grid

1.	Ⓐ Ⓑ Ⓒ Ⓓ Ⓔ	13.	Ⓐ Ⓑ Ⓒ Ⓓ Ⓔ	25.	Ⓐ Ⓑ Ⓒ Ⓓ Ⓔ	37.	Ⓐ Ⓑ Ⓒ Ⓓ Ⓔ
2.	Ⓐ Ⓑ Ⓒ Ⓓ Ⓔ	14.	Ⓐ Ⓑ Ⓒ Ⓓ Ⓔ	26.	Ⓐ Ⓑ Ⓒ Ⓓ Ⓔ	38.	Ⓐ Ⓑ Ⓒ Ⓓ Ⓔ
3.	Ⓐ Ⓑ Ⓒ Ⓓ Ⓔ	15.	Ⓐ Ⓑ Ⓒ Ⓓ Ⓔ	27.	Ⓐ Ⓑ Ⓒ Ⓓ Ⓔ	39.	Ⓐ Ⓑ Ⓒ Ⓓ Ⓔ
4.	Ⓐ Ⓑ Ⓒ Ⓓ Ⓔ	16.	Ⓐ Ⓑ Ⓒ Ⓓ Ⓔ	28.	Ⓐ Ⓑ Ⓒ Ⓓ Ⓔ	40.	Ⓐ Ⓑ Ⓒ Ⓓ Ⓔ
5.	Ⓐ Ⓑ Ⓒ Ⓓ Ⓔ	17.	Ⓐ Ⓑ Ⓒ Ⓓ Ⓔ	29.	Ⓐ Ⓑ Ⓒ Ⓓ Ⓔ	41.	Ⓐ Ⓑ Ⓒ Ⓓ Ⓔ
6.	Ⓐ Ⓑ Ⓒ Ⓓ Ⓔ	18.	Ⓐ Ⓑ Ⓒ Ⓓ Ⓔ	30.	Ⓐ Ⓑ Ⓒ Ⓓ Ⓔ	42.	Ⓐ Ⓑ Ⓒ Ⓓ Ⓔ
7.	Ⓐ Ⓑ Ⓒ Ⓓ Ⓔ	19.	Ⓐ Ⓑ Ⓒ Ⓓ Ⓔ	31.	Ⓐ Ⓑ Ⓒ Ⓓ Ⓔ	43.	Ⓐ Ⓑ Ⓒ Ⓓ Ⓔ
8.	Ⓐ Ⓑ Ⓒ Ⓓ Ⓔ	20.	Ⓐ Ⓑ Ⓒ Ⓓ Ⓔ	32.	Ⓐ Ⓑ Ⓒ Ⓓ Ⓔ	44.	Ⓐ Ⓑ Ⓒ Ⓓ Ⓔ
9.	Ⓐ Ⓑ Ⓒ Ⓓ Ⓔ	21.	Ⓐ Ⓑ Ⓒ Ⓓ Ⓔ	33.	Ⓐ Ⓑ Ⓒ Ⓓ Ⓔ	45.	Ⓐ Ⓑ Ⓒ Ⓓ Ⓔ
10.	Ⓐ Ⓑ Ⓒ Ⓓ Ⓔ	22.	Ⓐ Ⓑ Ⓒ Ⓓ Ⓔ	34.	Ⓐ Ⓑ Ⓒ Ⓓ Ⓔ		
11.	Ⓐ Ⓑ Ⓒ Ⓓ Ⓔ	23.	Ⓐ Ⓑ Ⓒ Ⓓ Ⓔ	35.	Ⓐ Ⓑ Ⓒ Ⓓ Ⓔ		
12.	Ⓐ Ⓑ Ⓒ Ⓓ Ⓔ	24.	Ⓐ Ⓑ Ⓒ Ⓓ Ⓔ	36.	Ⓐ Ⓑ Ⓒ Ⓓ Ⓔ		

AP CALCULUS BC PRACTICE TEST 2

SECTION I, PART A

Time: 55 Minutes
28 Questions

NO GRAPHING CALCULATOR IS ALLOWED ON THIS PORTION OF THE EXAM

Directions: Solve the following problems, using available space for scratchwork. After examining the form of the choices, decide which one is the best of the choices given and fill in the corresponding oval on the answer sheet. No credit will be given for anything written in the test book. Do not spend too much time on any one problem.

In this test:

(1) The domain of a function f is the set of all real numbers x for which $f(x)$ is a real number, unless otherwise specified.

(2) The inverse of a trigonometric function f may be indicated using the inverse function notation f or with the prefix "arc" (e.g., $\sin^{-1} x = \arcsin x$).

1. $\lim_{x \to a} \dfrac{\ln x - \ln a}{x - a} =$

(A) 0
(B) a
(C) 1
(D) $\dfrac{1}{a}$
(E) undefined

2. Let $f(x) = \int_1^{x^2} (3t^2 - 5t + 7)\,dt$. Let $P(x)$ be a second degree Taylor polynomial for f centered at $x_0 = -1$. Find a formula for $P(x)$.

(A) $P(x) = 5(x + 1) + 5(x + 1)^2$
(B) $P(x) = -10(x + 1) + 7(x + 1)^2$
(C) $P(x) = 5(x+1) + \dfrac{5}{2}(x+1)^2$
(D) $P(x) = -10(x + 1) + 14(x + 1)^2$
(E) $P(x) = -11(x + 1) + 14(x + 1)^2$

GO ON TO THE NEXT PAGE

3. $\lim_{x \to \infty} \frac{2x^3 - 5x + 4}{x^3 + 3x^2 - 1} =$

(A) -4

(B) 0

(C) 1

(D) 2

(E) ∞

4. $\frac{d}{dx}\left(-2x^3 + e^{2x} + 5\right) =$

(A) $-2x^3 + e^2x$

(B) $-\frac{1}{2}x^4 + \frac{1}{2}e^{2x} + 5x + C$

(C) $-6x^2 + e^2x$

(D) $-6x^2 + 2e^2x + 5x$

(E) $-6x^2 + 2e^2x$

5. The position vector for a particle moving in the xy-plane is given by $p(t) = \left\langle 3t^2, 2\sqrt{t} \right\rangle$. What is the velocity vector of the particle at time $t = 4$?

(A) $\left\langle 24, \frac{1}{2} \right\rangle$

(B) $\langle 48, 4 \rangle$

(C) $\left\langle 64, \frac{32}{3} \right\rangle$

(D) $\frac{\sqrt{2305}}{4}$

(E) $4\sqrt{145}$

6. $\int x \sin\left(x^2\right) dx =$

(A) $\frac{1}{2}x^2 \cos\left(x^2\right) + C$

(B) $-\frac{1}{2}x^2 \sin\left(x^2\right) + C$

(C) $\frac{1}{2}\cos\left(x^2\right) + C$

(D) $2x^2 \cos(x^2) + \sin(x^2) + C$

(E) $-\frac{1}{2}\cos\left(x^2\right) + C$

7. $\sum_{k=1}^{\infty} \frac{2^{k+3}}{5^k} =$

(A) $\frac{2}{5}$

(B) 8

(C) $\frac{16}{3}$

(D) $\frac{16}{5}$

(E) 12

8. $\lim_{x \to 0} \frac{\tan x}{x} =$

(A) 0

(B) $\frac{1}{2}$

(C) 1

(D) π

(E) ∞

GO ON TO THE NEXT PAGE

9. Which of the following series converge?

I. $\sum_{n=1}^{\infty} \frac{n}{n^3+1}$

II. $\sum_{n=1}^{\infty} \frac{2^n}{e^n}$

III. $\sum_{n=1}^{\infty} \frac{n}{n^2+1}$

(A) I only
(B) II only
(C) III only
(D) I and II only
(E) I and III only

10. $\int_1^5 \sqrt{1+x^4}\,dx$ could represent the length of what function from $x = 1$ to $x = 5$?

(A) $f(x) = 1 + x^4$
(B) $f(x) = \sqrt{1+x^4}$
(C) $f(x) = x^2$
(D) $f(x) = \frac{1}{3}x^3 + 2$
(E) $f(x) = x - 3$

11. Let $f(x) = \begin{cases} \frac{x^3-1}{x-1}, & \text{for } x \neq 1 \\ -3, & \text{for } x = 1 \end{cases}$.

Then which of the following statements are true?

I. $\lim_{x \to 1} f(x)$ exists.
II. f is continuous at $x = 1$.
III. f is differentiable at $x = 1$.

(A) I only
(B) II only
(C) III only
(D) I and II only
(E) I, II, and III

12. Suppose f is a continuous function. Use the table below to approximate $\int_{-1}^{9} f(x)dx$ using a right Riemann sum.

x	−1	1	5	7	9
$f(x)$	7	2	−3	−5	1

(A) −16
(B) −10
(C) 2
(D) 6
(E) 20

13. Suppose $y(4) = 2$ and $\frac{dy}{dt} = t - y$. Then use Euler's method with two steps to approximate the value of $y(3)$.

(A) −2
(B) −1
(C) −0.25
(D) 0.25
(E) 1.25

GO ON TO THE NEXT PAGE

14. The graph of the piecewise linear function f is shown in the graph below. Rank the following values from lowest to highest:

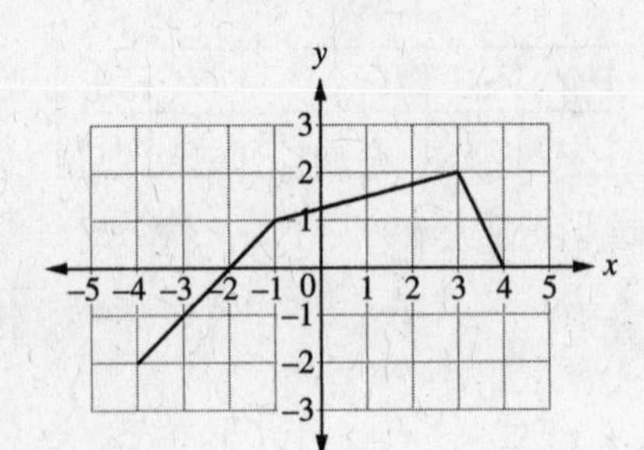

$a = f'(-2)$, $b = f'(2)$, $c = \int_{-4}^{-1} |f(x)|\,dx$, $d = \frac{1}{4}\int_{-1}^{3} f(x)\,dx$

(A) $a < b < c < d$

(B) $b < a < d < c$

(C) $b < a < c < d$

(D) $b < d < a < c$

(E) $a < d < b < d$

15. Find the slope of the tangent line to the graph of $x^2 - 2xy - y^2 = x - 1$ at the point (3,1).

(A) -2

(B) $\frac{1}{4}$

(C) $\frac{3}{8}$

(D) $\frac{3}{4}$

(E) 1

16. Which of the following series are convergent only for $-1 \le x < 1$?

I. $\sum_{k=1}^{\infty} \frac{x^k}{\sqrt{k}}$

II. $\sum_{k=1}^{\infty} \frac{x^k}{k}$

III. $\sum_{k=1}^{\infty} \frac{x^k}{k^2}$

(A) I only

(B) II only

(C) III only

(D) I and II only

(E) I and III only

17. The series $\sum_{k=1}^{\infty} \frac{(-1)^{k-1}}{k!}$ is to be approximated by $\sum_{k=1}^{n} \frac{(-1)^{k-1}}{k!}$ for some positive integer n. What is the lowest possible value of n that is guaranteed to approximate $\sum_{k=1}^{\infty} \frac{(-1)^{k-1}}{k!}$ with error less than $\frac{1}{1000}$?

(A) $n = 4$

(B) $n = 5$

(C) $n = 6$

(D) $n = 7$

(E) $n = 8$

GO ON TO THE NEXT PAGE

18. $\int \frac{5}{x^2+x-6}\,dx =$

(A) $5\ln\left|x^2+x-6\right|+C$

(B) $\frac{5}{2x+1}+C$

(C) $\frac{5x}{x^3+x^2-6x}+C$

(D) $\ln|x+3|-\ln|x-2|+C$

(E) $\ln|x-2|-\ln|x+3|+C$

19. Let $f(x)=\int_{-1}^{x}\sin^2 t\,dt$. Then $f''\left(\frac{\pi}{3}\right)=$

(A) $\frac{1}{4}$

(B) $\frac{1}{2}$

(C) $\frac{\sqrt{3}}{2}$

(D) $\frac{3}{4}$

(E) 1

Use the table below for problems 20 and 21.

x	f(x)	f′(x)	g(x)	g′(x)
1	3	−1	4	−3
2	−1	−2	2	2
3	0	1	1	−5
4	5	2	−4	4

20. Let $h(x)=f(g(x))$. Evaluate $h'(1)$.

(A) 12

(B) 5

(C) −2

(D) −6

(E) −13

21. Let $k(x)=f(x)\cdot g(x)$. Evaluate $k'(1)$.

(A) 12

(B) 5

(C) −2

(D) −6

(E) −13

22. Find all asymptotes for the graph of $f(x)=\frac{e^x}{1+e^x}$.

(A) $y=0$ and $y=1$

(B) $y=0$ only

(C) $y=\frac{1}{2}$ only

(D) $y=1$ only

(E) f has no asymptotes.

23. The graph of $r=1+2\sin\theta$ is shown below. Which of the following integrals represents the area of the inside loop?

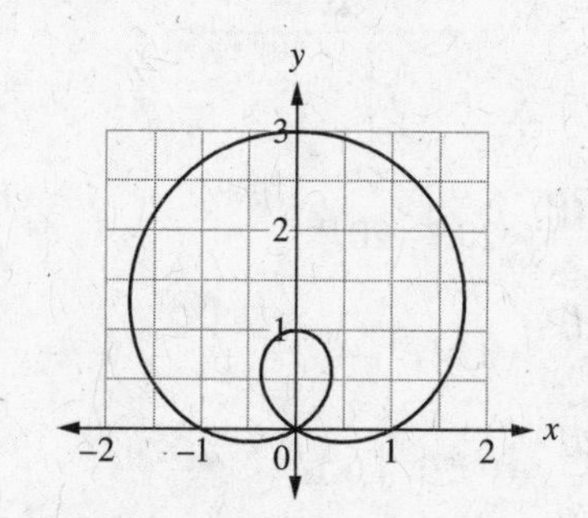

(A) $\frac{1}{2}\int_{0}^{2\pi}(1+2\sin\theta)^2\,d\theta$

(B) $\frac{1}{2}\int_{\frac{7}{6}\pi}^{\frac{3}{2}\pi}(1+2\sin\theta)^2\,d\theta$

(C) $\frac{1}{2}\int_{0}^{\pi}(1+2\sin\theta)^2\,d\theta$

(D) $\frac{1}{2}\int_{\pi}^{2\pi}(1+2\sin\theta)^2\,d\theta$

(E) $\frac{1}{2}\int_{\frac{7}{6}\pi}^{\frac{11}{6}\pi}(1+2\sin\theta)^2\,d\theta$

GO ON TO THE NEXT PAGE

24. $\int x \sec^2 x \, dx =$

(A) $\frac{1}{2}x^2 \tan x + C$

(B) $x \tan x + \ln|\cos x| + C$

(C) $x \tan x - \ln|\cos x| + C$

(D) $x \tan x + \frac{1}{2}x^2 \sec^2 x + C$

(E) $x \tan x - \frac{1}{2}x^2 \sec^2 x + C$

25. Which of the differential equations below could be used to model the logistic growth shown in the graph?

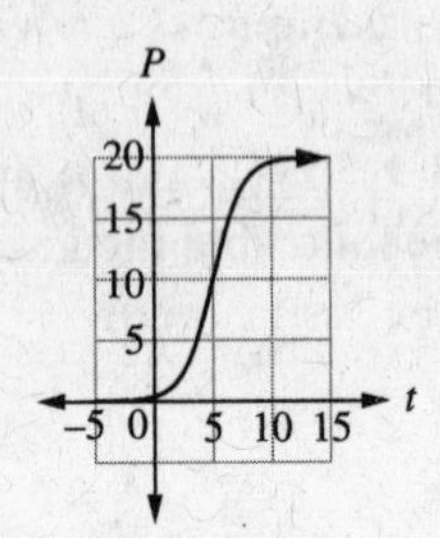

(A) $\frac{dP}{dt} = 20P - 20P^2$

(B) $\frac{dP}{dt} = 0.01P^2 - 0.2P$

(C) $\frac{dP}{dt} = 0.2P^2 + 0.1P$

(D) $\frac{dP}{dt} = 0.2P + 0.01P^2$

(E) $\frac{dP}{dt} = 0.2P - 0.01P^2$

26. The region in the first quadrant bordered by $f(x) = 2 - x^2$, the x-axis, the y-axis, and $x = 1$ is rotated around the x-axis. Find the volume of the solid.

(A) $\frac{5}{3}\pi$

(B) $\frac{2}{5}\pi$

(C) $\frac{19}{5}\pi$

(D) $\frac{43}{15}\pi$

(E) $\frac{19}{15}\pi$

27. $\lim_{x \to \infty} \frac{\tan\left(\frac{\pi}{4} + h\right) - 1}{h} =$

(A) 0

(B) $\frac{1}{2}$

(C) 1

(D) 2

(E) π

GO ON TO THE NEXT PAGE

28. Which differential equation below describes the slope field shown at right?

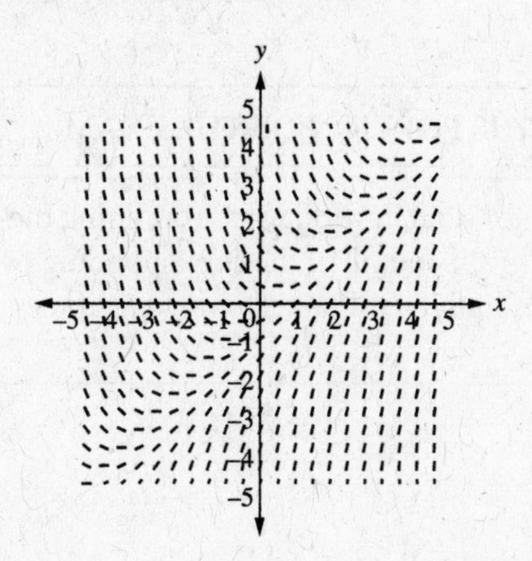

(A) $\frac{dy}{dx} = x - y$

(B) $\frac{dy}{dx} = x + y$

(C) $\frac{dy}{dx} = 1 + y$

(D) $\frac{dy}{dx} = y - x$

(E) $\frac{dy}{dx} = x + 1$

IF YOU FINISH BEFORE TIME IS CALLED, YOU MAY CHECK YOUR WORK ON THIS SECTION ONLY. DO NOT TURN TO ANY OTHER SECTION IN THE TEST. **STOP**

SECTION I, PART B

Time: 50 Minutes
17 Questions

A GRAPHING CALCULATOR IS ALLOWED ON THIS PORTION OF THE EXAM

Directions: Solve the following problems, using available space for scratchwork. After examining the form of the choices, decide which one is the best of the choices given and fill in the corresponding oval on the answer sheet. No credit will be given for anything written in the test book. Do not spend too much time on any one problem.

In this test:

(1) The exact numerical value of the correct answer does not always appear among the choices given. When this happens, select from among the choices the number that best approximates the numerical value.

(2) The domain of a function *f* is the set of all real numbers *x* for which $f(x)$ is a real number, unless otherwise specified.

(3) The inverse of a trigonometric function f may be indicated using the inverse function notation f^{1} or with the prefix "arc" (e.g., $\sin^{-1} x = \arcsin x$).

29. Let $f(-4) = -2, f'(-4) = 5, f''(-4) = 8$ and $f'''(-4) = -12$. Then a third degree Taylor polynomial for *f* centered at $x = -4$ is

(A) $P_3(x) = -2 + 5(x+4) + 4(x+4) -2(x+4)$
(B) $P_3(x) = -2 + 5(x+4) + 4(x+4)^2 -2(x+4)^3$
(C) $P_3(x) = -2 + 5(x+4) + 8(x+4)^2 -1$
$(x+24)^3$
(D) $P_3(x) = -2 + 5x + 4x^2 - 2x^3$
(E) $P_3(x) = -2 + 5x + 8x^2 - 12x^3$

30. The velocity of a particle is given by the graph shown below. The graph consists of line segments and a quarter-circle. Find the distance traveled by the particle from time $t = 0$ to time $t = 8$.

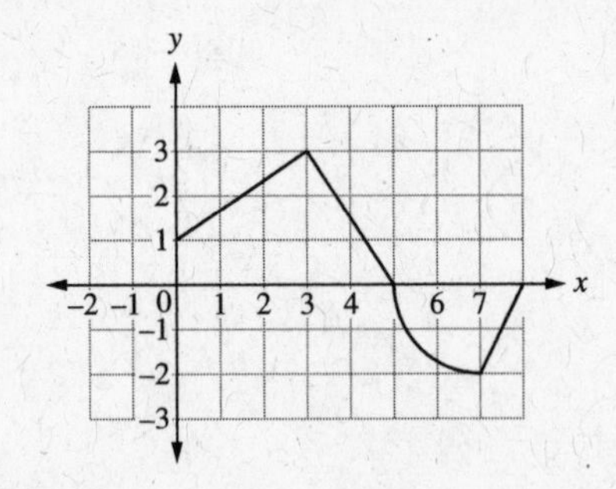

(A) $10 + \pi$
(B) $8 + \pi$
(C) $8 - \dfrac{\pi}{4}$
(D) $10 - \pi$
(E) $8 - \pi$

GO ON TO THE NEXT PAGE

31. A water tank has 1200 liters of water in it. A pump begins to drain the tank at a rate of $D(t)=6\sqrt{t+2}$ liters per minute. To the nearest minute, approximately how long does it take the pump to empty the tank?

(A) 12 minutes
(B) 31 minutes
(C) 43 minutes
(D) 55 minutes
(E) 64 minutes

32. The graph of $y=f'(x)$ is shown below. The domain of f is (0, 4). Which of the following statements is false?

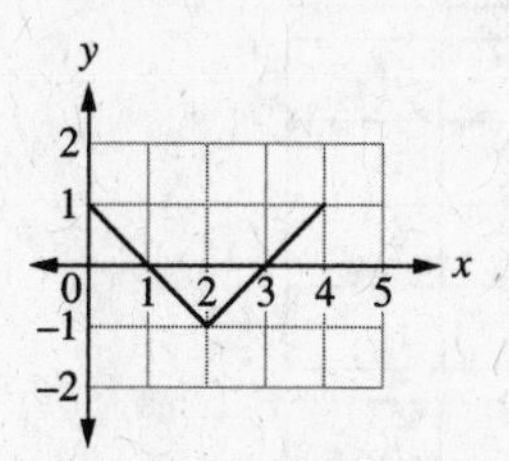

(A) f has a local maximum at $x=1$.
(B) f is concave down on the interval (1, 3).
(C) f is increasing on the intervals $(0,1)\cup(3,4)$.
(D) f has a point of inflection when $x=2$.
(E) f is continuous in the interval (0, 4).

33. The velocity of a particle is given by $v(t)=2t-3\cos(t^2)-3$, for $t\geq 0$. If the position of the particle at time $t=2$ is -4, then what is the position of the particle at time $t=3$?

(A) −7.082
(B) −3.057
(C) −2.724
(D) 1.276
(E) 9.418

34. If $\sum_{k=1}^{\infty} b_k$ converges and $0<a_k<b_k$ for all k, then which of the following statements is false?

(A) $\lim_{k\to\infty} a_k=0$.
(B) $\sum_{k=1}^{\infty}(-1)^k a_k$ converges.
(C) $\sum_{k=1}^{\infty}(-1)^k b_k$ converges.
(D) $\sum_{k=1}^{\infty}(a_k+b_k)b_k$ converges.
(E) $\sum_{k=1}^{\infty} a_k$ diverges.

35. On the graph of $y=f(x)$, the slope at any point (x, y) is equal to the value of y. If $f(2)=1$, then what is the value of $f(4)$?

(A) 2
(B) e
(C) 4
(D) 6
(E) e^2

36. The acceleration of a particle moving along an axis is $a(t)=3\sin(t^2+2)$ at any time t. If the velocity of the particle at time $t=2$ is $v(2)=-1$, then what is the velocity of the particle at time $t=7$?

(A) −1.269
(B) −0.487
(C) 0.513
(D) 1.011
(E) 1.385

GO ON TO THE NEXT PAGE

37. Find the length of the parametric curve given by $x(t) = 2\cos(3t)$ and $y(t) = 4 - e^t$ for $-1 \le t \le 3$.

(A) 19.560

(B) 28.155

(C) 87.651

(D) 184.098

(E) 276.738

38. A particle moves along the curve $y = x^2 - 3x + 5$. When $x = 4$, $\frac{dx}{dt} = 2$. Find the value of $\frac{dx}{dt}$.

(A) 2

(B) 8

(C) 10

(D) 15

(E) 20

39. Suppose f is a continuous, differentiable function, which is negative and increasing for all x. Let $g(x) = \int_7^x f(t)\,dt$. Then which of the tables below could be for the function g?

(A)

x	$g(x)$
3	4
4	2
5	1

(B)

x	$g(x)$
3	4
4	2
5	0

(C)

x	$g(x)$
3	4
4	3
5	1

(D)

x	$g(x)$
3	1
4	2
5	4

(E)

x	$g(x)$
3	1
4	3
5	4

GO ON TO THE NEXT PAGE

40. Find the area between one period of the curve $y = \dfrac{1}{2+\sin(\pi x)}$ and the x-axis.

(A) 0.385
(B) 0.770
(C) 1.155
(D) 1.540
(E) 1.925

41. The infinite series for $f(x) = \dfrac{e^x - 1}{x}$ can be written as

I. $\displaystyle\sum_{k=0}^{\infty} \frac{x^{k+1}}{k!}$

II. $\displaystyle\sum_{k=0}^{\infty} \frac{x^k}{(k+1)!}$

III. $\displaystyle\sum_{k=1}^{\infty} \frac{x^k}{(k+1)!}$

(A) I only
(B) II only
(C) III only
(D) I and II only
(E) II and III only

42. Find the area in the first quadrant bordered by $r = 2 + 2\cos\theta$.

(A) 0.712
(B) 2.571
(C) 5.685
(D) 8.712
(E) 17.425

43. $\displaystyle\lim_{x\to 3} \frac{\tan(\pi x)}{\cos\left(\frac{\pi}{2}x\right)} =$

(A) −2
(B) −1
(C) 1
(D) 2
(E) 3

44. What is the domain of the function $f(x) = \displaystyle\sum_{k=1}^{\infty} \frac{(x-3)^k}{k}$?

(A) $2 \le x < 4$
(B) $2 < x \le 4$
(C) $2 \le x \le 4$
(D) $-1 < x < 1$
(E) $-1 \le x < 1$

45. Let $f(x) = \int_{-2}^{\sin x} (t^2 - 3t + 2)\,dt$. At how many points in the interval $-3 < x < 6$ does f have a local maximum?

(A) 0
(B) 1
(C) 2
(D) 3
(E) 4

IF YOU FINISH BEFORE TIME IS CALLED, YOU MAY CHECK YOUR WORK ON THIS SECTION ONLY. DO NOT TURN TO ANY OTHER SECTION IN THE TEST. **STOP**

SECTION II

Time: 1 hour and 30 minutes
6 Problems
50 percent of total grade

Directions: Solve the following problems, using available space for scratchwork. Show how you arrived at your answer.

You may wish to look over the problems before starting to work on them; on the actual test, it is not expected that everyone will be able to complete all parts of all problems. All problems are given equal weight, but the individual parts of a particular problem are not necessarily given equal weight. You should not spend too much time on any one problem.

- You should write out all your work for each part. On the actual test, you will do this in the space provided in the test booklet. Be sure to write clearly and legibly. If you make a mistake, you can save time by crossing it out rather than trying to erase it. Erased or crossed-out work will not be graded.
- Show all your work. Clearly label any functions, graphs, tables, or other objects that you use. On the actual exam, you will be graded on the correctness and completeness of your methods as well as your answers. Answers without any supporting work may not receive credit.
- Justifications (i.e., the request that you "justify your answer") require that you give mathematical (non-calculator) reasons.
- Work must be expressed in standard mathematical notation, not calculator syntax.
- Unless otherwise specified, answers (numeric or algebraic) need not be simplified.
- If you use decimal approximations in calculations, the readers of the actual exam will grade you on accuracy. Unless otherwise specified, your final answers should be accurate to three places after the decimal point.
- Unless otherwise specified, the domain of function f is the set of all real numbers x for which $f(x)$ is a real number.

GO ON TO THE NEXT PAGE

PART A

Time: 30 Minutes
2 Problems

A GRAPHING CALCULATOR IS ALLOWED ON THIS PORTION OF THE EXAM

1. Let R be the region shown in the graph with $f(x) = 3\cos\left(\frac{\pi}{2}x\right)$ as the upper boundary and $g(x) = \sqrt{x} - 3$ as the lower boundary.

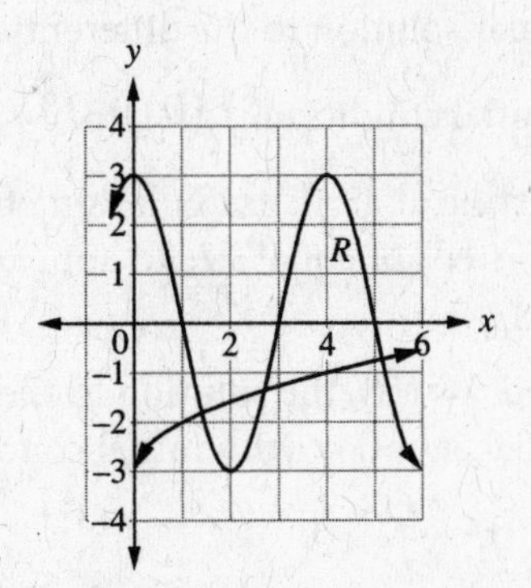

(a) Find the area of region R.
(b) Region R is rotated around the horizontal line $y = -2$. Find the volume of the resulting solid.
(c) Region R is the base of a solid. Cross-sections perpendicular to the x-axis are rectangles with height three times the base. Find the volume of the solid.

2. A hospital patient is receiving a drug on an IV drip. The rate at which the drug enters the body is given by $E(t) = \frac{4}{1+e^{-t}}$ cubic centimeters per hour. The rate at which the body absorbs the drug is given by $D(t) = 3^{\sqrt{t}-1}$ cubic centimeters per hour. The IV drip starts at time $t = 0$ and continues for 8 hours until time $t = 8$.

(a) How many cubic centimeters of the drug have entered the patient's bloodstream after 2 hours?
(b) Is the amount of drug in the body increasing or decreasing at time t = 6? Justify your answer.
(c) At what time is the amount of drug in the patient's bloodstream a maximum? Justify your answer.
(d) Find $E'(3)$ Using correct units, interpret your answer in the context of the problem.

GO ON TO THE NEXT PAGE

PART B

Time: 60 Minutes
4 Problems

NO GRAPHING CALCULATOR IS ALLOWED ON THIS PORTION OF THE EXAM

Note: If you have extra time, you can go back and work on Part A of Section II, but you cannot use a calculator to complete your work at this time.

3. The graphs of the polar curves $r = 2 + \cos\theta$ and $r = -3\cos\theta$ are shown on the graph below. The curves intersect when $\theta = \frac{2}{3}\pi$ and $\theta = \frac{4}{3}\pi$. Region R is in the second quadrant, bordered by each curve and the y-axis.

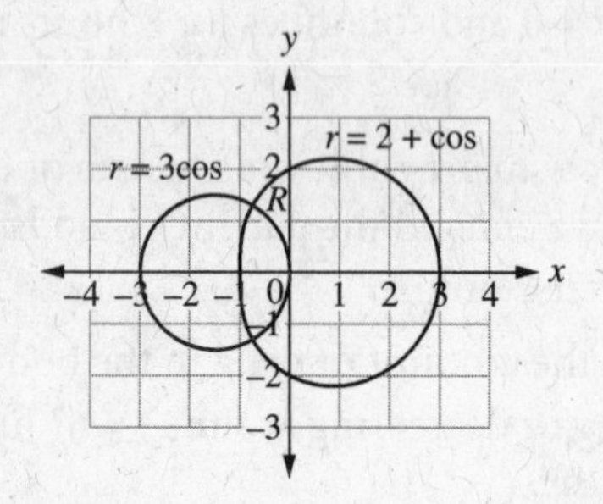

(a) Set up but do not evaluate a formula to find the area of R.

(b) A particle is moving along the curve $r = 2 + \cos\theta$ so that at time t, $\theta = t$. Find the velocity vector of the particle at time $t = \frac{\pi}{2}$.

(c) Set up but do not evaluate a formula to find the length of $r = 2 + \cos\theta$ between $\theta = \frac{2}{3}\pi$ and $\theta = \frac{4}{3}\pi$.

4. Let $\frac{dy}{dt} = 2t\left(y^2 + 1\right)$. Also let $y = f(t)$ be the particular solution to the differential equation with initial condition $f(0) = \sqrt{3}$.

(a) Starting at $t = 0$, use Euler's method with two steps of equal size to approximate $f(1)$.

(b) Find $y = f(t)$, the solution to the differential equation with initial condition $f(0) = \sqrt{3}$.

(c) Find the second degree Taylor polynomial for $y = f(t)$, centered at $t = 0$.

GO ON TO THE NEXT PAGE

5. The velocity of particle A moving along the x-axis is given by $v_A(t) = -t + 2\sin\left(\frac{\pi}{4}t\right)$ for $0 \le t \le 8$.

The particle is at position $x = -3$ at time $t = 0$. A second particle is also moving along the x-axis with position $x_B(t)$. The function $x_B(t)$ is a twice-differentiable function. Values of $x_B(t)$ at various times are given below.

t	0	1	3	6	8
$x_B(t)$	4	3	−2	−1	4

(a) Set up but do not evaluate an integral expression that gives the position of particle A at time $t = 4$.

(b) Set up but do not evaluate an integral expression that gives the total distance traveled by particle A for $0 \le t \le 8$.

(c) Find the acceleration of particle A at time t. Is the speed of particle A increasing, decreasing, or neither at time $t = 3$? Justify your answer.

(d) Does a point in the interval $0 \le t \le 8$ exist such that $v_B(t) = 0$? Justify your answer.

6. Let $f(x) = \frac{e^x - e^{-x}}{x}$.

(a) Find the first three nonzero terms and the general term of the Taylor series representation for f centered at $x = 0$.

(b) Use the Taylor series found in part (a) to evaluate $\lim_{x\to 0} \frac{e^x - e^{-x}}{x}$

(c) Let $g(x) = \frac{1 - e^{-x}}{x}$. Find the first three nonzero terms and the general term of the Taylor series representation for g centered at $x = 0$.

(d) The first four terms of the series for g are used to approximate $g\left(\frac{1}{10}\right)$ Show the error is less than $\frac{1}{1{,}000{,}000}$.

IF YOU FINISH BEFORE TIME IS CALLED, YOU MAY CHECK YOUR WORK ON THIS SECTION ONLY. DO NOT TURN TO ANY OTHER SECTION IN THE TEST. **STOP**

AP Calculus BC Practice Test 2 **Answer Key**

1. D
2. B
3. D
4. E
5. A
6. E
7. C
8. C
9. D
10. D
11. A
12. A
13. C
14. B
15. C
16. D
17. C
18. E
19. C
20. D
21. E
22. A
23. E
24. B
25. E
26. D
27. D
28. A
29. B
30. A
31. C
32. B
33. C
34. E
35. E
36. B
37. B
38. C
39. A
40. C
41. B
42. D
43. D
44. A
45. B

ANSWERS AND EXPLANATIONS

1. D

This is the definition of the derivative of $f(x) = \ln x$, evaluated at $x = a$.

$$\lim_{x\to a}\frac{\ln x - \ln a}{x-a} = \frac{d}{dx}(\ln x)\Big|_{x=a} = \frac{1}{x}\Big|_{x=a} = \frac{1}{a}$$

OR

Since the limit approaches the indeterminate form $\frac{0}{0}$, we can use L'Hôpital's rule:

$$\lim_{x\to a}\frac{\ln x - \ln a}{x-a} = \lim_{x\to a}\frac{\frac{1}{x}}{1} = \frac{1}{a}$$

2. B

Since $f(x) = \int_1^{x^2}(3t^2 - 5t + 7)\,dt$, then $f(-1) = 0$. $f'(x) = (3x^4 - 5x^2 + 7)\cdot 2x = 6x^5 - 10x^3 + 14x$, and $f'(-1) = -10$. $f''(x) = 30x^4 - 30x^2 + 14$, and $f''(-1) = 14$. Therefore,

$$P(x) = -10(x+1) + \frac{14}{2!}(x+1)^2 = -10(x+1) + 7(x+1)^2$$

3. D

Since the function is rational and the degrees of the numerator and denominator are equal, then the limit is equal to the ratio of the leading coefficients.

4. E

$$\frac{d}{dx}\left(-2x^3 + e^{2x} + 5\right) = -6x^2 + 2e^{2x}$$

5. A

Velocity is the derivative of position, so we differentiate each part of $p(t)$ and evaluate each of them at $t = 4$.

$$v(t) = \frac{d}{dt}(p(t)) = \left\langle 6t, \frac{1}{\sqrt{t}}\right\rangle \Rightarrow v(4) = \left\langle 24, \frac{1}{2}\right\rangle$$

6. E

Letting $u = x^2$, $du = 2x\,dx$, then

$$\int x\sin(x^2)\,dx = \frac{1}{2}\int \sin u\,du = -\frac{1}{2}\cos u + C = -\frac{1}{2}\cos(x^2) + C$$

7. C

We look for a way to write both the numerator and denominator as a power of *k*, then use the formula for the sum of an infinite geometric series to evaluate.

$$\sum_{k=1}^{\infty}\frac{2^{k+3}}{5^k}=\sum_{k=1}^{\infty}\frac{2^3\cdot 2^k}{5^k}=8\cdot\sum_{k=1}^{\infty}\left(\frac{2}{5}\right)^k=8\cdot\frac{\frac{2}{5}}{1-\frac{2}{5}}=8\cdot\frac{2}{3}=\frac{16}{3}$$

8. C

A commonly used limit is $\lim_{x\to 0}\frac{\sin x}{x}=1$, so using the fact that $\tan x=\frac{\sin x}{\cos x}$, we get

$$\lim_{x\to 0}\frac{\tan x}{x}=\lim_{x\to 0}\left(\frac{\sin x}{x}\cdot\frac{1}{\cos x}\right)=1\cdot 1=1$$

OR

Since the limit approaches the indeterminate form $\frac{0}{0}$, we can use L'Hôpital's rule:

$$\lim_{x\to 0}\frac{\tan x}{x}=\lim_{x\to 0}\frac{\sec^2 x}{1}=1$$

9. D

I. $\sum_{n=1}^{\infty}\frac{n}{n^3+1}$: $\frac{n}{n^3+1}<\frac{n}{n^3}=\frac{1}{n^2}$. $\sum_{n=1}^{\infty}\frac{1}{n^2}$ converges: *p*-series, $p = 2 > 1$. Therefore $\sum_{n=1}^{\infty}\frac{n}{n^3+1}$ converges by the basic comparison test.

II. $\sum_{n=1}^{\infty}\frac{2^n}{e^n}$: $\sum_{n=1}^{\infty}\frac{2^n}{e^n}=\sum_{n=1}^{\infty}\left(\frac{2}{e}\right)^n$ is a convergent geometric series because $r=\frac{2}{e}<1$.

III. $\sum_{n=1}^{\infty}\frac{n}{n^2+1}$: Let $a_n=\frac{n}{n^2+1}$ and let $b_n=\frac{1}{n}$. $\sum_{n=1}^{\infty}b_n$ is the harmonic series and therefore diverges.

$$\lim_{n\to\infty}\frac{a_n}{b_n}=\lim_{n\to\infty}\frac{n^2}{n^2+1}=1.$$

Therefore $\sum_{n=1}^{\infty}\frac{n}{n^2+1}$ diverges by the limit comparison test.

10. D

$$\text{arclength} = \int_a^b \sqrt{1+\left(f\ (x)\right)^2}\, dx,$$

so

$$f'(x) = x^2 \Rightarrow f(x) = \frac{1}{3}x^3 + C$$

11. A

$\lim\limits_{x\to 1} f(x) = \lim\limits_{x\to 1} \dfrac{x^3-1}{x-1} = \lim\limits_{x\to 1}\left(x^2+x+1\right) = 3$, so choice (I) is true. However f is not continuous at $x = 1$ because $f(1) = -3 \neq \lim\limits_{x\to 1} f(x)$. Since f is not continuous at $x = 1$, then f is not differentiable at $x = 1$ either.

12. A

$$\int_{-1}^{9} f(x)dx \approx f(1)\cdot 2 + f(5)\cdot 4 + f(7)\cdot 2 + f(9)\cdot 2 = -16$$

13. C

$y(3.5) \approx y(4) + y'(4)(-0.5$ $\qquad$ $y'(4) = 4 - 2 = 2$

$y(3.5) \approx 2 + 2(-0.5) = 1$

$y(3) \approx y(3.5) + y'(3.5)(-0.5)$ $\qquad$ $y'(3.5) = 3.5 - 1$

$y(3) \approx 1 + 2.5(-0.5) = -0.25$

14. B

$a = f'(-2) = 1$

$b = f'(2) = \dfrac{1}{4}$

$c = \displaystyle\int_{-4}^{-1} \left|f(x)\right| dx = \frac{5}{2}$

$d = \dfrac{1}{4}\displaystyle\int_{-1}^{3} f(x)\, dx = \frac{1}{4}\ \frac{1}{2}(1+2)\cdot 4\ = \frac{3}{2}$

Therefore, $b < a < d < c$

15. C

$$\frac{d}{dx}\left(x^2 - 2xy - y^2\right) = \frac{d}{dx}(x-1)$$

$$2x - 2x\frac{dy}{dx} - 2y - 2y\frac{dy}{dx} = 1 \quad (-2x-2y)\frac{dy}{dx} = 1 - 2x + 2y \quad \frac{dy}{dx} = \frac{1-2x+2y}{-2x-2y}$$

Therefore,

$$m = \left.\frac{1-2x+2y}{-2x-2y}\right|_{(x,y)=(3,1)} = \frac{3}{8}$$

16. D

Using the ratio test, one can find that all three series converge for $-1 < x < 1$. Test the endpoints of the interval on each series:

I. $x = 1$: $\sum_{k=1}^{\infty} \frac{1}{k^{\frac{1}{2}}}$ diverges because it is a *p*-series, $p = \frac{1}{2} < 1$.

$x = -1$: $\sum_{k=1}^{\infty} \frac{(-1)^k}{k^{\frac{1}{2}}}$ converges because of the alternating series test.

Therefore, $-1 \le x < 1$.

II. $x = 1$: $\sum_{k=1}^{\infty} \frac{1}{k}$ diverges because it is the harmonic series.

$x = -1$: $\sum_{k=1}^{\infty} \frac{(-1)^k}{k}$ converges because it is the alternating harmonic series.

Therefore, $-1 \le x < 1$.

III. $x = 1$: $\sum_{k=1}^{\infty} \frac{1}{k^2}$ converges because it is a *p*-series, $p = 2 > 1$.

$x = -1$: $\sum_{k=1}^{\infty} \frac{(-1)^k}{k^2}$ converges because it is absolutely convergent.

Therefore, $-1 \le x \le 1$. Since the series converges at the endpoint $x = 1$, Item III exceeds the range given in the problem and cannot be included as a correct answer.

17. C

For an alternating series, the error when summing the first n terms is less than the absolute value of the $(n + 1)^{st}$ term. So $\frac{1}{(n+1)!} < \frac{1}{1000}$, or $(n + 1)! > 1000$. $6! = 720$ and $7! = 5040$, so $n + 1 = 7$ and $n = 6$.

18. E

Using partial fractions, $\frac{5}{x^2+x-6}=\frac{1}{x-2}-\frac{1}{x+3}$.

Therefore,

$$\int \frac{5}{x^2+x-6}dx=\int\left(\frac{1}{x-2}-\frac{1}{x+3}\right)dx=\ln|x-2|-\ln|x+3|+C$$

19. C

Using the Fundamental Theorem of Calculus, we get: $f'(x)=\sin^2 x$. Differentiating again we get: $f''(x)=2\sin x\cos x$

Therefore,

$$f''\left(\frac{\pi}{3}\right)=2\cdot\frac{\sqrt{3}}{2}\cdot\frac{1}{2}=\frac{\sqrt{3}}{2}$$

20. D

Using the chain rule, $h'(x)=f'(g(x))\cdot g'(x)$, so $h'(1)=f'(4)\cdot g'(1)=2\cdot-3=-6$

21. E

Using the product rule, $k'(x)=f(x)\cdot g'(x)+g(x)\cdot f'(x)$, so $k'(1)=3\cdot-3+4\cdot-1=-13$

22. A

We must take two limits: the limit of f as x approaches both infinity and negative infinity.

$$\lim_{x\to\infty}\frac{e^x}{1+e^x}=\lim_{x\to\infty}\frac{1}{e^{-x}+1}=1$$

and $\lim_{x\to-\infty}\frac{e^x}{1+e^x}=0$. Therefore, $y=0$ and $y=1$.

OR

Since the limit approaches the indeterminate form $\frac{\infty}{\infty}$, we can use L'Hôpital's rule:

$$\lim_{x\to\infty}\frac{e^x}{1+e^x}=\lim_{x\to\infty}\frac{e^x}{e^x}=\lim_{x\to\infty}1=1.$$

23. E

The graph crosses through the origin when

$$1+2\sin\theta=0 \quad \sin\theta=-\frac{1}{2} \quad \theta=\frac{7}{6}\pi \text{ and } \theta=\frac{11}{6}\pi.$$

Therefore, the area is represented by

$$\frac{1}{2}\int_{\frac{7}{6}\pi}^{\frac{11}{6}\pi}(1+2\sin\theta)^2\,d\theta.$$

24. B

Using integration by parts,

$u = x \qquad dv = \sec^2 x\,dx$

$du = dx \qquad v = \tan x$

$$\int x\sec^2 x\,dx = x\tan x - \int \tan x\,dx = x\tan x + \ln|\cos x| + C.$$

25. E

Factoring each of the choices, the only one that results in a logistic differential equation with horizontal asymptotes at $P = 0$ and $P = 20$ is $\frac{dP}{dt} = 0.2P - 0.01P^2 = 0.01P(20-P)$.

Remember, our zeros here correspond to regions where the slope of the given curve is zero, or flat.

26. D

The graph is shown below. When rotated around the x-axis, the volume is given by

$$V = \int_0^1 \pi\left(2-x^2\right)^2 dx = \pi\int_0^1\left(4-4x^2+x^4\right)dx = \pi\left(4x-\frac{4}{3}x^3+\frac{1}{5}x^5\right)\Bigg|_0^1 = \frac{43}{15}\pi.$$

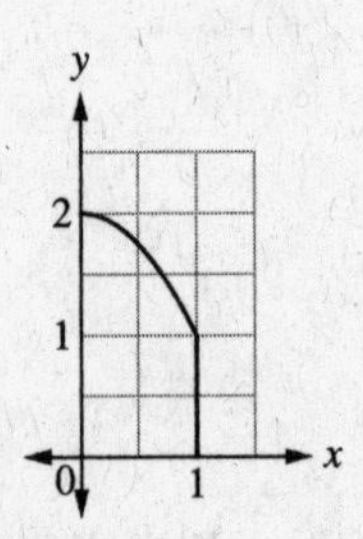

27. D

This is the definition of the derivative of $f(x) = \tan x$, evaluated at $x = \frac{\pi}{4}$

$$\lim_{h\to 0}\frac{\tan\left(\frac{\pi}{4}+h\right)-1}{h} = \frac{d}{dx}(\tan x)\Big|_{x=\frac{\pi}{4}} = \sec^2\left(\frac{\pi}{4}\right) = \left(\sqrt{2}\right)^2 = 2.$$

28. A

Notice the slopes of the segments are 0 along the line $y = x$. Below the line $y = x$, $x > y$, and notice the slopes are positive. Above the line $y = x$, $x < y$, and notice the slopes are negative. The only differential equation that meets all these criteria is (A).

29. B

$$P_3(x) = -2 + 5(x+4) + \frac{8}{2!}(x+4)^2 + \frac{-12}{3!}(x+4)^3 = -2 + 5(x+4) + 4(x+4)^2 - 2(x+4)^3$$

30. A

$$\text{distance} = \int_0^8 |v(t)|\, dt = 10 + \pi$$

31. C

Let the amount of water in the tank at time t be given by $f(t) = 1200 - \int_0^t 6\sqrt{x+2}\, dx$. Graphing this function shows an x-intercept at $t \approx 43.$ minutes.

32. B

f is concave down when f' is decreasing, on the interval (0,2). All the other statements are true.

33. C

The position at time $t = 3$ is given by the initial position at time $t = 2$ (–4) plus the change in position away from the initial position:

$$-4 + \int_2^3 v(t)\, dt \approx -2.724.$$

34. E

$\sum_{k=1}^{\infty} a_k$ must converge by the basic comparison test. All the other statements are true.

35. E

We solve the differential equation:

$$\frac{dy}{dx} = y \Rightarrow \frac{dy}{y} = dx \Rightarrow \ln y = x + C.$$

Using the point (2, 1), $C = -2$ and $y = f(x) = ex^{-2} \Rightarrow f(4) = e^2$.

36. B

The velocity at time $t = 7$ is given by the initial position at time $t = 2$ (–1) plus the change in velocity from the initial velocity:

$$v(7) = -1 + \int_2^7 a(t)\,dt \approx -0.487.$$

37. B

$x'(t) = -6\sin(3t)$ and $y'(t) = -e^t$

$$\text{arclength} = \int_{-1}^{3} \sqrt{(-6\sin(3t))^2 + (-e^t)^2}\,dt \approx 28.155$$

38. C

$$\frac{d}{dt}(y) = \frac{d}{dt}(x^2 - 3x + 5) \Rightarrow \frac{dy}{dt} = (2x-3)\frac{dx}{dt} \Rightarrow \frac{dy}{dt} = (2\cdot 4 - 3)\cdot 2 = 10$$

39. A

Since $g(x) = \int_7^x f(t)\,dt$, then $g'(x) = f(x)$ and $g''(x) = f'(x)$. The only choice that shows g decreasing and concave up is choice A.

40. C

Graphing the function, we see that the period is 2.

$$\int_0^2 \frac{1}{2 + \sin(\pi x)}\,dx \approx 1.155.$$

41. B

$$e^x = 1 + x + \frac{x^2}{2!} + \frac{x^3}{3!} + \frac{x^4}{4!} + \cdots \Rightarrow e^x - 1 = x + \frac{x^2}{2!} + \frac{x^3}{3!} + \frac{x^4}{4!} + \cdots \Rightarrow \frac{e^x - 1}{x} = 1 + \frac{x}{2!} + \frac{x^2}{3!} + \frac{x^3}{4!} + \cdots$$

II is the only choice that fits this series.

42. D

Graphing $r = 2 + 2\cos\theta$ in polar mode, we find the region in the first quadrant is defined when $0 \le \theta \le \frac{\pi}{2}$. Therefore,

$$\int_0^{\frac{\pi}{2}} \frac{1}{2}(2 + 2\cos\theta)^2\,d\theta \approx 8.712.$$

43. D

The limit approaches $\frac{0}{0}$, so we use L'Hôpital's rule:

$$\lim_{x\to 3}\frac{\tan(\pi x)}{\cos\left(\frac{\pi}{2}x\right)} = \lim_{x\to 3}\frac{\pi\sec^2(\pi x)}{-\frac{\pi}{2}\sin\left(\frac{\pi}{2}x\right)} = -2\cdot\lim_{x\to 3}\frac{\sec^2(\pi x)}{\sin\left(\frac{\pi}{2}x\right)} = -2\cdot\frac{(-1)^2}{-1} = 2$$

44. A

Let $a_k = \frac{(x-3)^k}{k}$, $a_{k+1} = \frac{(x-3)^{k+1}}{k+1}$. Then

$\lim_{k\to\infty}\left|\frac{a_{k+1}}{a_k}\right| = \lim_{k\to\infty}\left|\frac{(x-3)^{k+1}}{k+1}\cdot\frac{k}{(x-3)^k}\right| = |x-3|\cdot\lim_{k\to\infty}\frac{k}{k+1} = |x-3|$ $|x-3| < 1 \Rightarrow 2 < x < 4$. Testing the endpoints,

If $x = 4$, then $f(4) = \sum_{k=1}^{\infty}\frac{1}{k}$ which is divergent because this is the harmonic series.

If $x = 2$, then $f(2) = \sum_{k=1}^{\infty}\frac{(-1)^k}{k}$ which is convergent because this is the alternating harmonic series. Therefore the domain of f is: $2 \le x < 4$.

45. B

$f'(x) = (\sin^2 x - 3\sin x + 2)\cdot\cos x$

Graphing f' on the domain $-3 < x < 6$ there is only one place where the graph changes from positive to negative. Therefore the graph of f has only one local maximum.

FREE-RESPONSE ANSWERS

1. Solving $f(x) = g(x)$ for $2 < x < 6$ reveals intersections at $x \approx 2.7013188$ and $x \approx 5.1563171$. For clarity, let $a = 2.7013188$ and $b = 5.1563171$ in the solutions below.

   (a) $A = \int_a^b (f(x) - g(x))\,dx \approx 6.075$

   (b) The volume of a representative slice of width Δx is given by $\pi\left((f(x)+2)^2 - (g(x)+2)^2\right)\cdot\ x$.

   So the volume of the entire solid is $\int_a^b \pi\left((f(x)+2)^2 - (g(x)+2)^2\right)dx \approx 96.926$.

   (c) For a representative slice of the solid of width Δx, the base has length $f(x) - g(x)$. The height of the slice is $3(f(x) - g(x))$. So the volume of the slice is $3(f(x) - g(x))^2 \cdot \Delta x$. So the volume of the entire solid is $\int_a^b 3(f(x) - g(x))^2\,dx \approx 56.733$.

2. (a) Notice that the units of $E(t)$ are cubic centimeters per hour. Integrating $E(t)$ over the interval 0 to 2 hours gives us cubic centimeters:

   $\int_0^2 E(t)\,dt \approx 5.735 \text{ cm}^3$.

   (b) $E(6) - D(6) \approx -0.926 \text{ cm}^3/\text{hr}$. Therefore the amount of drug in the body is decreasing at time $t = 6$ hours.

   (c) Solving $E(t) - D(t) = 0$ results in $t \approx 5.09078034$.

   Let

   $f(t) = \int_0^t (E(x) - D(x))\,dx$ represent the amount of drug in the patient's body at time t.

   Then $f(0) = 0$, $f(5.0907834) \approx 7.321$ and $f(8) \approx 2.647$.

   So the amount of the drug in the patient's bloodstream is maximized at time $t \approx 5.091$ hours.

   OR

   Since $t \approx 5.091$ is the only relative extrema on $[0,8]$ and $f(t)$ changes sign from positive to negative at $t \approx 5.091$, then by the Extreme Value Theorem, the amount of the drug in the patient's bloodstream is maximized at time $t \approx 5.091$ hours.

   (d) $E'(3) \approx 0.181$. This means that the rate at which the drug is entering the patient's bloodstream is increasing at a rate of 0.181 cubic centimeters per hour per hour.

3. (a) First determine the angles that sweep out the area of region R. $r = 2 + \cos\theta$ intersects the positive y-axis at $\theta = \frac{\pi}{2}$ radians. $r = -3\cos\theta$ is at the origin at $\theta = \frac{\pi}{2}$ radians. The curves intersect each other in the second quadrant at $\theta = \frac{2}{3}\pi$ radians. Therefore, since $A = \frac{1}{2}\int_\alpha^\beta r^2\,d\theta$,

$$A=\frac{1}{2}\int_{\frac{\pi}{2}}^{\frac{2}{3}\pi}\left((2+\cos\theta)^2-(-3\cos\theta)^2\right)d\theta.$$

(b) $x = r\cos\theta = (2+\cos\theta)\cos\theta$ and $y = r\sin\theta = (2+\cos\theta)\sin\theta$. The velocity vector is $\left\langle \frac{dx}{d\theta}, \frac{dy}{d\theta}\right\rangle$ evaluated at $\theta = t = \frac{\pi}{2}$.

$$\frac{dx}{d\theta}=(2+\cos\theta)\cdot -\sin\theta+\cos\theta\cdot -\sin\theta=-2\sin\theta-2\sin\theta\cos\theta$$

$$\frac{dy}{d\theta}=(2+\cos\theta)\cdot\cos\theta+\sin\theta\cdot -\sin\theta=2\cos\theta+\cos^2\theta-\sin^2\theta$$

Evaluating at $\theta = t = \frac{\pi}{2}$ gives $\frac{dx}{d\theta}=-2$ and $\frac{dy}{d\theta}=-1$.

Therefore the velocity vector is $\langle -2,-1\rangle$.

(c) Using $\frac{dx}{d\theta}$ and $\frac{dy}{d\theta}$ as defined in part (b),arclength $=\int_{\frac{2}{3}\pi}^{\frac{4}{3}\pi}\sqrt{\left(\frac{dx}{d\theta}\right)^2+\left(\frac{dy}{d\theta}\right)^2}\,d\theta.$

4. (a) Remember when using Euler's method, each value you calculate is based on the previous value.

$f(0.5)\approx f(0)+f(0)\cdot 0.5 \quad f(0)=2.0(3+1)=0$

$f(0.5)\approx\sqrt{3}+0\cdot 0.5=\sqrt{3}$

$f(1)\approx f(0.5)+f(0.5)\cdot 0.5\ f(0.5)=2\cdot 0.5(3+1)=4$

$f(1)\approx\sqrt{3}+4\cdot 0.5=2+\sqrt{3}$

(b) Separating the variables,

$\frac{1}{y^2+1}dy=2t\,dt$

$\int\frac{1}{y^2+1}dy=\int 2t\,dt$

$\arctan y=t^2+C$

Substituting $\left(0,\sqrt{3}\right)$ gives $\arctan\sqrt{3}=0^2+C$, so $C=\frac{\pi}{3}$.

Therefore, $\arctan y=t^2+\frac{\pi}{3}$, so $y=\tan\left(t^2+\frac{\pi}{3}\right)$.

(c) Since $\frac{dy}{dt}=2t\left(y^2+1\right)$, $\frac{d^2y}{dt^2}=2t\cdot 2y\frac{dy}{dt}+2\left(y^2+1\right)$. Evaluating $\frac{d^2y}{dt^2}$ at $\left(0,\sqrt{3}\right)$ gives $f''(0) = 8$.
Since we already know that $f'(0) = 0$, $f(t)\approx\sqrt{3}+0t+\frac{8}{2!}t^2=\sqrt{3}+4t^2$.

5. (a) The position at time $t = 4$ is the initial position plus the change away from the initial position:

$$x_A(4)=-3+\int_0^4\left(-t+2\sin\left(\frac{\pi}{4}t\right)\right)dt.$$

(b) Do not forget to use absolute value; otherwise, you find the displacement of the particle.

$$d = \int_0^8 \left| -t + 2\sin\left(\frac{\pi}{4}t\right) \right| dt$$

(c) $a_A(t) = -1 + 2\cos\left(\frac{\pi}{4}t\right) \cdot \frac{\pi}{4} = -1 + \frac{\pi}{2}\cos\left(\frac{\pi}{4}t\right)$

$$v_A(3) = -3 + 2\sin\left(\frac{3}{4}\pi\right) = -3 + \sqrt{2} < 0$$

$$a_A(3) = -1 + \frac{\pi}{2}\cos\left(\frac{3}{4}\pi\right) = -1 - \frac{\sqrt{2}}{4}\pi < 0$$

Since $v_A(3)$ and $a_A(3)$ have the same sign, then the particle is speeding up at time $t = 3$.

(d) Since $x_B(t)$ is differentiable, then it is also continuous. So by the Mean Value Theorem, there must be at least one point in $0 \le t \le 8$ where $v_B(t) = 0$ because $x_B(0) = x_B(8) = 4$.

6. (a) Start with the series for e^x and build from there.

$$e^x = 1 + x + \frac{x^2}{2!} + \frac{x^3}{3!} + \cdots$$

$$e^{-x} = 1 - x + \frac{x^2}{2!} - \frac{x^3}{3!} + \cdots$$

$$f(x) = \frac{e^x - e^{-x}}{x} = \frac{2x + \frac{2x^2}{3!} + \frac{2x^5}{5!} + \cdots}{x} = 2 + \frac{2x^2}{3!} + \frac{2x^4}{5!} + \cdots + \frac{2x^{2k}}{(2k+1)!} + \cdots, \text{ for k} = 0, 1, 2, \cdots$$

(b) $\lim_{x \to 0} \frac{e^x - e^{-x}}{x} = \lim_{x \to 0} \left(2 + \frac{2x^2}{3!} + \frac{2x^4}{5!} + \cdots + \frac{2x^{2k}}{(2k+1)!} + \cdots\right) = 2$

(c) $e^{-x} = 1 - x + \frac{x^2}{2!} - \frac{x^3}{3!} + \cdots$

$$\frac{1 - e^{-x}}{x} = \frac{1 - \left(1 - x + \frac{x^2}{2!} - \frac{x^3}{3!} + \cdots\right)}{x} = 1 - \frac{x}{2!} + \frac{x^2}{3!} - \cdots + (-1)^k \frac{x^k}{(k+1)!} + \cdots,$$

for $k = 0, 1, 2, \ldots$

(d) The error is less than the absolute value of the fifth term, evaluated at $x = \frac{1}{10}$.

$$\left| -\frac{x^4}{5!} \right|_{x=\frac{1}{10}} = \frac{\left(\frac{1}{10}\right)^4}{120} = \frac{1}{1{,}200{,}000} < \frac{1}{1{,}000{,}000}$$

AP CALCULUS BC PRACTICE TEST 2 CORRELATION CHART

Multiple Choice Question	Review Chapter: Subsection of Chapter
1	Chapter 12: Computation of Derivatives
2	Chapter 26: Taylor Polynomial Approximation
3	Chapter 8: Limits Approaching Infinity
4	Chapter 12: Derivative of x^n Chapter 12: Exponential and Logarithmic Functions
5	Chapter 10: Vector Functions Chapter 12: Derivatives of Parametric, Polar, and Vector Functions Chapter 16: Interpreting the Derivative As a Rate of Change
6	Chapter 21: When *F* Is Complicated: The Substitution Game
7	Chapter 25: The Geometric Series with Applications
8	Chapter 7: Some Important Limits
9	Chapter 24: To Converge or Not to Converge Chapter 25: The Harmonic Series Chapter 25: Comparing Series to Test for Convergence or Divergence
10	Chapter 18: Applications of Integrals
11	Chapter 11: The Relationship Between Differentiability and Continuity
12	Chapter 23: Approximating the Area Under the Curve Using Riemann Sums
13	Chapter 16: Differential Equations Via Euler's Method
14	Chapter 14: Corresponding Characteristics of the Graphs of f and f' Chapter 17: Finding the Area Under a Curve
15	Chapter 12: Implicit Differentiation Chapter 13: The Slope of a Curve at a Point
16	Chapter 25: The Ratio Test for Convergence and Divergence
17	Chapter 25: Alternating Series with Error Bound
18	Chapter 21: Antiderivatives by Simple Partial Fractions

Multiple Choice Question	Review Chapter: Subsection of Chapter
19	Chapter 12: Chain Rule Chapter 20: What's So Fundamental? The Connection Between Integrals and Derivatives
20	Chapter 12: Chain Rule
21	Chapter 12: Rules for Computing Derivatives
22	Chapter 8: Horizontal Asymptotes
23	Chapter 10: Polar Functions Chapter 19: Antiderivatives—The Indefinite Integral
24	Chapter 21: Integration by Parts
25	Chapter 22: Solving Logistic Differential Equations and Using Them in Modeling
26	Chapter 18: Area Bounded by Curves
27	Chapter 12: Computation of Derivatives
28	Chapter 16: Differential Equations
29	Chapter 26: Taylor Polynomial Approximation
30	Chapter 17: Rates of Change and the Definite Integral Chapter 18: Accumulated Change
31	Chapter 17: Rates of Change and the Definite Integral
32	Chapter 14: Corresponding Characteristics of the Graphs of f and f' Chapter 15: Concavity and the Sign of f''
33	Chapter 18: Distance Traveled by a Particle Along a Line
34	Chapter 25: Comparing Series to Test for Convergence or Divergence
35	Chapter 22: Differential Equations Chapter 22: Separable Differential Equations
36	Chapter 22: Accumulation Functions and Antiderivatives
37	Chapter 10: Parametric Functions
38	Chapter 16: Related Rates
39	Chapter 11: The Relationship Between Differentiability and Continuity Chapter 14: Corresponding Characteristics of the Graphs of f and f' Chapter 15: Concavity and the Sign of f''

Multiple Choice Question	Review Chapter: Subsection of Chapter
40	Chapter 17: Finding the Area Under a Curve
41	Chapter 26: Maclaurin Series for Special Functions Chapter 26: Shortcuts Using Taylor/Maclaurin Series
42	Chapter 10: Polar Functions Chapter 17: Finding the Area Under a Curve
43	Chapter 16: L'Hôpital's Rule
44	Chapter 25: The Properties of Series Chapter 25: The Harmonic Series
45	Chapter 14: Critical Points and Local Extrema Chapter 20: What's So Fundamental? The Connection Between Integrals and Derivatives

AP CALCULUS BC PRACTICE TEST 2 CORRELATION CHART

Free Response Question	Review Chapter: Subsection of Chapter
1(a)	Chapter 18: Area Between Two Curves
1(b)	Chapter 18: Solids of Revolution
1(c)	Chapter 18: Volumes of Solids
2(a)	Chapter 17: Rates of Change and the Definite Integral Chapter 18: Accumulated Change
2(b)	Chapter 16: Interpreting the Derivative As a Rate of Change
2(c)	Chapter 14: Critical Points and Local Extrema
2(d)	Chapter 6: Evaluating a Derivative at a Point Chapter 16: Interpreting the Derivative As a Rate of Change
3(a)	Chapter 10: Polar Functions Chapter 17: Finding the Area Under a Curve
3(b)	Chapter 12: Derivatives of Parametric, Polar, and Vector Functions Chapter 16: Interpreting the Derivative As a Rate of Change
3(c)	Chapter 10: Polar Functions Chapter 19: Antiderivatives—The Indefinite Integral
4(a)	Chapter 16: Differential Equations Via Euler's Method
4(b)	Chapter 22: Separable Differential Equations
4(c)	Chapter 26: Taylor Polynomial Approximation
5(a)	Chapter 22: Accumulation Functions and Antiderivatives
5(b)	Chapter 18: Distance Traveled by a Particle Along a Line
5(c)	Chapter 16: Interpreting the Derivative As a Rate of Change
5(d)	Chapter 14: The Mean Value Theorem
6(a)	Chapter 26: Taylor Polynomial Approximation
6(b)	Chapter 26: Shortcuts Using Taylor/Maclaurin Series
6(c)	Chapter 26: Taylor Polynomial Approximation
6(d)	Chapter 26: How Close Is Close Enough? The Lagrange Error Bound

AP Calculus BC Practice Test 3
Answer Grid

1. (A) (B) (C) (D) (E)
2. (A) (B) (C) (D) (E)
3. (A) (B) (C) (D) (E)
4. (A) (B) (C) (D) (E)
5. (A) (B) (C) (D) (E)
6. (A) (B) (C) (D) (E)
7. (A) (B) (C) (D) (E)
8. (A) (B) (C) (D) (E)
9. (A) (B) (C) (D) (E)
10. (A) (B) (C) (D) (E)
11. (A) (B) (C) (D) (E)
12. (A) (B) (C) (D) (E)
13. (A) (B) (C) (D) (E)
14. (A) (B) (C) (D) (E)
15. (A) (B) (C) (D) (E)
16. (A) (B) (C) (D) (E)
17. (A) (B) (C) (D) (E)
18. (A) (B) (C) (D) (E)
19. (A) (B) (C) (D) (E)
20. (A) (B) (C) (D) (E)
21. (A) (B) (C) (D) (E)
22. (A) (B) (C) (D) (E)
23. (A) (B) (C) (D) (E)
24. (A) (B) (C) (D) (E)
25. (A) (B) (C) (D) (E)
26. (A) (B) (C) (D) (E)
27. (A) (B) (C) (D) (E)
28. (A) (B) (C) (D) (E)
29. (A) (B) (C) (D) (E)
30. (A) (B) (C) (D) (E)
31. (A) (B) (C) (D) (E)
32. (A) (B) (C) (D) (E)
33. (A) (B) (C) (D) (E)
34. (A) (B) (C) (D) (E)
35. (A) (B) (C) (D) (E)
36. (A) (B) (C) (D) (E)
37. (A) (B) (C) (D) (E)
38. (A) (B) (C) (D) (E)
39. (A) (B) (C) (D) (E)
40. (A) (B) (C) (D) (E)
41. (A) (B) (C) (D) (E)
42. (A) (B) (C) (D) (E)
43. (A) (B) (C) (D) (E)
44. (A) (B) (C) (D) (E)
45. (A) (B) (C) (D) (E)

AP CALCULUS BC PRACTICE TEST 3

SECTION I, PART A

Time: 55 Minutes
28 Questions

NO GRAPHING CALCULATOR IS ALLOWED ON THIS PORTION OF THE EXAM

Directions: Solve the following problems, using available space for scratchwork. After examining the form of the choices, decide which one is the best of the choices given and fill in the corresponding oval on the answer sheet. No credit will be given for anything written in the test book. Do not spend too much time on any one problem.

In this test:

(1) The domain of a function f is the set of all real numbers x for which $f(x)$ is a real number, unless otherwise specified.

(2) The inverse of a trigonometric function f may be indicated using the inverse function notation f^{-1} or with the prefix "arc" (e.g., $\sin^{-1} x = \arcsin x$).

1. Given $f(x)=\dfrac{2x-5}{5x+2}$, $f'(x) =$

(A) $\dfrac{-29}{(5x+2)^2}$

(B) $\dfrac{-21}{(5x+2)^2}$

(C) $\dfrac{29}{(5x+2)^2}$

(D) $\dfrac{8x-1}{(5x+2)^2}$

(E) $\dfrac{20x-21}{(5x+2)^2}$

2. A particle moves along a straight line so that for time $t \geq 0$, its velocity is given by $v(t) = \sin(3t)$. If the position of the particle at time $t = \dfrac{\pi}{3}$ is 7, what is the particle's postion at time $t = 0$?

(A) $-\dfrac{2}{3}$

(B) 5

(C) 7

(D) $\dfrac{19}{3}$

(E) $\dfrac{23}{3}$

GO ON TO THE NEXT PAGE

3. What is the value of $\sum_{n=0}^{\infty}\left(-\frac{3}{4}\right)^n$?

(A) $-\frac{3}{7}$

(B) $\frac{4}{7}$

(C) $\frac{7}{4}$

(D) 4

(E) None. The series diverges.

4. $\lim_{h\to 0}\frac{\sin\left(\frac{\pi}{3}+h\right)-\sin\left(\frac{\pi}{3}\right)}{h}=?$

(A) $\cos x$

(B) $\cos\left(\frac{\pi}{3}+h\right)$

(C) $\cos\left(\frac{\pi}{3}+h\right)-\frac{\sqrt{3}}{2}$

(D) $-\frac{1}{2}$

(E) $\frac{1}{2}$

5. The length of the curve $y=x^3$ for $x=0$ to $x=6$ is given by

(A) $\int_0^6\sqrt{1+3x^2}\,dx$

(B) $\int_0^6\sqrt{1+x^3}\,dx$

(C) $\int_0^6\sqrt{1+3x^4}\,dx$

(D) $\int_0^6\sqrt{1+9x^4}\,dx$

(E) $\int_0^6\sqrt{1+9x^6}\,dx$

6. $\int\frac{\sec^2 x}{1+\tan x}dx=$

(A) $\ln|1=\tan x|+C$

(B) $\tan x-\ln|\tan x|+C$

(C) $\ln\left(1+\frac{1}{\tan x}\right)+C$

(D) $\tan x - x + C$

(E) $\frac{1}{1+\tan^2 x}+C$

7. Let $y=f(x)$ be the solution to the differential equation $\frac{dy}{dx}=y-x+2$ with initial condition $f(2)=-3$. Using Euler's method with 2 steps of equal size, what is the approximation for $f(1.6)$?

(A) −10.4

(B) −4.52

(C) −1.96

(D) 0.48

(E) 1.88

x	1	5	9	13
$f(x)$	5	k	11	7

8. The function f is continuous on the closed interval [1,13] and has values given in the table above. The trapezoidal approximation for $\int_1^{13}f(x)dx$ found with 3 subintervals with equal length is 72. What is the value of k?

(A) −8

(B) $-\frac{32}{5}$

(C) 1

(D) 2

(E) 45

GO ON TO THE NEXT PAGE

9. The function g is twice-differentiable. $g''(x)$ never equals zero. If $g(2) = 5$, $g'(2) = \frac{1}{4}$, and $g''(2) = 2$, which of the following could be the value of $g(4)$?

(A) 4

(B) 4.5

(C) 5.25

(D) 5.5

(E) 6.5

10. A function h has a Maclaurin series given by

$$x^2 - \frac{x^4}{2} + \frac{x^6}{3} + \ldots + (-1)^n \frac{x^{2n}}{n} + \ldots$$

Which of the following is an expression for $h(x)$?

(A) $\cos x$

(B) $\ln(1 + x^2)$

(C) $\arctan(x^2)$

(D) $e^x - \sin x$

(E) $e^x + \sin x$

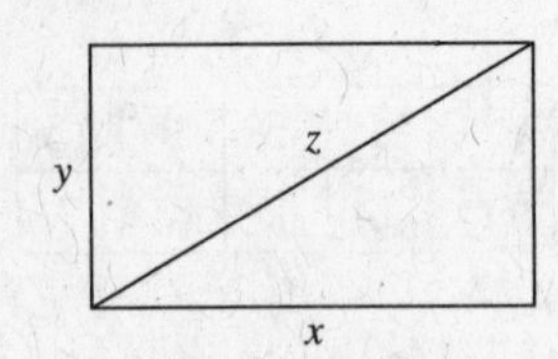

11. The sides and diagonal of the rectangle shown above are strictly increasing with time. When $x = 12$ and $y = 5$, $\frac{dx}{dt} = 2\frac{dy}{dt}$, and $\frac{dz}{dt} = k\frac{dy}{dt}$. What is the value of k at that moment?

(A) $\frac{6}{13}$

(B) $\frac{12}{13}$

(C) $\frac{29}{26}$

(D) $\frac{313}{169}$

(E) $\frac{29}{13}$

12. If $f'(x) = \cos(2x)$ and $f\left(\frac{\pi}{4}\right) = 2$, then $f\left(\frac{\pi}{12}\right) =$

(A) 1

(B) $\frac{3}{2}$

(C) $\frac{7}{4}$

(D) $\frac{5}{2}$

(E) 3

13. For time $t > 0$, the position of a particle moving in the xy-plane is given parametrically by $x = \frac{1}{2t+1}$ and $y = 6t - t^2$. What is the acceleration vector at time $t = 2$?

(A) $\langle -2, 2 \rangle$

(B) $\left\langle \frac{-2}{25}, 2 \right\rangle$

(C) $\langle 0, 2 \rangle$

(D) $\left\langle \frac{8}{125}, -2 \right\rangle$

(E) $\left\langle \frac{2}{25}, 2 \right\rangle$

14. $\int \frac{4}{x^2 - 1} dx =$

(A) $2\arctan\left(\frac{1}{2x}\right) + C$

(B) $4\ln|x^2 - 1| + C$

(C) $2\ln|x - 1| + 2\ln|x + 1| + C$

(D) $2\ln\left|\frac{x-1}{x+1}\right| + C$

(E) $2\ln\left|\frac{x+1}{x-1}\right| + C$

GO ON TO THE NEXT PAGE

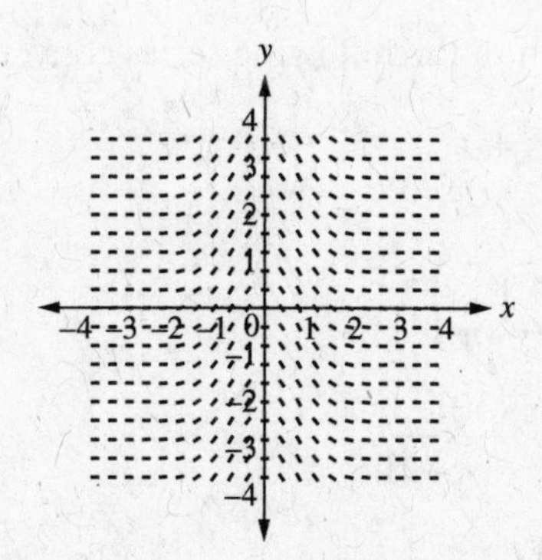

15. The slope field for a differential function is shown above. Which of the following could be a solution to the differential equation with the initial condition $y(0) = 2$?

    (A) $y = 2 - x^2$

    (B) $y = \sqrt{4 - x^2}$

    (C) $y = \dfrac{2}{2 + x^2}$

    (D) $y = 2e^{-}x^2$

    (E) $y = \dfrac{2}{x^2}$

16. If $f'(x) = |x + 1|$, which of the following could be the graph of f?

    (A)

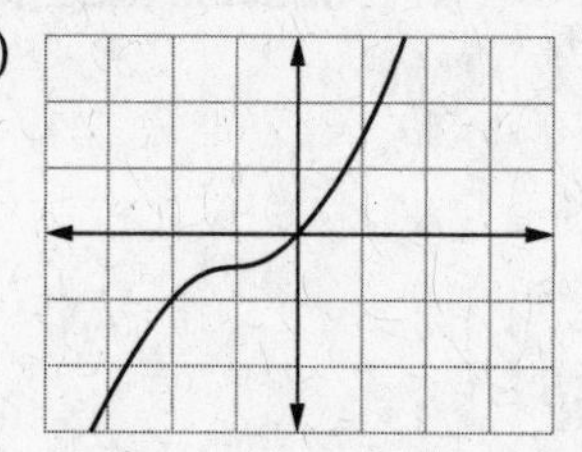

    (B)

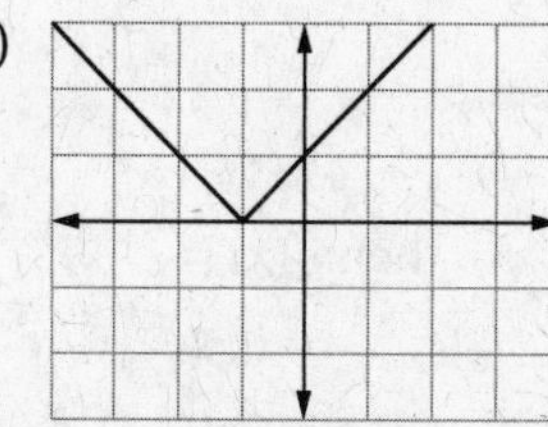

    (C)

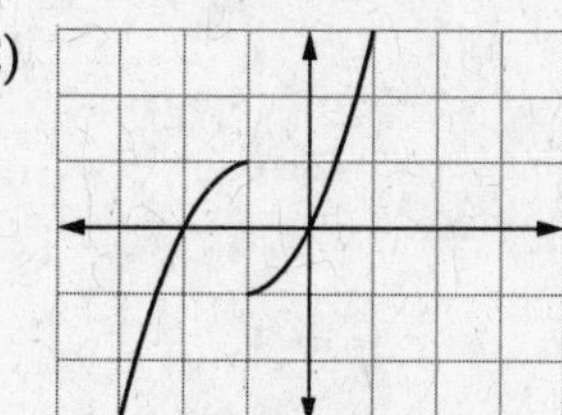

    (D)

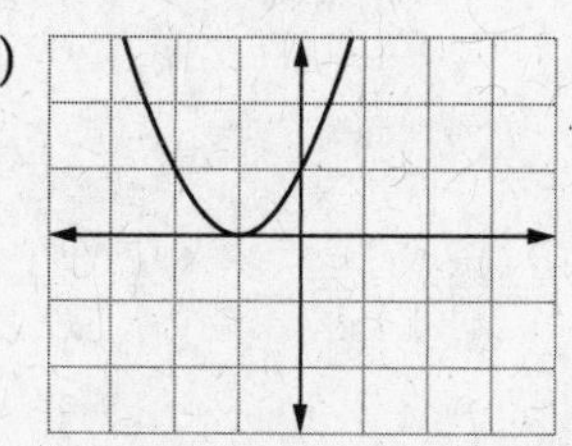

    (E)

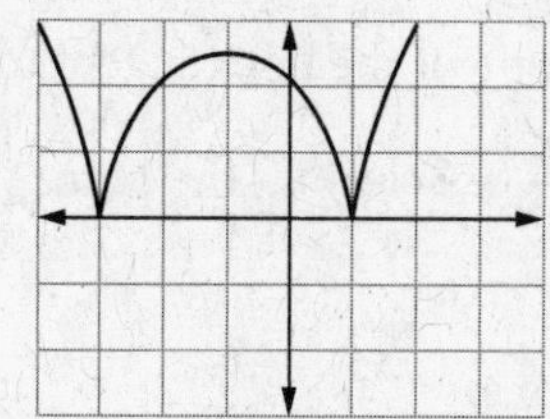

GO ON TO THE NEXT PAGE

17. The radius of convergence for the power series $\sum_{n=1}^{\infty}\frac{(x+4)^{2n}}{n}$ is 1. What is the interval of convergence?

(A) $3 \leq x < 5$
(B) $-1 \leq x < 1$
(C) $-1 < x < 1$
(D) $-5 < x < -3$
(E) $-5 \leq x < -3$

18. If $f(x) = \arcsin(x^3)$, then $f'(x) =$

(A) $\frac{1}{\sqrt{1-x^6}}$
(B) $\frac{-3x^2}{\sqrt{1-x^5}}$
(C) $\frac{3x^2}{\sqrt{1-x^5}}$
(D) $\frac{-3x^2}{\sqrt{1-x^6}}$
(E) $\frac{3x^2}{\sqrt{1-x^6}}$

19. What is the slope of the line tangent to the curve $y+3=\frac{x^3}{3}-3\cos y$ at the point $\left(2, \frac{\pi}{2}\right)$.

(A) -6
(B) -2
(C) 1
(D) 3
(E) 4

20. Which of the following series converge?

I. $\sum_{n=1}^{\infty}\frac{1}{\sqrt[3]{n}}$

II. $\sum_{n=1}^{\infty}\left(\frac{2}{\pi}\right)^n$

III. $\sum_{n=1}^{\infty}\frac{5(n+1)}{(n+2)!}$

(A) I only
(B) II only
(C) III only
(D) I and II only
(E) II and III only

21. The function given by $h(x)=8\sqrt{x}-2x-3$ attains its minimum value at $x=$

(A) -3
(B) 0
(C) 2
(D) 4
(E) 5

x	$g(x)$	$g'(x)$
1	1	3
6	-3	7

22. The function g has a continuous derivative. The table above gives selected values of g and its derivative for $x = 1$ and $x = 6$. If $\int_1^6 g(x)dx = 10$, what is the value of $\int_1^6 xg\,(x)dx$?

(A) -109
(B) $-36\frac{1}{3}$
(C) -29
(D) -19
(E) 19

GO ON TO THE NEXT PAGE

23. What is the slope of the line tangent to polar curve $r = 5\theta$ when $\theta = \frac{\pi}{2}$?

(A) $-\frac{\pi}{2}$

(B) $-\frac{2}{\pi}$

(C) 0

(D) $\frac{\pi}{2}$

(E) 5

24. Let $f(x) = \sqrt{x}$. If the rate of change of f at $x = k$ is 3 times the rate of change at $x = 4$, then $k = ?$

(A) $\frac{1}{2\sqrt{3}}$

(B) $\frac{1}{\sqrt{3}}$

(C) $\frac{4}{9}$

(D) $\frac{2}{3}$

(E) $\frac{3}{4}$

25. $\int_1^4 \frac{1}{(x-3)^2}dx =$

(A) $-\frac{3}{2}$

(B) $-\frac{1}{2}$

(C) $\frac{1}{2}$

(D) $\frac{3}{2}$

(E) None. The integral is divergent.

26. The coefficients of the power series $\sum_{n=0}^{\infty} a_n(x-3)^n$ are $a_0 = 7$ and $a_n = \frac{n+2}{4n-1}$ for $n \geq 1$. The radius of convergence for this series is

(A) 0

(B) $\frac{1}{4}$

(C) 1

(D) 4

(E) Infinite

27. Let f be a function defined by $f(x) = \int_3^{x^2} \sqrt{t^2 + t}dt$. $f'(2) =$

(A) $\sqrt{12}$

(B) $2\sqrt{6}$

(C) $4\sqrt{6}$

(D) $2\sqrt{20}$

(E) $4\sqrt{20}$

28. The function $h(x)$ is given by $h(x) = \cos\left(\frac{x+3}{x^3+1}\right)$. Which of the following statements is true.

I. The graph of h has a horizontal asymptote at $y = 0$.

II. The graph of h has a horizontal asymptote at $y = 1$.

III. The graph of h has a vertical asymptote at $x = 0$.

(A) I only

(B) II only

(C) III only

(D) I and III only

(E) II and III only

IF YOU FINISH BEFORE TIME IS CALLED, YOU MAY CHECK YOUR WORK ON THIS SECTION ONLY. DO NOT TURN TO ANY OTHER SECTION IN THE TEST.

STOP

SECTION I, PART B

Time: 50 Minutes
17 Questions

A GRAPHING CALCULATOR IS ALLOWED ON THIS PORTION OF THE EXAM

Directions:: Solve the following problems, using available space for scratchwork. After examining the form of the choices, decide which one is the best of the choices given and fill in the corresponding oval on the answer sheet. No credit will be given for anything written in the test book. Do not spend too much time on any one problem.

In this test:

(1) The exact numerical value of the correct answer does not always appear among the choices given. When this happens, select from among the choices the number that best approximates the numerical value.

(2) The domain of a function f is the set of all real numbers x for which $f(x)$ is a real number, unless otherwise specified.

(3) The inverse of a trigonometric function f may be indicated using the inverse function notation f^{-1} or with the prefix "arc" (e.g., $\sin^{-1} x = \arcsin x$).

29. If $\int_0^2 f(x)dx = 6$ and $\int_0^8 f(x)dx = -4$, then $\int_2^8 (2f(x)+3)dx =$

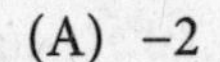

(A) −2
(B) 10
(C) 14
(D) 20
(E) 34

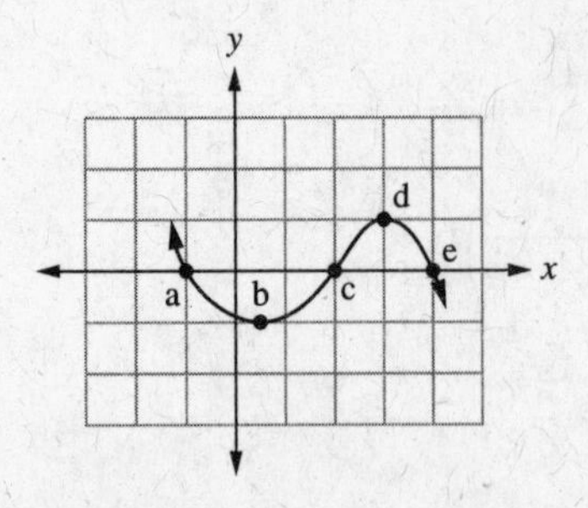

30. The figure above shows the graph of a polynomial function f. For which point will $f'(x) > f(x) > f''(x)$?

(A) a
(B) b
(C) c
(D) d
(E) e

GO ON TO THE NEXT PAGE

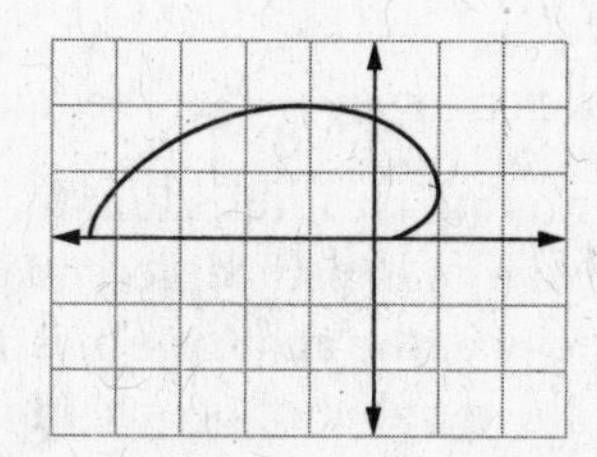

31. The graph above shows the polar curve $r = 3\theta + \sin\theta$ for $0 \le \theta \le \pi$. Let R be the region bounded by the polar curve and the x-axis. What is the area of R?

(A) 8.402
(B) 16.804
(C) 20.923
(D) 56.720
(E) 113.439

32. Let g be a differentiable function such that $g(5) = -2$ and $g'(5) = 4$. Which of the following statements could be false?

(A) $\lim_{x\to 5^-} g(x) = \lim_{x\to 5^+} g(x)$
(B) $\lim_{x\to 5} g(x) = 4$
(C) $\lim_{x\to 5} g(x) = -2$
(D) $\lim_{x\to 5} \dfrac{g(x)+2}{x-5} = 4$
(E) $\lim_{h\to 0} \dfrac{g(5+h)+2}{h} = 4$

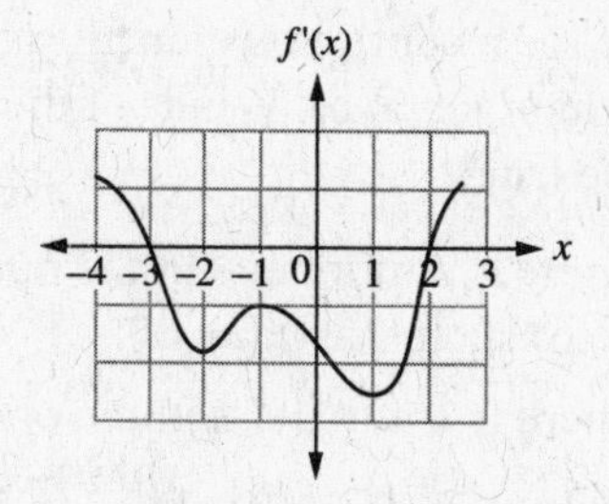

33. The figure above shows the graph of $f'(x)$, the derivative of some function f on the interval [-4, 2]. If $g'(x) = 2f'(x)$, how many points of inflection does the graph of g have on the interval [-4, 2]?

(F) One
(G) Two
(H) Three
(I) Four
(J) Five

34. Let R be the region in the first quadrant bounded by the x-axis, $y = \ln(x + 3)$, and the line $x = 4$. The region R is the base of a solid. For the solid, each cross-section perpendicular to the x-axis is a quarter-circle. What is the volume of the solid?

(K) 2.589
(L) 8.038
(M) 10.234
(N) 16.076
(O) 32.151

GO ON TO THE NEXT PAGE

35. Let $h(x)$ be a continuous function on the closed interval [a, b]. Which of the following must be true?

(A) There is a number c in the open interval (a, b) such that $h'(c) = 0$.

(B) There is a number c in the open interval (a, b) such that $h'(c) = \frac{h(b)-h(a)}{b-a}$.

(C) There is a number c in the open interval (a, b) such that $h(c) = 0$.

(D) There is a number c in the open interval (a, b) such that $h(c) \le h(x)$ for all x in [a, b].

(E) There is a number c in the open interval (a, b) such that $h(a) < h(c) < h(b)$.

x	1.4	1.8	2.0	2.2
$h(x)$	16.8	16.5	16	15

36. The function h is differe ntiable for all x. Selected values of $h(x)$ are shown in the table above. If h and $h'(x)$ are strictly decreasing on the interval $0 \le x \le 4$, which of the following could be the value for $h'(2)$?

(A) −1

(B) −2

(C) −2.5

(D) −4

(E) −5

37. The rate of change, $\frac{dP}{dt}$, of the number of people entering an auditorium for a concert is modeled by a logistic differential equation. The capacity of the auditorium is 1600. At 6:30 pm the number of people in the auditorium is 400 and is increasing at the rate of 600 people per hour. Which of the following differential equations describes the situation?

(A) $\frac{dP}{dt} = P(1600-P)+400$

(B) $\frac{dP}{dt} = \frac{1}{600}(1600-P)+400$

(C) $\frac{dP}{dt} = \frac{1}{600}P(1600-P)$

(D) $\frac{dP}{dt} = \frac{1}{800}P(1600-P)$

(E) $\frac{dP}{dt} = 600P(1600-P)$

x	$g(x)$	$g'(x)$	$g''(x)$	$g'''(x)$
5	0	0	−4	3

38. The third derivative of the function g is continuous on the interval (3, 8). Values of g and its first three derivatives at $x = 5$ are shown in the table above. What is $\lim_{x\to 5} \frac{f(x)}{(x-5)^2}$?

(A) −4

(B) −2

(C) 0

(D) $\frac{3}{2}$

(E) The limit does not exist.

GO ON TO THE NEXT PAGE

39. The first five terms of the Taylor expansion for $f(x)$ about $x = 1$ is given by $7-\frac{3}{2}(x-1)+\frac{7}{2}(x-1)^2-6(x-1)^3+2(x-1)^4$.

What is $f'''(1)$?

(A) −36
(B) −18
(C) −6
(D) 0
(E) 6

40. The function f has derivatives of all orders for all real numbers. Let $f^{(4)}(x) = e^{\tan x}$. If the third-degree Taylor polynomial for f centered at $x = 0$ is used to approximate $f(0.5)$, What is the Lagrange Error Bound for the maximum error on the interval [0, 1]?

(A) 0.012
(B) 0.023
(C) 0.042
(D) 0.198
(E) 0.677

41. The rate at which gravel is processed is given by $R(t)=6\sqrt{4+3\cos\left(t^2\right)}$, where $R(t)$ is measured in tones per hour and t is measured in hours. During the time interval $\sqrt{0 \le t \le 5}$ what is the average rate at which gravel is being processed, in tons per hour?

(A) 2.410
(B) 10.043
(C) 10.612
(D) 12.051
(E) 60.256

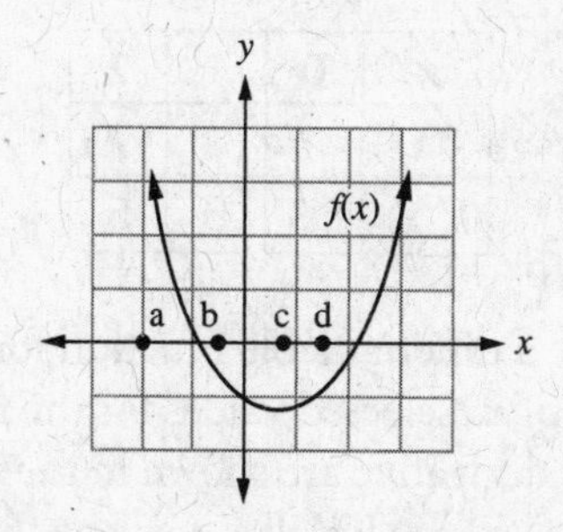

42. The figure above shows the graph of a function f. Which of the following has the smallest value?

(A) $f(a)$
(B) $f'(a)$
(C) $\frac{f(b)-f(a)}{b-a}$
(D) $f(b) - f(d)$
(E) $f'(c)$

43. The nth derivative of a function f at $x = 0$ is given by $f^{(n)}(0)=(-1)^n\frac{n+2}{(n+1)3^n}$ for $n \ge 0$.

Which of the following gives the first four nonzero terms for the Maclaurin series for f?

(A) $2-\frac{1}{2}x+\frac{4}{27}x^2-\frac{5}{108}x^3+\ldots$
(B) $2-\frac{1}{2}x+\frac{2}{27}x^2-\frac{5}{648}x^3+\ldots$
(C) $2+\frac{1}{2}x+\frac{4}{27}x^2+\frac{5}{108}x^3+\ldots$
(D) $2+\frac{1}{2}x+\frac{2}{27}x^2+\frac{5}{648}x^3+\ldots$
(E) $2-2x+\frac{4}{7}x^2-\frac{648}{5}x^3+\ldots$

GO ON TO THE NEXT PAGE

x	0	1	2
$f(x)$	4	2	7
$f'(x)$	−3	8	5

44. Let f be a differentiable function for all values of x. Selected values for the function and its derivative are shown in the table above. $f'(x) > 0$ for all x. Let g be the inverse function of f. What is the value of $g'(2)$?

(A) $\frac{1}{8}$

(B) $\frac{1}{5}$

(C) 1

(D) 5

(E) 8

45. Let f be a function with first derivative given by $f'(x) = \cos(x^2)$ for $-2 \le x \le 2$. For what value of x does f attain its maximum value on the closed interval $-2 \le x \le 2$?

(A) −1.772

(B) −1.253

(C) 0

(D) 1.253

(E) 1.772

IF YOU FINISH BEFORE TIME IS CALLED, YOU MAY CHECK YOUR WORK ON THIS SECTION ONLY. DO NOT TURN TO ANY OTHER SECTION IN THE TEST.

STOP

SECTION II

Time: 1 hour and 30 minutes
6 Problems
50 percent of total grade

Directions: Solve the following problems, using available space for scratchwork. Show how you arrived at your answer.

You may wish to look over the problems before starting to work on them; on the actual test, it is not expected that everyone will be able to complete all parts of all problems. All problems are given equal weight, but the individual parts of a particular problem are not necessarily given equal weight. You should not spend too much time on any one problem.

- You should write out all your work for each part. On the actual test, you will do this in the space provided in the test booklet. Be sure to write clearly and legibly. If you make a mistake, you can save time by crossing it out rather than trying to erase it. Erased or crossed-out work will not be graded.
- Show all your work. Clearly label any functions, graphs, tables, or other objects that you use. On the actual exam, you will be graded on the correctness and completeness of your methods as well as your answers. Answers without any supporting work may not receive credit.
- Justifications (i.e., the request that you "justify your answer") require that you give mathematical (non-calculator) reasons.
- Work must be expressed in standard mathematical notation, not calculator syntax.
- Unless otherwise specified, answers (numeric or algebraic) need not be simplified.
- If you use decimal approximations in calculations, the readers of the actual exam will grade you on accuracy. Unless otherwise specified, your final answers should be accurate to three places after the decimal point.
- Unless otherwise specified, the domain of function f is the set of all real numbers x for which $f(x)$ is a real number.

GO ON TO THE NEXT PAGE

PART A

Time: 30 Minutes
2 Problems

A GRAPHING CALCULATOR IS ALLOWED ON THIS PORTION OF THE EXAM

1. For $0 \le t \le 5$, where t is measured in hours, the rate at which wheat is delivered to a holding area of a mill is given by $H(t) = 650 + 450\sin\left(\frac{t^2}{4}\right)$ bushels per hour. The mill grinds wheat at a constant rate of 575 bushels/hour. At time $t = 0$, the holding area has 500 bushels of wheat.

   (a) How many bushels of wheat are delivered to the mill during the given time interval? Round your answer to the nearest bushel.

   (b) Write an expression for $W(t)$, the total amount of wheat in the holding area at any time t, where $0 \le t \le 5$.

   (c) Find the rate at which the number of bushels of wheat in the holding area is changing when $t = 4$ hours. Using correct units, interpret this value.

   (d) For $0 \le t \le 5$, what is the maximum number of bushels of wheat in the holding area? Justify your answer.

2. A particle moving along a curve in the xy-plane has position $(x(t), y(t))$ for $t > 0$, where $\frac{dx}{dt} = \left(\frac{4}{t} - 2\right)^{\frac{1}{4}}$ and $\frac{dy}{dt} = 2te^{-t}$. At time $t = 2$, the particle is at $(-4, 3)$.

   (a) Find the speed of the particle when $t = 2$.

   (b) What is the y-coordinate of the particle when $t = 5$

   (c) Write an equation for the line tangent to the graph of the particle when $t = 2$.

   (d) Is there a time t at which the particle is furthest left? If so, explain why and give the value of t and the x-coordinate at this time. If not, explain why not.

GO ON TO THE NEXT PAGE

PART B

Time: 60 Minutes
4 Problems

NO GRAPHING CALCULATOR IS ALLOWED ON THIS PORTION OF THE EXAM

Note: If you have extra time, you can go back and work on Part A of Section II, but you cannot use a calculator to complete your work at this time.

t	0	2	3	6	7	8
$H(t)$	70	83	85	81	78	75

3. To monitor the thermal pollution of a river, a biologist takes measurements of the temperature at selected times as shown in the table above. The temperature of the river can be modeled by a differentiable function, $H(t)$, where t is measured in hours and $H(t)$ is measured in °F.

   (a) Use the data in the table to approximate the rate at which the temperature is changing when $t = 4.5$ hours. Show the computations that lead to your answer.

   (b) Write an integral expression in terms of $H(t)$ for the average water temperature of the river during the 8 hours indicated by the data. Use a right Riemann sum on the five given intervals to approximate the average temperature of the river.

   (c) Suppose that $H'(3) = -\frac{1}{2}$ and $H''(t) < 0$ for all $t > 0$. Write the equation for the tangent line to the graph of H at $t = 2$ and use it to approximate $H(4.5)$. Is this estimate greater than or less than the actual value of $H(4.5)$? Give a reason for your answer.

   (d) Using the values in the table, evaluate $\int_0^8 (4 - 6H(t))dt$.

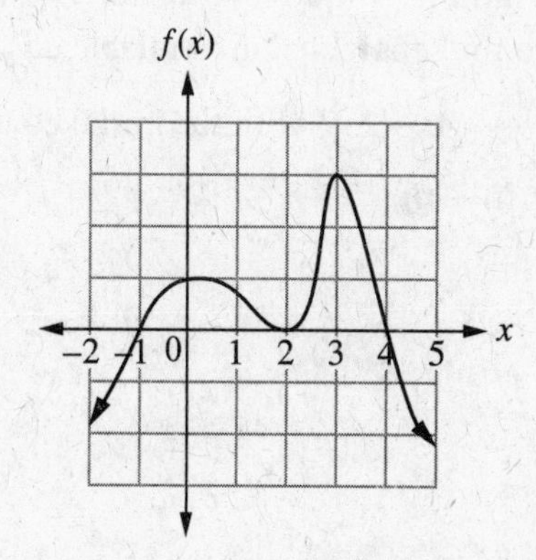

4. Let $f(x)$ be a function whose graph is shown above. Let $g(x) = \int_{-1}^{x} f(t)dt$. The area of the regions bounded by the graph of f and the x-axis $[-2, -1]$, $[-1, 2]$, $[2, 4]$, and $[4, 5]$ is 2, 5, 9, and 3 respectively. The graph of f has horizontal tangents at $x = 1$, $x = 2$, and $x = 3$.

   (a) Write the equation for the line tangent to the graph of g at $x = 2$.

   (b) For what value(s) of x does g have a relative maximum? Justify your answer.

   (c) For what values of x is the graph of g concave downward? Justify your answer.

   (d) Evaluate $\int_{-2}^{5} f(x)dx$.

GO ON TO THE NEXT PAGE

5. Let f be a function satisfying $f'(x) = 4x - 2x\ f(x) \lim_{x \to \infty} f(x)$. for all $x \geq 0$. $\lim_{x \to \infty} f(x) = 4$ and $f(0) = 10$.

   (a) Find the value of $\int_0^{\infty} (4x - 2x\ f(x))dx$.

   (b) Use Euler's method beginning at $x = 0$ to approximate $f(1)$ using 2 intervals of equal width.

   (c) Find the unique solution for $y = f(x)$ to the separable differentiable equation $\frac{dy}{dx} = 4x - 2xy$ with the initial condition $f(0) = 10$.

6. Let f be a function with derivative of all orders at $x = 2$. The values of f and its first three derivatives at $x = 2$ are shown below.

   $f(2) = 4$

   $f'(2) = 0$

   $f''(2) = -3$

   $f'''(2) = 12$

   $f^{(4)}(2) = 3$

   (a) Write the third-degree Taylor polynomial for f about $x = 2$ and use it to approximate $f(1.5)$.

   (b) Does f have a relative extrema at $x = 2$? If so, determine if the extrema is a relative maximum or relative minimum. If not, explain why not.

   (c) It is known that $f^{(4)}(x)$ is strictly increasing on [0, 2]. Use the Lagrange Error Bound to show that the third-degree Taylor polynomial approximates $f(1.5)$ within 0.01.

   (d) Let $h(x) = \int_2^x f(t)dt$. Write the third-degree Taylor polynomial for $h(x)$ about $x = 2$.

IF YOU FINISH BEFORE TIME IS CALLED, YOU MAY CHECK YOUR WORK ON THIS SECTION ONLY. DO NOT TURN TO ANY OTHER SECTION IN THE TEST. **STOP**

AP Calculus BC Practice Test 3 **Answer Key**

1. C
2. D
3. B
4. E
5. D
6. A
7. C
8. C
9. E
10. B
11. E
12. C
13. D
14. D
15. D
16. A
17. D
18. E
19. B
20. E
21. B
22. C
23. B
24. C
25. E
26. C
27. E
28. B
29. A
30. C
31. D
32. B
33. C
34. B
35. D
36. D
37. D
38. B
39. A
40. B
41. D
42. B
43. B
44. A
45. D

ANSWERS AND EXPLANATIONS

1. C

$$f' = \frac{(5x+2)(2)-(2x-5)(5)}{(5x+2)^2} = \frac{29}{(5x+2)^2}$$

2. D

$$\text{Position} = 7 + \int_{\frac{\pi}{3}}^{0} \sin(3t)dt = 7 + \frac{-2}{3} = \frac{19}{3}$$

3. B

This is a geometric series with first term of 1.

$$S = \frac{a_0}{1-r} = \frac{1}{1-\frac{-3}{4}} = \frac{4}{7}$$

4. E

This is the limit formula for the derivative of $y = \sin(x)$ evaluated at $x = \frac{\pi}{3}$.

$$\frac{d}{dx}(\sin x) = \cos x$$

$$\cos\left(\frac{\pi}{3}\right) = \frac{1}{2}$$

5. D

The formula for arc length of a curve is $\int_{x_1}^{x_2} \sqrt{1+(f(x))^2}\, dx$

In this case the result is $\int_0^6 \sqrt{1+(3x^2)^2}\, dx = \int_0^6 \sqrt{1+9x^4}\, dx$

6. A

Using u-substitution let

$$u = 1 + \tan x$$
$$du = \sec^2 x dx$$
$$dx = \frac{du}{\sec^2 x}$$

This gives

$$\int \frac{\sec^2 x}{1 + \tan x} dx \Rightarrow \int \frac{du}{u} = \ln|u| + C$$

Substituting back, the answer is $\ln|1 + \tan x| + C$

7. C

1st iteration:
$$m = y - x + 2\big|_{(2,-3)} = -3 - 2 + 2 = -3$$
$$T : y = -3(x-2) - 3\big|_{x=1.8} = -3(1.8-2) - 3 = -2.4$$

2nd iteration:
$$m = y - x + 2\big|_{(1.8,-2.4)} = -2.4 - 1.8 + 2 = -2.2$$
$$T : y = -2.2(x-1.8) - 2.4\big|_{x=1.6} = -2.2(1.6-1.8) - 2.4 = -1.96$$

8. C

$$4\left(\frac{1}{2}\right)(5+k) + 4\left(\frac{1}{2}\right)(k+11) + 4\left(\frac{1}{2}\right)(11+7) = 72$$
$$\Rightarrow 2k + 34 = 36$$
$$\Rightarrow k = 1$$

9. E

Since $g'(2) = \frac{1}{4}$, g must be increasing. Since $g''(2) = 4$, g must be concave upward. The equation for the tangent line at (2, 5) is $y = \frac{1}{4}(x-2) + 5$. When $x = 4$, $y = 6$. The value of the function must be greater than 6.

10. B

The Maclaurin series for $\ln(1+x) = x - \frac{x^2}{2} + \frac{x^3}{3} + \ldots + (-1)^n \frac{x^n}{n} + \ldots$ Substitute x^2 for x.

11. E

Let

$$x^2 + y^2 = z^2$$
$$\Rightarrow 2x\frac{dx}{dt} + 2y\frac{dy}{dt} = 2z\frac{dz}{dt}$$
$$\Rightarrow x\frac{dx}{dt} + y\frac{dy}{dt} = z\frac{dz}{dt}$$
$$\Rightarrow 12\left(2\frac{dy}{dt}\right) + 5\frac{dy}{dt} = 13\left(k\frac{dy}{dt}\right)$$
$$\Rightarrow 29 = 13k$$
$$\Rightarrow k = \frac{29}{13}$$

12. C

$$f'(x) = \cos(2x)$$
$$f(x) = \frac{1}{2}\sin(2x) + C$$
$$f\left(\frac{\pi}{4}\right) = \frac{1}{2}\sin\left(2\cdot\frac{\pi}{4}\right) + C = 2 \Rightarrow C = \frac{3}{2}$$
$$\therefore f(x) = \frac{1}{2}\sin(2x) + \frac{3}{2}$$
$$f\left(\frac{\pi}{12}\right) = \frac{1}{2}\sin\left(2\cdot\frac{\pi}{12}\right) + \frac{3}{2} = \frac{7}{4}$$

13. D

$$x(t) = (2t+1)^{-1} \qquad y(t) = 6t - t^2$$
$$x'(t) = -2(2t+1)^{-2} \qquad y'(t) = 6 - 2t$$
$$x''(t) = 8(2t+1)^{-3} \qquad y''(t) = -2$$
$$x''(2) = \frac{8}{125} \qquad y''(2) = -2$$
$$\vec{a}(t) = \left\langle \frac{8}{125}, -2 \right\rangle$$

14. D

By partial fractions,

$$\frac{4}{x^2-1}=\frac{2}{x-1}-\frac{2}{x+1}$$

$$\int\frac{4}{x^2-1}\,dx=\int\frac{2}{x-1}-\frac{2}{x+1}\,dx$$
$$=2\ln|x-1|-2\ln|x+1|$$
$$=2\ln\left|\frac{x-1}{x+1}\right|+C$$

15. D

(A) cannot be correct. It is a parabola.

(B) cannot be correct. It has a restricted domain from [–2, 2].

(C) cannot be correct. It does not equal 2 when $x = 0$.

(D) is the correct choice.

(E) cannot be correct. It has a vertical asymptote at $x = 0$.

16. A

An absolute value is always non-negative (but generally positive). If the derivative is positive, the function must be increasing for all values of x except when $f'(x) = 0$. This narrows the choice down to (A) or (C). (C) is discontinuous, so the derivative fails to exist at $x = -1$. That leaves (A) as the correct response.

17. D

Using the Ratio Test we find that the series converges as follows:

$$\lim_{n\to\infty}\left|\frac{\frac{(x+4)^{2(n+1)}}{n+1}}{\frac{(x+4)^{2n}}{n}}\right|=\lim_{n\to\infty}\left|(x+4)^2\right|<1$$
$$\Rightarrow -1<x+4<1 \Rightarrow -5<x<-3$$

The series does not converge at either endpoint because of the p-rule.

18. E

$$f(x) = \arcsin\left(x^3\right)$$

$$f'(x) = \frac{1}{\sqrt{1-\left(x^3\right)^2}} \cdot \frac{d}{dx}\left(x^3\right)$$

$$f'(x) = \frac{3x^2}{\sqrt{1-x^6}}$$

19. B

$$y+3 = \frac{x^3}{3} - 3\cos(y)$$

$$\Rightarrow y' = 3x^2 + 3\sin(y) \cdot y'$$

$$y' = \left.\frac{x^2}{1-3\sin(y)}\right|_{2,\frac{\pi}{2}} = \frac{4}{1-3} = -2$$

20. E

I. does not converge by the *p*-series rule.

II. converges because it is a geometric series with a ratio between −1 and 1.

III. converges because the factorial in the denominator makes the ratio test equal zero for all values of *x*.

21. B

The function is strictly increasing for all $x > 0$. This means the minimum must occur at the left endpoint where $x = 0$.

22. C

Using integration by parts,

$$\int_1^6 xg'(x)\,dx =$$

$$xg(x) - \int_1^6 g'(x)\,dx$$

$$xg(x) - g(x)\Big|_1^6 = 6(-3) - 1(-1) - 10 = -29$$

23. B

$$x = 3\theta \cos\theta$$

$$\frac{dx}{d\theta} = 3\theta(-\sin\theta) + \cos\theta \cdot 3$$

$$y = 3\theta \sin\theta$$

$$\frac{dy}{d\theta} = 3\theta\cos\theta + \sin\theta \cdot 3$$

$$\frac{dy}{dx} = \frac{\frac{dy}{d\theta}}{\frac{dx}{d\theta}} = \left.\frac{3\theta\cos\theta + \sin\theta \cdot 3}{3\theta(-\sin\theta) + \cos\theta \cdot 3}\right|_{x=\frac{\pi}{2}} = \frac{-1}{\frac{\pi}{2}} = \frac{-2}{\pi}$$

24. C

$$f(x) = \sqrt{x}$$

$$f'(x) = \frac{1}{2\sqrt{x}}$$

$$f'(4) = \frac{1}{2\sqrt{4}} = \frac{1}{4}$$

$$f'(k) = 3\left(\frac{1}{4}\right) = \frac{3}{4}$$

$$\frac{1}{2\sqrt{k}} = \frac{3}{4} \Rightarrow k = \frac{4}{9}$$

25. E

$$\int_1^4 \frac{1}{(x-3)^2}\,dx = \lim_{k\to 3^-}\int_1^k \frac{1}{(x-3)^2}\,dx + \lim_{k\to 3^+}\int_k^4 \frac{1}{(x-3)^2}\,dx$$

$$= \lim_{k\to 3^-}\left.\frac{-1}{(x-3)}\right|_1^k + \lim_{k\to 3^+}\left.\frac{-1}{(x-3)}\right|_k^4$$

$$= \lim_{k\to 3^-}\frac{-1}{(k-3)} - \frac{-1}{(1-3)} = \infty$$

Since the left-hand integral is divergent, the entire problem must diverge.

26. C

Using the ratio test

$$\lim_{n\to\infty}\left|\frac{\frac{(n+1)+2}{4(n+1)-1}\cdot(x-3)^{n+1}}{\frac{n+2}{4n-1}\cdot(x-3)^{n}}\right| = \lim_{n\to\infty}|(x-3)|<1$$

$$\Rightarrow -1<(x-3)<1$$

$$\Rightarrow -2<x<4$$

So the radius of convergence is 1.

27. E

$$f(x)=\int_1^{x^2}\sqrt{t^2+t}\,dt$$

$$\Rightarrow f(x)=F(t)\Big|_1^{x^2}=F\left(x^2\right)-F(1)$$

$$f'(x)=F'\left(x^2\right)(2x)-0$$

$$=f\left(x^2\right)(2x)-0$$

$$=\sqrt{\left(x^2\right)^2+x^2}\,(2x)$$

$$f'(2)=\sqrt{16+4}(2\cdot 2)=4\sqrt{20}$$

28. B

As $x\to\infty, \frac{x+3}{x^3+1}\to 0$ $\cos(0)=1$

29. A

$$\int_2^8(2f(x)+3)\,dx=\int_0^8(2f(x)+3)\,dx-\int_0^2(2f(x)+3)\,dx$$

$$=2\int_0^8 f(x)\,dx+24-\left(2\int_0^2 f(x)\,dx+6\right)$$

$$=2(-4)+24-(2(6)+6)$$

$$=-2$$

30. C

The function is increasing at $x=c$ so $f'(c)>0$. $f(c)=0$. The function is concave downward when $x=c$ so $f''(c)<0$. C is the only point where all three conditions are met.

31. D

$$A = \frac{1}{2}\int_0^{\pi} (3\theta + \sin\theta)^2 \, d\theta = 56.7204$$

32. B

This is the definition for the derivative of g at $x = 5$. We don't know much about $g'(x)$. We know that it exists at $x = 5$, but we don't know that the derivative itself is continuous or differentiable there.

33. C

Relative maximum and minimum points in the graph of the first derivative correspond to points of inflection in the original function.

34. B

This is a volume of a solid with a known cross-section. The volume is always $\int_a^b A(x)\,dx$, where $A(x)$ is the area of the cross-section.

$$V = \int_0^4 \frac{1}{4}\pi r^2 dx$$

$$\Rightarrow \frac{1}{4}\pi \int_0^4 (\ln(x+2))^2 \, dx = 8.038$$

35. D

(A) is not true because there is no guarantee the function has a root in the interval.

(B) is not true because there is no guarantee the function has a critical value in the interval.

(C) is not true because the conditions for the Mean Value Theorem require the function to be differentiable. This is not stated here.

(D) is true because the function does have a maximum value by the Extreme Value Theorem.

(E) is not true because the statement requires the function to be increasing for all values of x in the interval.

36. D

If $h'(x)$ is decreasing, the function must be concave downward. That means that $\frac{16-16.5}{2-1.8} > h'(2) > \frac{15-16}{2.2-2}$. Only choice (D) meets these conditions.

37. D

A logistic growth curve follows the differential equation $\frac{dP}{dt} = k(P)(M-P)$, where M is the maximum carrying capacity of the system. Only (D) meets this condition and the initial condition that the population is increasing at the rate of 600 people per hour.

38. B

$\lim_{x\to 5}\frac{g(x)}{(x-5)^2}=\frac{0}{0}$. Apply L'Hôpital's Rule. $\lim_{x\to 5}\frac{g'(x)}{2(x-5)}=\frac{0}{0}$. Apply L'Hôpital's Rule again. $\lim_{x\to 5}\frac{g''(x)}{2}=\frac{-4}{2}=-2$.

39. A

$$\frac{f'''(1)}{3!}=-6$$
$$\Rightarrow f'''(1)=-6\cdot 3!=-36$$

40. B

On the interval [0, 1], $e^{\tan x}$ is an increasing function. It will reach its maximum value at on the interval [0, 0.5] at $x = 0.5$. The maximum error $E<\left|\frac{f^{(4)}(0)}{4!}(0-0.5)^4\right|=0.023$

41. D

$$Average=\frac{1}{b-a}\int_a^b f(x)\,dx$$
$$Average=\frac{1}{5-0}\int_0^5 6\sqrt{4+3\cos\left(t^2\right)}\,dt=\frac{1}{5}(60.256)=12.051$$

42. B

$f(a)$ is a positive value.

$f'(a)$ has a large negative value.

$\frac{f(b)-f(a)}{b-a}$ is the slope of the secant line through a and b. This is a shallower value than $f'(a)$.

$f(b) - f(d)$ has a positive value.

$f'(c) = 0$.

Therefore, $f'(a)$ has the smallest value.

43. B

Coefficients are found as follows.

n	$f^{(n)}(0)$	$\frac{f^{(n)}(0)}{n!}$
0	2	2
1	$-\frac{1}{2}$	$-\frac{1}{2}$
2	$\frac{4}{27}$	$\frac{2}{27}$
3	$-\frac{5}{108}$	$-\frac{5}{648}$

44. A

$$g'(2) = \frac{1}{f'(g(2))} = \frac{1}{f'(1)} = \frac{1}{8}$$

45. D

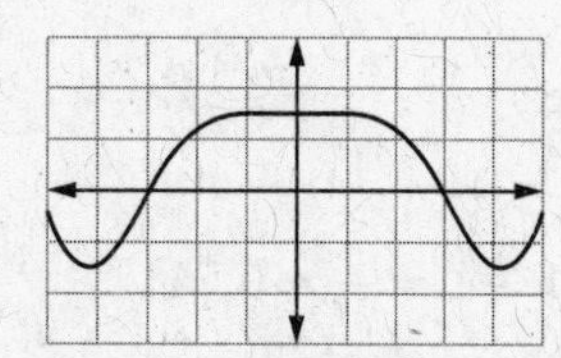

The graph of $f'(x)$ is shown above. $f(x)$ will have a relative maximum when the sign of $f'(x)$ changes from positive to negative. This occurs when $x = 1.253$.

FREE-RESPONSE ANSWERS

1. (a) $\text{Amount} = \int_0^5 H(t)\,dt$

$= \int_0^5 \left(650 + 450\sin\left(\frac{t^2}{4}\right)\right) dt = 3637.465 \text{ bushels}$

(b) $W(t) = 500 + \int_0^t (H(t) - 575)\,dt$

$W'(t) = H(t) - 575$

$W'(4) = H(4) - 575 = -265.561$ Bushels / hour

At time $t = 4$ hours, the number of bushels of wheat is decreasing at a rate of 265.561 bushels/hour.

(c) $W'(t) = H(t) - 575 = 0$

$t = 3.3638, 4.946$

Values of t	$W(t)$
0	500
3.638	1074.720
4.946	760.851
5	762.461

Maximum value = 1074.720 bushels

2. (a) $\text{Speed} = \sqrt{\left(\frac{dx}{dt}\right)^2 + \left(\frac{dy}{dt}\right)^2}\Bigg|_{t=2} = \sqrt{0 + \left(\frac{4}{e^2}\right)^2} = \frac{4}{e^2} \approx 0.541$

(b) $y = 3 + \int_2^5 \left(2te^{-t}\right) dt = 3.731$

(c) $m = \frac{\frac{dy}{dt}}{\frac{dx}{dt}} = \frac{2te^{-t}}{\frac{4}{t} - 2}\Bigg|_{t=2} = \textit{undefined}$

Equation: $x = -4$

(d) $\frac{dx}{dt} = \frac{4}{t} - 2 = 0 \Rightarrow t = 2$

$x''(t) = \frac{-4}{t^2}$

$x''(2) = -\frac{1}{4}$

By the second derivative test, the critical value is relative maximum. So, there is no value for t that gives a relative minimum,

3. (a) $H'(4.5) \approx \frac{H(6)-H(3)}{6-3} = \frac{81-85}{6-3} = -\frac{4}{3}$ °F/ hour

(b) $\text{Average} = \frac{1}{8-0}\int_0^8 H(t)\,dt$

$= \frac{1}{8}(2(83)+1(85)+3(81)+1(78)+1(75)) = \frac{647}{8} = 80\frac{7}{8}$ °F

(c) $T: y-85 = -\frac{1}{2}(x-3) \Rightarrow y = -\frac{1}{2}(x-3)+85$

$y = -\frac{1}{2}(4.5-3)+85 = 84\frac{1}{4}$

Since $H''(x) < 0$, the function is concave down so the approximation is an overestimate of the actual value.

(d)

$$\int_0^8 (4-6K(x))\,dx = \int_0^8 4\,dx - 6\int_0^8 K(x)\,dx$$
$$= 4x\big|_0^8 - 6K(x)\big|_0^8 = (32-0)-6(75-70)$$
$$= 2$$

4. (a) Point $g(2) = \int_{-1}^{2} f(x) = 5$ Slope $g'(x) = f(x) \Rightarrow g'(2) = f(2) = 0$ Tangent: $y=5$

(b) g has a relative maximum when the sign of $g'(x)$ (which is $f(x)$) changes from positive to negative. This occurs when $x = 4$.

(c) g is concave downward whenever $g''(x)$ is negative. This implies that $f'(x) < 0$ or that $f(x)$ is decreasing, This occurs when $1 < x < 2$ and $3 < x < 5$.

(d)

$$\int_{-2}^{5} f(x)\,dx = \int_{-2}^{-1} f(x)\,dx + \int_{-1}^{5} f(x)\,dx$$
$$= -2+5+9-3 = 9.$$

5. (a) $4x-2x \cdot f(x)=f'(x)$

$$\Rightarrow \int_0^{\infty}(4x-2x \cdot f(x))=\int_0^{\infty} f'(x)\,dx$$

$$=\lim_{b\to\infty}\int_0^b f'(x)\,dx=\lim_{b\to\infty} f(x)\Big|_0^b=\lim_{b\to\infty} f(b)-f(0)=4-10=-6$$

(b)

Orig x	Orig y	dx	$dy=(4x-2\,xf(x))dx$	New x	New
0	10	0.5	$(4(0)-2(0)(1))(0.5)=0$	0.5	10
0.5	10	0.5	$(4(0.4)-2(0.5)(10))(0.5)=-4$	1.0	6

(c) $\dfrac{dy}{dx}=4x-2xy$

$\dfrac{dy}{dx}=2x(2-y)$

$\displaystyle\int\frac{dy}{2-y}=\int 2x\,dx$

$-\ln|2-y|=x^2+C$

$-\ln|2-10|=0^2+C\Rightarrow C=-\ln(8)$

$\ln|2-y|=\ln(8)-x^2$

$y=2-8e^{-x^2}$

6. (a)

$$T_2(x) = f(2) + \frac{f'(2)}{1!}(x-2)^1 + \frac{f''(2)}{2!}(x-2)^2 + \frac{f'''(2)}{3!}(x-2)^3$$

$$= 4 - 0(x-2) - \frac{3}{2}(x-2)^2 + \frac{12}{6}(x-3)^3$$

$$T_2\left(\frac{3}{2}\right) = 4 - \frac{3}{2}\left(\frac{3}{2}-2\right)^2 + \frac{12}{6}\left(\frac{3}{2}-2\right)^3 = \frac{31}{8}$$

(b)

$$f'(x) = 0 + 3(x-2) + 6(x-2)^2 + \ldots + \frac{f^{(n)}(2)}{n!}(x-2)^n + \ldots$$

$f'(2) = 0$

$f''(2) = -3$

Since $f'(2) = 0$ and $f''(2) < 0$,

there is a relative maximum

at $x = 2$.

(c)

$$Error < \left|\frac{f^4{}_{\max[1,2]}(z)}{4!}(x-2)^4\right|$$

$$Error < \left|\frac{3}{4!}\left(\frac{3}{2}-2\right)^4\right| \Rightarrow Error < \frac{1}{128}$$

which is less than 0.01.

(d)

$$P_3(x) = \int_2^x \left(4 - \frac{3}{2}(t-2)^2\right)dt$$

$$\Rightarrow 4t - \frac{1}{2}(t-2)^3\Big|_2^x = 4x - \frac{1}{2}(x-2)^3 - 8 = 4(x-2) - \frac{1}{2}(x-2)^3$$

AP CALCULUS BC PRACTICE TEST 3 CORRELATION CHART

Multiple Choice Question	Review Chapter: Subsection of Chapter
1	Chapter 12: Rules for Computing Derivatives
2	Chapter 18: Distance Traveled by a Particle Along a Line
3	Chapter 25: The Geometric Series with Applications
4	Chapter 11: Physical Interpretation of the Derivative: Speed
5	Chapter 10: Polar Functions
6	Chapter 21: When *F* Is Complicated: The Substitution Game
7	Chapter 16: Differential Equations Via Euler's Method
8	Chapter 23: Trapezoidal Approximation
9	Chapter 14: Corresponding Characteristics of the Graphs of f and f' Chapter 15: Concavity and the Sign of f''
10	Chapter 26: Maclaurin Series for Special Functions
11	Chapter 16: Related Rates
12	Chapter 22: Initial Conditions—What Is C?
13	Chapter 10: Parametric Functions Chapter 16: Interpreting the Derivative As a Rate of Change
14	Chapter 21: Antiderivatives by Simple Partial Fractions
15	Chapter 16: Differential Equations Chapter 22: Differential Equations
16	Chapter 14: Corresponding Characteristics of the Graphs of f and f'
17	Chapter 26: From Series to Power Series
18	Chapter 12: Chain Rule
19	Chapter 12: Implicit Differentiation
20	Chapter 24: To Converge or Not to Converge Chapter 25: The Ratio Test for Convergence and Divergence
21	Chapter 16: Optimization
22	Chapter 21: Integration by Parts
23	Chapter 10: Polar Functions Chapter 12: Derivatives of Parametric, Polar, and Vector Functions
24	Chapter 16: Related Rates
25	Chapter 21: Antiderivatives by Improper Integrals

Multiple Choice Question	Review Chapter: Subsection of Chapter
26	Chapter 26: From Series to Power Series
27	Chapter 12: Chain Rule Chapter 20: What's So Fundamental? The Connection Between Integrals and Derivatives
28	Chapter 14: Critical Points and Local Extrema Chapter 16: Graphing Using Derivatives
29	Chapter 17: Properties of the Definite Integral
30	Chapter 14: Corresponding Characteristics of the Graphs of f and f' Chapter 15: Concavity and the Sign of f'
31	Chapter 6: Computing a Definite Integral Chapter 10: Polar Functions Chapter 17: Finding the Area Under a Curve
32	Chapter 11: The Relationship Between Differentiability and Continuity
33	Chapter 15: Inflection Points
34	Chapter 18: Volumes of Solids
35	Chapter 14: The Mean Value Theorem
36	Chapter 14: Corresponding Characteristics of the Graphs of f and f' Chapter 15: Concavity and the Sign of f'
37	Chapter 22: Solving Logistic Differential Equations and Using Them in Modeling
38	Chapter 16: L'Hôpital's Rule
39	Chapter 26: Taylor Polynomial Approximations
40	Chapter 26: How Close Is Close Enough? The Lagrange Error Bound
41	Chapter 18: Average Value of a Function
42	Chapter 11: A Geometric Interpretation: The Derivative As a Slope
43	Chapter 26: Taylor Polynomial Approximation
44	Chapter 12: The Derivative of Inverse Functions
45	Chapter 14: Critical Points and Local Extrema

AP CALCULUS BC PRACTICE TEST 3 CORRELATION CHART

Free Response Question	Review Chapter: Subsection of Chapter
1(a)	Chapter 22: Accumulation Functions and Antiderivatives
1(b)	Chapter 22: Accumulation Functions and Antiderivatives
1(c)	Chapter 18: Accumulated Change Chapter 20: Fundamental Theorem of Calculus (FTC)
1(d)	Chapter 14: Critical Points and Local Extrema Chapter 14: Absolute Extrema Chapter 16: Optimization
2(a)	Chapter 16: Interpreting the Derivative As a Rate of Change
2(b)	Chapter 22: Differential Equations
2(c)	Chapter 12: Derivatives of Parametric, Polar, and Vector Functions Chapter 13: The Slope of a Curve at a Point
2(d)	Chapter 15: Concavity and the Sign of f''
3(a)	Chapter 13: Approximating Derivatives Using Tables and Graphs
3(b)	Chapter 18: Average Value of a Function
3(c)	Chapter 13: Local Linear Approximation
3(d)	Chapter 20: What's So Fundamental? The Connection Between Integrals and Derivatives Chapter 21: Computing Antiderivatives Directly
4(a)	Chapter 13: The Slope of a Curve at a Point Chapter 14: Corresponding Characteristics of the Graphs of f and f'
4(b)	Chapter 14: Critical Points and Local Extrema Chapter 14: Corresponding Characteristics of the Graphs of f and f'
4(c)	Chapter 15: Concavity and the Sign of f''
4(d)	Chapter 17: Properties of the Definite Integral
5(a)	Chapter 21: Antiderivatives by Improper Integrals
5(b)	Chapter 16: Differential Equations Via Euler's Method
5(c)	Chapter 22: Separable Differential Equations
6(a)	Chapter 26: Taylor Polynomial Approximation
6(b)	Chapter 14: Critical Points and Local Extrema Chapter 26: Taylor Polynomial Approximation
6(c)	Chapter 26: How Close Is Close Enough? The Lagrange Error Bound
6(d)	Chapter 26: Taylor Polynomial Approximation